WILDLIFE DISEASE AND HEALTH IN CONSERVATION

Wildlife Management and Conservation

Paul R. Krausman, Series Editor

Wildlife Disease and Health in Conservation

EDITED BY DAVID A. JESSUP AND ROBIN W. RADCLIFFE

Published in Association with *THE WILDLIFE SOCIETY*

JOHNS HOPKINS UNIVERSITY PRESS | BALTIMORE

Printed in the United States of America on acid-free paper
9 8 7 6 5 4 3 2 1

Johns Hopkins University Press
2715 North Charles Street
Baltimore, Maryland 21218
www.press.jhu.edu

Library of Congress Cataloging-in-Publication Data

Names: Jessup, David A., editor. | Radcliffe, Robin W., editor.
Title: Wildlife disease and health in conservation / edited by David A. Jessup and Robin W. Radcliffe.
Description: Baltimore, Maryland : Johns Hopkins University Press, 2023. | Series: Wildlife management and conservation | Includes bibliographical references and index.
Identifiers: LCCN 2022038715 | ISBN 9781421446745 (hardcover ; acid-free paper) | ISBN 9781421446752 (ebook)
Subjects: LCSH: Wildlife diseases. | Wildlife conservation.
Classification: LCC SF996.4 .W553 2023 | DDC 639.9/6—dc23/eng/20220928
LC record available at https://lccn.loc.gov/2022038715

A catalog record for this book is available from the British Library.

Sponsors and Supporting Organizations

Animal and Plant Health Inspection Service
U.S. DEPARTMENT OF AGRICULTURE

College of Veterinary Medicine

Additional support was provided by Drs. Brant Schumaker of University of Wyoming, Michelle Verant of National Park Service, and Tonie Rocke of USGS-NWHC.

Contents

Preface

This book explores selected diseases that are very damaging to the health of people and animals (wild and domestic), or threatening to wildlife conservation efforts. Each chapter provides the wildlife professional with cutting-edge scientific information. Several diseases have only recently emerged (2017–2022). Extensive references are included at the end of each chapter to help those whose responsibility it is to plan for, respond to, or make policy, as well as those who conduct research on these serious health challenges. Traditional exhaustive taxonomy-based academic tomes often do not make this level of information available until 5–10 years after initial research and response efforts. Most chapters here are about relatively newer, invasive, and controversial wildlife health challenges, although a few classic diseases are included because they provide warnings and lessons we should have learned. We hope that this book is sufficiently compelling that it will hold the interest of wildlife decision makers whose primary expertise may not be biological. Like it or not, science is often trumped by real or perceived social, political, legal, or financial barriers or, perhaps more often, science is just not well enough understood—or trusted—as a basis to act upon alone.

More than anything else, these chapters are a series of stories about wildlife health and disease challenges told by people who have "walked the walk." Authors were given latitude to tell the story they felt needed telling as long as the basic information was provided and referenced. Therefore, chapter formats and perspectives vary considerably. Some offer a close-up view of a disease, while others take a broader, more "30,000-foot" view. Some of the chapters are about a single serious disease in a single genus; others are about diseases that cross species barriers or affect whole ecosystems. Certain chapters emphasize diagnosing a serious wildlife disease, treatment, or management option; others are stories about achieving or enhancing health in wildlife, domestic animals, humans, and their ecosystems. We hope that the One Health message shines through clearly in all chapters, that you can "hear the authors' voices" and sense what kind of people they are, and that their words and work inspire the coming generation.

This book is more about preventive medicine than about treating sick wild animals. This is because most wildlife health problems are more effectively dealt with on the population or ecosystem level rather than through care for individual animals. Although some clinical wildlife disease treatment information is provided, this book is largely for conservation biologists and practitioners of conservation medicine (a very descriptive term that seems to have, at least temporarily, lost favor). There are other good clinically oriented texts and veterinary books, which are cited at the conclusion of the chapters. We hope that readers will see that effectively treating disease in wildlife, and trying to maintain health of wildlife populations, often requires the practitioner to use a different set of skills and treatments. Although authors were asked to focus on biological and biomedical aspects of each disease or health problem, they were encouraged to comment on some nonbiological factors and to try to include legal, social, financial, or political perspectives. So, some of the "prescriptions" provided to improve health borrow from those disciplines as much as from biomedicine.

Here is a quick review of the book's five sections. The intimacy of connections between disease-causing agents, their host organisms, and their environment are perhaps most easily envisioned for health problems of marine and aquatic species. Animals surrounded by, and highly dependent on, their fluid medium have relatively little ability to avoid, escape, or neutralize agents of disease, and most are exquisitely sensitive to changes in their environment. Diseases in marine and aquatic environments are often multifactorial, have consequences up and down food chains, and may profoundly alter ecosystems. The aquatic section of the book (Part I) includes vertebrate and invertebrate examples, the latter being as yet a relatively poorly understood or explored area of wildlife and biomedical sciences. Several examples are provided of marine and aquatic health problems that have terrestrial origins, reflecting the fact politely stated, that "stuff rolls downhill." The story of sea star wasting syndrome reflects the power of a public plea for answers to catalyze research quickly, even in the face of very limited funding, jurisdictional confusion, and little or no previous experience with such a disease. The results of collaboration and cooperation are obvious in this story but not always universal in life sciences. Coral diseases and coral reef health are very closely tied to anthropogenic activities; unfortunately and ironically, if past is prologue, the current largely anthropogenic death or destabilization of the nearshore oceans and the ecological processes they support could cause the death, impoverishment, and displacement of millions of humans. Few if any wild fisheries remain healthy or as abundant as they once were. Many anadromous fisheries in the continental United States are completely, or mostly, dependent on intense management and stocking operations, which in turn are extremely dependent on intensive disease diagnostic efforts and treatment. The global advancement of chythrid fungus—a disease process only first described in 2005, to a point where it has destroyed many species of amphibians and threatens others—is another cautionary tale. Oil spills, although they receive a great deal of press, are much more confined in space and time than, say, coral disease. The manner in which these clearly man-made insults are responded to and cleaned up, how animals are treated and damage mitigated, and how it is all organized and paid for by the responsible party provides an excellent example of what humankind might be able to accomplish in other areas where we are clearly responsible for upsetting the natural balance and causing disease. The slow recovery of southern (California) sea otters (*Enhydra lutris nereis*) has been influenced by morbidity and mortality due to several forms of pollution, much of it "pathogen pollution." The large collaborative efforts to understand the problems; to mitigate and change behaviors; to integrate government, university, aquarium, and private nonprofit research and response capacities allowed high quality science to be linked with public outreach and fundraising to support it; and to develop social, political, and legal support to help promote the sea otters' recovery. We hope this section of the book provides impetus for all to recognize that marine and aquatic environments, and the animals that live in them, are every bit as susceptible to diseases and environmental perturbations that upset healthy relationships as those that occur on land and in closer contact with people and domestic animals.

Diseases affecting wild ungulates are important and foundational perhaps because they are large and visible; are a primary source of food, income, and recreation; and are now becoming more appreciated for the important ecological services they provide. Part II of the book starts out with the grandaddy of them all, rinderpest, and provides insight into the small ruminant version that has killed so many of the rare saiga antelope (*Saiga tatarica*) across the steppes of Asia in recent years. This chapter should also challenge the reader to recognize that, despite nearly 200 years of dealing with some of these diseases, we may have yet to act on some of the fundamental lessons they provide. The chapter on the three viruses that cause "hemorrhagic disease" in North American wild cervids provides an example of a similarly fast-moving and frequently fatal viral disease but one that has received far less investment because of their relatively minor impact on livestock economics and none on human health. These diseases can, and do, pose major concerns to wildlife managers and were foundational to establishment of wildlife disease research in the southeastern United States. Unlike the aforementioned viral diseases of wild ungulates, chronic wasting disease moves slowly (except when and where people move captive wildlife or their tissues) but the prion that causes it is nearly indestructible. No treatment or effective management is on the horizon and the long-term projected consequences are chilling for conservationists. This chapter also provides a clear example of the downside of ignoring one of the basic principles of the North American Model of Wildlife Conservation: commercialization and confinement of wildlife are fundamentally dangerous (Radcliffe and Jessup 2022).* Two bacterial diseases in this chapter, brucellosis and epizootic pneumonia of bighorn sheep (*Ovis canadensis*), demonstrate that diseases of livestock can have profound impacts on the management of wildlife. These bacteria are very different, but both play tricks on the hosts' immune response, can create carrier animals, and there are as yet no effective wildlife vaccines or effective population level treatments for either. With both diseases, there is a similar underlying political and financial story—the conflict between livestock interests and conservationists over access to, and use of, publicly owned western rangelands. The last chapter in this section deals with a global scourge that is poised to invade North America, having ravaged Eastern Europe and much of Asia for the last decade. This scourge is a virus that is highly infectious; lethal; resistant to disinfection in blood, in tissues (even commercial meat products), on boots, on tires, and in soils; and can be harbored in and transmitted by insects. That disease, African swine fever (ASF), is in two Caribbean nations at the time of this writing. With the widespread and purposeful movement and stocking of wild pigs across much of the United States over the past three decades, this is a horror story waiting to happen.

There are many, many diseases that primarily involve rodents, small carnivores, bats, and other small mammals. In Part III, we have chosen just five (in four chapters) to demonstrate the important challenges they present to wildlife management and how diseases can threaten wildlife species survival, ecosystem health and stability, human health, and social expectations. Perhaps no endangered species recovery story in North America has been followed more closely than that of the black-footed ferret (*Mustela nigripes*) in the Great Plains. But few know that almost as many efforts to re-establish them have failed as have succeeded, and two diseases (plague and distemper) are serious and fundamental problems. In North America, rabies has been reduced to periodic flareups from chronic infections in several small native carnivore and bat populations, but it provides an excellent example of how social expectations and persistent investment in efforts to reduce a disease in wildlife, after essential elimination from owned animals, can progress. The chapter also explains why rabies will likely never be eliminated. Unlike ASF, rabbit hemorrhagic disease (RHD) has recently entered and become epidemic in North America. Although it was kept at bay for decades, or limited to domestic rabbits, it has now arrived and jumped from domestic to wild lagomorphs. It is another highly infections, resistant, and lethal viral disease that is likely to have profound and long-term

* Radcliffe, R. W., and D. A. Jessup. 2022. Wildlife health and the North American model of wildlife conservation. Journal of Zoo and Wildlife Medicine 53:493–503.

ecological consequences (as rabbits are ecological engineers of various habitats and feed many carnivores) and may be impossible to eradicate in the foreseeable future. Closing this section, white-nose syndrome (WNS) is an insidious fungal disease that appears to have come from Europe where it causes little problem to bats. But for more than a decade, it has been a cause of major morbidity and mortality in hibernating, cave dwelling and social bats species of North America. The effort that has gone into understanding this disease and what can be done about it is another example of scientific collaboration and cooperation, and of hope on the horizon.

We may think of birds, with the power of flight, as being on the opposite end of the spectrum from aquatic animals, less bound by their environment, but as the reader will see in Part IV, freedom is relative and not a barrier to some serious diseases. Large waterfowl die-offs from the beginning to the middle of the 20th century became not only a subject of increased biological concern but helped galvanize social and political support to investigate them and, where possible, reduce impacts by improved wildlife management. How waterfowl diseases helped establish the wildlife health professions is touched on in that chapter. One of those foundational diseases, lead poisoning, is still with us today, prominently as a major barrier to recovery of the California condor (*Gymnogyps californianus*) and a consistent cause of mortality in other raptors and carrion-feeding birds. As depressing as this story is, the bright side is that major changes in hunting to help ameliorate the problem are underway. Like SARS-CoV-2, avian influenza (AI) viruses mutate, recombine, and change pathogenicity. In the months since the chapter on AI was written, multiple H5N1 outbreaks have occurred in wild birds, including cranes in Israel, raptors in North America, and migratory waterfowl on several continents. Recently, it has also caused pinniped and cetacean die-offs on both coasts of South America. Avian malaria has a large global distribution; however, as the chapter on this disease will convey, it is most destructive on tropical islands with endemic, rare, and sensitive species. Advances in vector control and bioengineering may offer hope for recovery of iconic Hawaiian native forest birds if we are not too late. The chapter on monitoring gyrfalcon (*Falco rusticolus*) health as a means of understanding health effects of climate change makes the argument for proactive science in the Arctic where climate effects are most apparent.

The last section of this book (Part V) provides examples of four diseases that easily cross species barriers, can infect human beings, and are causes of serious worldwide social concern. The first of these is tuberculosis (TB) of the variety that originated or evolved in cattle. Like brucellosis, it has been beaten back by sanitary measures in most developed nations but now remains in some wildlife populations to the consternation of all. The narrative of how bovine TB advanced through Kruger National Park is fascinating. Toxoplasmosis is perhaps the single most successful protozoal parasite at moving around the world. It causes a wide variety of disease signs and mortality in wild and domestic animals, and in human beings, with recent revelations of potential links to various neurodegenerative diseases reawakening public health concerns. Perhaps no disease frightens people more than Ebola, despite the fact that it is endemic in relatively small areas in central Africa. Unfortunately, these places are where chimpanzees and gorillas, our nearest relatives, live. This story is included in no small part because it shows what integrating conservation into communities can do to protect both people and animals and enrich their lives. No current book on wildlife health or disease would be complete without a chapter dealing with the cornonavirus diseases—SARS, Middle East respiratory syndrome (MERS), and COVID-19—that have recently held human society in thrall. Each of these diseases apparently originated in wild animals and are dangerous, often lethal, to human beings. Ironically, SARS warned us that coronaviruses in small mammals could jump species into humans, where they were very infectious, and could jump oceans when those affected humans travel with potentially catastrophic consequences. But after a short, aggressive and successful effort to combat its spread, the world went back to "business as usual." Then MERS warned us that a similar coronavirus of camels could be even more lethal (50% mortality rate). But, it is the third of these coronaviruses that turned our world upside down, albeit as a much less lethal, but far more infectious disease. Lessons do not get more important than that.

Few, if any, of these diseases are good news or likely to disappear soon, but where people have stepped up to deal with them effectively, one may see hope for a better world. Remember, none of these diseases pose as great an imminent threat to your health (with perhaps one or two exceptions) as a distracted driver in the lane next to you. All of them (except one or two) will shape our world in the future. It is not the purpose of this book to scare people. The purpose of this book is to provide knowledge about wildlife health and disease, examples of mistakes, but also of wise and health-promoting decisions and actions. Knowledge is power—power to act, to not be afraid, and to make a difference.

It has become popular in the last decade to state that 70% of emerging diseases are from wildlife. That is not really true, and it engenders fear and confusion. Several of these "70% diseases" are reemerging in wildlife but had clear origins in domestic animals and human civilization (TB, brucellosis, rabies, morbilliviruses, and plague); others are largely, or in part, the result of global commerce and our ability to move rapidly about the globe (several fish and amphibian diseases, some strains of bluetongue, ASF, RHD, WNS, and coronavirus diseases); and others are due to human intrusion into and exploitation of wildlife habitats (some coral diseases, rabies, plague, coronaviruses, and Ebola) or are directly the result of human activities, most often with prior knowledge of risk (lead poisoning, avian malaria, epizootic pneumonia of bighorn sheep, oil spills, and Arctic climate change). The common denominator: "'We have met the enemy, and he is us'" (*Pogo*, Walt Kelly, 1970).*

People need to be less afraid of wildlife (the vast majority mean us no harm) and their diseases and be more afraid of what we are doing as individuals, as societies, and as a global human race in pursuit of luxury, profit, power, and sometimes, just because we can. On the bright side, and as demonstrated several places in this book, we retain a good deal of power to reduce the damage done, reverse alterations to natural systems or let them heal, and protect ourselves and nature from the worst consequences of health problems or diseases we have unleashed.

* Kelly created a poster with the phrase for the first Earth Day in 1970.

Acknowledgments

Some people consume and retain information presented in text form, others are more visual learners, and most can do both. In this book we have adopted a strong visual approach, including charts and graphs, color photographs, and, in a fit of nostalgia, hand-drawn illustrations. We hope you agree that they are lovely—perhaps "works of art for disease wonks"—but kidding aside, we hope that memorable images help solidify knowledge gained, encourage your interest in and commitment to wildlife, and maybe provide some inspiration when it is needed. We feel that the art alone may make the book uniquely useful, even to folks who do not have biological science degrees or veterinary training. It may also make some of the content accessible to those outside the English-speaking world. We would be most gratified if this book were appreciated for its aesthetic as well as its intellectual qualities. We particularly want to thank Dr. Laura Donohue, our illustrator and now young veterinarian, whose captivating and thoughtful drawings make this book special.

The authors and coauthors of the chapters in this book willingly spent significant time providing their unique insights. In a number of cases, the chapters they wrote constitute a magnum opus near then end of a long career. It was a privilege working with one and all. Paul Payson, our copyeditor, did a superlative job and significantly improved the structure and clarity of the writing.

We also want to thank the sponsoring organizations, agencies, and individuals whose logos appear in the first pages of this book. Their sponsorship not only made it possible to provide a full-color illustrated book at an affordable price, but it also highlights the solidarity between and cooperation among like minded organizations. The vision of these sponsors and financial donors made it possible to produce something truly unique. Thank you, one and all.

Contributors

Carter T. Atkinson, BS, MS, PhD
US Geological Survey
Pacific Island Ecosystems Research Center
PO Box 44
Hawai'i National Park, HI 96718

Riley F. Bernard, BS, MS, PhD
Department of Zoology and Physiology
University of Wyoming
1000 East University Avenue
Laramie, WY USA 82071

Thomas E. Besser, BS, DVM, PhD, DACVM
Department of Veterinary Microbiology and Pathology
Washington State University
PO Box 647040
Pullman, WA USA 99164

Courtney F. Bowden, MS
US Department of Agriculture
Animal and Plant Health Inspection Service
Wildlife Services, National Wildlife Research Center
4101 Laporte Avenue
Fort Collins, CO USA 80521

Vienna R. Brown, MPH, PhD
US Department of Agriculture
Animal and Plant Health Inspection Service
Wildlife Services, National Feral Swine Damage Management Program
4101 Laporte Avenue
Fort Collins, CO USA 80521

Katherine E. Carr, BS
Veterinary Medical Teaching Hospital
School of Veterinary Medicine
University of California, Davis
1 Garrod Drive
Davis, CA USA 95616

E. Davis Carter, BS, MS, PhD
Center for Wildlife Health and One Health Initiative
University of Tennessee
2505 E.J. Chapman Drive
Knoxville, TN USA 37996

E. Frances Cassirer, BS, MS, PhD
Idaho Department of Fish and Game
3316 16th Street
Lewiston, ID USA 83501

Sonja A. Christensen, PhD
Department of Fisheries and Wildlife
Michigan State University
480 Wilson Road Room 13
East Lansing, MI USA 48824

Janna M. E. Freeman, BS
EpiCenter for Disease Dynamics
One Health Institute
School of Veterinary Medicine
University of California, Davis
1089 Veterinary Medicine Drive
Davis, CA 95616

Joseph K. Gaydos, VMD, PhD
The SeaDoc Society
Karen C. Drayer Wildlife Health Center—Orcas Island Office
942 Deer Harbor Road
Eastsound, WA 98245

Thomas Gidlewski, MS, VMD
US Department of Agriculture
Animal and Plant Health Inspection Service
Wildlife Services, National Wildlife Science Program
4101 Laporte Avenue
Fort Collins, CO USA 80521

Kirsten V. Gilardi, DVM, DACZM
Karen C. Drayer Wildlife Health Center
School of Veterinary Medicine
University of California, Davis
1089 Veterinary Medicine Drive
Davis, CA USA 95616

Matthew J. Gray, BS, MS, PhD
Center for Wildlife Health and One Health Initiative
University of Tennessee
2505 E.J. Chapman Drive
Knoxville, TN USA 37996

Rebecca H. Hardman, BS, MS, DVM, PhD
Florida Fish and Wildlife Conservation Commission
100 8th Avenue SE
St. Petersburg, FL USA 33701

Michelle G. Hawkins, VMD, DABVP (Avian Practice)
Department of Medicine and Epidemiology and California Raptor Center
School of Veterinary Medicine
University of California, Davis
1 Garrod Drive
Davis, CA 95616

Michael T. Henderson, MS
The Peregrine Fund
5668 West Flying Hawk Lane
Boise, ID 83709

David A. Jessup, BS, DVM, MPVM, DACZM
Karen C. Drayer Wildlife Health Center
School of Veterinary Medicine
University of California, Davis
1089 Veterinary Medicine Drive
Davis, CA USA 95616

Christine K. Johnson, VMD, MPVM, PhD
Karen C. Drayer Wildlife Health Center
School of Veterinary Medicine
University of California, Davis
1089 Veterinary Medicine Drive
Davis, CA USA 95616

Richard A. Kock, MA, VET MB, VMD-MRCVS
Royal Veterinary College
Hawkshead Lane North Mymms Hatfield
Hertfordshire
United Kingdom AL97TA

Julianna B. Lenoch, DVM, MPH, DACVPM
US Department of Agriculture
Animal and Plant Health Inspection Service
Wildlife Services, National Wildlife Disease Program
4101 Laporte Avenue
Fort Collins, CO USA 80521

Debra L. Miller, BS, MS, DVM, PhD
Center for Wildlife Health and One Health Initiative
University of Tennessee
2505 E.J. Chapman Drive
Knoxville, TN USA 37996

Melissa A. Miller, DVM, PhD
California Department of Fish and Wildlife
Marine Wildlife Veterinary Care and Research Center Santa Cruz, CA
Karen C. Drayer Wildlife Health Center
School of Veterinary Medicine
University of California, Davis
1089 Veterinary Medicine Drive
Davis, CA USA 95616

Michael W. Miller, DVM, PhD
Colorado Parks & Wildlife
4330 Laporte Avenue
Fort Collins, CO USA 80521-2153

Michele A. Miller, DVM, PhD, MPH, DECZM (ZHM)
Faculty of Medicine and Health Sciences
Department of Science and Innovation - National Research Foundation Centre of Excellence for Biomedical Tuberculosis Research
South African Medical Research Council Centre for Tuberculosis Research
Stellenbosch University
PO Box 241
Cape Town 8000, South Africa

Diego Montecino-Latorre, BVS, MPVM, PhD
Wildlife Conservation Society—Health Program
2300 Southern Boulevard
Bronx, NY USA 10460

Thomas Müller, DVM, PhD
Friedrich-Loeffler-Institut
Federal Research Institute for Animal Health
Insel Riems, Germany

Robin W. Radcliffe, DVM, DACZM
Cornell Conservation Medicine Program
College of Veterinary Medicine
Cornell University
602 Tower Road
Ithaca, NY USA 14853

Andrew M. Ramey, PhD
US Geological Survey
Alaska Science Center
4210 University Drive
Anchorage, AK USA 99508

Tonie E. Rocke, MS, PhD
US Geological Survey
National Wildlife Health Center
6006 Schroeder Road
Madison, WI USA 53711

J. Jeffrey Root, Ph.D.
US Department of Agriculture
Animal and Plant Health Inspection Service
Wildlife Services, National Wildlife Research Center
4101 Laporte Avenue
Fort Collins, CO USA 80521

Olivia Rozanski, MS, VMD
School of Veterinary Medicine
University of Pennsylvania
3800 Spruce Street
Philadelphia, PA USA 19104

Mark G. Ruder, DVM, PhD
Southeastern Cooperative Wildlife Disease Study
College of Veterinary Medicine
University of Georgia
589 D. W. Brooks Drive
Athens, GA USA 30602

Charles E. Rupprecht, VMD, MS, PhD
College of Forestry, Wildlife & Environment
Auburn University
602 Duncan Drive
Auburn, AL USA 36849

Brant Schumaker, DVM, MPVM, PhD
Division of Kinesiology and Health
University of Wyoming
1000 East University Avenue
Laramie, WY USA 82071

Karen Shapiro, MPVM, DVM, PhD
Department of Pathology, Microbiology and Immunology
School of Veterinary Medicine
University of California, Davis
1 Shields Avenue
Davis, CA USA 95616

Tierra Smiley Evans, DVM, PhD
One Health Institute
School of Veterinary Medicine
University of California, Davis
1089 Veterinary Medicine Drive
Davis, CA USA 95616

Marcela M. Uhart, DVM
Karen C. Drayer Wildlife Health Center
School of Veterinary Medicine
University of California, Davis
1089 Veterinary Medicine Drive
Davis, CA USA 95616

Elizabeth VanWormer, DVM, PhD
School of Veterinary Medicine and Biomedical Sciences
School of Natural Resources
University of Nebraska, Lincoln
3310 Holdrege Street
Lincoln, NE USA 68583

Michelle L. Verant, DVM, MPH, PhD
National Park Service
Natural Resources Stewardship and Science
Biological Resources Division—Wildlife Health
1201 Oakridge Drive, Suite 200
Fort Collins, CO USA 80525

E. Scott P. Weber III, VMD, MSc, DACVPM, CertAquatic Vet
Department of Medicine and Epidemiology
2108 Tupper Hall
School of Veterinary Medicine
University of California, Davis
1275 West Health Sciences Drive
Davis, CA USA 95616

Lisa L. Wolfe, MS, DVM
Colorado Parks and Wildlife
4330 Laporte Avenue
Fort Collins, CO USA 80521-2153

Krystal M. T. Woo, DVM
Lindsay Wildlife Experience
1931 First Avenue
Walnut Creek, CA USA 94597

Thierry M. Work, MS, DVM, MPVM
US Geological Survey
National Wildlife Health Center
Honolulu Field Station
300 Ala Moana Boulevard
Honolulu, HI USA 96850

Michael Ziccardi, DVM, MPVM, PhD, Hon. ACZM
Oiled Wildlife Care Network
Karen C. Drayer Wildlife Health Center
One Health Institute, School of Veterinary Medicine
University of California, Davis
1089 Veterinary Medicine Drive
Davis, CA USA 95616

WILDLIFE DISEASE AND HEALTH IN CONSERVATION

Introduction

DAVID A. JESSUP

Aldo Leopold's warning "The role of disease in wildlife conservation has probably been radically underestimated"* certainly seems prescient today. Over the past several decades, the traditional wildlife management perspective that disease is just another form of compensatory loss and of no significance to wildlife populations has been repeatedly challenged. We have seen that diseases can reduce wildlife populations, sometimes profoundly enough to affect harvestable surplus (of hunted or fished species), contribute to endangerment of sensitive species, alter food webs, and even affect ecosystem sustainability. In addition, wildlife diseases have radically altered conservation priorities through effects on domestic animal and human health. With the 2019 coronavirus pandemic, our perspective of wildlife health, and our world, has likely changed forever. A wildlife disease disrupted global economies, strained relationships between nations, and challenged laws and social norms. This is precisely what the One World–One Health† concept tells us—the health of animals, people, and environments are inexorably bound together.

Our first reaction as *Homo sapiens* is to look at threats to our well-being, whether health, wealth, or convenience and then "blame the messenger." If wildlife are involved in a health crisis, killing them is almost a knee-jerk reflex. But wildlife and their diseases are not trying to threaten us; they too are just trying to survive. The conditions for disease emergence include host, agent, and environmental factors, all of which may be affected by anthropogenic actions. Importantly, to have serious effects, the disease process must be able to propagate in susceptible host population(s) through time and space (Figure I.1). A portion of the efficacy of preventive medicine comes from our ability to use time and space against diseases of concern (isolation, social distancing, quarantine zones, fencing, etc.).

Some combination of contagiousness or high infectivity rate (R_0), persistence (survival in an infectious form) in the host or the environment, and relative pathogenicity (case fatality rate in people, domestic animals, or wildlife) largely determines whether

* Leopold, A. 1933. Game management (page 325). C. Scribner's Sons, New York, New York, USA, and London, England.

† Cook, R. A., W. B. Karesh, and S. A. Osofsky. 2004. One World, One Health: building interdisciplinary bridges to health in a globalized world. Conference summary. One World-One Health Symposium hosted by the Wildlife Conservation Society and The Rockefeller University, 29 September 2004, New York, New York, USA. http://www.oneworldonehealth.org/sept2004/owoh_sept04.html.

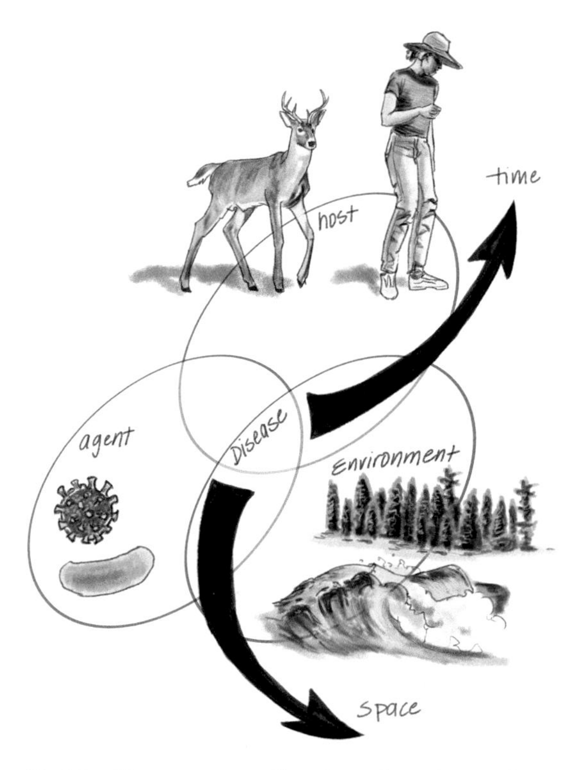

Fig. I.1. Diseases may manifest themselves where and when host, pathogen, and environmental factors come together. To become a wildlife health concern, this relationship must cause morbidity and/or mortality and persist through space and time. If understood and acted upon, these concepts may enable treatment or prevention, or allow management of some diseases, and even avoidance in the future. Illustration by Laura Donohue.

particular wildlife diseases will become a serious societal concern. Many disease-causing organisms, and most parasites, have coevolved with their hosts to some degree, limiting their pathogenicity, replication, or infectivity. But what balance may have existed can be upset by disease agents entering new hosts and new ecosystems, by hosts coming in contact with an existing disease they have no experience with, by changes in host immunity or agent pathogenicity, or by alterations in established host and agent ecological relationships. Often it is human activities that upset these balances, and you will find examples of each of these scenarios in the chapters of this book. Optimizing the health of wildlife and that of the ecosystems on which they (and we) depend (One Health) is one of the great challenges facing wildlife conservation, management and medicine in the 21st century.

In the past decade, we have seen a number of instances where wildlife health and disease problems have transcended the perceived borders between biology and the social, legal, financial, and political realms. Few would have believed, 10–20 years ago, that this would become the "new normal." Even the epidemiologists and microbe hunters who warned us of the medical consequences of species-jumping pathogens probably did not expect that a disease of apparent wildlife origin would change the world's economy and political order in 2020 and beyond.

Historically, health has been defined as the absence of injury or disease, and disease as any perturbation of normal physiologic function. Although these definitions are simplistic, they provide a sufficient base for further discussion. We have tried to use the use the word "disease" when referring to manifestation of illness and the organisms that cause morbidity and mortality in free-living wild animals as well as those organisms that also infect domestic animals or people (zoonotic diseases). We have tried to use the word "health" when referring to the more complex effects of disease on the resilience of free-living wild animal populations and ecosystems. "Health" is a relative term that implies balance and sustainability over time but, like all dynamic processes, is ever changing.

We do not mean to imply that health can be achieved by eliminating all diseases. Some, perhaps many, diseases are relatively benign and part of normal life processes. Some may provide protection against more severe pathogens; others may serve to check potentially destructive population increases. The "mission" of most disease-causing agents is simply to replicate themselves, the recognized disease being a byproduct of that process—and, often, not killing the host is advantageous. The wildlife diseases discussed in this book are some of the more deadly, infectious, or persistent, and particularly, are those with conservation implications or human health connections. There are four primary reasons for this:

1. Human actions are, and have been, the cause of some of the worst wildlife health problems we face.
2. What people have done can (and arguably should be) undone to restore health, ecological processes, and sustainability.
3. Human-induced alterations in health may be easier to treat using the medical, legal, social, financial, and political "medicines" available.
4. If we understand the biology and other facets of health problems we have created, we may be able to avoid recreating them in the future.

> But man is part of nature, and his war against nature is inevitably a war against himself . . . we in this generation must come to terms with nature, and I think we're challenged as mankind has never been challenged before to prove our maturity and our mastery, not of nature, but of ourselves.
>
> *(Rachel Carson, 1963)**

We have been repeatedly asked the questions: why this book, and why now? The answer to the first question should be apparent to anyone who opens and thumbs through the pages of this book. Art and science are not mutually exclusive and the manner in which they may reinforce each other is insufficiently appreciated. We have attempted to use both art and science to inform and inspire the reader, something that is seldom done with educational texts and serious science books anymore. Recognizing that striking illustrations are often what our brains retain most strongly, that they can reveal both complex and nearly invisible activities, and that they can illustrate processes that transit space and time, we felt it was imperative to use them now. Anatomic illustration may also reduce some of the revulsion and fear associated with pictures or descriptions of disease and death. We have used a "field notes" style of illustration in part to remind readers that good One Health work relies on a strong grounding in field biology and ecology.

The answer to the second question also seems obvious. At the time of the writing of this Introduction, we are two and a half years into a global pandemic of apparent wildlife origin, the most expensive pandemic in recorded history, with total numbers of attributed deaths approaching those of the 1918–1920 global influenza pandemic. The second global epidemic of monkeypox in a decade is now spreading in the human population. And this all comes on the heels of global health concerns over avian influenza, Zika, yellow fever, and dengue. There are many other examples of the challenges we face: African swine fever has crossed most of the globe over a decade and is now just off US shores, wild rabbit populations are dying of a new hemorrhagic disease, fungal diseases are killing large numbers of bats and causing multiple amphibian species extinctions, to name a few. Never has the importance of wildlife disease and managing wildlife for healthy and sustainable populations been more evident.

Science provides the knowledge and tools, and art inspires people to act. We hope this book inspires the reader to think about wildlife health and disease in new and thoughtful ways and do something about it. We hope it inspires you to think deeply about the sustainability of natural systems and what we can, and should, do to conserve them; to think long and hard about life more generally, the challenges our societies face to make things better or worse, and actions to mitigate those we can; and to think harder still about our place in Nature, the spirit that is in Nature,† and what we stand to lose if we do not find a better balance.

* Carson, Rachel. 1963. The silent spring of Rachel Carson. CBS Reports interview. https://www.youtube.com/watch?v=kVxMuQgRuzs.

† Jessup D. A. 1992. Spirit in Nature. Journal of Zoo and Wildlife Medicine 23:153–158.

I | Wildlife Disease and Health in Marine and Aquatic Species

1 Coral Reef Ecosystem Health

Thierry M. Work

Introduction

Currently, coral reefs occupy just 0.1% of the Earth's surface, mainly concentrated between the Tropics of Cancer and Capricorn, yet account for most of the marine biodiversity (Knowlton et al. 2010). Estimates suggest that 32% of all known marine species reside on coral reefs and that another 74% of estimated marine biodiversity in coral reefs remains to be discovered (Fisher et al. 2015). Given their disproportionate contribution to marine biodiversity globally, coral reefs have the well-deserved moniker of "rainforests of the seas." Corals support this biodiversity by dint of their myriad colony morphologies that form the three-dimensional structure of coral reefs. Consequently, this structure allows for the existence of ecological niches, akin to a tropical forest, providing habitat for numerous mollusks, crustacea, larval fish, and other biota that constitute reef ecosystems (Hughes et al. 2002). As a result, corals have effects on habitat at multiple scales from the microscopic all the way to continental, the latter including the formation of massive barrier reefs and lagoons. The propensity of such small organisms (a typical coral polyp is about 10–20 mm diameter) to form such large structures and affect species composition

A PACIFIC CORAL REEF SYSTEM

Illustration by Laura Donohue.

on a grand scale makes corals the archetypal ecosystem engineers (Wild et al. 2011).

Corals are invertebrates in the order Cnidaria and have been on the Earth for around 270 million years (Stanley and Fautin 2001). The fundamental unit of corals is the polyp, which is a mouth surrounded by tentacles, and corals can vary from large single polyps to colonies of thousands of polyps, all of which are interconnected. Individual polyps are composed of three tissue layers: (1) epidermis (with stinging cells), separated from (2) gastrodermis by a connective tissue layer called the mesoglea, and (3) the calicodermis, which is responsible for secreting the calcium carbonate skeleton that forms individual corallites, which make up the structure of the coral colony and, on a larger scale, coral reefs. Within the cells of the gastrodermis are unicellular dinoflagellates called zooxanthellae (an endosymbiont) (Work and Meteyer 2014) (Figure 1.1). Endosymbionts contain chloroplasts and photosynthesize. It is this coral host–endosymbiont symbiosis that allows corals to thrive in clear, nutrient-poor waters in tropical latitudes. The coral host provides habitat and protection to the endosymbionts, and endosymbionts provide critical nutrients to the coral host by their photosynthetic activities. Endosymbionts also serve as a photoprotective agent to protect coral host cells from damaging UV rays prevalent in tropical climates, and the endosymbionts are largely responsible for the vibrant colors seen in coral reefs. Finally, endosymbionts play an important role in modulating calcification in corals and, thus, accretion of the skeleton that comprises the 3-dimensional structure of the coral colony (Davy and Allemand 2012). This 3-dimensional framework is critical to coral reefs in that it provides a bevy of niches and habitats that enable the rich marine biodiversity for which coral reefs are renowned (Graham and Nash 2013).

Although coral reefs are marine phenomena, these habitats are closely linked to processes that occur on land. For example, mangroves and coastal wetlands are often associated with coral reef ecosystems and play an important role in coral reef health. Mangroves can provide nursery habitat for many coral reef associated organisms and, because of their shading ability, are thought to play an important role in providing refugia for corals from the effects of climate change (Yates et al. 2014). Wetlands and seagrass beds serve an important role in filtering of nutrients before they reach coral reefs and, thus, are important in maintaining water clarity needed for endosymbionts in corals to photosynthesize and grow (Figure 1.2). In the same vein, coral reefs provide structural integrity and wave attenuation that protect seagrass and coastal mangrove ecosystems from storm-surge damage (van de Koppel et al. 2015). Because of this close association, events that affect coastal processes can have direct effects on coral reefs as outlined later in this chapter.

Coral reefs provide substantial ecosystem services to humans. Numerous coastal communities globally depend on healthy coral reefs for protection, food security, and income (Moberg and Folke 1999). Coral reefs provide protection to coastal inhabitants in the form of barriers that can attenuate effects of hurricanes or typhoons (Fernando et al. 2008). Coral reefs also serve as important sources of tourism income for many coastal communities. For instance, in Hawaii alone in 2004, tourism associated with coral reef resources was worth in excess of $10 billion (USD) (Cesar and van Beukering 2004); globally, the value was estimated to be $36 billion (Spalding et al. 2017). Numerous human communities also depend on coral reefs for sources of protein and food security. For example, the estimated value of coral reef associated fisheries for six countries in the coral triangle (Indonesia, Malaysia, Papua New Guinea, Philippines, Solomon Islands, and Timor Leste) is about $1.2 billion (USD) (Cruz-Trinidad et al. 2014).

Coral reefs are normally resilient ecosystems having survived millions of years and undergone multiple extinction events (Stanley and Fautin 2001). Although corals have disappeared and reappeared, these changes have occurred over millennia, and it is the current rapid pace of coral disappearance that is of concern (Stanley and van de Schootbrugge

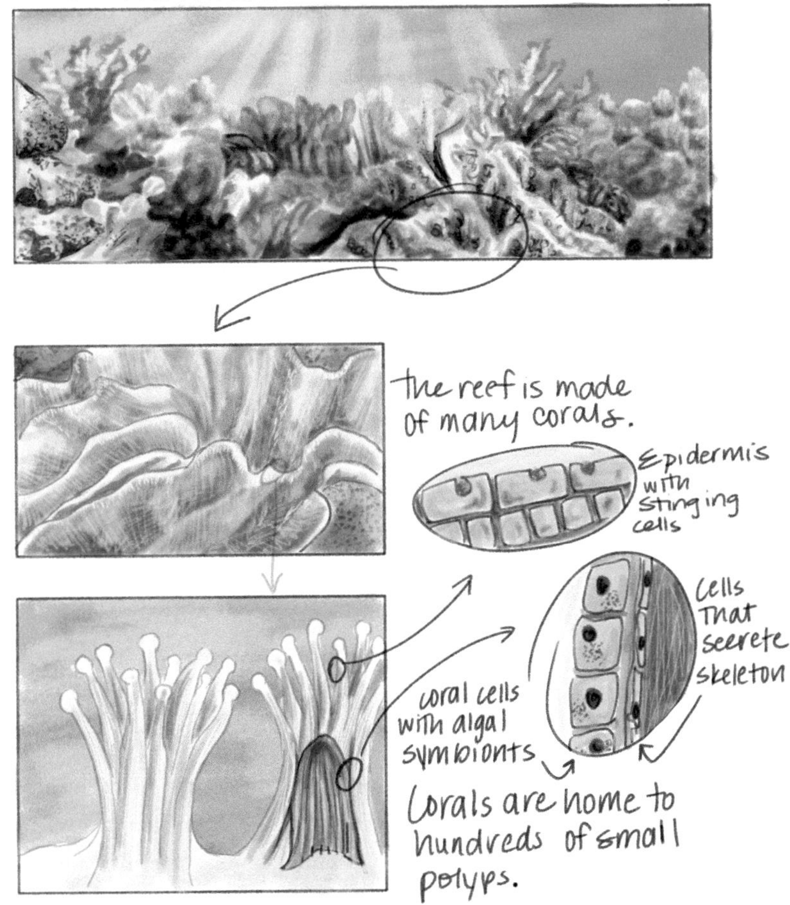

Fig. 1.1. Coral reefs influence processes from the macroscopic to microscopic scale. Top panel shows a barrier reef parallel to and protecting a coastline from storm surges. Second panel shows three-dimensional complexity of coral reefs providing habitat to various organisms. Bottom panels illustrate anatomy of corals. Coral polyps have three tissue layers, including epidermis with cnidae separated from gastrodermis by a connective tissue mesoglea. The gastrodermis contains intracellular photosynthesizing endosymbionts, and gastrodermal cells line the gastrovascular cavity (digestive tract). The calicodermis is the cell layer responsible for formation of the coral skeleton. Adapted from Work and Meteyer 2014. Illustration by Laura Donohue.

2018). The long lag time for coral reef regeneration lends a sense of urgency to understanding threats to coral reefs and devising ways to make these ecosystems more resistant to perturbation. One issue complicating efforts to restore coral reefs is a lack of a clear definition of what exactly constitutes a "healthy coral reef." Various attempts have been made to define "ecosystem health" mainly based on abilities of ecosystems to maintain structure, resilience, and vigor (Rapport 1989, Costanza and Mageau 1999). Those definitions have been criticized as too imprecise and not amenable to definitive and

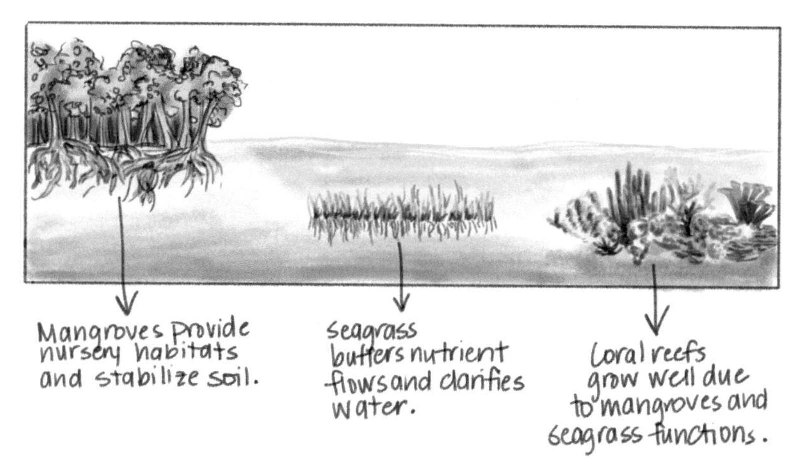

Fig. 1.2. Mangroves and wetlands are important to coral reef health. Coral reef ecosystems are intimately associated with processes that occur on land with mangroves providing critical nursery habitat and soil stabilization, and seagrasses buffering flows of nutrients both of which help ensure clarity of waters necessary for coral growth. Modified from Ogden 1988. Illustration by Laura Donohue.

measurable end points that would allow for policy-making decisions (Suter 1993, Downs et al. 2005). One metric of overall coral reef health that has been proposed is whether a coral reef is accreting or losing calcium carbonate over decades (Perry et al. 2008). This concept is appealing and useful locally, because coral reefs fundamentally are defined by their skeletons and geomorphic structures. However, measuring calcium deposition on a broad enough scale to incorporate both reef accretion and erosion by biological and physical means is presently impractical for widespread use (Dee et al. 2020).

Coral reefs face four major threats to their long-term existence on Earth—climate change, land-based pollution, overfishing, and disease—all of which have the potential to lead to rapid loss of corals over landscape scales. All these threats are directly due or related to human activities or inputs. As such, by modifying how we treat tropical coastal resources, humans have the power to either extirpate or preserve coral reefs for future generations, and action is needed sooner rather than later.

Climate Change

Increase in atmospheric CO_2 is leading to warming of oceans worldwide, and increased CO_2 saturation in seawater is reducing the pH of seawater and its buffering capacity (Kroeker et al. 2010). This poses two problems for corals. The first is elevated temperature. The polyp/endosymbionts mutually beneficial relationship depends on certain temperature-mediated homeostatic conditions that, when exceeded, lead to expulsion or death of endosymbionts from coral tissue and loss of photosynthetic pigment. This results in transparent tissues overlying white skeletons, a phenomenon termed bleaching (Coles and Brown 2003). Bleaching can have various degrees of severity. In mild cases, animals merely become pale or white but are able to recolonize their tissues with endosymbionts within weeks to months depending on environmental conditions and species affected (Rodrigues and Grottoli 2007). When recolonization does not occur, colonies die outright or often subsequently succumb to tissue loss diseases (Miller et al. 2009). There is now evidence that with increases in global warming, episodes of widespread bleaching have been on the rise leading to substantial loss of corals. The most recent example is the destruction of large swathes of the Great Barrier Reef in Australia by repeated bleaching events leading to loss of coral recruits and an uncertain outlook for the ability of the reefs to recover in the longer term (Dietzel et al. 2020). Moreover, the intervals between bleaching events are becoming shorter over time giving reefs less opportunity to recover (Hughes et al. 2018).

A second problem is ocean acidification. Coral calcification is an active metabolic process where calicodermis cells concentrate carbonate from seawater in extracellular compartments and deposit it in the form of aragonite, the chemical form of calcium carbonate that comprises coral skeletons (Tambutté et al. 2011). Calcification is an energy-intensive process, which is aided by endosymbiont energy production, and decreased pH of seawater encourages dissolution of the skeleton requiring increased energy on the part of the coral to compensate (Tambutté et al. 2011). Experimental studies examining oceanic conditions prior to industrialization indicate that oceans are already more acidic and that corals are currently operating under calcification regimes that are 25% lower than preindustrial conditions. With projected climate change models, coral reef calcification is likely to decline by 130% making them functionally extinct (Anthony 2016). Conceivably, corals could compensate for increased acidity by metabolically increasing calcification. There is also experimental evidence that coral polyps can survive acidic conditions by no longer accreting skeleton but then redepositing skeleton when environmental conditions are suitable (Fine and Tchernov 2007). However, in the wild in presence of predators and other limiting factors, such a strategy would likely be unsuccessful for long-term survival of corals, and absence of coral reef skeletal structure would not provide the diversity of habitats that make a functional reef. In aggregate, increased sea-surface temperatures are projected to lead to increased coral mortality, decreased growth, and decreased reproduction with concurrent shift from coral- to algae-dominated ecosystems (Hoegh-Guldberg 1999). Ocean acidification will also likely affect other calcifying organisms on coral reefs, such as mollusks, crustacea, and crustose or branching coralline algae while promoting growth of non-calcifying organisms that photosynthesize such as seagrasses (Kroeker et al. 2010). Finally, increased numbers and intensities of storms and hurricanes in tropical regions are predicted with climate change (Knutson et al. 2010). Hurricanes often lead to destruction of coral reefs, but given adequate time, they can recover. Thus, increased frequencies of storms will leave shorter time for recovery and could adversely affect longer term survival of coral reef systems. Although various nature-based (Seddon et al. 2020) and technological solutions (Creutzig et al. 2016) to address climate change have been proposed, their ability to help mitigate coral reef damage are unproven, and no plans for implementation currently exist.

Land-based Pollution and Degradation

Human populations are projected to increase, and a majority of this population growth is expected to occur in coastal zones mostly concentrated in the tropics (Neumann et al. 2015), where coral reefs are most abundant. Projected sea level rises and increased frequencies of hurricanes and coastal storms will likely exceed the economic capacity of coastal communities to recover from such events in the long term (Dinan 2017). Increases in human populations on coastlines will likely increase pressures on coral reef ecosystems that can be broadly compartmentalized into two categories: (1) habitat loss and (2) eutrophication. As coastal communities face increased storms and rising sea levels, they will have the choice of either moving to higher ground or fortifying coast lines. Seawalls and jetties may immediately delay effects of sea level rise; however, their constructions may lead to loss of coastal mangroves and wetlands (Galbraith et al. 2002) that, as stated previously, play an important role in filtering land-based pollutants and serving as nurseries for various organisms associated with coral reefs. Recent research shows that seagrasses also might help mitigate effects of ocean acidification by raising pH of seawater (Unsworth et al. 2012). Since 2009, there has been a global decrease in coastal wetlands of 33%, and future losses are projected to be concentrated in areas where coral reefs are located (Hu et al. 2017). Wetlands can be considered the "kidneys" of the coasts, filtering land-based detritus and pollutants before they reach coral reefs (van de Koppel et al. 2015), so their depletion would lead to increased deposits of land-based pollutants and nutrients on nearby coral reefs.

Land-based pollution affects corals in various ways and has been reviewed by Fabricius (2005). Sedimentation can be detrimental to corals because it blocks light thereby inhibiting photosynthesis of endosymbiotic algae. Sediments also smother and promote mucus production in corals as they attempt to slough particles deposited on their surfaces, and this process can be highly energy intensive. Sediments are also often laced with heavy metals and other contaminants that can be directly harmful to corals by killing tissues or adversely affecting reproduction (Rogers 1990). Land-based pollution also leads to increased nutrient concentrations in seawater that promote growth of biota harmful to corals. For instance, increased nutrients promote the growth of burrowing sponges that overgrow corals and erode coral skeletons (Schönberg et al. 2017), promote the growth of rapidly growing macroalgae or ascidians that can smother corals (Shenkar et al. 2008), and promote overgrowth of cyanobacteria (Charpy et al. 2012), which can cause diseases such as black band disease in corals (Richardson 1997).

Coral reefs are occasionally decimated by the predatory starfish, the crown-of-thorns (*Acanthaster planci*), that periodically multiply to very large numbers and consume corals. There is now accumulating evidence that population explosions of this starfish and the consequent damage to corals are associated with nutrient influxes from rivers that promote growth of phytoplankton and survival of starfish larvae that feed on those plankton (Birkeland 1982, Fabricius 2005, Fabricius et al. 2008). Zhao et al. (2021) noted the possibility that the net effects of nutrient enrichment in coastal waters and the adverse effects of nutrients on coral calcification could lead to a state of coral reefs that would transition from net calcium accretion to net dissolution. To decrease nutrient inputs into coastal ecosystems, Zhao et al. (2021) propose solutions including more efficient use of fertilizers in agriculture to reduce runoff, improved sewage management, restoration of riparian wetlands, making coastal aquaculture more sustainable by reducing nutrient outflows or moving such operations inland, and better overall management of land-sea nutrient flow in areas of high human habitation.

Overfishing

Coral reef ecosystems maintain their integrity and resilience in part by having a suitable complement of functional groups of organisms of which fishes play an important role. For instance, coral reefs removed from human influence are characterized by an abundance of top-level predators such as sharks and barracudas (Friedlander and DeMartini 2002). The presence of these large fish indicates that sufficient resources exist to support the biomass. An example of such a contrast is the main Hawaiian Islands where few top-level predators remain, and the protected northwestern Hawaiian Islands that are replete with sharks and jacks (Friedlander and DeMartini 2002) (Figure 1.3). Grazing fish such as parrotfish and surgeonfish play important roles in controlling macroalgae which due to their comparatively faster growth rates tend to overgrow corals (Jackson et al. 2001). Overfishing of parrotfish and angelfish has also led to overgrowth of reefs by sponges (Loh et al. 2015).

Pauly (1988) defined three types of overfishing: (1) growth overfishing or harvesting fishes before they have time to grow, (2) recruitment overfishing or harvesting fishes faster than they can reproduce, and (3) Malthusian overfishing or fishing that is a matter of survival leading to few incentives to manage a resource sustainably. It is the third type of overfishing that predominates in tropical oceans and, thus, poses such a threat to coral reefs. A reason why top-level predators are the first fishes to disappear in overfished areas is the human tendency to fish down the food chain where larger carnivores are targeted first followed by smaller lower trophic level fishes (grazers) and then finally remaining invertebrates

Fig. 1.3. Contrast between healthy and unhealthy coral reef. *Left*, healthy coral reef with abundant live coral and top-level predators such as large jacks and sharks, on the Johnston Atoll, Central Pacific. *Right,* degraded coral reef with no fish, few live corals, and algae-overgrown coral skeletons in St. Kitts, Caribbean. Photographs by T. M. Work.

(Pauly et al. 1998). The effects of overfishing on coral reefs make themselves known years after depletion of fish stocks and were summarized by Jackson (2001) including loss of top-level predators, loss of ecosystem engineers (e.g., corals), and increased microbial diseases secondary to pollution. Additional evidence of food web cascade effects of overfishing include increases of bioeroding sea urchins after overfishing of triggerfish in Africa. The resulting imbalance led to a transition from coral reef domination to seagrass beds (Valentine and Heck 2005).

Experience has shown that, if fisheries are properly managed, fish stocks can rebound rapidly. New ways of incentivizing fishermen to exploit fishes sustainably are emerging and showing success. These include rules-based fisheries where individual quotas of catches are assigned to particular groups or spatial quotas in the form of territorial use rights fisheries where local communities are enabled in enforcing sustainable traditional fisheries. These measures coupled with the establishment of no-take marine reserves have shown promise in aiding recovery of certain fisheries (Barner et al. 2015). No-take marine reserves, in particular, have been suggested as a potential solution not only to the depletion of fisheries, but also to the recovery of coral reef ecosystems (Halpern 2003). However, marine reserves would be most effective if established with clear goals and with concordance and support of local peoples (Hilborn et al. 2004). Finally, modulating or reducing demand for fish protein from developed economies that drives international offshore fisheries would likely reduce incentives for unsustainable fishing (Finkbeiner et al. 2017).

Disease

Disease in corals is one of the most pressing and salient threats to coral reef systems (Peters 2015), particularly in the Caribbean where this phenomenon has been most comprehensively documented and serves as a case study. In the late 1970s, coral reefs in the Caribbean were dominated by *Acropora* sp., highly branching corals that provided critical 3-dimensional structure to reefs. In the late 1970s and the 1980s, *Acropora* began dying in large areas of the Caribbean from tissue loss of unknown cause, termed white band disease (Gladfelter 1982). These die-offs were followed in the early 1980s by unexplained mortalities of the sea urchin *Diadema antillarum* that originated in Panama and within a year spread throughout the Caribbean, wiping out >90% of populations (Lessios 2016). Because urchins play a vital role in controlling macroalgae, die-offs of urchins led to overgrowth of algae on reefs that were already ailing; the end result was that, by the late 1980s, *Acropora* corals in some regions had declined between 40% and 100%, mainly due to white band disease followed by algal overgrowth. The remaining live corals consisted of mounding and encrusting corals with an altered and diminished 3-dimensional structure and associated biota (Aronson and Precht 2001, Peters 2015). In the Caribbean, there is currently a trend of flattening of coral reefs with attendant loss of biodiversity (Newman et al. 2015).

Various environmental cofactors have been associated with coral disease. Thermal anomalies leading to bleaching events have precipitated subsequent coral disease outbreaks as shown by longitudinal coral monitoring efforts in the US Virgin Islands (USVI), A widespread bleaching event in 2005 affected multiple coral species leading to the loss of almost half of coral cover in the region (Muller et al. 2008, Rogers et al. 2009). This was followed by tissue loss disease, termed white plague, affecting bleached corals more severely than unbleached (Muller et al. 2008, Rogers et al. 2009) (Figure 1.4).

Tissue-loss diseases continue to limit recovery of corals at certain sites in the USVI (Rogers and Muller 2012). Pollution can also exacerbate coral disease. For instance, associations between sewage-derived nitrogen and unspecified disease in *Porites* spp. have been documented in Guam (Redding et al. 2013), and there is a positive relationship between proximity to human habitations and presence of growth anomalies in corals (Aeby et al. 2011).

Historically, during disease outbreaks in corals and urchins, few efforts were made to determine why they were dying, in part due to absence of a systematic approach to understand causes of death. Although field surveys to document declines of corals coupled with deep sequencing of bacterial communities are frequently done, there has been relatively little effort to examine tissues at the microscopic level to document cellular damage (Work and Meteyer 2014), an approach that is commonly used to investigate disease in other animals (Work et al. 2008). Confusing the picture, some investigators have attempted to infer causation based on appearance of gross lesions without demonstrating suspect agents

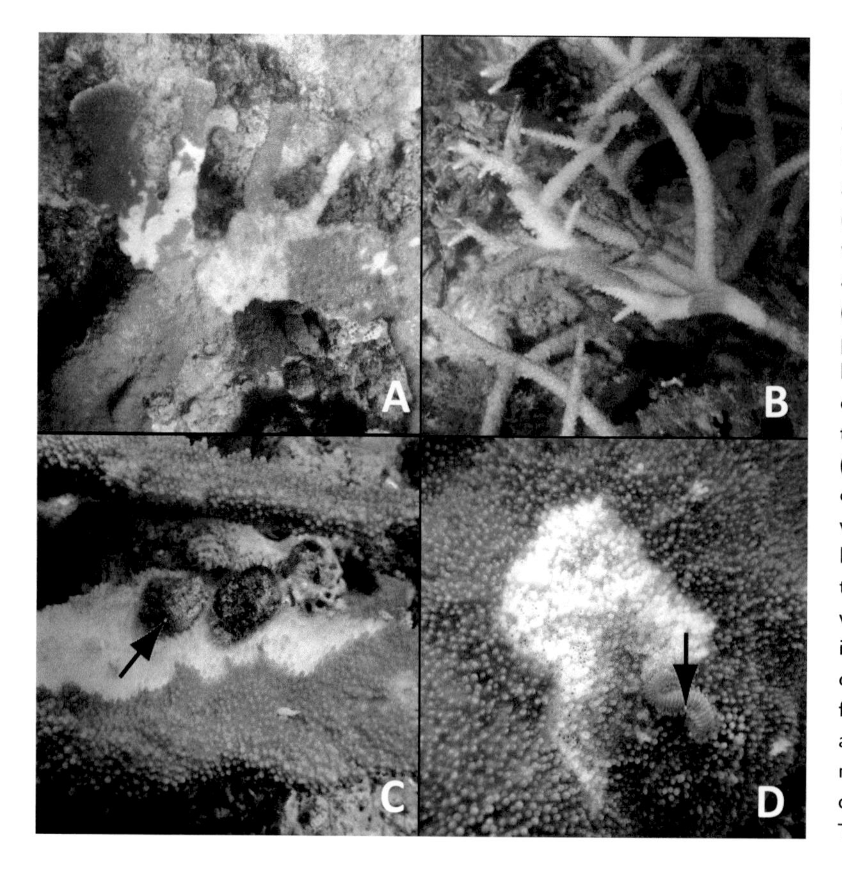

Fig. 1.4. Lesions encountered during the white plague outbreak in US Virgin Islands in 2004 that affected mainly *Acropora* spp. (A) *Acropora palmata* with white plague manifesting as acute to subacute tissue loss. (B) *A. cervicornis* with white plague also manifesting as tissue loss. (C) Snail (arrow) predation on *A. palmata*; note white area of tissue loss with scalloped edges. (D) Fireworm (arrow) predation on *A. palmata*. In the absence of a visible cause, differentiating tissue loss that can be explained as due to fireworm or other predation versus unexplained (white plague) in the field with certainty can be difficult. This highlights the need for additional exams at the tissue and cellular level to develop more robust case definitions of coral disease. Photographs by T. M. Work.

were causing cellular pathology (Work and Aeby 2006, Rogers 2010). In the case of the sea urchin die-off, urchin tissues were not available to examine, so an opportunity was lost to verify causation of an outbreak. This illustrates the need for a systematic veterinary approach and for the veterinary and nonveterinary scientific communities to cooperate to understand and deal with health and disease in coral reef organisms (Work et al. 2008). Subsequent studies have shown that *Diadema* are more likely to succumb to bacterial infections than other urchin species, indicating their lack of resistance may have played a role in their decline (Beck et al. 2014).

Like all organisms, coral have a limited host response repertoire and respond grossly to insults in three basic ways: (1) discoloration, where tissues are altered from normal baseline colors (e.g., bleaching); (2) growth anomalies, where skeleton or polyps adopt morphologies such as gigantism or skeletal aberrations; and (3) tissue loss, which can be graded as rapid (acute) leaving bare white skeleton or gradual (subacute) manifested by a border of white bare skeleton progressing to overgrowth of bare skeleton by algae (Figure 1.5) (Work and Aeby 2006). In some cases such as bleaching or predation by fish, gross lesions can have an explanation; however in most cases, the cause of a lesion cannot be determined based on gross morphology alone without additional laboratory diagnostics guided by histology (Work and Meteyer 2014).

Microscopically, coral tissues are analogous to mucus membranes; unlike vertebrates and some invertebrates like molluscs, corals do not manifest a prominent inflammatory response, which results in few clues to aid the diagnostician. Like other animals, because corals have a limited gross-response repertoire, a gross lesion can have multiple different microscopic manifestations. For instance, tissue loss in *Montipora* corals in Hawaii can be associated with invasion by ciliates, algae, or even corals that parasitize other corals (Work et al. 2012).

As of this writing, remnant coral reefs in the Caribbean are undergoing another massive wave of mortality that originated near Miami Harbor in 2014 (Precht et al. 2016) and has now spread throughout the Florida reef tract (Muller et al. 2020), the USVI (Brandt et al. 2021), and the Yucatan peninsula in Mexico (Alvarez-Filip et al. 2019). This disease, called stony coral tissue loss disease (SCTLD), affects over 20 species of corals (Aeby et al. 2019), leads to rapid tissue loss, and has led to the functional extirpation of at least one coral species, *Dendrogyra cylindrus* in Florida (Neely et al. 2021). Unlike the *Acropora* die-offs in the wider Caribbean in 1970s (Aronson and Precht 2001) and the 2005 bleaching and disease event in the USVI (Rogers et al. 2009), substantial efforts are being devoted to the examination of coral tissues microscopically, which has led to the hypothesis that SCTLD appears to be a dysfunction between the host coral and its symbiotic algae (Landsberg et al. 2020). To date, no infectious agents visible on light microscopy have been associated with SCTLD (Landsberg et al. 2020), although bacterial microbiomes are altered in affected corals (Meyer et al. 2019), which may be affecting their immune responses and thereby increasing susceptibility (Meiling et al. 2021). Based on light-microscopic findings, viruses or toxins may be involved in the pathogenesis of SCTLD. Recent studies using electron microscopy that allow for more detailed examination of structures at the subcellular level show that a virus targeting the endosymbionts leading to endosymbiont death followed by coral host cell death might be the root cause of SCTLD (Work et al. 2021). The findings of Landsberg et al. (2020) and Work et al. (2021) highlight the utility and importance of systematically investigating coral disease based on careful examination of animal tissues at the cellular level to illuminate fundamental causes of coral disease. Pursuant to these findings, the State of Florida is realigning research priorities to focus on the potential role of viruses as a cause of SCTLD.

Future Directions

Coral reef ecosystems are complex. They provide an excellent example of investigating and applying

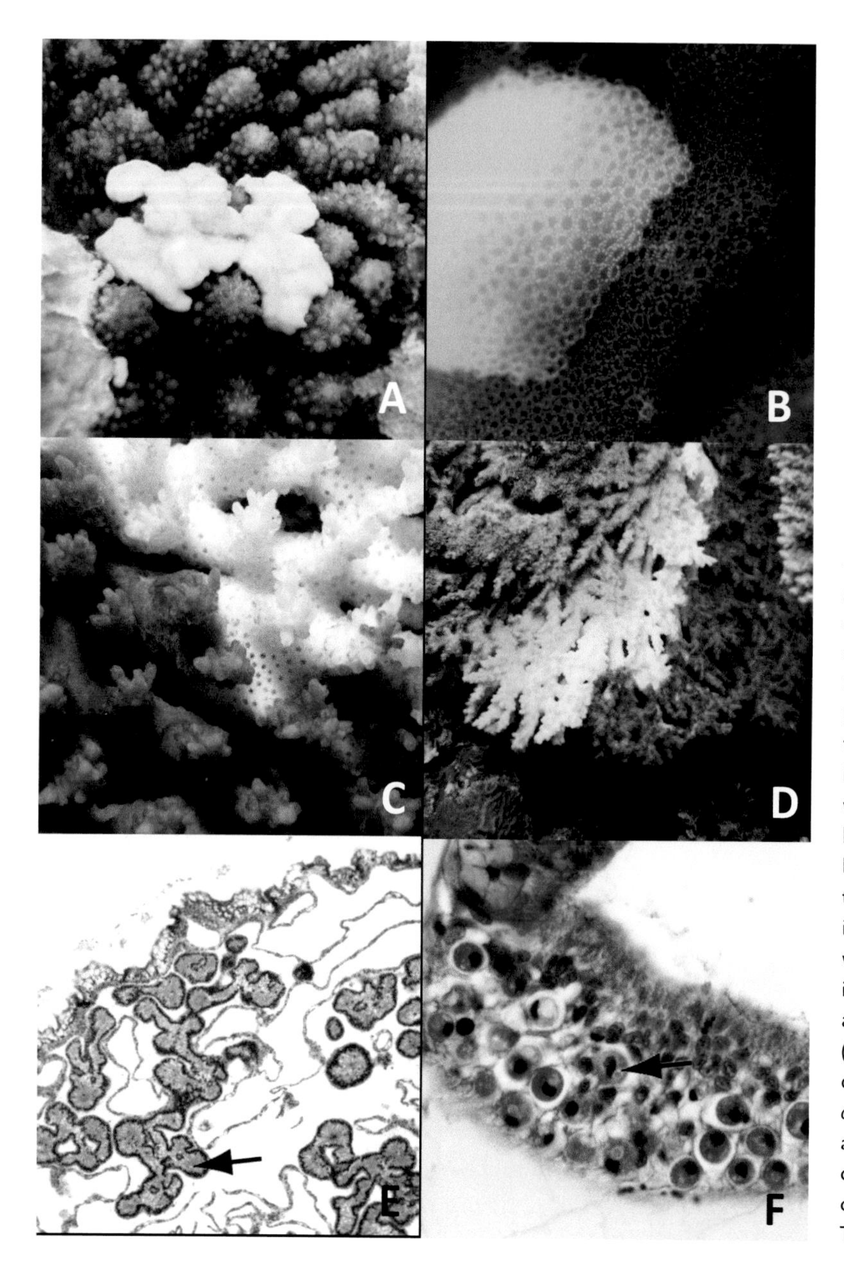

Fig. 1.5. Examples of principal gross lesions in corals and selected histology. (A) Growth anomaly in *Acropora digitifera*; note smooth skeleton bereft of polyp structure covered by white tissue. (B) discoloration (bleaching) in massive *Porites* sp.; note white polyps. (C) Acute tissue loss in *Acropora* sp.; note white bare skeleton with absence of tissues. (D) Subacute tissue loss in *Acropora* sp.; note band of white bare skeleton separating intact tissue (right) from algae-covered skeleton (left). (E) Photomicrograph of parasitic coral (arrow) invading *Montipora capitata* which grossly manifested acute tissue loss. (F) Endosymbionts (arrow) within gastrodermis of *Favites* sp. Photographs by T. M. Work.

health research at the ecosystem, as opposed to individual animal or even animal population level. This complexity would benefit from collaborative approaches, because problems facing coral reefs are also complex. Addressing climate change is a global endeavor. Addressing overfishing and land-based pollution are more regional and would benefit from expertise of not only fisheries scientists, but also sociologists, development experts, and local community leaders. Addressing coral mortality events would benefit from more engagement from the veterinary community and a partnership with biologists and ecologists. Managers are also striving to restore formerly degraded coral reefs by outplanting corals from nurseries (Ladd et al. 2018), but such efforts may be hampered by emerging diseases such as

Table 1.1. Threats to coral reefs and examples of near- and long-term actions that could aid in their survival

Threats	Near-term action	Long-term action
Climate change	Protect marine biodiversity hotspots (Queirós et al. 2016), coral reef restoration (Hein et al. 2021)	Carbon taxes (Andrew et al. 2010), carbon capture and sequestration (Kazemifar 2021), nature-based solutions (e.g., aforrestation; Seddon et al. 2020)
Pollution	Modernizing wastewater infrastructure (Larsen 2011)	Reducing nitrogen inputs, coastal wetland and mangrove restoration (Kroon et al. 2014)
Overfishing	Marine protected areas, more local control over fisheries resource (Barner et al. 2015)	Address (reduce) demand for fish in developed world, better governance of fisheries, address economic inequities promoting overfishing, harness technology to reduce bycatch (Finkbeiner et al. 2017)
Disease	Using systematic biomedical approaches based on cell pathology to develop more concrete case definitions of coral disease (Work et al. 2008, Work and Meteyer 2014)	Development of animal models for coral disease (Neff 2020) and additional tools to investigate disease (Pollock et al. 2011)

SCTLD. Coral reefs have survived for millennia and are likely to continue to do so if solutions to help them thrive can be found. Given the financial, social, and political implications of losing coral reefs on a large scale, it is useful to contemplate actions that could be taken in the near and longer term to reduce human impacts on these ecosystems (Table 1.1).

Summary

About half of the Earth's population resides in the tropics (Kummu and Varis 2011), a region that also harbors most of its biodiversity including coral reefs. This region also contains the highest density of human coastal populations (Neumann et al. 2015). Given the outsize role that coral reefs play in the economic health and food security of adjoining human coastal communities, the lives and health of these human communities will be highly dependent on whether coral reefs can continue to provide the necessary ecosystem services. This in turn will depend on whether or not humans decide to conserve these ecosystems. Perhaps there is no more profound example of One Health's admonition that human, animal, and ecosystem health are inexorably linked.

Disease is a significant threat to the long-term survival of coral reefs. By contributing to knowledge of what causes disease in coral reefs, wildlife health professionals can play an important role in long-term recovery of these ecosystems. This can be done by applying sound diagnostic medicine in partnership with other disciplines (coral ecology, biology, and social sciences) to better understand and manage what are clearly complex issues. Unfortunately, given the rate of decline of some coral reef ecosystems, time is of the essence. If we are to have any hope of reversing the situation, we must act sooner than later and using an all-hands-on-deck approach to find creative solutions to what currently seem to be an intractable problem.

ACKNOWLEDGMENTS

Esther Peters and Caroline Rogers provided constructive comments on earlier versions of this chapter. Any use of trade, firm, or product names is for descriptive purposes only and does not imply endorsement by the US Government.

LITERATURE CITED

Aeby, G. S., B. Ushijima, J. E. Campbell, S. Jones, G. J. Williams, J. L. Meyer, C. Häse, and V. J. Paul. 2019. Pathogenesis of a tissue loss disease affecting multiple species of corals along the Florida reef tract. Frontiers in Marine Science 6:668.

Aeby, G. S., G. J. Williams, E. C. Franklin, J. Haapkyla, C. D. Harvell, S. Neale, C. A. Page, L. Raymundo, B. Vargas-Ángel, B. L. Willis, et al. 2011. Growth anomalies on the coral genera *Acropora* and *Porites* are strongly associated with host density and human population size across the Indo-Pacific. PLOS One 6:e16887.

Alvarez-Filip, L., N. Estrada-Saldívar, E. Pérez-Cervantes, A. Molina-Hernández, and F. J. González-Barrios. 2019. A rapid spread of the stony coral tissue loss disease outbreak in the Mexican Caribbean. PeerJ 7:e8069.

Andrew, J., M. A. Kaidonis, and B. Andrew. 2010. Carbon tax: challenging neoliberal solutions to climate change. Critical Perspectives on Accounting 21:611–618.

Anthony, K. R. N. 2016. Coral reefs under climate change and ocean acidification: challenges and opportunities for management and policy. Annual Review of Environment and Resources 41:59–81.

Aronson, R. B., and W. F. Precht. 2001. White-band disease and the changing face of Caribbean coral reefs. Hydrobiologia 460:25–38.

Barner, A. K., J. Lubchenco, C. Costello, S. D. Gaines, A. Leland, B. Jenks, S. Murawski, E. Schwaab, and M. Spring. 2015. Solutions for recovering and sustaining the bounty of the ocean combining fishery reforms, rights-based fisheries management, and marine reserves. Oceanography 28:252–263.

Beck, G., R. Miller, and J. Ebersole. 2014. Mass mortality and slow recovery of *Diadema antillarum*: could compromised immunity be a factor? Marine Biology 161:1001–1013.

Birkeland, C. 1982. Terrestrial runoff as a cause of outbreaks of *Acanthaster planci* (Echinodermata: Asteroidea). Marine Biology 69:175–185.

Brandt, M. E., R. S. Ennis, S. S. Meiling, J. Townsend, K. Cobleigh, A. Glahn, J. Quetel, V. Brandtneris, L. M. Henderson, and T. B. Smith. 2021. The emergence and initial impact of stony coral tissue loss disease (SCTLD) in the United States Virgin Islands. Frontiers in Marine Science 8. doi: 10.3389/fmars.2021.715329.

Cesar, H. S. J., and P. J. H. van Beukering. 2004. Economic valuation of the coral reefs of Hawai'i. Pacific Science 58:231–242.

Charpy, L., B. E. Casareto, M. J. Langlade, and Y. Suzuki. 2012. Cyanobacteria in coral reef ecosystems: a review. Journal of Marine Biology 2012: 259571.

Coles, S. L., and B. E. Brown. 2003. Coral bleaching—capacity for acclimatization and adaptation. Advances in Marine Biology 46:183–223.

Costanza, R., and M. Mageau. 1999. What is a healthy ecosystem? Aquatic Ecology 33:105–115.

Creutzig, F., B. Fernandez, H. Haberl, R. Khosla, Y. Mulugetta, and K. C. Seto. 2016. Beyond technology: demand-side solutions for climate change mitigation. Annual Review of Environment and Resources 41:173–198.

Cruz-Trinidad, A., P. M. Aliño, R. C. Geronimo, and R. B. Cabral. 2014. Linking food security with coral reefs and fisheries in the coral triangle. Coastal Management 42:160–182.

Davy, S. K., and D. Allemand. 2012. Cell biology of cnidarian-dinoflagellate symbiosis. Microbiology and Molecular Biology Reviews 76:229–252.

Dee, S., M. Cuttler, M. O'Leary, J. Hacker, and N. Browne. 2020. The complexity of calculating an accurate carbonate budget. Coral Reefs 39:1525–1534.

Dietzel, A., M. Bode, S. R. Connolly, and T. P. Hughes. 2020. Long-term shifts in the colony size structure of coral populations along the Great Barrier Reef. Proceedings of the Royal Society B: Biological Sciences 287:20201432.

Dinan, T. 2017. Projected increases in hurricane damage in the United States: the role of climate change and coastal development. Ecological Economics 138:186–198.

Downs, C. A., C. M. Woodley, R. Richmond, L. Lanning, and R. Owen. 2005. Shifting the paradigm of coral reef 'health' assessment. Marine Pollution Bulletin 51:486–494.

Fabricius, K. E. 2005. Effects of terrestrial runoff on the ecology of corals and coral reefs: review and synthesis. Marine Pollution Bulletin 50:125–146.

Fabricius, K. E., K. Okaji, and G. De'ath. 2008. Three lines of evidence to link outbreaks of the crown-of-thorns seastar *Acanthaster planci* to the release of larval food limitation. Coral Reefs 29:593–605.

Fernando, H. J. S., S. P. Samarawickrama, S. Balasubramanian, S. S. L. Hettiarachchi, and S. Voropayev. 2008. Effects of porous barriers such as coral reefs on coastal wave propagation. Journal of Hydro-environment Research 1:187–194.

Fine, M., and D. Tchernov. 2007. Scleractinian coral species survive and recover from decalcification. Science 315:1811.

Finkbeiner, E. M., N. J. Bennett, T. H. Frawley, J. G. Mason, D. K. Briscoe, C. M. Brooks, C. A. Ng, R. Ourens, K. Seto, S. Switzer Swanson, et al. 2017. Reconstructing overfishing: moving beyond Malthus for effective and equitable solutions. Fish and Fisheries 18:1180–1191.

Fisher, R., R. A. O'Leary, S. Low-Choy, K. Mengersen, N. Knowlton, R. E. Brainard, and M. J. Caley. 2015. Species richness on coral reefs and the pursuit of convergent global estimates. Current Biology 25:500–505.

Friedlander, A. M., and A. E. DeMartini. 2002. Contrasts in density, size, and biomass of reef fishes between the northwestern and the main Hawaiian islands: the effects of fishing down apex predators. Marine Ecology Progress Series 230:253–264.

Galbraith, H., R. Jones, R. Park, J. Clough, S. Herrod-Julius, B. Harrington, and G. Page. 2002. Global climate change and sea level rise: potential losses of intertidal habitat for shorebirds. Waterbirds: The International Journal of Waterbird Biology 25:173–183.

Gladfelter, W. B. 1982. White-band disease in *Acropora palmata*: implications for the structure and growth of shallow reefs. Bulletin of Marine Science 32:639–643.

Graham, N. A. J., and K. L. Nash. 2013. The importance of structural complexity in coral reef ecosystems. Coral Reefs 32:315–326.

Halpern, B. S. 2003. The impact of marine reserves: do reserves work and does reserve size matter? Ecological Applications 13:117–137.

Hein, M. Y., T. Vardi, E. C. Shaver, S. Pioch, L. Boström-Einarsson, M. Ahmed, G. Grimsditch, and I. M. McLeod. 2021. Perspectives on the use of coral reef restoration as a strategy to support and improve reef ecosystem services. Frontiers in Marine Science 8:618303.

Hilborn, R., K. Stokes, J.-J. Maguire, T. Smith, L. W. Botsford, M. Mangel, J. Orensanz, A. Parma, J. Rice, J. Bell, et al. 2004. When can marine reserves improve fisheries management? Ocean & Coastal Management 47:197–205.

Hoegh-Guldberg, O. 1999. Climate change, coral bleaching and the future of the world's coral reefs. Marine and Freshwater Research 50:839–866.

Hu, S., Z. Niu, Y. Chen, L. Li, and H. Zhang. 2017. Global wetlands: potential distribution, wetland loss, and status. Science of the Total Environment 586:319–327.

Hughes, T. P., K. D. Anderson, S. R. Connolly, S. F. Heron, J. T. Kerry, J. M. Lough, A. H. Baird, J. K. Baum, M. L. Berumen, T. C. Bridge, et al. 2018. Spatial and temporal patterns of mass bleaching of corals in the Anthropocene. Science 359:80–83.

Hughes, T. P., D. R. Bellwood, and S. R. Connolly. 2002. Biodiversity hotspots, centres of endemicity, and the conservation of coral reefs. Ecology Letters 5:775–784.

Jackson, J. B. C. 2001. What was natural in the coastal oceans? Proceedings of the National Academy of Sciences USA 98:5411–5418.

Jackson, J. B. C., M. X. Kirby, W. H. Berger, K. A. Bjorndal, L. W. Botsford, B. J. Bourque, R. H. Bradbury, R. Cooke, J. Erlandson, J. A. Estes, et al. 2001. Historical overfishing and the recent collapse of coastal ecosystems. Science 293:629–637.

Kazemifar, F. 2021. A review of technologies for carbon capture, sequestration, and utilization: cost, capacity, and technology readiness. Greenhouse Gases: Science and Technology 12:200–230.

Knowlton, N., R. E. Brainard, R. Fisher, M. Moews, L. Plaisance, and M. J. Caley. 2010. Coral reef biodiversity. Pages 65–78 *in* A. D. McIntyre, editor. Life in the world's oceans: diversity, distribution, and abundance. John Wiley & Sons, Chichester, England.

Knutson, T. R., J. L. McBride, J. Chan, K. Emanuel, G. Holland, C. Landsea, I. Held, J. P. Kossin, A. K. Srivastava, and M. Sugi. 2010. Tropical cyclones and climate change. Nature Geoscience 3:157–163.

Kroeker, K. J., R. L. Kordas, R. N. Crim, and G. G. Singh. 2010. Meta-analysis reveals negative yet variable effects of ocean acidification on marine organisms. Ecology Letters 13:1419–1434.

Kroon, F. J., B. Schaffelke, and R. Bartley. 2014. Informing policy to protect coastal coral reefs: insight from a global review of reducing agricultural pollution to coastal ecosystems. Marine Pollution Bulletin 85:33–41.

Kummu, M., and O. Varis. 2011. The world by latitudes: a global analysis of human population, development level and environment across the north–south axis over the past half century. Applied Geography 31:495–507.

Ladd, M. C., M. W. Miller, J. H. Hunt, W. C. Sharp, and D. E. Burkepile. 2018. Harnessing ecological processes to facilitate coral restoration. Frontiers in Ecology and the Environment 16:239–247.

Landsberg, J. H., Y. Kiryu, E. C. Peters, P. W. Wilson, N. Perry, Y. Waters, K. E. Maxwell, L. K. Huebner, and T. M. Work. 2020. Stony coral tissue loss disease in Florida is associated with disruption of host–zooxanthellae physiology. Frontiers in Marine Science 7:576013.

Larsen, T. A. 2011. Redesigning wastewater infrastructure to improve resource efficiency. Water Science and Technology 63:2535–2541.

Lessios, H. A. 2016. The great *Diadema antillarum* die-off: 30 years later. Annual Review of Marine Science 8:267–283.

Loh, T. L., S. E. McMurray, T. P. Henkel, J. Vicente, and J. R. Pawlik. 2015. Indirect effects of overfishing on Caribbean reefs: sponges overgrow reef-building corals. PeerJ 3:e901.

Meiling, S. S., E. M. Muller, D. Lasseigne, A. Rossin, A. J. Veglia, N. MacKnight, B. Dimos, N. Huntley, A. M. S. Correa, T. B. Smith, et al. 2021. Variable species responses to experimental stony coral tissue loss disease (SCTLD) exposure. Frontiers in Marine Science 8:670829.

Meyer, J. L., J. Castellanos-Gell, G. S. Aeby, C. C. Häse, B. Ushijima, and V. J. Paul. 2019. Microbial community shifts associated with the ongoing stony coral tissue loss disease outbreak on the Florida reef tract. Frontiers in Microbiology 10:2244.

Miller, J., E. Muller, C. Rogers, R. Waara, A. Atkinson, K. R. T. Whelan, M. Patterson, and B. Witcher. 2009. Coral disease following massive bleaching in 2005 causes 60% decline in coral cover on reefs in the US Virgin Islands. Coral Reefs 28:925.

Moberg, F., and C. Folke. 1999. Ecological goods and services of coral reef ecosystems. Ecological Economics 29:215–233.

Muller, E. M., C. S. Rogers, A. S. Spitzack, and R. van Woesik. 2008. Bleaching increases likelihood of disease on *Acropora palmata* (Lamarck) in Hawksnest Bay, St John, US Virgin Islands. Coral Reefs 27:191–195.

Muller, E. M., C. Sartor, N. I. Alcaraz, and R. van Woesik. 2020. Spatial epidemiology of the stony-coral-tissue-loss disease in Florida. Frontiers in Marine Science 7:163.

Neely, K. L., C. L. Lewis, K. S. Lunz, and L. Kabay. 2021. Rapid population decline of the pillar coral *Dendrogyra cylindrus* along the Florida reef tract. Frontiers in Marine Science 8:656515.

Neff, E. P. 2020. The quest for an animal model of coral health and disease. Lab Animal 49:37–41.

Neumann, B., A. T. Vafeidis, J. Zimmermann, and R. J. Nicholls. 2015. Future coastal population growth and exposure to sea-level rise and coastal flooding—a global assessment. PLOS One 10:e0118571.

Newman, S. P., E. H. Meesters, C. S. Dryden, S. M. Williams, C. Sanchez, P. J. Mumby, and N. V. C. Polunin. 2015. Reef flattening effects on total richness and species responses in the Caribbean. Journal of Animal Ecology 84:1678–1689.

Ogden, J. C. 1988. The influence of adjacent systems on the structure and function of coral reefs. Pages 123–129. *in* J. H. Choat, D. Barnes, M. A. Borowitzka, J. C. Coll, P. J. Davies, P. Flood, B. G. Hatcher, D. Hopley, P. A. Hutchings, D. Kinsey, et al., editors. Proceedings of the 6th International Coral Reef Symposium. Volume 1. Plenary addressess and status review, 8–12 August 1988, Townsville, Australia.

Pauly, D. 1988. Some definitions of overfishing relevant to coastal zone management in Southeast Asia. Tropical Coastal Area Management 3:14–15.

Pauly, D., V. V. Christensen, J. Dalsgaard, R. Froese, and F. Torres, Jr. 1998. Fishing down marine food webs. Science 279:860–863.

Perry, C. T., T. Spencer, and P. S. Kench. 2008. Carbonate budgets and reef production states: a geomorphic perspective on the ecological phase-shift concept. Coral Reefs 27:853–866.

Peters, E. C. 2015. Diseases of coral reef organisms. Pages 147–178 *in* C. Birkeland, editor. Coral reefs in the Anthropocene. Springer Science+Business Media,, Dordrecht, The Netherlands.

Pollock, J. F., P. J. Morris, B. L. Willis, and D. G. Bourne. 2011. The urgent need for robust coral disease diagnostics. PLOS Pathogens 7:e1002183.

Precht, W. F., B. E. Gintert, M. L. Robbart, R. Fura, and R. van Woesik. 2016. Unprecedented disease-related coral mortality in southeastern Florida. Scientific Reports 6:31374.

Queirós, A. M., K. B. Huebert, F. Keyl, J. A. Fernandes, W. Stolte, M. Maar, S. Kay, M. C. Jones, K. G. Hamon, G. Hendriksen, et al. 2016. Solutions for ecosystem-level protection of ocean systems under climate change. Global Change Biology 22:3927–3936.

Rapport, D. J. 1989. What constitutes ecosystem health? Perspectives in Biology and Medicine 33:120–132.

Redding, J. E., R. L. Myers-Miller, D. M. Baker, M. Fogel, L. J. Raymundo, and K. Kim. 2013. Link between sewage-derived nitrogen pollution and coral disease severity in Guam. Marine Pollution Bulletin 73:57–63.

Richardson, L. L. 1997. Occurrence of the black band disease cyanobacterium on healthy corals of the Florida Keys. Bulletin of Marine Science 61:485–490.

Rodrigues, L. J., and A. G. Grottoli. 2007. Energy reserves and metabolism as indicators of coral recovery from bleaching. Limnology and Oceanography 52:1874–1882.

Rogers, C. S. 1990. Responses of coral reefs and reef organisms to sedimentation. Marine Ecology Progress Series 62:185–202.

Rogers, C. S. 2010. Words matter: recommendations for clarifying coral disease nomenclature and terminology. Diseases of Aquatic Organisms 91:167–175.

Rogers, C. S., and E. M. Muller. 2012. Bleaching, disease and recovery in the threatened scleractinian coral *Acropora palmata* in St. John, US Virgin Islands: 2003–2010. Coral Reefs 31:807–819.

Rogers, C. S., E. Muller, T. Spitzack, and J. Miller. 2009. Extensive coral mortality in the US Virgin Islands in 2005/2006: a review of the evidence for synergy among thermal stress, coral bleaching and disease. Caribbean Journal of Science 45:204–214.

Schönberg, C. H. L., J. K.-H. Fang, and J. L. Carballo. 2017. Bioeroding sponges and the future of coral reefs. Pages 179–372 *in* J. L. Carballo, and J. J. Bell, editors. Climate change, ocean acidification and sponges: impacts across multiple levels of organization. Springer International Publishing, Cham, Switzerland.

Seddon, N., A. Chausson, P. Berry, C. A. J. Girardin, A. Smith, and B. Turner. 2020. Understanding the value and limits of nature-based solutions to climate change and other global challenges. Philosophical Transactions of the Royal Society B Biological Sciences 375:20190120.

Shenkar, N., O. Bronstein, and Y. Loya. 2008. Population dynamics of a coral reef ascidian in a deteriorating environment. Marine Ecology Progress Series 367:163–171.

Spalding, M., L. Burke, S. A. Wood, J. Ashpole, J. Hutchison, and P. zu Ermgassen. 2017. Mapping the global value and distribution of coral reef tourism. Marine Policy 82:104–113.

Stanley, G., and B. van de Schootbrugge. 2018. The evolution of the coral–algal symbiosis and coral bleaching in the geologic past. Pages 9–26 *in* M. J. H. van Oppen, and J. M. Lough, editors. Coral bleaching: patterns, processes, causes and consequences. Springer International Publishing, Cham, Switzerland.

Stanley, G. D., and D. G. Fautin. 2001. The origins of modern corals. Science 291:1913–1914.

Suter, G. 1993. A critique of ecosystem health concepts and indexes. Environmental Toxicology and Chemistry 12:1533–1539.

Tambutté, S., M. Holcomb, C. Ferrier-Pagès, S. Reynaud, É. Tambutté, D. Zoccola, and D. Allemand. 2011. Coral biomineralization: from the gene to the environment. Journal of Experimental Marine Biology and Ecology 408:58–78.

Unsworth, R. F. K., C. J. Collier, G. M. Henderson, and L. J. McKenzie. 2012. Tropical seagrass meadows modify seawater carbon chemistry: implications for coral reefs impacted by ocean acidification. Environmental Research Letters 7:024026.

Valentine, J. F., and K. L. Heck. 2005. Perspective review of the impacts of overfishing on coral reef food web linkages. Coral Reefs 24:209–213.

van de Koppel, J., T. van der Heide, A. H. Altieri, B. K. Eriksson, T. J. Bouma, H. Olff, and B. R. Silliman. 2015. Long-distance interactions regulate the structure and resilience of coastal ecosystems. Annual Review of Marine Science 7:139–158.

Wild, C., O. Hoegh-Guldberg, M. S. Naumann, M. F. Florencia Colombo-Pallotta, M. Ateweberhan, W. K. Fit, R. Iglesias-Prieto, C. Palmer, J. C. Bythell, J. C. Ortiz, et al. 2011. Climate change impedes scleractinian corals as primary reef ecosystem engineers. Marine and Freshwater Research 62:205–215.

Work, T., and C. Meteyer. 2014. To understand coral disease, look at coral cells. EcoHealth 11:610–618.

Work, T. M., and G. S. Aeby. 2006. Systematically describing gross lesions in corals. Diseases of Aquatic Organisms 70:155–160.

Work, T. M., L. L. Richardson, T. R. Reynolds, and B. L. Willis. 2008. Biomedical and veterinary science can increase our understanding of coral disease. Journal of Experimental Marine Biology and Ecology 362:63–70.

Work, T. M., R. Russell, and G. S. Aeby. 2012. Tissue loss (white syndrome) in the coral *Montipora capitata* is a dynamic disease with multiple host responses and potential causes. Proceedings of the Royal Society of London. Series B: Biological Sciences 279:4334–4341.

Work, T. M., T. M. Weatherby, J. H. Landsberg, Y. Kiryu, S. M. Cook, and E. C. Peters. 2021. Viral-like particles are associated with endosymbiont pathology in Florida corals affected by stony coral tissue loss disease. Frontiers in Marine Science 8:750658.

Yates, K. K., C. S. Rogers, J. J. Herlan, G. R. Brooks, N. A. Smiley, and R. A. Larson. 2014. Diverse coral communities in mangrove habitats suggest a novel refuge from climate change. Biogeosciences 11:4321–4337.

Zhao, H., M. Yuan, M. Strokal, H. C. Wu, X. Liu, A. Murk, C. Kroeze, and R. Osinga. 2021. Impacts of nitrogen pollution on corals in the context of global climate change and potential strategies to conserve coral reefs. Science of the Total Environment 774:145017.

2 Sea Star Wasting Disease

Joseph K. Gaydos

Introduction

While the concept of wildlife conservation emerged from a history of harvest and population management, it also has gained considerable traction from humanity's recent, and often arms-length, fascination with wildlife that are seen as majestic, cuddly, or deemed intelligent. Sea stars, also called starfish (though they are invertebrates, not fish), are none of those. To quote Marie De Santis from her book *Starfish Detectives* (2018), starfish "don't roll over, suckle furry young, or blink with soulful eyes and [they are] an economic no-show to boot. Yet, the starfish is unquestionably a star! A seashore ambassador par excellence!" The public came unhinged with worry and demanded an explanation in 2013 when sea stars began losing limbs, shriveling up, and dying en masse, proving that effective conservation ambassadors don't have to be large, charismatic megafauna. This, and the need to understand what was happening, motivated scientists along the western coast of North America to collaborate on an underfunded mission to understand, and hopefully address, this unknown epizootic (De Santis 2018).

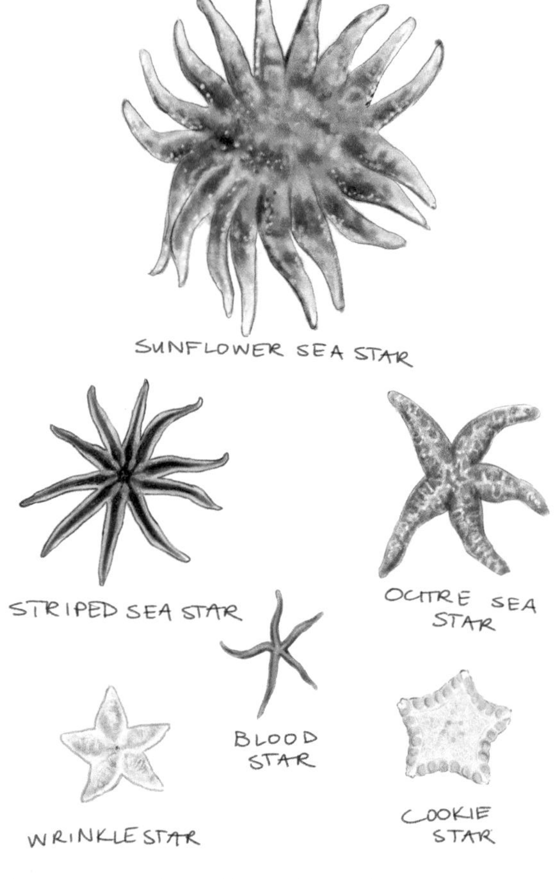

Six Sea stars of the Pacific Northwest. Illustration by Laura Donohue.

The northwestern coast of North America is a hotspot for sea star diversity and endemism. At least once during the lowest of the two daily low tides, numerous sea star species are revealed to people inquisitive enough to look for them over the huge swaths of exposed shoreline and small pools recently covered in much deeper water. Sea star species diversity within the intertidal zone pales in comparison, however, to what you can find in the subtidal. Dozens and dozens more species with even greater diversity in size, color, number of limbs, and dietary preferences can be found for those intrepid enough to don snorkel or SCUBA gear and explore the part of the ocean never exposed by tides. Many, like the classic and once nearly ubiquitous ochre star (*Pisaster ochraceous*), have 5 arms. Others can have 6, 10, 11, 12, 16, or even 24 arms. They can range in size from one that would be framed beautifully by your fingertip, to the largest, the sunflower star (*Pycnopodia helianthoides*), that can grow to be 1.2 m across and weigh 5 kg.

The sea star wasting disease (SSWD) epizootic was first identified in 2013 in Washington State (USA) and, soon after, in British Columbia (Canada) and California (USA). Not surprising, the disease was first observed in the ochre star and the large sunflower star, the two most abundant species once commonly found in the intertidal and subtidal zones, respectively. Since the 1970s, outbreaks of SSWD have been reported in the northeastern Pacific, California, and northern New England (Menge 1979, Dungan et al. 1982, Eckert et al. 2000, Bates et al. 2009, Hewson et al. 2019). Unlike these outbreaks, which all occurred in single species and localized areas, the epizootic that began in 2013 involved 23 species of sea stars (Table 2.1) and eventually spanned from Alaska (USA) to Baja California (Mexico) (MARINe, 2021). It has been called the largest documented marine disease epizootic of a noncommercial species (Hewson et al. 2014, Eisenlord et al. 2016, Harvell et al. 2019).

Table 2.1. Species affected by sea star wasting disease

Species	Common name	Effect	References
Asterias amurensis	Northern Pacific star	U	www.seastarwasting.org
Astrometis sertulifera	Fragile rainbow star	U	Hewson et al. 2014; www.seastarwasting.org
Astropecten polyacanthus	Sand star	U	www.seastarwasting.org
Crossaster papposus	Rose star	U	www.seastarwasting.org
Dermasterias imbricata	Leather star	L	Montecino-Latorre et al. 2016
Evasterias troschelii	Mottled star	M	Hewson et al. 2014, Montecino-Latorre et al. 2016, Kay et al. 2019
Henricia spp.	Blood stars	M	Montecino-Latorre et al. 2016
Leptasterias hexactis	Six-armed star	M	Montecino-Latorre et al. 2016, Jaffe et al. 2019
Linckia columbiae	Fragile star	U	www.seastarwasting.org
Lophaster furcilliger vexator	Crested star	U	www.seastarwasting.org
Luidia foliolata	Sand star	M	Hewson et al. 2014; www.seastarwasting.org
Mediaster aequalis	Vermillion star	L	Hewson et al. 2014, Montecino-Latorre et al. 2016
Orthasterias koehleri	Rose star	H	Hewson et al. 2014, Montecino-Latorre et al. 2016
Patiria miniata	Bat star	M	Hewson et al. 2014; www.seastarwasting.org
Pisaster brevispinus	Giant pink star	H	Heswon et al. 2014, Montecino-Latorre et al. 2016
Pisaster giganteus	Giant star	M	Hewson et al. 2014, www.seastarwasting.org
Pisaster ochraceus	Ochre star	H	Hewson et al. 2014, Eisenlord et al. 2016, Miner et al. 2018
Pteraster militaris	Wrinkled star	U	www.seastarwasting.org
Pteraster tesselatus	Slime star	U	www.seastarwasting.org
Pycnopodia helianthoides	Sunflower star	H	Hewson et al. 2014, Schultz et al. 2016; Montecino-Latorre et al. 2016, Harvell et al. 2019, Hamilton et al. 2021
Solaster dawsoni	Sun star	H	Montecino-Latorre et al. 2016
Solaster stimpsoni	Stimpson's star	H	Hewson et al. 2014, Montecino-Latorre et al. 2016
Styasterias forreri	Velcro star	U	Hewson et al. 2014; www.seastarwasting.org

Effects on the species are classified as high (H), medium (M), low (L), and unknown (U).

Clinical Signs, Pathology, and Etiology

Signs and Gross Lesions

The first signs of SSWD are lethargy, limb curling, and grip loss. Soon after, superficial white lesions appear on their spiny skin and increase in depth and diameter. Arms often detach from the central disc and there can be progressive general loss of body turgor, or "wasting," before the animal dies (Figure 2.1) (Hewson et al. 2014, Menge et al. 2016).

In some cases, signs in diseased stars progressed rapidly from "losing grip" to "disintegration" in as little as four hours (Menge et al. 2016). Time-lapse videos of sea stars moving about while their arms progressively fell off, photographs of piles of ossicles contrasted against vibrant scenes of rocks covered with sea stars that were taken just weeks prior, and a few well-placed quotes, spurred the media to call this a "sea star zombie apocalypse." While zombie apocalypse is not a scientific term, it captured public attention and spurred elected officials and management agencies to invest in addressing the epizootic.

Etiology

Identifying the etiology of SSWD was important. Pet owners wanted to know if their dog was going to lose its tongue after eating an affected star. Parents wanted to know if their kid was going to lose a finger after picking one up, and if so, which finger! Was this outbreak caused by a toxin that could affect anyone or an infectious agent specific to echinoderms like sea stars, sea urchins, and sea cucumbers? Relatively quickly (from a scientific perspective) after the outbreak was first identified, field observations, experimental infections, and molecular studies implicated a novel densovirus (*Parvoviridae*) and proposed the name sea star–associated densovirus (SSaDV; Hewson et al. 2014). Densoviruses are single-stranded DNA viruses. Experiments showed that the disease could be transmitted to healthy sea stars with virus-sized particles from affected animals. Replicating densovirus was present in diseased tissues, and there was an association between viral load and disease. The authors even found that SSaDV, or a closely related densovirus, was present in ethanol-preserved museum specimens from as far back as 1942. They hypothesized that the virus may have existed for so long but not caused an epizootic earlier because (1) additional unknown environmental factors were not present; (2) populations of adult stars, especially sunflower stars, had only recently reached unprecedentedly high numbers facilitating a density-dependent opportunity; and (3) a genetic mutation in the virus, perhaps a modification of capsid structure, may have led to increased SSaDV virulence. The public applauded the quick discovery and assumed the case was closed. It was far from closed.

Little was known about sea star pathogens and maybe even less about sea star immune systems. As scientists began to probe the SSaDV etiology theory, alternate hypotheses emerged. While Hewson et al. (2014) showed that injection of viral-sized particles from affected sunflower stars caused disease in unaffected animals, later attempts to replicate experimental infection in other species failed to reproduce disease in ochre stars, giant pink stars (*Pisaster brevispinus*), and mottled stars (*Evasterias troschelii*) (Hewson et al. 2018). As it turns out, a multiomic study (Jackson et al. 2021) showed that there are a multitude of densoviruses that occur in sea stars on the western coast of North America, with SSaDV being just one of them. In addition to densoviruses, investigators examined the possibility that RNA viruses could also be culprits but found no association between virus and disease in the two species tested, sunflower stars and ochre stars (Hewson et al. 2018). The same scientists also examined microcystin toxins, a potent toxin produced by blue-green algae, as a putative cause for wasting disease and found no significant relationship in average microcystin concentration between asymptomatic and SSWD-affected stars (Hewson et al. 2018). It has also been suggested

Fig. 2.1. Progression of sea star wasting disease. Illustration by Laura Donohue.

that wasting disease could be caused by remineralized microbial organic matter that clog a sea star's respiratory surfaces, reducing the amount of oxygen available to the animal (Aquino et al. 2021). The scientists demonstrated that hypoxic or anaerobic conditions were present in affected sea stars from 2013 and 2014 and concluded that wasting in sea stars is influenced by microorganisms inhabiting the animal–water interface. This could be true, even with a viral etiology, but alone does not explain how the disease can be transmitted from affected to non-affected animals by virus-sized fractions or how the disease occurred over the entire western coast of North America.

While the etiology (or etiologies) of this massive epizootic is still in question, there is growing recognition that the triad including host immunity, a possible pathogen, and environmental interaction have been at play with elevated sea temperatures likely playing an important role. Looking specifically at wasting disease in ochre stars in Oregon, Menge et al. (2016) did not find that warmer water temperatures were associated with epizootic onset or exacerbated disease. Studying ochre star mortality from Alaska to San Diego, California (USA), Miner et al. (2018) did not find strong evidence to support that elevated seawater temperatures contributed to the initial emergence of SSWD, but they did show that sustained, anomalously high seawater temperatures in 2014 and 2015 may have exacerbated the disease's impact as the epizootic progressed. Further laboratory and field experiments in the same species, showed that elevated water temperature was directly associated with exacerbated disease (Eisenlord et al. 2016). Experimental work on ochre stars supported this conclusion, showing that while cooler temperatures did not stop disease progression, disease progression was slowed in stars that were housed in cooler water (Kohl et al. 2016). In documenting the range-wide near extirpation of sunflower stars, Harvell et al. (2019) established that peak declines in nearshore waters were associated with anomalous elevations in sea surface temperatures. Limited data on deep water temperatures prevented an analysis of the role that elevated water temperatures played in sunflower star mortality in deep offshore waters, but deep-water trawl data showed precipitous sea star declines there as well (Harvell et al. 2019). This effectively squashed some scientists' optimism that there might have been deep-water refugia where sunflower stars did not die from wasting disease. The relationship between SSWD and elevated ocean temperatures is supported by analysis of another, and even more comprehensive data set ranging from Baja California to the Aleutian Islands (USA) showing a latitude-related difference in disease related to water temperature. Range-wide sunflower star density dramatically dropped by 94.3%; however, the epizootic occurred more rapidly and with even more extensive mortality in the southern half of sunflower star range compared to the northern half, largely attributed to warmer water temperatures (Hamilton et al. 2021). The mechanisms by which temperature exacerbates wasting disease are not understood, but elevated water temperatures and consequent compromise in a star's ability to thermoregulate are associated with lower celomic fluid volume, higher metabolic demands, and metabolic stress. It is hypothesized that these factors collectively lead to immunosuppression (discussed below).

The high density of some species of sea stars prior to disease onset has been suggested as a possible precipitating factor for the epizootic but is a difficult hypothesis to prove. One attempt to evaluate this idea did not show a density-dependent phenomena. Instead, the investigators showed that at the onset of the epizootic in Oregon, wasting disease was noted almost uniformly along the coast even though ochre star densities varied widely (Menge et al. 2016).

Pathogenesis and Pathology

It is not really understood if there is a relationship between size of a star and susceptibility to SSWD. Experimentally, both large and small stars of the same species were found to die from the disease, but it seemed that disease either occurred more often, or maybe was easier to recognize, in larger ochre stars compared to smaller ones (Eisenlord et al. 2016). This could be due to larger animals developing the disease faster once exposed or small individuals dying faster once they developed symptoms, both of which have been noted in experimental infections. Ochre stars come in a dramatic variety of colors—so much so that some people mistake them for different species when learning to identify them (Harley et al. 2006). Could ochre star color influence susceptibility to SSWD? One study found a disproportionally higher level of wasting disease in orange-colored ochre stars compared to other color morphs (Menge et al. 2016). Gene flow between ochre stars of different colors is high, and it is not really known if color polymorphism has an underlying genetic component (Harley et al. 2006); therefore, it is unclear if there is a causal connection linking color to susceptibility to wasting disease or if it is just a coincidence. Experimental injection of purple and orange ochre stars with a suite of noninfectious organic substances causing dermal ulceration that mimicked lesions seen in field cases of SSWD showed that orange-colored stars developed more severe lesions. However, the orange stars lived longer than purple stars (Work et al. 2021), lending support to the hypothesis that orange morphs may be less susceptible to disease or other causes of inflammation.

Sea Star Immunity

Sea stars have complex immune systems, and the coelomocytes found within the coelomic fluid have cellular and humoral immune functions. Experimental infection of sunflower stars using virus-sized fractionated filtrates from wasting sea stars gives some insight into how sea stars mount an immune response to the disease (Fuess et al. 2015). The coelomocyte transcriptomic response to infection is characterized by increased expression of genes associated with immune pathways. These include the Toll pathway, complement cascade, melanization, and arachidonic acid metabolism (Fuess et al. 2015). Postinfection expression of genes involved in G-coupled protein processes, Wnt signaling, cell adhesion, and neural processes likely contribute to the clinical signs of wasting disease including the loss of arms and melting or deflation of the body. Metatranscriptomic analysis comparing wasting to healthy stars is consistent with this observation, revealing that genes expressed in affected animals were related to structural changes and disintegration, enhanced protein degradation and autocatalytic protein modification, decreased energy metabolism, and changes in cellular signaling (Gudenkauf and Hewson 2015). Most recently, the creation of a high-quality reference genome for ochre stars has permitted the connection of SSWD with an environmental driver (elevated temperature), individual animal difference (age, size, and genotype), and genes related to the star's response (immune response, phagocytosis, cell death and wound healing, apoptosis, muscle contraction, and heat shock protein) (Ruiz-Ramos et al. 2020).

Diagnosis

Without knowing the etiology (or etiologies) responsible for the massive multispecies epizootic that started in 2013, diagnosing SSWD in individual animals is problematic. Further complicating diagnosis of SSWD is the fact that sea stars only have so many ways to react to a disease and die (Hewson et al. 2019). Gross lesions similar to SSWD lesions found in wild ochre stars were reproduced by injecting them with a suite of noninfectious organic substances including peptone, cells of the cultured microalga *Dunaliella tertiolecta*, and particulate organic matter from seawater (Work et al. 2021). Control animals (wild-caught stars brought into captivity) also developed similar, yet less-severe lesions, possibly from the stress of captivity, suggesting again that stars have little variability in how they respond to disease. Recall that multiple single-species, localized SSWD epizootics occurred prior to the multispecies, widespread outbreak (Menge 1979, Dungan et al. 1982, Eckert et al. 2000, Bates et al. 2009, Hewson et al. 2019). There is little understanding of what caused those outbreaks and how their etiology may be similar to, or vary from, the 2013 outbreak. Consequently, finding a single star on the beach or in a captive collection with clinical signs ranging from lethargy, limb curling, grip loss, superficial white dermal lesions, arm detachment, or loss of body turgor does not mean it is dying of the same disease that, since 2013, has killed 23 different species and untold millions of stars from Mexico to Alaska. To examine the role that densovirus(es) may play in this epidemic or in a diseased star from a captive collection, one could collect and test the tube feet, pyloric ceca, coelomic fluid, gonads, or body wall from live or dead animals. The choice of sample site is important as the microbial communities differ with regard to the tissue sample origin, and there seems to be a tissue tropism hierarchy for SSaDV (Jackson et al. 2018). There are published methods for quantitative polymerase chain reaction (qPCR) to identify SSaWD densovirus and similar densoviruses, including primers for nonstructural protein 1 and viral gene products 1 and 4 (Hewson et al. 2014, Jackson et al. 2021). Methods for thermal cycling conditions are also published for qPCR, but there are no commercial labs testing for the densoviruses of echinoderms at this time.

Treatment

Without knowledge of the underlying cause(s) of SSWD, it is difficult to treat affected animals in a captive setting, much less in wild populations. Anecdotal evidence of varying success in treating wasting captive animals exists, including treatment of individual animals showing signs of SSWD with antimicrobials. This was the impetus for a study that determined that a one-time intracoelomic injection of enrofloxacin (5mg/kg) or immersion in an enrofloxacin solution (5mg/L for 6 hours) produced vascular system fluid drug concentrations expected to exceed the minimum inhibitory concentration for many bacterial pathogens (Rosenberg et al. 2016). We still don't know if antimicrobial treatment is effective for treating wasting stars, as it has not been studied and published. It does appear that magnesium chloride ($MgCl_2$) applied to sea stars in a captive setting can slow the physical progression of wasting disease in *Leptasterias* spp. (Jaffe et al. 2019). The authors of the study hypothesized $MgCl_2$ may play a role in eliminating secondary bacterial infections that could complicate underlying viral infection. If this mechanism is true, this could explain anecdotal positive responses that affected stars have had to antimicrobial therapy.

A Multispecies Epizootic

Relative to single-host pathogens, there are little data to help us understand the dynamics and impact of pathogens that infect multiple host species. If SSWD is caused by one or more pathogens, it represents a grand opportunity to collect empirical data to help us better understand and eventually model rates of transmission. Such data would help define the role that physiological, immunological, and genetic backgrounds of different host species play in determining species susceptibility and how biodiversity is, or is not, affected by these pathogens (Dobson 2004). The SSWD epizootic that began in 2013 has affected more than 23 species of stars on the western coast of North America with varying degrees of morbidity and mortality. By far, the most affected and most well-studied species are the two stars that were most common in the intertidal and subtidal zones—the ochre and the sunflower star. Both are considered highly susceptible to the disease and demonstrated large-scale population declines from the epizootic (see Table 2.1). Comparatively speaking, little is known about the other 21 species affected by the disease. Complicating matters further, while scientists were able to identify SSWD in multiple species, understanding the population-level effects in those species has been more challenging. Like many pathogens that affect multiple species, there were winners and losers from the outbreak, in the ecological sense. In the northern Gulf of Alaska, SSWD did not affect all species uniformly and ultimately caused a shift in sea star assemblages with lower density and lower diversity (Konar et al. 2019). While complicated by preepizootic density and detectability, it was clear that, in the Salish Sea, some subtidal species, including rose stars (*Orthasterias koehleri*), giant pink stars, sunflower stars, sun stars (*Solaster dawsoni*), and Stimpson's stars (*Solaster stimpsoni*), were hit hard and populations declined after epizootic onset (Montecino-Latorre et al. 2016). For other species shown to be susceptible to SSWD in the wild and in captivity, including leather stars (*Dermasterias imbricata*), six-armed stars (*Leptasterias hexactis*), and *Mediaster* spp., wild populations of these species have not declined postoutbreak (Montecino-Latorre et al. 2016). In fact, the abundance of leather stars actually increased after 2013 possibly due to reduced predation by other star species like sunflower stars, or from increased prey availability after competing star species declined. Kay et al. (2019) evaluated the post-outbreak changes in population density of two sympatric species, ochre stars and mottled stars in Burrard Inlet, British Columbia (Canada) and found that ochre star populations were reduced in number and in size after 2013, while mottled star populations increased markedly. Both species were susceptible to SSWD, but experimentally, mottled star populations took longer to reach thresholds of 50% infected and 50% mortality than did ochre stars. However, changes in population densities postepizootic were not simply a result of differences in disease susceptibility with a consequent release of competition between two species. For example, dramatic declines in sunflower stars and ochre stars may have even opened feeding opportunities for *Evasterias* in the intertidal zone. Clearly, a multispecies outbreak like SSWD has both direct and indirect ecological effects that may benefit susceptible species like leather stars and mottled stars.

Ecosystem Impacts

The two most affected species, the ochre star and sunflower star, are keystone species (Paine 1966, Duggins 1983). A keystone species has a disproportionately large influence on the structure of the ecosystem, and the ecosystem will change without them. As ochre stars (and more importantly six-armed *Leptasterias* stars) disappeared from the intertidal zone, black turban snails doubled in number and the zones at which some species occurred in the rocky intertidal changed (Gravem and Morgan 2017). Perhaps more interesting from a biodiversity perspective, as SSWD decimated sunflower stars, kelp-grazing green urchins and red urchins began to appear in greater abundance (Montecino-Latorre et al. 2016, Schultz et al. 2016, Burt et al. 2018). This makes sense because sunflower stars prey on urchins and modify urchin behavior. In places where urchin-eating sea otters were absent, kelp density also declined (Schultz et al. 2016, Burt et al. 2018). More recently, in Monterey Bay and other parts of California sea otter range a lack of nutrient upwelling resulted in urchins being so nutrient and calorie poor that otters ceased taking them as prey items. With a paucity of sea stars this led to serious damage to and depletion of areas of kelp forest. Overstory kelp are important ecosystem engineers in that they create habitat and food for numerous commercially exploited and nonexploited fish and invertebrates. In northern California, increased ocean temperatures combined with the SSWD-associated loss of sunflower stars contributed to the collapse of the economically and socially important red abalone and red sea urchin fisheries (Rogers-Bennett and Catton 2019).

The 2013 SSWD epizootic along the western coast of North America has demonstrated that sea stars are not simply economic "no shows." Combined with other factors like climate change and ocean warming, SSWD is having a real impact on coastal ecosystems, fisheries, and economies.

Looking Forward

Time will reveal how the 2013 epizootic reshapes intertidal and subtidal communities along the western coast of North America, including which sea star populations will shrink, disappear, recover or remain dominant. Early evidence might suggest that the SSWD epizootic may have selected for ochre stars with some form of resistance. Comparison of pre- and post-outbreak adults revealed significant allele frequency changes at three loci, which has the potential to persist in subsequent generations and influence the resilience of this species to ongoing or future outbreaks (Schiebelhut et al. 2018). The future doesn't look as bright for sunflower stars, which have been extirpated throughout much of their range and were just recently listed as critically endangered by the International Union for the Conservation of Nature (Gravem et al. 2021).

Politics and Recovering a Critically Endangered Species

In 2016, the first paper was published documenting the severe depletion of sunflower stars in the Salish Sea (Montecino-Latorre et al. 2016). This news came on the heels of unpublished reports of major declines in this species throughout most of its range from California to Alaska. About this time, scientists began to wonder if the species should be considered for listing under the US Endangered Species Act (ESA). In November 2016, my colleagues and I submitted a proposal to the US National Oceanographic and Atmospheric Administration (NOAA) through the National Marine Fisheries Service for the species to be considered for listing as a "Species of Concern." The Species of Concern program was designed to identify imperiled species and fund work to recover them before they had to be listed as threatened or endangered under the US ESA. Following the inauguration of a new presidential administration in January 2017, the Species of Concern program was discontinued, and the petition was dead in the water. Scientists met again and discussed the pros and cons of proposing the species for listing as threatened or endangered under the US ESA. Scientists decided that the effort was not worth the benefit, given that the current administration was not favorable to reviewing new species proposed for ESA listing, and even a listing would not guarantee the availability of federal funding for sunflower star recovery. An alternative strategy was identified to propose listing through the International Union for the Conservation of Nature (IUCN) as a mechanism to document the decline and express the need for recovery efforts. With funding from The Nature Conservancy, two postdoctoral students led authorship on a proposal that resulted in the 2021 listing of this star as critically endangered under the IUCN Red List (Gravem et al. 2021). In August 2021, the Center for Biological Diversity petitioned the Secretary of Commerce and NOAA through the National Marine Fisheries Service to list the sunflower star as threatened or endangered under the US ESA. At the time of this writing, this proposal is still under review and consideration. Nonetheless, a dedicated group of scientists and conservationists continue to work on sunflower star recovery without much dedicated funding.

Summary

Science is linked to conservation action by more than just peer-reviewed publications and often requires scientists to collaborate with personnel from government agencies and prioritize networking and engaging with stakeholders before, during, and after their research is conducted (LeFlore et al. 2021). While ample mystery still surrounds the 2013 SSWD epizootic, it is an example of how scientists, natural resource managers, and nongovernmental organizations can work together and act quickly to understand and address an emerging wildlife health concern. Indeed, large wildlife health and conservation problems, particularly those involving complicated ecological relationships, can only be resolved or managed by cooperative efforts. Wildlife veterinarians and health professionals frequently find themselves facing situations where care and treatment of individual animals is of limited value and even trying to manage disease in a population is ineffective. It is at this point that the "patient" must encompass the whole ecosystem, or at times, even a portion of the planet.

LITERATURE CITED

Aquino, C. A., R. M. Besemer, C. M. DeRito, J. Kocian, I. R. Porter, P. T. Raimondi, J. E. Rede, L. M. Schiebelhut, J. P. Sparks, J. P. Wares, et al. 2021. Evidence that microorganisms at the animal–water interface drive sea star wasting disease. Frontiers in Microbiology 11:610009.

Bates, A. E., B. J. Hilton, and C. D. G. Harley. 2009. Effects of temperature, season and locality on wasting disease in the keystone predatory sea star *Pisaster ochraceus*. Diseases of Aquatic Organisms 86:245–251.

Burt, J. M., M. T. Tinker, D. K. Okamoto, K. W. Demes, K. Holmes, and A. K. Salomon. 2018. Sudden collapse of a mesopredator reveals its complementary role in mediating rocky reef regime shifts. Proceedings of the Royal Society B Biological Sciences 285:20180553.

De Santis, M. 2018. Starfish detectives: sirens in a changing sea. CreateSpace Independent Publishing Platform. San Bernardino, California, USA.

Dobson, A. 2004. Population dynamics of pathogens with multiple host species. American Naturalist 164:S64–S78.

Duggins, D. 1983. Starfish predation and the creation of mosaic patterns in a kelp-dominated community. Ecology 64:1610–1619.

Dungan, M. L., T. E. Miller, and D. A. Thompson. 1982. Catastrophic decline of a top carnivore in the Gulf of California rocky intertidal zone. Science 216:989–991.

Eckert, G. L., J. M. Engle, and D. J. Kushner. 2000. Sea star disease and population declines at the Channel Islands. Pages 390–393 *in* D. R. Browne, K. L. Mitchell, and H. W. Chaney, editors. Proceedings of the Fifth California Islands Symposium. Santa Barbara Museum of Natural History, Santa Barbara, California, USA.

Eisenlord, M.E., M. L. Groner, R. M. Yoshioka, J. Elliott, J. Maynard, S. Fradkin, M. Turner, K. Pyne, N. Rivlin, R. van Hooidonk, et al. 2016. Ochre star mortality during the 2014 wasting disease epizootic: role of population size structure and temperature. Philosophical Transactions Royal Society B Biological Sciences 371:20150212.

Fuess, L. E., M. E. Eisenlord, C. J. Closek, A. M. Tracy, R. Mauntz, S. Gignoux-Wolfsohn, M. N. Moritsch, R. Yoshioka, C. A. Burge, C. D. Harvell, et al. 2015. Up in arms: immune and nervous system response to sea star wasting disease. PLoS One 10:e0133053.

Gravem, S. A., W. N. Heady, V. R. Saccomanno, K. F. Alvstad, A. L. M. Gehman, T. N. Frierson, and S. L. Hamilton. 2021. *Pycnopodia helianthoides*. International Union for the Conservation of Nature, Red List of Threatened Species 2021, Gland, Switzerland.

Gravem, S. A., and S. G. Morgan. 2017. Shifts in intertidal zonation and refuge use by prey after mass mortalities of two predators. Ecology 98:1006–1015.

Gudenkauf, B. M., and I. Hewson. 2015. Metatranscriptomic analysis of *Pycnopodia helianthoides* (Asteroidea) affected by sea star wasting disease. PLoS One 10:e0128150.

Hamilton, S. L., R. Saccomanno, W. N. Heady, A. L. Gehman, S. I. Lonhart, R. Beas-Luna, F. T. Francis, L. Lee, L. Rogers-Bennett, A. K. Salomon, et al. 2021. Disease-driven mass mortality event leads to widespread extirpation and variable recovery potential of a marine predator across the eastern Pacific. Proceedings of the Royal Society B Biological Sciences 288:20211195.

Harley, C. D. G., M. S. Pankey, J. P. Wares, R. K. Grosberg, and M. J. Wonham. 2006. Color polymorphism and genetic structure in the sea star *Pisaster ochraceus*. Biological Bulletin 211:248–262.

Harvell, C. D., D. Montecino-Latorre, J. M. Caldwell, J. M. Burt, K. Bosley, A. Keller, S. F. Heron, A. K. Salomon, L. Lee, O. Pontier, et al. 2019. Disease epidemic and a marine heat wave are associated with the continental-scale collapse of a pivotal predator (*Pycnopodia helianthoides*). Science Advances 5:eaau7042.

Hewson, I., K., S. I. Bistolas, E. M. Quijano Cardé, J. B. Button, P. J. Foster, and J. M. Flanzenbaum. 2018. Investigating the complex association between viral ecology, environment, and Northeast Pacific sea star wasting. Frontiers in Marine Science 5:77.

Hewson, I., J. B. Button, B. M. Gudenkauf, B. Miner, A. L. Newton, J. K. Gaydos, J. Wynne, C. L. Groves, G. Hendler, M. Murray, et al. 2014. Densovirus associated with sea-star wasting disease and mass mortality. Proceedings of the National Academy of Sciences 111:17278–17283.

Hewson, I., B. Sullivan, E. W. Jackson, Q. Xu, H. Long, C. Lin, E. M. Quijano Cardé, J. Seymour, N. Siboni, M. R. L. Jones, and M. A. Sewell. 2019. Perspective: something old, something new? Review of wasting and other mortality in Asteroidea (Echinodermata). Frontiers in Marine Science 6:406.

Jackson, E. W., C. Pepe-Ranney, S. J. Debenport, D. H. Buckley, and I. Hewson. 2018. The microbial landscape of sea stars and the anatomical and interspecies variability of their microbiome. Frontiers in Microbiology 9:1829.

Jackson, E. W., R. C. Wilhelm, M. R. Johnson, H. L. Lutz, I. Danforth, J. K. Gaydos, M. Hart, and I. Hewson. 2021. Diversity of sea star-associated densoviruses and transcribed endogenized viral elements of densovirus origin. Journal of Virology 95(1): e01594-20.

Jaffe, N., R. Eberl, J. Bucholz, and C. S. Cohen. 2019. Sea star wasting disease demography and etiology in the brooding sea star *Leptasterias* spp. PLoS One 14:e0225248.

Kay, S. W. C., A.-L. Gehman, and C. D. G. Harley. 2019. Reciprocal abundance shifts of the intertidal sea stars, *Evasterias troschelii* and *Pisaster ochraceus*, following sea star wasting disease. Proceedings of the Royal Society B Biological Sciences 286:20182766.

Kohl, W. R., T. I. McClure, and B. G. Miner. 2016. Decreased temperature facilitates short-term sea star wasting disease survival in the keystone intertidal sea star *Pisaster ochraceus*. PLoS One 11:e0153670.

Konar, B., T. J. Mitchell, K. Iken, H. Coletti, T. Dean, D. Esler, M. Lindeberg, B. Pister, and B. Weitzman. 2019. Wasting disease and static environmental variables drive sea star assemblages in the Northern Gulf of Alaska. Journal of Experimental Marine Biology and Ecology 520:151209.

Multi-Agency Rocky Intertidal Network (MARINe). 2021. Sea star wasting syndrome. http://www.seastarwasting.org. Accessed 26 October 2021.

Menge, B. A. 1979. Coexistence between the seastars *Asterias vulgaris* and *A. forbesi* in a heterogeneous environment: a non-equilibrium explanation. Oecologia 41:245–272.

Menge, B. A., E. B. Cerny-Chipman, A. Johnson, J. Sullivan, S. Gravem, and F. Chan. 2016. Sea star wasting disease in the keystone predator *Pisaster ochraceus* in Oregon: insights into differential population impacts, recovery, predation rate, and temperature effects from long-term research. PLoS One 11:e0153994.

Miner, C. M., J. L. Burnaford, R. F. Ambrose, L. Antrim, H. Bohlmann, C. Blanchette, J. M. Engle, S. C. Fradkin, R. Gaddam, C. D. G. Harley, et al. 2018. Large-scale impacts of sea star wasting disease (SSWD) on intertidal sea stars and implications for recovery. PLoS One 13: e0192870.

Montecino-Latorre, D., M. E. Eisenlord, M. Turner, R. Yoshioka, C. D. Harvell, C. V. Pattengill-Semmens, J. D. Nichols, and J. K. Gaydos. 2016. Devastating transboundary impacts of sea star wasting disease on subtidal asteroids. PLoS One 11:e0163190.

LeFlore, M., D. Bunn, P, Sebastian, and J. K. Gaydos. 2021. Improving the probability that small-scale science will benefit conservation. Conservation Science and Practice 4:e571.

Paine, R. T. 1966. Food web complexity and species diversity. American Naturalist 100: 65–75.

Rogers-Bennett, L., and C. A. Catton. 2019. Marine heat wave and multiple stressors tip bull kelp forest to sea urchin barrens. Scientific Reports 9:15050.

Rosenberg, J., M. Haulena, B. E. Phillips, C. A. Harms, G. A. Lewbart, L. L. Lahner, and M. G. Papich. 2016. Population pharmacokinetics of enrofloxacin in purple sea stars (*Pisaster ochraceus*) following an intracoelomic injection or extended immersion. American Journal of Veterinary Research 77:1266–1275.

Ruiz-Ramos, D. V., L. M. Schiebelhut, K. J. Hoff, J. P. Wares, and M. N. Dawson. 2020. An initial comparative genomic autopsy of wasting disease in sea stars. Molecular Ecology 29:1087–1102.

Schiebelhut, L. M., J. B. Puritz, and M. N. Dawson. 2018. Decimation by sea star wasting disease and rapid genetic change in a keystone species, *Pisaster ochraceus*. Proceedings of the National Academy of Sciences USA 115:7069–7074.

Schultz, J. A., R. N. Cloutier, and I. M. Côte. 2016. Evidence for a trophic cascade on rocky reefs following sea star mass mortality in British Columbia. Peer J 4:e1980.

Work, T. M., T. M. Weatherby, C. M. DeRito, R. M. Besemer, and I. Hewson. 2021. Sea star wasting disease pathology in *Pisaster ochraceus* shows a basal-to-surface process affecting color phenotypes differently. Diseases of Aquatic Organisms 145:21–33.

3 Wild Fisheries

One Health Challenges Critical to Understanding and Protecting Fish in Our Aquatic Environs

E. Scott P. Weber III, Olivia Rozanski

Introduction

Water is crucial for all life and is essential to support growth and production of plant and animal agriculture, aquaculture, and wild fisheries. The quality and availability of natural water resources—freshwater, brackish, and marine—are critical factors for the health of our planet. Native fish, particularly anadromous species that spend portions of their lives in both fresh and salt water, enhance the stability, sustainability, and health of the environment. Indigenous peoples on several continents use native fish for subsistence and have depended on certain fish species as cultural, ceremonial, historical, and economic sine qua nons. The reliance on native and anadromous fish by indigenous tribes in North America occurs along every coastline and follows rivers and their tributaries to their sources of origin found deep in the continent's interior (Smith 2011). Fish species historically important and celebrated by indigenous peoples include, but are not limited to, salmon and sturgeon species of the Pacific Northwest; salmon, cod, sturgeon, and striped bass in the Atlantic region; goodeids and pupfish in desert regions; and several sturgeon species along inland lakes, rivers, and their tributaries in the upper Midwest and Mountain West (see figure on the right).

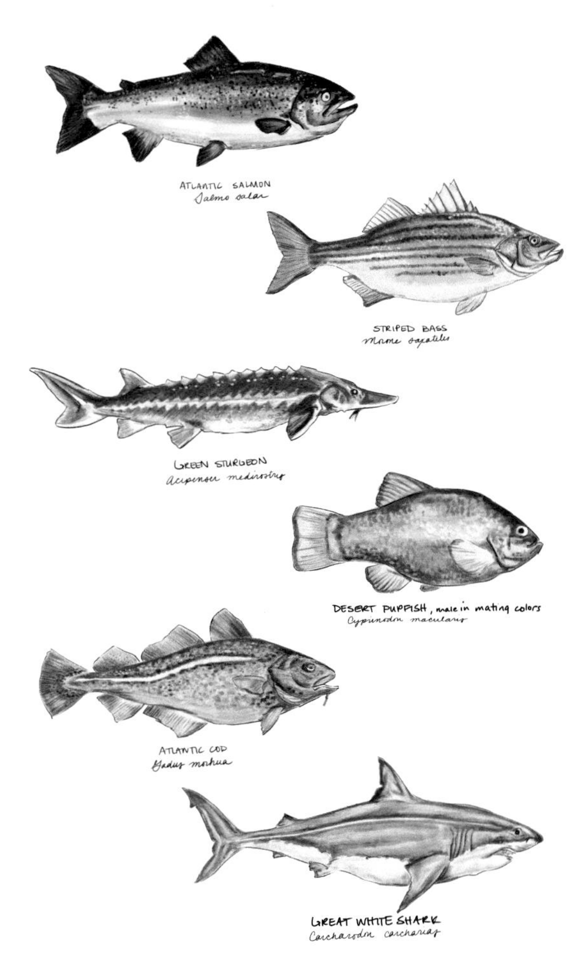

Six native fish species protected in North America for various reasons. Illustration by Laura Donohue.

Fish are the largest and most diverse classes of vertebrates, representing half of all taxonomic phyla. The 36,058 species of fish currently identified are divided into 4 classes, 59 orders, 490 families and approximately 4,300 genera with 300–500 new species added annually since 2002 (Eschmeyer et al. 2021). An obscure deep-water genus of fish called the bristlemouths (*Cyclothone spp.*) from the family Gonostomatidae are the most abundant vertebrates on the planet, numbering in the trillions, but only a few marine biologists and fishermen have even seen them (McClain et al. 2001, Wienerroither et al. 2009). Our knowledge of aquatic organisms and aquatic environments is rudimentary when compared with our understanding of terrestrial species. This makes aquatic animal health a challenging and exciting field.

Fish have adapted to some of the harshest environments on the planet, perhaps in part to survive in the harsh conditions that may be less hospitable to infectious pathogens and parasites (Tobler et al. 2007). Fish can thrive in toxic conditions where other vertebrate species cannot survive, such as the cave dwelling *Poecilia mexicana* that resides in environments with toxic sulfur levels (Jourdan et al. 2014). Native pupfish and goodeids in North America are critically endangered due to habitat loss in desert regions as water is diverted for agriculture and urban and suburban development (Minckley and Deacon 1968; Bailey et al. 2007), yet these species are remarkable in their survivability (Figure 3.1). The Julimes pupfish (*Cyprinodon julimes*) can tolerate temperature ranges from –1.9°C to 45.5°C (Minckley and Minckley 1986), while another species, the Devils Hole pupfish (*Cyprinodon diabolis*), has a habitat restricted to a narrow subterranean fissure in Death Valley California (Martin et al. 2016).

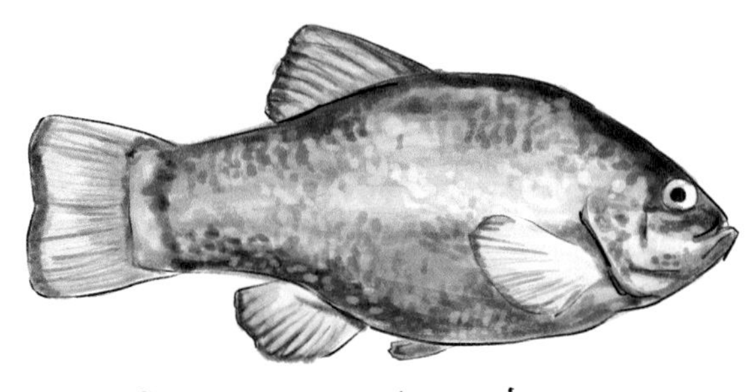

Fig. 3.1. Desert pupfish (*Cyprinodon macularius*). Illustration by Laura Donohue.

The loss of fish biodiversity in temperate regions such as California in the 20th century illustrates how native fish populations are critical for maintaining healthy ecosystems (Moyle and Williams 1990). Scientists and veterinarians have limited knowledge of natural history, husbandry, and diseases of most fish, except those species that are commercially important as pets, laboratory research models, used for aquaculture, or caught as sportfish or food fish. The most pressing concerns for fish conservation, as with other poikilotherms, includes climate change, infectious diseases, invasive species, environmental pollution, overfishing, and habitat loss or alteration (Gibbons et al. 2000). This chapter introduces these threats as related to fish health and explores the complicated acclimatization of fish to this ever-changing environment.

Infectious Disease

Study of infectious disease outbreaks in wild fisheries is challenging. Diseases in wild fish can be caused by pathogens from microbes to metazoan parasites. Diseases that threaten aquatic environments include existing, reemerging, and newly emerging pathogens. These pathogens may be species specific, involve multiple hosts, have complex life cycles, cross-infect different vertebrate taxa, and may be zoonotic. Recent outbreaks that are reportable for international trade include parasites like the monogenean trematode *Gyrodactylus salaris* in salmonid fish, and viruses like infectious hematopoietic necrosis virus (IHNV), salmonid alphavirus (SAV), spring viremia of carp (SVCV), and viral hemorrhagic septicemia virus (VHSV) (WOAH 2021, Weber 2012). These diseases and others pose a major risk to dwindling wild fish stocks and disease outbreaks in wild stocks can involve several species and be more severe in waters adjacent to infected fish farms.

Excellent references for investigating wild fish mortality events are available, with a comprehensive field manual produced by US Fish and Wildlife Service (USFWS) (Meyer and Barclay 1990) providing the foundation for current protocols. Acquiring appropriate environmental and historical information when investigating large fish mortality events is vital as many cases are complicated by carry-over effects, biotic factors, and multiple stressors (La and Cooke 2011). A definitive diagnosis may not be reached if inadequate information is gathered, or inappropriate or insufficient samples are taken. Modeling of environmental factors during a fish kill in Minnesota demonstrated that maximum nighttime land surface temperature was the most critical factor in a series of wild fish mortalities (Phelps et al. 2019), a finding emphasizing the importance of climate change to wild fisheries. Woo et al. (2020) recently published a comprehensive review of climate change and infectious diseases of fish.

For wild fish mortality investigations, it is important to take representative and replicate environmental samples, collect fresh dead specimens or euthanize moribund fish for diagnostic evaluation, and conduct necropsy and histopathology on representative samples. Fish samples are often frozen during mortality events for convenience, but necropsies on fresh dead fish provide more accurate results and help distinguish primary from secondary pathogens. Multiple sections should be taken from each tissue or lesion to provide serial samples for formalin fixation, alcohol preservation, freezing, and fresh tissue testing. This sectioning preserves a specific tissue sample-set to allow comparisons for histopathology, molecular diagnostics, and toxicologic investigations. Fresh tissues and cultures can be submitted immediately for microbiology and virology to specialized aquatic animal health laboratories that provide growth conditions specific to fish pathogens, such as lower incubation temperatures for microbial swabs, and maintenance of fish species–specific cell lines for virus isolation.

The most common clinical signs for many fish disease outbreaks include erosions, swelling, or fusing of the gill filaments; hemorrhage on the fins, skin, or in the eyes; skeletal abnormalities; darkening of the skin; enophthalmos, exophthalmos, corneal opacity, or ocular inclusions; dermal ulceration; ventral hemorrhagic petechiation; white, black, or translucent fecal casts; rectal or cloacal prolapse; and muscle erosions and necrosis. Internal gross necropsy findings may include enlarged spleen and liver; hemorrhagic mottling of the organs; hepatic lipidosis; large or pinpoint granulomas or masses; dark or light pigmented mottling of the organs; internal masses; edematous or cystic gonads; black or white cysts in the organs or muscles; gastric dilation or gastrointestinal intussusception; and fluid-filled swimbladder (Figure 3.2). Unfortunately, many fish diseases are not associated with pathognomonic lesions, so clinical samples from a mortality event should be supported with additional diagnostic data.

Emerging and Reemerging Viral Pathogens

Viral diseases are a threat to wild fisheries. Viruses are readily transmitted from and between commercially raised and wild fish, and because many viral diseases have increases or decreases in morbidity and mortality based on ideal thermal ranges for the pathogen, climate change and global movement of animals can allow viruses to be found in new geographic areas, to infect new host species, or to cause more severe morbidity and greater mortality. Viruses can be easily translocated via movement of ill, subclinical, or dead animals or animal products. Most mitigation strategies for viral disease outbreaks in fish include prevention using strict testing protocols for exported or imported fish, fry, or reproductive products; careful disinfection and cleaning of equipment used for handling or transport of fish or fish products; careful management of water associated with shipment and proper treatment of water discharge from culture facilities, when applicable; avoidance of feeding cultured fish any live foods or unprocessed wild fishery product; and promptly recording and investigating all unusual mortality

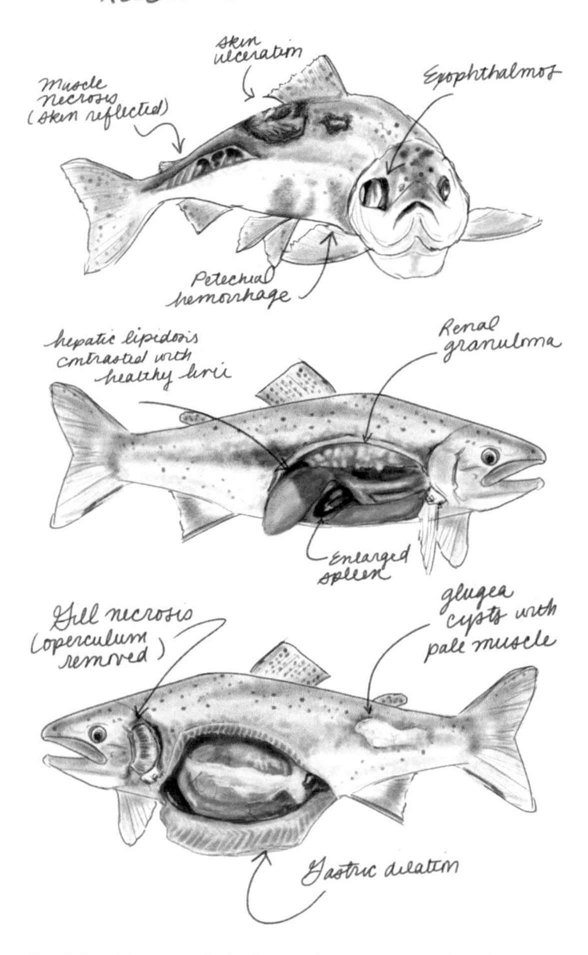

Fig. 3.2. Gross pathological lesions associated with several different infectious and noninfectious etiologies illustrated in an Atlantic salmon (*Salmo salar*). Illustration by Laura Donohue.

events in wild or cultured fish. If preventive measures fail to limit the spread of a viral infection, treatment or mitigation is seldom an option for fish viral disease outbreaks nor is treatment or mitigation practical for infected wild fish populations. While many viral pathogens are species or family specific, some such as infectious pancreatic necrosis virus have expanded their host range as demonstrated by outbreaks in nine wild marine fish associated with Atlantic salmon (*Salmo salar*) farms off the Shetland Islands (Wallace et al. 2008). One of the most significant reemerging pathogens is VHSV, a novirhabdovirus, that has infected over 140 species of both freshwater and marine fish in the Northern Hemisphere (Escobar et al. 2018), with 4 genotypes and 9 subtypes now recognized. The ranaviruses infect a wide range of wild and domestic fish from tropical marine Banggai cardinalfish (*Pterapogon kauderni*) (Weber et al. 2009) to temperate brackish water threespine stickleback (*Gasterosteus aculeatus*) (Waltzek et al. 2012) to cold water marine lumpfish (*Cyclopterus lumpus*) (Stagg et al. 2020), and these viruses can jump to other vertebrate classes with transmission of FV3-like ranaviruses between fish, amphibians, and chelonians (Brenes et al. 2014) (See Chapter 4).

Koi Herpes Virus

Common carp (*Cyprinus carpio*) are a major wild and farmed food source and a key species in wild aquatic ecosystems worldwide. Koi herpes virus (KHV) is caused by cyprinid herpesvirus CyHV-3 and has been responsible for mass mortalities in wild and farmed populations of native and introduced carp across the world. Koi herpes virus disease was first described following a series of outbreaks in the United States and Israel (Hedrick et al. 2000a). Typical clinical signs of KHV disease include anorexia, dermal ulceration, enophthalmos, gasping at the water surface, or positioning their body near rapid water flow areas like waterfalls and filter outlets. Most fish have extensive erosions and destruction of the filament structure of the gills (Figure 3.2). In naïve populations, KHV causes high morbidity approaching 80–100% and mortality of 70–80% (Walster 1999, Steinhagen et al. 2016). Asymptomatic carriers and latency in previously exposed fish can allow virus spread in wild populations, though outbreaks are only recognized when large numbers of carp die as seasonal water temperatures enter the infective range for the virus (St-Hilaire et al. 2005). Control of KHV requires effective testing methods and appropriate biosecurity measures to prevent movement of infected fish. Viral DNA can be detected using

Taqman® real-time polymerase chain reaction (PCR) for (Bercovier et al. 2005) and antibodies to KHV exposure by enzyme-linked immunosorbent assay (ELISA) (Adkison et al. 2005). Because previously exposed or recovered fish may be asymptomatic carriers, a combination of tests is needed to screen for the pathogen and antibodies to aid control efforts. Mortality events have occurred in wild *C. carpio* in several countries through incidental, accidental, or intentional release of infected varieties of domestic *C. carpio*. The first mortality event in Canada was recorded in the Kawartha Lakes region, Ontario, in 2007 with subsequent outbreaks in southwestern Ontario in 2008 (Garver et al. 2010). Following 27 mass mortalities across Minnesota, Iowa, Pennsylvania, and Wisconsin, USA, in 2017–2018, wild *C. carpio* were screened for three internationally reportable cyprinid viruses (KHV; carp edema virus, CEV; and SVCV) using real-time quantitative PCR (qPCR) (Padhi et al. 2019). Of the cases examined, KHV, CEV, and KHV–CEV coinfection was diagnosed in 13, 2, and 4 of these mortality events, respectively (Padhi et al. 2019). Based on KHV's high mortality, Australia considered using the virus as a biocontrol for invasive *C. carpio*, but epidemiological data suggests that release of KHV into thermally suitable waterways would result in short-term mortality but would not reduce carp numbers below their reproductive threshold (Becker et al. 2019).

Carp Edema Virus

Carp edema virus, a newly emerging virus of wild carp, is a large, double-stranded DNA virus in the poxvirus family (Poxviridae). The disease was first described in Japan in freshwater koi and characterized by fish laying down on the bottom substrate, and fish recovered when exposed to a bath treatment using NaCl (Hosoya and Suzuki 1976). On gross necropsy, the gills of ill and deceased fish appeared edematous and swollen, in sharp contrast to the necrotic gills associated with KHV (Ono et al. 1986: Oyamatsu, et al. 1997) (Figure 3.2). Recent koi outbreaks have occurred in South Korea (Kim et al. 2020) and from Chiang Mai province in Thailand (Pikulkaew et al. 2020). Outbreaks of CEV in commercial common carp and koi have been reported from several European countries posing socioeconomic threats to aquaculture industries and wild fisheries (Way and Stone 2013; Lewisch et al. 2015; Way et al. 2017). A mortality event of 340 adult wild *C. carpio* in Italy was attributed to CEV (Marsella et al. 2021). Several wild outbreaks of CEV have been reported in the United States including a large mortality of over 500 wild spawning *C. carpio* in New Jersey (Lovy et al. 2018). Review of host fish species at an outbreak at Lake Swartout in Minnesota suggests that CEV has a narrow species range similar to CyHV-3, and predators could translocate live virus in their feces to uninfected areas after feeding on ill or deceased CEV-infected *C. carpio* (Tolo et al. 2021).

Mycobacteriosis

Atypical mycobacteria are in the same family as those causing chronic diseases like tuberculosis in humans, birds, and other mammals. Mycobacteriosis causes significant complications for managing collections of fish, and it has been identified in hundreds of freshwater and marine ornamentals, laboratory animals, aquaculture-reared fish, and wild populations. A new species of mycobacteria was identified in 1997 in wild striped bass (*Morone saxatilis*) from the Chesapeake Bay (Figure 3.3). The bass exhibited ulcerative dermatitis with granulomatous lesions identified in various organs at necropsy (Heckert et al. 2001) (see Figure 3.2). Morbidity and mortality associated with mycobacteria infections in Atlantic menhaden (*Brevoortia tyrannus*) were also observed in surveys from several Chesapeake Bay tributaries, with some waterways having prevalence as high as 57% (Kane et al. 2007).

Quantitative assessment for *Mycobacteria* spp. can help assess the environmental health of aquatic ecosystems as this pathogen favors poor water quality conditions including warm polluted waters (Dailloux

Fig. 3.3. Striped bass (*Morone saxatilis*). Illustration by Laura Donohue.

et al. 1992). Australian researchers identified *Mycobacteria* spp. while conducting surveys for myxozoas and they noted mycobacteria infections persisted in a large range of water-quality conditions (Delghandi et al. 2020). Low pH of streams at northern latitudes have historically been associated with acid rain caused by industrial pollution. In Finland, a drop in environmental pH enhanced the growth of *Mycobacteria* spp., while decreasing growth of other heterotrophic bacteria in both brook water (Iivanainen et al. 1993) and sediment (Iivanainen et al. 1999). Some *Mycobacteria* spp. favor warming temperatures; others, like *M. chelonae*, exhibit thermoresistance (Groner et al. 2018, Schulze-Röbbecke and Buchholtz. 1992), and elevated temperatures may inhibit *M. shottsii* and *M. pseudoshottsii* found in the Chesapeake Bay (Gauthier et al. 2021). A further complication in the Chesapeake Bay is the identification of *Mycobacteria* spp. from the northern snakehead (*Channa argus*), an invasive fish species, that has expanded its range throughout the Potomac catchment area (Densmore et al. 2016). Mycobacteria has also been detected in Pacific estuaries and caused mortality in both captive and wild endangered delta smelt (*Hypomesus transpacificus*) (Baxa et al. 2015). From 2007 until 2012 *Mycobacteria* spp. outbreaks were diagnosed annually in delta smelt captured for breeding programs.

Mycobacteriosis is nearly impossible to eradicate in captive and wild populations given its ubiquitous nature and environmental persistence. Despite significant advances in molecular diagnostics, definitive diagnosis by culture can take weeks. Some fish that test negative may harbor infection but not shed bacteria, and some fish may only test positive because infected material is passing though the gastrointestinal tract. Therefore, fecal swabs and screening for acid-fast organisms by cytology, fecal analysis, or biopsy can only be presumptive for positive animals showing clinical signs of infection. Treatment of clinically infected fish is highly discouraged because of the lack of antemortem testing, the empirical length of treatment, limited pharmacokinetic data supporting effective antibiotics in most fish, the risk for atypical mycobacterial zoonoses, and potential development of antimicrobial resistance when using antibiotics that are critical for human and veterinary medicine.

Epizootic Ulcer Syndrome

Several fish fungal pathogens have been identified over the last 20 years. The water mold *Aphanomyces invadans*, which causes epizootic ulcerative syndrome (EUS), poses a significant risk to wild fresh and brackish water fish in tropical and subtropical regions. The disease was first identified in the early 1970s in wild ayu (*Plecoglossus altivelis*) (Egusa and Masuda 1971), a goldfish (*Carassius auratus*) from Japan (Miyazaki and Egusa 1972), and in estuarine wild mullet (*Mugil cephalus*) in Australia (McKenzie and Hall 1976). Epizootic ulcerative syndrome causes red spots on the dermis of fish that progress to dermal ulceration and muscular necrosis (Figure 3.2). *Aphanomyces invadens* may be part of a syndrome associated with another yet undiscovered primary pathogen, rather than the sole causative agent of EUS (Roberts et al. 1993). Based on surveys of clinically ill wild Australian bony bream (*Nematalosa erebi*), EUS has expanded its range in the Murray–Darling River drainage (Go et al. 2012). Similarly, EUS was identified in the Chobe–Zambezi rivers, and subsequent surveys revealed it spread southward in Zimbabwe and infected at least seven new species (Sibanda et al. 2018). India has experienced similar outbreaks of EUS in both cultured and wild species, with the pathogen expanding its host range to over 94 species (Pradhan et al. 2014). EUS poses an additional threat to wild fisheries because of warming water temperatures in temperate regions and increased globalization of fish and the fishery products trade. Many tropical and subtropical fish in the ornamental fish trade can be clinically infected or carriers of EUS, and several of these fish species can also become invasives if released into suitable environments, exposing naïve wild fish populations to EUS.

Migrating and Emigrating Myxozoas

Many fish myxozoan parasites have a complicated multistage life cycle that includes a fish host and an intermediate oligochaete worm. Habitat changes and rising global temperatures allow these parasitic diseases to infect a variety of temperate species across several continents. Two infectious disease examples are whirling disease (WD) caused by *Myxobolus cerebralis* and proliferative kidney disease (PKD) caused by *Tetracapsuloides bryosalmonae*. The initial dissemination of both parasites into new geographic areas followed widespread export and stocking of brown trout (*Salmo trutta*) for recreational fishing in freshwater streams. Clinical signs associated with WD include skeletal deformities of the spine and skull, blackening of the tail, and swimming in a circling pattern (Figure 3.4).

Eggs of brown trout were imported from Germany to New York in 1883 by aquaculturist Fred Mather. Mather wanted to please anglers by introducing an attractive faster growing fish with a tolerance for

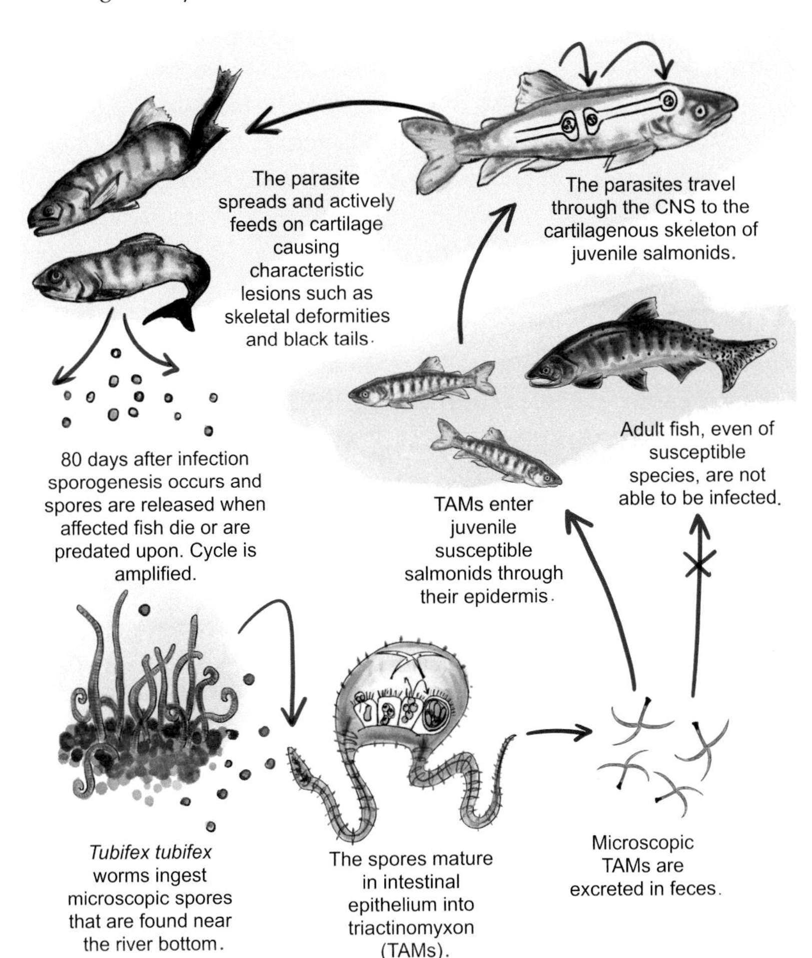

Fig. 3.4. The life cycle of whirling disease in both an intermediate host, the sludge worm (*Tubifex tubifex*), and the final host, brook trout (*Salvelinus fontinalis*). Illustration by Laura Donohue.

slightly warmer temperatures than native brook trout (*Salvelinus fontinalis*) (Mather 1887). The eggs were hatched, and fish were raised in New York and Michigan. Under the US Fish Commission, 4,900 brown-trout fry were released into Michigan's Baldwin River in 1884 as the first wild trout introduction. To create a more robust American stock, breeding fish and additional eggs from Scotland were imported the following year. Based on the success of these early introductions and favorability of both size and gaminess for anglers, *S. trutta* has been stocked in 45 states in the United States and has achieved self-sustaining populations in 34 of the 50 states (USGS 2022).

Proliferative kidney disease is marked by anemia causing pale gills, exophthalmia, and enlarged, gray or mottled kidneys and spleen (Figure 3.2). Warming temperatures have allowed PKD to expand into northern regions and impact Alaskan populations of salmonids (Gorgoglione et al. 2020).

Whirling disease was largely confined to relatively small regions of Europe prior to World War II and then expanded throughout the continent after the war. Hoffman (1969) suggested that unrestricted transfers of rainbow trout (*Oncorhynchus mykiss*) throughout Europe coupled with rearing salmonids in earthen ponds were to blame. The ravages of war on the European continent may have contributed to creating more suitable habitat for the host oligochaetes, which thrive in poor water quality conditions and areas of high silt and detritus. Following World War II, environmental disturbances of streams and rivers in North America were likely factors in WD's spread into wild fisheries. The first recorded epizootic in North America was among *S. fontinalis* at Benner Springs Fish Research Station, Centre County, Pennsylvania, in 1956 (Hoffman 1962). By 1968, WD had been identified in 10 states from the east to west coast (Bartholomew and Reno 2002).

The importance of salmonids for sportfishing, commercial fisheries, and aquaculture have resulted in significant funding for WD research and disease management in hatcheries and production facilities. This support has made possible the molecular diagnostic testing necessary to quickly identify infected and asymptomatic fish (Baxa et al. 2002). Research on the intermediate oligochaete host revealed new information about the worm life cycle; based on these findings, modifications in aquaculture were made to eliminate earthen ponds and raceways and facilitate drainage so the ponds would not accumulate sediment that supports the intermediate host (Hedrick, et al. 1998) (Figure 3.4). Raceways, equipment, and transport trucks are now routinely cleaned and disinfected, and ultraviolet irradiation and other safe disinfectant methods were developed to inactivate the waterborne infective stages of the parasite, both in rearing facilities and discharge or transport water (Hedrick et al. 2000*b*). Through a German–US collaboration, a WD-resistant rainbow trout strain was successfully reared in European and US commercial and state hatcheries for stocking streams to crossbreed with and help restore wild rainbow trout populations in the US Rocky Mountain states (Hedrick et al. 2003, Schisler et al. 2006).

As part of stream restoration after forest fires, surveying of streams for both the oligochaete host and infective stage of *M. cerebralis* was initiated by a collaborative team in California. Yellow Creek in the Humbug Valley, Plumas County, was selected for the survey because the stream had undergone changes in the sediment layers and structure following the Storrie Forest fire, making stream conditions more amenable to harboring the oligochaete worm that carries WD (Figure 3.4). Sediment samples from 16 sites were analyzed in this pilot study for oligochaetes using molecular diagnostics. To assess areas of the stream that supported infective stages of the parasite for fish, susceptible fingerling rainbow trout were placed in fish traps in six locations for the entire summer. Examination of sediments and fingerling fish at the University of California Davis Aquatic Animal Health Laboratory revealed site differences in oligochaete worm populations and infective stages of *M. cerebralis* in the trout fingerlings (Weber et al. 2013). This labor-intensive work was only accomplished through the participation and collaboration of a diverse stakeholder team and provided a baseline on the status of WD that may be used to better predict how changes in habitat, alterations to stream substrate composition, and water quality conditions may affect and influence forestry-management decisions and environmental-restoration projects for maintaining healthy wild and stocked salmonid populations.

Fish, both wild and hatchery raised, provide a major source of protein globally, offer rural economic opportunities, provide diverse recreational activities, and are a source of cultural enrichment and biodiversity. Disease outbreaks in wild fisheries are no longer isolated events restricted to pathogen and host studies. In the last several decades, researchers have advanced diagnostic capabilities for identifying pathogens and understanding taxonomic and genetic phylogeny. Metagenomics and other molecular techniques have helped to identify previously undescribed microbes from host species during disease outbreaks. Sometimes these advanced molecular diagnostics are used to investigate mortality events without fully describing the pathological significance, infectious disease ecology of the agents, or the effects of environmental changes to aquatic systems. Cooperative and collaborative approaches, like that described for WD screening and mitigation, are needed for all infectious diseases that pose a threat to wild fisheries. Improving fish health requires a concerted team approach lead by aquatic animal and fisheries professionals and supported by a diverse group of stakeholders. As wild fishery disease outbreaks become more complicated, the political, economic, and social implications of management decisions become increasingly critical for successful plan implementation. Health-management strategies need to be communicated effectively to a range of audiences: infectious-disease experts, anglers, the public, industry, environmentalists, and policy makers. This approach can ensure that evidence-based fish health decisions will be made with scientific rigor and practical approaches based on ethical considerations. Effectively addressing One Health concerns associated with wild fisheries will enable experts to adapt plans to ever changing events and environments, while maintaining political and public support. The end goal is to promote and support the preservation and protection of healthy wild fisheries by enabling broad and effective stewardship of our diverse aquatic ecosystems in perpetuity.

Invasive Species

Invasive species are becoming an increasingly more common problem with globalization of the world economy. Aquatic environments are particularly threatened by invasive plants, animals, and pathogens (NEANS 2021). Aquatic invasives are often introduced by stocking of game species, accidental release of aquaculture animals, intentional release of nonnative pet or collection species, hitchhiking on boats and equipment or in ballast water, accidental escapes due to natural or human facilitated disasters, and environmental alterations to the landscape by dam and canal construction or coastal development and industrial projects. The problems that invasive species impose are multifactorial. Sometimes they are a threat due to direct predation, such as the northern snakehead fish and red lionfish (*Pterois volitans* and *Pterois miles*). Severe damage to both native vegetation and numerous native fish species in North American waters has resulted from invasion of Asian carp including bighead carp (*Hypophthalmichthys nobilis*), silver carp (*Hypophthalmichthys molitrix*), black carp (*Mylopharyngodon piceus*) and grass carp (*Ctenopharyngodon idella*), in the Mississippi River and its tributaries. Invasive species can be carriers of infectious disease, such as KHV, VHS, WD, and infectious salmon anemia or parasites like *Sphaerothecum destruens*, to previously unexposed native populations. Introduced prey species can also supplant normal prey (Spikmans et al. 2020). This has led to thiaminase toxicity to native and stocked Salmonidae, including endangered Atlantic salmon in the northeastern Unites States and Canada after alewife

(*Alosa pseudoharengus*) became the predominate prey item (Honeyfield et al. 2002). The best way forward to help mitigate invasive aquatic introductions is active surveillance and ecological investigation of the impacts of introduced species, coupled with communication and education efforts to raise public vigilance, and ensure the threat remains at the forefront of government conservation priorities.

Environmental Pollution

Pollution and habitat destruction are serious challenges to protecting our diverse aquatic environments and their inhabitants. Chemicals, including flame retardants, household cleaners, plastics, insecticides, pesticides, herbicides, heavy metals, and pharmaceuticals, are polluting streams, lakes, rivers, and oceans. The Mississippi River "dead zone" in the Gulf of Mexico, which happens yearly is a glaring example. On a scale never seen before, a flotilla of plastic debris known as the Great Pacific Garbage Patch, has accumulated off the coast of Hawaii and is now the size of Texas. Marine plastic pollution has been a focus for researchers and conservationists for four decades beginning with the first description of waterborne plastic waste (Coleman and Wehle 1984), from the National Marine Fisheries Service oceanic surveys of plastics (Day et al. 1990), and by the discovery of the great garbage patch flotilla in 1997 by Charles Moore (Moore 2003). Other pollutants include pharmaceutical compounds, endocrine disruptors, and more localized problems like oil spills and other accidental chemical releases. The following sections focus on antimicrobial resistance (AMR), endocrine disruptors, and plastics pollution.

Environmental Antimicrobial Resistance and Endocrine Disruptors

Medical, industrial, and agricultural industries generate antimicrobial and pharmaceutical wastes that are pervasive and are found even in habitats once considered pristine. Wastewater testing data show increased frequencies and greater quantities of antibiotics and antibiotic-resistant microbes in marine ecosystems (Felis et al. 2020). Advances in molecular diagnostics and analytical chemistry have improved identification and monitoring of even small amounts of chemical or antibiotic contamination in aquatic animal populations and environments (Ng and Gin 2019). A survey of lakes in China detected 39 antibiotics with a high proportion of antibiotic resistance (Liu et al. 2018). In North America, standard sewage treatment involves the aeration, mixing, and holding of sewage for days to weeks; this water includes antibiotics flushed down toilets or disposed of in other way, providing an ideal media and incubation period for microbes to develop AMR in wastewater holding tanks.

Antimicrobial resistance is a global problem. A survey of apex marine predatory fish off the US gulf coast, Cape Cod Bay, Florida Keys, and Belize revealed the persistence and environmental dissemination of AMR in wild fish populations at nearly all study sites (Blackburn et al. 2010). Swiss researchers discovered AMR in all raw fish and shellfish samples imported for human consumption that were surveyed for bacterial indicator organisms (Boss et al. 2016). A comprehensive monitoring program in 7 European countries detected 17 antibiotics in effluents of wastewater treatment plants, including medically important ciprofloxacin, cefalexin, and azithromycin, with the highest average concentrations measured in Ireland, Portugal, and Spain (Rodriguez-Mozaz et al. 2020).

Marine plastic pollution is also contributing to the evolution of AMR in marine microbial communities. The cooccurrence of antibiotic resistance genes (ARGs) and metal resistance genes (MRGs) in microbial communities in ocean environments was analyzed using metagenomic data on plastic particles from marine polluted areas, showing ARGs and MRGs present and significantly higher in this plastic-polluted microbiota as compared with unpolluted seawater biota from the North Pacific Gyre (Yang et al. 2019). Multidrug-resistance genes and multimetal-resistance genes were also the predominate classes of genes found in the plastic microbiota from highly polluted areas. Although most ARG contamination is from human wastewater or terrestrial agricultural effluent, aquaculture also contributes to antimicrobial resistance. For nearly 30 years, researchers have known antibiotics such as quinolones, oxalinic acid, and flumequine could accumulate in crustaceans, blue mussels (*Mytilis edulis*), and the gastrointestinal tract of saithe (*Pollachius virens*), mackerel (*Scomber scombrus*), ballan wrasse (*Labrus bergylta*), haddock (*Melanogrammus aeglefinus*), ling (*Molva molva*), cod (*Gadus morhua*), pollack (*Pollachius pollachius*), flounder (*Platichthys flesus*), salmon (*Salmo salar*), and crab (*Cancer pagurus*) proximate to aquaculture facilities that had treated using medicated feeds (Ervik et al. 1994). Samuelsen et al. (1992) observed the presence of both the salmonid pathogen *Aeromonas salmonicida* and oxalinic acid in the GI tract of wild and farmed fish, mollusks, and crabs. In one of the largest metaanalyses of the scientific literature spanning 40 countries, AMR associated with aquaculture correlated by multiantibiotic resistance index was linked to a comparable multiantibiotic resistance index from human clinical bacteria (Reverter et al. 2020). The study concluded that a pooling effect of AMR resistance in the environment may be attributable to the commingling of multiple sources of environmental contamination between human medical, agricultural, industrial, and aquaculture antibiotic usage. The highest AMR levels for aquaculture were observed from warmer climates and in economically vulnerable countries (Reverter et al. 2020).

Endocrine disruptors and their effects on aquatic organisms also result from current wastewater management practices. One of the first studies detecting estrogenic chemicals was in wastewater from sewage treatment plants that discharged into British rivers (Desbrow et al. 1998). Investigating wild populations of riverine fish (roach, *Rutilus rutilus*), Jobling et al. (1998) discovered that estrogenic chemicals caused widespread sexual disruption and high incidences of intersexuality in wild populations of roach in rivers throughout the United Kingdom. Similarly, researchers in the United States found a high prevalence of organic wastewater contaminants from residential, industrial, and agriculture sources in most streams sampled (Kolpin et al. 2002). Some of the most detected compounds in this study included coprostanol (fecal steroid), cholesterol (plant and animal steroid), *N*, *N*-diethyltoluamide (insect repellant), caffeine (stimulant), triclosan (antimicrobial disinfectant), tri(2-chloroethyl) phosphate (fire retardant), and 4-nonylphenol (nonionic detergent metabolite) (Kolpin et al. 2002). Kidd et al. (2007) conducted an ecotoxicological study in Canada exposing fathead minnow (*Pimephales promelas*) to low concentrations (5–6 ng/L) of 17α-ethynylestradiol. Altered reproductive physiology in the minnows included male feminization and disrupted oogenesis in females that resulted in a complete population collapse of this species in a lake. Although most early research was related to identifying estrogenic endocrine disruptors in the environment and their effects on a variety of wildlife species, androgenic effects related to growth-promoting veterinary pharmaceuticals have been identified and found to alter behavior in wild-caught female eastern mosquitofish (*Gambusia holbrooki*) (Bertram et al. 2018).

US Geological Survey (USGS) mobile laboratories are assessing both the presence and long-term effects of endocrine disruptors in the Shenandoah River watershed (Barber et al. 2019*b*). Barber et al. (2019*a*) showed endocrine disruption in a model organism (fathead minnow) along several sites on the Shenandoah by measuring vitellogenin induction in adult males. Another study compared endocrine disruptor levels from environmental samples with wild largemouth bass (*Micropterus salmoides*) or smallmouth bass (*Micropterus dolomieu*) samples from watersheds in 19 National Wildlife Refuges (NWRs) in the northeastern United States (Iwanowicz et al. 2016). Male *M. dolomieu* had intersex gonadal tissues from 60% to 100% of specimens examined at all sample sites (Iwanowicz et al. 2016). Solutions to remediate environmental endocrine disruptors are also being investigated. Polymide-6 particles have been used to selectively remove and recover estrogenic endocrine disruptors from wastewater (Tizaoui et al. 2017), and hopefully this research can be practically scaled for large wastewater treatment facilities to help reduce discharge into vulnerable and protected waterways.

Fig. 3.5. Great white shark (*Carcharodon carcharias*). Illustration by Laura Donohue.

Plastic Pollution

Microplastics are widespread in the aquatic environment. One of the earliest studies on ingestion of polystyrene spherules, investigated fish from southern New England waters and discovered 8 of 14 species had ingested plastic (Carpenter et al. 1972). The highest concentrations of microplastic pollution (>10 up to 10^4 items/m^3) was found in the Mediterranean Sea, the North Sea, the Black Sea, and the South China Sea (Waldschläger et al. 2020). Azevedo-Santos et al. (2019) identified over 427 mainly carnivorous species from various aquatic habitats around the world that had ingested plastic. Analysis of records for 555 species of pelagic and benthic deep-sea fish with documented records for nearly 172,000 individuals, found that the incidence for plastic ingestion had doubled over the last decade, increasing to 26% (Savoca et al. 2021).

Microplastics leach persistent organic pollutants (POPs). Well-known POPs such as phthalates, bisphenol A (BPA), and brominated flame retardants are a few plastic additives that have been proven to have harmful impacts on living organisms (Baršienė et al. 2006; Canesi and Fabbri 2015; Balbi et al. 2016; Wang et al. 2020). The most leached POPs are polycyclic aromatic hydrocarbons (PAHs) (Antunes et al. 2013; Mai et al. 2018, Romeo et al. 2015). Microplastics are not the only source of marine PAH contamination. PAHs are also introduced into the marine environment through oil spills, with an estimated 700 million liters of crude oil dumped accidentally or intentionally into the oceans each year (Kleindienst et al. 2015). Through the ingestion of trace metals, POPs, and PAHs by marine microbial and invertebrate populations, biomagnification of these metals and compounds may occur in the food chain with apex predators like the great white shark (*Carcharodon carcharias*) and humans being at risk of accumulating high levels of these toxins (Figure 3.5) (Croteau et al. 2005; Kelly et al. 2007; LeBlanc 1995; Marsili et al. 2016).

Further research is needed to determine the toxic effects of microplastics and their secondary contaminants such as POPs, PAHs, ARGs, MRGS, and heavy metals in wild, aquaculture, or mariculture-reared fish. Understanding the effects of plastic pollution in our aquatic environments and addressing this problem is a One Health challenge of enormous proportion.

Summary

During the early 1990s, commercial and recreational fishermen in the communities of Tomales and Bodega Bay were reeling from the closure of commercial coho salmon (*Oncorhynchus kisutch*) trolling and shortened fishing seasons in the area. These happened shortly after Brown et al. (1994) published an assessment of California *O. kisutch* that estimated California's spawning numbers were below 5000 and that populations for many historic breeding areas numbered less than 100 individuals, genetically unsustainable numbers. In 1994, collapse of the New England ground fishery, including Atlantic cod was imminent (Figure 3.6). By 1998, the wild

Fig. 3.6. Atlantic cod (*Gadus morhua*). Illustration by Laura Donohue.

Fig. 3.7. Atlantic salmon (*Salmo salar*). Illustration by Laura Donohue.

Atlantic salmon population along the pristine River Deveron in Scotland collapsed as they did elsewhere in the United Kingdom, Ireland, Europe, and the northeastern United States (Figure 3.7). By the end of the 1990s, the great American shad (*Alosa sapidissima*) runs along the Delaware river that were highly anticipated cultural events each spring for Native American tribes and American colonists had practically ceased (Hardy 1999). These events inspired one author (SW) of this chapter to dedicate a career to fish and aquatic ecosystem health.

In the future, climate change is likely to make conditions less amenable to native species and conversely favor some invasive ones. Increases in ambient water temperatures of streams, rivers, and oceans can profoundly affect overall water quality as many chemicals become more toxic at warmer temperatures. This is exacerbated in dammed rivers and streams that are subject to considerable annual water level fluctuations—all conditions unsuitable for breeding, ova hatching, and larval fish growth, but supportive for the establishment of many aquatic pathogens in the aquatic microbiome (Duda et al. 2021). The green sturgeon (*Acipenser medirostris*), a species that has evolved and persisted for 200 million years, is an example of a fish that has experienced a population collapse as dammed rivers in California have prevented their access to adequate breeding habitat (Figure 3.8) (Mayfield and Cech 2004, Adams et al. 2007). These disruptive climate and environmental changes can allow for new pathogens to emerge or allow existing pathogens to expand their geographic ranges or alter their species specificity by infecting compromised native fish species and further exacerbating restoration efforts.

One Health and Conservation Medicine concepts are critical to understanding the interconnectivity of aquatic and terrestrial environments. The terminal migration of salmon, American shad, and some other anadromous species up into natal streams provides (or provided) a source of protein for many animals and a massive influx of nitrogen and other soil enriching organic matter to most of the Northern Hemisphere forests to benefit native plants and animals. This is an irreplaceable component of the ecological cycle of life (Ackert 2012). Another example is wolf (*Canis lupus*) reintroduction and subsequent reduction of elk (*Cervus canadensis*) grazing along streams that allowed natural restoration of aquatic habitats in Yellowstone and Isle Royale National Parks (Smith and Peterson 2021). Although complicated, wolf predation altered ungulate behavior since wolves could more effectively prey on ungulates along rivers and streams coursing through valleys;

Fig. 3.8. Green sturgeon (*Acipenser medirostris*). Illustration by Laura Donohue.

fewer ungulates feeding in these areas allowed natural vegetation to stabilize the riverbanks and allow for other species such as beavers (*Castor canadensis*) to reestablish where they previously had been extirpated in Yellowstone Park, returning a stabilizing effect on streambanks (Smith and Peterson 2021).

Aquatic animal health requires the collaboration of interdisciplinary teams including modelers, environmental scientists, environmental engineers, land developers, fisheries and wildlife biologists, aquatic animal veterinarians, public policy makers, and communications experts to solve complex issues facing entire aquatic ecosystems. Unfortunately, this occurs far too infrequently, and aquatic animal health professionals too seldom are able to translate fisheries disease research into remediation strategies. To be successful, we need to

- investigate wild fisheries health threats and disease outbreaks using advanced diagnostics and evidence-based science;
- create expansive wildlife corridors using riparian watersheds linking rivers and streams that lead to bays and estuaries like the Chesapeake Bay;
- recognize the importance of diverse aquatic habits, and preserve and protect exiting marshes, swamps, and dunes along coastal waterways from development; and
- better understand the consequence of dams, diversions and water policies prior to development, construction or implementation of new projects and policies.

Successful wetland restoration and protection not only benefits native wildlife but helps clean waterways, buffers flooding of inland communities, restores diverse habitats and local groundwater tables, and promotes species diversity of both plants and animals. Substantial socioeconomic and environmental-health benefits result from such restoration. Understanding diseases in aquatic organisms, from infectious to nutritional and genetic to toxins, is exciting for health professionals, but confirmed diagnosis should not be our endpoint. The opportunities in wild fisheries are limitless when professional skills are integrated into diverse teams to solve complex environmental problems to save, preserve and protect these diverse and unique natural habitats.

DEDICATION

This chapter is dedicated to John J. "Tiger" Thouron, O.B.E. Tiger directed and expanded the Thouron award established by his parents Sir John Rupert Hunt Thouron and Esther DuPont, to support graduate study and foster Anglo-American friendship, with more than 700 Thouron scholars crossing the Atlantic during his lifetime. He was also a sportsman and conservationist, managing a bird preserve in Pennsylvania and working on the Deveron, Bogie, and Isla Trust in Scotland to ensure the health of rivers and streams. This chapter recognizes the work of the Alliance for the Chesapeake Bay (Ryan Davis), Chesapeake Bay Foundation (Ashley E. Spotts), Pheasants Forever, Inc. and Quail Forever (Julia Smith), and the USDA CREP Program in returning 30 acres of our property at White Marsh Hollow into contiguous wildlife habitat and riparian watershed critical to the Chesapeake Bay watershed.

LITERATURE CITED

Ackert, L. 2012. Archimedes: new studies in the history and philosophy of science and technology. Volume 34. Sergei Vinogradskii and the cycle of life: from the thermodynamics of life to ecological microbiology, 1850–1950. Springer Science and Business Media, Dordrecht, The Netherlands.

Adams, P. B., C. Grimes, J. E. Hightower, S. T. Lindley, M. L. Moser, and M. J. Parsley. 2007. Population status of North American green sturgeon, *Acipenser medirostris*. Environmental Biology of Fishes 79:339–356.

Adkison, M. A., O. Gilad, and R. P. Hedrick. 2005. An enzyme linked immunosorbent assay (ELISA) for detection of antibodies to the koi herpesvirus (KHV) in the serum of koi *Cyprinus carpio*. Fish Pathology 40:53–62.

Antunes, J. C., J. G. L. Frias, A. C. Micaelo, and P. Sobral. 2013. Resin pellets from beaches of the Portuguese coast and adsorbed persistent organic pollutants. Estuarine, Coastal and Shelf Science 130:62–69.

Azevedo-Santos, V. M., G. R. L. Gonçalves, P. S. Manoel, M. C. Andrade, F. P. Lima, and F. M. Pelicice. 2019.

Plastic ingestion by fish: a global assessment. Environmental Pollution 255:2018–2019.

Bailey, N. W., C. Macias Garcia, and M. G. Ritchie. 2007. Beyond the point of no return? A comparison of genetic diversity in captive and wild populations of two nearly extinct species of Goodeid fish reveals that one is inbred in the wild. Heredity 98:360–367.

Balbi, T., S. Franzellitti, R. Fabbri, M. Montagna, E. Fabbri, and L. Canesi. 2016. Impact of bisphenol A (BPA) on early embryo development in the marine mussel *Mytilus galloprovincialis*: effects on gene transcription. Environmental Pollution 218:996–1004.

Baršienė, J., J. Šyvokienė, and A. Bjornstad. 2006. Induction of micronuclei and other nuclear abnormalities in mussels exposed to bisphenol A, diallyl phthalate and tetrabromodiphenyl ether-47. Aquatic Toxicology 78(Supplement):S105–S108.

Barber, L. B., J. L. Rapp, C. Kandel, S. H. Keefe, J. Rice, P. Westerhoff, D. W. Bertolatus, and A. M. Vajda. 2019*a*. Integrated assessment of wastewater reuse, exposure risk, and fish endocrine disruption in the Shenandoah River watershed. Environmental Science and Technology 53:3429–3440.

Barber, L.B., A. M. Vajda, D. W. Bertolatus, J. E. Dietze, M. L. Hladik, L. R. Iwanowicz, J. R. Jasmann, A. Jastrow, S. H. Keefe, D. W. Kolpin, et al. 2019*b*. Assessment of endocrine disruption in the Shenandoah River watershed—chemical and biological data from mobile laboratory fish exposures and other experiments conducted during 2014, 2015, and 2016. US Geological Survey data release. https://doi.org/10.5066/F7QF8S22. Accessed 11 Aug 2021.

Bartholomew, J. L., and P. W. Reno. 2002. The history and dissemination of whirling disease. Pages 3–24 *in* J. L. Bartholomew, and J. C. Wilson, editors. Whirling disease: reviews and current topics. American Fisheries Society Symposium 29, Bethesda, Maryland, USA.

Baxa, D. V., M. El-Matbouli, K. B. Andree, M. Caffara, S. J. Grésoviac, C. S. Friedman, and R. P. Hedrick. 2002. In situ hybridization: a detection tool for fish pathogens and its application on recent advances on whirling disease research. Pages 293–300 *in* Diseases in Asian aquaculture IV: proceedings of the fourth symposium on diseases in Asian aquaculture. Fish Health Section, Asian Fisheries Society, Quezon City, Philippines.

Baxa, D. V., A. Javidmehr, S. M. Mapes, and S. J. Teh. 2015. Subclinical mycobacterium infections in wild delta smelt. Austin Journal of Veterinary Science and Animal Husbandry 2:1004.

Becker, J. A., M. P. Ward, and P. M. Hick. 2019. An epidemiologic model of koi herpesvirus (KHV) biocontrol for carp in Australia. Australian Zoologist 40:25–35.

Bercovier, H., Y. Fishman, R. Nahary, S. Sinai, A. Zlotkin, M. Eyngor, O. Gilad, A. Eldar, and R. P. Hedrick. 2005. Cloning of the koi herpesvirus (KHV) gene encoding thymidine kinase and its use for a highly sensitive PCR based diagnosis. BMC Microbiology 5:13.

Bertram, M. G., M. Saaristo, J. M. Martin, T. E. Ecker, M. Michelangeli, C. P. Johnstone, and B. B. M. Wong. 2018. Field-realistic exposure to the androgenic endocrine disruptor 17β-trenbolone alters ecologically important behaviours in female fish across multiple contexts. Environmental Pollution 243:900–911.

Blackburn, J. K., M. A. Mitchell, M. C. H. Blackburn, A. Curtis, and B. A. Thompson. 2010. Evidence of antibiotic resistance in free-swimming, top-level marine predatory fishes. Journal of Zoo and Wildlife Medicine 41:7–16.

Boss, R., G. Overesch, and A. Baumgartner. 2016. Antimicrobial resistance of *Escherichia coli*, Enterococci, *Pseudomonas aeruginosa*, and *Staphylococcus aureus* from raw fish and seafood imported into Switzerland. Journal of Food Protection 79:1240–1246.

Brenes, R., D. L. Miller, T. B. Waltzek, R. P. Wilkes, J. L. Tucker, J. C. Chaney, R. H. Hardman, M. D. Brand, R. R. Huether, and M. J. Gray. 2014. Susceptibility of fish and turtles to three ranaviruses isolated from different ectothermic vertebrate classes. Journal of Aquatic Animal Health 26:118–126.

Brown, L. R., P. B. Moyle, and R. M. Yoshiyama. 1994. Historical decline and current status of coho salmon in California. North American Journal of Fisheries Management 14:237–261.

Canesi, L., and E. Fabbri. 2015. Environmental effects of BPA: focus on aquatic species. Dose-Response 13(3):1559325815598304.

Carpenter, E. J., S. J. Anderson, G. R. Harvey, H. P. Miklas, and B. B. Peck. 1972. Polystyrene spherules in coastal waters. Science 178:749–750.

Coleman, F. C., and D. H. S. Wehle. 1984. Plastic pollution: a worldwide oceanic problem. Parks 9:9–12.

Croteau, M. N., S. N. Luoma, and A. R. Stewart. 2005. Trophic transfer of metals along freshwater food webs: evidence of cadmium biomagnification in nature. Limnology and Oceanography 50:1511–1519.

Dailloux, M., C. Laurain, M. Weber, and P. H. Hartemann. 1992. Water and nontuberculous mycobacteria. Water Research 33:2219–2228.

Day, R. H., D. G. Shaw, and S. E. Ignell. 1990. The quantitative distribution and characteristics of neuston plastic in the North Pacific Ocean, 1985–88. Pages 247–266 *in* L. S. Shomura and M. L. Godfrey, editors. Proceedings of the second international conference on marine debris. US National Oceanic and

Atmospheric Administration, Technical Memorandum NOAA-TM-N MFS-SWFSC-154, Washington, D.C., USA.

Delghandi, M. R., K. Waldner, M. El-Matbouli, and S. Menanteau-Ledouble. 2020. Identification *Mycobacterium spp.* in the natural water of two Austrian rivers. Microorganisms 8:130.

Densmore, C. L., L. R. Iwanowicz, A. Henderson, D. D. Iwanowicz, and J. S. Odenkirk. 2016. Mycobacterial infection in northern snakehead (*Channa argus*) from the Potomac River catchment. Journal of Fish Diseases 39:771–775.

Desbrow, C., E. J. Routledge, G. C. Brighty, J. P. Sumpter, and M. Waldock. 1998. Identification of estrogenic chemicals in STW effluent. 1. Chemical fractionation and in vitro biological screening. Environmental Science and Technology 32:1549–1558.

Duda, J. J., C. E. Torgersen, S. J. Brenkman, R. J. Peters, K.T. Sutton, H. A. Connor, P. Kennedy, S. C. Corbett, E. Z. Welty, A. Geffre, et al. 2021. Reconnecting the Elwha River: spatial patterns of fish response to dam removal. Frontiers in Ecology and Evolution 9:765488.

Egusa, S., and N. Masuda. 1971. A new fungal disease of *Plecoglossus altivelis*. Fish Pathology 6:41–46.

Ervik, A., B. Thorsen, V. Eriksen, B. T. Lunestad, and O. B. Samuelsen. 1994. Impact of administering antibacterial agents on wild fish and blue mussels *Mytilus edulis* in the vicinity of fish farms. Diseases of Aquatic Organisms 18:45–51.

Eschmeyer, W. N., R. Fricke, and R. Van der Laan, editors. 2021. Catalog of fishes: genera, species, references. http://researcharchive.calacademy.org/research/ichthyology/catalog/fishcatmain.asp. Accessed 16 Nov 2021.

Escobar, L. E., J. Escobar-Dodero, and N. B. Phelps. 2018. Infectious disease in fish: global risk of viral hemorrhagic septicemia virus. Reviews in Fish Biology and Fisheries 28:637–655.

Felis, E., J. Kalka, A. Sochacki, K. Kowalska, S. Bajkacz, M. Harnisz, and E. Korzeniewska. 2020. Antimicrobial pharmaceuticals in the aquatic environment—occurrence and environmental implications. European Journal of Pharmacology 866:172813.

Garver, K. A., L. Al-Hussinee, L. M. Hawley, T. Schroeder, S. Edes, V. LePage, E. Contador, S. Russell, S. Lord, R. M. Stevenson, and B. Souter. 2010. Mass mortality associated with koi herpesvirus in wild common carp in Canada. Journal of Wildlife Diseases 46:1242–1251.

Gauthier, D. T., A. N. Haines, and W. K. Vogelbein. 2021. Elevated temperature inhibits *Mycobacterium shottsii* infection and *Mycobacterium pseudoshottsii* disease in striped bass *Morone saxatilis*. Diseases of Aquatic Organisms 144:159–174.

Gibbons, J. W., D. E. Scott, T. J. Ryan, K. A. Buhlmann, T. D. Tuberville, B. S. Metts, J. L. Greene, T. Mills, Y. Leiden, S. Poppy, et al. 2000. The global decline of reptiles, déjà vu amphibians: reptile species are declining on a global scale. Six significant threats to reptile populations are habitat loss and degradation, introduced invasive species, environmental pollution, disease, unsustainable use, and global climate change. BioScience 50:653–666.

Go, J., I. Marsh, M. Gabor, V. Saunders, R. L. Reece, J. Frances, C. Boys, and L. J. Gabor. 2012. Detection of *Aphanomyces invadans* and epizootic ulcerative syndrome in the Murray-Darling drainage. Australian Veterinary Journal 90:513–514.

Gorgoglione, B., C. Bailey, and J. A. Ferguson. 2020. Proliferative kidney disease in Alaskan salmonids with evidence that pathogenic myxozoans may be emerging north. International Journal for Parasitology 50:797–807.

Groner, M. L., J. M. Hoenig, R. Pradel, R. Choquet, W. K. Vogelbein, D.T. Gauthier, and M. A. M. Friedrichs. 2018. Dermal mycobacteriosis and warming sea surface temperatures are associated with elevated mortality of striped bass in Chesapeake Bay. Ecology and Evolution 8:9384–9397.

Hallett, S. L., A. Hartigan, A., and S. D. Atkinson. 2015. Myxozoans on the move: dispersal modes, exotic species and emerging diseases. Pages 343–362 *in* B. Okamura, A. Gruhl, and J. L. Bartholomew, editors. Myxozoan evolution, ecology and development. Springer International, Cham, Switzerland.

Hardy, C. 1999. Fish or foul: a history of the Delaware River basin through the perspective of the American shad, 1682 to the present. Pennsylvania History: A Journal of Mid-Atlantic Studies 66:506–534.

Heckert, R. A., S. Elankumaran, A. Milani, and A. Baya. 2001. Detection of a new *Mycobacterium* species in wild striped bass in the Chesapeake Bay. Journal of Clinical Microbiology 39:710–715.

Hedrick, R. P., M. A. Adkison, M. El-Matbouli, and E. MacConnell. 1998. Whirling disease: re-emergence among wild trout. Immunological Reviews 166:365–376.

Hedrick R. P., O. Gilad, S. Yun, J. V. Spangenberg, G. D. Marty, R. W. Nordhausen, M. J. Kebus, H. Bercovier, and A. Eldar. 2000*a*. A herpesvirus associated with mass mortality of juvenile and adult koi, a strain of common carp. Journal of Aquatic Animal Health 12:44–57.

Hedrick, R. P., T. S. McDowell, G. D. Marty, G. T. Fosgate, K. Mukkatira, K. Myklebust, and M. El-Matbouli. 2003. Susceptibility of two strains of rainbow trout (one with suspected resistance to whirling disease) to *Myxobolus cerebralis* infection. Diseases of Aquatic Organisms 55:37–44.

Hedrick, R. P., T. S. McDowell, G. D. Marty, K. Mukkatira, D. B. Antonio, K. B. Andree, Z. Bukhari, and T. Clancy. 2000*b*. Ultraviolet irradiation inactivates the waterborne infective stages of *Myxobolus cerebralis*: a treatment for hatchery water supplies. Diseases of Aquatic Organisms 42:53–59.

Hoffman, G. L. 1962. Whirling disease of trout. US Fish and Wildlife Service, Fishery Leaflet 508, Washington, D.C., USA.

Hoffman, G. L. 1969. Intercontinental and transcontinental dissemination and transfaunation of fish parasites with emphasis on whirling disease (*Myxosoma cerebralis*) and its effects on fish. Pages 69–81 *in* S. F. Snieszko, editor. Symposium on diseases of fisheries and shellfishes. American Fisheries Society, Special Publication No. 5, Bethesda, Maryland, USA.

Honeyfield, D. C., J. P. Hinterkopf, and S. B. Brown. 2002. Isolation of thiaminase-positive bacteria from alewife. Transactions of the American Fisheries Society 131:171–175.

Hosoya, H., and M. Suzuki. 1976. Effect of NaCl solution bath on a disease of juvenile colorcarp accompanied by mass mortality, arising at the rainy season. Reports of the Niigata Prefecture Inland Fish Experimental Station 4:69–71. (In Japanese.)

Iivanainen, E. K., P. J. Martikainen, P. K. Väänänen, and M.-L. Katila.1993. Environmental factors affecting the occurrence of mycobacteria in brook waters. Applied and Environmental Microbiology 59:398–404.

Iivanainen, E., P. J. Martikainen, P. K. Väänänen, and M.-L. Katila. 1999. Environmental factors affecting the occurrence of mycobacteria in brook sediments. Journal of Applied Microbiology 86:673–681.

Iwanowicz, L. R., V. S. Blazer, A. E. Pinkney, C. P. Guy, A. M. Major, K. Munney, S. Mierzykowski, S. Lingenfelser, A. Secord, K. Patnode, et al. 2016. Evidence of estrogenic endocrine disruption in smallmouth and largemouth bass inhabiting northeast US national wildlife refuge waters: a reconnaissance study. Ecotoxicology and Environmental Safety 124:50–59.

Jobling, S., M. Nolan, C. R. Tyler, G. Brighty, and J. P. Sumpter.1998. Widespread sexual disruption in wild fish. Environmental Science and Technology 32:2498–2506.

Jourdan, J., D. Bierbach, R. Riesch, A. Schießl, A. Wigh, L. Arias-Rodriguez, J. R. Indy, S. Klaus, C. Zimmer, and M. Plath. 2014. Microhabitat use, population densities, and size distributions of sulfur cave dwelling *Poecilia mexicana*. PeerJ 2:e490.

Kane, A. S., C. B Stine, L. Hungerford, M. Matsche, C. Driscoll, and A. M. Baya. 2007. Mycobacteria as environmental portent in Chesapeake Bay fish species. Emerging Infectious Diseases 13:329.

Kelly, B. C., M. G. Ikonomou, J. D. Blair, A. E. Morin, and F. A. P. C. Gobas. 2007. Food web–specific biomagnification of persistent organic pollutants. Science 317:236–239.

Kidd, K. A., P. J. Blanchfield, K. H. Mills, V. P. Palace, R. E. Evans, J. M. Lazorchak, and R. W. Flick. 2007. Collapse of a fish population after exposure to a synthetic estrogen. Proceedings of the National Academy of Sciences USA 104:8897–8901.

Kim, S. W., S. S. Giri, S. G. Kim, J. Kwon, W. T. Oh, and S. C. Park. 2020. Carp edema virus and cyprinid herpesvirus-3 coinfection is associated with mass mortality of koi (*Cyprinus carpio haematopterus*) in the Republic of Korea. Pathogens 9:222.

Kleindienst, S., J. H. Paul, and S. B. Joye. 2015. Using dispersants after oil spills impacts on the composition and activity of microbial communities. Nature Reviews Microbiology 13:388–396.

Kolpin, D. W., E. T. Furlong, M. T. Meyer, E. M. Thurman, S. D. Zaugg, L. B. Barber, and H. T. Buxton. 2002. Pharmaceuticals, hormones, and other organic wastewater contaminants in US streams, 1999–2000: a national reconnaissance. Environmental Science and Technology 36:1202–1211.

La, V. T., and S. J. Cooke. 2011. Advancing the science and practice of fish kill investigations. Reviews in Fisheries Science 19:21–33.

LeBlanc, G. A. 1995. Trophic-level differences in the bioconcentration of chemicals: implications in assessing environmental biomagnification. Environmental Science and Technology 29:154–160.

Lewisch, E., B. Gorgoglione, K. Way, and M. El-Matbouli. 2015. Carp edema virus/Koi sleepy disease: an emerging disease in Central-East Europe. Transboundary Emerging Diseases 62:6–12.

Liu, X., S. Lu, W. Guo, B. Xi, and W. Wang. 2018. Antibiotics in the aquatic environments: a review of lakes, China. Science of the Total Environment 627:1195–1208.

Lovy, J., S. E. Friend, L. Al-Hussinee, and T. B. Waltzek. 2018. First report of carp edema virus in the mortality of wild common carp *Cyprinus carpio* in North America. Diseases of Aquatic Organisms 131:177–186.

Mai, L., L. J. Bao, L. Shi, L. Y. Liu, and E. Y. Zeng. 2018. Polycyclic aromatic hydrocarbons affiliated with microplastics in surface waters of Bohai and Huanghai Seas, China. Environmental Pollution 241:834–840.

Marsili, L., D. Coppola, M. Giannetti, S. Casini, M. C. Fossi, J. H. Van Wyk, E. Sperone, S. Tripepi, P. Micarelli, and S. Rizzuto. 2016. Skin biopsies as a sensitive non-lethal technique for the ecotoxicological studies of great white shark (*Carcharodon carcharias*) sampled in South Africa. Expert Opinion on Environmental Biology Journal 5:1.

Marsella, A., T. Pretto, M. Abbadi, R. Quartesan, L. Cortinovis, E. Fiocchi, A. Manfrin, and A. Toffan. 2021. Carp edema virus-related mortality in wild adult common carp (*Cyprinus carpio*) in Italy. Journal of Fish Diseases 44:939–947.

Martin, C. H., J. E. Crawford, B. J. Turner, and L. H. Simons. 2016. Diabolical survival in Death Valley: recent pupfish colonization, gene flow and genetic assimilation in the smallest species range on Earth. Proceedings of the Royal Society B Biological Sciences 283:20152334.

Mather, F. 1887. Brown trout in America. Bulletin of the United States Fish Commission 7:21–22.

Mayfield, R. B., and J. J. Cech Jr. 2004. Temperature effects on green sturgeon bioenergetics. Transactions of the American Fisheries Society 133:961–970.

McClain, C., M. Fougerolle, M. Rex, and J. Welch. 2001. MOCNESS estimates of the size and abundance of a pelagic gonostomatid fish *Cyclothone pallida* off the Bahamas. Journal of the Marine Biological Association of the United Kingdom 81:869–871.

McKenzie, R. A., and W. T. K. Hall. 1976. Dermal ulceration of mullet (*Mugil cephalus*). Australian Veterinary Journal 52:230–231.

Meyer, F. P., and L. A. Barclay, editors. 1990. Field manual for the investigation of fish kills. US Fish and Wildlife Service Resource Publication 177, Washington, D.C., USA.

Minckley, W. L., and J. E. Deacon. 1968. Southwestern fishes and the enigma of "endangered species": man's invasion of deserts creates problems for native animals, especially for freshwater fishes. Science 159:1424–1432.

Minckley, W. L., and C. O. Minckley. 1986. *Cyprinodon pachycephalus*, a new species of pupfish (Cyprinodontidae) from the Chihuahuan desert of northern México. Copeia 1986:184–192.

Miyazaki, T., and S. Egusa. 1972. Studies on mycotic granulomatosis in freshwater fish I. Mycotic granulomatosis in goldfish. Fish Pathology 7:15–25.

Moore, C. 2003. Trashed: across the Pacific Ocean, plastics, plastics, everywhere. Natural History Magazine, November 2003. https://www.naturalhistorymag.com/features/172720/trash-revisited. Accessed 18 Nov 2021.

Moyle, P. B., and J. E. Williams. 1990. Biodiversity loss in the temperate zone: decline of the native fish fauna of California. Conservation Biology 4:275–284.

Ng, C., and K. Y. H. Gin. 2019. Monitoring antimicrobial resistance dissemination in aquatic systems. Water 11:71.

Northeast Aquatic Nuisance Species Panel (NEANS). 2021. NEANS homepage. https://www.northeastans.org. Accessed 13 Oct 2021.

Ono, S. I., A. Nagai, and N. Sugai. 1986. A histopathological study on juvenile colorcarp, *Cyprinus carpio*, showing edema. Fish Pathology 21:167–175.

Oyamatsu, T., N. Hata, K. Yamada, T. Sano, and H. Fukuda. 1997. An etiological study on mass mortality of cultured colorcarp juveniles showing edema. Fish Pathology 32:81–88.

Padhi, S. K., I. Tolo, M. McEachran, A. Primus, S. K. Mor, and N. B. D. Phelps. 2019. Koi herpesvirus and carp oedema virus: infections and coinfections during mortality events of wild common carp in the United States. Journal of Fish Diseases 42:1609–1621.

Phelps, N. B., I. Bueno, D. A. Poo-Muñoz, S. J. Knowles, S. Massarani, R. Rettkowski, L. Shen, H. Rantala, P. L. F. Phelps, and L. E. Escobar. 2019. Retrospective and predictive investigation of fish kill events. Journal of Aquatic Animal Health 31:61–70.

Pikulkaew, S., K. Phatwan, W. Banlunara, M. Intanon, and J. K. Bernard. 2020. First evidence of carp edema virus infection of koi *Cyprinus carpio* in Chiang Mai Province, Thailand. Viruses 12:1400.

Pradhan, P. K., G. Rathore, N. Sood, T. R. Swaminathan, M. K. Yadav, D. K. Verma, D. K. Chaudhary, R. Abidi, P. Punia, and J. K. Jena. 2014. Emergence of epizootic ulcerative syndrome: large-scale mortalities of cultured and wild fish species in Uttar Pradesh, India. Current Science 106:1711–1718.

Reverter, M., S. Sarter, D. Caruso, S. C. Avarre, M. Combe, E. Pepey, L. Pouyaud, S. Vega-Heredía, H. De Verdal, and R. E. Gozlan. 2020. Aquaculture at the crossroads of global warming and antimicrobial resistance. Nature Communications 11:1870.

Roberts, R. J., L. G. Willoughby, and S. Chinabut. 1993. Mycotic aspects of epizootic ulcerative syndrome (EUS) of Asian fishes. Journal of Fish Diseases 16: 169–183.

Rodriguez-Mozaz, S., I. Vaz-Moreira, S. V. Della Giustina, M. Llorca, D. Barceló, S. Schubert, T. U. Berendonk, I. Michael-Kordatou, D. Fatta-Kassinos, J. L. Martinez, et al. 2020. Antibiotic residues in final effluents of European wastewater treatment plants and their impact on the aquatic environment. Environment International 140:105733.

Romeo, T., M. D'Alessandro, V. Esposito, G. Scotti, D. Berto, M. Formalewicz, S. Noventa, S. Giuliani, S. Macchia, D. Sartori, et al. 2015. Environmental quality assessment of Grand Harbour (Valletta, Maltese Islands): a case study of a busy harbour in the Central Mediterranean Sea. Environmental Monitoring and Assessment 187:747.

Samuelsen, O. B., B. T. Lunestad, B. Husevåg, T. Hølleland, and A. Ervik. 1992. Residues of oxolinic acid in wild fauna following medication in fish farms. Diseases of Aquatic Organisms, 12:111–119.

Savoca, M., A. McInturf, and E. Hazen. 2021. Plastic ingestion by marine fish is widespread and increasing. Global Change Biology 27:2188–2199.

Schisler, G. J., K. A. Myklebust, and R. P. Hedrick. 2006. Inheritance of *Myxobolus cerebralis* resistance among F1-generation crosses of whirling disease resistant and susceptible rainbow trout strains. Journal of Aquatic Animal Health 18:109–115.

Schulze-Röbbecke, R., and K. Buchholtz. 1992. Heat susceptibility of aquatic mycobacteria. Applied and Environmental Microbiology 58:1869–1873.

Sibanda, S., D. M. Pfukenyi, M. Barson, B. Hang'ombe, and G. Matope. 2018. Emergence of infection with *Aphanomyces invadans* in fish in some main aquatic ecosystems in Zimbabwe: a threat to national fisheries production. Transboundary and Emerging Diseases 65:1648–1656.

Smith, B. D. 2011. The subsistence economies of Indigenous North American societies: a handbook. Smithsonian Institution Scholarly Press, Washington, D.C., USA.

Smith, D. W., and R. O. Peterson. 2021. Intended and unintended consequences of wolf restoration to Yellowstone and Isle Royale National Parks. Conservation Science and Practice 3:e413.

Spikmans, F., P. Lemmers., H. L. M. op den Camp, F. Kappen, A. Blaakmeer, G. van der Velde, F. van Langevelde, R. S. E. W. Leuven, and T. A. van Alen. 2020. Impact of the invasive alien topmouth gudgeon (*Pseudorasbora parva*) and its associated parasite *Sphaerothecum destruens* on native fish species. Biological Invasions 22:587–601.

Stagg, H. E. B., S. Guðmundsdóttir, N. Vendramin, N. M. Ruane, H. Sigurðardóttir, D. H. Christiansen, A. Cuenca, P. E. Petersen, E. S. Munro, V. L. Popov, et al. 2020. Characterization of ranaviruses isolated from lumpfish *Cyclopterus lumpus* L. in the North Atlantic area: proposal for a new ranavirus species (European North Atlantic Ranavirus). Journal of General Virology 101:198–207.

St-Hilaire, S., N. Beevers, K. Way, R. M. Le Deuff, P. Martin, and C. Joiner. 2005. Reactivation of koi herpesvirus infections in common carp *Cyprinus carpio*. Diseases of Aquatic Organisms 67:15–23.

Steinhagen, D., V. Jung-Schroers, and M. Adamek. 2016. Impact of cyprinid herpesvirus 3 (koi herpesvirus) on wild and cultured fish. CAB Reviews 11:041.

Tizaoui, C., S. Ben Fredj, and L. Monser. 2017. Polyamide-6 for the removal and recovery of the estrogenic endocrine disruptors estrone, 17β-estradiol, 17α-ethinylestradiol and the oxidation product 2-hydroxyestradiol in water. Chemical Engineering Journal 328:98–105.

Tobler, M., I. Schlupp, F. J. García de León, M. Glaubrecht, and M. Plath. 2007. Extreme habitats as refuge from parasite infections? Evidence from an extremophile fish. Acta Oecologica 31:270–275.

Tolo, I. E., S. K. Padhi, P. J. Hundt, P. G. Bajer, S. K. Mor, and N. B. D. Phelps. 2021. Host range of carp edema virus (CEV) during a natural mortality event in a Minnesota lake and update of CEV associated mortality events in the USA. Viruses 13:400.

US Geological Survey (USGS). 2022. *Salmo trutta* Linnaeus, 1758. https://nas.er.usgs.gov/queries/factsheet.aspx?SpeciesID=931/. Accessed 24 Aug 2022.

Waldschläger, K., S. Lechthaler, G. Stauch, and H. R. Schüttrumpf. 2020. The way of microplastic through the environment—application of the source-pathway-receptor model. Science of the Total Environment 71:136584.

Wallace, I. S., A. Gregory, A. G. Murray, E. S. Munro, and R. S. Raynard. 2008. Distribution of infectious pancreatic necrosis virus (IPNV) in wild marine fish from Scottish waters with respect to clinically infected aquaculture sites producing Atlantic salmon, *Salmo salar* L. Journal of Fish Diseases 31:177–186.

Walster, C. 1999. Clinical observations of severe mortalities in koi carp, *Cyprinus carpio*, with gill disease. Fish Veterinary Journal 3: 54–58.

Waltzek T. B., G. D. Marty, M. E. Alfaro, W. R. Bennett, K. A. Garver, M. Haulena, E. S. Weber III, and R. P. Hedrick. 2012. Systemic iridovirus from threespine stickleback *Gasterosteus aculeatus* represents a new megalocytivirus species (family Iridoviridae). Diseases of Aquatic Organisms 98:41–56.

Wang, X., Q. Zhu, X. Yan, Y. Wang, C. Liao, and G. Jiang. 2020. A review of organophosphate flame retardants and plasticizers in the environment: analysis, occurrence and risk assessment. Science of the Total Environment 731:139071.

Way, K., O. Haenen, D. Stone, M. Adamek, S. M. Bergmann, L. Bigarré, N. Diserens, M. El-Matbouli, M. C. Gjessing, V. Jung-Schroers, and E. Leguay. 2017. Emergence of carp edema virus (CEV) and its significance to European common carp and koi *Cyprinus carpio*. Diseases of Aquatic Organisms, 126:155–166.

Way, K., and D. Stone. 2013. Emergence of carp edema virus-like (CEV-like) disease in the UK. CEFAS Finfish News 15:32--34.

Weber, E. S., K. Malm, D. V. Baxa, S. C. Yun, and L. Campbell. 2013. Yellow Creek whirling disease study: investigating the presence and potential severity of whirling disease in Yellow Creek. Pages 1–14 *in* Final Report submitted to US Forest Service and California Department of Fish and Game, Sacramento, California, USA.

Weber, E. S., III, T. B. Waltzek, D. A. Young, E. L. Twitchell, A. E. Gates, A. Vagelli, G. R. Risatti, R. P. Hedrick, and F. Salvatore Jr. 2009. Systemic iridovirus infection in the Banggai cardinalfish (*Pterapogon kauderni* Koumans 1933). Journal of Veterinary Diagnostic Investigation 21:306–320.

Weber, E. S. P., III. 2012. Emerging infectious diseases in aquaculture and fisheries. Pages 314–327 *in* A. A. Aguirre, R. S. Ostfield, and P. Daszak, editors. New directions in conservation medicine: applied cases of ecological health. Oxford University Press, New York, New York, USA.

Wienerroither, R., F. Uiblein, F. Bordes, and T. Moreno. 2009. Composition, distribution, and diversity of pelagic fishes around the Canary Islands, Eastern Central Atlantic. Marine Biology Research 5:328–344.

Woo, P. T. K., J. Leong, and K. Buchmann, eds. 2020. Climate change and infectious fish diseases. Commonwealth Agricultural Bureaux International, Wallingoford, UK.

World Organisation for Animal Health (WOAH). 2021. OIE animal diseases portal. OIE aquatic disease manual. https://www.oie.int/en/what-we-do/animal-health-and-welfare/animal-diseases/. Accessed 22 Oct 2021.

Yang, Y., G. Liu, W. Song, C. Ye, H. Lin, Z. Li, and W. Liu. 2019. Plastics in the marine environment are reservoirs for antibiotic and metal resistance genes. Environment International 123:79–86.

4 Ranavirosis and Chytridiomycosis

The Impact on Amphibian Species

Debra L. Miller, E. Davis Carter, Rebecca H. Hardman, Mathew J. Gray

Introduction

Global amphibian populations are in decline from multiple threats (Collins and Storfer 2003, Wake and Vredenburg 2008). Although habitat destruction and degradation are playing a role in population declines, emerging infectious diseases are responsible for the most recent significant losses in global amphibian biodiversity (Scheele et al. 2019). Many pathogens can cause disease in amphibians, but for the purpose of this chapter, we will concentrate on a virus, *Ranavirus*, and two chytrid fungi, *Batrachochytrium dendrobatidis* (Bd) and *B. salamandrivorans* (Bsal), that are listed as notifiable pathogens by the World Organization for Animal Health (OIE). Susceptibility to these pathogens varies among host species and often among pathogen species, strains, and isolates (Hoverman et al. 2010, Dang et al. 2017). Additionally, other factors (e.g., environmental temperature, contaminants, concurrent infections, and life stage) can influence host susceptibility and likelihood of pathogen transmission. Importantly, the expanding distribution of these pathogens has been (and continues to be) facilitated by humans, including transport on contaminated equipment (e.g., hiking boots, waders, kayaks, and boats) and through wildlife trade networks (Kolby et al. 2014, Fitzpatrick et al. 2018, Casais et al. 2019). These three pathogens have been detected in amphibian trade (Schloegel et al. 2009, Kolby et al. 2014, Fitzpatrick et al. 2018); in the case of the chytrid fungi, release of infected human-managed amphibians into the wild has occurred with negative impacts on populations (Walker et al. 2008, Martel et al. 2020). For most countries, importation of wildlife does not require an animal health certificate verifying uninfected status of OIE-notifiable pathogens (Grant et al. 2017), unlike domesticated animals, thus providing few barriers to the global translocation of pathogens. Recent surveys in the United States indicate that pet amphibian businesses are in support of a clean trade program (Cavasos et al. 2021). Moreover, US pet consumers are willing to pay more for certified pathogen-free amphibians (Cavasos et al. 2021). Future steps needed include identifying the structure of a clean-trade program and garnering government and industry support for implementation.

When considering diseases, the intersection of host, pathogen, and environment provides a conceptual framework for understanding how and why they emerge (see Figure I.1). This diagram demonstrates that disease results not just from the pathogen, host, or environment but from a combination of conditions that occurs at the intersection

The class Amphibia contains a wide variety of species, which are found on all continents except Antarctica, and belong to one of three orders (Anura, frogs and toads; Caudata, salamanders; Gymnophiona, caecilians) whose general body forms are shown here. Illustration by Laura Donohue.

of all three (in certain situations, perhaps two of the three). Disease is the etiologic outcome of pathogen infection and negative impacts on a host that leads to morbidity and possibly mortality. Hence, disease describes what is occurring. For free-living wildlife however, it is necessary to also investigate why and how morbidity and mortality events are occurring if there is any hope of mitigating outbreaks. For example, disease outbreaks could occur due to introduction of a novel pathogen, such as Bsal outbreaks in Europe (Stegen et al. 2017). Alternatively, elevated levels of stressors in the environment could decrease immune function and result in emergence of endemic pathogens, such as ranavirus outbreaks associated with deicing runoff (Hall et al. 2020). Below are disease descriptions of three OIE-notifiable amphibian pathogens that have been associated with both local and widespread die-off events. In discussing these intriguing pathogens and the devastation they cause, it should continually be asked "why and how" so that solutions can be developed to limit and halt their spread.

Ranavirus

Etiology

Ranaviruses are large double-stranded DNA viruses in the family Iridoviridae that form icosahedral capsids. They are capable of infecting reptiles, amphibians, and fish, whereas the other genera within the family infect only invertebrates (*Iridovirus* and *Chloriridovirus*) or only fish (*Megalocytivirus* and *Lymphocystivirus*) (Chinchar et al. 2017). Although there are reports of many different ranaviruses, there are currently seven accepted species including frog virus 3 (FV3), *Ambystoma tigrinum* virus (ATV), common midwife toad virus (CMTV), epizootic haematopoietic necrosis virus, European North Atlantic virus, Santee-Cooper ranavirus, and Singapore grouper iridovirus (ICTV 2022). The most notable (and perhaps dangerous) attribute of *Ranavirus* is found within FV3 and CMTV because they can be transmitted among vertebrate classes, namely reptiles, fish (see Chapter 3), and amphibians (Brenes et al. 2014, von Essen et al. 2020). Hence, ranaviruses can have devastating effects on aquatic ecosystems across multiple trophic levels.

Ranaviruses can be found on every continent except Antarctica (Duffus et al. 2015), but they have been most devastating to wild populations of amphibians in North America and Europe. In North America, FV3 and ATV are the most commonly reported ranaviruses infecting amphibians, although there is evidence of CMTV infection in some cases

(Claytor et al. 2017, Vilaça et al. 2019). The history of FV3 is particularly interesting as it dates back to the 1960s when Dr. Allan Granoff (at St. Jude's Children's Hospital in Memphis Tennessee) was studying neoplasms caused by viruses (Granoff et al. 1966). While researching cancer in children, Dr. Granoff used the ranid herpesvirus-1 model of northern leopard frogs (*Lithobates pipiens*), where frogs infected as tadpoles develop renal carcinoma as adults. During this investigation he discovered FV3, and it was believed to not be lethal for amphibians at the time. Subsequently, a polymerase chain reaction (PCR) test for ranavirus was developed and retrospective testing of previous cases by Green et al. (2002) found that, of 44 amphibian mortality events occurring between 1996 and 2001, more than half (57%) were due in part to ranaviruses. Regardless of whether ranaviruses have been recently expanding their range or if diagnostic tests and awareness have simply increased detection (Gray and Chinchar 2015), it is clear that ranaviruses can negatively impact amphibian populations. For example, Wheelwright et al. (2014) reported mortality of thousands of wood frog (*Lithobates sylvaticus*) tadpoles in a matter of days. Stage-structured models also demonstrate that populations of susceptible host species can go extinct in a matter of years due to ranavirus outbreaks (Earl and Gray 2014, Earl et al. 2016). Thus, population-level impacts of ranaviruses can be severe (Price et al. 2014).

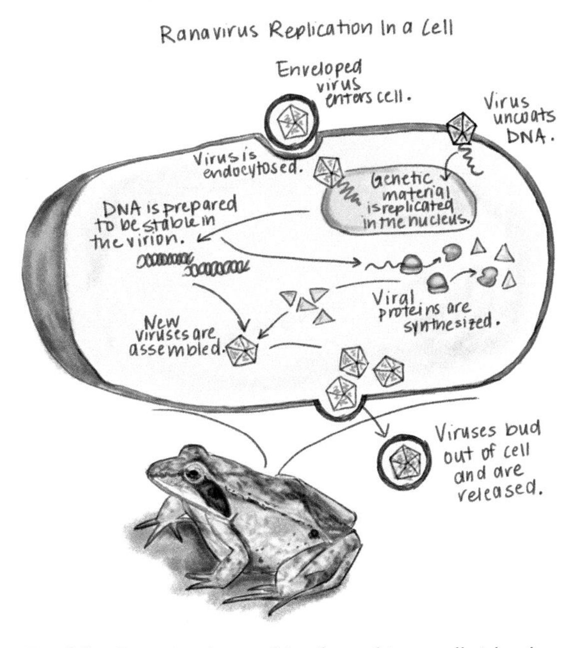

Fig. 4.1. Ranavirus is capable of attacking a cell either by entering the host cell through endocytosis or as a naked virion that injects viral DNA into the cell. Viral DNA is replicated in the nucleus and assembled into paracrystalline arrays in the cytoplasm. Virions then exit the cell via budding through the plasma membrane. Illustration by Laura Donohue.

Pathogenesis and Pathology

Ranaviruses attack cells by invading them via one of two pathways (Figure 4.1). Enveloped virions (virions encased in a plasma membrane) enter a host cell through receptor-mediated endocytosis in one pathway, and naked virions (virions lacking the aforementioned membrane) interact with the host cell plasma membrane and inject viral DNA directly into the cell's cytoplasm in the other (Jancovich et al. 2015). Either way, viral DNA is then replicated and transported from the nucleus to the cytoplasm where viral assembly occurs. The accumulating virions can form paracrystalline arrays, which are observable using transmission electron microscopy. Enveloped virions are subsequently formed as they bud through the cell plasma membrane (Jancovich et al. 2015).

Although there are no pathognomonic lesions for ranavirosis, common lesions are seen regardless of the host (amphibian, reptile, or fish) or *Ranavirus* species (Figure 4.2) (Miller et al. 2015). Ranaviruses most notably target endothelial cells (i.e., those lining blood vessels), epithelial cells (i.e., those lining the respiratory tract, gastrointestinal tract, and renal tubules), and hepatocytes (Miller et al. 2015). The commonly observed lesions are hemorrhage, necrosis, and edema in those tissues or organs. Hemorrhages are often severe, and this pathogen has been referred to as the "Ebola of the amphibian world" (Gray and Miller 2013). Behavioral changes that may be seen include difficulty swimming or righting

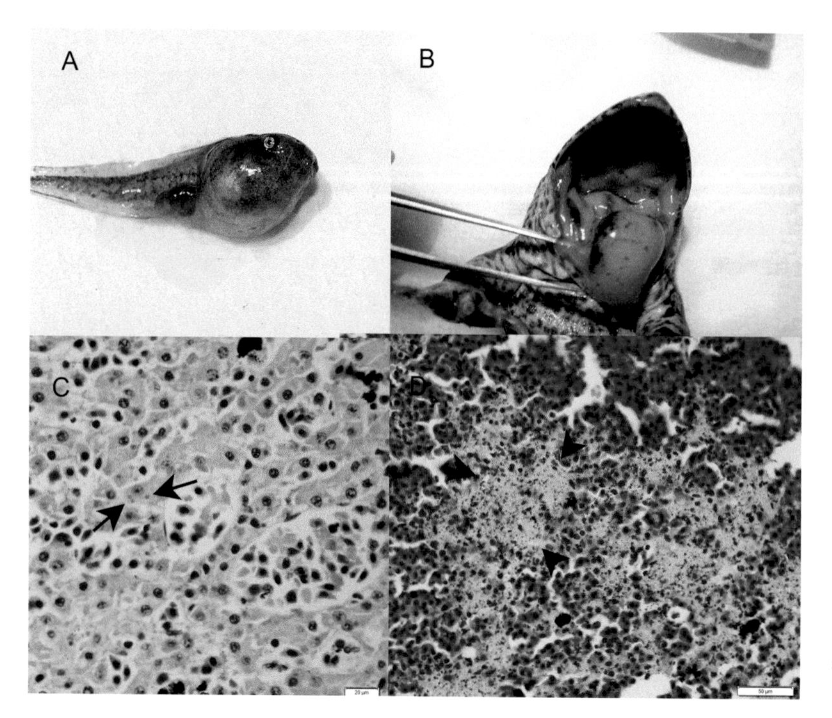

Fig. 4.2. Examples of lesions observed in amphibians infected with ranavirus. (A) Tadpole with hemorrhages and edema. (B) Adult anuran with numerous oral hemorrhages. (C) Inclusion bodies (arrows) within hepatocytes in the liver of an adult anuran infected with ranavirus. (D) Necrosis (arrowheads) in the spleen of an adult anuran infected with ranavirus.

themselves, impaired ambulation, and labored breathing. The latter is most often noted in turtles due to necrosis within the oral cavity, nares, and lungs. Ultimately, death is due to massive organ damage, blood loss, or inability to breathe. Often secondary (opportunistic) organisms (e.g., parasites, bacteria, and fungi) take advantage of the weakened host and contribute to death (Miller et al. 2015).

Ecology and Epidemiology

Morbidity and mortality events have impacted free-living amphibian (and reptile and fish) populations worldwide. Price et al. (2014) documented multispecies declines at several locations in northern Spain due to CMTV. Populations at sites that were exposed to ranaviruses were more likely to experience declines over a 14-year longitudinal period, and model predictions illustrated that population extirpation was possible (Bosch et al. 2021). Earl and Gray (2014) predicted that isolated wood frog populations could go extinct in as little as 5 years if ranavirus outbreaks occurred yearly. Furthermore, the introduction of ranaviruses to amphibian populations of threatened or endangered species could have drastic consequences for population stability (Earl et al. 2016). Thus, for highly susceptible host species, ranaviruses can have substantial population impacts (Hartmann et al. 2022).

Interestingly, patterns of mortality observed in Europe and North America differ in the life stage impacted. In North America, ranavirosis is often observed in larval and metamorphic individuals, while in Europe, mortality has been observed in adults (Brunner et al. 2015). Ranavirus outbreaks in both regions are often observed seasonally (Brunner et al. 2015; Green et al. 2002), which has been attributed to several factors. First, amphibian species tend to be more susceptible to ranaviruses at warmer temperatures (Brand et al. 2016, Price et al. 2019). One hypothesis is that increased host susceptibility is a result of increased viral replication rates at warmer temperatures. Second, the timing of outbreaks may result from subclinically infected adults introducing the pathogen to aquatic environments during breeding (Brunner et al. 2004). Third, during larval development, amphibian susceptibility to infection shifts (Haislip et al. 2011). Although the most sensitive stage of larval development may be species specific, it is clear that clinical disease and mortality often coincide with larval stages at or approaching metamorphosis (Cullen et al. 1995, Warne et al. 2011). Susceptibility to ranaviruses may peak at this stage due to immune system shifts (Rollins-Smith 1998, Rollins-Smith 2001). It is likely that a combination of abiotic and biotic factors at a site contribute to the likelihood of a ranavirus outbreak (Brunner et al. 2015).

Management

Limited treatments or preventative measures have been developed for ranavirosis in amphibians, and currently supportive care is useful only in rehabilitation or similar human-managed systems. Some antiviral treatments have been successful in treating ranavirus infections in turtles, and vaccine development for fish ranaviruses is making promising advancements (Gray et al. 2017, Yao et al. 2022). Vaccines have been developed for *Andrias davidianus* ranavirus (strain of CMTV) and appear to provide enhanced protective immunity for Chinese giant salamanders (*Andrias davidianus*); however, efficacy against other ranaviruses and in other host species is unknown (Zhou et al. 2017, Chen et al. 2018). Ultimately, treatment and prevention strategies developed in human-managed settings may provide limited protection for wild amphibian populations. Currently, the best approach for managing disease in wild populations is identifying strategies that limit severity or frequency of a ranavirus outbreak. For example, reducing contact rates between individuals by decreasing host density or increasing habitat structure possibly can lower pathogen transmission (Malagon et al. 2020). Dewatering wetlands for a couple of months is another option in the wild to inactivate ranavirus at outbreak sites (Gray et al. 2017).

Overall, our understanding of ranaviruses has improved over time and has armed natural resource organizations with the ability to conduct pathogen surveillance programs (Gray et al. 2017). The continual identification of potential sources for transmission is helping to mitigate human-facilitated spread of the pathogen in rehabilitation programs, wet markets, and water-sporting events. An important recent tool is data sharing. The Ranavirus Reporting System is publicly available and can be easily accessed online at https://brunnerlab.shinyapps.io/GRRS_Interactive/ (Brunner et al. 2021).

Chytrid: *Batrachochytrium dendrobatidis*

Etiology

The amphibian chytrid fungus Bd has been the most devastating pathogen to vertebrate biodiversity on a global scale. This fungus causes the disease chytridiomycosis that has led to many population declines and extinction events (Scheele et al. 2019). It was first isolated in 1999; however, the spread of the pathogen likely preceded its description by many years (O'Hanlon et al. 2018, Longcore et al. 1999). The introduction of Bd to naive amphibian populations has resulted in population declines in over 500 species of amphibians, nearly half of which are now extinct or threatened with extinction (Scheele et al. 2019). Over the course of 20 years, Bd has marched through Costa Rica and Panama into South America at a rate of around 20–200 km/year, decimating amphibian species in its wake (Lips et al. 2006, Lips et al. 2008). Like ranaviruses, Bd occurs globally on every continent where amphibians exist; however, in contrast to ranavirus, the hardest hit species have tended to be in the tropical regions of Australia and South America (Scheele et al. 2019).

Pathogenesis and Pathology

Chytrids are aquatic fungi within the class Chytridiomycetes. Rather than hyphae, chytrids exist as infectious zoospores that are free swimming or as reproductive zoosporangia that are nonmotile (Figure 4.3). Similar to ranaviruses, chytrid fungi

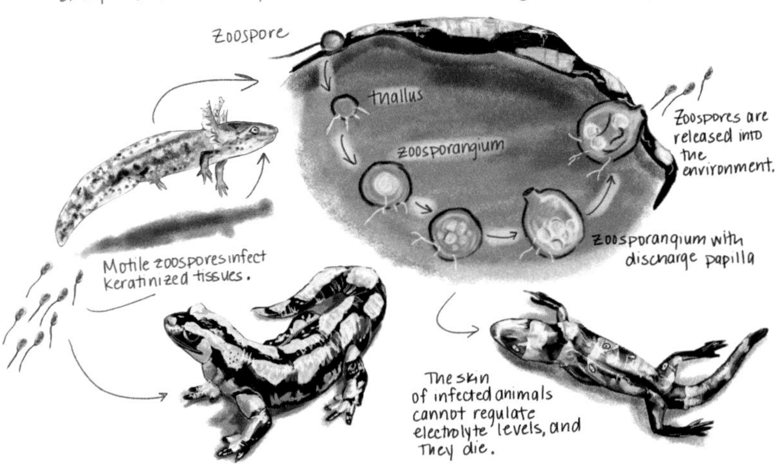

Fig. 4.3. Life cycle for *Batrachochytrium dendrobatidis* (Bd) and *Batrachochytrium salamandrivorans* (Bsal). Encysted spores of Bsal and zoospores may exist in the environment. Encysted spores are infectious and more resistant than zoospores to environmental degradation. Illustration by Laura Donohue.

of the genus *Batrachochytrium* are unique within their taxonomic group for being pathogens of vertebrates (Powell 1993). For *Batrachochytrium* species (Bd and Bsal), zoosporangia are found in keratinized skin of their hosts (Figure 4.4); therefore, host life stage is a factor in disease development. Keratin is found within the tooth rows of amphibian larvae, and infections are localized to this area until late stages of larval development when limbs emerge (i.e., keratinized digits) and larval skin transitions to its postmetamorphic state (McMahon and Rohr 2015). Chytrid invasion of the limited keratinized regions of larvae make mortality less frequent. With enough damage to the keratinized tooth rows, feeding may be impaired resulting in smaller body size, slower growth rates, or even starvation (Parris and Cornelius 2004, Venesky et al. 2009, Alvarado-Rybak et al. 2021). In contrast, juveniles and adults are covered in keratinized skin and thus can be heavily infected.

Amphibian skin is a very important organ. It is not only a mechanical protector, but it provides glandular secretions (lubrication and chemical deterrents from predators) and houses a rich microbiome that contributes to protection from would-be invaders (Rollins-Smith 2020). Beyond its protective qualities, amphibian skin is important for maintaining osmotic balance and for respiration, the latter being especially important in lungless amphibians (Plethodontidae;

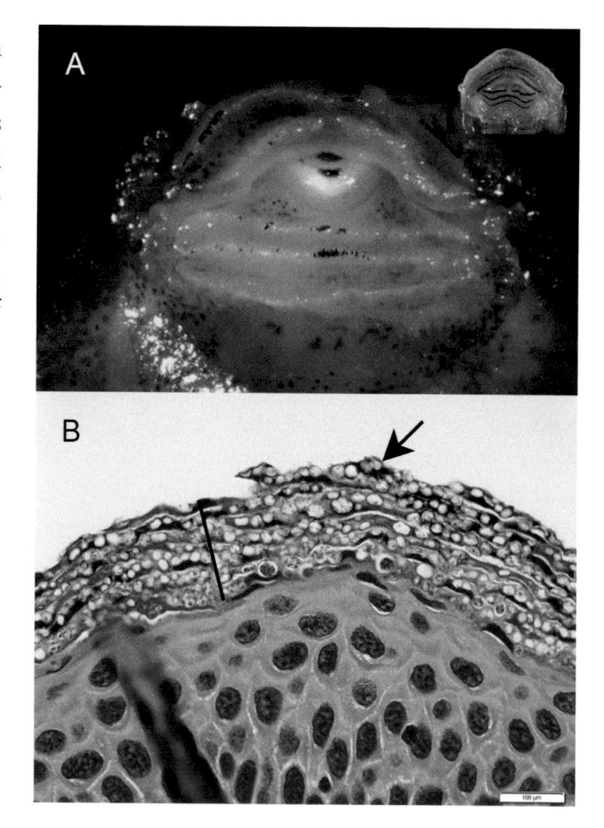

Fig. 4.4. Examples of lesions observed in amphibians infected with *Batrachochytrium dendrobatidis.* (A) Loss of keratin in tadpoles is demonstrated by the loss of pigmented tooth rows. Inset of uninfected tadpole for comparison. (B) Epidermis from the toe of an adult amphibian infected with chytrid (arrow) throughout the entire keratin layer (black line shows the extent of the keratin layer of the epidermis).

Wells 2007). Destruction of this vital organ via chytrid invasion can be lethal due to depletion of electrolytes, especially decreased plasma sodium and potassium, ultimately leading to cardiac arrest (Voyles et al. 2009).

In the case of Bd, excessive shedding and proliferation of the epidermis is observed in adults. These animals are often described as having no fear of humans because they do not attempt to flee when approached. This behavior may be due to electrolyte imbalances impairing muscle contractions causing decreased locomotion (Voyles et al. 2009). It is not uncommon for an animal to die in a statue posture. Based on field surveillance, anurans and caudates can both be infected by Bd with a larger proportion of anurans being threatened by Bd chytridiomycosis than caudates (Scheele et al. 2019).

Treatment

Antifungal medications can be used for treatment of amphibians managed in human care (Berger et al. 2010). Among the various antifungal treatments are itraconazole as a 0.005% or 0.01 % bath for 5 min/day for up to 10 days (Jones et al. 2012); and terbinafine as a 0.005% or 0.01% bath for 5 min/day for 5 days or 6 treatments over 10 days (Bowerman et al. 2010). Heat treatment can be used to raise body temperature of amphibians to 30°C for 10 days if the species can withstand this temperature and duration (Chatfield and Richards-Zawacki 2011, Bletz 2013). It is important to remember that, similar to mammals and birds, one size does not fit all when developing therapeutic options. The success of available treatment strategies likely relates to the severity of infection and tolerance of the host species (Wilber et al. 2017); thus, it is important to research current therapeutic options specific to a species (and a pathogen) before instituting any. Species preferring cooler temperatures may be less tolerant of heat treatment than tropical species, and larval stages or metamorphs may be more sensitive to chemotherapeutics than adults. Alternative therapeutics may include application of plant-derived fungicides to the environment or in a water bath with an infected amphibian (Silva et al. 2019). Proactive treatment can help protect species in human-managed breeding colonies or provide a boost of protection for animals being released into the wild through repatriation (head-start) programs. Use of slow-release implants to deliver antifungals when animals are released into new environments (Hardman et al. 2021) are being tested. Another preventative treatment is microbiome enhancement with beneficial bacterial species (Woodhams et al. 2016). The short generation time of many amphibians might allow for human-managed breeding programs to develop amphibian populations that are more resistant or tolerant to disease (Woodhams et al. 2011).

Management

Good biosecurity measures are essential to prevent the spread of pathogens within the laboratory or among field sites (Gray et al. 2018). Several disinfectants have been identified that can be safely used for biosecurity with 5 min of 1% Virkon S® or 1 min of either 4% sodium hypochlorite or 70% ethanol being effective against ranavirus, Bd, and Bsal (Bryan et al. 2009, Van Rooij et al. 2017). For molecular studies, it is important to also consider whether the disinfectant simply inactivates the pathogen (e.g., ethanol) or also denatures the DNA (e.g., bleach solutions) to prevent false positives (Cashins et al. 2008). For example, forceps and scalpel blades used to collect tissue samples, as well as materials used to collect environmental DNA, should be cleaned with 1:1 bleach solution (commercial bleach:water) prior to their next usage (Goldberg et al. 2016).

Chytrid: *Batrachochytrium salamandrivorans*

Etiology

In 2013, it became clear that Bd was not the only chytrid of concern to the amphibian world (Martel et al. 2013). A wild population of fire salamanders (*Salamandra salamandra*) in the Netherlands began to die in large numbers (Spitzen-van der Sluijs et al. 2013), and the pathogen looked similar to Bd but resulted in different gross lesions. Similar to Bd-infected amphibians, excessive skin sloughing was observed. Unlike Bd infections however, target lesions (areas of ulceration surrounded by a rim of pigment or hemorrhage) occurred throughout the entire body, but were still limited to the skin (Figure 4.5). These "target" lesions gave the infected fire salamanders the appearance of having been blasted by a miniature shotgun. After isolating the organism, successfully culturing it, and conducting experimental trials to fulfill Koch's postulates, Martel et al. (2013) confirmed that a new organism was the cause of the die-offs and molecularly characterized the isolate as *B. salamandrivorans*—the chytrid fungus that "devours" salamanders.

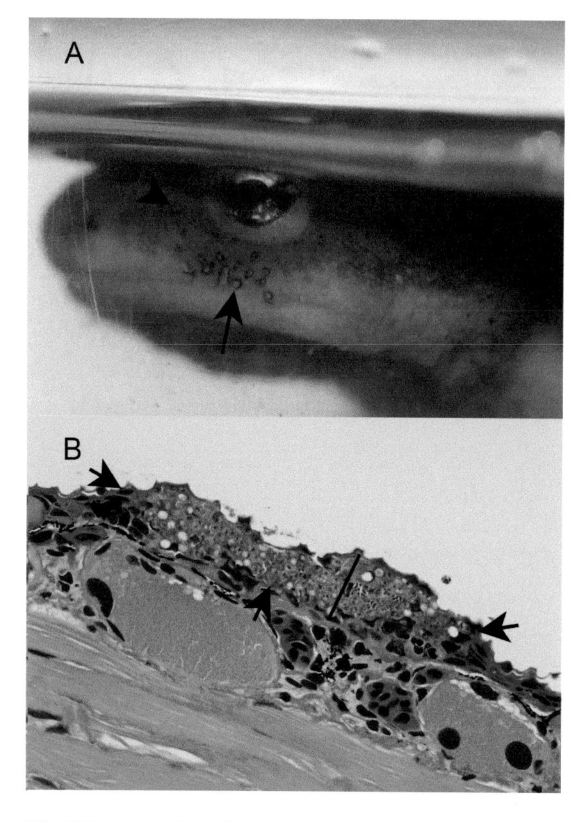

Fig. 4.5. Examples of lesions observed in amphibians infected with *Batrachochytrium salamandrivorans* (Bsal). (A) Target lesions (arrow and arrowhead) are commonly seen in Bsal but can be hard to notice in heavily pigmented areas (arrowhead). (B) Microscopically, these target lesions appear as craters (arrows) filled with Bsal organisms and, in this case, consuming the entire epidermis (black line shows the extent of the epidermis).

Ecology and Epidemiology

Like Bd, Bsal is endemic in Southeast Asia (Martel et al. 2014). Unlike ranaviruses and Bd, Bsal is currently not thought to be globally distributed and was introduced to Europe only recently. After thorough investigation, it is thought that Bsal was introduced to the Netherlands via pet trade networks (Martel et al. 2014). Since the original die-off, biologists have implemented surveillance throughout Europe and other parts of the globe in a race to avoid catastrophic species loss, as was caused by Bd. Unfortunately, Bsal is now detected throughout Europe in both wild and human-managed populations (Cunningham et al. 2015, Fitzpatrick et al 2018, Martel et al. 2020), but extensive sampling efforts have yet to detect Bsal in North America (Klocke et al. 2017, Waddle et al. 2020, Hill et al. 2021). With North America being a hotspot of global caudate diversity, government agencies and researchers have been on high alert (Gray et al. 2015). In response, the North American Bsal Task Force (https://www.salamanderfungus.org/) was launched in 2016 (Grant et al. 2017), with the goal of preventing the invasion of North America by Bsal and developing response strategies if introduction occurs. The Appalachian Mountains region of eastern North America is home to at least half of the salamander families that exist in the world with very high species richness and abundance. Several studies have identified susceptible species in the region (Martel et al. 2014, Carter et al. 2019, Friday et al. 2020). Alarmingly, the thermal climate within the Appalachian region falls within the preferred environmental suitability of Bsal, and ample host species create ideal conditions for disease emergence (Yap et al. 2015, Richgels et al. 2016, Carter et al. 2021).

Management and Treatment

Due to the threat of Bsal invasion, the US Fish and Wildlife Service implemented a rule in 2016 (USFWS 2016) using the Lacey Act that banned the importation of 201 salamander species into the United States that could serve as possible hosts (Grant et al. 2017). While this regulatory response decreased risk by reducing US salamander imports from 5% to <1% of annual amphibian imports (Gray et al. 2015, Can et al. 2019, Grear et al. 2021), it did not eliminate the threat. Import bans can lead to an increase in illegal wildlife trade (Rivalan et al. 2007). In addition, it is now known that Bsal can infect frogs (Nguyen et al. 2017, Stegen et al. 2017, Towe et al. 2021), which comprise 99% of global amphibian trade (Can et al. 2019, Scheffers et al. 2019). Around 250,000 fire-bellied toads are imported into the United States each year, and Bsal infection prevalence of this species in one German pet store was 8% (Nguyen et al. 2017). Thus, it is highly likely that thousands of Bsal-positive amphibians have entered the US pet trade without detection. Similarly, it is likely that salamanders infected with Bsal entered the United States prior to the 2016 trade ban (Yuan et al. 2018). To date, Bsal surveillance in amphibian trade has been limited to one study (Klocke et al. 2017). In cooperation with the pet amphibian industry, there is urgent need to increase surveillance for Bsal in captivity and devise strategies to contain it if a detection occurs (https://tiny.utk.edu/pijac).

Identifying effective Bsal-management options is in its infancy (Thomas et al. 2019). Reducing host density and increasing habitat structure can decrease host contact rates (Malagon et al. 2020); however, across a range of host densities and infection prevalence treatments, Bsal transmission was density independent (Tompros et al. 2021). Thus, manipulating host density may not be an effective management option. Reducing environmental persistence of Bsal or increasing host tolerance to infection might be a reasonable option (Islam et al. 2021, Wilber et al. 2021). Recently completed studies suggest that very small amounts of plant-derived fungicides (e.g., curcumin, allicin, and thymol) have strong anti-Bsal qualities (Tompros 2021). Also, several bacterial species have Bsal-inhibitory qualities making them good candidates for bioaugmentation (Bletz et al. 2018, Bates et al. 2019). Regardless of treatment options, the high invasion potential of Bsal ($R_0 > 4$; where R_0 is the infectivity rate, the average number of susceptible individuals that become infected by one infected individual) into susceptible populations (Tompros et al. 2021) emphasizes the importance of preventing it from being introduced (Canessa et al. 2018, Thomas et al. 2019).

A key discovery is that, although caudate species are most affected by Bsal, anuran species can be infected, possibly functioning as resistant, reservoir, susceptible, or amplification hosts (Stegen et al. 2017; Towe et al. 2021). Some anuran species can die from Bsal chytridiomycosis, for example, the Cuban tree frog (*Osteopilus septentrionalis*; Towe et al. 2021). This frog is highly invasive and has established wild populations in 14 countries and four US states (Cottrell and Ventosa 2018). Cuban tree frogs are also in the pet trade and documented to frequently "hitchhike" on many materials in transshipment including plants and cars, increasing the probability of accidental introduction of Bsal or other pathogens (Johnson 2017, Hedges et al. 2018). Some larval salamander species can become infected with Bsal (M. J. Gray, Center for Wildlife Health and One Health Initiative, University of Tennessee, unpublished data) and may serve as reservoirs similar to Bd.

There is great variation in how species respond to Bsal infection (Martel et al. 2014, Carter et al. 2019), and disease progression measured by infection load is not always linear. As with Bd, Bsal zoospores detected on the skin may decline temporarily with skin sloughing or other tolerance mechanisms eliminate infected tissues. Infected animals can appear to have cleared infection with diagnostic tests, only to have infections build once again. Ultimately, it appears that skin sloughing and reductions in lesions do not infer much protection from developing Bsal chytridiomycosis, at least in very susceptible species (Wilber et al. 2021). This is similar to what is seen in snakes with snake fungal disease (caused by *Ophidiomyces ophidiicola*) (Lorch et al. 2016) and bats with white-nose syndrome (caused by *Pseudogymnoascus destructans*) (Magnino et al. 2021).

The discovery of Bsal was fortuitous as it could easily have been missed or dismissed as a variant of Bd if the investigators had not conducted complete pathological evaluations of the fire salamander mortality events (Martel et al. 2013). This emphasizes the importance of complete pathological evaluations in morbidity and mortality events (Thomas et al. 2018). All too often biologists rely on pathogen-specific tests, such as PCR for testing diseased animals. This may be done for economic reasons, but the ability to diagnose the cause of the disease is severely hampered if tests are performed for one or two potential etiologies (Duffus et al. 2017). A negative pathogen-specific test (e.g., PCR) does not reveal the cause of a disease, and a positive test means the pathogen is present but not that it is the cause of disease. With amphibians, the best evidence of causation comes from histopathology, followed by additional diagnostic testing (Miller et al. 2015, Thomas et al. 2018).

Although this chapter has focused on describing current knowledge of three notifiable amphibian pathogens, it is worth noting that there are other pathogens (known and unknown) as well as parasites and toxins that cause amphibian mortality (Green and Converse 2005). The oldest known pathogen discussed in this chapter has only been known since 1966, while the most deadly wildlife pathogen known (Bd) was first described in 1999, and its chytrid relative, Bsal, about a decade later. It is very likely that new disease etiologies will emerge as enzootic pathogens and parasites spread from their native ranges to other areas around the globe and infect naive species. As Powell (1993) stated, "The best studied disease systems aren't the only systems worth studying."

Summary

There are multiple contributors to diseases that lead to organism compromise and death, such as anthropogenic stressors, opportunistic pathogens, and their interaction. For example, it only takes a rise of 2°C to increase the likelihood of disease due to ranavirus (Brand et al. 2016), and this is well within the present-day projected effects of climate change. Likewise, inadequate vegetation buffers and increased sedimentation and run-off of contaminants is severely impacting many aquatic species. Opportunistic pathogens, even resident organisms within a host's natural microbiome that might otherwise be beneficial, can be problematic if allowed to multiply unchecked and invade the weakened host. One hypothesis for how Bsal kills is that secondary bacterial infections cause septicemia due to widespread skin destruction (Bletz et al. 2018). There is no shortage of fungi, parasites, bacteria, and viruses within almost every ecosystem waiting for an opportunity to take advantage of animals with compromised innate and adaptive immune systems. Even with respect to the pathogens discussed in this chapter, animals can become coinfected and increase the likelihood of death (Longo et al. 2019). Furthermore, a weakened host is susceptible to things other than infectious agents, such as predation and environmental contaminants.

The joining together of researchers, biologists, veterinarians, industry, and others to solve these problems has been a great accomplishment of this work—a real One Health approach (Gibbs and Gibbs 2013). Cooperation has helped avoid duplication of effort; increased sharing of critical resources and ideas; maximized the number of answered questions; standardized diagnostic testing; and led to the development of guidelines, treatments, and management solutions. A One Health approach is efficient and effective and necessary for managing wildlife and zoonotic diseases in the 21st century.

It is said that Earth may be experiencing the sixth mass extinction event in 540 million years (Barnosky et al. 2011, Kolbert 2014), and a fairly new recognized player in this event is movement of animals (or in some cases, their parts and pathogens) across the globe. For amphibians, this means ranaculture (e.g., frog leg production) and the pet trade. For amphibians, trade is an open transmission pathway and one that has received limited investigation. New efforts are underway by coauthor M. J. Gray and colleagues to find the common ground with wildlife-trade stakeholders to save amphibians and maybe other life forms by using amphibians as our guide (https://tiny.utk.edu/pijac).

Much remains to be learned about these pathogens and the diseases they cause. Furthermore, we must remember that disease is a part of nature and may help to maintain ecosystem stability and population health. Personal acceptance of the One Health concept suggests it is our responsibility as good stewards of life on Earth to promote a balanced approach to health and sustainability in nature.

LITERATURE CITED

Alvarado-Rybak, M., P. Acuña, A. Peñafiel-Ricaurte, T. R. Sewell, S. J. O'Hanlon, M. C. Fisher, A. Valenzuela-Sánchez, A. A. Cunningham, and C. Azat. 2021. Chytridiomycosis outbreak in a Chilean giant frog (*Calyptocephalella gayi*) captive breeding program: genomic characterization and pathological findings. Frontiers in Veterinary Science 8:733357.

Bates, K. A., J. M. G. Shelton, V. L. Mercier, K. P. Hopkins, X. A. Harrison, S. O. Petrovan, and M. C. Fisher. 2019. Captivity and infection by the fungal pathogen *Batrachochytrium salamandrivorans* perturb the amphibian skin microbiome. Frontiers in Microbiology 10:1834.

Barnosky, A. D., N. Matzke, S. Tomiya, G. O. U. Wogan, B. Swartz, T. B. Quental, C. Marshall, J. L. McGuire, E. L. Lindsey, K. C. Maguire, et al. 2011. Has the Earth's sixth mass extinction already arrived? Nature 471:51–57.

Berger, L., R. Speare, A. Pessier, J. Voyles, L. F. Skerratt. 2010. Treatment of chytridiomycosis requires urgent clinical trials. Diseases of Aquatic Organisms 92:165–174.

Bletz, M. 2013. Probiotic bioaugmentation of an anti-Bd bacteria, *Janthinobacterium lividum*, on the amphibian, *Notophthalmus viridescens*: transmission efficacy and persistence of the probiotic on the host and non-target effects of probiotic addition on ecosystem components. Thesis, James Madison University, Harrisonburg, Virginia, USA.

Bletz, M. C., M. Kelly, J. Sabino-Pinto, E. Bales, S. Van Praet, W. Bert, F. Boyen, M. Vences, S. Steinfartz, F. Pasmans, and A. Martel. 2018. Disruption of skin microbiota contributes to salamander disease. Proceedings of the Royal Society B Biological Sciences 285:20180758.

Bosch, J., A. Mora-Cabello de Alba, S. Marquínez, S. J. Price, B. Thumsová, and J. Bielby. 2021. Long-term monitoring of amphibian populations of a national park in northern Spain reveals negative persisting effects of *Ranavirus*, but not *Batrachochytrium dendrobatidis*. Frontiers in Veterinary Science 8:645491.

Bowerman, J., C. Rombough, S.R. Weinstock, and G.E. Padgett-Flohr. 2010. Terbinafine hydrochloride in ethanol effectively clears *Batrachochytrium dendrobatidis* in amphibians. Journal of Herpetological Medicine and Surgery 20:24–28.

Brand, M. D., R. D. Hill, R. Brenes, J. C. Chaney, R. P. Wilkes, L. Grayfer, D. L. Miller, and M. J. Gray. 2016. Water temperature affects susceptibility to ranavirus. EcoHealth 13:350–359.

Brenes, R., M. J. Gray, T. B. Waltzek, R. P. Wilkes, and D. L. Miller. 2014. Transmission of ranavirus between ectothermic vertebrate hosts. PLoS One 9:e92476.

Brunner, J., D. Olson, M. Gray, D. Miller, and A. Duffus. 2021. Global patterns of ranavirus detections. FACETS 6:912–924.

Brunner, J., D. Schock, E. Davidson, and J. Collins. 2004. Intraspecific reservoirs: complex life history and the persistence of a lethal ranavirus. Ecology 85:560–566.

Brunner, J. L., A. Storfer, M. J. Gray, and J. T. Hoverman. 2015. Ranavirus ecology and evolution: from epidemiology to extinction. Pages 71–104 *in* M. J. Gray and V. G. Chinchar, editors. Ranaviruses: lethal pathogens of ectothermic vertebrates. Springer International Publishing, Cham, Switzerland.

Bryan, L. K., C. A. Baldwin, M. J. Gray, and D. L. Miller. 2009. Efficacy of select disinfectants at inactivating *Ranavirus*. Diseases of Aquatic Organisms 84:89–94.

Can, Ö. E., N. D'Cruze, and D. W. Macdonald. 2019. Dealing in deadly pathogens: taking stock of the legal trade in live wildlife and potential risks to human health. Global Ecology and Conservation 17:e00515.

Canessa, S., C. Bozzuto, E. Grant, S. Cruickshank, M. Fisher, J. Koella, S. Lötters, A. Martel, F. Pasmans, B. Scheele, et al. 2018. Decision-making for mitigating wildlife diseases: from theory to practice for an emerging fungal pathogen of amphibians. Journal of Applied Ecology 55:1987–1996.

Carter, E. D., M. C. Bletz, M. Le Sage, B. LaBumbard, L. A. Rollins-Smith, D. C. Woodhams, D. L. Miller, and M. J. Gray. 2021. Winter is coming—temperature affects immune defenses and susceptibility to *Batrachochytrium salamandrivorans*. PLoS Pathogens 17:e1009234.

Carter, E. D., D. L. Miller, A. C. Peterson, W. B. Sutton, J. P. W. Cusaac, J. A. Spatz, L. Rollins-Smith, L. Reinert, M. Bohanon, L. A. Williams, et al. 2019. Conservation risk of *Batrachochytrium salamandrivorans* to endemic lungless salamanders. Conservation Letters 13:e12675.

Casais, R., A. R. Larrinaga, K. P. Dalton, P. Domínguez Lapido, I. Márquez, E. Bécares, E. D. Carter, M. J. Gray, D. L. Miller, and A. Balseiro. 2019. Water sports could contribute to the translocation of ranaviruses. Scientific Reports 9:2340.

Cashins, S. D., L. F. Skerratt, and R. A. Alford. 2008. Sodium hypochlorite denatures the DNA of the amphibian chytrid fungus *Batrachochytrium dendrobatidis*. Diseases of Aquatic Organisms 80:63–67.

Cavasos, K., C. N. Poudyal, M. Gray, A. Warwick, J. Brunner, J. Piovia-Scott, N. Fefferman, C. M. Bletz, J. Lockwood, and J. Jones. 2021. Executive Summary: Amphibian Consumer and Business Survey. One Health Initiative, Department of Forestry, Wildlife and Fisheries, University of Tennessee, Knoxville, USA. https://onehealth.tennessee.edu/wp-content/uploads/sites/78/2021/12/PIJAC-Exec-Summary.pdf. Accessed 23 Feb 2022.

Chatfield, M. W., and C. L. Richards-Zawacki. 2011. Elevated temperature as a treatment for *Batrachochytrium dendrobatidis* infection in captive frogs. Diseases of Aquatic Organisms 94:235–238.

Chen, Z.-Y., T. Li, X.-C. Gao, C.-F. Wang, and Q.-Y. Zhan. 2018. Protective immunity induced by DNA vaccination against ranavirus infection in Chinese giant salamander *Andrias davidianus*. Viruses 10:52.

Chinchar, V. G., P. Hick, I. A. Ince, J. K. Jancovich, R. Marschang, Q. Qin, K. Subramaniam, T. B. Waltzek, R. Whittington, and T. Williams. 2017. ICTV virus taxonomy profile: Iridoviridae. Journal of General Virology 98:890–891.

Claytor, S. C., K. Subramaniam, N. Landrau-Giovannetti, V. G. Chinchar, M. J. Gray, D. L. Miller, C. Mavian, M. Salemi, S. Wisely, and T. B. Waltzek. 2017. Ranavirus phylogenomics: signatures of recombination and inversions among bullfrog ranaculture isolates. Virology 511:330–343.

Collins, J. P., and A. Storfer. 2003. Global amphibian declines: sorting the hypotheses. Diversity and Distributions 9:89–98.

Cottrell, V., and E. Ventosa. 2018. Invasive species compendium: Osteopilus septentrionalis (Cuban tree frog). https://www.cabi.org/isc/datasheet/71203. Accessed 23 Feb 2022.

Cullen, B. R., L. Owens, and R. J. Whittington. 1995. Experimental infection of Australian anurans (*Limnodynastes terraereginae* and *Litoria latopalmata*) with Bohle iridovirus. Diseases of Aquatic Organisms 23:83–92.

Cunningham, A. A., K. Beckmann, M. Perkins, L. Fitzpatrick, R. Cromie, J. Redbond, M. F. O'Brien, P. Ghosh, J. Shelton, and M. C. Fisher. 2015. Emerging disease in UK amphibians. Veterinary Record 176:468.

Dang, T. D., C. L. Searle, and A. R. Blaustein. 2017. Virulence variation among strains of the emerging infectious fungus *Batrachochytrium dendrobatidis* (Bd) in multiple amphibian host species. Diseases of Aquatic Organisms 124:233–239.

Duffus, A. L. J., H. M. A. Fenton, M. J. Gray, and D. L. Miller. 2017. Investigating amphibian and reptile mortalities: a practical guide for wildlife professionals. Herpetological Review 48:550–557.

Duffus, A. L. J., T. B. Waltzek, A. C. Stöhr, M. C. Allender, M. Gotesman, R. J. Whittington, P. Hick, M. K. Hines, and R. E. Marschang. 2015. Distribution and host range of ranaviruses. Pages 9–57 *in* M. J. Gray and V. G. Chinchar, editors. Ranaviruses: lethal pathogens of ectothermic vertebrates. Springer International Publishing, Cham, Switzerland.

Earl, J. E., J. C. Chaney, W. B. Sutton, C. E. Lillard, A. J. Kouba, C. Langhorne, J. Krebs, R. P. Wilkes, R. D. Hill, D. L. Miller, et al. 2016. Ranavirus could facilitate local extinction of rare amphibian species. Oecologia 182:611–623.

Earl, J. E., and M. J. Gray. 2014. Introduction of ranavirus to isolated wood frog populations could cause local extinction. EcoHealth 11:581–592.

Fitzpatrick, L. D., F. Pasmans, A. Martel, and A. A. Cunningham. 2018. Epidemiological tracing of *Batrachochytrium salamandrivorans* identifies widespread infection and associated mortalities in private amphibian collections. Scientific Reports 8:13845.

Friday, B., C. Holzheuser, K. R. Lips, and A. V. Longo. 2020. Preparing for invasion: assessing risk of infection by chytrid fungi in southeastern plethodontid salamanders. Journal of Experimental Zoology Part A Ecological and Integrative Physiology 333:829–840.

Gibbs, S. E., and E. P. Gibbs. 2013. The historical, present, and future role of veterinarians in One Health. Pages 31–47 *in* J. S. Mackenzie, M. Jeggo, P. Daszak, and J. A. Richt, editors. Current topics in microbiology and immunology. Volume 365. One Health: the human–animal–environment interfaces in emerging infectious diseases. Springer, Berlin, Heidelberg, Germany.

Goldberg, C. S., C. R. Turner, K. Deiner, K. E. Klymus, P. F. Thomsen, M. A. Murphy, S. F. Spear, A. McKee, S. J. Oyler-McCance, R. S. Cornman, et al. 2016. Critical considerations for the application of environmental DNA methods to detect aquatic species. Methods in Ecology and Evolution 7:1299–1307.

Granoff, A., P. E. Came, and D. C. Breeze. 1966. Viruses and renal carcinoma of *Rana pipiens*. I. The isolation and properties of virus from normal and tumor tissue. Virology 29:133–148.

Grant, E. H. C., E. Muths, R. A. Katz, S. Canessa, M. J. Adams, J. R. Ballard, L. Berger, C. J. Briggs, J. T. H. Coleman, M. J. Gray, et al. 2017. Using decision analysis to support proactive management of emerging infectious wildlife diseases. Frontiers in Ecology and the Environment 15:214–221.

Gray, M. J., and V. G. Chinchar. 2015. Introduction: history and future of ranaviruses. Pages 1–7 *in* M. J. Gray and V. G. Chinchar, editors. Ranaviruses: lethal pathogens of ectothermic vertebrates. Springer International Publishing, Cham, Switzerland.

Gray, M. J., A. L. Duffus, K. H. Haman, R. N. Harris, M. C. Allender, T. A. Thompson, M. R. Christman, A. Sacerdote-Velat, L. A. Sprague, and J. M. Williams. 2017. Pathogen surveillance in herpetofaunal populations: guidance on study design, sample collection, biosecurity, and intervention strategies. Herpetological Review 48:334–351.

Gray, M. J., J. P. Lewis, P. Nanjappa, B. Klocke, F. Pasmans, A. Martel, C. Stephen, G. Parra Olea, S. A. Smith, A. Sacerdote-Velat, et al. 2015. *Batrachochytrium salamandrivorans*: the North American response and a call for action. PLoS Pathogens 11:e1005251.

Gray, M. J., and D. L. Miller. 2013. The rise of ranavirus: an emerging pathogen threatens ectothermic vertebrates. The Wildlife Professional 2013(Spring):51–55.

Gray, M. J., J. A. Spatz, E. D. Carter, C. M. Yarber, R. P. Wilkes, and D. L. Miller. 2018. Poor biosecurity could lead to disease outbreaks in animal populations. PLoS One 13:e0193243.

Grear, D. A., B. A. Mosher, K. L. D. Richgels, and E. H. C. Grant. 2021. Evaluation of regulatory action and surveillance as preventive risk-mitigation to an emerging global amphibian pathogen *Batrachochytrium salamandrivorans* (Bsal). Biological Conservation 260:109222.

Green, D. E., and K. A. Converse. 2005. Diseases of frogs and toads. Pages 89–117 *in* S.K. Majumdar, J. E. Huffman, F. J. Brenner, and A. L. Panah, editors. Wildlife diseases: landscape epidemiology, spatial distribution, and utilization of remote sensing technology. Pennsylvania Academy of Science, Easton, USA.

Green, D. E., K. A. Converse, and A. K. Schrader. 2002. Epizootiology of sixty-four amphibian morbidity and mortality events in the USA, 1996–2001. Annals of the New York Academy of Sciences 969:323–339.

Haislip, N. A., M. J. Gray, J. T. Hoverman, and D. L. Miller. 2011. Development and disease: how susceptibility to an emerging pathogen changes through anuran development. PLoS One 6:e22307.

Hall, E. M., J. L. Brunner, B. Hutzenbiler and E. J. Crespi. 2020. Salinity stress increases the severity of ranavirus epidemics in amphibian populations. Proceedings of the Royal Society B Biological Sciences 287:20200062.

Hardman, R. H., S. Cox, S. D. Reinsch, H. C. Schwartz, S. Skeba, D. McGinnity, M. J. Souza, and D. L. Miller. 2021. Efficacy of subcutaneous implants to provide continuous plasma terbinafine in hellbenders (*Cryptobranchus allaeganiensis*) for future prophylactic use against chytridiomycosis. Journal of Zoo and Wildlife Medicine 52:300–305.

Hartmann, A. M., M. L. Maddox, R. J. Ossiboff, and A. V. Longo. 2022. Sustained ranavirus outbreak causes mass mortality and morbidity of imperiled amphibians in Florida. EcoHealth 19:8–14.

Hedges, B., L. Díaz, B. Ibéné, R. Joglar, R. Powell, F. Bolaños, and G. Chaves. 2010. The IUCN Red List of Threatened Species. *Osteopilus septentrionalis*. https://www.iucnredlist.org/fr/species/55811/11368202. Accessed 23 Feb 2022.

Hill, A. J., R. H. Hardman, W. B. Sutton, M. S. Grisnik, J. H. Gunderson, and D. M. Walker. 2021. Absence of *Batrachochytrium salamandrivorans* in a global hotspot for salamander biodiversity. Journal of Wildlife Diseases 5:553–560.

Hoverman, J. T., M. J. Gray, and D. L. Miller. 2010. Anuran susceptibilities to ranaviruses: role of species identity, exposure route, and a novel virus isolate. Diseases of Aquatic Organisms 89:97–107.

International Committee on the Taxonomy of Viruses (ICTV). 2022. Genus: Ranavirus. https://ictv.global/report/chapter/iridoviridae/iridoviridae/ranavirus. Accessed 23 Feb 2022.

Islam, R. M., M. J. Gray, and A. Peace. 2021. Identifying the dominant transmission pathway in a multi-stage infection model of the emerging fungal pathogen *Batrachochytrium salamandrivorans* on the eastern newt. Pages 193–216 *in* M. I. Teboh-Ewungkem and G. A. Ngwa, editors. Mathematics of planet Earth. Volume 7. Infectious diseases and our planet. Springer International Publishing, Cham, Switzerland.

Jancovich, J. K., Q. Qin, Q.-Y. Zhang, and V. G. Chinchar. 2015. Ranavirus replication: molecular, cellular, and immunological events Pages 105–139 *in* M. J. Gray and V. G. Chinchar, editors. Ranaviruses: lethal pathogens of ectothermic vertebrates. Springer International Publishing, Cham, Switzerland.

Johnson, S. A. 2017. The Cuban treefrog (*Osteopilus septentrionalis*) in Florida. Institute of Food and Agricultural Sciences, University of Florida, Gainesville, USA. https://edis.ifas.ufl.edu/publication/UW259. Accessed 22 Aug 2022.

Jones, M. E. B., D. Paddock, L. Bender, J. L. Allen, M. D. Schrenzel, and A. P. Pessier. 2012. Treatment of chytridiomycosis with reduced-dose itraconazole. Diseases of Aquatic Organisms 99:243–249.

Klocke, B., M. Becker, J. Lewis, R. C. Fleischer, C. R. Muletz-Wolz, L. Rockwood, A. A. Aguirre, and B. Gratwicke. 2017. *Batrachochytrium salamandrivorans* not detected in US survey of pet salamanders. Scientific Reports 7:13132.

Kolbert, E. 2014. The sixth extinction: an unnatural history. Henry Holt and Company, New York, New York, USA.

Kolby, J. E., K. M. Smith, L. Berger, W. B. Karesh, A. Preston, A. P. Pessier, and L. F. Skerratt. 2014. First evidence of amphibian chytrid fungus (*Batrachochytrium dendrobatidis*) and ranavirus in Hong Kong amphibian trade. PLoS One 9:e90750.

Lips, K. R., F. Brem, R. Brenes, J. D. Reeve, R. A. Alford, J. Voyles, C. Carey, L. Livo, A. P. Pessier, and J. P. Collins. 2006. Emerging infectious disease and the loss of biodiversity in a Neotropical amphibian community. Proceedings of the National Academy of Sciences USA 103:3165–3170.

Lips, K. R., J. Diffendorfer, J. R. Mendelson III, and M. W. Sears. 2008. Riding the wave: reconciling the roles of disease and climate change in amphibian declines. PLoS Biology 6:e72.

Longcore, J., A. Pessier, and D. Nichols. 1999. *Batrachochytrium dendrobatidis* gen. et sp. nov., a chytrid pathogenic to amphibians. Mycologia 91:219–227.

Longo, A., R. C. Fleischer, and K. Lips. 2019. Double trouble: co-infections of chytrid fungi will severely impact widely distributed newts. Biological Invasions 21:2233–2245.

Lorch, J. M., S. Knowles, J. S. Lankton, K. Michell, J. L. Edwards, J. M. Kapfer, R. A. Staffen, E. R. Wild, K. Z. Schmidt, A. E. Ballmann, et al. 2016. Snake fungal disease: an emerging threat to wild snakes. Philosophical Transactions of the Royal Society of London Series B Biological Sciences 371:20150457.

Magnino, M. Z., K. A. Holder, and S. A. Norton. 2021. White-nose syndrome: a novel dermatomycosis of biologic interest and epidemiologic consequence. Clinics in Dermatology 39:290–303.

Malagon, D. A., L. A. Melara, O. F. Prosper, S. Lenhart, E. D. Carter, J. A. Fordyce, A. C. Peterson, D. L. Miller, and M. J. Gray. 2020. Host density and habitat structure influence host contact rates and *Batrachochytrium salamandrivorans* transmission. Scientific Reports 10:5584.

Martel, A., M. Blooi, C. Adriaensen, P. Van Rooij, W. Beukema, M. C. Fisher, R. A. Farrer, B. R. Schmidt, U. Tobler, K. Goka, et al. 2014. Recent introduction of a chytrid fungus endangers western Palearctic salamanders. Science 346:630–631.

Martel, A., A. Spitzen-van der Sluijs, M. Blooi, W. Bert, R. Ducatelle, M. C. Fisher, A. Woeltjes, W. Bosman, K. Chiers, F. Bossuyt, et al. 2013. *Batrachochytrium salamandrivorans* sp. nov. causes lethal chytridiomycosis in amphibians. Proceedings of the National Academy of Sciences USA 110:15325–15329.

Martel, A., M. Vila-Escale, D. Fernández-Giberteau, A. Martinez Silvestre, S. Canessa, S. Praet, P. Pannon, K. Chiers, A. Ferran, M. Kelly, et al. 2020. Integral chain management of wildlife diseases. Conservation Letters 13:e12707.

McMahon, T. A., and J. R. Rohr. 2015. Transition of chytrid fungus infection from mouthparts to hind limbs during amphibian metamorphosis. EcoHealth 12:188–193.

Miller, D. L., A. P. Pessier, P. Hick, and R. J. Whittington. 2015. Comparative pathology of ranaviruses and diagnostic techniques. Pages171–208 *in* M. J. Gray and V. G. Chinchar, editors. Ranaviruses: lethal pathogens of ectothermic vertebrates. Springer International Publishing, Cham, Switzerland.

Nguyen, T. T., T. V. Nguyen, T. Ziegler, F. Pasmans, and A. Martel. 2017. Trade in wild anurans vectors the urodelan pathogen *Batrachochytrium salamandrivorans* into Europe. Amphibia-Reptilia 38:554–556.

O'Hanlon, S. J., A. Rieux, R. A. Farrer, G. M. Rosa, B. Waldman, A. Bataille, T. A. Kosch, K. A. Murray, B. Brankovics, M. Fumagalli, et al. 2018. Recent Asian origin of chytrid fungi causing global amphibian declines. Science 360:621–627.

Parris, M. J., and T. O. Cornelius. 2004. Fungal pathogen causes competitive and developmental stress in larval amphibian communities. Ecology 85:3385–3395.

Powell, M. J. 1993. Looking at mycology with a Janus face: a glimpse at Chytridiomycetes active in the environment. Mycologia 85:1–20.

Price, S. J., T. W. J. Garner, R. A. Nichols, F. Balloux, C. Ayres, A. Mora-Cabello de Alba, and J. Bosch. 2014. Collapse of amphibian communities due to an introduced ranavirus. Current Biology 24:2586–2591.

Price, S. J., W. T. M. Leung, C. J. Owen, R. Puschendorf, C. Sergeant, A. A. Cunningham, F. Balloux, T. W. J. Garner, and R. A. Nichols. 2019. Effects of historic and projected climate change on the range and impacts of an emerging wildlife disease. Global Change Biology 25:2648–2660.

Richgels, K. L. D., R. E. Russell, M. J. Adams, C. L. White, and E. H. C. Grant. 2016. Spatial variation in risk and consequence of *Batrachochytrium salamandrivorans* introduction in the USA. Royal Society Open Science 3:150616.

Rivalan, P., V. Delmas, E. Angulo, L. S. Bull, R. J. Hall, F. Courchamp, A. M. Rosser, and N. Leader-Williams. 2007. Can bans stimulate wildlife trade? Nature 447:529–530.

Rollins-Smith, L. A. 1998. Metamorphosis and the amphibian immune system. Immunological Reviews 166:221–230.

Rollins-Smith, L. A. 2001. Neuroendocrine-immune system interactions in amphibians: implications for understanding global amphibian declines. Immunological Research 23:273–280.

Rollins-Smith, L. A. 2020. Global amphibian declines, disease, and the ongoing battle between *Batrachochytrium* fungi and the immune system. Herpetologica 76:178–188.

Scheele, B. C., F. Pasmans, L. F. Skerratt, L. Berger, A. Martel, W. Beukema, A. A. Acevedo, P. A. Burrowes, T. Carvalho, A. Catenazzi, et al. 2019. Amphibian fungal panzootic causes catastrophic and ongoing loss of biodiversity. Science 363: 1459–1463.

Scheffers, B. R., B. F. Oliveira, I. Lamb, and D. P. Edwards. 2019. Global wildlife trade across the tree of life. Science 366:71–76.

Schloegel, M., L., A. M. Picco, A. Kilpatrick, A. Davies, A. Hyatt, and P. Daszak. 2009. Magnitude of the US trade in amphibians and presence of *Batrachochytrium dendrobatidis* and ranavirus infection in imported North American bullfrogs (*Rana catesbeiana*). Biological Conservation 142:1420–1426.

Silva, S., L. Matz, M. M. Elmassry, and M. J. San Francisco. 2019. Characteristics of monolayer formation in vitro by the chytrid *Batrachochytrium dendrobatidis*. Biofilm 1:100009.

Spitzen-van der Sluijs, A., F. Spikmans, W. Bosman, M. de Zeeuw, T. van der Meij, E. Goverse, M. Kik, F. Pasmans, and A. Martel. 2013. Rapid enigmatic decline drives the fire salamander (*Salamandra salamandra*) to the edge of extinction in the Netherlands. Amphibia-Reptilia 34:233–239.

Stegen, G., F. Pasmans, B. R. Schmidt, L. O. Rouffaer, S. Van Praet, M. Schaub, S. Canessa, A. Laudelout, T. Kinet, and C. Adriaensen. 2017. Drivers of salamander extirpation mediated by *Batrachochytrium salamandrivorans*. Nature 544:353–356.

Thomas, V., M. Blooi, P. Van Rooij, S. Van Praet, E. Verbrugghe, E. Grasselli, M. Lukac, S. Smith, F. Pasmans, and A. Martel. 2018. Recommendations on diagnostic tools for *Batrachochytrium salamandrivorans*. Transboundary and Emerging Diseases 65:e478–e488.

Thomas, V., Y. Wang, P. Van Rooij, E. Verbrugghe, V. Baláž, J. Bosch, A. A. Cunningham, M. C. Fisher, T. W. J. Garner, M. J. Gilbert, et al. 2019. Mitigating *Batrachochytrium salamandrivorans* in Europe. Amphibia-Reptilia 40:265–290.

Tompros, A. 2021. Management strategies to reduce invasion potential of *Batrachochytrium salamandrivorans*. Thesis, University of Tennessee, Knoxville, USA.

Tompros, A., A. D. Dean, A. Fenton, M. Q. Wilber, E. D. Carter, and M. J. Gray. 2021. Frequency-dependent transmission of *Batrachochytrium salamandrivorans* in eastern newts. Transboundary and Emerging Diseases 69:731–741.

Towe, A. E., M. J. Gray, E. D. Carter, M. Q. Wilber, R. J. Ossiboff, K. Ash, M. Bohanon, B. A. Bajo, and D. L Miller. 2021. *Batrachochytrium salamandrivorans* can devour more than salamanders. Journal of Wildlife Diseases 57:942–948.

US Fish and Wildlife Service (USFWS). 2016. Injurious wildlife species: listing salamanders due to risk of salamander chytrid fungus. Federal Register https://www.federalregister.gov/documents/2016/01/13/2016-00452/injurious-wildlife-species-listing-salamanders-due-to-risk-of-salamander-chytrid-fungus/. Accessed 23 Feb 2022.

Van Rooij, P., F. Pasmans, Y. Coen, and A. Martel. 2017. Efficacy of chemical disinfectants for the containment of the salamander chytrid fungus *Batrachochytrium salamandrivorans*. PLoS One 12: e0186269.

Venesky, M. D., M. J. Parris, and A. Storfer. 2009. Impacts of *Batrachochytrium dendrobatidis* infection on tadpole foraging performance. EcoHealth 6:565–575.

Vilaça, S. T., J. F. Bienentreu, C. R. Brunetti, D. Lesbarrères, D. L. Murray, and C. J. Kyle. 2019. Frog virus 3 genomes reveal prevalent recombination between ranavirus lineages and their origins in Canada. Journal of Virology 93:e00765-19.

von Essen, M., W. T. M. Leung, J. Bosch, S. Pooley, C. Ayres, and S. J. Price. 2020. High pathogen prevalence in an amphibian and reptile assemblage at a site with risk factors for dispersal in Galicia, Spain. PLoS One 15:e e0236803.

Voyles, J., S. Young, L. Berger, C. Campbell, W. F. Voyles, A. Dinudom, D. Cook, R. Webb, R. A. Alford, L. F. Skeratt, and R. Speare. 2009. Pathogenesis of chytridiomycosis, a cause of catastrophic amphibian declines. Science 326:582–585.

Waddle, J. H., D. A. Grear, B. A. Mosher, E. H. C. Grant, M. J. Adams, A. R. Backlin, W. J. Barichivich, A. B. Brand, G. M. Bucciarelli, D. L. Calhoun, et al. 2020. *Batrachochytrium salamandrivorans* (Bsal) not detected in an intensive survey of wild North American amphibians. Scientific Reports 10:13012.

Wake, D. B., and V. T. Vredenburg. 2008. Are we in the midst of the sixth mass extinction? A view from the world of amphibians. Proceedings of the National Academy of Sciences USA 105(Supplement 1):11466–11473.

Walker, S. F., J. Bosch, T. Y. James, A. P. Litvintseva, J. A. Oliver Valls, S. Piña, G. García, G. A. Rosa, A. A. Cunningham, S. Hole, et al. 2008. Invasive pathogens threaten species recovery programs. Current Biology 18:R853–R854.

Warne, R., E. Crespi, and J. Brunner. 2011. Escape from the pond: stress and developmental responses to ranavirus infection in wood frog tadpoles. Functional Ecology 25:139–146.

Wells, D. K. 2007. The ecology and behavior of amphibians. University of Chicago Press, Chicago, Illinois, USA.

Wheelwright, N. T., M. J. Gray, R. D. Hill, and D. L. Miller. 2014. Sudden mass die-off of a large population of wood frog (*Lithobates sylvaticus*) tadpoles in Maine, USA, likely due to ranavirus. Herpetological Review 45:240–242.

Wilber, M. Q., E. D. Carter, M. J. Gray, and C. J. Briggs. 2021. Putative resistance and tolerance mechanisms have little impact on disease progression for an emerging salamander pathogen. Functional Ecology 35:847–859.

Wilber, M. Q., R. A. Knapp, M. Toothman, and C. J. Briggs. 2017. Resistance, tolerance and environmental transmission dynamics determine host extinction risk in a load-dependent amphibian disease. Ecology Letters 20:1169–1181.

Woodhams, D. C., M. Bletz, J. Kueneman, and V. McKenzie. 2016. Managing amphibian disease with skin microbiota. Trends in Microbiology 24:161–164.

Woodhams, D. C., J. Bosch, C. J. Briggs, S. Cashins, L. R. Davis, A. Lauer, E. Muths, R. Puschendorf, B. R. Schmidt, B. Sheafor, et al. 2011. Mitigating amphibian disease: strategies to maintain wild populations and control chytridiomycosis. Frontiers in Zoology 8:8.

Yao, J.-Y., C.-S. Zhang. X.-M. Yuan, L. Huang, D.-Y. Hu, Z. Yu, W.-L. Yin, L.-Y. Lin, X.-Y. Pan, G.-L. Yang, et al. 2022. Oral vaccination with recombinant *Pichia pastoris* expressing iridovirus major capsid protein elicits protective immunity in largemouth bass (*Micropterus salmoides*). Frontiers in Immunology 13:852300.

Yap, T. A., M. S. Koo, R. F. Ambrose, D. B. Wake, and V. T. Vredenburg. 2015. Averting a North American biodiversity crisis. Science 349:481–482.

Yuan, Z., A. Martel, J. Wu, S. Praet, S. Canessa, and F. Pasmans. 2018. Widespread occurrence of an emerging fungal pathogen in heavily traded Chinese urodelan species. Conservation Letters 11:e12436.

Zhou, X., X. Zhang, Y. Han, Q. Jia, and H. Gao. 2017. Vaccination with recombinant baculovirus expressing ranavirus major capsid protein induces protective immunity in Chinese giant Ssalamander, *Andrias davidianus*. Viruses 9:195.

5 Oil Spills and Their Management

MICHAEL ZICCARDI

Introduction

Oil spills can cause serious wildlife health problems to both individual animals as well as wildlife populations (particularly when they occur in nearshore marine environments), but they are unlike most other wildlife "diseases." The primary acute toxic effects of oil on wildlife are often inextricably linked to the secondary effects that oiling causes (such as hypothermia, dehydration, and malnutrition). Exposure to petroleum compounds, although obviously not infectious and of limited geographic scope, can also have extremely long-term ecosystem-level effects (like some persistent pathogens). For example, more than 30 years after the 1989 *Exxon Valdez* oil spill (EVOS), measurable petroleum contamination in intertidal invertebrates and health effects in sea otters (*Enhyda lutris kenyoni*) and other keystone species studied as part of the postspill Nearshore Vertebrate Project continue to be demonstrated (Figure 5.1).

Oil spills are one of the most visible and clearly anthropogenic wildlife health problems that occur today and a quintessential One Health issue (e.g., a health problem affecting humans, animals, and plants and the environment in which they exist). They typically engender a great deal of public attention

The desired outcome to oiled wildlife response—a successful release of a normal animal back to a clean environment. Illustration by Laura Donohue.

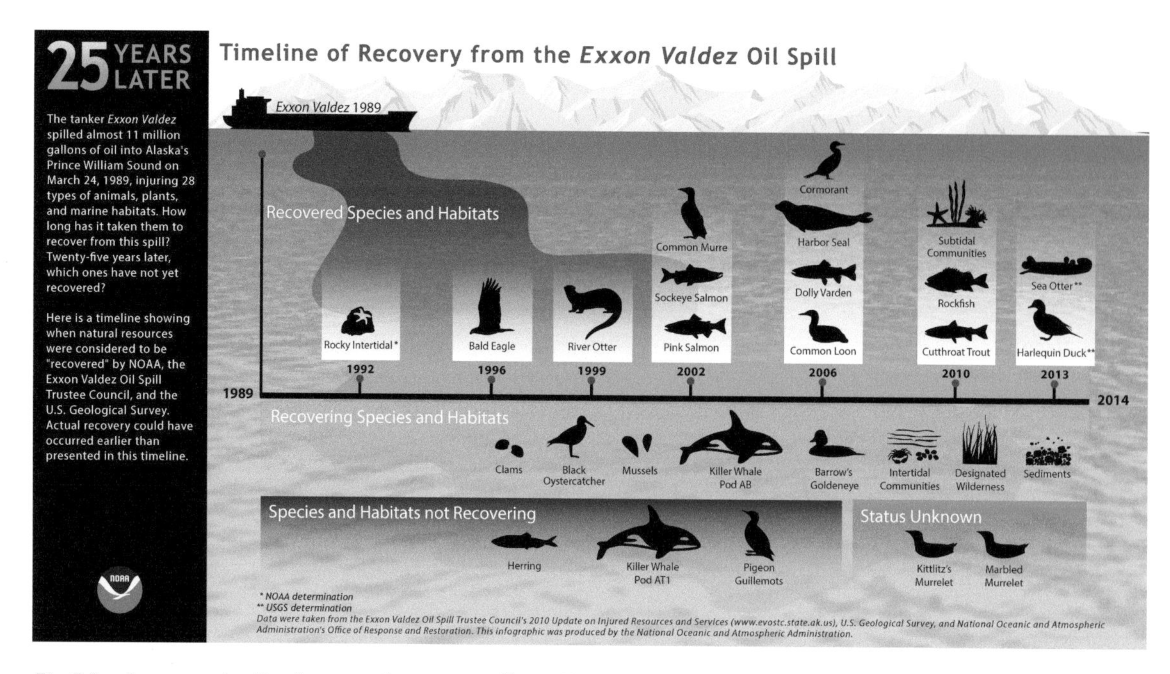

Fig. 5.1. Recovery timeline for natural resources affected by the *Exxon Valdez* oil spill in Prince William Sound, Alaska, USA, as determined by the National Oceanic and Atmospheric Administration (NOAA), with data taken from the *Exxon Valdez* Oil Spill Trustee Council's 2010 update on injured resources and services, US Geological Survey, and NOAA's Office of Response and Restoration. Infographic credit: NOAA.

and concern that often generate substantial anger against the "spiller" and a demand for wildlife care and accountability. Saving oiled wildlife, which in itself is an animal welfare issue, is just part of the picture. Gathering data and scientific knowledge of potential future use is a secondary concern. Additionally, there are serious legal and financial issues with most spills, including but not limited to, both civil and potentially criminal charges with fines and penalties levied against the "responsible party." This requires documentation of affected animals (live and dead) and maintaining carcasses as part of a legal chain of custody. In the United States, the polluter is not only financially responsible for the cost of response to the spill (of which wildlife recovery and care is one part), but also for restoration projects to address damages to the natural resource. This Natural Resource Damage Assessment (NRDA) process is legally distinct from the response effort, determining the overarching injuries to not only the environment and animals, but also to humans (e.g., recreational loss and financial damages), and can take many years, if not decades, to be settled.

One debate that often occurs during large-scale spills where wildlife are affected and a rescue and rehabilitation effort is mounted is whether the effort is "worth it." Some biologists argue that funds allocated to recovering and caring for individual oiled animals would be better spent on buying habitat or funding research and that those few animals that are "successfully" returned to the environment will never return to "normal." What is misleading in this argument is that oiled wildlife care actually adds to, rather than competes for, settlement funds that may purchase wildlife habitat or fund long-term studies. Information gleaned from the wildlife response feeds into both the damage assessment as well as the investigative processes and assures that the responsible party pays for the damage done in the final legal and financial settlements. Additionally, funds used for wildlife response are not deducted from what the responsible party must pay to restore wildlife habitats damaged by spilled petroleum or to help mitigate present and future loss of wildlife. At least in the United States, they are two entirely separate legal processes.

For biologists, veterinarians, and conservationists, an oil spill response is not just about improving our ability to save oiled wildlife (although that has been vastly improved over the last 40 years) or better understanding the effects of oil on wildlife and various ecosystems, which has dramatically improved following both the EVOS incident and the *Deepwater Horizon* (DWH) spill (see Box 5.1). It's about comprehensively and simultaneously trying to optimize response, mitigate damage, improve knowledge and techniques and, to the extent possible, restore wildlife populations and the ecosystems they depend on in the midst of a disaster. In some ways, the cooperation and degree of sophistication often found in oil spill response, including the wildlife efforts in many regions, is far ahead of other wildlife health management efforts: before oil spills occur in North America, approximately what will be done, who will do it, and how and who will pay for it (including emergency funds should no spiller come forward) are already determined. Comprehensive and collaborative programs (such as the California-based Oiled Wildlife Care Network) have also been developed to provide well-trained personnel and volunteers, immediately available well-designed facilities and to focus research efforts between disasters—a model that rarely exists for other wildlife health emergencies. Although laws, social customs, and species biology vary widely across the globe, such response networks have been established in North America, Japan, New Zealand, South Africa, and portions of Europe. Therefore, while this chapter describes the multiple effects that petroleum exposure can have on wildlife species (with a focus on the critical response elements needed for an effective wildlife response effort), the reader might consider whether a similar approach might be developed for other serious, fairly predictable, and recurring wildlife health problems.

General Oil Toxicity

Oils are extremely complex mixtures, varying widely depending on where they are extracted and the extent of processing, among other factors (Neff 1990). Petroleum can contain thousands of organic and inorganic compounds; most are different types of hydrocarbons (up to 98%) but may include sulfur, nitrogen, oxygen, heavy metals, and trace elements (NRC 2003). Aromatic hydrocarbons are the most acutely toxic elements of petroleum products, with the smaller monocyclic aromatics (e.g., benzene, toluene, ethylbenzene, and xylene) linked with carcinogenicity, organ damage, and even death in vertebrates (Neff 1979). While these compounds are readily bioavailable due to high water solubility, they are often not found in high concentrations due to extreme volatility (ATSDR 1997). Several polycyclic (or polynuclear) aromatic hydrocarbons (PAHs) are also carcinogenic (as well as cause reproductive failure in laboratory animals), but they are relatively insoluble in water (ATSDR 1995).

When considering the toxicity of petroleum compounds to wildlife, many factors influence the negative effects, including (but not limited to) mode and duration of exposure, constituents of the product, species sensitivity, and age or health status of the individual (Jessup and Leighton 1996). Additionally, fully understanding the underlying health-related effects of oil exposure go far beyond that seen with immediate morbidity and mortality and should include effects on behavior such as foraging, migration, reproduction, and distribution (Friend et al. 1999). The following sections will attempt to detail some of the specific effects documented in the main vertebrate wildlife species of veterinary concern (birds, sea turtles, and marine mammals) categorized by the main exposure routes.

Specific Oil Effects

External Exposure

Typically, the most numerous and visible higher vertebrates affected by coastal oil spills are marine

Box 5.1. *Deepwater Horizon:* Knowledge from catastrophe

On 20 April 2010, a blowout on the *Deepwater Horizon* (DWH) oil rig—an ultradeep water offshore drilling rig operating at a depth of approximately 1500 meters in the Gulf of Mexico—caused a massive explosion, killing 11 crewmen (Figure 5.2). The rig sank two days later, causing the exploratory well to gush uncontained for 87 days until it was successfully capped. While there were several estimates as to the total volume of crude oil that was released, the US government's estimate of 796 million liters (5 million barrels) is considered the most accurate, making it the largest marine oil spill in history.

A massive cleanup and containment effort was begun almost immediately—trying to limit the spread of what became more than a 147,600 square kilometer slick. The massive size of the slick and its significant distance from nearest shoreline made the use of traditional cleanup techniques, such as skimmers and ships to gather spilled oil from the surface, problematic and the calm waters of the Gulf made so-called "alternative response techniques" more attractive. More than 6.8 million liters of the chemical dispersant Corexit (EC9500A and EC9527A) was applied, attempting to drive the

Fig. 5.2. Select images from the 2010 *Deepwater Horizon* oil spill in the Gulf of Mexico, USA, including a controlled burn, an oiled sea turtle being recovered, a stranded bottlenose dolphin, and the consistency of the oil. Photograph credit: US National Oceanic and Atmospheric Administration's Office of Response and Restoration and M. Ziccardi.

surface slick into the water column and dilute the rising oil from the wellhead before it surfaced. However, their massive use caused significant concerns from the environmental community due to potential toxicity. Similarly, more than 400 controlled burns were successfully ignited during the cleanup, eventually removing an estimated 220,000–310,000 barrels of oil from the Gulf (Allen et al. 2011). However, the potential of entraining sea turtles in the burning slick, as well as the toxic pyrogenic compounds sinking to the sea floor or rising in the air, raised concerns. In all, more than 47,000 responders were working at the height of the response, attempting to stem the spread of oil to the shorelines (Ramseur and Hagerty 2014); however, oil came ashore in the marshlands of Louisiana in May and the beaches of Mississippi, Alabama, and Florida in June.

The spread of the huge slick also caused serious concern for offshore wildlife populations, including (but not limited to) 28 species of marine mammals, 5 of the world's 7 sea turtle species, and large rafts of seabirds. The fears of wildlife impacts increased when the oil entered the lush and fragile habitat of the Gulf's marshes and estuaries, as well as the sandy beaches that act as nesting grounds for several sea turtle species. Large-scale bird, marine mammal, and sea turtle recovery and rehabilitation efforts were quickly mobilized, eventually encompassing rehabilitation operations that stretched from the Texas–Louisiana border all the way into Florida. These were supported by on-water rescue operations that extended from nearshore bird and mammal recovery to juvenile sea turtle rescue efforts occurring far offshore, and sea turtle nest translocation efforts that sent more than 14,000 eggs to the Atlantic seaside for eventual hatching and release. In all, more than 3,000 live and 8,500 dead birds, 536 live and 613 dead sea turtles, and 12 live and 191 dead marine mammals were collected during the active operational period of the response (DHNRDA Trustees 2018). However, during the subsequent Natural Resource Damage Assessment process, the total numbers of impacted animals were estimated at 102,400 birds and 202,600 sea turtles killed, and between 2% and 17% of 16 different marine mammal populations were presumed lost.

While the DWH spill was not the first such incident to occur in United States waters, the concerns of the public and its visibility in the media were comparable to that of the 1989 *Exxon Valdez* oil spill (EVOS) in Prince William Sound, Alaska. In that incident, approximately 42 million liters (or just 5% of the DWH) spilled into the Sound; however, the acute effects on wildlife in the area were just as, if not more, severe, with an estimated 4,000 sea otter deaths from acute oiling (Garrott et al. 1993), and upwards of 435,000 seabird deaths in its aftermath (Figure 5.3) (Ford et al. 1991).

As a consequence of EVOS, the knowledge of the effects of oil on wildlife made a tremendous step forward (Jessup and Leighton 1996), and more organized and professional animal care procedures and protocols were developed (Williams and Davis 1995, OWCN 2014). Since the DWH incident, additional information has surfaced on the potential effects of oil on wildlife species—both from acute external exposure as well as low-level internal exposure (presumably from inhalation and ingestion) (Takeshita et al. 2021). Similarly, more effective and integrated oiled wildlife response management strategies were developed both during DWH (Ziccardi et al. 2015) as well as from international consortia as a consequence of DWH (e.g., the Global Oiled Wildlife Response System) (Ziccardi et al. 2021).

(continued)

Fig. 5.3. Select images from the 1989 *Exxon Valdez* oil spill in Prince William Sound, Alaska, USA, including an live oiled seabird, the vessel offloading fuel, a dead oiled sea otter, and the collection of a dead oiled bird. Credit: *Exxon Valdez* Oil Spill Trustee Council.

birds. In these species, the microscopic alignment of barbicels and barbules of the feathers (akin to pieces of Velcro mating to one another; Figure 5.4) forms a barrier that excludes water and traps a critical layer of air next to the skin, providing insulation in species with high normal body temperatures (Albers 2003). Exposure to oil causes this microstructure to be (at least temporarily) destroyed (Jenssen and Ekker 1988), allowing water to penetrate to the skin, leading to hypothermia and an inability to remain buoyant (Stephenson and Andrews 1997, O'Hara and Morandin 2010). Birds can temporarily offset hypothermia somewhat by increasing metabolic effort to maintain core body temperatures (Jenssen 1994, Cunningham et al. 2017), but this typically results in rapid starvation as energy stores (e.g., fat and muscle) are depleted (Fry and Lowenstine 1985, Bursian and Dean 2017). Additionally, when marine birds lose buoyancy, they cannot forage, compounding existing dehydration and starvation that leads to drowning or increased predation (Vermeer and Vermeer 1975). Species, degree of oiling, individual animal condition, and environmental conditions, among other factors, can directly impact the timing of injuries associated with external petroleum exposure (IEc 2015).

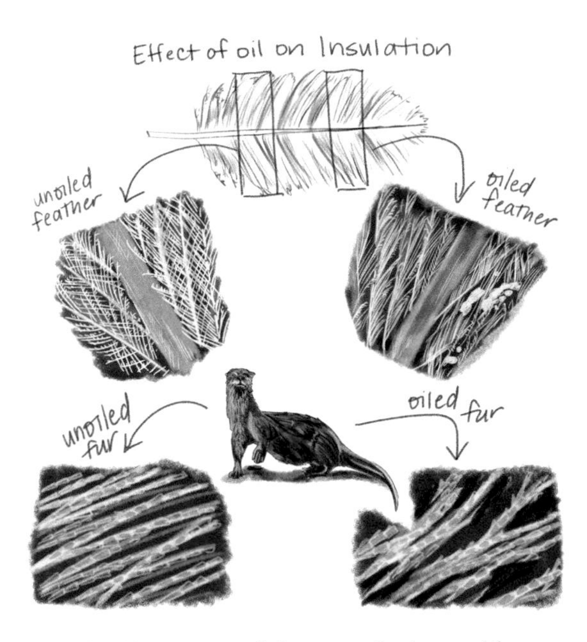

Fig. 5.4. Microstructural changes to feathers and fur upon petroleum exposure. Illustration by Laura Donohue.

Effects similar to birds can be seen with several heavily furred aquatic mammals (e.g., sea otters and fur seals), as the density and alignment of interlocking hair bundles trap an insulating air layer (Tarasoff 1972) (Figure 5.4). Oil exposure allows water to seep under the pelage (Davis et al. 1988), decreasing both buoyancy and insulation. To maintain this layer, sea otters must continually groom themselves; when oiled, such grooming is relatively ineffective and can lead to substantial ingestion of oil (Siniff et al. 1982). Sea otters have an extremely high metabolic rate (estimated at 2.4 times that of a comparable terrestrial mammal) (Costa and Kooyman 1982) and surface to volume ratio of their pelage. They must eat up to 25% of their body weight per day (Kenyon 1969), which can lead to additional petroleum exposure from ingesting prey species (Neff et al. 1987).

Oil on the skin, mucus membranes, and other sensitive tissues in vertebrates can also cause both acute and chronic physical damage to the epidermis and underlying layers, including irritation, burning, and permanent damage (depending on the compound's makeup, weathering, and other physical properties) (Mazet et al. 2002). The skin of birds is particularly sensitive, being extremely thin and fragile over most of the body surface (Bauck et al. 1997); highly refined products (e.g., gasoline) can cause chemical burns even if cleaned off promptly (Helm et al. 2014). External effects to vertebrates also depend on how easily the oil adheres to the animal, with unfeathered or unfurred species (e.g., cetaceans) being more resistant to oil sticking to their rubbery skin (St. Aubin et al. 1985).

The presence of oil on the surface of animals can also significantly impact their locomotion—flying, swimming, or ambulation on land. Sea turtles were regularly observed trapped in oil during the DWH oil spill (Stacy 2012), and seal pups encased in oil have been seen to drown in previous events (Davis and Anderson 1976). In experimental studies following DWH, lightly oiled western sandpipers (*Calidris mauri*) were observed to have decreased takeoff speed and angles that resulted in greater wingbeat frequencies and amplitudes (Maggini et al. 2015), and oiled homing pigeons (*Columba livia*) were observed to have 1.6 times greater return time, with longer distances, slower speeds, and more changes in elevation (Perez et al. 2017). These movement changes, in concert with heat loss and decreased nutrition, can rapidly deplete energy stores, leading to mortality even with light oiling.

Reproductive success in bird species can also be significantly affected by external oil exposure. Avian embryos can be directly damaged by the transfer of as little as 1–20 μL of oil from laying females to eggs (Hartung 1967), penetrating through the egg pores and either killing the developing embryo or, in some cases, causing deformities (Hoffman 1979, Szaro 1979, Finch et al. 2011). Physical oiling of eggs can also occur after laying if adults have external oiling present, allowing toxic components (likely PAHs) to penetrate the pores and, at sufficient doses, cause systemic organ failure in the developing embryo (Couillard and Leighton 1990).

Inhalation

For vertebrates that live at the air–water interface, the inhalation of volatiles from oil spills has caused direct damage to respiratory tissues (e.g., lungs or air sacs) (Harr et al. 2017), leading to decreased blood oxygenation; alterations to diving, swimming, or flying capabilities; and subsequent foraging impacts. In birds, due to gas exchange across thin respiratory tissues, smaller PAHs can also transfer more readily to the bloodstream (Brown et al. 1997), leading to higher potential for systemic effects. Because marine mammals and sea turtles make large inhalations prior to diving, significant exposure to volatile compounds and inhalation of oil droplets can easily occur. During DWH, asphyxiation by oil was documented as the primary cause of sea turtle death, and most turtles collected had significant amounts of oil in the mouth and esophagus (Stacy 2012). A decreased respiratory rate was observed in most live animals, and pneumonia was seen in several (Stacy et al. 2017).

Neurological deficits have been observed in live oiled animals (Massey 2006), although it is unclear if the deficits noted were directly related to inhalation of volatiles causing direct narcotic-type effects on the central nervous system (ATSDR 1997) or from other biomedical causes (e.g., hypoglycemia or liver dysfunction). In any case, direct morphological changes have been observed in nervous tissues (Peterson et al. 2003), and alterations in behavior can lead to lack of avoidance of predators, inability to forage, decreased reproductive efforts, decreased migratory habits, and other secondary but significant effects. During EVOS, harbor seals were observed exhibiting abnormally tame or lethargic behavior. This might be explained by brain lesions found in some oiled harbor seals (*Phoca vitulina*) and Steller sea lions (*Eumetopias jubatus*), likely caused by the toxic systemic effects of inhaled hydrocarbons (Spraker et al. 1994).

The most notable respiratory effects in oil spills appear to be those seen in marine mammals. Direct respiratory damage a key finding in sea otters affected in EVOS: interstitial pulmonary emphysema was observed in 73% of heavily contaminated, 45% of moderately contaminated, and 15% of lightly contaminated dead animals (Lipscomb et al. 1993), and subcutaneous emphysema was regularly diagnosed in live otters (Williams and Davis 1995). Coastal bottlenose dolphins (*Tursiops truncatus*) in heavily oiled Barataria Bay, Louisiana, following the DWH spill were even more impacted. While previous experimental studies suggested that cetaceans would avoid surface slicks (St. Aubin et al. 1985), this appears not to have been the case in the bay, and animals in the area were severely affected. In 2011, 43% of the evaluated dolphins were considered "unhealthy" (with 17% given a poor or grave prognosis) and were five times more likely to have moderate to severe lung disease (e.g., bacterial pneumonia) when compared to a control group (Schwacke et al. 2013). Follow-up studies found impaired stress responses (due to damage to the adrenal system), various inflammatory conditions, and reproductive dysfunction that lasted for at least 4 years after the incident (Venn-Watson et al. 2015, De Guise et al. 2017, Kellar et al. 2017).

Ingestion

Preening or grooming of oil from feathers and fur is one of the first responses to external oiling (Figure 5.5). These cleaning behaviors, in addition to fouling of (and bioconcentration in) food items, can lead to significant ingestion of petroleum. For example, in experimental studies conducted after the DWH spill, double-crested cormorants (*Phalacrocorax auritus*) having oil that coated 20% of their bodies ingested approximately 13 g of oil by day 8 of the study (Cunningham et al. 2017).

Fig. 5.5. Internal effects of oil exposure in seabirds. Illustration by Laura Donohue.

Ingestion of oil can cause a host of negative effects in vertebrate species. Initially, direct damage can occur to the tissues of the gastrointestinal tract, leading to diarrhea and dehydration and wasting even in the presence of increased food and water uptake (Balseiro et al. 2005, Cunningham et al. 2017), with gastric erosions or ulcers and degeneration of intestinal villi in severe cases (Fry and Lowenstine 1985, Lipscomb et al. 1993). Even if this initial insult is corrected in rehabilitation centers, "refeeding syndrome" can occur, where sudden electrolyte imbalances can lead to seizures, hemolytic anemia, cardiac failure, and death (Orosz 2013, Fravel et al. 2016).

When petroleum compounds are absorbed into the bloodstream via the gastrointestinal tract, a host of negative effects can occur. Briefly, these include the following.

LIVER

"First pass" metabolism by the liver can remove toxic components (Troisi et al. 2006) but can also produce reactive intermediate compounds, including oxygen radicals, that can damage the liver and impair its function (Harvey 1991). This hepatic damage can disrupt plasma proteins (leading to immune dysfunction) (Briggs et al. 1996), decrease carbohydrate metabolism (leading to altered nutrition) (Hazelwood 1986), and decrease clotting factors (leading to increased bleeding when injured) (Hochleithner et al. 2006).

RED BLOOD CELLS

Regenerative and nonregenerative anemias are observed in oil-affected animals (Rebar et al. 1995, Tseng 1999), although the precise cause of such anemias is unclear. Experimental studies have shown destructive anemias associated with oral oil exposure in birds (Leighton 1985, Troisi et al. 2007) and in mink (*Mustela vison*) as a model for sea otters (Schwartz et al. 2004), caused by oxidative damage to the cell membranes leading to the denaturing of hemoglobin (and the formation of Heinz bodies in birds). Hemolytic anemias, however, are not universally seen in "natural" exposures and likely are just one component of spill-related anemias.

WHITE BLOOD CELLS

Both the overall numbers and composition of white blood cells, or leukocytes, can be seriously affected by oil absorption. This is likely due to shifts in production from leukocytes to erythrocytes, malnutrition, stress, and damage from reactive-intermediate compounds (Rocke et al. 1984, Briggs et al. 1996, Bursian et al. 2017). If this immunocompromise is significant, it can lead to significant morbidity and mortality due to secondary infections.

KIDNEYS

Damage to the renal system is manifested by changes in kidney function due to glomerulonephritis (Fry and Lowenstine 1985, Harr et al. 2017) or severe dehydration due to gastrointestinal effects (Leighton 1986). Decreases in kidney function can lead to electrolyte imbalances (thereby affecting fluid volume and blood pressure) (Tseng and Ziccardi 2019) and to a decreased capacity to eliminate metabolic waste, reduced hemostasis, and generalized debilitation (Echols 2006).

HEART

While there are significant data related to cardiac-associated pathology in fish species following oil exposure (Incardona et al. 2014), this issue is an emerging concern in birds and marine mammals. In bird studies following DWH, the ingestion of oil was linked to changes in cardiovascular function, including flaccid heart musculature, increased ejection velocities and volumes, and decreased perfusion or blood pressure (Harr et al. 2017). It is currently unclear if these changes are related to direct toxic effects on heart muscle (Ou and Ramos 1992), alterations in myocardial conduction (Brette et al. 2014), or secondary to changes in other organ systems (Leighton 1986).

ENDOCRINE

Oil exposure can have both direct or indirect effects on the adrenal gland, leading to chronic effects on plasma corticosterone levels (Rattner and Eastin 1981, Lattin et al. 2014). Additional data suggest direct effects of PAHs (or their metabolism) on the thyroid gland (Rattner et al. 1984, Jenssen et al. 1990) as well as damage to the hypothalamus–pituitary–adrenal axis (Mohr et al. 2010, Schwacke et al. 2013). A study on dead stranded dolphins in Barataria Bay showed chronic adrenal insufficiency, thin adrenal cortices, and increased risk of bacterial pneumonias from the subsequent immunodeficiency (Venn-Watson et al. 2015).

REPRODUCTIVE

In addition to the external effects of oil on incubating eggs, ingestion can lead to direct alterations in reproductive function, including embryo or fetal mortality, teratogenesis, failed hatching or births, and increased chick or pup abandonment (well summarized in birds by Leighton 1993). Changes in behavior in multiple oiled adult bird species have been observed in field studies, primarily consisting of cessation of breeding effort, abandonment of nests, and disruption of pair bonds (Fry et al. 1986, Butler et al. 1988). Following the DWH spill, only 20% of pregnant dolphins in Barataria Bay produced viable calves, and the estimated annual survival rate was low (86.8%) compared with a pregnancy success rate of 83% and survival rate of 95.1% in Sarasota Bay, Florida, at the same time (Lane et al. 2015). Additionally, perinate dolphins collected during this time were found to more likely have died *in utero* (or very soon after birth), to have fetal distress, to have pneumonia not associated with lungworm infection, and to have a higher incidence of *Brucella* sp. infections identified via lung polymerase chain reaction assays (Colegrove et al. 2016).

Wildlife Response to Oil Spills

Since the EVOS in 1989, oiled wildlife response has evolved from an activity conducted separately by smaller charitable organizations (or financially supported through donations and volunteer efforts) to one that is fully integrated into, and supported by, the overall spill response effort under a structured Incident Command System (ICS). The wildlife response is typically a relatively small, albeit publicly visible, fraction of the overall efforts during an oil spill response. For example, of the tens of thousands of responders that were actively working on the incident at the height of the DWH incident, active wildlife responders made up less than 1% of the total workforce. It is difficult to contain public desire to help and is both politically and socially important to channel these energies. Volunteer use also helps keep cost of wildlife care down; they are usually a very small part of the overall disaster expenses, even in more wildlife-intensive spills varying from approximately 1%–5% of overall response costs (Jessup 1998, Massey et al. 2005). Because of the high sensitivity of the public to such accidents, oiled wildlife response care is a necessity in most regions, with both the wildlife regulatory agencies and responsible party perceived to be doing a poor job—even if the oil cleanup process is successful—when inadequate wildlife care or excessive suffering is perceived. It is imperative that wildlife professionals understand the overarching spill response structure so that they can understand their roles and that their efforts are supported, acknowledged, and integrated within the ICS structure.

In the United States (as well in regions where the tenets of ICS are applied), oil spill response follows a standardized organizational chart, with an Incident Commander (or Unified Command when multiple stakeholders are involved) leading the effort, and four defined sections under them: Operations, Planning, Logistics, and Administration and Finance (FEMA 2017). Wildlife response typically falls under the Wildlife Branch within the Operations Section, but works closely with other sections, particularly the Environmental Unit within the Planning Section. The specific activities initiated by the Wildlife Branch during oil spills differ widely depending upon many factors, including the size of the spill, type of product spilled, time of year, species potentially affected, and location. However, in general, oiled wildlife response can be divided into three different response strategies: primary (e.g., "keeping oil away from wildlife"), secondary (e.g., "keeping wildlife away from the oil"), and tertiary (e.g., actions taken if primary and secondary actions fail—typical wildlife collection and rehabilitation practices).

Primary Response Actions

The prevention of oil spread using cleanup techniques and protection strategies to keep oil from entering sensitive habitats is one of the most efficient ways to help wildlife. While wildlife professionals don't typically conduct these operations, they often do take part in general reconnaissance (to determine numbers and locations of animals at risk to direct cleanup and protection strategies) and collection of oiled carcasses (to reduce secondary impacts to scavengers).

Secondary Response Actions

Secondary wildlife response tactics include deterrence (or hazing) and preemptive capture of wildlife. Birds and other animals can be hazed from risky areas using close-range and longer range visual or auditory devices designed to scare them from that area, deter them from entering certain locations via exclusion devices (e.g., fences and netting), or attract them to less risky regions through bait or environmental manipulation (Petras 2003, Gorenzel and Salmon 2008). Deterrence activities come with some risks to both the human responders as well as the animals, as some techniques use potentially dangerous and regulated materials. Also, deterrence actions are only effective when there are safe locations where animals can be directed, when species will not quickly return, and when the geographic area involved is small enough that it can be effectively controlled. Animals vary in responses to techniques, and habituation and desensitization also occurs. Hazing activities must take place only under the authority and oversight of trustee agencies since such actions are typically legally designated as "harassment" or "take."

The capture (and subsequent handling, transportation, short-term holding, and release) of unoiled, at-risk species has been done in select incidents, including the relocation of tens of thousands of African penguins (*Spheniscus demersus*) during the *Treasure* oil spill in South Africa (Wolfaardt et al. 2008), and the capture and long-term holding of New Zealand dotterels (*Charadrius obscurus aquilonius*) during the *M/V Rena* oil spill (Gartrell et al. 2014). Similar to deterrent use, preemptive capture can be very effective in certain situations (discrete oil releases, endangered species, or animals that tolerate capture and holding) but is logistically intensive, has the distinct possibility of secondary injury to animals in the field and facility, and must be done under the auspices of the appropriate trustee agency.

Tertiary Response Actions

Tertiary response efforts are those that are typically thought of as the "standard wildlife response efforts" during oil spill events, including the capture and care of oil-affected animals (Figure 5.6). These efforts are not always undertaken during spills because, in certain situations, the return of "normal" animals (e.g., behaviorally and medically healthy individuals that can contribute to population recruitment) after rehabilitation cannot be accomplished. This decision may be based on logistical or resource limitations, condition of the affected animals, and other factors, but a concerted effort directed at the elimination of animal suffering (e.g., humane euthanasia) should always be considered even when full capture and care cannot be undertaken. Tertiary response actions fall into several main categories.

RECOVERY

Wildlife search and collection, also known as wildlife recovery, focuses on the collection and capture of dead and live oiled animals. Ideally, most oiled wildlife are captured by trained and experienced wildlife responders and by government agency personnel. In every incident, however, oil-affected wildlife are collected by well-meaning members of the public; therefore, an effective means to report oiled animal observations coupled with messaging to the public to report oiled wildlife immediately (and not collect animals for both human and animal safety) should be established. Specific procedures for the capture of live oiled wildlife are beyond the scope of this chapter; but many excellent resources are available for unoiled wildlife capture (Geraci and Lounsbury 2005, Whitworth et al. 2007, Townsend et al. 2018). Trained recovery personnel must systematically cover areas where oiled wildlife are likely to be found, and they should have permits from appropriate wildlife management agencies. At a minimum, accurate records are required to document each collected animal (e.g., date, time, location, and name) and legal requirements (such as "chain of custody" procedures) must be followed if required. Prior to the capture of any oiled animal the following should be considered: (1) captures should only be attempted if they can be performed in a safe manner for personnel as well as the animals; (2) potential benefits of capture must outweigh potential negative consequences (e.g., a small amount of oil on the fur of most pinnipeds will not warrant the capture of the animal); (3) rescuers should not enter nesting grounds or rookeries where disturbance might cause disruptions; and (4) "downstream" wildlife response planning (e.g., transport and care facility) should be

Fig. 5.6. Steps in the professional response to oil-affected birds. Illustration by Laura Donohue.

in place. The collection of dead animals during oil spills is also important to reduce the potential for secondary contamination as mentioned above, to better determine the overall scope of impacts to wildlife populations, and to document the losses. If possible, dead oiled animals should be collected and transported to a facility for a full evaluation under controlled conditions (see below for processing details). While it is understood that only a fraction of impacted animals will be collected during spills, these data are invaluable to the NRDA process—a systematic and quantitative approach to develop comprehensive and accurate estimates of impacts to wildlife that leads to legal and financial settlement.

STABILIZATION

Ideally, live oiled animals should either be stabilized *in situ* or quickly taken to a field stabilization site (e.g., a temporary first-aid station established close to the recovery area or an existing wildlife facility that is not set up for full oiled animal care but has the infrastructure to provide a sheltered place where initial first aid can be administered). At this point, initial triage (or prioritization and sorting of treatment) can be done, including the determination of likelihood of success of rehabilitation and early humane euthanasia using veterinary-approved means when survival to release is deemed unlikely. For those animals where treatment is warranted, field stabilization typically involves restoring normal core temperature (e.g., by supplemental heat or cooling and placement of containers in sheltered areas), addressing life-threatening conditions, and administering fluids (orally or parenterally). Animals are then transported in an organized, safe manner to an animal-care facility where full rehabilitative care can begin.

PROCESSING

The protocol should include individually marking all live and dead animals with a unique identifier (via tagging, marking, or microchipping, depending on the species); recording of field data; collecting key demographic data (e.g., species and, when possible, age class and sex); digitally photographing each animal (which encompasses the oiled regions on the body and, if possible, the identification); collecting an external oil sample (either several body feathers or an external swab of fur); and beginning an oiled animal intake form. When processing live animals, it is important to collect oil samples and data while minimizing the animal handling time and remaining safe. For larger, more dangerous animals (e.g., sea otters), it may be more efficient to conduct processing procedures at the same time as intake (see below), as chemical or significant manual restraint may be required. While live animal processing necessarily takes priority over dead animal assessment, dead animals can provide more extensive information as to the extent and degree of oil effects, as well as possible changes to live animal rehabilitation protocols (Ziccardi et al. 2015). The steps of processing a dead animal are similar to those for a live animal with the exception that no intake form is generated, and a complete necropsy is performed, whenever possible and approved. In particular, dead pinnipeds, cetaceans, and sea turtles should be necropsied, as it may provide the only means to determine whether it was exposed and died from that exposure. General necropsy methods and techniques used during oil spills are similar to those utilized during nonspill periods, with differences largely in sampling strategies for PAH analyses and the need for full photographic documentation and appropriate chain-of-custody procedures (Ziccardi et al. 2015).

INTAKE

Intake procedures are similar to those used in normal wildlife rehabilitation practices, with an emphasis on petroleum exposure and a need for efficiency when hundreds of animals may be moving through the system (OWCN 2014). Key data to be collected include an oiling evaluation (degree, extent, percentage, areas, and signs), attitude, alertness, body condition, hydration (and body weight whenever feasible), morphometrics, rectal temperature, and findings from a complete whole-body examination (with particular emphasis on those systems that may be affected from oil exposure). Following examination, blood samples should be drawn for assessment of, at a minimum, anemia and serum and plasma protein levels. During intake, the examiner and medical staff develop a treatment plan for each animal and triage animals for care based on such factors as legal status (e.g., threatened, endangered, or species of special concern), age class, historical success of that species or age class in rehabilitation, medical status (e.g., severe wounds and fractures), and characteristics of the spill response (e.g., size, caseload, available resources, and product spilled).

PREWASH CARE

A common mistake in the rehabilitation of oiled birds and heavily furred mammals is to wash newly admitted animals before they are physiologically stable. Because the washing process is rigorous and stressful, some oiled animals may require extensive prewash care to address nutritional compromise; dehydration; hypothermia; and the stress of capture, transport, and handling. However, other animals and species (such as cetaceans, most pinnipeds, and sea turtles) may be immediately cleaned on intake if blood work and initial exam show no abnormalities beyond physical coating. Nutrition and hydration support should be provided (e.g., birds gavage-fed high-calorie nutritional slurries alternating with rehydrating solutions up to six to eight times daily, with volumes dependent on species, size, and health status) and regularly re-examined to determine when they are medically stable enough (at a minimum, good alertness or attitude and blood values) to move to cleaning.

CLEANING

The goal of cleaning oiled wildlife (e.g., washing, rinsing, and, in some instances, drying completely) is to remove all external contamination to allow the affected animal to regain normal function (e.g., natural waterproofing and buoyancy, thermoregulation, normal ambulation, and ability to feed). Each taxon has specific requirements for successful cleaning (Jessup et al. 2012, Ziccardi et al. 2015, Tseng and Ziccardi 2019). Standard wash procedures include the use of dishwashing detergent (diluted for birds and furred mammals, full strength in other species) in water heated to, and maintained at, physiologically normal body temperatures. This solution is manually agitated around, or massaged into, the oiled pelage, and rinsed off until oil is completely removed. Monitoring of core body temperature and close observation of animals for distress should be always done. Rinsing with fresh water heated to physiologically normal body temperatures using an adjustable high pressure (40–60 psi, 276–414 kPa) nozzle until all detergent is removed from the feathers or underfur is as critical as removing oil. For heavily furred mammals and birds, rinse water with only 2–5 grains of hardness is highly recommended, since minerals in the water will bind with microscopic amounts of detergent and cause "soap" or salt crystals to form within the structure of the feather or on the hair strand, thereby reducing their waterproofing capabilities. Large, aggressive, or densely furred animals almost always require sedation or anesthesia for cleaning, so appropriate chemical agents with concomitant monitoring equipment must be available. After animals are completely clean, birds and densely furred mammals are then allowed to dry using towels, pet groomer dryers, heat lamps, or ambient conditions in their pen, depending on species, to allow a trapped layer of air to return in the undercoat. As can be appreciated from these details, significant facility infrastructure and supply requirements are necessary to support effective cleaning of oiled wildlife. Concerns related to maximizing human health and safety (e.g., bite wounds, heat exhaustion, slips, trips, falls and chemical exposure) are paramount—the cleaning of oiled wildlife should only be attempted by professionals that have the experience and knowledge on how to conduct these activities safely.

PRERELEASE CONDITIONING

Animals should be moved into appropriate enclosures (typically outdoor pools, aviaries, or pens) that are of sufficient size to allow accurate determination of behavior, feeding habits, waterproofing status, and recapture when necessary. Pen and water quality must be excellent to prevent recontamination of feathers or fur by oily feces and fish waste, and animals should be continually observed for signs of inadequate waterproofing. Excellent nutritional support should be provided, and regular medical checks to chart the resolution of acute and chronic oil exposure (e.g., external or internal organ damage) should be done to plot progress. Specific criteria must be met before animals can be released; these have been developed to increase the likelihood that animals are healthy, are in good body condition, have appropriate exercise tolerance, and have the ability to perform the full range of behaviors required for survival in the wild. Such criteria are taxon specific but usually include, at a minimum, normal behavior; good body weight; waterproof; and normal physical examination, hematology, and serum chemistry values (OWCN 2014). For animals that do not meet release criteria, several options are available including additional rehabilitation, euthanasia, or placement in a permanent care facility. However, for those that meet prerelease criteria, a release plan should be developed and approved both by the Incident Command and the appropriate wildlife trustees prior to releasing them into an appropriate clean environment. Postrelease monitoring (at a minimum bands, tags, or brands but ideally radio or satellite telemetry) are critical to collect data on survival rates, behavior, and reproductive success.

As was stated in the introduction, one contentious issue surrounding the recovery and rehabilitation of oiled wildlife is whether the effort on individual animals is cost efficient, whether survival is poor, and whether animals return to normal ecological standing after oil exposure. There is considerable variability among postrelease survival studies in the literature (dependent on species differences, aspects of the spill, or details of rehabilitation methods). In general, larger more robust birds (e.g., pelicans, gulls, penguins, ducks, and geese) tend to do well, while small and delicate birds (e.g., plovers and sandpipers) do not fare as well. Despite this variation, more recent studies have clearly shown survival far in excess of estimates generated prior to 2003 (Sharp 1996), a return to reproductive success, and in several cases no difference between survival of oiled and rehabilitated and control birds (Henkel and Ziccardi 2018). Only with continued efforts to better understand the success of tertiary oiled wildlife response after release can these questions be answered with strong scientific proof.

Summary

Since the devastating impact of the 1989 EVOS to Prince William Sound, much has been learned by wildlife professionals on the effects of, and how best to respond to, oil exposure in wildlife. A similar leap in scientific knowledge has more recently occurred following the 2010 DWH spill, with greater and more detailed results continuing to emerge on how petroleum can affect animals at all trophic levels.

Over this time, two key truths have emerged as critical for appreciating oil spill response and effects on wildlife populations. First, to best mitigate the impacts to wildlife associated with spills, extensive and proactive planning is absolutely crucial for an effective and successful response. Previous practices of contacting the local wildlife rehabilitation organization to address affected animals at the time of a release is no longer considered "best practice" for appropriate contingency planning. As a consequence of the DWH spill, the importance of proper oiled wildlife planning efforts as an integral part of overall spill response has been internationally recognized. In fact, the International Petroleum Industry Environmental Conservation Association-International Association of Oil and Gas Producers (IPIECA-IOGP) Oil Spill Response Joint Industry Project (OSR-JIP) now includes oiled wildlife response involving specialist personnel as one of the 15 key capabilities necessary for effective preparedness and response (IPIECA-IOGP 2015). This recognition has led to the development of numerous globally acceptable planning and response documents, including animal-care standards and systems of readiness and response encompassing training, equipment, and personnel readiness (IPIECA-IOGP 2016, 2017). Oiled wildlife response and care is becoming the global expectation.

A second emerging trend is a better appreciation that "health"—of both the environment and the organisms contained within—is intertwined, with human, plant, domestic animal, and environmental health (One Health) describing this interconnectivity. Oil spills in aquatic habitats are clear examples of this connection, with direct negative effects on the animals and plants in that environment and often indirect effects on human populations (e.g., chronic adverse health effects on cleanup workers and psychosocial impacts from environmental disasters). Additionally, oil exposure has significant long-term chronic effects on key wildlife populations (such as bottlenose dolphins), as well on the plant and human communities that coexist in these regions (Shultz et al. 2015, Kwok et al. 2019, Takeshita et al. 2021). Only through strong, collaborative science where wildlife response professionals add key data on the overarching effects of oil on wildlife can we fully appreciate and mitigate the true health-related impacts of oiling events in the future.

LITERATURE CITED

Albers, P. H. 2003. Petroleum and individual polycyclic aromatic hydrocarbons. Pages 341–371 *in* D. J. Hoffman, B. A. Rattner, G. A. Burton, Jr, J. Cairns, Jr., editors. Handbook of ecotoxicology Second Edition. CRC Press, Boca Raton, Florida, USA.

Allen, A. A., D. Jaeger, N. J. Mabile, and D. Costanzo. 2011. The use of controlled burning during the Gulf of Mexico *Deepwater Horizon* MC-252 oil spill response. American Petroleum Institute, Washington, D.C., USA.

Agency for Toxic Substances and Disease Registry (ATSDR). 1995. Toxicological profile for polycyclic aromatic hydrocarbons. US Dept. of Health and Human Services, Public Health Service, Agency for Toxic Substances and Disease Registry, Chamblee, Georgia, USA.

Agency for Toxic Substances and Disease Registry (ATSDR). 1997. Toxicological profile for benzene (update). US Dept. of Health and Human Services, Public Health Service, Agency for Toxic Substances and Disease Registry, Chamblee, Georgia, USA.

Balseiro, A., A. Espi, I. Marquez, V. Perez, M. Ferreras, J. G. Marín, and J. M. Prieto. 2005. Pathological features in marine birds affected by the Prestige's oil spill in the north of Spain. Journal of Wildlife Diseases 41:371–378.

Bauck, L., S. Orosz, and G. Dorrestein. 1997. Avian dermatology. Pages 548–562 *in* R. B. Altman, S. L. Clubb, G. M. Dorrestein, and K. Quesenberry, editors. Avian medicine and surgery. WB Saunders, Philadelphia, Pennsylvania, USA.

Brette, F., B. Machado, C. Cros, J. P. Incardona, N. L. Scholz, and B. A. Block. 2014. Crude oil impairs cardiac excitation-contraction coupling in fish. Science 343:772–776.

Briggs, K. T., S. H. Yoshida, and M. E. Gershwin. 1996. The influence of petrochemicals and stress on the immune system of seabirds. Regulatory Toxicology and Pharmacology 23:145–155.

Brown, R. E., J. D. Brain, and N. Wang. 1997. The avian respiratory system: a unique model for studies of respiratory toxicosis and for monitoring air quality. Environmental Health Perspectives 105:188.

Bursian, S. J., and K. Dean, editors. 2017. The effects of exposure to *Deepwater Horizon* oil on avian flight and health. Ecotoxicology and Environmental Safety 146:1–133.

Bursian, S. J., K. M. Dean, K. E. Harr, L. Kennedy, J. E. Link, I. Maggini, C. Pritsos, K. L. Pritsos, R. Schmidt, and C. G. Guglielmo. 2017. Effect of oral exposure to artificially weathered *Deepwater Horizon* crude oil on blood chemistries, hepatic antioxidant enzyme activities, organ weights and histopathology in western sandpipers (*Calidris mauri*). Ecotoxicology and Environmental Safety 146:91–97.

Butler, R., A. Harfenist, F. Leighton, and D. Peakall. 1988. Impact of sublethal oil and emulsion exposure on the reproductive success of Leach's storm-petrels: short and long-term effects. Journal of Applied Ecology 25:125–143.

Colegrove, K. M., S. Venn-Watson, J. Litz, M. J. Kinsel, K. A. Terio, E. Fougeres, R. Ewing, D. A. Pabst, W. A. McLellan, S. Raverty, et al. 2016. Fetal distress and in utero pneumonia in perinatal dolphins during the northern Gulf of Mexico unusual mortality event. Diseases of Aquatic Organisms 119:1–16.

Costa, D. P., and G. L. Kooyman. 1982. Oxygen consumption, thermoregulation, and the effect of fur oiling and washing on the sea otter, *Enhydra lutris*. Canadian Journal of Zoology 60:2761–2767.

Couillard, C. M., and F. A. Leighton. 1990. The toxicopathology of Prudhoe Bay crude oil in chicken embryos. Fundamentals of Applied Toxicology 14:30–39.

Cunningham, F., K. Dean, K. Hanson-Dorr, K. Harr, K. Healy, K. Horak, J. Link, S. Shriner, S. Bursian, and B. Dorr. 2017. Development of methods for avian oil toxicity studies using the double-crested cormorant (*Phalacrocorax auritus*). Ecotoxicology and Environmental Safety 141:199–208.

Davis, J. E., and S. S. Anderson. 1976. Effects of oil pollution on breeding grey seals. Marine Pollution Bulletin 7:115–118.

Davis, R., T. Williams, J. Thomas, R. Kastelein, and L. Cornell. 1988. The effects of oil contamination and cleaning on sea otters (*Enhydra lutris*). II. Metabolism, thermoregulation, and behavior. Canadian Journal of Zoology 66:2782–2790.

Deepwater Horizon Natural Resource Damage Assessment (DHNRDA) Trustees. 2018. Deepwater Horizon oil spill: final programmatic damage assessment and restoration plan and final programmatic environmental impact statement. US National Oceanic and Atmospheric Administration, National Marine Fisheries Service, Silver Springs, Maryland, USA.

De Guise, S., M. Levin, E. Gebhard, L. Jasperse, L. B. Hart, C. R. Smith, S. Venn-Watson, F. Townsend, R. Wells, and B. Balmer. 2017. Changes in immune functions in bottlenose dolphins in the northern Gulf of Mexico associated with the *Deepwater Horizon* oil spill. Endangered Species Research 33:291–303.

Echols, M. 2006. Evaluating and treating the kidneys. Clinical Avian Medicine 2:451–492.

Federal Emergency Management Agency (FEMA). 2017. National incident management system. Federal Emergency Management Agency, Hyattsville, Maryland, USA.

Finch, B. E., K. J. Wooten, and P. N. Smith. 2011. Embryotoxicity of weathered crude oil from the Gulf of Mexico in mallard ducks (*Anas platyrhynchos*). Environmental Toxicology and Chemistry 30:1885–1891.

Ford, R., M. Bonnell, D. Varoujean, G. Page, B. Sharp, D. Heinemann, and J. Casey. 1991. Assessment of direct seabird mortality in Prince William Sound and the western Gulf of Alaska resulting from the *Exxon Valdez* oil spill. Ecological Consulting. Inc., Portland, Oregon, USA.

Fravel, V. A., W. Van Bonn, F. Gulland, C. Rios, A. Fahlman, J. L. Graham, and P. J. Havel. 2016. Intraperitoneal dextrose administration as an alternative emergency treatment for hypoglycemic yearling California sea lions (*Zalophus californianus*). Journal of Zoo and Wildlife Medicine 47:76–82.

Friend, M., J. C. Franson, and E. A. Ciganovich. 1999. Field manual of wildlife diseases: general field procedures and diseases of birds. US Geological Survey, Biological Resources Division Information and Technology Report 1999-001, Madison, Wisconsin, USA.

Fry, D. M., and L. J. Lowenstine. 1985. Pathology of common murres and Cassin's auklets exposed to oil. Archives of Environmental Contamination and Toxicology 14:725–737.

Fry, D. M., J. Swenson, L. Addiego, C. Grau, and A. Kang. 1986. Reduced reproduction of wedge-tailed shearwaters exposed to weathered Santa Barbara crude oil. Archives of Environmental Contamination and Toxicology 15:453–463.

Garrott, R. A., L. L. Eberhardt, and D. M. Burn. 1993. Mortality of sea otters in Prince William Sound following the *Exxon Valdez* oil spill. Marine Mammal Science 9:343–359.

Gartrell, B., R. Collen, J. Dowding, H. Gummer, S. Hunter, E. King, L. Laurenson, C. Lilley, K. Morgan, H. McConnell, et al. 2014. Captive husbandry and veterinary care of northern New Zealand dotterels (*Charadrius obscurus aquilonius*) during the *CV Rena* oil-spill response. Wildlife Research 40:624–632.

Geraci, J. R., and V. J. Lounsbury. 2005. Marine mammals ashore: a field guide for strandings. National Aquarium, Baltimore, Maryland, USA.

Gorenzel, P., and T. Salmon. 2008. Bird hazing manual: techniques and strategies for dispersing birds from spill sites. University of California, Agriculture and Natural Resources Publication 21638, Oakland, USA.

Harr, K. E., D. R. Reavill, S. J. Bursian, D. Cacela, F. L. Cunningham, K. M. Dean, B. S. Dorr, K. C. Hanson-Dorr, K. Healy, K. Horak, et al. 2017. Organ weights and histopathology of double-crested cormorants (*Phalacrocorax auritus*) dosed orally or dermally with artificially weathered Mississippi Canyon 252 crude oil. Ecotoxicology and Environmental Safety 146:52–61.

Hartung, R. 1967. Energy metabolism in oil-covered ducks. Journal of Wildlife Management 31:798–804.

Harvey, R. G. 1991. Polycyclic aromatic hydrocarbons: chemistry and carcinogenicity. Cambridge University Press, Cambridge, England.

Hazelwood, R. 1986. Carbohydrate metabolism. Pages 303–325 *in* P. D. Sturkie, editor. Avian physiology. Springer, New York, New York, USA.

Helm, R. C., H. R. Carter, R. G. Ford, D. M. Fry, R. L. Moreno, C. Sanpera, and F. S. Tseng. 2014. Overview of efforts to document and reduce impacts of oil spills on seabirds. Pages 431–453 in M. Fingas, editor. Handbook of oil spill science and technology. John Wiley and Sons, Inc., Hoboken, New Jersey, USA.

Henkel, L. A., and M. H. Ziccardi. 2018. Life and death: how should we respond to oiled wildlife? Journal of Fish and Wildlife Management 9:296–301.

Hochleithner, M., C. Hochleithner, and L. Harrison. 2006. Evaluating and treating the liver. Pages 441–449 *in* G. J. Harrison and T. L. Lightfoot, editors. Clinical avian medicine. Spix Publishing, Inc, Palm Beach, Florida, USA.

Hoffman, D. J. 1979. Embryotoxic and teratogenic effects of petroleum hydrocarbons in mallards (*Anas platyrhynchos*). Journal of Toxicology and Environmental Health 5:835–844.

Industrial Economics, Inc (IEc). 2015. Literature-based fate estimate of birds exposed to the *Deepwater Horizon*/ Mississippi Canyon oil spill: panel summary. Industrial Economics, Inc., Cambridge, Massachusetts, USA.

Incardona, J. P., L. D. Gardner, T. L. Linbo, T. L. Brown, A. J. Esbaugh, E. M. Mager, J. D. Stieglitz, B. L. French, J. S. Labenia, C. A. Laetz, et al. 2014. *Deepwater Horizon* crude oil impacts the developing hearts of large predatory pelagic fish. Proceedings of the National Academy of Sciences USA 111:E1510–E1518.

International Petroleum Industry Environmental Conservation Association-International Association of Oil and Gas Producers (IPIECA-IOGP). 2015. Tiered preparedness and response. IPIECA-OGP, London, England.

International Petroleum Industry Environmental Conservation Association-International Association of Oil and Gas Producers (IPIECA-IOGP). 2016. Wildlife response preparedness. IPIECA-OGP, London, England.

International Petroleum Industry Environmental Conservation Association-International Association of Oil and Gas Producers (IPIECA-IOGP). 2017. Key principles for the protection, care, and rehabilitation of oiled wildlife. IPIECA-IOGP, London, England.

Jenssen, B. M. 1994. Review article: Effects of oil pollution, chemically treated oil, and cleaning on thermal balance of birds. Environmental Pollution 86:207–215.

Jenssen, B. M., and M. Ekker. 1988. A method for evaluating the cleaning of oiled seabirds. Wildlife Society Bulletin 16:213–215.

Jenssen, B. M., M. Ekker, and K. Zahlsen. 1990. Effects of ingested crude oil on thyroid hormones and on the mixed function oxidase system in ducks. Comparative Biochemistry and Physiology C Comparative Pharmacology 95:213–216.

Jessup, D. A. 1998. Diversity-rehabilitation of oiled wildlife. Conservation Biology 12:1153–1155.

Jessup, D. A., and F. A. Leighton. 1996. Oil pollution and petroleum toxicity to wildlife. Pages 141–156 *in* A. Fairbrother, L. N. Locke, and G. L. Hoff, editors. Noninfectious diseases of wildlife. Iowa State University Press, Ames, USA.

Jessup, D. A., L. C. Yeates, S. Toy-Choutka, D. Casper, M. J. Murray, and M. H. Ziccardi. 2012. Washing oiled sea otters. Wildlife Society Bulletin 36:6–15.

Kellar, N. M., T. R. Speakman, C. R. Smith, S. M. Lane, B. C. Balmer, M. L. Trego, K. N. Catelani, M. N. Robbins, C. D. Allen, R. S. Wells, et al. 2017. Low reproductive success rates of common bottlenose dolphins *Tursiops truncatus* in the northern Gulf of Mexico following the *Deepwater Horizon* disaster (2010–2015). Endangered Species Research 33:143–158.

Kenyon, K. W. 1969. The sea otter in the eastern Pacific Ocean. US Fish and Wildlife Service, Bureau of Sport Fisheries and Wildlife, North American Fauna 68, Washington, D.C., USA.

Kwok, R. K., A. K. Miller, K. B. Gam, M. D. Curry, S. K. Ramsey, A. Blair, L. S. Engel, and D. P. Sandler. 2019. Developing large-scale research in response to an oil spill disaster: a case study. Current Environmental Health Reports 6:174–187.

Lane, S. M., C. R. Smith, J. Mitchell, B. C. Balmer, K. P. Barry, T. McDonald, C. S. Mori, P. E. Rosel, T. K. Rowles, T. R. Speakman, et al. 2015. Reproductive outcome and survival of common bottlenose dolphins sampled in Barataria Bay, Louisiana, USA, following the *Deepwater Horizon* oil spill. Proceedings of the Royal Society B Biological Sciences 282:20151944.

Lattin, C. R., H. M. Ngai, and L. M. Romero. 2014. Evaluating the stress response as a bioindicator of sub-lethal effects of crude oil exposure in wild house sparrows (*Passer domesticus*). PLoS One 9: e102106.

Leighton, F. A. 1985. Morphological lesions in red blood cells from herring gulls and Atlantic puffins ingesting Prudhoe Bay crude oil. Veterinary Pathology 22:393–402.

Leighton, F. A. 1986. Clinical, gross, and histological findings in herring gulls and Atlantic puffins that ingested Prudhoe Bay crude oil. Veterinary Pathology 23:254–263.

Leighton, F. A. 1993. The toxicity of petroleum oils to birds. Environmental Reviews 1:92–103.

Lipscomb, T., R. Harris, R. Moeller, J. Pletcher, R. Haebler, and B. Ballachey. 1993. Histopathologic lesions in sea otters exposed to crude oil. Veterinary Pathology Online 30:1–11.

Maggini, I., L. Kennedy, A. MacMillan, K. Elliot, R. MacCurdy, C. Pritsos, K. Dean, and C. Guglielmo. 2015. *Deepwater Horizon* avian toxicity testing phase 2: western sandpiper *Calidris mauri* thermoregulation, takeoff, and endurance flight studies. (W12, W15/17, and W16). US Fish and Wildlife Service, *Deepwater Horizon* Natural Resources Damage Assessment and Restoration Office, Fairhope, Alabama, USA.

Massey, J. G. 2006. Summary of an oiled bird response. Journal of Exotic Pet Medicine 15:33–39.

Massey, J. G., S. Hampton, and M. Ziccardi. 2005. A cost/ benefit analysis of oiled wildlife response. American Petroleum Institute, Washington, D.C., USA.

Mazet, J. A., S. H. Newman, K. V. Gilardi, F. S. Tseng, J. B. Holcomb, D. A. Jessup, and M. H. Ziccardi. 2002. Advances in oiled bird emergency medicine and management. Journal of Avian Medicine and Surgery 16:146–149.

Mohr, F., B. Lasley, and S. Bursian. 2010. Fuel oil-induced adrenal hypertrophy in ranch mink (*Mustela vison*): effects of sex, fuel oil weathering, and response to adrenocorticotropic hormone. Journal of Wildlife Diseases 46:103–110.

Neff, J. M. 1979. Polycyclic aromatic hydrocarbons in the aquatic environment. Applied Science Publishers, London, England.

Neff, J. M. 1990. Composition and fate of petroleum and spill-treating agents in the marine environment. Pages 1–33 *in* J. R. Geraci and D. J. St. Aubin, editors. Sea mammals and oil: confronting the risks. Elsevier, Amsterdam, The Netherlands.

Neff, J. M., R. E. Hillman, R. S. Carr, R. L. Buhl, and J. I. Lahey. 1987. Histopathologic and biochemical responses in arctic marine bivalve molluscs exposed to experimentally spilled oil. Arctic:220–229.

US National Research Council (NRC). 2003. Oil in the sea III: inputs fates and effects. National Academy Press, Washington, D.C., USA.

O'Hara, P. D., and L. A. Morandin. 2010. Effects of sheens associated with offshore oil and gas development on the feather microstructure of pelagic seabirds. Marine Pollution Bulletin 60:672–678.

Orosz, S. E. 2013. Critical care nutrition for exotic animals. Journal of Exotic Pet Medicine 22:163–177.

Ou, X., and K. S. Ramos. 1992. Proliferative responses of quail aortic smooth muscle cells to benzo[*a*]pyrene: implications in PAH-induced atherogenesis. Toxicology 74:243–258.

Oiled Wildlife Care Network (OWCN). 2014. Protocols for the care of oil-affected birds. University of California Davis, School of Veterinary Medicine, Oiled Wildlife Care Network, Davis, USA.

Perez, C. R., J. K. Moye, D. Cacela, K. M. Dean, and C. A. Pritsos. 2017. Low level exposure to crude oil impacts avian flight performance: the *Deepwater Horizon* oil spill effect on migratory birds. Ecotoxicology and Environmental Safety 146:98–103.

Peterson, C. H., S. D. Rice, J. W. Short, D. Esler, J. L. Bodkin, B. E. Ballachey, and D. B. Irons. 2003. Long-term ecosystem response to the *Exxon Valdez* oil spill. Science 302:2082–2086.

Petras, E. 2003. A review of marine mammal deterrents and their possible applications to limit killer whale (*Orcinus orca*) predation on Steller sea lions (*Eumetopias jubatus*). US National Oceanic and Atmospheric Administration, National Marine Fisheries Service, Alaska Fisheries Science Center, Seattle, Washington, USA.

Ramseur, J. L., and C. L. Hagerty. 2014. *Deepwater Horizon* oil spill: recent activities and ongoing developments. Congressional Research Service, Library of Congress, Washington, D.C., USA.

Rattner, B. A., and W. C. Eastin, Jr. 1981. Plasma corticosterone and thyroxine concentrations during chronic ingestion of crude oil in mallard ducks (*Anas platyrhynchos*). Comparative Biochemistry and Physiology C Comparative Pharmacology 68C:103–107.

Rattner, B. A., V. P. Eroschenko, G. A. Fox, D. M. Fry, and J. Gorsline. 1984. Avian endocrine responses to environmental pollutants. Journal of Experimental Zoology 232:683–689.

Rebar, A., T. Lipscomb, R. Harris, and B. E. Ballachey. 1995. Clinical and clinical laboratory correlates in sea otters dying unexpectedly in rehabilitation centers following the *Exxon Valdez* oil spill. Veterinary Pathology 32:346–350.

Rocke, T., T. Yuill, and R. Hinsdill. 1984. Oil and related toxicant effects on mallard immune defenses. Environmental Research 33:343–352.

Schwacke, L. H., C. R. Smith, F. I. Townsend, R. S. Wells, L. B. Hart, B. C. Balmer, T. K. Collier, S. De Guise, M. M. Fry, L. J. Guillette Jr, et al. 2013. Health of common bottlenose dolphins (*Tursiops truncatus*) in Barataria Bay, Louisiana, following the *Deepwater Horizon* oil spill. Environmental Science and Technology 48:93–103.

Schwartz, J. A., B. M. Aldridge, B. L. Lasley, P. W. Snyder, J. L. Stott, and F. C. Mohr. 2004. Chronic fuel oil toxicity in American mink (*Mustela vison*): systemic and hematological effects of ingestion of a low-concentration of bunker C fuel oil. Toxicology and Applied Pharmacology 200:146–158.

Sharp, B. E. 1996. Post-release survival of oiled, cleaned seabirds in North America. Ibis 138:222–228.

Shultz, J. M., L. Walsh, D. R. Garfin, F. E. Wilson, and Y. Neria. 2015. The 2010 *Deepwater Horizon* oil spill: the trauma signature of an ecological disaster. Journal of Behavioral Health Services and Research 42:58–76.

Siniff, D., T. D. Williams, A. M. Johnson, and D. L. Garshelis. 1982. Experiments on the response of sea otters *Enhydra lutris* to oil contamination. Biological Conservation 23:261–272.

Spraker, T. R., L. F. Lowry, and K. J. Frost. 1994. Gross necropsy and histopathological lesions found in harbor seals. Pages 281–311. *in* T. R. Loughlin, editor. Marine mammals and the Exxon Valdez. Academic Press, Cambridge, Massachusetts, USA.

St. Aubin, D., J. Geraci, T. Smith, and T. Friesen. 1985. How do bottlenose dolphins, *Tursiops truncatus*, react to oil films under different light conditions? Canadian Journal of Fisheries and Aquatic Sciences 42:430–436.

Stacy, B. 2012. Summary of findings for sea turtles documented by directed captures, stranding response, and incidental captures under response operations during the BP *Deepwater Horizon* (Mississippi Canyon 252) oil spill. National Oceanic and Atmospheric Administration Assessment and Restoration Division, Deepwater Horizon Sea Turtles Natural Resource Damage Assessment Technical Working Group Report, Seattle, Washington, USA.

Stacy, N. I., C. L. Field, L. Staggs, R. A. MacLean, B. A. Stacy, J. Keene, D. Cacela, C. Pelton, C. Cray, M. Kelley, et al. 2017. Clinicopathological findings in sea turtles assessed during the *Deepwater Horizon* oil spill response. Endangered Species Research 33:25–37.

Stephenson, R., and C. A. Andrews. 1997. The effect of water surface tension on feather wettability in aquatic birds. Canadian Journal of Zoology 75:288–294.

Szaro, R. C. 1979. Bunker C fuel oil reduces mallard egg hatchability. Bulletin of Environmental Contamination and Toxicology 22:731–732.

Takeshita, R., S. J. Bursian, K. M. Colegrove, T. K. Collier, K. Deak, K. M. Dean, S. De Guise, L. M. DiPinto, C. J. Elferink, A. J. Esbaugh, et al. 2021. A review of the toxicology of oil in vertebrates: what we have learned following the *Deepwater Horizon* oil spill. Journal of Toxicology and Environmental Health Part B 24:355–394.

Tarasoff, F. J. 1972. Anatomical adaptations in the river otter, sea otter, and harp seal with reference to thermal regulation. Pages 111–141 *in* R. J. Harrison, editor. Functional anatomy of marine mammals. Volume 2. Academic Press, Cambridge, Massachusetts, USA.

Townsend, F. I., C. R. Smith, and T. K. Rowles. 2018. Health assessment of bottlenose dolphins in capture–release studies. Pages 823–834 *in* F. M. D. Gulland, L. A. Dierauf, K. L. Whitman, editors. CRC handbook of marine mammal medicine. Third Edition. CRC Press, Boca Raton, Florida, USA.

Troisi, G. M., S. Bexton, and I. Robinson. 2006. Polyaromatic hydrocarbon and PAH metabolite burdens in oiled common guillemots (*Uria aalge*) stranded on the east coast of England (2001–2002). Environmental Science and Technology 40:7938–7943.

Troisi, G., L. Borjesson, S. Bexton, and I. Robinson. 2007. Biomarkers of polycyclic aromatic hydrocarbon (PAH)-associated hemolytic anemia in oiled wildlife. Environmental Research 105:324–329.

Tseng, F. S. 1999. Considerations in care for birds affected by oil spills. Seminars in Avian and Exotic Pet Medicine 8:21–31.

Tseng, F. S., and M. Ziccardi. 2019. Care of oiled wildlife. Pages 75–84 *in* S. M. Hernandez, H. W. Barron, E. A. Miller, R. F. Aguilar, and M. J. Yabsley, editors. Medical management of wildlife species: a guide for practitioners. John Wiley and Sons, New York, New York, USA.

Venn-Watson, S., K. M. Colegrove, J. Litz, M. Kinsel, K. Terio, J. Saliki, S. Fire, R. Carmichael, C. Chevis, W. Hatchett, et al. 2015. Adrenal gland and lung lesions in Gulf of Mexico common bottlenose dolphins (*Tursiops truncatus*) found dead following the *Deepwater Horizon* oil spill. PLoS One 10:e0126538.

Vermeer, K., and R. Vermeer. 1975. Oil threat to birds on the Canadian west coast. Canadian Field-Naturalist 89:278–298.

Whitworth, D., S. Newman, T. Mundkur, and P. Harris. 2007. Wild birds and avian influenza: an introduction to applied field research and disease sampling techniques. Food and Agriculture Organization of the United Nations, FAO Animal Production and Health Manual Number 5, Rome. Italy.

Williams, T. M., and R. W. Davis, editors. 1995. Emergency care and rehabilitation of oiled sea otters: a guide for oil spills involving fur-bearing marine mammals. University of Alaska Press, Fairbanks, USA.

Wolfaardt, A., L. Underhill, R. Altwegg, J. Visagie, and A. Williams. 2008. Impact of the *Treasure* oil spill on African penguins *Spheniscus demersus* at Dassen Island: case study of a rescue operation. African Journal of Marine Science 30:405–419.

Ziccardi, M., J. Bergeron, B. L. Chilvers, A. Grogan, C. Hebert, P. Kelway, H. Nijkamp, S. Regmann, V. Ruoppolo, L. Smith, et al. 2021. Creating a legacy of oiled wildlife response preparedness through the post-Macondo Oil Spill Response-Joint Industry Project. International Oil Spill Conference Proceedings 2021(1):687447.

Ziccardi, M. H., S. M. Wilkin, T. K. Rowles, and S. Johnson. 2015. Pinniped and cetacean oil spill response guidelines. US National Oceanic and Atmospheric Administration, National Marine Fisheries Service, NOAA Technical Memorandum NMFS-OPR-52, Silver Springs, Maryland, USA.

6 Prescriptions for Reducing Ocean Pollution and Saving Southern Sea Otters

David A. Jessup, Melissa A. Miller

Introduction

In this chapter, the medical terms "patient," "diagnosis," "treatment," "prognosis," and "prescription" (Rx in medical shorthand) will be used to emphasize the parallels between medical care of individuals; groups and populations; and traditional animal, population, and ecosystem management practices. By doing so, we encourage our readers to design and implement wildlife research, as well as the conservation efforts coming from it, to align with One World, One Health principles.

When animal disease and mortality impair recovery of threatened species, it's important wherever possible to determine the underlying cause(s) (diagnosis) of these events; whether these factors are associated with human activity (to guide and optimize medical response); and how these problem(s) are best mitigated at the individual, population, and ecosystem level (prescription and treatment). Although treatments at the various levels are not mutually exclusive, treatment of individual patients for many wild marine mammals is often inefficient and ultimately may not facilitate population recovery or prevent suffering and loss, whereas preventive medicine is often more efficient and effective. When significant wildlife health problems are found to be caused by people, the case for humans mitigating them (e.g., you broke it, you fix it), becomes more compelling and, in some cases, legally imperative. Treatment of complicated wildlife disease problems requires significant resources; substantial public support; and the application of social, political, financial, and legal remediation efforts in addition to conventional management methods. Forty years of collaborative effort toward diagnosing and treating human-associated health issues to advance the recovery of southern sea otters (*Enhydra lutris nereis*) in California will be presented as an example.

Brief Review of Patient History

During the 1990s, faltering recovery of the southern (or California) sea otter population required conservationists to reassess complex interactions between these animals and their environment in relation to nutrition, disease, toxins, predators, and anthropogenic causes of morbidity and mortality. Despite decades of legal protection and aggressive conservation and research efforts, California sea otter population recovery was hindered by periods of high mortality, including deaths of prime-aged adults and reproductive-age females. Between 1979 and 1993, sea otter carcasses were recovered and examined

Sea otters are a keystone species in the kelp forests of the northern Pacific Ocean, helping maintain the abundance and diversity of a large suite of kelp-dependent species by limiting the impacts of sea urchins and other kelp grazers. In turn, kelp forests buffer storm and tidal swell, sequester carbon, reduce ocean acidity, and provide abundant nearshore habitat. Illustration by Laura Donohue.

by biologists, while veterinary pathologists examined many of those recovered from 1994 through 2021, yielding one of the most extensive and detailed pathology datasets for wild marine mammals; this dataset currently contains necropsy records for >10,000 southern sea otters, which is triple the total living population. In some years during the 1990s to early 2000s, carcasses representing 10% or more of the entire population were recovered from beaches annually. This is not sustainable for a species with relatively low reproductive (less than one pup/adult female per year) and recruitment rates (less than 50% of pups survive 1 year).

During the 1980s, sea otter entanglement and mortality due to gill and drift net fisheries was documented (Wendell et al. 1986), and these fisheries were modified or phased out to reduce impacts over the following decade. From the late 1990s to the early 2000s, up to 50% of sea otter mortality in some years was attributed to infectious disease (Thomas and Cole 1996, Kreuder et al. 2003). Some appeared to be evolutionarily new sea otter pathogens, others were associated with anthropogenic activity, and pathogen flow from land to sea via freshwater runoff was documented for the first time. Classical examples of this "pathogen pollution" (Daszak et al. 2001) include sea otter infection with the land-based protozoan parasites *Toxoplasma gondii* and *Sarcocystis neurona*, which are linked to fecal egg stage (oocyst or sporocyst) shedding by terrestrial felids and marsupials, respectively. Numerous additional connections between sea otter mortality and exposure to land-based pathogens and pollutants have been reported since, including parasites originating from terrestrial rodents (Miller et al. 2020*a*; see Table 1), fungal pathogens, and facilitation of harmful agal blooms (HABs) (Jessup et al. 2004, 2007; Miller et al. 2010*a*, 2020*b*). Southern sea otter mortality due to marine HAB exposure may be influenced by nutrient loading from coastal freshwater and estuary sources, and biotoxins can also flow downstream and enter the ocean following freshwater HAB events (Kudela et al. 2008, Miller et al. 2010*b*). Many persistent organic pollutants (POPs) and contaminants of ecological concern (COECs) are detectable in the blood of live and tissues of dead southern sea otters, in some cases at concentrations 20–50 times higher than otters sampled from more pristine habitats, such as Alaska (Nakata et al. 1998; Kannan et al. 2006; Jessup et al. 2010). Some diseases appear to be more prominent in otters from urbanized coastlines, near freshwater outflows, or following storm events, implicating land–sea pathogen, toxin, and nutrient pollution as drivers of California sea otter morbidity and mortality (Miller et al. 2002, 2010*a*, 2010*b*; Jessup et al. 2004; Burgess et al. 2018, 2020; Shapiro et al. 2012, 2019). This concept was first summarized as "sea otters in a dirty ocean" by Jessup et al. (2007).

Although we will discuss examples of land-based pathogens and toxins implicated in sea otter death in this chapter, our primary goal is to provide information on One Health prescriptions that were used to treat the patient(s); in this case, California sea otter populations and the ecosystem upon which they (and we) depend. We hope this approach will demonstrate methods and actions that can help advance wildlife conservation efforts beyond diagnosis and prescription for sick individuals, and toward a more holistic approach for treatment and management of animal populations and ecosystems that can provide more lasting and sustainable benefits. All causes of sea otter morbidity and mortality described in this chapter have clear anthropogenic origins or influences and are not solely "natural" processes. In addition, many pathogens and pollutants, including parasites, bacteria, anthropogenic chemicals, and HAB toxins, are easily concentrated within filter feeding and detritivore invertebrates such as bivalves, worms, and crabs that southern sea otters commonly consume. Thus, certain diets may predispose otters (and humans and other animals) to the combined effects of biological and chemical pollution that have been concentrated through a common food source. Individual sea otters have clear dietary specializations (Estes et al. 2003) that females teach to their pups, so they may manifest the "you are what you eat" paradigm (Figure 6.1). Importantly, treatment at the population or ecosystem level (e.g., focusing on pollution of anthropogenic origin and from impaired watersheds flowing into the ocean) can address multiple forms of pollution at the same time, providing direct benefits to diverse species and entire ecosystems. However, these efforts are more easily accomplished in some areas than others, and the positive effects may take years or even decades to manifest. Diagnosing and treating causes of land–sea pollution requires assessment of key biological and ecological components to optimize and guide the efforts, but also requires adequate funding and local, regional, and state-wide outreach to inspire changes in policy, human behavior, and societal priorities.

Fig. 6.1. Sea otters are food specialists, of which three guilds (or specialties) have been identified, and females teach these food gathering skills to their pups. Some sea otter foods, particularly filter or detritus feeding invertebrates (like this Washington clam), may harbor and concentrate harmful toxins, pollutants, bacteria, and protozoa. Photograph by Joe Tomoleoni.

The Cause(s): Bacterial

Bacterial infections, including pathogens originating from feces, are a cause of sea otter illness and death (Thomas and Cole 1996, Kreuder et al. 2003, Jessup et al. 2007, W. A. Miller et al. 2005, 2006, M. A. Miller et al. 2020*b*). Further, nutrient-rich runoff or discharges containing feces, fertilizer, and wastewater can contribute to eutrophication and HABs (Howard et al. 2007, Cochlan et al. 2008, Kudela et al. 2008, Coombs 2008). Old or inadequate sewage infrastructure is one part of the problem. In some coastal communities, clay sewage pipes and other rudimentary sewage collection systems were installed decades ago; these systems cannot handle the needs of growing populations or larger and more diverse waste generated by restaurants and businesses. An example is Pacific Grove, California, located next to Monterey. Due to outdated municipal infrastructure, sewage releases to the ocean around Pacific Grove were common over the past 40 years and have repeatedly caused beach closures (NMSP 2006, Price 2018). The frequency and severity of these pollution events often peaks during or after heavy seasonal precipitation (Sercu et al. 2009), providing one example of the larger environmental issue. Current methods used to measure sewage spill-related impacts to the coastal economy from lost beach use and impaired tourism do not take into account threats to coastal marine wildlife health. Also, many residences in rural coastal counties are still connected to aging, leaky septic systems, and boats sometimes discharge untreated sewage directly into the ocean. Reminding citizens, local politicians, and municipal engineers that their decisions and priorities can have critical consequences for both their constituents and marine wildlife can catalyze support for improving old and inadequate infrastructure. Sea otters, other marine wildlife, and a healthy environment have significant political and social support along the central California coast, where they contribute to a vibrant economy whose income from tourism is second only to that from agriculture.

The Causes: Protozoal

In addition to sewage and nutrient discharges, many tons of feces from pets, agricultural animals, unhoused communities, and wildlife are deposited in fields, yards, streets, culverts and beaches annually (Dabritz et al. 2006, Sercu et al. 2009, Oates et al. 2017, Mladenov et al. 2020). Although land–sea pathogen flow can be difficult to trace with high precision in most cases, a unique quality of several protozoal parasites is that they have a narrow set of definitive (egg-shedding) hosts. The two most common protozoa infecting southern sea otters are *Toxoplasma gondii* and *Sarcocystis neurona*, which are shed only in the feces of terrestrial felids (cats) and marsupials (opossums), respectively (Figure 6.2). The tough, environmentally resistant infective egg stage of each parasite, called oocysts for *T. gondii* and sporocysts for *S. neurona*, are shed by the millions in feces of these terrestrial hosts and can survive for months to years in the environment under optimal conditions. (See Chapter 23 for more complete discussion of life cycles.) Dabritz et al. (2006) demonstrated that feral and free-roaming cats in three California coastal communities were a significant source of oocysts that wash into the ocean. Keeping cats indoors or in outdoor enclosures (catios) and disposing of feces properly has both wildlife and public health (One Health) implications (ABC 2022).

The infective dose for these parasites is also extremely low, as little as a single oocyst. Although consumption of protozoa encysted in the tissues of infected intermediate hosts (e.g., small mammals and birds) is another potential means of *T. gondii* or *S. neurona* transmission, this route of infection is negligible for southern sea otters, a species that consumes marine and estuarine invertebrates almost exclusively. Favored southern sea otter prey items include several species of bivalves, gastropods, and crustaceans that are also consumed by humans (see Chapter 23). Many invertebrates concentrate small particles in their tissues, including protozoal eggs, via their feeding activity. For example, Mediterranean mussels (*Mytilus gallopronvincialis*) efficiently accumulated *T. gondii* oocysts in a laboratory setting; exposed mussels remained polymerase chain reaction (PCR) positive for *T. gondii* for 21 days postexposure, and parasite infectivity from mussel tissue was confirmed via mouse bioassay (Arkush et al. 2003). Subsequent studies detected these parasites via PCR in other sea otter prey items, both in experimental studies and the coastal marine environment (M. A. Miller et al. 2008, Mazzillo et al. 2013, Michaels et al. 2016). Another study revealed that sea otter dietary choices significantly affect their risk of *T. gondii* and *S. neurona* infection; otters that consumed many small marine snails (*Tegula* spp.) had 12 times greater probability of *T. gondii* infection than those that rarely consumed them (Johnson et al. 2009). Also, otters living and foraging in certain coastal areas were four times more likely to become infected with *T. gondii* than otters living in other places. Sea otters feeding on prey from sandy-bottom habitats (such as clams and worms) near Monterey were more likely to become infected with *S. neurona*. Epidemiological research showed that 62% (66 of 107) freshly dead, beach-cast sea otters and 42% (49 of 116) live otters sampled along the California coast from 1997 through 2001 were seropositive for *T. gondii* (Miller et al. 2002), and otters sampled near areas of maximal freshwater runoff were three times more likely to be seropositive for *T. gondii* than otters sampled in areas of low flow. This research supported the hypothesis that ingestion of prey items containing infectious, environmentally resistant eggs shed in the feces of felids or opossums and transported via freshwater runoff into the marine ecosystem was the most plausible route of protozoal infection for sea otters (Figure 6.2). From this work, it became clear that sea otters were excellent sentinels for assessing the spread of terrestrial pathogens into marine ecosystems (Conrad et al. 2005).

Both parasites are important causes of sea otter morbidity and mortality. In the late 1990s, infection with *T. gondii* was first recognized as a cause of southern sea otter mortality and as a potential contributor to slow sea otter population recovery in California (Thomas and Cole 1996; Miller et al 2004). *Toxoplasma gondii* and *S. neurona* infections were considered the primary cause of death for 23% of fresh dead sea otters examined from 1998 through 2001 and contributed to the death of another 11% of animals (Kreuder et al. 2003). Although native felids, including bobcats (*Lynx rufus*) and mountain lions (*Puma concolor*) coevolved alongside sea otters in coastal California and are competent definitive hosts for *T. gondii*, the introduction of millions of domestic cats into California over the last century poses substantial added risk of *T. gondii* infection for southern sea otters. In addition to host and environmental factors, parasite genotype also plays an important role in disease outcome. An unusual genotype of *T. gondii* (Type X) predominates in infected southern sea otters and appears to be more pathogenic than other strains (Shapiro et al. 2019). Type X *T. gondii* was identified in feces shed by coastal-living cats (M. A. Miller et al. 2008; VanWormer et al. 2014), again emphasizing this connection.

Toxoplasma gondii infection is widespread in human populations globally (Aguirre et al. 2019); approximately 40 million people are infected, and it is a leading cause of foodborne illness in the United States (CDC 2022). Although healthy individuals usually experience mild flu-like symptoms, severe disease and death can occur in individuals with weakened immune systems or babies infected *in utero* (Dubey and Beattie 1998). In people with AIDS or on chronic immunosuppressive therapy, *T. gondii*

Fig. 6.2. The life cycles of *Toxoplasma gondii* and *Sarcocystic neurona* are very similar, but they have different definitive hosts: cats for the former and opossum for the latter. Both appear to enter marine environments via freshwater runoff, classic forms of pathogen pollution. Illustration by Laura Donohue.

infection can lead to encephalitis, myocarditis, and death (see Chapter 23). Chronic *T. gondii* infection and the associated host inflammatory response may also contribute to development of neurodegenerative disease (Brown et al. 2005; Torrey et al. 2021).

Sarcocystis neurona has a life cycle similar to *T. gondii*, but the Virginia opossum (*Didelphis virginiana*) is the definitive host (Dubey et al. 2001) (Figure 6.2). Opossums were brought to California as a game species in the early 20th century and are now widespread, particularly in urban and suburban areas. *Sarcocystis neurona* causes fatal neurologic, systemic, and cardiac disease in sea otters (Miller et al. 2010*a*); although this parasite is also pathogenic to horses and other animals, is not known to cause human disease. During April 2004, approximately 40 sea otters died over 1 month along the coast near Morro Bay, California; necropsy and histopathology identified *S. neurona* infection as the cause (Miller et al. 2010*a*). This epizootic followed a local severe storm event that flushed out regional creeks and streams and likely deposited *S. neurona* sporocysts from opossum feces into the ocean, where they were concentrated in invertebrate prey items. This was the first reported "point source" outbreak of a terrestrial-origin parasite described in marine mammals and provides an especially clear example of land–sea pathogen flow.

The Cause(s): Nutrients and Pollutants

Some of the most productive and intensively farmed land in the United States is adjacent to southern sea otter habitat. Heavy use of nitrogen-based fertilizers, phosphates, and other nutrients; light and porous soils; and seasonally heavy rainfall can result in significant nutrient pulses into nearshore streams, estuaries, and coastal waters. Elevated levels of urea (from fertilizers or animal and human waste) in water can enhance replication of the marine diatom *Pseudo-nitzschia australis* and its production of the potent neurotoxin domoic acid (Howard et al. 2007; Cochlan et al. 2008; Coombs 2008; Kudela et al. 2008). Domoic acid is the cause of amnesic shellfish poisoning in humans (Perl et al. 1990) and is associated with large-scale mortality events in marine birds and mammals (Kudela et al. 2008), including sea otters (Kreuder et al. 2005). As with the protozoa described above, invertebrates can concentrate domoic acid and other biotoxins in their tissues and retain these potent toxins for weeks to months. In a recent study, domoic acid intoxication was a primary cause of death for at least 9% (52 of 560) of southern sea otters and a contributing cause for 20% (112 of 549) (Miller et al. 2020*b*). These findings were conservative estimates that were primarily based on acute effects; the total impact of domoic acid toxicosis on sea otters is certainly higher. Toxic HAB events appear to be occurring more frequently over time, and domoic acid accumulation in invertebrates has repeatedly delayed the commercial Dungeness crab (*Metacarcinus magister*) harvest in California over the last 15 years due to human health risks. In addition, cyanotoxins produced by cyanobacteria (*Microcystis* spp.) in nutrient-enriched freshwater lakes, streams, and estuaries can reach the ocean, killing sea otters (Miller et al. 2010*b*, 2020*b*).

As noted, POPs like dieldrin, chlordane, DDT and their bioactive congeners, and PCBs and COECs like polyaromatic hydrocarbons (PAHs; products of hydrocarbon emissions and combustion, and from plastics) have been found at high concentrations in tissues of necropsied and blood of live, apparently healthy, southern sea otters (Nakata et al. 1998, Jessup et al. 2010). Although a few point sources of POPs and COECs are known, most originate from ill-defined nonpoint sources on land, and many are legacy pollutants that are remnants of 150 years of intense agricultural production along the California coast. Although it is illegal under state and federal law to discharge harmful pollutants into fresh or marine waters, it is challenging to enact or enforce laws or develop mitigation measures when no primary source or polluter is identifiable, especially when necropsy examinations cannot demonstrate a clear cause–effect relationship.

Multicausality: Complex Processes

Several complex, multicausal disease processes have been identified in southern sea otters. One example is cardiomyopathy syndrome; significant risk factors for development of this condition include domoic acid intoxication, *S. neurona* and *T. gondii* infection, and older sea otter age (Kreuder et al. 2005, Miller et al. 2020*b*); genetic factors may also play a role (Carter et al. 2022). End-lactation syndrome (ELS) in adult female sea otters is not a discrete disease but a physical expression of the high cost of maternal reproduction in this unique species (Chinn et al. 2016). Fatal ELS results from the high energetic cost of pup care and pressure due to food limitation and can be further exacerbated by mating trauma, secondary bacterial infections, and concurrent disease (Miller et al. 2020*b*). Great white sharks (*Carcharodon carcharias*) commonly kill sea otters but do not consume them; shark-associated sea otter deaths began to rise in the early 2000s and have now eclipsed many other causes of southern sea otter mortality (Tinker et al. 2015; Miller et al. 2020*b*).

Extensive research has confirmed that southern sea otters are facing a multitude of health threats in their environment, including many conditions with clear anthropogenic associations (Table 6.1). Based on decades of detailed necropsies, wild southern sea otters are commonly coping with three, four, or more significant disease processes at the same time (Kreuder et al. 2003; Miller et al. 2020*b*). When Miller et al. (2020*b*) examined all causes of southern sea otter deaths, infectious disease (e.g., pooled bacterial, fungal, viral, and parasite infections) was the single largest category (63%; 354 of 560), not including additional cases where pathogens invaded tissues following traumatic or other events (e.g., sequelae).

Protozoa provide one excellent example of the high frequency of concurrent infection and disease. Although the only known source for these parasites is protozoal eggs shed in the feces of terrestrial felids and marsupials (Figure 6.2), concurrent infection with *T. gondii* and *S. neurona* is extremely common in southern sea otters, representing at least 41% of 542 sampled otters in a recent study (Miller et al. 2021). Because the protozoal detection methods used in the study were relatively insensitive (histopathology and serology instead of more sensitive PCR assays), the actual proportion of concurrent infections is likely even higher. In addition, coinfection with both parasites was a primary or contributing cause of death for at least 14 otters in the study, and recent research suggests that concurrent *T. gondii* and *S. neurona* infection can result in increased disease severity in marine mammals (Gibson et al. 2011).

Another important consideration is that population threats and disease processes are not static in space or time; new threats emerge, while others can become endemic and less catastrophic over time. The health of individual wild animals, small groups, and populations is influenced by many factors; these factors are subject to periodic fluctuations and long-term trends, especially in relation to major external stressors such as pollution and climate change. This is why continuous or periodic population monitoring is critical for understanding health/disease in wildlife populations and ecosystems, especially for rare and endangered species. Single studies or short-term datasets only represent a brief snapshot in time of the complex, dynamic, and ever-changing interaction between hosts, pathogens, toxins, and their environment. Table 6.1 shows the anthropogenic-associated health problems identified in southern sea otters, potential treatments for individual sea otters, human health consequences, and One Health prescriptions for population and ecosystem levels to reduce health risks to southern sea otters and humans alike.

Prescription, Treatment, Mitigation, and Management

Comprehensive sea otter health research conducted from 1994 to the present day has provided ample evi-

Table 6.1. California sea otters serving as sentinels for several One Health issues, including many forms of pathogen pollution

Health problem (*Causative agent*)	Individual otter treatment	Population or ecosystem treatment	Potential sea otter and human health impacts
Toxoplasmosis (*Toxoplasma gondii*)	Antiprotozoal medications such as diclazuril and ponazuril, and supportive care (limited response)	Keeping "cats indoors, reduce hardscapes and runoff, vegetative buffers, dispose of cat feces in sanitary landfills.	Systemic toxoplasmosis, meningoencephalitis, chronic neurodegenerative disease, abortion, congenital defects, cardiomyopathy
Sarcocystosis (*Sarcocystis neurona*)	Antiprotozoal medications such as diclazuril and ponazuril, and supportive care (limited response)	Reduce hardscape, install vegetative buffers, minimize fecal deposition near waterways, enhance wetland filtration, control invasive species	Meningoencephalomyelitis, cardiomyopathy, myositis, reproductive failure in sea otters, not humans
Fecal bacterial exposure (*Escherichia coli, Salmonella enterica, Campylobacter* spp., etc.)	Antibiotics and supportive care	Improve sewage infrastructure, reduce hardscape-associated runoff, install vegetative buffers, minimize fecal deposition near waterways, enhance wetland filtration	Gastroenteritis, diarrhea, peritonitis, cholecystitis, sepsis
Domoic acid (Produced by *Pseudo-nitzchia* sp. marine diatoms)	Antiseizure medications, supportive care (limited response, and residual brain and cardiac damage is common)	Reduce nutrient flushing from land and contamination of watersheds, vegetative buffers, wetland filtration, increase porous surfaces in urban areas	Amnesic shellfish poisoning; acute, subacute, and chronic toxicosis; cardiomyopathy; reproductive effects; cognitive impairment
Persistent organic pollutants and contaminants of ecological concern	None	Prevent environmental contamination where possible; immobilize contaminated sediments using wetlands and vegetative buffers	Altered reproductive, neuroendocrine, and immune function; toxicity at high doses
Microcystin (Produced by cyanobacteria such as *Microcystis* sp.)	Oral bile acid sequestrants such as cholestyramine if caught early	Reduce phosphates and other nutrient loading in lakes, ponds, and streams	Hepatic and renal necrosis, coagulopathy, hepatic encephalopathy
Capillariasis (Hepatitis caused by *Capillaria hepatica*, a common rodent parasite)	Mebendazole, albendazole; treatment does not kill encysted eggs in rodent livers, which are the infective stage for sea otters and humans	Control rodent populations near waterways and coastlines, especially non-native rats that are common hosts; dispose of rodent carcasses by incineration, burial or in approved landfills	Hepatic necrosis, peritonitis, peritoneal adhesions

These conditions are often caused by or associated with anthropogenic activity, land-based runoff to nearshore marine waters. Treating these conditions at the population and ecosystem level versus individual level is far more cost effective and efficacious in reducing risk of exposure and disease for sea otters, other marine mammals, humans, and domestic animals.

dence that a high proportion of sea otter morbidity and mortality is associated with human activity. But what is to be done about it? In central California, treatment of stranded, injured, and diseased sea otters is often provided by Monterey Bay Aquarium (MBA) staff, assisted by other entities. Given finite resources and recognizing the limited value that individual animal care has for population recovery, treatment at the ecosystem (watershed and nearshore ocean) level became a core focus of the groups that eventually coalesced into the Southern Sea Otter Alliance. Through the 1980s and into the 1990s, there were three distinct groups working on sea otter conservation and management in California; those employed by US Fish and Wildlife Service (USFWS), the California Department of Fish and

Game (CDFG), and the MBA. The federal program, which began after southern sea otters were listed as threatened under the Endangered Species Act (ESA) in 1977, was responsible for population recovery and monitoring the small population of otters translocated to San Nicholas Island. The CDFG program attempted to balance the interests of sea otters (state listed as threatened) and those of fishers and shell fishers (particularly those harvesting abalone and sea urchin) and to determine causes of sea otter death. The MBA program focused on study of otter mother–pup dynamics and rehabilitation of injured and orphaned animals. Although individuals in these programs could be collegial, differences in program priorities sometimes led to dispute and criticism, which was occasionally aired in public.

In 1998, concurrent with changes in personnel and a nadir in funding, the situation changed: leaders of the three programs met, devised a way to divide up the work equitably, and pledged mutual support to optimize sea otter conservation efforts. Although USFWS maintained statutory authority, their sea otter research and management program had been transferred to US Geologic Survey (USGS). The USGS program took the lead on yearly population counts and field research efforts. The CDFG (California Department of Fish and Wildlife, CDFW, since 2013) took responsibility for postmortem examination of dead sea otters, pollution response, aerial support of counts, and on-water work in support of USGS and MBA research efforts. The MBA focused on research on tagged otters in Monterey Bay, improving telemetry, advancing animal care, optimizing rehabilitated otter release success, and public outreach. The USGS sea otter program was housed at the University of California Santa Cruz, Marine Science Institute, which added connections and support from university faculty. The CDFW program was closely aligned with the University of California-Davis School of Veterinary Medicine faculty and brought along those resources. All parties agreed to support each other's research efforts, share permits, resources, equipment and personnel, animal data, and samples; to work together for funding; to meet yearly to review findings; and to have an open public research forum. This became the Southern Sea Otter Alliance, which provided two decades of collaborative research that was translated into policy and actions to optimize sea otter population recovery. *Rx: Use all your resources (as in integrative medicine).*

It is hard to overstate the importance of public outreach to successful large-scale conservation efforts. Indeed, without public support few wildlife health and conservation efforts will succeed. Although government and university researchers can publish (and to a lesser extent publicize) their findings, getting this information into the minds of the public and motivating them to care enough to pressure politicians to act seldom results from research efforts alone. The Southern Sea Otter Alliance had the advantage of the MBA with its sophisticated outreach efforts, including staff and volunteers communicating with tens of thousands of people monthly, many of whom were captivated by seeing exhibit sea otters. Recall that this all took place in the early days of modern media when direct personal contact was critical.

One particularly noteworthy interaction took place when an MBA volunteer told visitors that toxoplasmosis, caused by a parasite originating from cat feces, was killing sea otters (one of the exhibit otters had chronic neurologic deficits due to protozoal damage). Among those visitors was a state senator who represented the Sacramento area and his son. He quickly gathered like-minded politicians, including those representing the Monterey Bay area and other locations with sea otters, and they consulted with scientists working on sea otter health. Research scientists from the University of California-Davis working with the Southern Sea Otter Alliance were called to testify before the legislature. The result was passage of a bill that stated the problem (toxoplasmosis), mandated that cat litter sold in California have a label warning of potential health risks and proper disposal of used cat litter (to closed landfills, not down the toilet or street drain), and established a tax checkoff fund to support sea otter research. This annual research funding has steadily climbed over 15 years from about $150,000 to nearly $280,000. Much of this support has been used for health or population recovery related projects and to fund critical staff needs. The sea otter toxoplasma story and knowledge of other health risks to sea otters from human-associated pollution and actions are now well known to the public and North American wildlife health and conservation experts. *Rx: Inform and involve the public and decision makers (client patient education).*

In California, the entities responsible for investigation of and actions to control chemical pollutants entering state waters are the state and regional Water Quality Control Boards. With their help, data from sea otter health investigations was aligned with detailed water quality monitoring information, resulting in major advances in the ability to map environmental risks for specific sea otter diseases and pollutants (M. A. Miller et al. 2002, 2009; W. A. Miller et al. 2006). This collaboration also facilitated the identification of potential nonpoint sources of POP discharge to the coastal environment (Jessup et al. 2010). Two agricultural drainages that empty into Monterey Bay near Moss Landing after rainfall events (Moro Cojo and Tembladero Slough) were identified as having the highest levels of DDT, DDE, and DDD in the entire state. The Water Quality Board provided funding for laboratory testing of sea otter blood and tissue for these pollutants, funded pollution-related research, and ensured that the regulatory community was aware of all study results. Unfortunately, these legacy pesticides were widely used during the 1940s through the 1970s, and point sources were not identified. However, when it came time to dredge the Moss Landing Harbor, the dredge spoils, heavily laden with DDTs, other POPs and COECs, were not pumped to the ocean but instead were placed into a clay lined pit near the harbor, which was capped and paved to create a large boat launch ramp and parking lot; these spoils remain entombed to this day, slowly degrading but not leaching into the Pacific Ocean or being picked up by the filter feeding marine invertebrates that sea otters and people eat. *Rx: Collaborate and treat what you can when you can (triage).*

Clean water is a precious resource in California; it is needed to support a large human population and enable the state's valuable agricultural production, much which is centered on coastal farms. Near the coast, if water is pumped from wells too aggressively, salt water can intrude into underground freshwater aquifers and degrade them. One method of preventing this is to pump water back into the aquifer; but where to get the water? In the Watsonville area sewage is tertiary-treated and filtered, then pumped back underground to slow or halt saltwater influx (Figure 6.3). This substantial investment in state-of-the-art water recycling reduces the potential for fecal bacteria and other pathogens to sicken humans or otters and degrade freshwater and marine habitat, while also remediating aquifer depletion. *Rx: Some clouds have silver linings (supportive care for the patient).*

Vegetative buffers and wetlands (both natural or man-made) are critical tools that can effectively reduce ocean contamination by pathogens and pollutants that are harmful to sea otters or people (W. A. Miller et al. 2008, Shapiro et al. 2010, Hogan et al. 2013) (Figure 6.3). Protozoal oocysts can adhere to microalgae and biofilms that grow on marsh plants, sedges, reeds, and sea grasses and keep them from being flushed toward the ocean (Figure 6.3). Vegetative buffers can also retain particulate matter that contains pesticides, nutrients, and fecal bacteria that might flow to the ocean, be taken up by filter-feeding bivalves, and consumed by sea otters (or people) (Figure 6.3). Irrigation drainage canals are often treated with herbicides to keep them bare and facilitate rapid downstream water flow. However, vegetative buffers can help sequester and reduce various forms of pollution originating from agricultural and urban centers (Figure 6.3). Reconstruction of wetlands and replacement of impervious surfaces (e.g., asphalt and cement) with more porous surfaces such as decomposed granite, gravel,

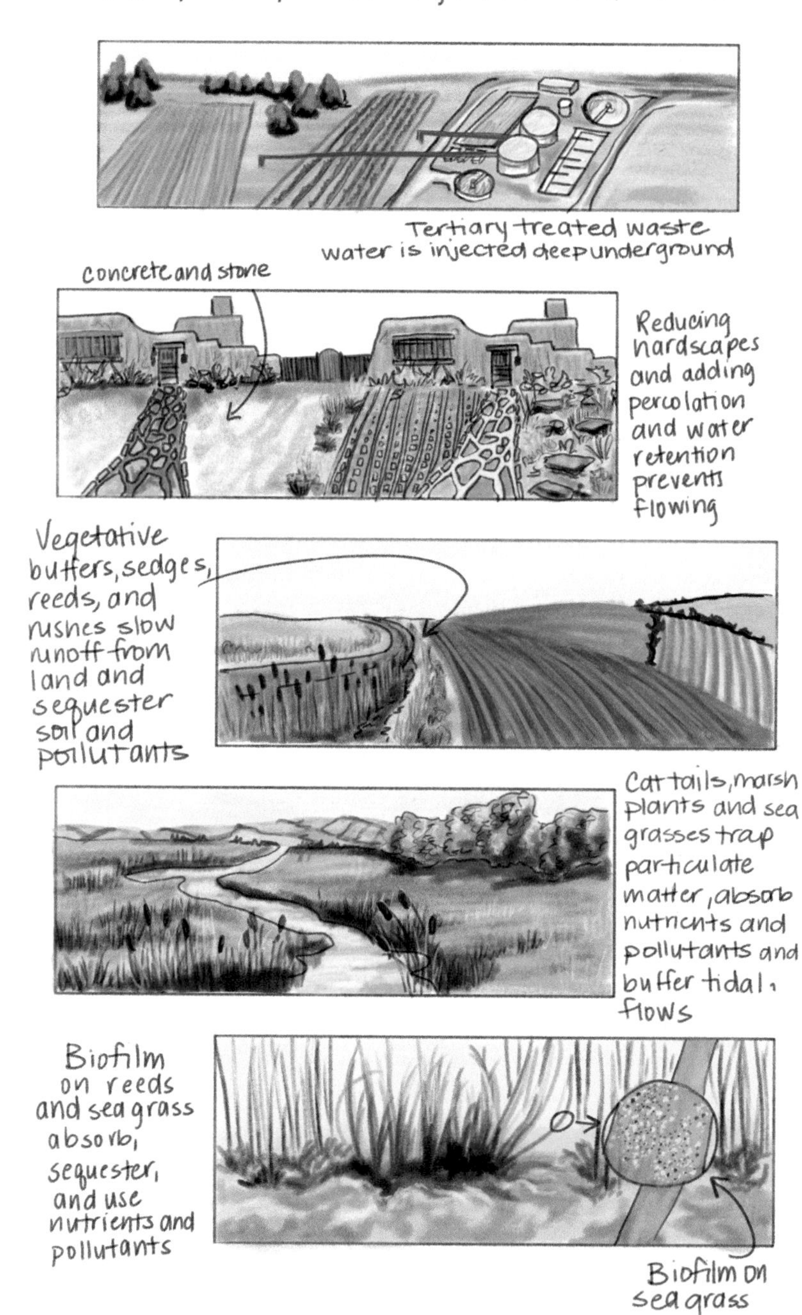

Fig. 6.3. Various methods can be used to reduce the pathogens and pollutants washed from land into the ocean. These include tertiary treatment and deep injection of wastewater, permeable landscaping, vegetative buffers, and marshland rehabilitation. Illustration by Laura Donohue.

and vegetated landscaping can reduce seasonal "flushing" events that are associated with sea otter disease (e.g., Miller et al. 2010*a*). Cities surrounding Monterey Bay have warning signs on street drains that lead to the ocean, and most cities are striving to reduce storm runoff by decreasing hardscape and promoting landscaping practices that encourage groundwater percolation (Figure 6.3). This helps reduce seasonal pulses of runoff into the ocean and replenishes local aquifers. Rehabilitating marshlands, developing vegetative buffers, and improved terraforming can also be used as mitigation measures by communities and businesses as an alternative to fines for pollution violations. Just as planting trees and restoring native soils and ecosystems can help sequester carbon to reduce global climate change, installing porous surfaces and vegetative buffers are a relatively simple, low-tech method to help mitigate pathogen and chemical pollution, recharge aquifers, and clean up streams and oceans. *Rx: Some relatively simple basic environmental treatments can be widely beneficial (preventive medicine).*

The Pajaro River empties into the Pacific Ocean just south of Watsonville, California. In February 2007, sea otters were found dead near its mouth, and necropsy revealed acute hepatic necrosis and icterus suggestive of exposure to a severe hepatotoxin (Miller et al. 2010*b*) (Figures 6.4A–C). Water samples collected upriver contained the cyanobacterial

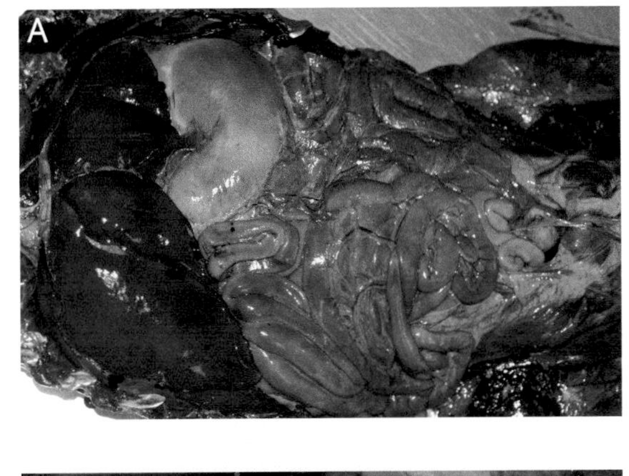

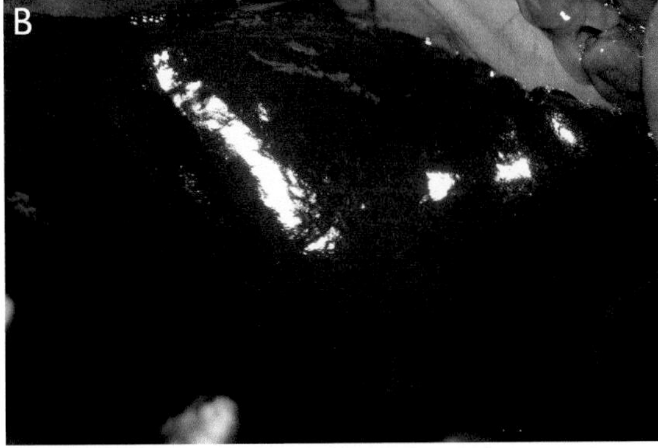

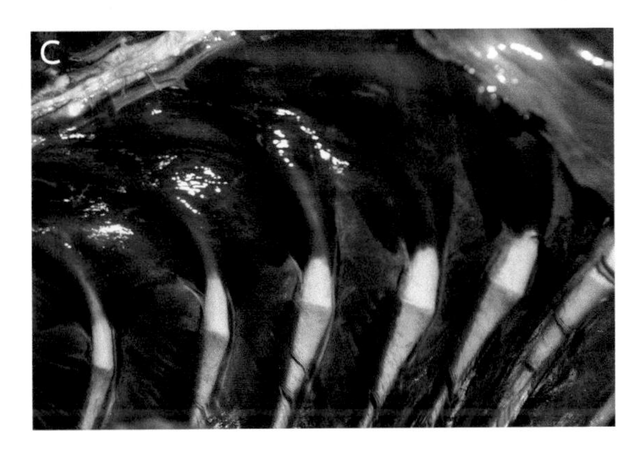

Fig. 6.4. The deaths of southern sea otters from acute liver failure were traced back to cyanobacterial blooms in a small lake draining into Monterey Bay: (A) diffusely swollen and mottled liver and icteric (yellow bile pigment stained) gastric serosa; (B) multiple pale foci in the liver suggestive of hepatic necrosis, and diffuse icterus of the duodenal serosa (upper right corner); and (C) severe icterus of costochondral junctions of the ribs and sternebrae from a sea otter with suspected fatal microcystin toxicosis. (D) Sample of cyanobacterial bloom material containing extremely high concentrations of cyanobacteria and microcystin toxin from Pinto Lake, which drains into Monterey Bay via the Pajaro River. Photographs by Melissa Miller and Robert Ketley.

toxin microcystin, a potent hepatotoxin (Figure 6.4D). Further upstream, a small water body (Pinto Lake) was found to be covered with a dense, malodorous, bright green scum, dominated by *Microcystis* sp. cyanobacteria, which are major microcystin producers. Water and scum samples collected from Pinto Lake (Figure 6.4D) contained some of the highest concentrations of microcystin ever recorded in the United States, exceeding by at least three orders of magnitude the US Environmental Protection Agency alert levels for drinking water or recreational water contact. Phosphate enrichment of water is one of several factors that can facilitate development of cyanobacterial super-blooms (HABs) with toxin production. Nutrient sources for Pinto Lake included historical clearing of adjacent redwood forests, local agricultural practices, and aging septic systems. Although the lake was a valued community resource for recreational boating and fishing, it had become a toxic health hazard. After months of coordinated effort by city, county, state, and federal resources, Pinto Lake was recognized as a federal priority for pollution mitigation, and California state water quality scientists established total daily maximum loads for microcystin discharges from Pinto Lake, a policy action that facilitated a robust state response. The resulting human resources and funding allowed the community to initiate both short- and long-term mitigation measures which appear to be helping to address this issue over time. In addition to improving community health and safety, these efforts have reduced the risk of additional sea otter deaths. In this case, investigation of dying sea otters along the coast helped identify a related and even more serious human and animal health problem upstream. The combined efforts of policymakers, water quality and biological scientists, and the local community helped to elicit federal and state resources to address the problem. In addition, CDFW scientists helped to spread the word about potential effective treatments for microcystin intoxication in exposed animals; this resulted in the successful recovery of a dog with severe microcystin poisoning, and collaborative community remediation of the contaminated lake where the dog was exposed (Rankin et al. 2013). *Rx: One Health can motivate community action and expenditures (improving patient lifestyle).*

Globally, but particularly in the towns and cities from Morro Bay to Monterey Bay, sea otters and ocean health matter. This is not only because otters are cute, but increasingly because people and communities recognize that they are valuable in other ways, were nearly lost forever, and have not yet recovered. It should be noted that these communities also have viable—if somewhat reduced—fisheries, and clean, fresh seafood is valued. Employees of many sea otter research organizations and nongovernmental organizations travel to local schools and colleges to educate students about sea otter health and their importance to estuaries, oceans, and people. In 2018–2019, tourism generated $2.98 billion (USD) in income to the economy of Monterey Bay-area counties, second only to agriculture, which generated $3.9 billion; a good portion of this revenue came from people who want to see sea otters. The MBA is the region's top tourist attraction, and it and four major university marine science stations around Monterey Bay employ hundreds of people who are acutely aware that sea otters are an ideal sentinel for marine pollution and potential human health risks, a keystone species, and a culturally and financially important living asset. *Rx: Educated and motivated people can be part of the solution not just a source of problems (outreach and promotion of community involvement in health).*

Summary

Efforts to understand and mitigate causes of southern sea otter morbidity and mortality, and the development of comprehensive management plans to foster population recovery, have taken 40 years to develop. This accomplishment required the combined resources and sustained collaboration of the Southern Sea Otter Alliance, made up of experts from the USGS; USFWS; CDFW; MBA; The Marine Mammal Center; University of California-Davis,

Wildlife Health Center; University of California Santa Cruz, Institute for Marine Science; and several advocacy organizations. Forty years of consistent population surveys, carcass recovery and examination, and studies on human-associated pathogens and toxins have provided a clear understanding of the ever-evolving challenges for southern sea otter health and population recovery. This work required merging of ecological and epidemiological datasets with those developed by water-quality resource managers, biologists, epidemiologists, pathologists, and clinical veterinarians. By pooling data and expertise gained from study of live-stranded, dead, captive, and free-ranging sea otters, scientists have gained a unique, "cradle to grave" perspective regarding sea otter biology and health threats. This work has also enabled the development of predictive models that can help optimize conservation decisions and predict and mitigate future threats (Tinker et al. 2021). The southern sea otter story is a classic example of what widely collaborative and cooperative wildlife health programs can accomplish, and it is a quintessential example of One Health thinking in action.

It may sound cynical, but people can be motivated by recognizing that there are strong personal and financial benefits for maintaining healthy wildlife and ecosystems. Southern sea otter health problems are a classic example of One Health—the concept that human, animal, and environmental health are inexorably linked (Jessup et al. 2004, 2007). Most of the pathogens and toxins that sicken or kill sea otters can also affect other marine wildlife, pets, livestock, and humans. Sea otters are a keystone species; they protect and promote the growth of kelp forests by feeding on kelp grazers like sea urchins (*Strongylocentrotus* spp.) (Estes and Duggins 1995) (see figure on page 91). Kelp forests and the immense biological diversity that they support are more likely to survive over the coming years with the help of sea otters than without them. In turn, kelp forests buffer storm surges, reduce coastal flooding, and slow beach erosion and are major marine carbon sequestering macroorganisms, comparable to tropical rain forests. Where sea otters enhance kelp forests, more carbon dioxide can be sequestered, thus modulating ocean acidification and climate change (Wilmers et al. 2012).

Sea otters also benefit coastal estuaries by promoting the survival and proliferation of eelgrass; in a manner similar to vegetative buffers, wetlands, and kelp forests, eelgrass proliferation can slow and trap sediments that carry pathogens and toxins and help reverse the impacts of estuarine eutrophication (Hughes et al. 2013). These same habitats can help conserve and protect coastal land and buffer the effects of sea level rise due to climate change. Although seldom considered in the context of wildlife conservation priorities, these ecosystem services could easily

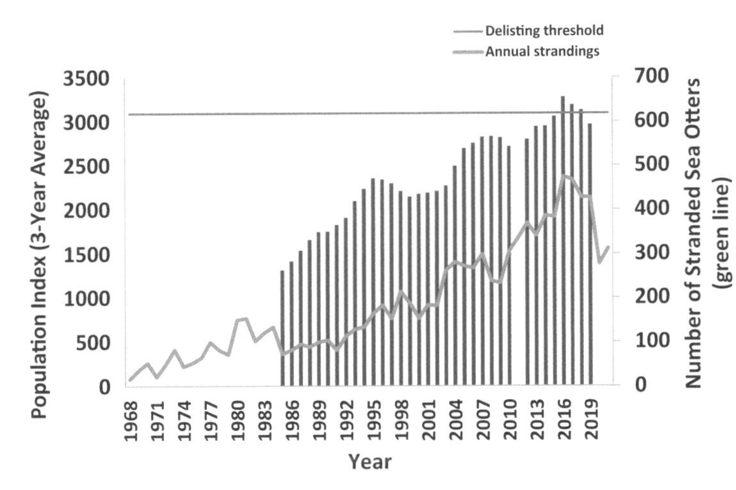

Fig. 6.5. Three-year running average (blue bars) of annual sea otter counts off California's coast. The population exceeded the Endangered Species Act delisting criteria of 3,090 individuals in 2016, 2017, and 2018. The green line shows the number of dead sea otters recovered from California beaches by year. Note this number rises in proportion to increasing population until about 2016. Both trends are positive indicators of improving sea otter survival. Graph by Colleen Young, California Department of Fish and Wildlife.

be worth hundreds of millions of dollars, and they provide considerable local, regional, and global benefits to humans and other animals. To summarize, efforts to treat or mitigate anthropogenic threats (including pathogens and toxins), that reduce sea otter health and abundance bring both direct and measurable, and less tangible but priceless benefits to Californians and, by example, the entire world. The treatments prescribed (or described) above include identifying anthropogenic-associated causes of morbidity and mortality (particularly those with One Health implications); optimizing public awareness; seeking knowledge, social mandates. and political and legal support to mitigate them; and translating scientific knowledge and public will into specific actions to improve the long-term health and sustainability of nearshore ocean environments, upon which so many species, including sea otters and people, depend.

Did the patients (otters and ocean) recover?

The primary criteria for removing southern sea otters from threatened status under the ESA is that they reach a population count of 3,090 for three consecutive years. As shown in Figure 6.5, this occurred in 2016–2018, and the USFWS is currently studying potential delisting. Also beginning in 2016, the absolute number of stranded sea otters began to drop after years of closely tracking sea otter population numbers. This trend preceded by several years the social changes wrought by the coronavirus pandemic (reduced beach visitation), but 2020 and 2021 numbers might also reflect effects from the global pandemic. Unfortunately, the COVID pandemic also caused the cancellation of yearly range-wide sea otter counts in 2020 and 2021, complicating the delisting deliberations.

Although the prognosis would seem to be encouraging, it is impossible to say if the "prescriptions" described above have saved a specified number of sea otters or to what extent each "treatment" contributed to southern sea otter population recovery or cleaner watersheds and oceans. That is the nature of preventive medicine. We have few ways to quantify health improved, or diseases prevented, or lives saved. Some reduction in sea otter and public health risks are likely given the 40 years of collaborative research, management, and mitigation efforts, so One Health principals have been served. However, we must not become complacent; new health challenges like climate change, ocean acidification, emerging pathogens, and other anthropogenic events are over the horizon.

> In the end, we will conserve only what we love, we will love only what we understand, and we will understand only what we are taught.
>
> *Baba Daum*

LITERATURE CITED

American Bird Conservancy (ABC). 2022. Cats indoors. ABC, The Plains, Virginia, USA. https://abcbirds.org/program/cats-indoors/. Accessed 10 Feb 2021.

Aguirre, A. A., T. Longcore, M. Barbieri, H. Dabritz, D. Hill, P. N. Klein, C. Lepczyk, E. L. Lilly, R. McLeod, J. Milcarsky, et al. 2019. The One Health approach to toxoplasmosis: epidemiology, control, and prevention strategies. EcoHealth 16:378–390.

Arkush, K. D., M. A. Miller, C. M. Leutenegger, I. A. Gardner, A. E. Packham, A. R. Heckeroth, A. M. Tenter, B. C. Barr, and P. A. Conrad. 2003. Molecular and bioassay-based detection of *Toxoplasma gondii* oocyst uptake by mussels (*Mytilus galloprovincialis*). International Journal of Parasitology 33:1087–1097.

Brown, A. S., C. A. Schaefer, C. P. Quesenberry Jr, L. Liu, V. P. Babulas, and E. S. Susser. 2005. Maternal exposure to toxoplasmosis and risk of schizophrenia in adult offspring. American Journal of Psychiatry 162:767–773.

Burgess, T., M. T. Tinker, M. A. Miller, J. Bodkin, M. Murray, J. Saarinen, P. Conrad, and C. Johnson. 2018. Defining the risk landscape in the context of pathogen pollution: *Toxoplasma gondii* in sea otters along the Pacific Rim. Royal Society Open Science 5:171178.

Burgess, T. L., M. T. Tinker, M. A. Miller, W. A. Smith, J. L. Bodkin, M. J. Murray, L. M. Nichol, J. M. Saarinen, S. Larson, J. A. Tomoleoni, et al. 2020. Spatial epidemiological patterns reveal mechanisms of land-sea transmission for *Sarcocystis neurona* in a coastal marine mammal. Scientific Reports 10:3683.

Carter, N.H., M. A. Miller, M. E. Moriarty, M. T. Tinker, R. B. Gagne, C. K. Johnson, M. J. Murray, M. M. Staedler, B. Bangoura, S. Larson, et al. 2022. Investigating associations among relatedness, genetic diversity, and causes of mortality in southern sea otters (*Enhydra lutris nereis*). Journal of Wildlife Diseases 58:63–75.

Centers for Disease Control and Prevention (CDC). 2022. Parasites - toxoplasmosis (*Toxoplasma* infection). https://www.cdc.gov/parasites/toxoplasmosis/index.html . Accessed 14 Mar 2022.

Chinn, S., M. A. Miller, M. Staedler, M. Tinker, M. D. Harris MD, F. Batac, E. Dodd, and L. Henkel. 2016. The high cost of motherhood: end-lactation syndrome (ELS) in southern sea otters (*Enhydra lutris nereis*). Journal of Wildlife Diseases 52:307–318.

Cochlan, W. P., J. Herndon, and R. M. Kudela. 2008. Inorganic and organic nitrogen uptake by the toxigenic diatom *Pseudo-nitzschia australis* (Bacillariophyceae). Harmful Algae 8:111–118.

Conrad, P. A., M. A. Miller, C. Kreuder, E. R. James, J. Mazet, H. Dabritz, D. A. Jessup, F. D. Gulland, and M. E. Griggs. 2005. Transmission of *Toxoplasma*: clues from the study of sea otters as sentinels of *Toxoplasma gondii* flow into the marine environment. International Journal of Parasitology 35:1155–1168.

Coombs, A. 2008. Urea pollution turns tides toxic. Nature News. doi: 10.1038/news.2008.1190.

Dabritz, H. A., E. R. Atwill, and I. A. Gardner, et al. 2006. Outdoor fecal deposition by free-roaming cats and attitudes of cat owners and non-owners toward stray pets, wildlife, and water pollution. Journal of the American Veterinary Medical Association, 229:74–81.

Dabritz, H., and P. A. Conrad. 2010. Cats and *Toxoplasma*: implications for public health. Zoonoses and Public Health 57(1):34–52.

Daszak P. A., A. A. Cunningham, and A. D. Hyatt. 2001. Anthropogenic environmental change and the emergence of infectious diseases in wildlife. Acta Tropica 78:103–116.

Dubey, J. P., and C. P. Beattie. 1998. Toxoplasmosis of animals and man. CRC Press, Boca Raton, Florida, USA.

Dubey, J. P., A. C. Rosypal, N. J. Thomas, D. S. Lindsay, J. F. Stanek, S. M. Reed, and W. J. A. Saville. 2001. *Sarcocystis neurona* infections in sea otters (*Enhydra lutris*): evidence of natural infection with *Sarcocystis* and transmission of infection to opossums (*Didelphis virginiana*). Journal of Parasitology 87:1387–1393.

Estes, J. A., and D. O. Duggins. 1995. Sea otters and kelp forests in Alaska: generality and variation in a community ecological paradigm. Ecological Monographs 65:75–100.

Estes, J. A., M. L. Riedman, M. M. Staedler, M. T. Tinker, and B. E. Lyon. 2003. Individual variation in prey selection by sea otters: patterns, causes, and implications. Journal of Animal Ecology 72:144–155.

Gibson, A. K., S. Raverty, D. M. Lambourn, J. Huggins, S. L. Magargal, and M. E. Grigg. 2011. Polyparasitism is associated with increased disease severity in *Toxoplasma gondii*-infected marine sentinel species. PLoS Neglected Tropical Diseases 5:e1142.

Hogan, J. N., M. E. Daniels, F. G. Watson, S. C. Oates, M. A. Miller, P. A. Conrad, K. Shapiro, D. Hardin, C. Dominik, A. Melli, et al. 2013. Hydrologic and vegetative removal of *Cryptosporidium parvum*, *Giardia lamblia*, and *Toxoplasma gondii* surrogate microspheres in coastal wetlands. Applied and Environmental Microbiology 79:1859–1865.

Howard, M. D. A., W. P. Cochlan, N. Ladizinsky, and R. M. Kudela. 2007. Nitrogenous preference of toxigenic *Pseudo-nitzschia australis* (Bacillariophyceae) from field and laboratory experiments. Harmful Algae 6:206–217.

Hughes, B. B., R. Eby, E. Van Dyke, M. T. Tinker, C. I. Marks, K. S. Johnson, and K. Wasson. 2013. Recovery of a top predator mediates negative eutrophic effects on seagrass. Proceedings of the National Academy of Sciences USA 110:15313–15318.

Jessup, D. A., C. K. Johnson, J. E. Estes, D. C. Brenner, W. M. Jarman, S. Reese, E. Dodd, M. T. Tinker, and M. H. Ziccardi. 2010. Persistent organic pollutants in the blood of free-ranging sea otters (*Enhydra lutris* spp.) in Alaska and California. Journal of Wildlife Diseases 46:1215–1233.

Jessup D. A., M. Miller, J. Ames, M. Harris, C. Kreuder, P. A. Conrad, and J. A. K. Mazet. 2004. Southern sea otter (*Enhydra lutris nereis*) as a sentinel of marine ecosystem health. EcoHealth 1:239–245.

Jessup, D. A., M. A. Miller, C. K. Johnson, P. A. Conrad, M. T. Tinker, J. Estes, J. A. K. Mazet. 2007. Sea otters in a dirty ocean. Journal of the American Veterinary Medical Association 231:1648–1652.

Johnson, C. K., M. T. Tinker, J. A. Estes, P. A. Conrad, M. Staedler, M. A. Miller, D. A. Jessup, and J. A. K. Mazet. 2009. Prey choice and habitat use drive sea otter pathogen exposure in a resource-limited coastal system. Proceedings of the National Academy of Sciences USA 106:2242–2247.

Kannan, K., E. Perrotta, and N. J. Thomas. 2006. Association between perfluorinated compounds and pathological conditions in southern sea otters. Environmental Science and Technology 40:4943–4948.

Kreuder, C., M. A. Miller, D. A. Jessup, L. J. Lowenstine, M. D. Harris, J. A. Ames, T. E. Carpenter, P. A. Conrad, and J. A. Mazet. 2003. Patterns of mortality in the southern sea otter (*Enhydra lutris nereis*) from 1998–2001. Journal of Wildlife Diseases 39:495–509.

Kreuder, C., M. A. Miller, L. J. Lowenstine, P. A. Conrad, T. E. Carpenter, D. A. Jessup, and J. A. K. Mazet. 2005. Evaluation of cardiac lesions and risk factors associated with myocarditis and dilated cardiomyopathy in southern sea otters (*Enhydra lutris nereis*). American Journal of Veterinary Research 66:289–299.

Kudela, R. M., J. Q. Lane, and W. P. Cochlan. 2008. The potential role of anthropogenically derived nitrogen in the growth of harmful algae in California, USA. Harmful Algae 8:103–110.

Mazzillo, F., K. Shaprio, and M. Silver. 2013. A new pathogen transmission mechanism in the ocean: the case of sea otter exposure to the land-parasite *Toxoplasma gondii*. PLoS One 8:e82477.

Michaels, L., D. Rejmanek, B. Aguilar, P. Conrad, and K. Shapiro. 2016. California mussels (*Mytilus californianus*) as sentinels for marine contamination with *Sarcocystis neurona*. Parasitology 143:762–769.

Miller, M. A., B. A. Byrne, S. S. Jang, E. M. Dodd, E. Dorfmeier, M. D. Harris, J. Ames, D. Paradies, K. Worcester, D. A. Jessup, et al. 2009. Enteric bacterial pathogen detection in southern sea otters (*Enhydra lutris nereis*) is associated with coastal urbanization and freshwater runoff. Veterinary Research 41:01.

Miller, M. A., P. A. Conrad, M. Harris, B. Hatfield, G. Langlois, D. A. Jessup, S. L. Magargal, A. E. Packham, S. Toy-Choutka, A. C. Melli, et al. 2010*a*. A protozoal-associated epizootic impacting marine wildlife: mass-mortality of southern sea otters (*Enhydra lutris nereis*) due to *Sarcocystis neurona* infection. Veterinary Parasitology 172:183–194.

Miller, M. A., P. J. Duignan, E. Dodd, F. Batac, M. Staedler, J. A. Tomoleoni, M. Murray, H. Harris, and C. Gardiner. 2020*a*. Emergence of a zoonotic pathogen in a coastal marine sentinel: *Capillaria hepatica* (syn. *Calodium hepaticum*)-associated hepatitis in southern sea otters (*Enhydra lutris neries*). Frontiers in Marine Science 7:335.

Miller, M. A., I. A. Gardner, C. Kreuder, D. Paradies, K. R. Worcester, D. A. Jessup, E. Dodd, M. D. Harris, J. A. Ames, A. E. Packham, et al. 2002. Coastal freshwater runoff is a risk factor for *Toxoplasma gondii* infection of southern sea otters (*Enhydra lutris nereis*). International Journal of Parasitology 32:997–1006.

Miller, M. A., M. E. Grigg, C. Kreuder, E. R. James, A. C. Melli, P. R. Crosbie, D. A. Jessup, J. C. Boothroyd, D. Brownstein, and P. A. Conrad. 2004. An unusual genotype of *Toxoplasma gondii* is common in California sea otters (*Enhydra lutris nereis*) and is a cause of mortality. International Journal of Parasitology 34:275–284.

Miller, M. A., R. M. Kudela, A. Mekebri, D. Crane, S. C. Oates, M. T. Tinker, M. Staedler, W. A. Miller, S. Toy-Choutka, C. Dominik, et al. 2010*b*. Evidence for a novel marine harmful algal bloom: cyanotoxin (microcystin) transfer from land to sea otters. PLoS One 5(9):e12576.

Miller, M. A., W. A. Miller, P. A. Conrad, E. R. James, A. C. Melli, C. M. Leutenegger, H. A. Dabritz, A. E. Packham, D. Paradies, M. Harris, et al. 2008. Type X *Toxoplasma gondii* in a wild mussel and terrestrial carnivores from coastal California: new linkages between terrestrial mammals, runoff, and toxoplasmosis of sea otters. International Journal of Parasitology 38:1319–1328.

Miller, M. A., M. E. Moriarty, L. Henkel, M. T. Tinker, T. L. Burgess, F. I. Batac, E. Dodd, C. Young, M. D. Harris, D A. Jessup, et al. 2020b. Predators, disease, and environmental change in the nearshore ecosystem: mortality in southern sea otters (*Enhydra lutris nereis*) from 1998–2012. Frontiers in Marine Science 7:582.

Miller, W. A., D. J. Lewis, M. D. A. Pereira, M. Lennox, P. A. Conrad, K. W. Tate, and E. R. Atwill. 2008. Farm factors associated with reducing *Cryptosporidium* loading in storm runoff from dairies. Journal of Environmental Quality 37:1875–1882.

Miller, W. A., M. A. Miller, I. A. Gardner, E. R. Atwill, J, Ames, D. Jessup, A. Melli, D. Paradies, K. Worcester, P. Olin, et al. 2005. New genotypes and factors associated with *Cryptosporidium* detection in mussels (*Mytilus* spp.) along the California coast. International Journal of Parasitology 35:1103–1113.

Miller, W. A., M. T. Miller, I. A. Gardner, E. R. Atwill, B. A. Byrne, S. Jang, M. Harris, J. Ames, D. Jessup, D. Paradies, et al. 2006. *Salmonella* spp., *Vibrio* spp., *Clostridium perfringens*, and *Plesiomonas shigelloides* detected in freshwater and marine invertebrates from coastal California ecosystems. Microbial Ecology 52: 198–206.

Mladenov, N., M. E. Verbyla, A. M. Kinoshita, R. Gersberg, J. Calderon, F. Pinongcos, M. Garcia, and M. Gil. 2020. San Diego River contamination study: increasing preparedness in the San Diego River watershed for potential contamination. Final Report. San Diego State University, San Diego, California, USA. https://sdrc.ca.gov/wp-content/uploads/2020/10/SDSU_Final-Report_v6-1.pdf. Accessed 11 Feb 2022.

Moriarty, M. E., M. T. Tinker, M. A. Miller, J. A. Tomoleoni, M. Staedler, J. A. Fujii, F. I. Batac, E. M. Dodd, R. M. Kudela, V. Zubkousky-White, et al. 2021. Exposure to domoic acid is an ecological driver of cardiac disease in southern sea otters. Harmful Algae 101:101973.

Moriarty, M. E., M. A. Miller, M. J. Murray, P. J. Duignan, C. T. Gunther-Harrington, C. L. Field, L. M. Adams, T. L. Schmitt, and C. K. Johnson. 2021. Exploration of serum cardiac troponin as a biomarker of cardiomyopathy in southern sea otters (*Enhydra lutris nereis*). American Journal of Veterinary Research 82:529–537.

Nakata, H., K. Kannan, L. Jing, N. Thomas, S. Tanabe, and J. P. Giesy. 1998. Accumulation pattern of pesticides and polychlorinated biphenyls in southern sea otters (*Enhydra lutris nereis*) found stranded along coastal California, USA. Environmental Pollution 103:45–53.

National Marine Sanctuary Program (NMSP). 2006. Part V—Water quality issues. Pages 207–285 *in* Monterey Bay National Marine Sanctuary—Proposed Action Plans. National Marine Sanctuary Program, National Ocean Service, National Oceanic and Atmospheric Administration, Department of Commerce, Silver Springs, Maryland. https://nmssanctuaries.blob.core.windows.net/sanctuaries-prod/media/archive/jointplan/reptoad/mb_pdf/waterquality.pdf. Accessed 16 Feb 2022.

Oates, S. C., M. A. Miller, D. Hardin, C. Dominik, D. Jessup, and W. A. Smith. 2017. Estimating daily relative dog abundance, fecal density and loading rates on two marine recreational beaches in Monterey County. Marine Pollution Bulletin 125:451–458.

Perl, T. M., L. Bédard, T. Kosatsky, J. C. Hockin, E. C. D. Todd, and R. S. Remis. 1990. An outbreak of toxic encephalopathy caused by eating mussels contaminated with domoic acid. New England Journal of Medicine 322:1775–1780.

Price, K. 2018. Addressing the problem of sewage spills. Monterey Herald, Feb 24, 2018. https://www.montereyherald.com/2018/02/24/addressing-the-problem-of-sewage-spills/. Accessed 18 Feb 2021.

Rankin KA, Alroy KA, Kudela RM, Murray MJ, Miller, MA. 2013. Treatment of cyanobacterial (microcystin) toxicosis using oral cholestyramine: case report of a dog from Montana. Toxins. 5: 1051–1063.

Sercu, B., L. C. Van De Werfhorst, J. Murray, and P. A. Holden. 2009. Storm drains are sources of human fecal pollution during dry weather in three urban Southern California watersheds. Environmental Science and Technology 43:293–298.

Shapiro, K., P. A. Conrad, J. A. K. Mazet, W. W. Wallender, W. A. Miller, and J. L. Largier. 2010. Effect of estuarine wetland degradation on transport of *Toxoplasma gondii* surrogates from land to sea. Applied and Environmental Microbiology 76:6821–6828.

Shapiro, K, M. Miller, and J. Mazet. 2012. Temporal association between land-based runoff events and California sea otter (*Enhydra lutris nereis*) protozoal mortalities. Journal of Wildlife Diseases 48:394–404.

Shapiro, K., E. VanWormer, A. Packham, E. Dodd, P. A. Conrad, and M. Miller. 2019. Type X strains of *Toxoplasma gondii* are virulent for southern sea otters (*Enhydra lutris nereis*) and present in felids from nearby watersheds. Proceedings of the Royal Society B Biological Sciences 286:20191334.

Tinker, M. T., B. B. Hatfield, M. D. Harris, and J. A. Ames. 2015. Dramatic increase in sea otter mortality from white sharks in California. Marine Mammal Science 32:309–326.

Tinker, M. T., L.P. Carswell, J.A. Tomoleoni, B.B. Hatfield, M. D. Harris, M. A. Miller, M. E. Moriarty, C.K. Johnson, C. Young, L.A. Henkel, et al. 2021, An integrated population model for southern sea otters. US Geological Survey Open-File Report 2021-1076, Reston, Virginia, USA.

Thomas, N. J., and R. A. Cole. 1996. Risk of disease and threats to the wild population. Endangered Species Update 13(12):23–27.

Torrey, E. F. 2021. Parasites, pussycats, and psychosis: the unknown dangers of human toxoplasmosis. Springer, Cham, Switzerland.

VanWormer, E., M. A. Miller, P. A. Conrad, M. E. Grigg, D. Rejmanek, T. E. Carpenter, and J. A. K. Mazet. 2014. Using molecular epidemiology to track *Toxoplasma gondii* from terrestrial carnivores to marine hosts: implications for public health and conservation. PLoS Neglected Tropical Diseases 8:e2852.

Wendell, F., R. Hardy, and J. A. Ames. 1986. An assessment of the accidental take of sea otters, *Enhydra lutris*, in gill and trammel nets. US National Oceanic and Atmospheric Administration Marine Resources Technical Report 54, Washington, D.C., USA

Wilmers, C. C., J. A. Estes, M. Edwards, K. L. Laidre, and B. Konar. 2012. Do trophic cascades affect the storage and flux of atmospheric carbon? An analysis of sea otters and kelp forests. Frontiers in Ecology and the Environment 10:409–415.

Wildlife Disease and Health in Wild Ungulates

7 Great Viral Plagues of Large Mammals

Lessons for Humanity from Morbillivirus Evolution—Emergence, Decline, and Relationships with Nature

Richard A. Kock

Introduction to Morbilliviruses: Rinderpest, Peste des Petits Ruminants, Distemper, and Measles

The morbilliviruses are part of the Paramyxovirus family, negative-strand RNA enveloped viruses in the order Mononegavirales. Vertebrates serve as natural hosts (Figure 7.1). That is the science, now for the story.

Rinderpest (cattle plague)

A well-documented historic example of the emergence and spread of a disease among domestic and wild ruminants is the morbillivirus of cattle plague. It became known as rinderpest* in the late 19th century after its severe impact on Afrikaans farmers in South Africa, the endpoint of a particularly devastating epidemic, which tore through Africa over a decade. This began during the period of colonization; the rinderpest virus (RPV) entered Africa with cattle from India used to feed the troops, and the disease virtually halted the expansion of European cultures from the Cape to Egypt due to the loss of cattle and animal resources. Entire indigenous

* Afrikaans word for "cattle plague."

Saiga antelope on the steppes of Asia, perhaps a pre-Pleistocene progenitor host of the morbilliviruses. Illustration by Laura Donohue.

Fig. 7.1. Morbillivirus hosts around the world. Illustration by Laura Donohue.

human communities were wiped out as a consequence of losing their main food source(s) or having to seek new livelihoods as their cattle died. Rinderpest also had a profound effect on wildlife populations, which were severely diminished, and distributions changed forever.

> We found that this dreadful epidemic had swept off all the cattle and wild buffalo, and much of the other game beside. The vultures and hyenas were too surfeited to devour the putrid carcasses, which lay under almost every tree near water. (Lugard 1893)

The virus first emerged, along with domestication of cattle, in Eurasia around 9000 years ago, perhaps adapting from other wild artiodactyl species' viruses (Figure 7.2). Once the virus was established in cattle in Asia, it was the likely cause of recorded early European animal plagues, including the cattle epizootic of AD 376 to 386 (Cáceres 2011). It became a disease described in the English language as "Steppe Murrain" or plague of the grasslands. Much of the history of cattle plague stems from the belief that the virus emerged from the vast regions beyond civilization, where the foreign invaders from the Khan Dynasties wrought havoc. There was even a cattle breed, herded with the Mongol armies, apparently resistant to the disease, believed to have spread the plague, weakening their enemies through devastating epidemics (Scott 2000).

This is an old story worth revisiting, in the context of the current pandemic of SARS CoV-2, which has ravaged human populations on six continents since 2019 and is thought to have animal origins. It is postulated that earlier human epidemics caused by a human morbillivirus (HMV), measles, diverged from rinderpest virus. Measles is familiar to every parent where the infection persists, especially in children. It is part of being human. What isn't understood, but highly topical, is the pathway for a natural animal virus to adapt to and infect humans. This is believed to have occurred during the genesis of livestock farming in the Middle East (Furuse et al. 2010). Uhl et al. (2019) postulate that the progenitor of HMV originated around 376 BCE and that animal domestication drove virus cross-species events starting around 900 CE. Genomic analysis suggests divergence of these two viruses as early as the 6th century BCE (Düx et al. 2020), but emergence may have extended over long periods and multiple events making historical precision challenging. The growing association between human and animal communities and emerging infections predates, by centuries, current human pandemic diseases but only recently have we begun to understand the links.

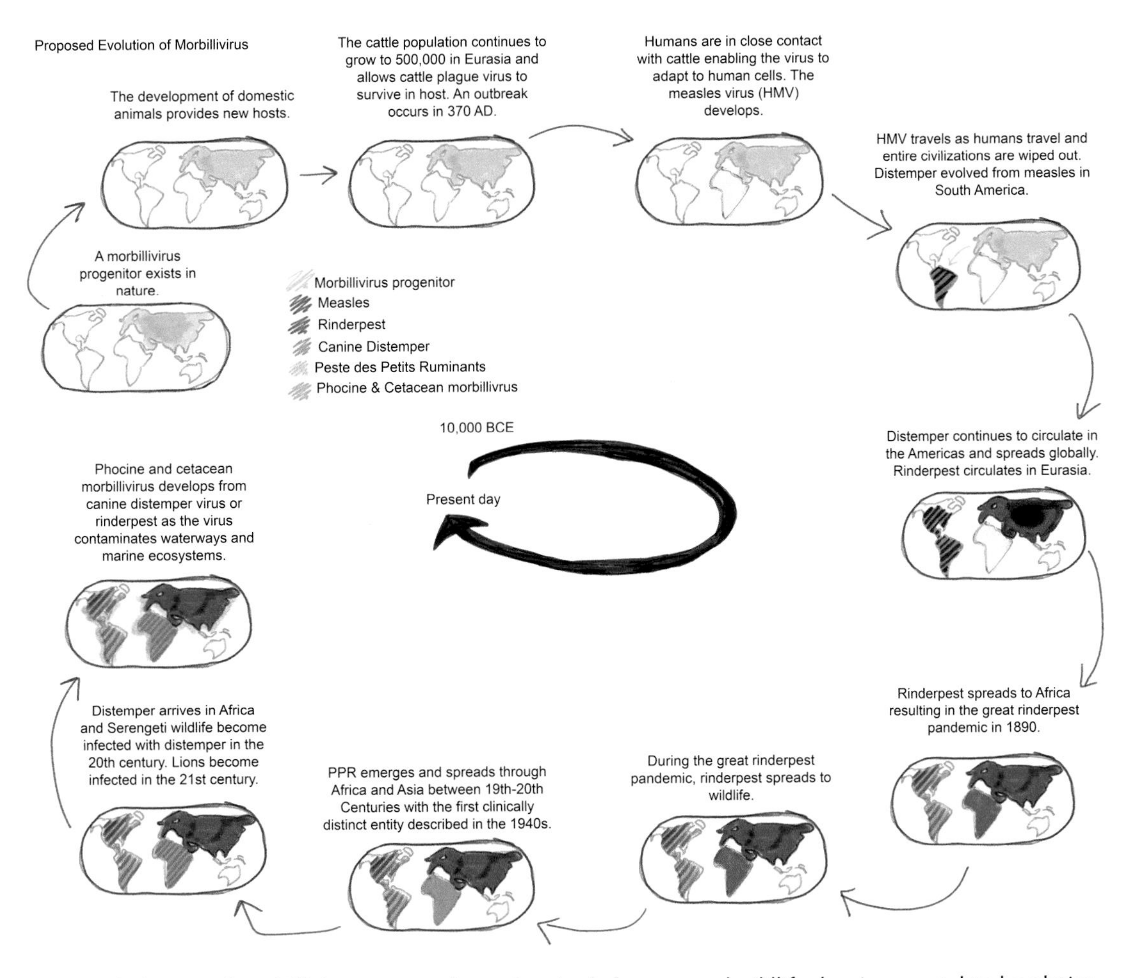

Fig. 7.2. The history of morbilliviruses among domestic animals, humans, and wildlife showing a postulated evolutionary story from the emergence of rinderpest in cattle through to the contemporary marine mammal morbilliviruses in the 21st century. Illustration by Laura Donohue.

Rinderpest virus became a scourge of the cattle-owning dynasties of Europe and, from the 17th century, was recognised as a unique syndrome. Pope Clement XI's physician understood this disease to be due to a contagion, well before our current understanding and nomenclature of pathogens developed (Scott 2000). As a result, more regulation was imposed on cattle herds, and it led to the establishment of the first veterinary professionals in France (1761 CE) and Britain (1791 CE). The final stage in the RPV story unfolded in the 21st century through a programme of global vaccination that eliminated the virus from cattle and from nature, the first success of this kind in the veterinary field and only the second infection of mammals to be eliminated after smallpox of humans.

Peste des Petits Ruminants

Based on molecular studies, peste des petits ruminants (PPR) likely emerged in West Africa around the same time (mid-19th century) as RPV became established in Africa, but it took about a century to be recognised as a distinct clinical entity and virus, gain a foothold, and spread along with growing small livestock populations in Asia and across Africa

(Libeau et al. 2014). Ironically, presence of RPV may have led to social preferences for small livestock, and reduced susceptibility of livestock and wildlife surviving RPV to PPR was not appreciated. Even as we celebrated the eradication of RPV its cousin virus, PPR, may be benefiting. The pandemic emergence of PPR and the close association between sheep, goats, and humans poses the question: Are we on the cusp of another emergent morbillivirus? We could answer this important question if we knew the precise mechanisms for these host switches and adaptations. A similar pattern of emergence is known with canine distemper (CDV), a morbillivirus disease of dogs and other domestic and wild species. CDV is closely related to HMV based on molecular studies, and viral codon evidence suggests CDV may have previously infected humans; measles had invaded South America prior to the emergence of canine distemper and could have facilitated spillover and adaptation to dogs and its first reports in the 1700s BCE, whilst another trail is suggested from South American bat morbilliviruses (Uhl et al 2019, Quintero-Gil et al. 2019). Because of the increasing domestic dog population in the 18th century, CDV expanded its range globally; it is now found in a wide range of mammals and may even be considered a candidate, due to its ability to infect primates, to be a future human pathogen.

Emerging Disease from Human–Animal Associations

As we gained understanding of morbilliviruses in people and domestic animals, we knew little of their importance in wildlife or the impacts among diverse species. The emergence of disease remains a narrow concern to the affected community, with focus on treatment and control primarily within humans and domestic animals. As science evolves, germ, genetic, and evolutionary theory progresses, and pieces of the epidemiological puzzle of diseases begin to fall into place. Surprisingly, it has taken events such as SARS CoV-2 in the 21st century for humanity to reconsider infectious diseases and heed warnings from history and science about the risk of evolution and emergence of contagion between animals and humans. Humans and domestic animals seem to be particularly cursed with unpleasant pathogens arising from their close association. *Homo sapiens* is home to about 1400 pathogens, of which 900 have suggested animal roots. Domestic animals fall foul of catastrophic mortalities from a myriad of organisms, seemingly with little resistance, from diseases that threaten animal-based food systems.

Ecologists once considered disease in nature as normal, and of little consequence to biodiversity, but regrettably this comfortable position is no longer tenable as significant wildlife disease outbreaks are a more frequent occurrence (Tompkins et al. 2015, Kock et al. 2018). Contemporary history teaches us that the evolution of "disease" is no accident. Humans along with their domestic animals and crops continue to settle in previously undeveloped environments. It is inevitable that microorganisms will step out of their niche in nature, to occupy a large and open, monotypic, playing field of hosts. Demands on domesticated animals for high productivity or desired traits in behaviour and stress of intense management have reduced capacities for resistance and immunity against disease agents and further accelerated the emergence of adapted organisms. We have compensated for this by adapting to disease threats, by generating biosecurity measures, and by attempting to isolate herds with vaccination and various medical tools and pharmacopeia to artificially support them in the face of infections. Much of this knowledge derived from early medical science focused on human diseases, which emerged from the consequences of population growth, sendentarization, and population clustering.

With the advent of animal diseases impacting human assets like cattle, a separate animal health discipline was born, veterinary medicine. Ironically, human and nonhuman animal disciplines followed a separate disciplinary training route, and there was some loss of integrative thinking on diseases and their emergence. For example, over the last 60–70 years, human medical education has increasingly ignored zoonoses (which for day-to-day medical practitioners are rare) and taken our relationships with the animal kingdom for granted. In extensive livestock-owning communities, zoonoses are endemic and ignored or neglected. For example, there is an absence of a word in the Masai vocabulary in East Africa to explain zoonosis (Queenan et al. 2017). This is not a sign of ignorance but most likely a result of their close association with animals, adaptation to animal pathogens, and tolerance to infection. At no point in human history, even to the present day, has society seriously questioned whether domesticating animals was a good or bad thing. It was just one of many opportunities taken in support of human expansion and development. Thus, the long list of zoonoses has strong links to our dependency on animals and is demonstrated by the fact that of the 900 animal-origin human pathogens, perhaps 675 or so have some association with wildlife species in their origins (Taylor et al. 2001, Jones et al. 2008) directly or through a domestic animal bridge or vector. The vast majority of human infections acquired day to day from animals comes from the use of domestic animals for food (Johnson et al. 2020). This has been further compounded by industrialization of livestock creating pathogen factories (Gilbert et al. 2017). In summary, about 43% of the broadly defined "emerging infectious diseases" of humans originate in some way from wildlife, but evidence suggests that this is actually on the decline, probably along with biodiversity loss, whilst pandemic diseases of humans are increasing (Kock and Cáceres-Escobar 2022). The majority of emerging pathogens with wildlife origins are from peridomestic ("human-adapted") wildlife (Johnson et al. 2020). Hence, the human–domestic domain is the key area of concern for the interface between wildlife species and domestics. We have little to directly fear from most free-ranging wild animals, but we need to better understand the risks of putting ourselves and animals in closer contact. This is especially true as we expand farmed species, destroy habitats, disturb wild species, and alter their environments. It should also be remembered that the main human health challenges are not from the rare direct transmission events but from their spread via human to human contact, as occurs with mycobacterium tuberculosis, influenza, malaria, and the contemporary Ebola and SARS CoV-2 virus. Directly transmitted infections from wildlife are extremely rare (Kock 2014) whatever the evolutionary histories and relationships.

Morbilliviruses and Wildlife

Rinderpest Virus

Few animal diseases have more history than cattle plague, and yet the early literature does not mention wildlife at all. It is only in more contemporary science arising from the great 19th century disease events in Egypt and subsequently the Great 1890s RPV Pandemic in Africa, that made wildlife a focus of attention. The increased interest in wildlife may simply have been a result of the huge mortalities, affecting a range of species, among some of the largest aggregations of ruminant animals on earth, and vividly described in diaries during the early colonial period (Lugard 1893). The Asian story seems much more closely tied to cattle and has benefited from ancient Chinese and Roman texts on plagues but with no attention to wildlife (Decker 2020). One possible explanation for the paucity of wildlife disease reports from Asia is that other than the large herds of saiga antelope (*Saiga tatarica*) and a variety of gazelles in Central Asia, there were no large aggregations of wild ruminants regularly observed and reported by scholars. The absence of Asian wildlife accounts may also stem from the illiteracy of hunting migrants and armies in extremely remote and unchartered territories far from centres of emerging civilization. Gazelles appear to be resistant to RPV virus, with only very recent historical accounts of Mongolian gazelle (*Procapra gutturosa*) mortality being proven due to incursion of RPV into Mongolia from China in the early part of the 20th century (Sodnomdarjaa et al. 2013). Apparently, a field expedition from the People's Commissariat of the Soviet Union experimentally proved the susceptibility of gazelle to RPV in 1938 after a mass die-off (Kurchenko 1995, Maidar 1958). In these early publications however, nothing was said about the susceptibility of saiga antelope on the great Asian steppes, which have proven to be highly susceptible to the PPR virus (Pruvot et al. 2020).

The fact that RPV did not invade the African continent until mid-19th century may be due to limited cattle populations and trade connections between Asia and sub-Saharan Africa. Areas where most of the wild ruminant populations lived, on the great savannahs of East Africa, were not impacted until colonial times. There is, however, evidence for epidemics among indigenous African breeds as early as the 1700s in West Africa (Curasson 1942), but without spilling into wildlife populations, until the large colonial import of domestic European and Asian breeds. The great RPV pandemic of the 1890s appears to have started with expeditions into what is now modern-day Ethiopia that brought infected cattle from India during colonization. The interest in African resources and colonial expansion was rapid and extensive, often described as the scramble for Africa by the great European powers of the time. Movement of cattle is considered central to the epidemiological spread of RPV, but wildlife populations no doubt contributed to vectoring the virus across landscapes and provided opportunity for the bridging of disease between cattle and wildlife. This may have been critical for the jump into southern Africa by way of the large dry miombo forests of southern Tanzania where no cattle occurred due to trypanosomiasis, a parasitic disease transmitted by tsetse fly (*Glossina* spp.) associated with African bushlands and forest.

Research in the 20th century showed that, although African buffalo (*Syncerus caffer*) were not critical to maintaining the virus, they were biologically competent to do so (Kock et al. 2008). Likewise, the annual presence of yearling disease in wildebeest (*Connochaetes* spp.) in the Serengeti suggests long-term maintenance in wildlife populations after the great plague, which was ecologically significant. Plant and animal species distributions were found to be changed after the Great RPV Pandemic, showing how significant RPV was to wildlife ecology (Holdo et al 2009, Kingdon 1990). The loss of entire populations over this time frame is undeniable, and the recovery of wildlife populations after its elimination, equally so. The recovery of the Serengeti herds is the best example. Since the elimination of the last vestiges of RPV virus in Kenya around 2003, it is remarkable to witness the stability of the wild ruminant populations in that country (KWS 2021), despite growing human populations, poaching, regular droughts, and loss of habitat. Certainly, over the period of the 1990s when several epidemics affected wildlife in Kenya and Tanzania without any reports in cattle, it became clear that the disease could manifest severely in wildlife alone (R. A. Kock et al. 1999). In species like African buffalo, 60% mortalities were estimated in populations in naïve herds, while the disease in partially immune herds was at times cryptic with only mild disease signs. Ultimately, control of the disease and its elimination from cattle resulted in its eradication, while the disease in wildlife helped its persistence in East Africa, its last stronghold. In the end, the wildlife populations were not large enough to maintain the virus after it was eliminated from cattle and it burned out. But it was only through working with both livestock and wildlife in the eradication campaign that the cryptic virus was tracked and eliminated, showing how collaboration between distinct domestic and wildlife veterinary disciplines was fruitful (Kock 2008).

Clinical expression, epidemiology, and pathology of rinderpest in affected wildlife species are broadly similar to the cattle disease (Figure 7.3). Whilst whole populations of species like African buffalo can suffer epidemics lasting several years, many associated artiodactyl species show few detected clinical

Fig. 7.3. The general pathology of the morbilliviruses. Illustration by Laura Donohue.

infections or no signs during these epidemics suggesting resistance to infection due to innate immunity, or they are refractory to infection for other cellular, genetic, or ecological reasons. In these species (e.g., gazelles), there is some evidence of being refractory to infection during epidemics with negative serology, despite close contact with buffalo (Kock et al. 2006*a*). The majority of cases reported with sufficient detail for pathological analysis are from African buffalo (subfamily Bovinae: which includes cattle and domestic buffalo and which were the main domestic hosts of rinderpest) (N. D. Kock et al. 1999). African buffalo are a close wild relative to cattle. The other species in Africa most affected were the so-called bovine or tragelaphine antelopes such as kudu (*Tragelaphus* spp.), and live and dead animals in these species presented with only blindness and major ocular pathology, which suggest that the eye was the portal of entry. There are anomalies from these general observations, including disease in giraffe (*Camelopardalis* spp.), warthog (*Phacochoerus* spp.), and eland (*Taurotragus* spp.), where clinical disease can be severe and cause death (N. D. Kock et al. 1999, R. A. Kock et al. 1999) or where population suppression was evident as with wildebeest (Holdo et al. 2009). Clinical and pathological data on rinderpest in wildlife in Asia other than anecdote is lacking, although gazelles were blamed for introduction of disease to cattle (Sodnomdarjaa et al. 2013).

It is challenging to make bold assumptions on the role of a disease in conservation, but RPV probably did represent an existential threat to the wild ruminants of the continent. Although the last surviving strains appeared to only mildly affect cattle, some wildlife species, such as buffalo and Tragelaphine antelopes, remained highly susceptible to the end. The history of RPV and its impact on wildlife in Asia will remain something of an enigma.

Peste des Petits Ruminants

As RPV passed into history, PPR (goat plague) virus was increasing. After the last case of RPV was diagnosed in the Meru National Park, Kenya, in African buffalo in December 2001 (Kock et al. 2006), the emergence of PPR in buffalo was identified, serologically, within 3 years in Uganda (and confirmed there in livestock in 2008) and in antelope in Sudan, where it has persisted (Kock 2006; Fernandez Aguilar et al. 2020). The virus subsequently spread to Kenya in 2006 (Spiegel and Havas 2019) and Tanzania in 2007 (Mdetele et al. 2021), and the virus has now been serologically demonstrated in a range of wildlife species, subsequent to or associated with livestock outbreaks in East Africa (Fernandez Aguilar et al. 2020, Mahapatra et al. 2015, Jones et al. 2021). Despite the obvious risk and evident immune-ecological association between the viruses, recognition of this relationship was surprisingly limited among veterinary services and professionals, and nothing was done to stop the spread of PPR. The virus finally reached the China Sea by 2013, Central Asia by 2014, and Mongolia by 2016, with the declaration in 2015 of PPR being the subject of a global eradication campaign by the Food and Agriculture Organization of the United Nations and the World Organisation for Animal Health (OIE) (Fine et al. 2020). The PPR virus might have evolved from the RPV virus, and it was only a matter of time for its impact on wildlife to be observed, given its likely adaptation to so many hosts predicated by its progenitor virus (Figure 7.4).

Most early reports describe disease in a diversity of captive wildlife in the Middle East (e.g., Furley et al. 1987) followed by observations elsewhere in Near East and Asian wildlife, mostly in free-ranging wild caprine and ovine populations in mountainous regions (Munir 2013, ul-Rahman et al. 2018). Perhaps one reason for the slow recognition of PPR spread in Africa was that clinical PPR in wildlife was not obvious or mostly absent, and national veterinary services were reluctant to declare infection without laboratory confirmation, given its novelty and the trade implications. The African history is not without suspected cases, with disease declared to the OIE by Sudan as early as 2007. These outbreaks were not confirmed until 2017 however, when virus was detected in Dorcas gazelles (*Gazella dorcas*) in Dinder National Park on the Ethiopian border (Asil et al. 2019). Unfortunately, this event lacked important epidemiology, pathology, and other details to enable confirmation of an outbreak in free-ranging wildlife, which would be the first in Africa. There are frequent reports of disease in captive Dorcas gazelle, which are traded in Sudan, so this suggests there might be some confusion in the reporting of this outbreak, which included only three cases for diagnostics.

Many African species in captivity have shown disease due to PPR, so it remains a challenge to understand why free-ranging wildlife are apparently resistant. Possible explanations include cross protection from another circulating morbillivirus (seropositives detected in wildlife to PPR tests may not be specific to PPR), innate resistance of wildlife species in the majority of African ecologies, or virus-specific factors (e.g., lineages) in this region, although virulence seems independent of lineage with PPR. The low seroprevalence in wild populations reported suggests the lack of apparent disease is not a result of immunosterilisation of wild populations (Fernandez Aguilar et al. 2020, Jones et al. 2021).

In Asian wildlife (and in captivity), the situation is contradictory to the Africa story: a number of

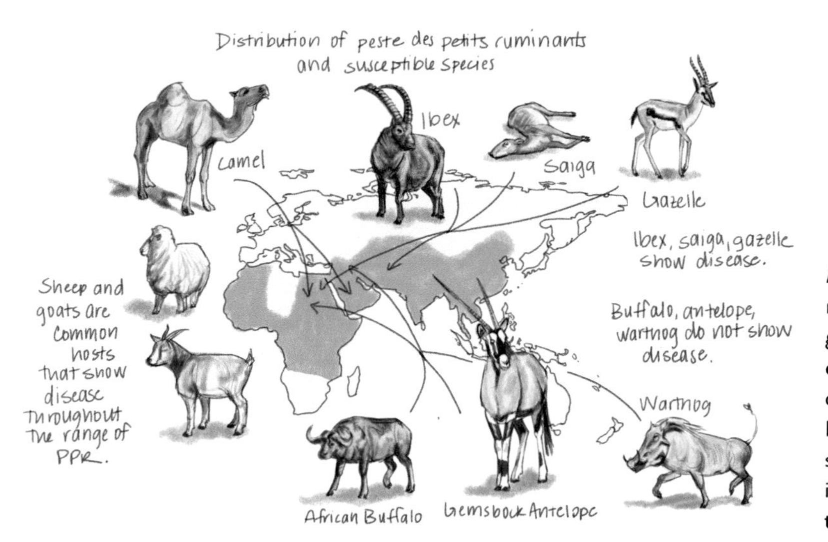

Fig. 7.4. Peste des petits ruminants is often described as a goat plague, and over the last decade has become a serious disease of wildlife in Asia. Infection in Africa is usually subclinical except in captivity or in stressed populations. Illustration by Laura Donohue.

severe wildlife epidemics were reported that were spatially and temporally widespread, including the near extinction of a subspecies of antelope, saiga (*Saiga tatarica mongolica*) in Mongolia (Pruvot et al. 2020; Hoffmann et al. 2012; Abubakar et al. 2011; Marashi et al. 2017). The majority of the epidemics observed are in wild caprine (*Capra ibex* and *Capra aegagrus*) and ovine (*Ovis ammon*) species with a scattering of cases in gazelles, bharal (*Pseudois nayaur*), and other small wild ruminants. The worry about PPR in Asia is that almost all wild ungulates are under severe pressure, with high levels of competition from domestic animals, degraded land, poor nutrition, and low population resilience: scattered, isolated small herds survive in remote and inaccessible landscapes and are often hunted. The added burden of PPR might well lead to extirpation of subpopulations and eventually species extinction. Many of these Asian species were on the International Union for Conservation of Nature red list of threatened and endangered species even before PPR arrived. The risk of wildlife vectoring virus across mountainous terrain and national borders also potentially creates a serious problem, complicating vaccination programmes in livestock and eradication efforts, and creating further threats from livestock keepers persecuting wildlife for fear of epidemics arising from them. The suspicion of persistence of PPR in the southern Gobi region despite government reports of elimination after the 2016 epidemic in Mongolia (WOAH 2022), might fit in this category.

Clinically affected and reported cases including mortality and pathology are in the *Capra* and *Gazella* species in Asia (ul-Rahman et al. 2018) but also in saiga antelope (Pruvot et al. 2020) and a wide spectrum of species of this genus in captivity, including African species (e.g., *Oryx* and *Gazella* genera). Ocular lesions in gazelles were reported, which echo some findings in some species with rinderpest.

The clinical presentation of PPR in both captive and the free-ranging Asian wildlife species is similar to that seen in goats and sheep; occasionally, epidemics circulate among wildlife that show similar mortality patterns to naïve domestic animal events, which can result in 90% mortality rates (Pruvot et al. 2020). Oculonasal discharge and diarrhoea are common signs with respiratory and enteric pathologies dominating and short periods of clinical disease prior to death or recovery.

These results and observations suggest, in the context of different species, ecologies and geographies, some form of innate immunity, cross protection with other morbillivirus, such as CDV, or viral incompetence in various hosts and locations, preventing disease expression and wider circulation among exposed species of wildlife. In antibody-positive wildlife populations in eastern Africa without observed clinical disease, there appear to be low titers, based on available but unvalidated serological tests used (Jones et al. 2021). This result may in part confirm the theory of complete or incomplete species cellular or humoral barriers to viral infection and disease under different phylogenies and ecologies. The virus is naturally most adapted to the abundant domestic goat and sheep hosts, where it apparently first evolved (i.e., a result of phylogenetically determined cell viral entry capacities) and, thereon, explaining disease expressed in the closest wild relatives after spillover. Some ecological factors may complicate this explanation, with apparent anomalies and outlier susceptibilities, with species in different habitats or management conditions and species such as the saiga antelope. The saiga, not closely related to the main domestic hosts but is a more ancient species, has been difficult to place phylogenetically. It is a true antelope, so it is related to progenitors of the *Capra* genus.

Canine Distemper

Canine distemper virus (CDV) provides a more contemporary example of morbillivirus evolution and growing threat to wildlife. The CDV virus appears to have emerged in South America in dogs around the 18th century and possibly from human measles infections, which were rampant amongst indigenous communities at the time of colonization by the Spanish. CDV was introduced to Europe and subsequently spread globally, with domestic dogs filling the vital role of maintenance host. This was at the same time that use of dogs in various roles resulted in their numbers increasing to hundreds of millions of individuals. Effective vaccines were developed for domestic dogs in the 1950s and 1960s and are widely applied with good effect. But CDV shows wide species adaptability and CDV can be found in both terrestrial and marine environments, demonstrating the high plasticity of these RNA viruses (Beineke et al. 2015) and its origins and evolution remain open to debate (Quintero-Gil et al. 2019). The molecular evidence suggests phocine distemper virus (PDV) diverged from CDV in the 17th century (Stokholm et al. 2021). Their evolution is by no means over. As genomic techniques explore deeper associations, related bat morbilliviruses have been described in South America (Drexler et al. 2012) and what appears to be spillover of CDV into seals and further evolution of these viruses in a range of water mammals (PDV) (Kennedy et al. 2019). Cetacean morbilliviruses (CEMV), first characterized in the 1980s, seems to share an association, based on signalling lymphocyte activation molecule receptors, with ruminant viruses (Van Bressem et al. 2014), suggesting a more ancient geneology.

The CDV epidemic in the lions of the Greater Serengeti ecosystem (Roelke-Parker et al. 1996, Kock et al. 1998) was unusual as the virus had largely been silent for decades except in domestic dogs. The emergence of CDV in Serengeti lions was later attributed to climate impacts, altered tick and blood parasite loads, and immunologic resistance to the virus (Munson et al. 2008). What is certain is that these viruses have been involved in major epidemic disease in wildlife species, occasionally threatening critically endangered species such as Ethiopian wolves (Gordon et al. 2015) and Amur tigers (Gilbert et al. 2020), where vaccination has been advocated as a last-ditch effort to save the species. Infection of nonhuman primates has raised the possibility of reinfection of humans with a CDV variant. This may further complicate efforts to eradicate measles, which has proven difficult to date. The World Health Organization launched an effort in 2001 to eliminate HMV (Holzman et al. 2020), but progress has slowed with increasing vaccine hesitancy, reports of adverse reactions, and a general shift among some communities against vaccine programmes, especially in so-called "developed countries." It should be remembered that CDV and measles viruses are close enough genetically that modified-live CDV vaccines for young puppies manufactured in the 1970s and 1980s in North America included measles virus to help assure immune response might override declining maternally derived antibodies and enhance protection. The two lyophilized live viruses were mixed together in the reconstituted product and then injected into dogs where they could further replicate.

Summary

The Emergence of Pandemics (Plagues), Exemplified by the History of Morbilliviruses and Their Resolution

Plagues are not an accident and truly are a human story. These historic events and their complex epidemiology have been driven by the expansion and growth of people together with their domesticated, companion and food, animals. The viruses spread by populations migrating over thousands of years, across and between continents, with conquering armies, political ideologies, and fundamentally through human development of animal-based food systems. The morbilliviruses all share similar transmission pathways and pathologies and now affect a large part of the mammal taxon, emerging from perhaps quite modest beginnings in a nondescript wild species on the steppes of Asia at the dawn of civilization. Rinderpest virus may well be the progenitor of all these modern viruses, a portal for an ancient wild Asian morbillivirus through cattle, and plausibly the pathway for HMV, PPR, and CDV emergence. Further, human colonization was a major boost for RPV, HMV, and CDV. The dramatic RPV plague in Africa affecting the most diverse community of ungulates on earth at the end of the 19th century had a massive impact on the ecology and species diversity of eastern Africa, and a lasting impact on veterinary thinking, especially in control options. PDV and CEMV emerged more recently, perhaps through "effluent flows" from domestic animals, with the increasing pollution of the seas with human and animal waste. The consequences of agriculture and these human–domestic animal associations and subsequent spread of these viruses, along with demographic changes in their populations has been dramatic but poorly appreciated. Even with over 96% of the earth's mammal biomass now attributed to humans and domestic animals (Bar-on et al. 2018), the morbillivirus story may not be complete.

The last chapter of the story may be the growing invasion of these viruses back into wildlife populations, as the intensity of the interface with the domestic domain increases. It is clear that, without swift action and major change in our animal-based food systems, biodiversity as a whole will not recover. Individual susceptible wildlife species that no longer have the resilience to prevent these viruses from having an impact will rapidly diminish. The eradication of RPV (over 50 years of veterinary effort), the primary source of all of this harm, was clearly one step in the right direction for affected species, but it is unlikely that the same outcome can be achieved for all the spawned morbilliviruses or others on the horizon. The stalling of HMV eradication and the global emergence of CDV and its relatives suggests that, if we are to resolve these existential threats, development strategies must change. "Nature abhors a vacuum" (Aristotle), and the only likely solution is restoration of biodiversity and complex ecologies, with the inherent stabilities, generated by "fair" competition rebalancing ecosystems and microbial communities. The last pockets of resilience to morbilliviruses, demonstrated in places like the Serengeti, shows us the right pathway to recovery: through sustaining biodiversity and restoration of ecosystems. Conservationists need to understand this history so they can bring this knowledge into their narratives and, through lobbying, influence change in political landscapes. Unfortunately, we are distracted by other issues, such as the dominance of the climate change narrative, which is also rooted in very similar drivers, yet the importance of biodiversity to climate change is rarely discussed.

Therefore, the most obvious conservation, veterinary, and public health (One Health) strategy to resolve plagues and pandemics would be to seek a gradual reduction of domestic animal numbers globally and support a natural decline in human population, thereby freeing up space for wild mammal species and achieving ecosystem stability in the fullness of time.

This chapter explores the interface and concerns between livestock and wildlife health, but in truth, the problem is the interface created by humans. Indeed, agriculture was more of a convenience, an accident of experience rather than a necessity for human survival (Kock et al. 2012). This article argues that agriculture's opening up of animal pathogen factories is an unacceptable and existential risk. Morbilliviruses, arguably the most transmissible and most pathogenic and threatening of RNA viruses, need large animal populations to sustain them, so the solution is simple: reduce those populations. Vaccination of domestic species for these diseases makes sense as an interim measure, given that individual immune systems gain strong memories from artificial and natural infection and mount effective resistance for life. On the other hand, evidence suggests that disparate small wildlife populations are unable to maintain the viruses for long, so despite their bad reputation, are not a significant part of the morbillivirus equation and just need to be protected from exposure to prevent extinction.

Conclusion

This story begs the question as to the fate of humanity, if this trend in pathogen emergence, which was set in motion millennia ago, continues. Are the coronaviruses just another warning? Are they the result of opportunity for these plastic viruses, the attraction of humanity and domestic animals and their abundance? Are they a long-time wildlife threat, as old and new narratives would have us believe? Are these plagues just nature trying to rebalance?

LITERATURE CITED

Abubakar, M., Z. I. Rajput, M. J. Arshed, G. Sarwar, and Q. Qurban Ali. 2011. Evidence of peste des petits ruminants virus (PPRV) infection in Sindh ibex (*Capra aegagrus blythi*) in Pakistan as confirmed by detection of antigen and antibody. Tropical Animal Health Production 43:745–747.

Asil, R. M., M. Ludlow, A. Ballal, S. Alsarraj, H. Wegdan, W. H. Ali, B. A. Mohamed, S. M. Mutwakil, and N. A. Osman 2019. First detection and genetic characterization of peste des petits ruminants virus from Dorcas gazelles "Gazella dorcas" in the Sudan, 2016–2017. Archive of Virology 164:2537–2543.

Bar-On, Y. M., R. Phillips and R. Milo. 2018. The biomass distribution on Earth. Proceedings of the National Academy of Sciences USA 115:6506–6511.

Beineke, A., W. Baumgärtner, and P. Wohlsein, 2015. Cross-species transmission of canine distemper virus—an update. One Health 1:49–59.

Cáceres, S.B. 2011. The long journey of cattle plague. Canadian Veterinary Journal 52:1140.

Curasson, G. 1942. La peste bovine. Pages 28–302 *in* Traité de pathologie exotique vétérinaire et comparée, Tome I. Maladies à ultra virus. Vigot Frères, Paris, France.

Decker, M. 2020. Animal and zoonotic diseases in the ancient and late antique Mediterranean: three case studies. https://h3de.org/2020/07/13/animal-and-zoonotic-diseases-in-the-ancient-and-late-antique-mediterranean-three-case-studies/. Accessed 20 Jan 2022.

Drexler, J. F., V. M. Corman, M. A. Muller, G. D. Maganga, P. Vallo, T. Binger, F. Gloza-Rausch, V. M. Cottontail, A. Rasche, S. Yordanov, et al. 2012. Bats host major mammalian paramyxoviruses. Nature Communications 3:796.

Düx, A., S. Lequime, L. V. Patrono, B. Vrancken, S. Boral, J. F. Gogarten, A. Hilbig, D. Horst, K. Merkel, B. Prepoint, et al. 2020. Measles virus and rinderpest virus divergence dated to the sixth century BCE. Science 368:1367–1370.

Fernandez Aguilar, X., M. Mahapatra, M. Begovoeva, G. Kalema-Zikusoka, M. Driciru, C. Ayebazibwe, D. S. Adwok, M. Kock, J.-P. K. Lukusa, J. Muro, et al. 2020. Peste des petits ruminants at the wildlife–livestock interface in the northern Albertine Rift and Nile Basin, East Africa. Viruses 12:293.

Fine, A. E., M. Pruvot, C. T. O. Benfield, A. Caron, G. Cattoli P. Chardonnet M. Dioli, T. Dulu M. Gilbert, R. Kock, et al. 2020. Eradication of peste des petits ruminants virus and the wildlife–livestock interface. Frontiers in Veterinary Science 7:50.

Furley, C. W., W. P. Taylor, and T. U. Obi. 1987. An outbreak of peste des petits ruminants in a zoological collection Veterinary Record 121:443–447.

Furuse, Y., A. Suzuki, and H. Oshitani. 2010. Origin of measles virus: divergence from rinderpest virus between the 11th and 12th centuries. Journal of Virology 7:52.

Gilbert, M., X. Xiao, and T.P. Robinson. 2017. Intensifying poultry production systems and the emergence of avian influenza in China: a 'One Health/Ecohealth' epitome. Archive of Public Health 75:48.

Gilbert, M., N. Sulikhan, O. Uphyrkina, M. Goncharuk, L. Kerley, E. Hernandez Castro, R. Reeve, T. Seimon, D. McAloose, I. V. Seryodkin, et al. 2020. Distemper, extinction, and vaccination of the Amur tiger. Proceedings of the National Academy of Sciences USA 117:31954–31962.

Gordon, C. H., A. C. Banyard, A. Hussein, M. K. Laurenson, J. R. Malcolm, J. Marino, F. Regassa, A.-M. E. Stewart, A. R. Fooks, and C. Sillero-Zubiri. 2015. Canine distemper in endangered Ethiopian wolves. Emerging Infectious Diseases 21:824–832.

Hoffmann, B., H. Wiesner, J. Maltzan, R. Mustefa, M. Eschbaumer, F. Arif, and M. Beer. 2012. Fatalities in wild goats in Kurdistan associated with peste des petits ruminants virus. Transboundary and Emerging Diseases, 59:173–176.

Holdo R. M., A. R. E. Sinclair, B. M. Bolker, A. P. Dobson, K. L. Metzger, M. E. Ritchie, and R. D. Holt. 2009. A disease-mediated trophic cascade in the Serengeti creates a major carbon sink. PLoS Biology 7:e210.

Holzmann, H., H. Hengel, M. Tenbusch, and H. W. Doerr. 2016. Eradication of measles: remaining challenges. Medical Microbiology and Immunology 205:201–208.

Johnson, C. K., P. L. Hitchens, P. S. Pandit, J. Rushmore, T. S. Evans, C. C. W. Young, and M. M. Doyle. 2020. Global shifts in mammalian population trends reveal key predictors of virus spillover risk. Proceedings of the Royal Society B Biological Sciences 287:2019–2736.

Jones, B. A., M. Mahapatra, D. Mdetele, J. Keyyu, F. Gakuya, E. Eblate, I. Lekolool, C. Limo, J. N. Ndiwa, P. Hongo, et al. 2021. Peste des petits ruminants virus infection at thewildlife–livestock interface in the Greater Serengeti Ecosystem, 2015–2019. Viruses 13:838.

Jones, K. E., N. G. Patel, M. A. Levy, A. Storeygard, D. Balk, J. L. Gittleman, and Peter Daszak. 2008. Global trends in emerging infectious diseases. Nature 451:990–993.

Kennedy, J. M., J. A. P. Earle, S. Omar, H. Abdullah, O. Nielsen, M. E. Roelke-Parker, and S. L. Cosby 2019. Canine and phocine distemper viruses: global spread and genetic basis of jumping species barriers. Viruses. 11:944.

Kenya Wildlife Service (KWS). 2021. National Wildlife Census Report. http://www.kws.go.ke/file/3550/download?token=L8M6oawm. Accessed 20 Jan 2022.

Kingdon, J. 1990. Island Africa: the evolution of Africa's rare animals and plants. Princeton University Press, Princeton, New Jersey, USA.

Kock, N. D., R. A. Kock, J. Wambua, and J. Mwanzia. 1999. Pathological changes in free ranging African ungulates during a rinderpest epizootic in Kenya 1993–1997 Veterinary Record 145:527–528.

Kock, R. A. 2006. Rinderpest and wildlife. Pages 144–162 *in* T. Barrett, P.-P. Pastoret, and W. Taylor, editors. Rinderpest and peste des petits ruminants virus. Plagues of large and small ruminants. Academic Press, London, England.

Kock, R. A. 2008. The role of wildlife in the epidemiology of rinderpest in east and central Africa 1994–2004: a study based on serological surveillance and disease investigation. Thesis, Cambridge University, Cambridge, England.

Kock, R. 2014. Drivers of disease emergence and spread: is wildlife to blame? Onderstepoort Journal of Veterinary Research 81(2):E1–E4.

Kock, R. A., R. G. Wallace, and R. G. Alders. 2012. Wildlife, wild food, food security and human society. Pages 7180 *in* Proceedings of the OIE Global Conference on Wildlife Animal Health and Biodiversity—Preparing for the Future. World Organisation for Animal Health, Paris, France.

Kock, R., and H. Cáceres-Escobar, editors. 2022. Situation analysis on the roles and risks of wildlife in the emergence of human infectious diseases. International Union for the Conservation of Nature, Gland, Switzerland.

Kock, R., W. S. K. Chalmers, J. Mwanzia, C. Chillingworth, J. Wambua, P. G. Coleman, and W. Baxendale. 1998. Canine distemper antibodies in lions in the Masai Mara. Veterinary Record 142:662–665.

Kock, R. A., M. Orynbayev, and S. Robinson, et al. 2018. Saigas on the brink: Multidisciplinary analysis of the factors influencing mass mortality events. Science Advances 17:eaao2314.

Kock, R. A., J. M. Wambua, J. Mwanzia, H. Wamwayi, E. K. Ndungu, T. Barrett, N. D. Kock, and P. B. Rossiter. 1999. Rinderpest epidemic in wild ruminants in Kenya 1993–7. Veterinary Record 145:275–283.

Kock, R. A., H. M. Wamwayi, P. B. Rossiter, G. Libeau, E. Wambwa, J. Okori, F. S. Shiferaw, and T. D. Mlengeya. 2006. Re-infection of wildlife populations with rinderpest virus on the periphery of the Somali ecosystem in East Africa. Rinderpest in East Africa: continuing re-infection of wildlife populations on the periphery of the Somali ecosystem. Preventive Veterinary Medicine 75:63–80.

Kurchenko, F. P. 1995. Prophylaxis and actions to eliminate rinderpest. Veterinariya-Moskva, 8:27–31. (In Russian.) Cited in International Atomic Energy Agency (IAEA). 1999. Support for rinderpest surveillance in West Asia. Fifth Coordination Meeting of the IAEA Regional Model Project RAW/5/004, 9–15 October 1999. IAEA, Vienna, Austria.

Libeau, G., A. Diallo, and S. Parida. 2014. Evolutionary genetics underlying the spread of peste des petits ruminants virus. Animal Frontiers 4:14–20.

Libeau, G., S. Guendouz, E. S. Swai, O. Nyasebwa, S. L. Koyie, H. Oyas, S. Parida, and R. Kock. 2021. Peste des petits ruminants virus infection at the wildlife–livestock interface in the greater Serengeti ecosystem, 2015–2019. Viruses 13:838.

Lugard, F.D. 1893. The rise of our East African empire. Vol. I. Nyasaland and eastern Africa. William Blackwood and Sons, Edinburgh, Scotland.

Mahapatra, M., K. Sayalel, M. Muniraju, E. Eblate, R. D. Fyumagwa, L. Shilinde, M. Mdaki, J. D. Keyyu, S. Parida, and R. Kock. 2015. Spillover of peste des petits ruminants virus from domestic to wild ruminants in the Serengeti ecosystem, northern Tanzania. Emerging Infectious Diseases 21:2230–2234.

Maidar, D. 1958. Development of veterinary services in Mongolia. PhD Thesis, Mongolian State University, Ulaanbaatar, Mongolia.

Mdetele, D.P., E. Komba, M. D. Seth, G. Misinzo, R. Kock, and B. A. Jones. 2021. Review of peste des petits ruminants occurrence and spread in Tanzania. Animals 11:1698.

Marashi M., S. Masoudi, M. K. Moghadam, H. Modirrousta, M. Marashi, M. Parvizifar, M. Dargi, M. Saljooghian, F. Homan, B. Hoffmann, et al. 2017. Peste des petits ruminants virus in vulnerable wild small ruminants, Iran, 2014–2016. Emerging Infectious Diseases 23:704–706.

Munir, M. 2013. Role of wild small ruminants in the epidemiology of peste des petits ruminants. Transboundary and Emerging Diseases 61:414–424.

Munson L., K. Terio, R. Kock, T. Mlengeya, M., Roelke, E. Dubovi, B. Summers, T. Sinclair, and C. Packer. 2008.

Climate extremes promote fatal co-infections during canine distemper epidemics in African lions. PLOS One Biology 3(6):e2545.

Pruvot, M., A. E. Fine, C. Hollinger, S. Strindberg, B. Damdinjav, B. Buuveibaatar, B. Chimeddorj, G. Bayandonoi, B. Khishgee, B. Sandag, et al. 2020. Outbreak of peste des petits ruminants in critically endangered Mongolian saiga and other wild ungulates. Emerging Infectious Diseases 26:51–62.

Queenan, K., P. Mangesho, M. Ole-Neselle, E. Karimuribo, M. Rweyemamu, R. A. Kock, and B. Häsler. 2017. Using local language syndromic terminology in participatory epidemiology: lessons for One Health practitioners among the Maasai of Ngorongoro, Tanzania. Preventive Veterinary Medicine 139A:42–49.

Quintero-Gil, C., S. Rendon-Marin, M. Martinez-Gutierrez, and J. Ruiz-Saenz 2019. Origin of canine distemper virus: consolidating evidence to understand potential zoonoses. Frontiers in Microbiology 10:1982.

Roelke-Parker, M. E., L. Munson, C. Packer, R. A. Kock, S. Cleaveland, M. Carpenter, S. J. O'Brien, A. Pospischil, R. Hofmann-Lehmann, H. Lutz, et al. 1996. A canine distemper virus epidemic in Serengeti lions (*Panthera leo*). Nature 379:441–445.

Scott, G. R. 2000. The murrain now known as rinderpest. Newsletter of the Tropical Agriculture Association, U.K., 20 (4):14–16.

Sodnomdarjaa, R., S. J. B. Tserendorj, and J. Bekhochir. 2013. History of rinderpest control and eradication in Mongolia. Mongolian Journal of Agricultural Sciences 11:172–190.

Spiegel, K. A., and K. A. Havas. 2019. The socioeconomic factors surrounding the initial emergence of peste des petits ruminants in Kenya, Uganda, and Tanzania from 2006 through 2008. Transboundary and Emerging Diseases. 66:627–633.

Stokholm, I., W. Puryear, K. Sawatzki, S. W. Knudsen, T. Terkelsen, P. Becher, U. Siebert, and M. T. Olsen. 2021. Emergence and radiation of distemper viruses in terrestrial and marine mammals. Proceedings of the Royal Society B Biological Sciences 288:1969.

Taylor, L. H., S. M. Latham, and M. E. Woolhouse. 2001. Risk factors for human disease emergence. Philosophical Transactions of the Royal Society of London B Biological Sciences 356:983–989.

Tompkins, D. M., S. Carver, M. E. Jones, M. Krkošek and L. F. Skerratt. 2015. Emerging infectious diseases of wildlife: a critical perspective. Trends in Parasitology 31:149–159.

Uhl, E. W., C. Kelderhouse, J. Buikstra, J. P. Blick, B. Bolon, and R. J. Hogan. 2019. New World origin of canine distemper: interdisciplinary insights. International Journal of Paleopathology 24:266–278.

Ul-Rahman, A., J. M. Wensman, M. Z. Abubakar, P. Shabbir, and P. Rossiter. 2018. Peste des petits ruminants in wild ungulates. Tropical Animal Health and Production 50:1815–1819.

Van Bressem, M.-F, P. J. Duignan, A. Banyard, M. Barbieri, K. M. Colegrove, S. De Guise, G. Di Guardo, A. Dobson, M. Domingo, D. Fauquier, et al. 2014. Cetacean morbillivirus: current knowledge and future directions. Viruses.6:5145–5181.

World Organisation for Animal Health (WOAH). 2022. Peste des petits ruminants. https://www.woah.org/en/disease/peste-des-petits-ruminants/. Accessed 20 Jan 2022.

8 Chronic Wasting Disease

Michael W. Miller, Lisa L. Wolfe

Introduction

Chronic wasting disease (CWD; Williams and Young 1980), the common name for the prion disease of "deer" and related species (cervids), occurs in North America, Asia, and Europe (Table 8.1, Figure 8.1). Once a mere scientific curiosity, CWD has emerged since the 1990s to be described by some—deservedly or otherwise—as a wildlife disease of critical concern. Unchecked epidemics of the more infectious forms of CWD can have implications that touch upon conservation, ecology, socioeconomics, and public health. Despite considerable gains in knowledge about CWD over the four decades since first described by Williams and Young, this disease nonetheless presents epidemiologic and management challenges for wildlife managers and other stewards of free-ranging and captive cervids.

This chapter focuses primarily on CWD in North American cervids (Williams 2005) because that is the most familiar, longest studied, and thus best-understood form. We note where possible comparative features of, similarities to, and differences from the two forms of CWD described in Scandinavia (Benestad et al. 2016, Pirisinu et al. 2018, Tranulis et al. 2021).

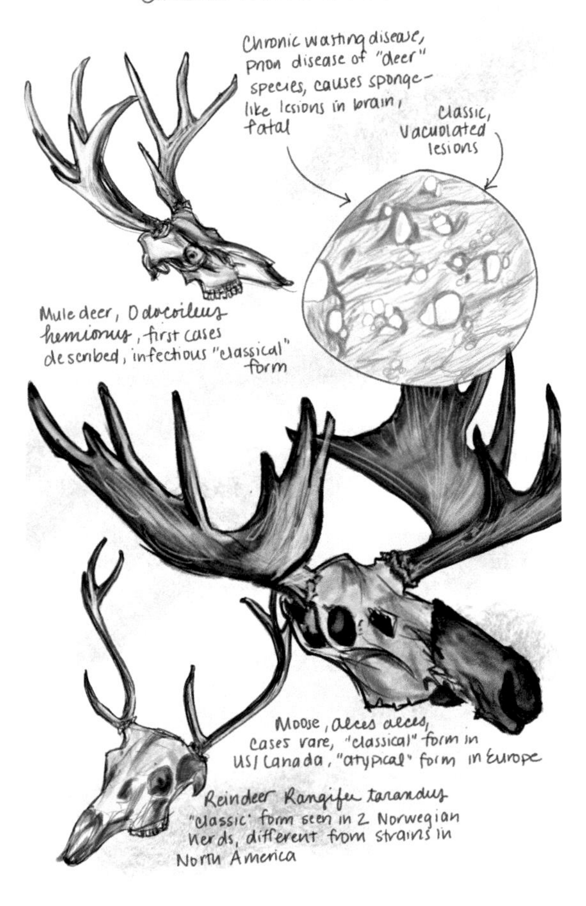

Illustration by Laura Donohue.

Table 8.1. Reported occurrences of classical and atypical chronic wasting disease (CWD) phenotypes in cervid species in North America, Asia, and Europe in captive (C) and free-ranging (F) settings

	Phenotype	
Continent and species	Classical	Atypical
North America		
Mule deer (*Odocoileus hemionus*)	C, F	
White-tailed deer (*Ododcoileus virginianus*)	C, F	
Wapiti (*Cervus canadensis*)	C, F	
Red deer (*C. elaphus*)	C	
Moose (*Alces alces*)	F	
Reindeer or caribou (*Rangifer tarandus*)	C	
Asia		
Wapiti (*C. canadensis*)	C	
Red deer (*C. elaphus*)	C	
Sika deer (*C. nippon*)	C	
Europe		
Red deer (*C. elaphus*)		F
Moose (*A. alces*)		F
Reindeer or caribou (*R. tarandus*)	F	

Classical CWD in Asia can be traced to North America (Sohn et al. 2002, 2020). The prion strains associated with classical CWD in North America and Europe are different from one another, and the strain associated with atypical CWD is distinct from either classical CWD strain (Tranulis et al. 2021). Empty cells indicate that no cases have been reported.

History

The history of CWD is not linear and all CWD cannot be traced to a common point of origin, despite widely held beliefs to the contrary. Williams and Young (1980, 1982) first described "chronic wasting disease" in the late 1970s from cases observed among captive mule deer (*Odocoileus hemionus*) and wapiti (*Cervus canadensis*) held in Colorado and Wyoming, USA. By the time it was formally described, this wasting syndrome already had been recognized for a decade or more by animal caretakers and veterinarians in a handful of research and zoo facilities in Colorado, Wyoming, and Ontario, Canada, that were connected through exchanges of live and presumably exposed cervids (Williams and Young 1980, 1992). For example, the Toronto Zoo imported mule deer from the Denver and Cheyenne Mountain Zoos (Williams and Young 1992, Dubé et al. 2006). Multiple research facilities in north-central Colorado and Wyoming also exchanged animals to varying degrees. As noted by Williams and Young (1992), "Movement of animals from Wyoming to Fort Collins was considerable, but only a few animals have moved in the other direction." Establishing a diagnostic profile for CWD led to the disclosure of additional cases in captivity and in the wild (Williams and Young 1992, Williams and Miller 2002). As awareness of and surveillance for CWD have grown, so have the number and geographic extent of cases (Williams et al. 2002, Miller and Fischer 2016, Sohn et al. 2020, Tranulis et al. 2021).

Part of CWD's apparent growth can be attributed to true expansion of its global footprint, but part is a function of a broadening surveillance effort finding foci that already had been present for some time. A common mistake made in recounting the history of CWD is equating the first time a case was detected in a location to the first time CWD had *ever* occurred in that location (Miller and Fischer 2016). In numerous instances, undetected, unrecognized, unconfirmed, or unreported cases of CWD clearly had occurred prior to the first confirmed case. For example, investigation of the "first" CWD case detected in Saskatchewan, Canada, in 1996 revealed an extensive outbreak among captive wapiti facilities in that province and in neighboring Alberta. This was traced back to imports from South Dakota, USA, in the 1980s (Argue et al. 2007), uncovering outbreaks involving captive wapiti facilities in several states that had gone undetected for a decade or more (Williams et al. 2002). Similarly, modeling estimated that the "new" free-ranging CWD focus first detected in Wisconsin, USA, in 2001 may have begun several decades earlier (Wasserberg et al. 2009), perhaps predating the western US outbreak or arising in parallel.

The story of CWD's emergence and spread is neither clean nor simple. Both human-assisted and natural sources of spread—occurring alone or in varied combinations and sequences—likely have driven CWD's distribution. Exposed animals apparently were moving in commerce in parts of the United States and

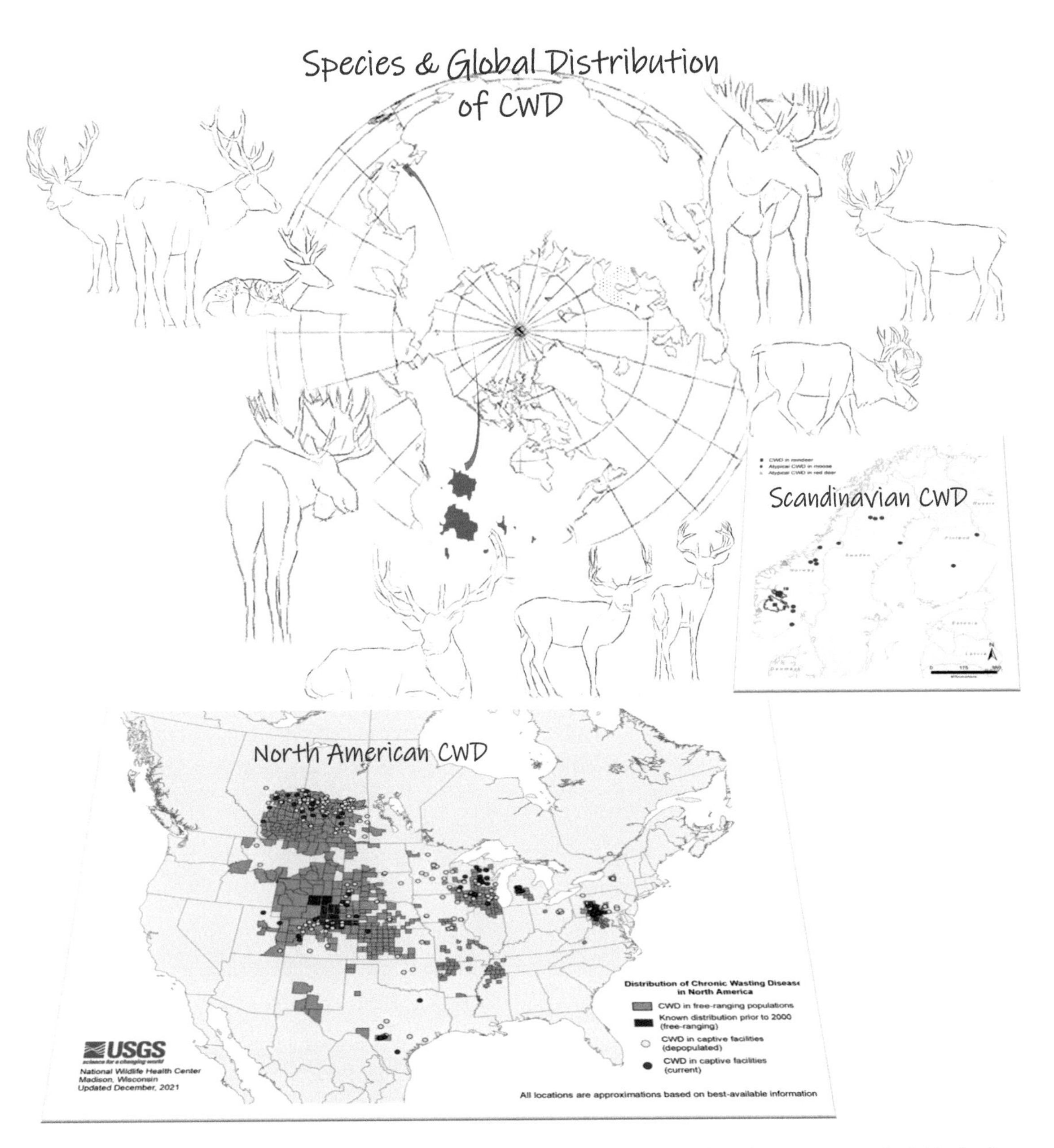

Fig. 8.1. Chronic wasting disease (CWD) in one or more forms has been diagnosed on three continents. Individuals of seven species (and hybrids) spanning four genera in the Family Cervidae have been infected. North America has the widest geographic and species distribution of CWD, with free-ranging foci of varied sizes scattered across the United States and western Canada involving mule deer (*Odocoileus hemionus*), white-tailed deer (*O. virginianus*), wapiti (*Cervus canadensis*), and occasionally moose (*Alces alces*), as well as cases in countless affected captive cervid facilities (US Geological Survey 2021). After receiving infected captive wapiti from North America, the Republic of Korea has reported outbreaks involving captive wapiti, red deer (*C. elaphus*), and sika deer (*C. nippon*; Sohn et al. 2020). Two unique forms of CWD also have been described in Scandinavia since 2016: a presumably infectious form affecting at least two Norwegian reindeer (*Rangifer tarandus*) herds and an "atypical" form presenting as scattered cases—primarily in moose but also red deer—from Norway, Sweden, and Finland (Tranulis et al. 2021). Maps derived from the US Geological Survey (2021) and Tranulis et al. (2021); cervid species silhouettes adapted from scaled two-dimensional images accessed at https://www.dimensions.com.

Canada by the 1970s (Williams and Young 1992, Williams et al. 2002). There were documented transfers from Colorado zoos to Ontario (Dubé et al. 2006) and from South Dakota to Saskatchewan to South Korea (Sohn et al. 2002, Kahn et al. 2004), as well as movements among commercial facilities within the United States (Williams et al. 2002). But countless other cervid transfers that remain mostly undocumented or unacknowledged also have occurred among exposed zoos, private brokers and collectors, and other entities. Such transfers may have laid part of the foundation for CWD's emergence elsewhere in the United States and Canada. Spillover from affected captive cervid facilities into the wild—including exposed cervids that escaped or were intentionally released—seems to best explain some but not all free-ranging outbreaks. Natural expansion through migratory and dispersal movements of affected cervids has driven spread in the wild in multiple locations. Yet critical details are and likely will remain incomplete.

Few CWD outbreaks have a clear, singular explanation for their origins (Miller and Fischer 2016). Even in North America, not all CWD cases have plausible, let alone documented, interconnections (Schwabenlander et al. 2013, Walther et al. 2019). The circumstances that gave rise to the index outbreak of CWD reported by Williams and Young (1980, 1992) were uncertain but do not appear unique (Box 8.1). Data from Europe show that similar circumstances likely have been replicated in other locations, before or since. Novel features of Scandinavian cases (Pirisinu et al. 2018, Nonno et al. 2020; see Etiology) suggest CWD on the European continent originated independently (Tranulis et al. 2021), belying the dogma that all occurrences of CWD somehow connect back to those "first" cases described by Williams and Young. Multiple starts for CWD in North American seem equally plausible.

Etiology

Chronic wasting disease belongs to a collection of unusual diseases called transmissible spongiform encephalopathies (TSEs; Hörnlimann et al. 2007).

Box 8.1. Origins of chronic wasting disease

The origins of the prion strains that cause chronic wasting disease (CWD) in North American or Scandinavian cervids are unknown. Several hypotheses on their origins have been proposed.

- Scrapie-associated prion spillover: Similarities between CWD and scrapie, historical geographic overlaps with CWD in North America and in Scandinavia, timing and circumstances of emergence, lack of cohesive epidemiological explanations for all foci, multiple distinct prion strains, and oral susceptibility of white-tailed deer (*Odocoileus virginianus*) to classical scrapie (Greenlee et al. 2015) support this hypothesis.
- A new or a reemerging disease of cervids: Relatively limited global distribution, *PRNP* polymorphisms, and poor infectivity in bovids and other noncervids support this hypothesis. Although unproven, overproduction of cellular prion protein and spontaneous misfolding has been suggested as a mechanism giving rise to sporadic prion disease in humans (Prusiner 1995) and cervids (Tranulis et al. 2021). Spontaneous (i.e., rare, one in a million) misfolding events would be difficult to distinguish from rare exogenous exposure events. Spontaneous misfolding events also would not be sufficient to drive epidemic patterns observed with classical CWD in cervids.
- An unknown source: Although unlikely, a prion strain could be originating in and spilling over from an as yet unrecognized domestic or free-ranging host species.

Considering multiple strains and two phenotypes of CWD have been observed, these potential origins are not mutually exclusive.

Prions, or proteinaceous infectious agents (Prusiner 1982, 1998), are widely accepted as causing CWD and other TSEs. Prions propagate like a self-perpetuating biotoxin rather than conventional bacteria or viruses. In essence, the malformed disease-associated prion protein (PrP^d) alters the shape (conformation) of the normal cell surface prion protein already found in the host (PrP^c) to make new PrP^d copies. Interestingly, this simple shape-shifting trick makes PrP^d more resistant to enzymes and helps avoid host defenses. The disease-causing protein is chemically identical to the normal PrP^c already present, and this unconventional propagation strategy is the main reason why standard diagnostic tests, vaccines, and treatments for bacterial and viral diseases will not work on CWD and other prion diseases.

Prion diseases affect a variety of mammalian species (Prusiner 1995). The "portrait" chapters in Hörnlimann et al. (2007) offer especially accessible overviews of these diseases. Prion diseases of other hoofed stock species include scrapie of domestic sheep (*Ovis aries*) and goats (*Capra hircus*) and bovine spongiform encephalopathy (BSE) of cattle (*Bos taurus*) and other species in the Family Bovidae. Scrapie and BSE each come in more than one form, broadly grouped as "classical" or "atypical" (see review by Greenlee and West Greenlee 2015). All prion diseases can be transmitted by consumption or inoculation of infected tissue, but CWD and scrapie also behave as infectious diseases and can be transmitted naturally like more common viral and bacterial diseases.

Chronic wasting disease has become, by convention, the generic name for TSEs of cervids. Yet CWD is not a singular entity. At least two disease forms (phenotypes) and at least four prion strains occur among the "CWD" cases encountered in various species and geographic locations (Table 8.1). A single phenotype of North American CWD is recognized. This form has been termed "classical CWD", akin to the "classical scrapie" term used to denote the oldest recognized form of that disease. Two or more prion strains have been associated with classical CWD in North America (Angers et al. 2010, Hannaoui et al. 2021). At least two additional, distinct prion strains occur among Scandinavian cervids with CWD (Nonno et al. 2020, Bian et al. 2021, Tranulis et al. 2021). One or more Scandinavian strains are associated with a novel phenotype called "atypical CWD", akin to "atypical scrapie" (Pirisinu et al. 2018, Tranulis et al. 2021). The rather wide natural host range and genetic variation within some host species may lead to further strain variation (Arifin et al. 2021, Bian et al. 2021, Hannaoui et al. 2021).

The origins of North American and Scandinavian CWD strains are unknown (Box 8.1). Several possibilities have been suggested (Williams and Miller 2003, EFSA 2017, Tranulis et al. 2021). More than one origin event now seems all but certain given the occurrence of distinct CWD-associated prion strains and disease phenotypes on separate continents.

Hosts and Susceptibility

Chronic wasting disease occurs in at least seven species (and hybrids) spanning four genera in the Family Cervidae (Table 8.1). Natural susceptibility in more cervid species seems likely, perhaps with the exception of fallow deer (*Dama dama*) that have shown resistance to prolonged natural exposure (Rhyan et al. 2011). As with other TSEs, CWD-associated prions can be transmitted to a variety of species via intracerebral (IC) inoculation with infectious tissue (Williams 2005; Mathiason 2017; EFSA 2017, 2019). Susceptibility to oral challenge is more limited and better predicts natural susceptibility (e.g., Harrington et al. 2008, Rhyan et al. 2011, Williams et al. 2018).

Similarities and differences in the amino acid sequences of PrP^c influence the natural host range of a prion strain (Goldmann 2008). A single gene (*PRNP*) encodes PrP^c, whose sequences are highly shared among cervid species (Robinson et al. 2012b, Robinson et al. 2019, Arifin et al. 2021). Unique biochemical properties of the PrP^c encoded by cervid species may enhance natural vulnerability to misfolding (Benestad and Telling 2018). Despite obvious physical and behavioral differences among susceptible cervid species their predominant or "wild type" PrP^c^s are essentially the same, varying by only one amino acid at most. These close relationships likely facilitate interspecies transmission where habitation or ranges overlap.

Within most host species, however, one or more natural variations in *PRNP* gene coding (polymorphisms) have been observed (Robinson et al. 2012*b*, Robinson et al. 2019, Arifin et al. 2021). These polymorphisms can result in amino acid substitutions within PrP^c that modulate prion disease course and relative infection risk. As a rule, individuals of the "wild-type" *PRNP* genotype and amino acid sequences tend to show higher relative infection probability and more rapid disease progression than those with the alternative *PRNP* coding. For example, white-tailed deer (*Odocoileus virginianus*) with the *PRNP* genotype encoding the wild-type amino acid polymorphism at position 96 were more likely to be infected and to have infections detected by biopsy than those of other genotypes (Robinson et al. 2012*a*, Thomsen et al. 2012, Seabury et al. 2020). Within host species, individuals carrying minor *PRNP* polymorphisms are relatively rare in nature, suggesting these genotypes may suffer from selective disadvantages in the absence of CWD (Jewell et al. 2005, Wolfe et al. 2014, Monello et al. 2017).

Similarities and differences in PrP^c amino acid sequences among species tend to be predictive of potential host ranges of prion strains. Although the *PRNP* gene is highly conserved among mammalian species, *PRNP* gene sequences vary across taxa (Stewart et al. 2012, Cullingham et al. 2020). The differences among species—called "species barriers"—play a role in limiting the natural host range of a prion strain. Species barriers to CWD infection are characterized by inefficient IC transmission on the first (primary) passage attempt and lack of primary oral transmission (Harrington et al. 2008). Substantive differences in *PRNP* gene coding between cervids and species outside the Family Cervidae may largely explain observed species barriers to natural CWD prion transmission (Williams et al. 2018, Fox et al. 2021, Wolfe et al. 2022). Of perhaps greatest interest and consequence, CWD has shown little natural propensity for causing disease in humans (Waddell et al. 2018). Classical BSE prions proved to be an exception to the general pattern, with a far wider host range (including humans) than initially believed (Houston and Andréoletti 2018). The known strains of CWD prions differ from classical BSE prions in cattle (Bruce et al. 1997, Race et al. 2002, Tamgüney et al. 2006, Tranulis et al. 2021). Nonetheless, concerns about CWD and other animal prion diseases understandably linger in the wake of BSE's unexpected capacity to cause disease in humans (Houston and Andréoletti 2018). The assessment of CWD's zoonotic potential continues (e.g., EFSA 2019, Pritzkow et al. 2022, Tranulis et al. 2021).

Demographics and Field Biology

The transmission and demographics of infection within affected cervid herds are best understood from studies of classical North American CWD in mule deer, white-tailed deer, and wapiti. Classical and atypical CWD differ in patterns of apparent transmissibility. Classical CWD is contagious within (and likely among) susceptible species. Prion shedding in saliva, feces, and urine occurs through much of the disease course and seems particularly important in transmission of classical CWD (Figure 8.2; Tamgüney et al. 2009, Mathiason 2017, Plummer et al. 2017). Routes involving semen, placental tissues, and milk could play seasonal roles (Mathiason 2017). The classical form in Scandinavian reindeer (*Rangifer tarandus*) is presumably contagious given its other similarities to classical North American CWD and classical scrapie (Benestad et al. 2016, Mysterud et al. 2019*a*). In contrast, the atypical CWD reported in Scandinavian moose (*Alces alces*) and red deer (*Cervus elaphus*) appears less likely to be contagious (EFSA 2019, Mysterud et al. 2021*a*, Tranulis et al. 2021), or perhaps less contagious than classical CWD or otherwise acquired.

Fig. 8.2. Classical chronic wasting disease (CWD) is an infectious prion disease. Key epidemiological features of classical CWD include a rather long, invariably fatal disease course and multiple transmission pathways. The general relationships between prion transmission sources (blue shading) and the disease course are shown for genetic "wild type" mule deer (*Odocoileus hemionus*) in a wildland setting. An infected host sheds prions throughout much of the ~2 year disease course, beginning well before clinical signs appear. Prions from excreta and carcass remains can persist in the environment long after the host has died, further extending the infectious period associated with a single case. Stylized mule deer timeline based on laboratory and field observations; see text for additional details and references. Similar patterns would be expected for classical CWD in free-ranging, wild-type white-tailed deer (*O. virginianus*), wapiti (*Cervus canadensis*), and perhaps reindeer (*Rangifer tarandus*). Illustration by Laura Donohue.

Prions persist in the environment. Experimentally, paddocks contaminated with residual prions were still infectious >2 years after excreta or >1 year after decomposed carcasses from classical CWD-infected mule deer had been deposited (Miller et al. 2004). Anecdotally, however, heavily contaminated environments can remain infectious for a decade or more, similar to the persistence reported for scrapie prions (Georgsson et al. 2006). Soil contamination and perhaps plant uptake may foster local intra- and interspecies transmission (Fig. 8.2; Miller et al. 2004, Zabel and Ortega 2017, Plummer et al. 2018). Ingesting infectious tissues or carcasses remains (Miller et al. 2004) could transmit both classical and atypical CWD.

Classical and atypical CWD infections occur predominantly among adult cervids. Cases tend to be rare among young-of-the-year and yearling age classes. CWD infection of younger age classes becomes more common with increasing overall prevalence (proportion or percent of animals infected within the herd). The average age of infected adult animals varies and likely declines over the course of an epidemic (Figure 8.3; Miller and Williams 2003, Fisher et al. 2022). In contrast, atypical CWD cases reported thus far among Scandinavian moose were all relatively aged individuals (Pirisinu et al. 2018, Ågren et al. 2021, Tranulis et al. 2021). Adult (≥2 years old) males tend to be infected with classical CWD at higher rates than adult females in both North American deer

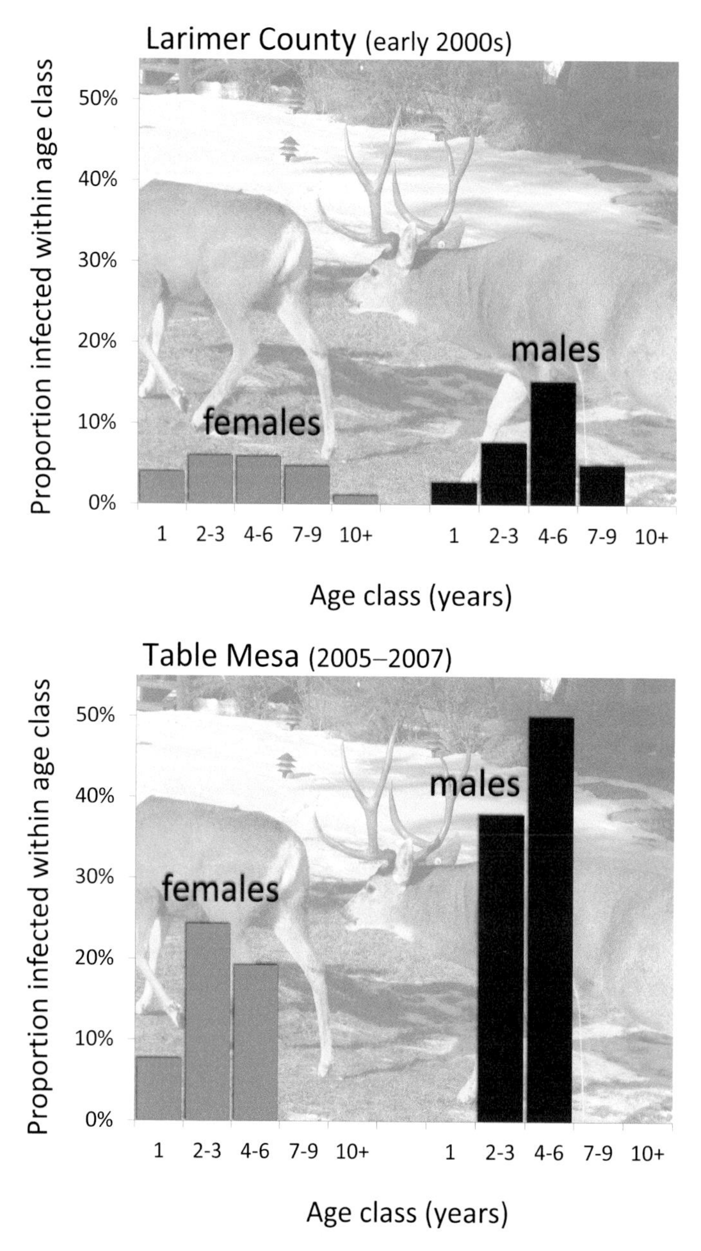

Fig. 8.3. Sex, age, and classical chronic wasting disease (CWD). Classical CWD epidemics in North American deer species (*Odocoileus* spp.) and Norwegian reindeer (*Rangifer tarandus*) show similar demographic features (e.g., Robinson et al. 2012*a*, Mysterud et al. 2019, Miller and Wolfe 2021). Two common patterns have emerged, as illustrated with data from mule deer (*O. hemionus*) herds in northcentral Colorado, USA: (1) infection rates tend to be approximately two times higher in males ("bucks") than in females ("does") from the same herd, and (2) "prime aged" adults show higher infection rates than very young or very old deer. As the overall rate of infection in a herd increases, individuals become infected and succumb at younger ages. Older aged animals become rare. In the examples shown, 25% of females and 4% of males were over 6 years old in the heavily hunted Larimer County herds (top panel), whereas only 6% of females and 2% of males were over 6 years old in the unhunted Table Mesa herd where overall infection rates were much higher (bottom panel).

species (Figure 8.3; Samuel and Storm 2016, Miller and Wolfe 2021) and in Norwegian reindeer (Mysterud et al. 2019*a*), although this pattern may eventually weaken in unchecked epidemics (e.g., Edmunds et al. 2016, Fisher et al. 2022). Sex differences seem less pronounced in wapiti (Monello et al. 2014) and in captive settings and unapparent thus far among the relatively few atypical CWD cases reported.

The temporal and spatial dynamics of classical CWD are relatively slow. Outbreaks in free-ranging cervid herds may be especially difficult to detect in the earliest phase because cases are rare and limited in spatial distribution (Heisey et al. 2010). Moreover, infected animals often succumb to predation or other causes of death before reaching end-stage disease in the wild and may be missed by surveillance focused too narrowly on a "wasting" presentation. (See Signs and Gross Lesions) Consequently, CWD often is well established in a herd or jurisdiction by the time it is first detected. A phase of linear to exponential growth in apparent prevalence, occurring over the span of a decade or more in free-ranging settings, has been described in unchecked epidemics of classical CWD (EFSA 2018, Miller et al. 2020, LaCava et al. 2021, Smolko et al. 2021). During this phase, apparent prevalence estimated from random sampling (e.g., via harvest surveys) provides a reasonable proxy for disease incidence (the rate of new infections per year or other timespan; Miller and Wolfe 2021). Apparent prevalence can reach high levels, with 1 in every 2–3 harvested adult males becoming infected (e.g., Rivera et al. 2019, LaCava et al. 2021). The expected dynamics beyond this growth phase are less certain because long-term field data remain limited. Patterns should become clearer in coming decades if recurring assessments continue in some of the longest known outbreak areas in the United States and western Canada.

Annual survival of animals infected with CWD is lower than in uninfected individuals of the same sex and age class (Monello et al. 2014, Edmunds et al. 2016, DeVivo et al. 2017). As a result, sustained high prevalence can compromise the performance and resilience of infected herds. The presence of *PRNP* genotypes with lower infection probabilities—especially heterozygotes—may provide some level of buffering against herd decline (Williams et al. 2014, LaCava et al. 2021). Nonetheless, herds heavily infected with CWD may have limited capacity to support sport harvest or recover from severe weather events or other perturbations (Monello et al. 2014, Williams et al. 2014, Edmunds et al. 2016, DeVivo et al. 2017). The vulnerability of infected animals to predation (Miller et al. 2008, DeVivo et al. 2017, Fisher et al. 2022) could help sustain predator densities even as prey abundance declines.

The spatial dynamics of classical CWD have been more difficult to discern, in part because surveillance approaches have confounded observed patterns of "spread" in numerous jurisdictions. In particular, the common practice of continuously establishing new surveillance areas on the periphery of known foci can give the illusion of spatial "spread" when in fact all that has grown with certainty is the footprint of surveillance effort. Human-assisted movements of infected animals also have confounded spatial dynamics, perhaps best illustrated by the high probability that spillover events from one or more of at least 40 CWD-infected commercial cervid facilities drove initial seeding of CWD into the wild in Saskatchewan (Kahn et al. 2004, Norbert et al. 2016). Where sufficient field data allow for analyses, social and habitat connectivity appear to explain observed patterns of CWD's natural expansion across landscapes occupied by infected cervids (Robinson et al. 2013, Kelly et al. 2014, Norbert et al. 2016, Mejía-Salazar et al. 2018, Smolko et al. 2021).

In contrast to classical CWD, atypical CWD in Scandinavian moose has thus far appeared as comparatively rare cases (~1 in ≥10,000 harvest samples) with scattered spatial distribution and limited clustering (Figure 8.1; Tranulis et al. 2021). The observed pattern seems more suggestive of a sporadic disease, although being sporadic does not preclude low-level transmissibility, exposure to an exogenous prion source through spillover from another host, foodborne exposure, or some other means.

Signs and Gross Lesions

Prion infection in cervids has a more ambiguous clinical and postmortem presentation than one might expect (Figure 8.4). The typical suspect case encountered in the field rarely matches the profile of the drooling, stumbling, walking skeleton image that has become the face of CWD. Moreover, sensationalized monikers like "zombie deer" and "mad deer" do disservice to CWD surveillance because distorting the suspect profile and focusing it too narrowly can impede disease detection and control efforts.

Fig. 8.4. The field presentation of chronic wasting disease (CWD) cases varies with disease stage and other factors. The earliest signs may be subtle, inconsistent, and especially difficult to judge in the field. As the disease progresses, behavioral changes and loss of body condition become more apparent but still may not be recognized by untrained observers. (The mule deer buck labeled as a "later clinical" case was described as apparently healthy by field technicians a few days before the photograph was taken.) End-stage CWD cases present the more familiar "suspect" image. However, by this stage affected animals may not last long in the field and are simply found dead. Necropsy findings tend to be relatively nonspecific. Given the vagaries of presenting signs and necropsy findings, those surveying cervids for CWD should retain a broad view of what qualifies as a "suspect" and rely on laboratory testing to screen for prion infection regardless of whether field observations and necropsy findings are more or less suggestive of CWD.

Behavioral changes and poor body condition are hallmarks of late-stage classical and atypical CWD (Williams 2005, Pirisinu et al. 2018, Ågren et al. 2021), but these signs are relatively nonspecific and easily confused with other causes of malnutrition or chronic disease. Infected animals can show additional but equally nonspecific signs, including "fuzzy" appearance or retained winter hair coat, nasal discharge and coughing, excess salivation, and dullness or unresponsiveness (Williams 2005). The earliest signs—less awareness of or response to adverse stimuli, straying from the herd—may be subtle, inconsistent, and especially difficult to judge in the field (Figure 8.4). Handling, immobilization, or stressful events may accentuate signs (e.g., Benestad et al. 2016). Diarrhea is *not* among the signs of CWD in cervids; itchiness (pruritus), as seen in scrapie, has not been reported. Concurrent health problems or death by accidents, predation, or harvest may confound the clinical picture. Other diseases that can appear similar to CWD in cervids include rabies, plague, and tuberculosis, which are all transmissible to humans. Given the vagaries of presenting signs, those surveying cervids for CWD should retain a broad view of what qualifies as a "CWD suspect" and rely on laboratory testing to identify true cases.

Affected animals do not recover from CWD. Clinical disease is progressive once signs appear and can last from a few weeks to perhaps a year or more. An inapparent incubation period lasting a year or more precedes the clinical phase (Williams 2005, Mathiason 2017). The recorded length of the clinical phase depends in part upon the timeframe within which the subtle early signs are first recognized. Secondary complications (e.g., aspiration pneumonia) can shorten survival, as can predation, accidents, or harvest. Some *PRNP* genotypes of deer (*Odocoileus* spp.) and wapiti can live longer with CWD than others (e.g., O'Rourke et al. 2007, Johnson et al. 2011, Wolfe et al. 2014). Prion strain also may influence the course of disease: all Scandinavian moose and red deer cases thus far have involved older aged animals (≥10 years; Pirisinu et al. 2018, Ågren et al. 2021, Tranulis et al. 2021).

As with clinical appearances, gross necropsy findings are nonspecific (Williams 2005). Moreover, postmortem findings can be misleading in cases with secondary complications or unrelated problems. Animals that die of uncomplicated, end-stage CWD are usually adults but minimally should be late yearlings, and their carcasses should be emaciated (Figure 8.4). Some cases will have a distended, watery rumen and/or a flaccid, distended esophagus; scant formed feces may be present. However, lesions of aspiration pneumonia, trauma, foreign body ingestion, parasitism, dental wear, and other CWD-related or unrelated conditions often muddy the postmortem picture, especially in free-ranging animals (Williams 2005, Pirisinu et al. 2018, Ågren et al. 2021). In some cases, the circumstances surrounding death (e.g., an unusual accident or location for finding a carcass) suggest behavioral anomalies even in the absence of clinical disease. Potential suspect cases should be screened for prion disease by laboratory testing regardless of whether necropsy findings are more or less suggestive of CWD.

Pathogenesis and Pathology

The predominant pattern of PrP^d deposition in classical CWD is consistent with natural infection via oral exposure. This pattern has been reproduced experimentally by oral inoculation in a number of susceptible cervid hosts (e.g., Fox et al. 2006, Balachandran et al. 2010, Sohn et al. 2020) and closely resembles the pathogenesis of classical scrapie in sheep (Williams 2005). Deposits can be detected in tonsil, retropharyngeal lymph node, and other gut-associated lymphoid tissues relatively early in the disease course. Somewhat later, PrP^d deposits become apparent in central nervous system (CNS) tissues. The extent and intensity of PrP^d deposition increases through end-stage disease, ultimately involving a long list of organs and multiple sites within the brain (Williams 2005). Timing and intensity of PrP^d deposition can be influenced by *PRNP* genotype (e.g., Fox et al. 2006, Thomsen et al. 2012). A subset of natural cases in classical CWD epidemics may have PrP^d deposits in CNS sites only; such cases may be the result of more direct exposure of nerves in the lining of the mouth, nasal passages, or pharynx (Spraker et al. 2004). In atypical CWD cases, PrP^d deposits are limited to the CNS, with an overall pattern and intensity distinct from classical CWD (Pirisinu et al. 2018, Vikøren et al. 2019).

The microscopic spongiform lesions specific for TSE develop later in the disease course, becoming prominent in multiple CNS sites in end-stage clinical cases (Williams 2005, Benestad et al. 2016, Pirisinu et al. 2018). The parasympathetic vagal nucleus in the dorsal portion of the medulla oblongata at the obex is the earliest and most reliable CNS site for demonstrating spongiform lesions, showing consistent involvement in susceptible species and all forms. Other histopathology lesions from intercurrent disease (e.g., aspiration pneumonia) may be present in some cases (Williams 2005).

Diagnosis

Diagnosing CWD requires laboratory confirmation because clinical signs and necropsy findings are not specific to prion disease. Most contemporary CWD tests demonstrate PrP^{d} in lymphoid and/or CNS tissues (Haley and Richt 2017). For the classical North American and Scandinavian reindeer forms, screening lymphoid tissue (especially retropharyngeal lymph node or tonsil) has proven sensitive for detecting infections throughout most of the disease course (Williams 2005, Haley and Richt 2017, EFSA 2018). In contrast, all atypical CWD cases in Scandinavian moose and red deer described thus far had only CNS involvement (Pirisinu et al. 2018, Vikøren et al. 2019, Ågren et al. 2021). Screening both brainstem and oropharyngeal lymphoid tissues (e.g., retropharyngeal lymph node and palatine tonsil) in suspect cases may provide the broadest sensitivity in detecting classical and atypical forms of CWD (EFSA 2019).

Immunohistochemistry (IHC), Western blot, and enzyme-linked immunosorbent assays using various monoclonal antibodies have been developed for CWD diagnostics (Haley and Richt 2017, EFSA 2018). Potential influence of *PRNP* genotypes on epitopes targeted by monoclonal antibodies should be considered in selecting and interpreting diagnostics (EFSA 2018). Ultrasensitive amplification assays—particularly real-time quaking-induced conversion (RT-QuIC)—have shown potential utility in research applications (Caughey et al. 2017). Routine histopathology of the parasympathetic vagal nucleus in the brainstem also can be used to demonstrate spongiform lesions in suspect cases. Reviews cited herein provide greater details about CWD diagnostics.

Diagnostics for detecting infected animals while still alive remain comparatively limited. Tonsil or rectal mucosa biopsies have been used most extensively to screen for cases using IHC (Wolfe et al. 2007, Spraker et al. 2009, Thomsen et al. 2012) or RT-QuIC (Haley et al. 2020*a*). Testing other prospective samples with RT-QuIC also has been explored (Haley et al. 2020*a*, Ferreira et al. 2021). *PRNP* polymorphisms can influence apparent diagnostic sensitivity of live animal testing and should be considered in interpreting negative results (e.g., Thomsen et al. 2012). Preclinical infections may be undetectable in cases lacking PrP^{d} in peripheral lymphoid tissues.

Outcome and Treatment

Death is the expected outcome of CWD infection (Williams 2005). No effective medications, preventive treatments, or vaccines are available (Wolfe et al. 2012, 2020; Pilon et al. 2013; Goñi et al. 2015; Wood et al. 2018).

Treatments for reducing or removing prions in contaminated environments offer little more promise than animal-oriented treatments. Sodium hypochlorite (bleach), sodium hydroxide (lye or caustic soda), hypochlorous acid, and a few other compounds can lower or eliminate prion infectivity (e.g., Hughson et al. 2016, Williams et al. 2019, Alcaron et al. 2021). Treating prion-contaminated soil can reduce its infectivity (e.g., Sohn et al. 2019, Booth et al. 2021). Whether such treatments can completely interrupt transmission appears more doubtful (Miller et al. 1998, Alcaron et al. 2021). Similar to recommendations for scrapie (e.g., Alcaron et al. 2021), environmental treatments aimed at CWD control seem best suited for lightly contaminated captive settings as an adjunct to other disease management practices, with sustained monitoring for recrudescence. Landscape-scale environmental treatment is impractical, and thus strategies for preventing heavy prion contamination or attraction to contaminated sites will remain more important approaches in free-ranging settings (e.g., Mejía-Salazar et al. 2018, Plummer et al. 2018).

Management

Attempts to control CWD in free-ranging cervids vary with respect to goals and approaches. Establishing realistic goals can be problematic, particularly when management responses are taken up before CWD distribution and occurrence are completely understood. Stemming epidemic growth seems more feasible than eradicating well-established CWD (Potapov et al. 2016, Miller et al. 2020). Early or preemptive interventions may offer the best chances for success (Belsare et al. 2021, Mysterud et al 2021*b*), but their effectiveness will be more difficult to "prove." Control (not elimination) may be more achievable in captive settings (Miller et al. 1998, Haley et al. 2020*b*).

Sustained culling (Manjerovic et al. 2014, Mysterud et al. 2019b) and hunting (Miller et al. 2020, Conner et al. 2021, Smolko et al. 2021) seem to be the most practical approaches for large-scale application in the wild. Either culling or harvest apparently can blunt epidemic growth when applied with sufficient intensity and duration (Manjerovic et al. 2014, Miller et al. 2020). Culling focused in the vicinity of prior cases yields higher removal of infected animals than more random harvest-based approaches, but the disparate outcomes seen in Illinois, USA (Manjerovic et al. 2014), versus Alberta, Canada (Smolko et al. 2021), illustrate the importance of sustained application in curbing epidemic growth. Campaigns to suppress or eradicate CWD via broad and local population reductions and selective removal of infected animals have met with varying degrees of effectiveness and public acceptance (Uehlinger et al. 2016; Mysterud et al. 2019*b*, 2021*b*; Rivera et al. 2019; Smolko et al. 2021). Local preferences and support ultimately dictate the viability of control options available to managers (e.g., Holsman et al. 2010, Uehlinger et al. 2016, Meeks et al. 2022, Smolko et al. 2021). Eliminating artificial point sources of exposure—for example, feeding and baiting stations, mineral licks, and grain leaks—can help curb exposure and transmission (Mejía-Salazar et al. 2018, Plummer et al. 2018, Mysterud et al. 2019*c*). Long-term or recurring assessments—ideally replicated and within an adaptive management framework—are needed to gage the effectiveness of attempts to control CWD (Conner et al. 2021, Smolko et al. 2021).

In terms of natural control, predation shows theoretical promise (Wild et al. 2011) with some empirical support. Mountain lions (*Puma concolor*), for example, selectively remove infected mule deer year-round (Krumm et al. 2010, Fisher et al. 2022) and in doing so may help reduce prion burdens in the environment (Baune et al. 2021, Fisher et al. 2022). Societal intolerance by humans holds predator abundance below levels likely needed to make this a stand-alone option for CWD control in most locations, but even limited numbers of predators could help reduce the chances of CWD becoming established (e.g., Belsare et al. 2021). Canids generally resist prion infection (Stewart et al. 2012), and mountain lions also appear resistant to natural exposure (Wolfe et al. 2022). Neither canid nor felid predators seem likely to propagate CWD prions, but a small fraction of ingested prions may evade digestion and be passed transiently after infected tissues are consumed (Nichols et al. 2015, Baune et al. 2021). Relying on natural selection of less susceptible *PRNP* genotypes seems ill-advised as a control strategy for free-ranging herds given evidence that these genotypes may have poorer overall fitness in the wild (Monello et al. 2017, LaCava et al. 2021, Fisher et al. 2022). However, selective breeding could have application in commercial cervid propagation (Haley et al. 2020*b*, 2021; Seabury et al. 2020).

Effectively containing CWD at any geographic or administrative scale depends on early detection of new foci and sources of infection. To this end, mortalities of qualifying "suspect" adult cervids should be examined for CWD wherever prion diseases occur in any species of domestic or wild ruminant. Surveillance in places lacking known sources of prion infection also could help clarify distribution of the atypical CWD form seen thus far only in Scandinavian moose and red deer. Surveys must be carefully designed to provide meaningful, biologically relevant data (see EFSA 2018 and references therein).

Without additional knowledge about the origin(s) and duration of established foci, and about the relative importance of known and assumed risk factors, it is difficult to identify the main factors driving CWD's geographic spread and emergence among captive and free-ranging cervid populations. But exposing susceptible animals to cervids from known infected or exposed herds or housing them on pastures or in facilities or regions with a history of CWD are substantial risks. Scrapie exposure (direct or indirect, past or recent) should be more objectively assessed as a potential risk factor, particularly when other risk factors cannot be identified or appear relatively implausible.

Financial, Legal, Political, and Social Factors

As if unconventional biology and insidious nature weren't enough to contend with, nonbiological factors also influence the motivation, rate, and degree of progress made toward understanding and addressing CWD. Perceptions about CWD vary widely and can change over time (e.g., Holsman et al. 2010, Pattison-Williams et al. 2020, Meeks et al. 2022). Those underestimating the diversity and potential adversity in social perspectives related to CWD and its control have seen their efforts to address this disease derailed. Sociopolitical interest in CWD has thus far been episodic in the United States and Canada; whether a similar pattern will emerge in Europe remains to be determined. The lack of sustained attention span erodes support for long-term programs and funding. Limitations on fiscal resources and regulatory authorities have in turn hampered surveillance, monitoring, and control. Devising regulations can be difficult when regulatory authorities are divided or incomplete or at cross purposes. In some countries and at more local levels of government, CWD has become a wedge issue, pitting wildlife and agricultural authorities or hunting and conservation constituencies against one another. For example, wildlife authorities may want to ban commercial cervid farming, while agriculture authorities are seeking to promote its expansion. Likewise, managing herds for trophy hunting to satisfy constituents may foster epidemic growth. Disease occurrence can impact cervid-related revenues and commerce. Questions of liability for disease occurrence and transfer remain largely uncharted territory.

Dealing with CWD is challenging under current conditions. Nonetheless, many if not all of CWD's fiscal, legal, and sociopolitical facets could become vastly more complicated overnight if CWD-associated prions were shown to cause disease in humans. Here's hoping that bridge will remain uncrossed.

Summary

Four decades after its first description in North America in the late 1970s, the emergence of new forms of "chronic wasting disease" in Scandinavia since 2016 underscores the limits of our scientific knowledge about this unconventional wildlife disease. Despite considerable progress in understanding CWD, it remains a diagnostic, epidemiologic, and management challenge for stewards of free-ranging and captive cervids. Clinical appearances can be subtle and nonspecific, diagnostic tools for live animals are limited, and key information on history and risk factors often is incomplete. No therapies or vaccines are available and there are few proven tools for managing affected populations. Management through harvest and culling seem to be the most viable paths forward; however, its longer term effectiveness remains an open question. But the consequences of unchecked epidemics can be substantial. Given recent events, those managing cervid species anywhere in the world should consider the potential for encountering prion disease when developing herd health monitoring and management programs. Where CWD occurs, those managing cervid resources should redouble efforts to find practicable and sustainable control strategies.

LITERATURE CITED

Ågren, E. O., K. Sörén, D. Gavier-Widén, S. L. Benestad, L. Tran, K. Wall, G. Averhed, N. Doose, J. Våge, and M. Nöremark. 2021. First detection of chronic wasting disease in moose (*Alces alces*) in Sweden. Journal of Wildlife Diseases 57:461–463.

Alcaron, P., F. Marco-Jimenez, V. Horigan, A. Ortiz-Pelaez, B. Rajanayagam, A. Dryden, H. Simmons, T. Konold, C. Marco, J. Charnley, et al. 2021. A review of cleaning and disinfection guidelines and recommendations following an outbreak of classical scrapie. Preventive Veterinary Medicine 193:e105388.

Angers, R. C., H.-E. Kang, D. Napier, S. Browning, T. Seward, C. Mathiason, A. Balachandran, D. McKenzie, J. Castilla, C. Soto, et al. 2010. Prion strain mutation determined by prion protein conformational compatibility and primary structure. Science 328:1154–1158.

Argue, C. K., C. Ribble, V. W. Lees, J. McLane, and A. Balachandran. 2007. Epidemiology of an outbreak of chronic wasting disease on elk farms in Saskatchewan. Canadian Veterinary Journal 48:1241–1248.

Arifin, M. I., S. Hannaoui, S. C. Chang, S. Thapa, H. M. Schatzl, and S. Gilch. 2021. Cervid prion protein polymorphisms: role in chronic wasting disease pathogenesis. International Journal of Molecular Sciences 22:e2271.

Balachandran, A., N. P. Harrington, J. Algire, A. Soutyrine, T. R. Spraker, M. Jeffrey, L. González, and K. I. O'Rourke. 2010. Experimental oral transmission of chronic wasting disease to red deer (*Cervus elaphus elaphus*): early detection and late stage distribution of protease-resistant prion protein. Canadian Veterinary Journal 51:169–178.

Baune, C., L. L. Wolfe, K. C. Schott, K. A. Griffin, A. G. Hughson, M. W. Miller, and B. Race. 2021. Reduction of chronic wasting disease prion seeding activity following digestion by mountain lions. mSphere 6:e00812-21.

Belsare, A. V., J. J. Millspaugh, J. R. Mason, J. Sumners, H. Viljugrein, and A. Mysterud. 2021. Getting in front of chronic wasting disease: model-informed proactive approach for managing an emerging wildlife disease. Frontiers in Veterinary Science 7:e608235.

Benestad, S. L., G. Mitchell, M. Simmons, B. Ytrehus, and T. Vikøren. 2016. First case of chronic wasting disease in Europe in a Norwegian free-ranging reindeer. Veterinary Research 47:e88.

Benestad, S. L., and G. C. Telling. 2018. Chronic wasting disease: an evolving prion disease of cervids. Pages 447–462 *in* M. Pocchiari and J. Manson, editors. Handbook of Clinical Neurology. Volume 153. Human prion diseases. Elsevier, Amsterdam, The Netherlands.

Bian, J., S. Kim, S. J. Kane, J. Crowell, J. L. Sun, J. Christiansen, E. Saijo, J. A. Moreno, J. DiLisio, E. Burnett, et al. 2021. Adaptive selection of a prion strain conformer corresponding to established North American CWD during propagation of novel emergent Norwegian strains in mice expressing elk or deer prion protein. PLoS Pathogens 17:e1009748.

Booth, C. J., S. S. Lichtenberg, R. J. Chappell, and J. A. Pedersen. 2021. Chemical inactivation of prions is altered by binding to the soil mineral montmorillonite. ACS Infectious Diseases 7:859–870.

Bruce, M. E., R. G. Will, J. W. Ironside, I. McConnell, D. Drummond, A. Suttie, L. McCardle, A. Chree, J. Hope, C. Birkett, et al. 1997. Transmissions to mice indicate that 'new variant' CJD is caused by the BSE agent. Nature 389:498–501.

Caughey, B., C. D. Orru, B. R. Groveman, A. G. Hughson, M. Manca, L. D. Raymond, G. J. Raymond, B. Race, E. Saijo, and A. Kraus. 2017. Amplified detection of prions and other amyloids by RT-QuIC in diagnostics and the evaluation of therapeutics and disinfectants. Progress in Molecular Biology and Translational Science 150:375–388.

Conner, M. M., M. E. Wood, A. Hubbs, J. Binfet, A. Holland, L. R. Meduna, A. Roug, J. P. Runge, T. D. Nordeen, M. J. Pybus, et al. 2021. The relationship between harvest management and chronic wasting disease prevalence trends in western mule deer herds. Journal of Wildlife Diseases 57:831–843.

Cullingham, C. I., R. M. Peery, A. Dao, D. I. McKenzie, and D. W. Coltman. 2020. Predicting the spread-risk potential of chronic wasting disease to sympatric ungulate species. Prion 14:56–66.

DeVivo, M. T., D. R. Edmunds, M. J. Kauffman, B. A. Schumaker, J. Binfet, T. J. Kreeger, B. J. Richards, H. M. Schätzl, and T. E. Cornish. 2017. Endemic chronic wasting disease causes mule deer population decline in Wyoming. PLoS One 12:e0186512.

Dubé, C., K. G. Mehren, I. K. Barker, B. L. Peart, and A. Balachandran. 2006. Retrospective investigation of chronic wasting disease of cervids at the Toronto Zoo, 1973–2003. Canadian Veterinary Journal 47:1185–1193.

Edmunds, D. R., M. J. Kauffman, B. A. Schumaker, F. G. Lindzey, W. E. Cook, T. J. Kreeger, R. G. Grogan, and T. E. Cornish. 2016. Chronic wasting disease drives population decline of white-tailed deer. PLoS One 11:e0161127.

European Food Safety Authority (EFSA), Panel on Biological Hazards, A. Ricci, A. Allende, D. Bolton, M. Chemaly, R. Davies, P. S. Fernández Escámez, R. Gironés, L. Herman, K. Koutsoumanis, R. Lindqvist, et al. 2017. Scientific opinion on chronic wasting disease (CWD) in cervids. EFSA Journal 15:e04667.

European Food Safety Authority (EFSA), Panel on Biological Hazards, A. Ricci, A. Allende, D. Bolton, M. Chemaly, R. Davies, P. S. Fernández Escámez, R. Gironés, L. Herman, K. Koutsoumanis, R. Lindqvist, et al. 2018. Scientific opinion on chronic wasting disease (II). EFSA Journal 16:e05132.

European Food Safety Authority (EFSA), Panel on Biological Hazards, K. Koutsoumanis, A. Allende, A. Alvarez-Ordoňez, D. Bolton, S. Bover-Cid, M. Chemaly, R. Davies, A. De Cesare, L. Herman, F. Hilbert, et al. 2019. Scientific opinion on the update on chronic wasting disease (CWD) III. EFSA Journal 17:e05863.

Ferreira, N. C., J. M. Charco, J. Plagenz, C. D. Orru, N. D. Denkers, M. A. Metrick, A. G. Hughson, K. A. Griffin, B. Race, E. A. Hoover, et al. 2021. Detection of chronic wasting disease in mule and white-tailed deer by RT-QuIC analysis of outer ear. Scientific Reports 11:e7702.

Fisher, M. C., R. A. Prioreschi, L. L. Wolfe, J. P. Runge, K. A. Griffin, H. M. Swanson, and M. W. Miller. 2022. Apparent stability masks underlying change in a mule deer herd with unmanaged chronic wasting disease. Communications Biology 5:e15.

Fox, K. A., J. E. Jewell, E. S. Williams, and M. W. Miller. 2006. Patterns of PrP^{CWD} accumulation during the course of chronic wasting disease infection in orally inoculated mule deer (*Odocoileus hemionus*). Journal of General Virology 87:3451–3461.

Fox, K. A., S. M. Muller, T. R. Spraker, M. E. Wood, and M. W. Miller. 2021. Opportunistic surveillance of captive and free-ranging bighorn sheep (*Ovis canadensis*) in Colorado, USA, for transmissible spongiform encephalopathies. Journal of Wildlife Diseases 57:338–344.

Georgsson, G., S. Sigurdarson, and P. Brown. 2006. Infectious agent of sheep scrapie may persist in the environment for at least 16 years. Journal of General Virology 87:3737–3740.

Goldmann W. 2008. PrP genetics in ruminant transmissible spongiform encephalopathies. Veterinary Research 39:e30.

Goñi, F., C. K. Mathiason, L. Yim, K. Wong, J. Hayes-Klug, A. Nalls, D. Peyser, V. Estevez, N. Denkers, J. Xu, et al. 2015. Mucosal immunization with an attenuated *Salmonella* vaccine partially protects white-tailed deer from chronic wasting disease. Vaccine 33:726–733.

Greenlee, J. J., and M. H. West Greenlee. 2015. The transmissible spongiform encephalopathies of livestock. ILAR Journal 56:7–25.

Greenlee, J., S. J. Moore, J. Smith, M. H. West Greenlee, and R. Kunkle. 2015. Scrapie transmits to white-tailed deer by the oral route and has a molecular profile similar to chronic wasting disease and distinct from the scrapie inoculum. Prion 9(Supplement 1):S62.

Haley, N. J., R. Donner, D. M. Henderson, J. Tennant, E. A. Hoover, M. Manca, B. Caughey, N. Kondru, S. Manne, A. Kanthasamay, et al. 2020*a*. Cross-validation of the RT-QuIC assay for the antemortem detection of chronic wasting disease in elk. Prion 14:47–55.

Haley, N., R. Donner, K. Merrett, M. Miller, and K. Senior. 2021. Selective breeding for disease-resistant *PRNP* variants to manage chronic wasting disease in farmed whitetail deer. Genes 12:e1396.

Haley, N. J., D. M. Henderson, R. Donner, S. Wyckoff, K. Merrett, J Tennant, E. A. Hoover, D. Love, E. Kline, A. D. Lehmkuhl, et al. 2020*b*. Management of chronic wasting disease in ranched elk: conclusions from a longitudinal three-year study. Prion 14:76–87.

Haley, N. J., and J. A. Richt. 2017. Evolution of diagnostic tests for chronic wasting disease, a naturally occurring prion disease of cervids. Pathogens 6:e35.

Hannaoui, S., E. Triscott, C. Duque Velásquez, S. C. Chang, M. I. Arifin, I. Zemlyankina, X. Tang, T. Bollinger, H. Wille, D. McKenzie, et al. 2021. New and distinct chronic wasting disease strains associated with cervid polymorphism at codon 116 of the Prnp gene. PLoS Pathogens 17:e1009795.

Harrington, R. D., T. V. Baszler, K. I. O'Rourke, D. A. Schneider, T. R. Spraker, H. D. Liggitt, and D. P. Knowles. 2008. A species barrier limits transmission of chronic wasting disease to mink (*Mustela vison*). Journal of General Virology 89:1086–1096.

Heisey, D. M., E. E. Osnas, P. C. Cross, D. O. Joly, J. A. Langenberg, and M. W. Miller. 2010. Linking process to pattern: estimating spatiotemporal dynamics of a wildlife epidemic from cross-sectional data. Ecological Monographs 80:221–240.

Holsman, R. H., J. Petchenik, and E. E. Cooney. 2010. CWD after "the fire": Six reasons why hunters resisted Wisconsin's eradication effort. Human Dimensions of Wildlife 15:180–193.

Hörnlimann, B., D. Riesner, and H. Kretzschmar, editors. 2007. Prions in humans and animals. Walter de Gruyter GmbH & Co. Berlin, Germany.

Houston, F., and O. Andréoletti. 2018. The zoonotic potential of animal prion diseases. *In* M. Pocchiari and J. Manson, editors. Human prion diseases. Handbook of Clinical Neurology 153:447–462.

Hughson, A. G., B. Race, A. Kraus, L. R. Sangaré, L. Robins, B. R. Groveman, E. Saijo1, K. Phillips, L. Contreras, V. Dhaliwal, et al. 2016. Inactivation of prions and amyloid seeds with hypochlorous acid. PLoS Pathogens 12:e1005914.

Jewell, J. E., M. M. Conner, L. L. Wolfe, M. W. Miller, and E. S. Williams. 2005. Low frequency of PrP genotype 225SF among free-ranging mule deer (*Odocoileus hemionus*) with chronic wasting disease. Journal of General Virology 86:2127–2134.

Johnson, C. J., A. Herbst, C. Duque-Velasquez, J. P. Vanderloo, P. Bochsler, R. Chappell, and D. McKenzie. 2011. Prion protein polymorphisms affect chronic wasting disease progression. PLoS One 6:e17450.

Kahn, S., C. Dubé, L. Bates, and A. Balachandran. 2004. Chronic wasting disease in Canada: part 1. Canadian Veterinary Journal 45:397–404.

Kelly, A. C., N. E. Mateus-Pinilla, W. Brown, M. O. Ruiz, M.R. Douglas, M. E. Douglas, P. Shelton, T. Beissel, and J. Novakofski. 2014. Genetic assessment of environmental features that influence deer dispersal: implications for prion-infected populations. Population Ecology 56:327–340.

Krumm, C. E., M. M. Conner, N. T. Hobbs, D. O. Hunter, and M. W. Miller. 2010. Mountain lions prey selectively on prion-infected mule deer. Biology Letters 6:209–211.

LaCava, M. E. F., J. L. Malmberg, W. H. Edwards, L. N. L. Johnson, S. E. Allen, and H. B. Ernest. 2021. Spatio-temporal analyses reveal infectious disease-driven selection in a free-ranging ungulate. Royal Society Open Science 8:e210802.

Manjerovic, M. B., M. L. Green, N. Mateus-Pinilla, and J. Novakofski. 2014. The importance of localized culling in stabilizing chronic wasting disease prevalence in white-tailed deer populations. Preventive Veterinary Medicine 113:139–145.

Mathiason, C. K. 2017. Scrapie, CWD, and transmissible mink encephalopathy. Progress in Molecular Biology and Translational Science 150:267–292.

Meeks, A., N. C. Poudyal, L. I. Muller, and C. Yoest. 2022. Hunter acceptability of chronic wasting disease (CWD) management actions in Western Tennessee. Human Dimensions of Wildlife 27:457–471.

Mejía-Salazar, M. F., C. L. Waldner, Y. T. Hwang, and T. K. Bollinger. 2018. Use of environmental sites by mule deer: a proxy for relative risk of chronic wasting disease exposure and transmission. Ecosphere 9:e02055.

Miller, M. W., and J. R. Fischer. 2016. The first five (or more) decades of chronic wasting disease: lessons for the five decades to come. Transactions of the North American Wildlife and Natural Resources Conference 81:110–120.

Miller, M. W., J. P. Runge, A. A. Holland, and M. D. Eckert. 2020. Hunting pressure modulates prion risk in mule deer herds. Journal of Wildlife Diseases 56:781–790.

Miller, M. W., H. M. Swanson, L. L. Wolfe, F. G. Quartarone, S. L. Huwer, C. H. Southwick, and P. M. Lukacs. 2008. Lions and prions and deer demise. PLoS One 3:e4019.

Miller, M. W., M. A. Wild, and E. S. Williams. 1998. Epidemiology of chronic wasting disease in captive Rocky Mountain elk. Journal of Wildlife Diseases 34:532–538.

Miller, M. W., and E. S. Williams. 2003. Horizontal prion transmission in mule deer. Nature 425:35–36.

Miller, M. W., E. S. Williams, N. T. Hobbs, and L. L. Wolfe. 2004. Environmental sources of prion transmission in mule deer. Emerging Infectious Diseases 10:1003–1006.

Miller, M. W., and L. L. Wolfe. 2021. Inferring chronic wasting disease incidence from prevalence data. Journal of Wildlife Diseases 57:718–721.

Monello, R. J., N. L. Galloway, J. G. Powers, S. A. Madsen-Bouterse, W. H. Edwards, M. E. Wood, K. I. O'Rourke, and M. A. Wild. 2017. Pathogen-mediated selection in free-ranging elk populations infected by chronic wasting disease. Proceedings of the National Academy of Sciences USA 114:12208–12212.

Monello, R. J., J. G. Powers, N. T. Hobbs, T. R. Spraker, M. K. Watry, and M. A. Wild. 2014. Survival and population growth of a free-ranging elk population with a long history of exposure to chronic wasting disease. Journal of Wildlife Management 78:214–223.

Mysterud, A., S. L. Benestad, C. M. Rolandsen, and J. Våge. 2021a. Policy implications of an expanded chronic wasting disease universe. Journal of Applied Ecology 58:281–285.

Mysterud A, K. Madslien, H. Viljugrein, T. Vikøren, R. Andersen, M. E. Güere, S. L. Benestad, P. Hopp, O. Strand, B. Ytrehus, et al. 2019*a*. The demographic pattern of infection with chronic wasting disease in reindeer at an early epidemic stage. Ecosphere 10:e02931.

Mysterud, A., O. Strand, and C. M. Rolandsen. 2019*b*. Efficacy of recreational hunters and marksmen for host culling to combat chronic wasting disease in reindeer. Wildlife Society Bulletin 43:683–692.

Mysterud, A., H. Viljugrein, C. M. Rolandsen, and A. V. Belsare. 2021*b*. Harvest strategies for the elimination of low prevalence wildlife diseases. Royal Society Open Science 8:e210124.

Mysterud, A., H. Viljugrein, E. J. Solberg, and C. M. Rolandsen. 2019c. Legal regulation of supplementary cervid feeding facing chronic wasting disease. Journal of Wildlife Management 83:1667–1675.

Nichols, T. A., J. W. Fischer, T. R. Spraker, Q. Kong, and K. C. VerCauteren. 2015. CWD prions remain infectious after passage through the digestive system of coyotes (*Canis latrans*). Prion 9:367–375.

Nonno, R., M. A. Di Bari, L. Pirisinu, C. D'Agostino, I. Vanni, B. Chiappini, S. Marcon, G. Riccardi, L. Tran, T. Vikøren, et al. 2020. Studies in bank voles reveal strain differences between chronic wasting disease prions from Norway and North America. Proceedings of the National Academy of Sciences USA 117:31417–31426.

Norbert, B. R., E. H. Merrill, M. J. Pybus, T. K. Bollinger, and Y. T. Hwang. 2016. Landscape connectivity predicts chronic wasting disease risk in Canada. Journal of Applied Ecology 53:1450–1459.

O'Rourke K. I., T. R. Spraker, D. Zhuang, J. J. Greenlee, T. E. Gidlewski, and A. N. Hamir. 2007. Elk with a long incubation prion disease phenotype have a unique PrP^{d} profile. NeuroReport 18:1935–1938.

Pattison-Williams, J. K., L. Xie, W. L. Adamowicz, M. Pybus, and A. Hubbs. 2020. An empirical analysis of hunter response to chronic wasting disease in Alberta. Human Dimensions of Wildlife 25:575–589.

Pilon, J. L., J. C. Rhyan, L. L. Wolfe, T. R. Davis, M. P. McCollum, K. I. O'Rourke, T. R. Spraker, K. C. VerCauteren, M. W. Miller, T. Gidlewski, et al. 2013. Immunization with a synthetic peptide vaccine fails to protect mule deer (*Odocoileus hemionus*) from chronic wasting disease. Journal of Wildlife Diseases 49:694–698.

Pirisinu, L., L. Tran, B. Chiappini, I. Vanni, M. A. Di Bari, G. Vaccari, T. Vikøren, K. Madslien, J. Våge, T. Spraker, et al. 2018. A novel type of chronic wasting disease detected in European moose (*Alces alces*) in Norway. Emerging Infectious Diseases 24:2210–2218.

Plummer, I. H., C. J. Johnson, A. R. Chesney, J. A. Pedersen, and M. D. Samuel. 2018. Mineral licks as environmental reservoirs of chronic wasting disease prions. PLoS One 13:e0196745.

Plummer, I. H., S. D. Wright, C. J. Johnson, J. A. Pedersen, and M. D. Samuel. 2017. Temporal patterns of chronic wasting disease prion excretion in three cervid species. Journal of General Virology 98:1932–1942.

Potapov, A., E. Merrill, M. Pybus, and M.A. Lewis. 2016. Chronic wasting disease: transmission mechanisms and the possibility of harvest management. PLoS ONE 11:e0151039.

Pritzkow, S., D. Gorski, F. Ramirez, G. C. Telling, S. L. Benestad, and C. Soto. 2022. North American and Norwegian chronic wasting disease prions exhibit different potential for interspecies transmission and zoonotic risk. The Journal of Infectious Diseases 225:542–551.

Prusiner, S. B. 1982. Novel proteinaceous infectious particles cause scrapie. Science 216:136–144.

Prusiner, S. B. 1995. The prion diseases. Scientific American 272:48–51,54–57.

Prusiner, S. B. 1998. Prions. Proceedings of the National Academy of Sciences USA 95:13363–13383.

Race, R. E., A. Raines, T.G.M. Baron, M.W. Miller, A. Jenny, and E.S. Williams. 2002. Comparison of abnormal prion protein glycoform patterns from transmissible spongiform encephalopathy agent-infected deer, elk, sheep, and cattle. Journal of Virology 76:12365–12368.

Rhyan, J. C., M. W. Miller, T. R. Spraker, M. McCollum, P. Nol, L. L. Wolfe, T. R. Davis, L. Creekmore, and K. I. O'Rourke. 2011. Failure of fallow deer (*Dama dama*) to develop chronic wasting disease when exposed to a contaminated environment and infected mule deer (*Odocoileus hemionus*). Journal of Wildlife Diseases 47:739–744.

Rivera, N. A., A. L. Brandt, J. E. Novakofski, and N. E. Mateus-Pinilla. 2019. Chronic wasting disease in cervids: prevalence, impact and management strategies. Veterinary Medicine: Research and Reports 10:123–139.

Robinson, A.L., H. Williamson, M. E. Güere, H. Tharaldsen, K. Baker, S. L. Smith, S. Pérez-Espona, J. Krojerová-Prokešová, J. M. Pemberton, W. Goldmann, et al. 2019. Variation in the prion protein gene (*PRNP*) sequence of wild deer in Great Britain and mainland Europe. Veterinary Research 50:e59.

Robinson, S. J., M. D. Samuel, C. J. Johnson, M. Adams, and D. I. McKenzie. 2012*a*. Emerging prion disease drives host selection in a wildlife population. Ecological Applications 22:1050–1059.

Robinson, S. J., M. D. Samuel, K. I. O'Rourke, and C. J. Johnson. 2012*b*. The role of genetics in chronic wasting disease of North American cervids. Prion 6: 153–162.

Robinson, S. J., M. D. Samuel, R. E. Rolley, and P. Shelton. 2013. Using landscape epidemiological models to understand the distribution of chronic wasting disease in the Midwestern USA. Landscape Ecology 28:1923–1935.

Samuel, M. D., and D. J. Storm. 2016. Chronic wasting disease in white-tailed deer: infection, mortality, and implications for heterogeneous transmission. Ecology 97:3195–3205.

Schwabenlander, M. D., M. R. Culhane, S. M. Hall, S. M. Goyal, P. L. Anderson, M. Carstensen, S. J. Wells, W. B. Slade, and A. G. Armién. 2013. A case of chronic wasting disease in a captive red deer (*Cervus elaphus*). Journal of Veterinary Diagnostic Investigation 25:573–576.

Seabury, C. M., D. L. Oldeschulte, E. K. Bhattarai, D. Legare, P. J. Ferro, R. P. Metz, C. D. Johnson, M. A. Lockwood, and T. A. Nichols. 2020. Accurate genomic predictions for chronic wasting disease in US white-tailed deer. G3 Genes|Genomes|Genetics 10:1433–1441.

Smolko, P., D. Seidel, M. Pybus, A. Hubbs, M. Ball, and E. Merrill. 2021. Spatio-temporal changes in chronic wasting disease risk in wild deer during 14 years of surveillance in Alberta, Canada. Preventive Veterinary Medicine 197:e105512.

Sohn, H.-J., J.-H. Kim, K.-S. Choi, J.-J. Nah, Y.-S. Joo, Y.-H. Jean, S.-W. Ahn, O.-K. Kim, D.-Y. Kim, and A. Balachandran. 2002. A case of chronic wasting disease in an elk imported to Korea from Canada. Journal of Veterinary Medical Science. 64:855–858.

Sohn, H.-J., G. Mitchell, Y. H. Lee, H. J. Kim, K.-J. Park, A. Staskevicus, I. Walther, A. Soutyrine, and A. Balachandran. 2020. Experimental oral transmission of chronic wasting disease to sika deer (*Cervus nippon*). Prion 14:271–277.

Sohn, H.-J., K.-J. Park, I.-S. Roh, H.-J. Kim, H.-C. Park, and H.-E. Kang. 2019. Sodium hydroxide treatment effectively inhibits PrP^{CWD} replication in farm soil. Prion 13:137–140.

Stewart, P., L. Campbell, S. Skogtvedt, K.A. Griffin, J. M. Arnemo, M. Tryland, S. Girling, M. W. Miller, M. A. Tranulis, and W. Goldmann. 2012. Genetic predictions of prion disease susceptibility in carnivore species based on variability of the prion gene coding region. PLoS One 7:e50623.

Spraker, T., A. Balachandran, D. Zhuang, and K. I. O'Rourke. 2004. Variable patterns of distribution of PrP^{CWD} in the obex and cranial lymphoid tissues of Rocky Mountain elk (*Cervus elaphus nelsoni*) with subclinical chronic wasting disease. Veterinary Record 155:295–302.

Spraker, T. R., K. C. VerCauteren, T. Gidlewski, D. A. Schneider, R. Munger, A. Balachandran, and K. I. O'Rourke. 2009. Antemortem detection of PrP^{CWD} in preclinical, ranch-raised Rocky Mountain elk (*Cervus elaphus nelsoni*) by biopsy of the rectal mucosa. Journal of Veterinary Diagnostic Investigation 21:15–24.

Tamgüney, G., K. Giles, E. Bouzamondo-Bernstein, P. J. Bosque, M. W. Miller, J. Safar, S. J. DeArmond, and S. B. Prusiner. 2006. Transmission of elk and deer prions to transgenic mice. Journal of Virology 80:9104–9114.

Tamgüney, G., M. W. Miller, L. L. Wolfe, T. M. Sirochman, D. V. Glidden, C. Palmer, A. Lemus, S. J. DeArmond, and S. B. Prusiner. 2009. Asymptomatic deer excrete infectious prions in faeces. Nature 461:529–532.

Thomsen, B. V., D. A. Schneider, K. I. O'Rourke, T. Gidlewski, J. McLane, R. W. Allen, A. A. McIsaac, G. B. Mitchell, D. P. Keane, T. R. Spraker, et al. 2012. Diagnostic accuracy of rectal mucosa biopsy testing for chronic wasting disease within white-tailed deer (*Odocoileus virginianus*) herds in North America: effects of age, sex, polymorphism at PRNP codon 96, and disease progression. Journal of Veterinary Diagnostic Investigation 24:878–887.

Tranulis, M. A., D. Gavier-Widén, J. Våge, M. Nöremark, S.-L. Korpenfelt, M. Hautaniemi, L. Pirisinu, R. Nonno, and S. L. Benestad. 2021. Chronic wasting disease in Europe: new strains on the horizon. Acta Veterinaria Scandinavica 63:e48.

Uehlinger, F. D., A. C. Johnston, T. K. Bollinger, and C. L. Waldner. 2016. Systematic review of management strategies to control chronic wasting disease in wild deer populations in North America. BMC Veterinary Research 12:1–16.

United States Geological Survey. 2021. Distribution of chronic wasting disease in North America. https://www.usgs.gov/media/images/distribution-chronic-wasting-disease-north-america-0. Accessed 10 Dec 2021.

Vikøren, T., J. Våge, K. I. Madslien, K. H. Roed, C. M. Rolandsen, L. Tran, P. Hopp, V. Veiberg, M. Heum, T. Moldal, et al. 2019. First detection of chronic wasting disease in a wild red deer (*Cervus elaphus*) in Europe. Journal of Wildlife Diseases 55:970–972.

Waddell, L., J. Greig, M. Mascarenhas, A. Otten, T. Corrin, and K. Hierlihy. 2018. Current evidence on the transmissibility of chronic wasting disease prions to human—a systematic review. Transboundary and Emerging Diseases 65:37–49.

Walther, I., A. Staskevicius, A. Soutyrine, and G. Mitchell. 2019. Diagnosis of CWD in a herd of farmed red deer. Prion 13:S59–S60.

Wasserberg, G., E. E. Osnas, R. E. Rolley, and M. D. Samuel. 2009. Host culling as an adaptive management tool for chronic wasting disease in white-tailed deer: a modeling study. Journal of Applied Ecology 46:457–466.

Wild, M. A., N. T. Hobbs, M. S. Graham, and M. W. Miller. 2011. The role of predation in disease control: a comparison of selective and nonselective removal on prion disease dynamics in deer. Journal of Wildlife Diseases 47:78–93.

Williams, A. L., T. J. Kreeger, and B. A. Schumaker. 2014. Chronic wasting disease model of genetic selection favoring prolonged survival in Rocky Mountain elk (*Cervus elaphus*). Ecosphere 5:e60.

Williams, E. S. 2005. Chronic wasting disease. Veterinary Pathology 42:530–549.

Williams, E. S., and M. W. Miller. 2002. Chronic wasting disease in deer and elk in North America. Revue Scientifique et Technique Office International des Epizooties 21:305–316.

Williams, E. S., and M. W. Miller. 2003. Transmissible spongiform encephalopathies in non-domestic animals: origin, transmission and risk factors. Revue Scientifique et Technique Office International des Epizooties 22:145–156.

Williams, E. S., M. W. Miller, and E. T. Thorne. 2002. Chronic wasting disease: implications and challenges for wildlife managers. Transactions of the North American Wildlife and Natural Resources Conference 67:87–103.

Williams, E. S., D. O'Toole, M. W. Miller, T. J. Kreeger, and J. E. Jewell. 2018. Cattle (*Bos taurus*) resist chronic wasting disease following oral inoculation challenge or ten years' natural exposure in contaminated environments. Journal of Wildlife Diseases 54:460–470.

Williams, E. S., and S. Young. 1980. Chronic wasting disease of captive mule deer: a spongiform encephalopathy. Journal of Wildlife Diseases 16:89–98.

Williams, E. S., and S. Young. 1982. Spongiform encephalopathy of Rocky Mountain elk. Journal of Wildlife Diseases 18:465–471.

Williams, E. S., and S. Young. 1992. Spongiform encephalopathies of Cervidae. Revue scientifique et technique Office international des Epizooties 11:551–567.

Williams, K., A. G. Hughson, B. Chesebro, and B. Race. 2019. Inactivation of chronic wasting disease prions using sodium hypochlorite. PLoS One 14:e0223659.

Wolfe, L. L, M. M. Conner, and M. W. Miller. 2020. Effect of oral copper supplementation on susceptibility in white-tailed deer (*Odocoileus virginianus*) to chronic wasting disease. Journal of Wildlife Diseases 56:568–575.

Wolfe, L. L., K. A. Fox, K. A. Griffin, and M. W. Miller. 2022. Mountain lions resist long-term dietary exposure to chronic wasting disease. Journal of Wildlife Diseases 58:40–49.

Wolfe, L. L., K. A. Fox, and M. W. Miller. 2014. "Atypical" chronic wasting disease in *PRNP* genotype 225FF mule deer. Journal of Wildlife Diseases 50:660–665.

Wolfe, L. L., D. A. Kocisko, B. Caughey, and M. W. Miller. 2012. Assessment of prospective preventive therapies for chronic wasting disease in mule deer. Journal of Wildlife Diseases 48:530–533.

Wolfe, L. L., T. R. Spraker, L. González, M. P. Dagleish, T. M. Sirochman, J. C. Brown, M. Jeffrey, and M. W. Miller. 2007. PrP^{CWD} in rectal lymphoid tissue of deer (*Odocoileus* spp.). Journal of General Virology 88:2078–2082.

Wood, M. E., P. Griebel, M. L. Huizenga, S. Lockwood, C. Hansen, A. Potter, N. Cashman, J. W. Mapletoft, and S. Napper. 2018. Accelerated onset of chronic wasting disease in elk (*Cervus canadensis*) vaccinated with a PrP^{Sc}-specific vaccine and housed in a prion contaminated environment. Vaccine 36:7737–7743.

Zabel, M., and A. Ortega. 2017. The ecology of prions. Microbiology and Molecular Biology Reviews 81:e00001-17.

9 Epizootic Bighorn Sheep Pneumonia Caused by *Mycoplasma ovipneumoniae*

THOMAS E. BESSER, E. FRANCES CASSIRER

Introduction

Bighorn sheep (*Ovis canadensis*), one of two wild sheep species native to North America, once numbered a million or more animals. Precipitous declines followed European settlement, and the species' future was threatened by the early 20th century (Buechner 1960). Despite recovery efforts, many populations remain small and isolated, and in select areas bighorn sheep still receive special protection under the United States Endangered Species Act (ECOS 1998, 1999). The causes of bighorn sheep declines and extirpations in the 19th century were poorly documented but included unregulated market hunting, competition for forage with livestock, and from the earliest published descriptions, outbreaks of infectious diseases. The most frequently mentioned disease prior to the mid-20th century was scabies, perhaps due to its visually dramatic nature. However, investigators also reported pneumonia as the cause of relatively early die-offs, and pneumonia outbreaks are recognized as a major factor limiting bighorn sheep numbers today (Box 9.1) (Brooks 1923, Grinnell 1928, Cassirer et al. 2018).

Bighorn sheep are highly social ungulates native to xeric habitats in rugged desert, canyon, and mountainous terrain in western North America from the

Illustration by Laura Donohue.

Box 9.1. Witnessing a catastrophic mortality event

Something odd was happening at the National Bison Range (NBR) at the end of July 2016: the 80-plus bighorn ewes and their lambs, which normally formed a large cohesive nursery group (see Figure 9.1) lasting through the summer, had abruptly fragmented into a number of dispersed groups by 1 August. Later in August, an unusual number of coughing ewes were seen, some of which appeared to be losing weight at a time when they're normally regaining condition lost during lactation. So, it was not a complete surprise when the first dead sheep were found in mid-September. By the time the pneumonia epizootic ended, 85% of the 223 bighorn sheep in the NBR population had died. All lambs born in 2017 and 2018 also died before reaching 3 months of age, many showing signs of pneumonia prior to their death. The start of this outbreak was preceded by repeated observations of young bighorn sheep from the herd on short-duration, off-refuge exploratory movements toward areas with domestic sheep operations, one of which was subsequently found to contain multiple ewes carrying the outbreak-associated genetic strain of the causative pathogen, *Mycoplasma ovipneumoniae*. This severe outbreak highlights the two key phases of epizootic pneumonia of bighorn sheep: an initial outbreak resulting in deaths of animals of all age classes, followed in subsequent years by repeated outbreaks largely limited to lambs of the year. Many similar epizootics have been reported over the past century, often following observed or suspected contacts with domestic sheep (*Ovis aries*) or goats (*Capra hircus*).

Besser et al. 2021, Cassirer et al. 2018

Canadian Rockies to northwestern Mexico and east to the badlands of the Dakotas and Nebraska. Although both sexes have horns, adults are sexually dimorphic and only males have curled "big horns" (see image on page 145). Populations or subpopulations are usually defined by space use of female groups, which may be interconnected by movements of males, particularly during the breeding season. Except for the period around breeding, which varies seasonally across latitudes, females and adult males use different and largely separate habitats. Females usually give birth to one lamb and, to evade predators, often raise their offspring communally in nursery groups (Figure 9.1), a social structure that promotes spread of infectious agents (Nunn et al. 2015). Individuals are relatively long lived, with documented lifespans up to 18 years for ewes and 13 years for rams, allowing for long duration and persistent infections that may shape the epidemiology of respiratory disease (Jorgenson et al. 1997).

What Causes Epizootic Bighorn Sheep Pneumonia?

An accurate understanding of the cause of a disease is important because it can lead to more effective prevention and control measures (Asokan and Asokan 2016). The causal agent of epizootic bighorn sheep pneumonia remained elusive for many years; however, recent evidence identifies the bacterium *Mycoplasma ovipneumoniae* as the primary spillover pathogen (defined below) (Besser et al. 2013, Cassirer et al. 2018). The bacterium was first linked to pneumonia in a wild sheep species in a case report of a pneumonia epizootic in Dall's sheep (*Ovis dalli*) at the Toronto Zoo (Black et al. 1988). Subsequently, *M. ovipneumoniae* was detected by bacteriologic cultures from bighorn sheep, a technically difficult procedure with poor sensitivity for this agent (Rudolph et al. 2007). The larger role of *M. ovipneumoniae* in epizootic pneumonia was only recognized after molecular diagnostic methods that do not require bacteriologic culture, first 16S metagenomics and then

Fig. 9.1. Part of the bighorn sheep (*Ovis canadensis*) nursery group at the National Bison Range prior to the 2016 pneumonia epizootic, as an example of the lamb group size and interaction that can promote pathogen transmission. Photograph by J. T. Hogg.

Box 9.2. Evidence that *Mycoplasma ovipneumoniae* causes epizootic bighorn sheep pneumonia

1. Strong statistical association with the disease: *M. ovipneumoniae* is detected in the vast majority of bighorn sheep pneumonia epizootics but not in most healthy bighorn populations.
2. One strain per epizootic: Single *M. ovipneumoniae* genetic strains are detected within epizootics, and the initial detection of the epizootic strain is temporally linked to the epizootic onset.
3. First pathogen to invade the lung tissues: *M. ovipneumoniae* predominates in lung tissues very early in the disease course within an individual bighorn sheep. Infections with multiple other bacterial species follow when *M. ovipneumoniae* has disabled lung defences.
4. Chronic carriers and lamb pneumonia: At the end of an all-ages epizootic, some surviving ewes become chronic carriers of *M. ovipneumoniae*, serving as the source of infection for lambs born in subsequent years. Removal or stochastic loss of the last chronic carrier ewe correlates with cessation of lamb pneumonia epizootics.
5. Accurate experimental replication: Key features of epizootic pneumonia are reproduced by inoculation with *M. ovipneumoniae* alone, including epizootic transmission within groups, a prolonged disease course, a single epizootic strain type, and polymicrobial pneumonia with lesions attributable to both *Mycoplasma* and secondary bacterial infections.

Niang et al. 1998; Nicholas et al. 2008; Besser et al. 2008, 2012*b*, 2013, 2021; USDA APHIS 2015; Cassirer et al. 2018; Weyand et al. 2018; Kamath et al. 2019; Felts 2020; Garwood et al. 2020; Spaan et al. 2021.

specific polymerase chain reaction (PCR) tests, were applied to investigation of the disease in free-living bighorn sheep (Besser et al. 2008, Besser et al. 2013).

Box 9.2 summarizes the lines of evidence supporting *M. ovipneumoniae* as the primary infectious agent initiating epizootic bighorn sheep pneumonia. No previous or current competing theory for the cause of bighorn sheep pneumonia is supported by any of these points. *Mycoplasma ovipneumoniae* induces pneumonia indirectly by impairing normal pulmonary defense mechanisms, resulting in polymicrobial infections with predominantly obligately anaerobic, opportunistic pathogens (Besser et al. 2012*b*, Besser et al. 2013). Further research is needed to determine the degree to which the species composition of the secondary bacterial infections drives variation in disease severity, the speed of progression of the disease within individual bighorn sheep, and the overall mortality rate of the epizootic event.

Other Causes of Bighorn Sheep Respiratory Disease

Pasteurella haemolytica, later reclassified as *Mannheimia haemolytica*, was long considered the cause of epizootic bighorn sheep pneumonia, based initially on its frequent isolation from the lung tissues of pneumonic bighorn sheep and its recognition as the most important respiratory pathogen of domestic cattle and sheep (Miller 2001, CAST 2008). The *M. haemolytica* theory was also supported by the ability of leukotoxic strains to rapidly kill bighorn sheep after experimental challenge (Foreyt et al. 1994) and by the demonstration that it was transmitted from domestic sheep or goats to bighorn sheep in direct contact (Rudolph et al. 2003, Lawrence et al. 2010). Despite the plausibility of the *M. haemolytica* hypothesis and many years of investigation, no evidence directly supportive of a primary causal role for this bacterium in epizootic bighorn sheep pneumonia has been reported. For example, *M. haemolytica* has not been statistically associated with the disease nor have epizootic-associated strain types been identified. Finally, the pneumonia that followed *M. haemolytica* challenge of bighorn sheep differs from the naturally occurring disease in its peracute time course and the microbiology and pathology of the pneumonic lungs (Foreyt et al. 1994, Besser et al. 2014). Similarly, *P. multocida* and *Bibersteinia trehalosi*, also frequently isolated from the lungs of pneumonic bighorn sheep, lack evidence supporting a primary causal role and so likely represent secondary infections.

Fatal pneumonia that was plausibly attributable primarily to *M. haemolytica* was reported in one of three captive bighorn sheep after prolonged (90 days) commingling with *M. ovipneumoniae*-free domestic sheep; a yearling bighorn ram died within 12 hours of the onset of observed illness. At necropsy, very high numbers of *M. haemolytica* were isolated from its lungs and histologic lesions characteristic of those induced by *M. haemolytica* were observed (Besser et al. 2012a). Other instances of *M. haemolytica*-associated pneumonia in captive bighorn sheep experimentally commingled with *M. ovipneumoniae*-free domestic sheep have subsequently been observed (Thomas Besser, Washington State University, unpublished data). Similar disease events occurring in free-ranging individuals or herds could potentially be distinguished from *M. ovipneumoniae*-associated epizootics by absence of *M. ovipneumoniae* (or absence of a newly introduced *M. ovipneumoniae* strain type), a peracute disease course, and absence of polymicrobial pneumonia (Foreyt et al. 1994, Bunch et al. 1999, CAST 2008, Dassanayake et al. 2010).

Several other infectious agents are known to cause respiratory diseases in bighorn sheep. Lungworms (*Protostrongylus stilesi*, *P. rushii*, and *Muellerius capillaris*) have been detected in bighorn sheep in moister Rocky Mountain habitats, where the snail intermediate host can survive, and once were considered an important contributor to epizootic respiratory disease (Forrester 1971, Hibler et al. 1982, Jenkins et al. 2007). Other studies, however, suggest that the widespread presence of lungworms in healthy bighorn sheep populations occupying suitable habitat (Hibler et al. 1982), the lack of observed beneficial effects of anthelminthic-based reductions in lungworm prevalence (Miller et al. 2000), and the inability of experimental challenge with lungworms to reproduce epizootic bighorn pneumonia (Samson et al. 1987), indicate that lungworms—while undoubtedly resulting in some burden of verminous pneumonia—are not associated with larger scale epizootics in bighorn lambs or adults (Besser et al. 2013). Respiratory viruses, including parainfluenza-3 virus (PI3), respiratory syncytial virus (RSV), and bovine viral diarrhea virus, are known to infect bighorn sheep (Parks et al. 1972, Parks and England 1974, Dunbar et al. 1985, Spraker et al. 1986, Elliott et al. 1994, Cassirer et al. 1996, Rudolph et al. 2007, Wolff et al. 2016*b*). Challenge of captive bighorn sheep with PI3 and RSV induced infection and seroconversion, clinical signs of anorexia and lethargy, and histologic evidence of pneumonia (Dassanayake et al. 2013). In the wild, repeated observations of coughing bighorn sheep at Lookout Mountain, Oregon, triggered euthanasia of two animals and capture and sampling of 24 additional animals; *M. ovipneumoniae* was not detected, but a lamb euthanized for sampling had bronchopneumonia with *P. multocida* infection and a positive RSV antigen test. It was concluded that this population experienced an epizootic of nonfatal respiratory disease, most likely associated with RSV infections (Julia Burco, Oregon Department of Fish and Wildlife, personal communication).

Bighorn Sheep Pneumonia as a Spillover Disease

Early 20th century observers had noted that bighorn sheep pneumonia epizootics frequently followed contact with livestock, especially domestic sheep (Warren 1910, Brooks 1923, Grinnell 1928, Shillinger 1937, CAST 2008). As early as 1937, the disease risk to bighorn sheep in direct contact with domestic sheep was confirmed experimentally when healthy captive bighorn sheep "gradually died off" after introduction of healthy domestic sheep (Shillinger 1937). Many similar experiments followed with the same fatal outcome for bighorn sheep (see reviews in Wehausen et al. 2011, Besser et al. 2012a). As a result, epizootic bighorn sheep pneumonia must be considered a spillover disease, that is, a disease resulting when a pathogen carried by a reservoir host (domestic sheep or goats for *M. ovipneumoniae*) infects a novel host (bighorn sheep). Spillover disease generally requires that the virulence mechanisms of the pathogen can harm the novel host, and spillover diseases become epizootic only if the pathogen is capable of persisting, amplifying, and being efficiently transmitted within the novel host species; all these processes are more likely in closely related hosts such as domestic and bighorn sheep (Alexander et al. 2018). Although they diverged from a common ancestor millions of years ago, bighorn and domestic sheep are closely related, can hybridize, and can produce viable and fertile offspring (Young and Manville 1960). Transmission of *M. ovipneumoniae* from domestic sheep or goats to bighorn sheep may require direct contact, facilitated by the social compatibility or attraction these species exhibit (Figure 9.2).

Mycoplasma ovipneumoniae is found at high prevalence within and among domestic sheep and goat populations (USDA APHIS 2015, Heinse et al. 2016), the same host species previously reported to be associated anecdotally and experimentally with pneumonia epizootics in bighorn sheep. The bacterium generally causes little overt disease in domestic small ruminants, although it has been associated with chronic nonprogressive pneumonia in domestic lambs, and epizootic pneumonia in domestic goats (Alley et al. 1999, Rifatbegovic et al. 2011). *Mycoplasma ovipneumoniae* is readily transmitted by direct contact, which is enabled by the high degree of social acceptance between bighorn sheep and domestic sheep and goats (Smith 1954, Young and Manville 1960). Therefore, *M. ovipneumoniae* reservoir hosts in and near bighorn ranges—such as domestic sheep grazed on public rangelands, goats used for weed control, and small ruminant operations on private lands within and adjacent to bighorn ranges—

Fig. 9.2. Bighorn sheep (*Ovis canadensis*) and domestic sheep (*Ovis aries*) and goats (*Capra hircus*) are highly social and freely interact, especially in the absence of others of their own species. Photograph by E. F. Cassirer.

provide diverse opportunities for spillover infection of bighorn sheep (Figure 9.3).

Cases of *M. ovipneumoniae* spillover to other Caprinae native to North America, including Dall's sheep, mountain goats (*Oreamnos americanus*), and muskoxen (*Ovibos moschatus*) have each been associated with respiratory disease epizootics (Black et al. 1988, Handeland et al. 2014, Wolff et al. 2019). In addition, cross-species transmission of *M. ovipneumoniae* from mountain goats was reported following translocation of naïve bighorn sheep into adjacent range (Wolff et al. 2016*a*).

Domestic goats carry strains of *M. ovipneumoniae* that are genetically distinct from those carried by domestic sheep (Maksimovic et al. 2016, Kamath et al. 2019). Among *M. ovipneumoniae* strains detected in bighorn sheep, domestic sheep strains of *M. ovipneumoniae* far outnumber domestic goat strains (85 versus 3 strain types, respectively, in Kamath et al. 2019). In experimental studies involving a small number of bighorn sheep, challenge with domestic goat clade *M. ovipneumoniae* induced clinical signs of respiratory disease but no fatalities, in contrast to the high mortality of bighorn sheep induced by domestic sheep clade strains in similar studies (Besser et al. 2012*a*, 2014, 2017). Fatal pneumonia in wild bighorn sheep due to spillover of a domestic goat *M. ovipneumoniae* strain has been observed however (Cassirer et al. 2017). Therefore, the relative virulence of domestic goat and sheep *M. ovipneumoniae* clades for bighorn sheep remains to be determined: more research is needed on the propensity of goat clade strains to spill over and infect bighorn sheep, the disease severity, and the persistence of infection in bighorn sheep following such spillovers.

The strain diversity of *M. ovipneumoniae* is sufficiently high that detecting the bighorn sheep epizootic strain type in a nearby herd or flock of domestic goats or sheep identifies it as the presumptive source of the spillover (Besser et al. 2021). More research is needed, however, to clarify the degree of local and regional dissemination of strain types among domestic sheep flocks to determine the strength of inference that can be based upon a detection of the

Fig. 9.3. The risk landscape for *Mycoplasma ovipneumoniae* spillover to bighorn sheep (*Ovis canadensis*). Illustration by Laura Donohue.

epizootic strain within a reservoir host herd or flock adjacent to the affected bighorn range (Kamath et al. 2019, Besser et al. 2021).

The host specificity of *M. ovipneumoniae* for Caprinae is relative, not absolute. Detection of *M. ovipneumoniae* in small numbers of cattle (*Bos taurus*), caribou (*Rangifer tarandus*), mule and white-tailed deer (*Odocoileus hemionus* and *O. virginianus*, respectively) has been reported (Wolfe et al. 2010, Highland et al. 2018, Lieske et al. 2022). A single report of *M. ovipneumoniae* in cattle was thought to be associated with close contact and transmission from bighorn sheep during a pneumonia epizootic (Wolfe et al. 2010). A single *M. ovipneumoniae* strain first detected in a yearling caribou with pneumonia (Rovani et al. 2019) was subsequently shown to be present in both other Alaskan caribou and in Dall's sheep, without any apparent disease association, over a period exceeding 10 years and across a range exceeding 500 km (Lieske et al. 2022). The degree of risk to bighorn sheep resulting from Cervidae or cattle shedding *M. ovipneumoniae* is probably low, but some uncertainty remains due to the currently unknown (or unreported) prevalence and duration of infection and to the unknown virulence in bighorn sheep of the strain(s) detected in non-Caprinae hosts for bighorn sheep (Wolfe et al. 2010, Highland et al. 2018, Lieske et al. 2022). The lack of social attraction and close contact between bighorn sheep and Cervidae or cattle may further reduce the impact of such reservoirs.

Initiation and Progression of *M. ovipneumoniae* Infection in Individual Bighorn Sheep

Mycoplasma ovipneumoniae does not survive for long in the environment, and transmission is thought to require direct contact or proximity with an infected animal (Nicholas et al. 2008). Transmission has been documented in bighorn sheep occupying pens separated by 10–30 m however, presumably by aerosolized droplets from coughing animals (Besser et al. 2014, Felts 2020). The speed and efficiency of direct contact transmission may be affected by the pathogen load carried by the infected animal, presence of clinical signs in the source animal that may increase exposure (such as nasal discharge, head shaking, or coughing), the frequency and duration of close contact, and perhaps also the age—or previous exposure-related susceptibility of the recipient animal. For example, when a susceptible bighorn sheep was placed in a small pen with an asymptomatic domestic sheep carrier, *M. ovipneumoniae* was first detected in the bighorn sheep 28 days later (Besser et al. 2014). Following experimental *M. ovipneumoniae* exposure, two to three weeks typically elapse before the first clinical signs (increased watery nasal discharge and other signs of upper respiratory infection) are observed (Besser et al. 2014, 2017). Progressive upper and lower respiratory tract clinical signs then appear over the following weeks to months (Besser et al. 2014, 2017). Animals that survive this progression may slowly recover over the subsequent months. Observed progression of all-age pneumonia epizootics in free-ranging bighorn sheep are generally consistent with a similar process of rapid transmission from the first symptomatic animals, followed by slower development of disease and mortality over the following days, weeks, and months (Besser et al. 2021). Disease progression is more rapid in lamb epizootics, where clinical signs are usually first observed at around three weeks of age, and death occurs a median of three weeks later (Cassirer et al. 2018).

The slow development of clinical signs following *M. ovipneumoniae* infection is thought to be associated with progressive infection of the ciliated respiratory passages from the nose to the middle ears, sinuses, and trachea and bronchi, which in turn triggers a cascade of secondary bacterial infections (Figure 9.4). *Mycoplasma ovipneumoniae* specifically interacts with motile cilia on the luminal surfaces of the epithelial cells lining the medium and larger airways, binding to and cross-linking respiratory cilia, impairing their ability to sweep out surface mucus containing inhaled

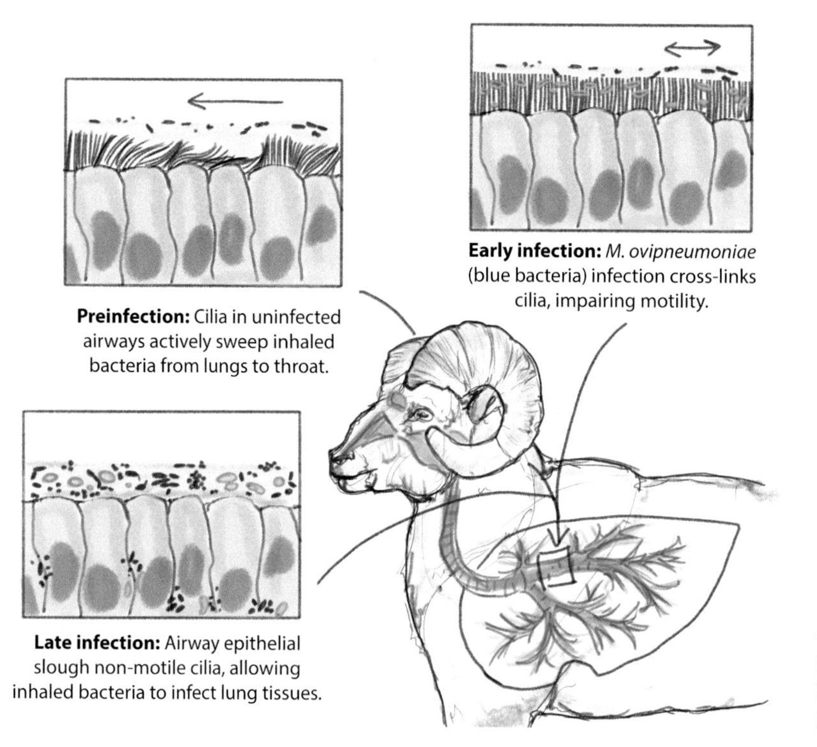

Fig. 9.4. Conceptual model of polymicrobial pneumonia induced by *Mycoplasma ovipneumoniae*. Illustration by Laura Donohue.

bacteria. Subsequently, the cilia are sloughed (Niang et al. 1998). These effects on ciliated epithelia forms the basis of a pathogenicity hypothesis for bighorn sheep: inactivation of ciliated epithelia enables inhaled bacteria to establish infections in the eustachian tubes, paranasal sinuses, conjunctivae, trachea, and bronchi and bronchioles. Invasion of *M. ovipneumoniae* to tissues beyond the respiratory tract and its draining lymph nodes has not been reported.

Clinical Signs Associated with *M. ovipneumoniae* Infection

Clinical signs observed in bighorn sheep acutely infected with *M. ovipneumoniae* primarily involve the respiratory tract, including increased lacrimation, serous to purulent nasal discharge, repetitive chewing and nose licking behaviors, ear drooping or paresis, and paroxysmal bouts of coughing. Coughing ranges from infrequent or intermittent to paroxysmal and prolonged; while prolonged observation may be required to observe spontaneous coughing in resting bighorn sheep, inducing the animals to move often promptly triggers increased coughing. Chronic pneumonia in bighorn sheep typically leads to progressive and severe loss of body condition. Bighorns with prolonged progressive pneumonia may eventually become insensitive to approaching humans and, terminally, may become ataxic. Bighorn sheep that survive pneumonia for 2–4 months usually show diminishing clinical signs (Besser et al. 2017, 2021), and by 6–12 months after initial infection, most appear clinically normal.

Necropsy Lesions and Microbiology

Epizootic bighorn sheep pneumonia presents as a bronchopneumonia but may show large variation in gross and microscopic lesions within and among individual animals. Gross lesions (Figure 9.5) of

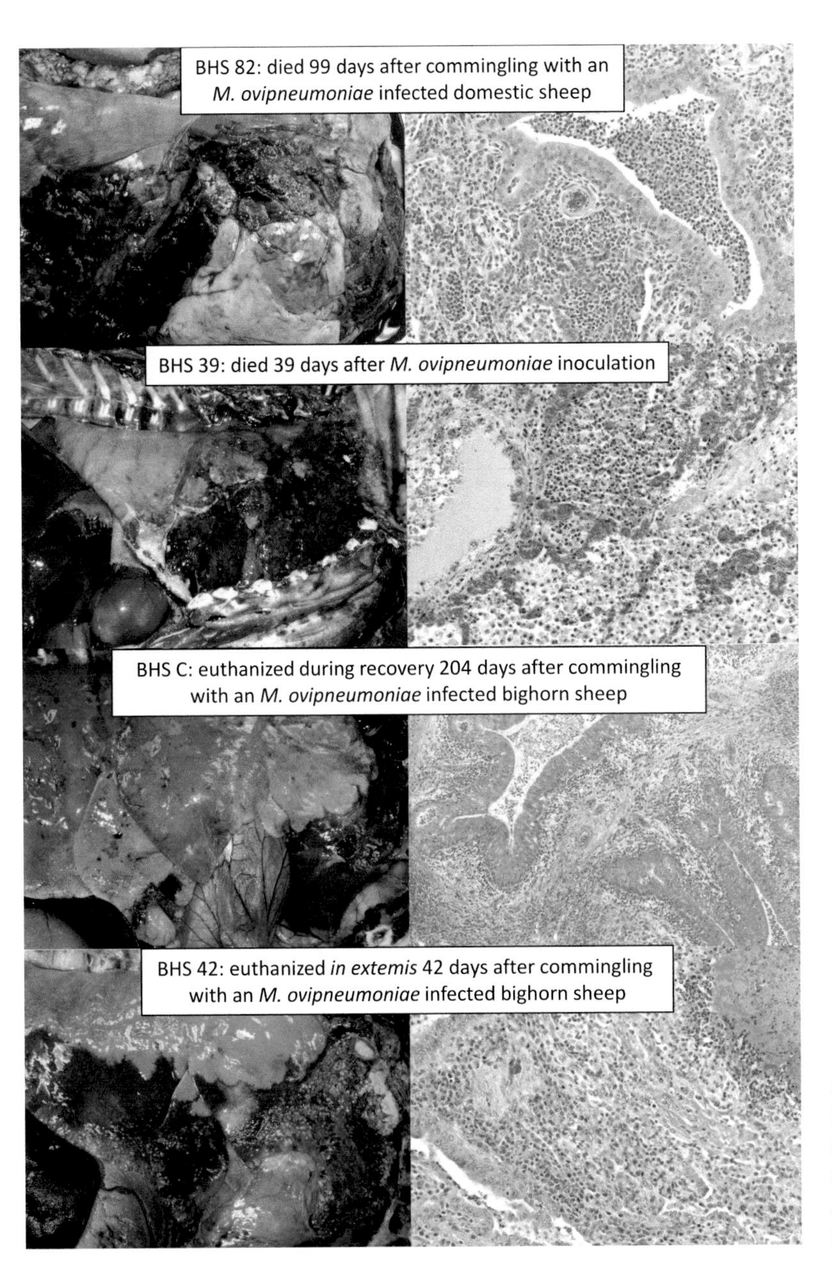

Fig. 9.5. Gross and histological lung lesions of four bighorn sheep (*Ovis canadensis*) at different times following *Mycoplasma ovipneumoniae* infection. Besser et al. (2014)

epizootic bighorn pneumonia includes consolidation (noninflation) of lung tissues especially in the anterior and ventral areas of the lung, which may affect anywhere from 5% to >50% of the lung volume. Consolidated tissues range from dark red to pale grey or white, with the latter often indicating microabscessation. In some animals, especially later in the disease course, larger abscesses may form. Serous or purulent sinusitis and, in lambs, otitis media and middle ear abscessation are also commonly observed. Additional gross lesions that may be seen include fibrinous or fibrous adhesions between the lung surface and the rib cage, turbid fluid in the pleural cavity, areas of hemorrhage, and deposition of fibrin, all likely associated with secondary (not caused by *M. ovipneumoniae*) bacterial infections. Microscopic lesions are similarly variable, including bronchiolitis, bronchiolar epithelial hyperplasia, bronchial lymphocytic cuffing, alveolar necrosis, streaming degenerative neutrophils (oat cells), and fibrin deposition. Some veterinary pathologists may interpret specific lesions as expected from (or typical of) infections with specific bacterial groups; however, due to the highly variable polymicrobial bacterial populations that characterize the later stages of this disease, these attributions should be considered tentative (Besser et al. 2013). Following recovery and apparent resolution of respiratory disease, some lesions, such as fibrous adhesions between the lung and the rib cage, and lung abscesses, may persist for years.

Sampling and Diagnostics

Real-time PCR tests on nasal swabs are a reliable method for detecting *M. ovipneumoniae* in both acutely and chronically infected bighorn sheep (Cassirer et al. 2018, Manlove et al. 2019). In acute infection, a broad range of other upper and lower respiratory tract samples, including conjunctivae, tympanic bullae, sinus cavity membranes, and bronchi, are all likely to be positive. Lung tissue samples are less reliable than are direct swab samples of ciliated larger airways. Most (70–95%) surviving bighorn sheep clear *M. ovipneumoniae* infection by a year or two after the epizootic. Carriage of *M. ovipneumoniae* in domestic sheep and goats is also reliably detected by nasal swab sampling (Heinse et al. 2016, USDA APHIS 2015). Strong age-related infection patterns are seen in populations of bighorn sheep, domestic sheep, and domestic goats infected with *M. ovipneumoniae*, with high prevalence of infection in young animals during most of the first year of life, followed by gradual clearance to average adult prevalence by 12–24 months of age (Besser et al. 2008, Plowright et al. 2017, Besser et al. 2019).

A competitive enzyme-linked immunosorbent assay (cELISA) test to detect antibody indicating exposure to *M. ovipneumoniae* is available through the Washington Animal Disease Diagnostic Laboratory (Ziegler et al. 2014). Infected herds of bighorn sheep generally show high cELISA seroprevalence (median, 0.67), while most uninfected herds are uniformly seronegative (Besser et al. 2008, 2013; Cassirer et al. 2018). Following infection and subsequent clearance of *M. ovipneumoniae*, bighorn sheep maintain antibody detectable by the cELISA for 6 months to 2 or more years. Testing by cELISA may be useful in initial screening for *M. ovipneumoniae* infection or exposure in previously unsampled bighorn sheep herds: seroprevalence of 10% or less in a sample of 10 or more animals strongly suggests a lack of herd infection or recent (within 2 years) exposure (Besser et al. 2013). Negative cELISA tests in sera of 3- to 12-month-old bighorn lambs indicate the absence of *M. ovipneumoniae* carriers within the ewe group (Garwood et al. 2020). Based on limited data, mountain goats seem to respond similarly to bighorn sheep in their cELISA responses to *M. ovipneumoniae* exposure. The cELISA test has less diagnostic value in domestic sheep due to their lower cELISA seroprevalence and higher PCR prevalence (USDA APHIS 2015), and the cELISA test has no known diagnostic value in domestic goats.

Genetic strain types of *M. ovipneumoniae* are distinguished by DNA sequences of conserved genes (multilocus sequence typing or MLST) (Cassirer et al. 2017, Einarsdottir et al. 2018). The MLST profiles of *M. ovipneumoniae* strains of domestic sheep origin are distinct from those carried by domestic goats, providing a tool to identify the likely source host of novel spillover strains (Kamath et al. 2019). Presence of identical MLST strains suggests *M. ovipneumoniae* transmission, whether within or between bighorn sheep populations or from domestic sheep and goat reservoirs to bighorn sheep (Justice-Allen et al. 2016, Kamath et al. 2019, Besser et al. 2021). Further, adult bighorn sheep immune to an endemic *M. ovipneumoniae* strain remain susceptible to novel strains, and new all-ages pneumonia epizootics have been reported following infection with new MLST *M. ovipneumoniae* strain types (Besser et al. 2008, Justice-Allen et al. 2016, Cassirer et al. 2017, Felts 2020). Future genetic studies with whole genome sequencing methods may provide additional power to identify strains and virulence genes as well as link gene alleles with specific biological properties.

Mycoplasma ovipneumoniae–Associated Epizootics

Acute all-age outbreaks are the most dramatic manifestation of this disease in free-ranging bighorn sheep (Figure 9.6). Most populations, particularly those in the United States, have experienced all-age pneumonia outbreaks (Monello et al. 2001, WAFWA Wild Sheep Working Group 2012, Cassirer et al. 2018), and several new outbreaks are reported every year. The number reported likely underestimates the true frequency because many bighorn sheep populations are not monitored annually. In some cases, no signs of disease are observed, but population surveys reveal drops in numbers consistent with major mortality events, which may subsequently be linked to detection of a newly introduced *M. ovipneumoniae* strain.

Infection spreads rapidly within populations during acute outbreaks, but mortality rates can be variable. Death rates in 82 pneumonia events documented between 1978 and 2015 ranged from 5% to 100% with a median of 48% (Cassirer et al. 2018). The sources of variation in severity among pneumonia epizootics are largely unknown but may include factors such as *M. ovipneumoniae* strain virulence, differences in the composition and relative numbers of secondary invading bacterial species, the degree of population substructure and its effect on transmission, and genetic or other host factors that may influence susceptibility to disease. Some of this reported variability may also be an artifact of the lack of visibility of some herds due to factors including the remoteness or accessibility of the population, infrequent monitoring, or the season in which the epizootic occurs. As serious as the immediate effects of all-age pneumonia outbreaks can be, the annual or periodic lamb pneumonia mortality that often follows can have more significant long-term consequences for populations. Low recruitment associated with pneumonia epizootics in lambs may last a few years or become a phase of the disease lasting decades (Manlove et al. 2016). While most survivors of initial all-age outbreaks recover, some may become chronic carriers of *M. ovipneumoniae* (Plowright et al. 2017). Chronic carrier ewes infect their lambs, which lack effective passive immune protection, and disease spreads rapidly within nursery groups (Besser et al. 2008, Manlove et al. 2017). Some chronic carriers exhibit persistent upper respiratory disease, but most are not identifiable by visible symptoms. Plowright et al. (2017) found that older animals were more likely to be chronic carriers and paranasal sinus tumors are also a cofactor suspected of increasing the likelihood of persistence of *M. ovipneumoniae* as well as *P. multocida* and other infectious agents (Fox et al. 2015). Host genotype may also influence carriage status (Martin et al. 2021). Rams may also chronically carry *M. ovipneumoniae* and could serve as a source of infection to ewe groups but, due to their life history, are unlikely to have a role in transmission of *M. ovipneumoniae* directly to young lambs (Cassirer et al. 2018).

Management of Pneumonia in Bighorn Sheep Populations

Proximity to domestic sheep or goats and relatively larger bighorn sheep population sizes have been identified as factors associated with greater risk of pneumonia outbreaks in several retrospective reviews (Singer et al. 2000, Monello et al. 2001, Sells et al. 2015, Cassirer et al. 2018). While increased risk of outbreaks associated with proximity to domestic sheep and goats seems logical for this spillover disease, the mechanism for the observed relationship with larger population size is unclear, as no consistent evidence has been found for predisposing environmental stressors or density-dependent effects on vital rates prior to pneumonia epizootics (Cassirer et al. 2018, Besser et al. 2021). Managing for smaller populations runs counter to restoration goals, and further investigation is needed to understand and address this relationship.

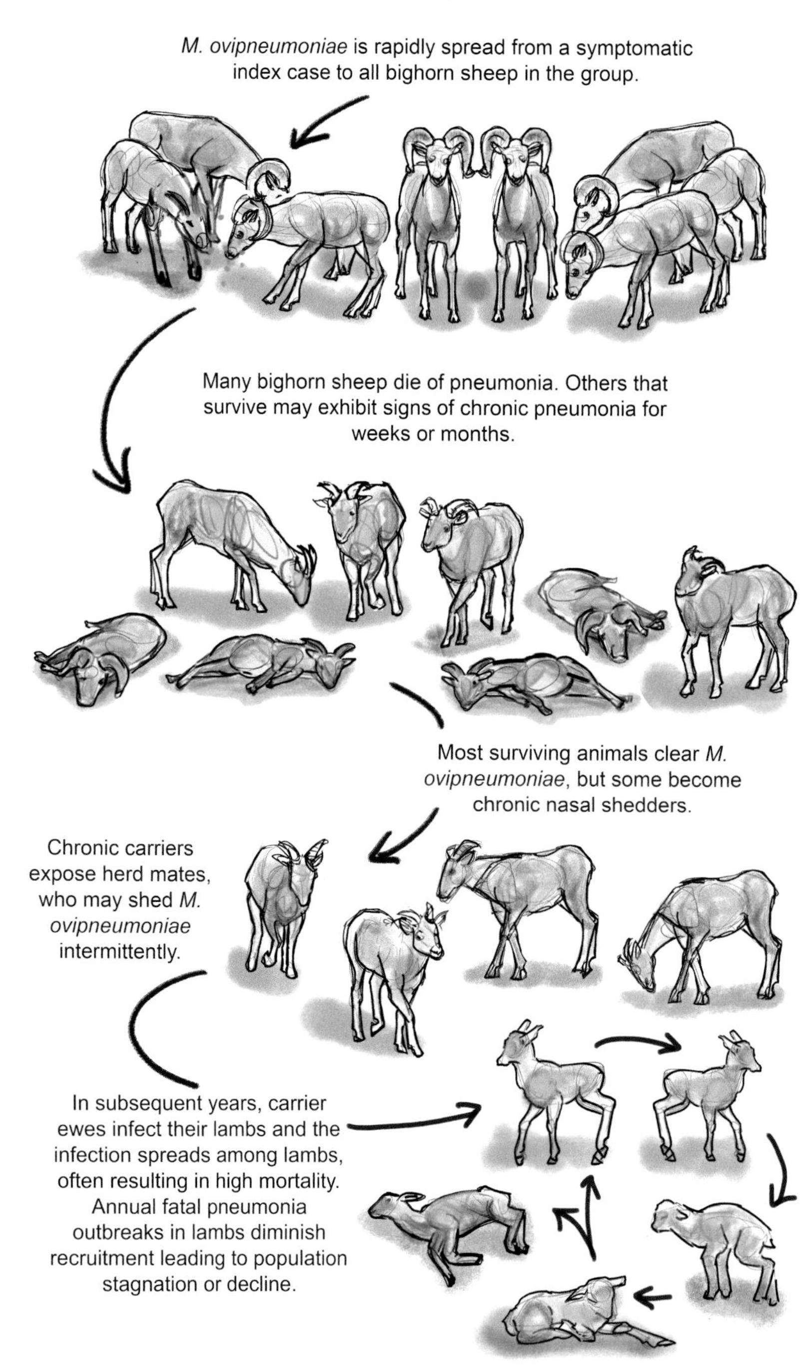

Fig. 9.6. Epizootic bighorn sheep pneumonia. Illustration by Laura Donohue.

State, provincial, and federal resource management agencies have developed guidelines intended to reduce the risk of spillover from domestic sheep and goats (WAFWA Wild Sheep Working Group 2012, Bureau of Land Management 2016). These guidelines include conducting risk assessments for interspecies contact; removal of bighorn sheep known or suspected to have contacted domestic sheep or goats; careful assessment of risk of pathogen transmission prior to conducting bighorn sheep translocations; and coordination among land management agencies, private landowners, and other stakeholders on management of domestic sheep and goats on or near ranges occupied by bighorn sheep. Guidelines for bighorn sheep translocations also include pretranslocation health screening and population performance assessments in source and recipient (or adjacent) bighorn sheep populations (WAFWA Wildlife Health Committee, 2014).

Minimizing interspecies contact has been a contentious component of management of public lands grazing allotments because the American Sheep Industry Association and the North American Pack Goat Association disagree with resource management agencies' assessment of risk (American Sheep Industry 2020, North American Pack Goat Association 2021). In response to several court decisions on the management of domestic sheep grazing allotments in bighorn sheep habitat, federal land management agencies and the US Geological Survey have collaborated on a model that is used to quantitatively assess the risk of contact between bighorn sheep and public lands grazing allotments (Carpenter et al. 2014; O'Brien et al. 2014, 2021).

Management efforts to stop or moderate the course of ongoing disease outbreaks following spillover have included nonselective removals and selective culling of more severely affected bighorn sheep in attempts to reduce onward pathogen transmission (Cassirer et al. 1996, Edwards et al. 2010, Bernatowicz et al. 2016, Ramsey et al. 2016). These interventions have not shown success at reducing severity or extent of epizootics. By the time deaths are detected, infection is likely widespread, and disease progression is well underway. Complete eradication of surviving populations following outbreaks has been necessary in some cases to prevent transmission of *M. ovipneumoniae* to adjacent unexposed populations (McFarlane and Aoude 2010, Bernatowicz et al. 2016).

Veterinary approaches to disease control have also been tried, including administration of antibiotics and vaccines that target Pasteurellaceae, use of anthelmintics to control lungworms, and supplementation with selenium to boost immune function (Miller et al. 2000, Cassirer et al. 2001, Goldstein et al. 2005, Coggins 2006, Cox 2006, Sirochman et al. 2012). These approaches have not produced any clear beneficial effects. Translocations that supplement infected bighorn populations with additional susceptible animals are likely counterproductive (Plowright et al. 2013, Almberg et al. 2022).

The most promising intervention targets elimination of chronic *M. ovipneumoniae* carriage from infected populations. Cessation of pneumonia outbreaks in lambs has occurred following the death, removal, or experimental isolation of *M. ovipneumoniae* carrier ewes (Cassirer et al. 2018, Felts 2020, Garwood et al. 2020, Besser et al. 2021, Spaan et al. 2021). Selective removal of chronic carriers of *M. ovipneumoniae* and depopulation and reintroduction have been identified as the two management actions that promote local eradication of *M. ovipneumoniae* (Almberg et al. 2022). Additional research and adaptive management are needed to improve the efficiency and effectiveness of selective removals for clearing persistent infection from wild sheep, particularly in larger populations and those where access is limited. This may involve better identifying social behaviors and demographic factors associated with persistence and transmission. Further work is needed to establish repeatability of treatment effects and to monitor longer term outcomes. From a practical standpoint, development of simple, reliable animal side diagnostic tests could help facilitate selective removals.

Domestic Sheep and Goats May Determine the Fate of Bighorn Sheep Populations

As a trigger for epizootic pneumonia, *M. ovipneumoniae* links the health of bighorn sheep populations to the presence of domestic sheep and goats. Even if prevention of spillover on public lands is eventually achieved, complete separation of wild sheep from domestic sheep and goats is unlikely to be attainable in many places without addressing increasing numbers of domestic sheep and goats on private lands in or near bighorn sheep ranges (USDA APHIS 2020, Thorne et al. 2021). Sheep and goat operations on private lands tend to hold smaller numbers of animals compared to public lands operations but cumulatively pose a significant spillover risk, given that a single contact between a susceptible bighorn and an animal infected with *M. ovipneumoniae* is sufficient to initiate an epizootic. Management of this risk is contingent on cooperation and active collaboration between domestic sheep and goat owners and wildlife managers and conservationists. Many owners of small flocks of domestic sheep and goats are unaware that their animals may pose a disease risk to bighorn sheep. Outreach and education efforts to domestic sheep and goat owners can improve their understanding of the importance of maintaining separation. An approach to reducing *M. ovipneumoniae* spillover to bighorn sheep that is particularly relevant to small domestic sheep and goat operations is operation-level reduction or elimination of *M. ovipneumoniae.* Operations with fewer than 100 animals are more likely to be naturally free of *M. ovipneumoniae* (USDA APHIS 2015), and plausible approaches to eliminate *M. ovipneumoniae*—including within-herd quarantine and segregation, test and treat, or test and cull—are more feasible in small operations than large ones. Elimination of *M. ovipneumoniae* may also offer direct health and productivity benefits to the domestic herds and flocks (Besser et al. 2019, Manlove et al. 2019). Antibiotic and vaccine therapies for domestic sheep and goats are currently problematic (Ziegler et al. 2014, Einarsdottir et al. 2018) but, if developed, could greatly increase the possibilities for eliminating *M. ovipneumoniae* from domestic herds and flocks. These efforts may be aided by ongoing research to better understand the genetic basis for *M. ovipneumoniae* carriage in domestic sheep (Mousel et al. 2021). While pilot projects have shown considerable promise, methods for elimination of *M. ovipneumoniae* from domestic sheep and goat operations are still in the research stage (Besser et al. 2019).

Summary

Research since 2005 has identified *M. ovipneumoniae* as the primary agent causing bighorn sheep pneumonia epizootics. The disease occurs when a previously naïve population experiences spillover infection with *M. ovipneumoniae* for the first time or when a new genetic strain of *M. ovipneumoniae* is introduced into a population previously infected with another strain. After the initial epizootic following spillover, chronic carrier ewes can infect lamb cohorts in subsequent years, significantly impairing recruitment. Identification and removal of chronic carrier ewes can end pneumonia epizootics in lambs and speed population recovery. Keeping wild sheep populations healthy—by stopping spillover and transmission—may be accomplished by preventing or minimizing direct contact between *M. ovipneumoniae* reservoir hosts and bighorn sheep, and by local elimination of *M. ovipneumoniae* from domestic sheep and goats on private lands in or adjacent to bighorn sheep ranges.

Eventual control of epizootic bighorn sheep pneumonia requires overcoming significant challenges but is nevertheless favored by several factors, including the availability of reliable diagnostic tests, the limited host range, the lack of an environmental reservoir of *M. ovipneumoniae*, and natural clearance of infection by most bighorn sheep. The broad base of public support and funding for restoration of this charismatic species increases the odds of future success.

ACKNOWLEDGEMENTS

The authors acknowledge that the ideas and concepts presented in this chapter are the result of discussions (and occasionally, constructive arguments) with many people in academia, wildlife agencies, and nongovernmental organizations knowledgeable and passionate about bighorn sheep and disease. We also gratefully acknowledge the generous financial support provided to our research by many diverse sources (including federal and state agencies as well as nongovernmental organizations). Finally, this research would not have been possible without the hands-on contributions of our many colleagues, co-workers, and volunteers, without whom collecting samples, running assays, analyzing data, and preparing research publications would not have been possible.

LITERATURE CITED

Alexander, K. A., C. J. Carlson, B. L. Lewis, W. M. Getz, M. V. Marathe, S. G. Eubank, C. E. Sanderson, and J. K. Blackburn. 2018. The ecology of pathogen spillover and disease emergence at the human-wildlife-environment interface. Nature Public Health 5:267–298.

Alley, M. R., G. Ionas, and J. K. Clarke. 1999. Chronic non-progressive pneumonia of sheep in New Zealand - a review of the role of *Mycoplasma ovipneumoniae*. New Zealand Veterinary Journal 47:155–160.

Almberg, E. S., K. R. Manlove, E. F. Cassirer, J. M. Ramsey, K. Lackey, J. A. Gude, and R. K. Plowright. 2022. Modeling management strategies for chronic disease in wildlife: predictions for the control of respiratory disease in bighorn sheep. Journal of Applied Ecology 59:693–703.

American Sheep Industry Association. 2020. Bighorn sheep in domestic sheep grazing allotments. https://www.sheepusa.org/wp-content/uploads/2022/03/ASI-Bighorn-Grazing-2022.pdf. Accessed 31 Dec 2021.

Asokan, G. V., and V. Asokan. 2016. Bradford Hill's criteria, emerging zoonoses, and One Health. Journal of Epidemiology and Global Health 6:125–129.

Bernatowicz, J., D. Bruning, E. F. Cassirer, R. B. Harris, K. Mansfield, and P. Wik. 2016. Management responses to pneumonia outbreaks in three Washington State bighorn herds: lessons learned and questions yet unanswered. Proceedings of the Biennial Symposium of the Nothern Wild Sheep and Goat Council 20:38–61.

Besser, T. E., E. F. Cassirer, M. A. Highland, P. Wolff, A. Justice-Allen, K. Mansfield, M. A. Davis, and W. Foreyt. 2013. Bighorn sheep pneumonia: sorting out the cause of a polymicrobial disease. Preventive Veterinary Medicine 108:85–93.

Besser, T. E., E. F. Cassirer, A. Lisk, D. D. Nelson, K. Manlove, P. C. Cross, and J. Hogg. 2021. Natural history of a bighorn sheep pneumonia epizootic: source of infection, course of disease, and pathogen clearance. Ecology and Evolution 11:14366–14382.

Besser, T. E., E. F. Cassirer, K. A. Potter, and W. J. Foreyt. 2017. Exposure of bighorn sheep to domestic goats colonized with *Mycoplasma ovipneumoniae* induces sub-lethal pneumonia. PLoS One 12:e0178707.

Besser, T. E., E. F. Cassirer, K. A. Potter, K. Lahmers, J. L. Oaks, S. Shanthalingam, S. Srikumaran, and W. J. Foreyt. 2014. Epizootic pneumonia of bighorn sheep following experimental exposure to *Mycoplasma ovipneumoniae*. PLoS One 9:e110039.

Besser, T. E., E. F. Cassirer, K. A. Potter, J. VanderSchalie, A. Fischer, D. P. Knowles, D. R. Herndon, F. R. Rurangirwa, G. C. Weiser, and S. Srikumaran. 2008. Association of *Mycoplasma ovipneumoniae* infection with population-limiting respiratory disease in free-ranging Rocky Mountain bighorn sheep (*Ovis canadensis canadensis*). Journal of Clinical Microbiology 46:423–430.

Besser, T. E., E. F. Cassirer, C. Yamada, K. A. Potter, C. Herndon, W. J. Foreyt, D. P. Knowles, and S. Srikumaran. 2012*a*. Survival of bighorn sheep (*Ovis canadensis*) commingled with domestic sheep (*Ovis aries*) in the absence of *Mycoplasma ovipneumoniae*. Journal of Wildlife Diseases 48:168–172.

Besser, T. E., M. A. Highland, K. Baker, E. F. Cassirer, N. J. Anderson, J. M. Ramsey, K. Mansfield, D. L. Bruning, P. Wolff, J. B. Smith, et al. 2012*b*. Causes of pneumonia epizootics among bighorn sheep, western United States, 2008–2010. Emerging Infectious Diseases 18:406–414.

Besser, T. E., J. Levy, M. Ackerman, D. Nelson, K. Manlove, K. A. Potter, J. Busboom, and M. Benson. 2019. A pilot study of the effects of *Mycoplasma ovipneumoniae* exposure on domestic lamb growth and performance. PLoS One 14:e0207420.

Black, S. R., I. K. Barker, K. G. Mehren, G. J. Crawshaw, S. Rosendal, L. Ruhnke, J. Thorsen, and P. S. Carman. 1988. An epizootic of *Mycoplasma ovipneumoniae* infection in captive Dall's sheep (*Ovis dalli dalli*). Journal of Wildlife Diseases 24:627–635.

Brooks, A. 1923. The Rocky Mountain sheep in British Columbia. Canadian Field-Naturalist 37:23–25.

Buechner, H. K. 1960. The bighorn sheep in the United States, its past, present, and future. Wildlife Monographs 4:3–174.

Bunch, T. D., W. M. Boyce, C. P. Hibler, W. R. Lance, T. R. Spraker, and E. S. Williams. 1999. Diseases of North American wild sheep. *In* R. V. P. R. Krausman, editor. Mountain sheep of North America. University of Arizona Press, Tucson, USA.

Bureau of Land Management. 2016. Management of domestic sheep and goats to sustain wild sheep. US Department of the Interior, Bureau of Land Management, Washington, D.C., USA. https://www.blm.gov/sites/blm.gov/files/uploads/ManagementofDomesticSheepandGoatstoSustainWildSheep_Manual1730.pdf. Accessed 14 Feb 2022.

Carpenter, T. E., V. L. Coggins, C. McCarthy, C. S. O'Brien, J. M. O'Brien, and T. J. Schommer. 2014. A spatial risk assessment of bighorn sheep extirpation by grazing domestic sheep on public lands. Preventive Veterinary Medicine 114:3–10.

Cassirer, E. F., K. R. Manlove, E. S. Almberg, P. L. Kamath, M. Cox, P. Wolff, A. Roug, J. Shannon, R. Robinson, R. B. Harris, et al. 2018. Pneumonia in bighorn sheep: risk and resilience. Journal of Wildlife Management 82:32–45.

Cassirer, E. F., K. R. Manlove, R. K. Plowright, and T. E. Besser. 2017. Evidence for strain-specific immunity to pneumonia in bighorn sheep. Journal of Wildlife Management 81:133–143.

Cassirer, E. F., L. E. Oldenburg, V. Coggins, P. Fowler, K. M. Rudolph, D. L. Hunter, and W. J. Foreyt. 1996. Overview and preliminary analysis of a bighorn sheep dieoff, Hells Canyon 1995–96. Proceedings of the Biennial Symposium of the Nothern Wild Sheep and Goat Council 10:78–86.

Cassirer, E. F., K. M. Rudolph, P. Fowler, V. L. Coggins, D. L. Hunter, and M. W. Miller. 2001. Evaluation of ewe vaccination as a tool for increasing bighorn lamb survival following pasteurellosis epizootics. Journal of Wildlife Diseases 37:49–57.

Coggins, V. L. 2006. Selenium supplementation, parasite treatment, and management of bighorn sheep at Lostine River, Oregon. Proceedings of the Biennial Symposium of the Nothern Wild Sheep and Goat Council 15:98–106.

Council for Agricultural Science and Technology (CAST). 2008. Pasteurellosis transmission risks between domestic and wild sheep. CAST, Volume CAST Commentary QTA2008-1, Ames, Iowa, USA.

Cox, M. 2006. Effects of mineral supplements on California bighorn sheep in northern Nevada. Proceedings of the Biennial Symposium of the Nothern Wild Sheep and Goat Council 15:107–120.

Dassanayake, R. P., S. Shanthalingam, C. N. Herndon, R. Subramaniam, P. K. Lawrence, J. Bavananthasivam, E. F. Cassirer, G. J. Haldorson, W. J. Foreyt, F. R. Rurangirwa, et al. 2010. *Mycoplasma ovipneumoniae* can predispose bighorn sheep to fatal *Mannheimia haemolytica* pneumonia. Veterinary Microbiology 145:354–359.

Dassanayake, R. P., S. Shanthalingam, R. Subramaniam, C. N. Herndon, J. Bavananthasivam, G. J. Haldorson, W. J. Foreyt, J. F. Evermann, L. M. Herrmann-Hoesing, D. P. Knowles, et al. 2013. Role of *Bibersteinia trehalosi*, respiratory syncytial virus, and parainfluenza-3 virus in bighorn sheep pneumonia. Veterinary Microbiology 162:166–172.

Dunbar, M. R., D. A. Jessup, J. F. Evermann, and W. J. Foreyt. 1985. Seroprevalence of respiratory syncytial virus in free-ranging bighorn sheep. Journal of the American Veterinary Medical Association 187:1173–1174.

Edwards, V. L., J. Ramsey, C. Jourdonnais, R. Vinkey, M. Thompson, N. Anderson, T. Carlsen, C. Anderson. 2010. Situational response to four bighorn sheep die-offs in western Montana. 2010. Proceedings of the Biennial Symposium of the Nothern Wild Sheep and Goat Council 17:29–50.

Einarsdottir, T., E. Gunnarsson, and S. Hjartardottir. 2018. Icelandic ovine *Mycoplasma ovipneumoniae* are variable bacteria that induce limited immune responses in vitro and in vivo. Journal of Medical Microbiology 67:1480–1490.

Elliott, L. F., W. M. Boyce, R. K. Clark, and D. A. Jessup. 1994. Geographic analysis of pathogen exposure in bighorn sheep (*Ovis canadensis*). Journal of Wildlife Diseases 30:315–318.

Environmental Conservation Online System (ECOS). 1998. Peninsular bighorn sheep (*Ovis canadensis nelsoni*). US Fish and Wildlife Service, Pacific Southwest Region, Sacramento, California, USA. https://ecos.fws.gov/ecp/species/4970. Accessed 14 Feb 2022.

Environmental Conservation Online System (ECOS). 1999. Sierra Nevada bighorn sheep (*Ovis canadensis sierrae*). US Fish and Wildlife Service, Pacific Southwest Region, Sacramento, California, USA. https://ecos.fws.gov/ecp/species/3646. Accessed 14 Feb 2022.

Felts, B. L. 2020. Epidemiological investigations of bighorn sheep respiratory disease and implications for management. Dissertation, South Dakota State University, Brookings USA. https://openprairie.sdstate.edu/etd/3932. Accessed 14 Feb 2022.

Foreyt, W. J., K. P. Snipes, and R. W. Kasten. 1994. Fatal pneumonia following inoculation of healthy bighorn sheep with *Pasteurella haemolytica* from healthy domestic sheep. Journal of Wildlife Diseases 30:137–145.

Forrester, D. J. 1971. Bighorn sheep lungworm–pneumonia complex. Pages 158–173 *in* J. W. Davis, and R. C. Anderson, editors. Parasitic diseases of wild mammals. Iowa State University Press, Ames, USA.

Fox, K. A., N. M. Rouse, K. P. Huyvaert, K. A. Griffin, H. J. Killion, J. Jennings-Gaines, W. H. Edwards, S. L. Quackenbush, and M. W. Miller. 2015. Bighorn sheep (*Ovis canadensis*) sinus tumors are associated with coinfections by potentially pathogenic bacteria in the upper respiratory tract. Journal of Wildlife Diseases 51:19–27.

Garwood, T. J., C. P. Lehman, D. P. Walsh, E. F. Cassirer, T. E. Besser, and J. A. Jenks. 2020. Removal of chronic *Mycoplasma ovipneumoniae* carrier ewes eliminates pneumonia in a bighorn sheep population. Ecology and Evolution 10:3491–3502.

Goldstein, E. J., J. J. Millspaugh, B. E. Washburn, G. C. Brundige, and K. J. Raedeke. 2005. Relationships among fecal lungworm loads, fecal glucocorticoid metabolites, and lamb recruitment in free-ranging Rocky Mountain bighorn sheep. Journal of Wildlife Diseases 41:416–425.

Grinnell, G. B. 1928. Mountain sheep. Journal of Mammalogy 9:1–9.

Handeland, K., T. Tengs, B. Kokotovic, T. Vikoren, R. D. Ayling, B. Bergsjo, O. G. Sigurethardottir, and T. Bretten. 2014. *Mycoplasma ovipneumoniae*—a primary cause of severe pneumonia epizootics in the Norwegian muskox (*Ovibos moschatus*) population. PLoS One 9:e106116.

Heinse, L., L. Hardesty, and R. Harris. 2016. Risk of pathogen spillover to bighorn sheep from small domestic sheep and goat flocks on private land in Washington. Wildlife Society Bulletin 40:625–633.

Hibler, C. P., T. R. Spraker, and E. T. Thorne. 1982. Protostrongylosis in bighorn sheep. Pages 208–213 *in* E. T. Thorne, N. Kinsgston, W. R. Jolley, and R. C. Bergstrom, editors. Diseeases of wildlife in Wyoming. Wyoming Game and Fish Department, Cheyenne, USA.

Highland, M. A., D. R. Herndon, S. C. Bender, L. Hansen, R. F. Gerlach, and K. B. Beckmen. 2018. *Mycoplasma ovipneumoniae* in wildlife species beyond subfamily Caprinae. Emerging Infectious Diseases 24:2384–2386.

Jenkins, E. J., A. M. Veitch, S. J. Kutz, T. K. Bollinger, J. M. Chirino-Trejo, B. T. Elkin, K. H. West, E. P. Hoberg, and L. Polley. 2007. Protostrongylid parasites and pneumonia in captive and wild thinhorn sheep (*Ovis dalli*). Journal of Wildlife Diseases 43:189–205.

Jorgenson, J. T., M. FestaBianchet, J. M. Gaillard, and W. D. Wishart. 1997. Effects of age, sex, disease, and density on survival of bighorn sheep. Ecology 78:1019–1032.

Justice-Allen, A., E. Butler, J. Pebworth, A. Munig, P. Wolff, and T. E. Besser. 2016. Investigation of pneumonia mortalities in a *Mycoplasma*-positive desert bighorn sheep population and detection of a different strain of *Mycoplasma ovipneumoniae*. Proceedings of the Biennial Symposium of the Nothern Wild Sheep and Goat Council 20:68–72.

Kamath, P. L., K. Manlove, E. F. Cassirer, P. C. Cross, and T. E. Besser. 2019. Genetic structure of *Mycoplasma ovipneumoniae* informs pathogen spillover dynamics between domestic and wild Caprinae in the western United States. Scientific Reports 9:15318.

Lawrence, P. K., S. Shanthalingam, R. P. Dassanayake, R. Subramaniam, C. N. Herndon, D. P. Knowles, F. R. Rurangirwa, W. J. Foreyt, G. Wayman, A. M. Marciel, et al. 2010. Transmission of *Mannheimia haemolytica* from domestic sheep (*Ovis aries*) to bighorn sheep (*Ovis canadensis*): unequivocal demonstration with green fluorescent protein-tagged organisms. Journal of Wildlife Diseases 46:706–717; Erratum, Journal of Wildlife Diseases, 46:1346.

Lieske, C., R. F. Gerlach, M. Frances, and K. B. Beckmen. 2022. Multilocus sequence typing of *Mycoplasma ovipneumoniae* detected in Dall's sheep (*Ovis dalli dalli*) and caribou (*Rangifer tarandus granti*) in Alaska, USA. Journal of Wildlife Diseases 58:625–630.

Maksimović, Z., C. De la Fe, J. Amores, A. Gómez-Martín, and M. Rifatbegović. 2016. Comparison of phenotypic and genotypic profiles among caprine and ovine *Mycoplasma ovipneumoniae* strains. Veterinary Record 180:180.

Manlove, K., M. Branan, K. Baker, D. Bradway, E. F. Cassirer, K. L. Marshall, R. S. Miller, S. Sweeney, P. C. Cross, and T. E. Besser. 2019. Risk factors and productivity losses associated with *Mycoplasma ovipneumoniae* infection in United States domestic sheep operations. Preventive Veterinary Medicine 168:30–38.

Manlove, K., E. F. Cassirer, P. C. Cross, R. K. Plowright, and P. J. Hudson. 2016. Disease introduction is associated with a phase transition in bighorn sheep demographics. Ecology 97:2593–2602.

Manlove, K. R., E. F. Cassirer, R. K. Plowright, P. C. Cross, and P. J. Hudson. 2017. Contact and contagion: probability of transmission given contact varies with demographic state in bighorn sheep. Journal of Animal Ecology 86:908–920.

Martin, A. M., E. F. Cassirer, L. P. Waits, R. K. Plowright, P. C. Cross, and K. R. Andrews. 2021. Genomic association with pathogen carriage in bighorn sheep (*Ovis canadensis*). Ecology and Evolution 11:2488–2502.

McFarlane, L., and A. Aoude. 2010. Status of Goslin Unit bighorn sheep pneumonia outbreak in Utah. Proceedings of the Biennial Symposium of the Nothern Wild Sheep and Goat Council 17:53.

Miller, M. W. 2001. Pasteurellosis. Page 558 *in* E. S. Williams and I. K. Barker, editors. Infectious diseases of wild mammals. Iowa State University Press, Ames, USA.

Miller, M. W., J. E. Vayhinger, D. C. Bowden, S. P. Roush, T. E. Verry, A. N. Torres, and V. D. Jurgens. 2000. Drug treatment for lungworm in bighorn sheep: reevaluation of a 20-year-old management prescription. Journal of Wildlife Management 64:505–512.

Monello, R. J., D. L. Murray, and E. F. Cassirer. 2001. Ecological correlates of pneumonia epizootics in bighorn sheep herds. Canadian Journal of Zoology 79:1423–1432.

Mousel, M. R., S. N. White, M. K. Herndon, D. R. Herndon, J. B. Taylor, G. M. Becker, and B. M. Murdoch. 2021.

Genes involved in immune, gene translation, and chromatic organization pathways associated with *Mycoplasma ovipneumoniae* presence in nasal secretions of domestic sheep. PLoS One 16:e0247209.

Niang, M., R. F. Rosenbusch, M. C. DeBey, Y. Niyo, J. J. Andrews, and M. L. Kaeberle. 1998. Field isolates of *Mycoplasma ovipneumoniae* exhibit distinct cytopathic effects in ovine tracheal organ cultures. Zentralblat Veterinarmedicin A 45:29–40.

Nicholas, R., R. Ayling, and L. McAuliffe. 2008. Mycoplasma diseases of ruminants: disease, diagnosis and control. Commonwealth Aagricultural Bureax International, Cambridge, Massachusetts, USA.

North American Pack Goat Association. 2021. Legal/land use documents. North American Pack Goat Association, Rye, Colorado, USA. https://www.napga.org/resources/. Accessed 31 Dec 2021.

Nunn, C. L., F. Jordan, C. M. McCabe, J. L. Verdolin, and J. H. Fewell. 2015. Infectious disease and group size: more than just a numbers game. Philosophical Transactions of the Royal Society B Biological Sciences 370:20140111.

O'Brien, J. M., T. E. Carpenter, C. S. O'Brien, and C. M Ccarthy. 2014. Incorporating foray behavior into models estimating contact risk between bighorn sheep and areas occupied by domestic sheep. Wildlife Society Bulletin 38:321–331.

O'Brien, J. M., A. Titolo, P. Cross, F. Quamen, and M. Woolever. 2021. Bighorn sheep risk tool: US Geological Survey software release. Bighorn Sheep Risk of Contact Tool · Wiki · BHS Risk Tool Group / Bighorn Sheep Risk Tool · GitLab (usgs.gov). US Geological Survey, Reston, Virginia, USA. https://code.usgs.gov/bhs-risk-tool-group. Accessed 14 Feb 2022.

Parks, J. B., and J. J. England. 1974. A serologic survey for selected viral infections of Rocky Mountain bighorn sheep. Journal of Wildlife Diseases 10:107–110.

Parks, J. B., G. Post, T. Thorne, and P. Nash. 1972. Parainfluenza-3 virus infection in Rocky Mountain bighorn sheep. Journal of the American Veterinary Medical Association 161:669–672.

Plowright, R. K., K. Manlove, E. F. Cassirer, P. C. Cross, T. E. Besser, and P. J. Hudson. 2013. Use of exposure history to identify patterns of immunity to pneumonia in bighorn sheep (*Ovis canadensis*). PLoS One 8:e61919.

Plowright, R. K., K. R. Manlove, T. E. Besser, D. J. Paez, K. R. Andrews, P. E. Matthews, L. P. Waits, P. J. Hudson, and E. F. Cassirer. 2017. Age-specific infectious period shapes dynamics of pneumonia in bighorn sheep. Ecology Letters 20:1325–1336.

Ramsey, J., K. Carson, E. Almberg, M. Thompson, R. Mowry, L. Bradley, J. Kolbe, and C. Jourdonnais. 2016. Status of western Montana bighorn sheep herds and discussion of control efforts after all-age die-offs. Proceedings of the Biennial Symposium of the Nothern Wild Sheep and Goat Council 20:19–37.

Rifatbegovic, M., Z. Maksimovic, and B. Hulaj. 2011. *Mycoplasma ovipneumoniae* associated with severe respiratory disease in goats. Veterinary Record 168:565.

Rovani, E. R., K. B. Beckmen, and M. A. Highland. 2019. *Mycoplasma ovipneumoniae* associated with polymicrobial pneumonia in a free-ranging yearling barren ground caribou (*Rangifer tarandus granti*) from Alaska, USA. Journal of Wildlife Diseases 55:733–736.

Rudolph, K. M., D. L. Hunter, W. J. Foreyt, E. F. Cassirer, R. B. Rimler, and A. C. S. Ward. 2003. Sharing of *Pasteurella* spp. between free-ranging bighorn sheep and feral goats. Journal of Wildlife Diseases 39:897–903.

Rudolph, K. M., D. L. Hunter, R. B. Rimler, E. F. Cassirer, W. J. Foreyt, W. J. DeLong, G. C. Weiser, and A. C. Ward. 2007. Microorganisms associated with a pneumonic epizootic in Rocky Mountain bighorn sheep (*Ovis canadensis canadensis*). Journal of Zoo and Wildlife Medicine 38:548–558.

Samson, J., J. C. Holmes, J. T. Jorgenson, and W. D. Wishart. 1987. Experimental infections of free-ranging Rocky Mountain bighorn sheep with lungworms (*Protostrongylus* spp.; Nematoda: Protostrongylidae). Journal of Wildlife Diseases 23:396–403.

Sells, S. N., M. S. Mitchell, J. J. Nowak, P. M. Lukacs, N. J. Anderson, J. M. Ramsey, J. A. Gude, and P. R. Krausman. 2015. Modeling risk of pneumonia epizootics in bighorn sheep. Journal of Wildlife Management 79:195–210.

Shillinger, J. E. 1937. Disease relationship between domestic animals and wildlife. Transactions of the North American Wildlife and Natural Resources Conference 2:298–302.

Singer, F. J., C. M. Papouchis, and K. K. Symonds. 2000. Translocations as a tool for restoring populations of bighorn sheep. Restoration Ecology 8:6–13.

Sirochman, M. A., K. J. Woodruff, J. L. Grigg, D. P. Walsh, K. P. Huyvaert, M. W. Miller, and L. L. Wolfe. 2012. Evaluation of management treatments intended to increase lamb recruitment in a bighorn sheep herd. Journal of Wildlife Diseases 48:781–784.

Smith, D. R. 1954. The bighorn sheep of Idaho; its status, life history and management. Idaho Department of Fish and Game, Boise, USA. https://www.arlis.org/docs/vol1/Susitna/33/APA3352.pdf. Accessed 14 Feb 2022.

Spaan, R. S., C. W. Epps, R. Crowhurst, D. Whittaker, M. Cox, and A. Duarte. 2021. Impact of *Mycoplasma ovipneumoniae* on juvenile bighorn sheep (*Ovis *canadensis*) survival in the northern Basin and Range ecosystem. Peer Journal 9:e10710.

Spraker, T. R., J. K. Collins, W. J. Adrian, and J. H. Olterman. 1986. Isolation and serologic evidence of a respiratory syncytial virus in bighorn sheep from Colorado. Journal of Wildlife Diseases 22:416–418.

Thorne, J. W., B. M. Murdoch, B. A. Freking, R. R. Redden, T. W. Murphy, J. B. Taylor, and H. D. Blackburn. 2021. Evolution of the sheep industry and genetic research in the United States: opportunities for convergence in the twenty-first century. Animal Genetics 52:395–408.

US Department of Agriculture Animal and Plant Health Inspection Service (USDA APHIS). 2015. *Mycoplasma ovipneumoniae* on US sheep operations. USDA APHIS, Fort Collins, Colorado, USA.

US Department of Agriculture Animal and Plant Health Inspection Service (USDA APHIS). 2020. How is the US goat industry growing? USDA APHIS, Fort Collins, Colorado, USA.

Warren, E. R. 1910. The mountain sheep. Pages 9–12 *in* The mammals of Colorado: an account of the several species found within the boundaries of the state, together with a record of their habits and distribution. G. P. Putnam's Sons, The Knickerbocker Press, New York, New York, USA, and London, England.

Wehausen, J. D., S. T. Kelley, and R. R. Ramey. 2011. Domestic sheep, bighorn sheep, and respiratory disease: a review of experimental evidence. California Fish and Game 97:7–24.

Western Association of Fish and Wildlife Agencies (WAFWA) Wild Sheep Working Group. 2012. Recommendations for domestic sheep and goat management in wild sheep habitat. https://wafwa.org/initiatives/wsi/. Accessed 31 Dec 2021.

Western Association of Fish and Wildlife Agencies (WAFWA) Wildlife Health Committee. 2014. Bighorn sheep herd health monitoring recommendations. https://wafwa.org/wp-content/uploads/2020/07/BHS-herd-health-monitoring_Final-1_3_2015.pdf. Accessed 10 Feb 2022.

Weyand, L. K., E. F. Cassirer, and T. E. Besser. 2018. Fatal pneumonia in bighorn sheep lambs: the critical role of *Mycoplasma ovipneumoniae* carrier ewes. Proceedings of the Biennial Symposium of the Nothern Wild Sheep and Goat Council 21:113.

Wolfe, L. L., B. Diamond, T. R. Spraker, M. A. Sirochman, D. P. Walsh, C. M. Machin, D. J. Bade, and M. W. Miller. 2010. A bighorn sheep die-off in southern Colorado involving a Pasteurellaceae strain that may have originated from syntopic cattle. Journal of Wildlife Diseases 46:1262–1268.

Wolff, P., M. Cox, C. McAdoo, and C. A. Anderson. 2016*a*. Disease transmission between sympatric mountain goats and bighorn sheep. Proceedings of the Biennial Symposium of the Nothern Wild Sheep and Goat Council 20:79.

Wolff, P. L., J. A. Blanchong, D. D. Nelson, P. J. Plummer, C. McAdoo, M. B. Cox, T. E., J. Munoz-Gutierrez, and C. A. Anderson. 2019. Detection of *Mycoplasma ovipneumoniae* in pneumonic mountain goat (*Oreamnos americanus*) kids. Journal of Wildlife Diseases 55:206–212.

Wolff, P. L., C. Schroeder, C. McAdoo, M. Cox, D. D. Nelson, J. F. Evermann, and J. F. Ridpath. 2016*b*. Evidence of bovine viral diarrhea virus infection in three species of sympatric wild ungulates in Nevada: life history strategies may maintain endemic infections in wild populations. Frontiers in Microbiology 7:292.

Young, S. P., and R. H. Manville. 1960. Records of bighorn hybrids. Journal of Mammalogy 41:523–525.

Ziegler, J. C., K. K. Lahmers, G. M. Barrington, S. M. Parish, K. Kilzer, K. Baker, and T. E. Besser. 2014. Safety and immunogenicity of a *Mycoplasma ovipneumoniae* bacterin for domestic sheep (*Ovis aries*). PLoS One 9:e95698.

10 Bovine Brucellosis in the Greater Yellowstone Area

BRANT SCHUMAKER

Introduction

It may seem strange to discuss bovine (cattle) brucellosis in a book on wildlife health. To be sure, bovine brucellosis is the correct name for this disease. The bacterial disease, caused by *Brucella abortus*, is primarily a cattle (*Bos taurus*) disease that has infected wildlife populations in the United States and globally. However, a successful eradication campaign in cattle in the United States has led attention to focus on the last remaining US endemic source of the disease—wild Rocky Mountain elk (*Cervus canadensis nelsoni*) and American bison (*Bison bison*) in the Greater Yellowstone Area (GYA) (Figure 10.1).

It is most commonly accepted that bovine brucellosis was introduced to North America when cattle were brought in by European settlers as a source of food. The first report of brucellosis in North American wild ruminants was in 1917 in bison residing in Yellowstone National Park (YNP) (Mohler 1917). It is presumed that cattle kept for park employees commingled with wild bison and transmitted the bacterium (Meagher and Meyer 1994). The first report of brucellosis in elk was in 1930 on the National Elk Refuge, where three of the nine animals sampled tested positive (Murie 1951). Subsequent transmis-

Illustration by Laura Donohue.

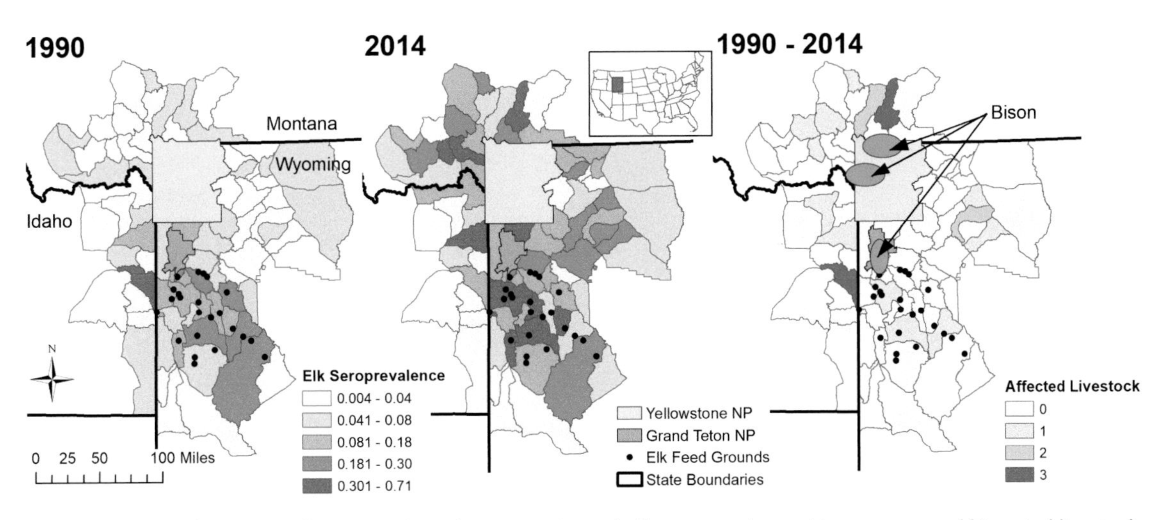

Fig. 10.1. Map of Greater Yellowstone Area showing estimated elk seroprevalence, bison ranges, and historical livestock premises (1990–2014) divided by elk hunt district. Gray-shaded areas are National Parks (NP). Map by P. C. Cross. Data used with permission from Brennan et al. (2017).

sion of brucellosis occurred between bison living in YNP and sympatric elk. Many of the unfed elk in Montana and Wyoming can trace the origins of their bacterial isolates back to elk from Wyoming feedgrounds (Kamath et al. 2016).

During the height of the Great Depression in 1934, bovine brucellosis was considered to be the most important cattle disease of the time due to disease-associated calf losses, infertility, decreased milk production, and lameness. In 1934 and 1935, brucellosis affected 11.5% of adult cattle. An eradication campaign began as part of an economic recovery program in order to reduce the US cattle population due to the Great Depression and concurrent drought conditions (Ragan 2002). In the late 1940s, the losses to the livestock industry from brucellosis were estimated at $100 million annually (Becton 1977). Funding from the US Congress for a state–federal cooperative Brucellosis Eradication Program (BEP) began in 1954. In 1957, around the time of the inception of the BEP, there were 123,964 herds found to be infected with brucellosis. This was considered a gross underestimate due to inadequate surveillance (Ragan 2002) and going from "farm to farm" for "Bangs disease" was a significant part of most large animal veterinary practices from the 1950s to the 1970s. By 1976, the nationwide incidence had fallen to less than 0.7%, corresponding to $20 million in losses annually (Becton 1977) (Figure 10.2).

Further attention paid to the disease in the last 50 years has led to eradication of brucellosis everywhere in the nation except for the GYA, where the disease remains endemic in the wild elk and bison populations. Now, in the GYA, the main concern is the persistent but intermittent spillover of the bacterium from wild ungulates back into livestock.

From 1990 to 2001, no infected domestic herds were identified in the GYA, and there were no affected cattle herds in the United States (Ragan 2002). For one six-month period beginning in February of 2008, every state, along with Puerto Rico and the US Virgin Island had achieved Brucellosis Class Free status for the first time in the history of the Brucellosis Eradication Program. It did not last, however; just six months later, two herds in Montana tested positive. Continuously since that time, there has been at least one active livestock quarantine in the GYA states: Idaho, Montana, and Wyoming. From April 2002 to August 2021, one domestic elk, seven domestic bison, and 30 cattle herds in the GYA were determined to be infected (National Academies of Sciences, Engineering, and Medicine, 2017; H. Hasel, D. Lawrence,

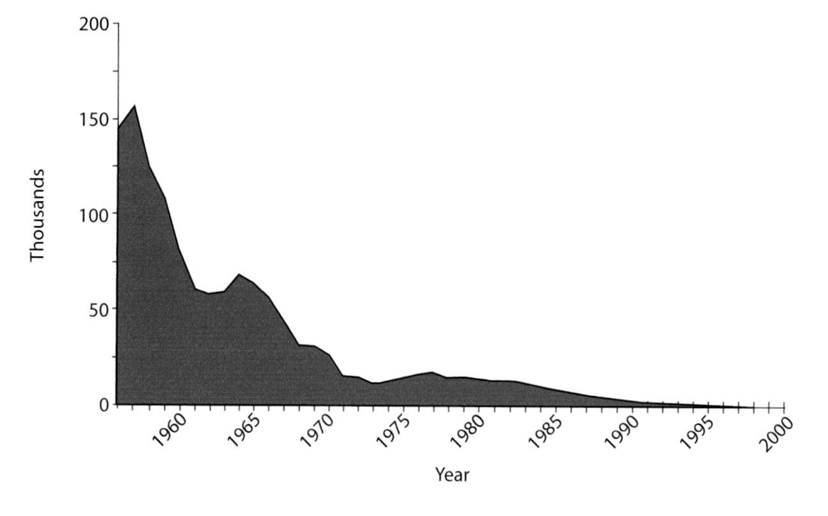

Fig. 10.2. Reduction in brucellosis-infected cattle herds in the Unites States in the latter half of the 20th century, between 1955 and 2000. Data from 1955 to 1969 include retesting of infected herds (1964 data are estimated), and data from 1970 to 2000 include only distinct infected herds. Used with permission from Ragan (2002).

and E. Liska, personal communication), all with brucellosis genotypes linked with elk.

Etiology

Bovine brucellosis is caused by a bacterium that has evolved to live inside the cells of the immune system, normally in charge of killing such bacteria. These immune system cells are called monocytes when they are in circulation or macrophages when they are in tissues. If the immune system is working normally, a monocyte or macrophage devours bacteria into an intracellular packet and then fuses that packet with another one containing enzymes to digest the newly captured bacteria. *Brucella* spp. persist in these immune system cells by blocking the fusion of the packet containing the enzymes with the packet engulfing itself and enjoys a protected "free ride" through the host body, transported by these cells. Because *Brucella* spp. are so good at evading the immune system, it is presumed that most infections are for the life of the wild or domestic animal host.

Demographics and Field Biology

In United States, bovine brucellosis primarily affects elk, bison, and cattle in the GYA. In Canada, a small pocket of bovine brucellosis exists in wild wood bison in and around Wood Buffalo National Park (Tessaro et al. 1990). In male bison, inflammation of the testicles and scrotal enlargement have been the primary clinical signs of brucellosis (Corner and Connell 1958; Tunnicliff and Marsh 1935). There may also be no clinical signs of the disease and exposure to the bacterium is only found after testing for antibody levels in blood samples. In female ungulates, brucellosis presents as spontaneous late-term abortion, premature birth, or birth of nonviable calves (Thorne et al. 1979). Retained placenta, while common in domestic cattle, is not typically observed in elk but occasionally seen in bison (Thorne et al. 1978*a*, Rhyan et al. 2001). Secondary infections of the carpus have been observed in free-ranging elk and were commonly seen in research animals. Sometimes these infections regressed spontaneously but they were also a cause of severe lameness in chronically infected elk (Thorne et al. 1978*a*, 1979). It is unknown to what degree domestic and wild ungulates have the capacity to clear an infection with *B. abortus*, in large part because of the lack of a gold-standard antemortem diagnostic test for infection, rather than exposure. Calves born to infected cow elk and bison often lose their serologic titers, and isolation of *B. abortus* from seronegative animals through bacterial culture can occur (Thorne 1982, Rhyan et al. 2009).

Rare cases in moose (*Alces alces*) reportedly produce a fatal infection characterized by generalized weakness, emaciation, and death (Corner and Connell 1958, Jellison et al. 1953, Fenstermacher and Olsen 1942). Field infections in white-tailed (*Odocoileus virginianus*) and mule deer (*Odocoileus hemionus*) have not been described, although a very small number of deer have tested positive for *Brucella* antibodies and artificial inoculations have shown that deer theoretically may be infected with the bacterium (Baker et al. 1962; Davis et al. 1984). Nine bighorn sheep (*Ovis canadensis*) naturally exposed to an aborted elk fetus subsequently developed inflammation of the placenta and lymph nodes, as well as abortion in females and inflammation of the testes and epididymis in males (Kreeger et al. 2004).

Antibodies to *Brucella* spp. have been detected in carnivores, including coyotes (*Canis latrans*). One infected coyote shed *B. abortus* in postpartum vaginal discharge for 11 days, suggesting that coyotes (and possibly other scavengers) may be minor contributors to spreading *B. abortus* in the environment, especially in areas with a high prevalence in domestic and wild ungulates (Davis et al. 1979). Clinical signs in wild carnivores have not been described (Thorne 1982).

The percentage of animals testing positive for *Brucella* antibodies in blood samples collected from bison living within YNP ranges from 40–60% (Hobbs et al. 2015). The historical trend in elk brucellosis prevalence has been that elk wintering on the National Elk Refuge or state-run feedgrounds in Wyoming had prevalences of brucellosis in the 10–40% range, whereas nonfed elk had much lower prevalences. In the 21st century however, large parcels of land, often owned by hobby ranchers who discourage hunting on their properties, have led to large populations of elk congregating on these parcels. The prevalence of brucellosis in some of these unfed herds of elk are now higher than many elk populations receiving supplemental feed (Scurlock and Edwards 2010). Whereas it was once viewed that closing elk feedgrounds alone would potentially "get rid of" brucellosis in the GYA, managers must now consider strategies to reduce densities of elk herds on private land away from feedgrounds.

Transmission of the *B. abortus* bacteria occurs after contact with infected reproductive tissues and fluids through ingestion, inhalation, or direct contact with mucous membranes. Opportunities for exposure to the bacteria within bison populations have been observed with groups of pregnant females being curious and sniffing or licking an aborted fetus or newly born calf of an infected cow, sometimes leading to spikes in infection and even "abortion storms" subsequent to this type of group exposure. In contrast, elk are less social during calving and sequester themselves during normal birthing events and the dam almost invariably meticulously cleans the birth site and eats fetal afterbirth to reduce predator attractants. Therefore, transmission is more likely when elk spontaneously abort their fetus on feedlines or pastures where other elk are grazing, where cattle feeds are spread or stored, or on spring pastures or lowland grazing allotments that elk may migrate through when cattle have been turned out for early grazing.

Close association of elk and cattle with aborting elk has been the primary driver of new brucellosis cases in wild and domestic ungulates (Thorne et al. 1979). Whole-genome sequencing of *Brucella* isolates paired with epidemiologic data has implicated elk as the primary source of new brucellosis cases in cattle and domestic bison (Kamath et al. 2016, Rhyan et al. 2013) over the last 20 years or more. In contrast, there have been no domestic herds infected with brucellosis since 1998 that were infected by contact with wild bison. This led the most recent National Academy of Science review to suggest future efforts focus primarily on elk, their management and development of more effective vaccines for them or for cattle (National Academies of Sciences, Engineering, and Medicine 2017).

Transmission between species is more likely during winter months, when harsh climates and forage scarcity create more pressure for commingling at sites with more abundant forage (Figure 10.3). During the summer months, animals disperse across abundant grazing areas leading to much fewer encounters. Abundant data from radio-collared cattle, elk, and deer in the Starkey Experimental Forest in eastern Oregon have shown a behavioral cascade of larger bodied ungulates displacing smaller species from preferred patches (Coe et al. 2001). Close interactions that do occur are rare but may be very important in driving disease transmission in the summer months (zu Dohna et al. 2014). Individual elk who interact more closely with herdmates may be also more likely to interact across species barriers, especially when cattle are herded onto new pastures on public grazing allotments (Cuzzocreo 2017). In recent years, hunters of shed antlers have been anecdotally implicated in driving elk off late spring–early summer pasture and creating more commingling events.

Signs and Gross Lesions

Bovine brucellosis presents primarily as third trimester abortions in host species (Figure 10.4). Elk normally give birth from May to June, but abortions can occur as early as February. Fluids and placental membranes in normal calves born to infected dams can harbor *Brucella* bacteria, and shedding can occur as late as 17 days postpartum making the elk transmission window extend from February to June (Thorne 1982). Calves born during subsequent pregnancies are often healthy, although weak calves and neonatal mortality may occur (Thorne 1982).

Gross lesions in male bison may include inflammation of the testes and distension of the scrotum (Corner and Connell 1958). Rare cases in moose show evidence of inflammation of the pleural space, pericardium, and abdominal wall; enlarged lymph nodes; and intermittent areas of necrosis of the spleen, liver, and kidneys (Corner and Connell 1958; Fenstermacher and Olsen 1942; Jellison et al. 1953). Bull moose may also exhibit inflammation of the

Fig. 10.3. Risks of interspecies transmission of *Brucella abortus* in the Greater Yellowstone Area. Illustration by Laura Donohue.

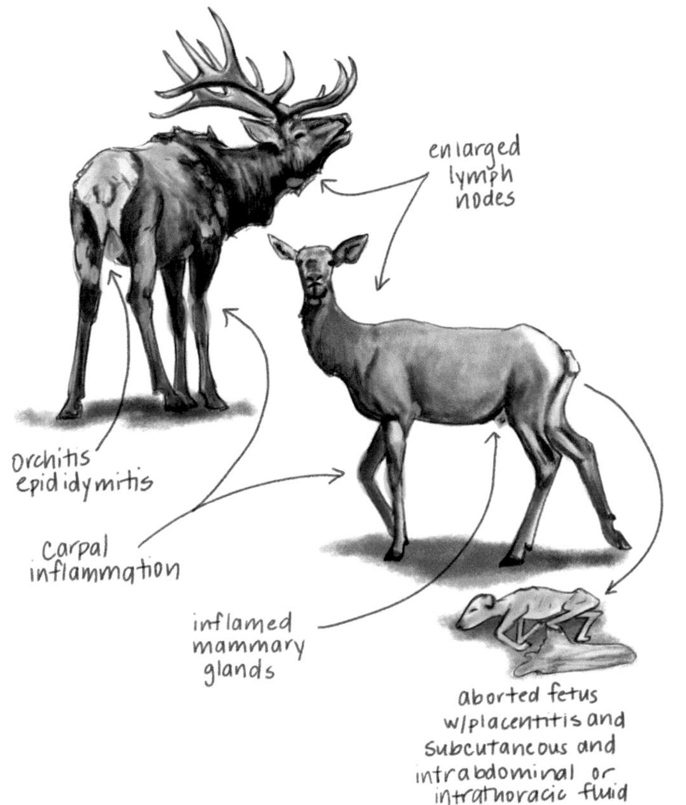

Fig. 10.4. Common clinical signs of bovine brucellosis in elk. Illustration by Laura Donohue.

testes and epididymis (Corner and Connell 1958; Fenstermacher and Olsen 1942). Infected elk often show few if any gross lesions other than enlargement of lymph nodes, especially associated with the head, mammary glands, and reproductive tract. Occasional inflammation to the carpi of infected elk is also observed. Aborted fetuses show nonspecific signs of fetal death and inflammation in the placenta (if still present). There may be clear red-tinged fluid under the skin of the fetus and within the abdominal and thoracic cavities. Fetal lungs may be edematous and meconium may be present in the stomach. Placentas are rarely found due to dams eating them. If present, they may be thickened with purulent exudate (Thorne 1982). Although retained placenta is a common sign of brucellosis in cattle, it does not occur frequently in elk (Thorne et al. 1978*a*, 1979).

Pathogenesis and Pathology

Brucella spp. invade the host ungulate through the mucous membranes of the eyes, nose, and mouth through investigating, ingesting, or inhaling the bacteria. The bacteria transmigrate through the epithelial cell lining of the respiratory or digestive tract where they are phagocytized by host macrophages. The bacteria localize in the lymph nodes in the head and then spread through the lymphatic system to the blood, establishing an intermittent bacteremia (National Academies of Sciences, Engineering, and Medicine 2017). Invasion of systemic lymph tissues causes a mild, acute lymphadenitis (Payne 1959).

During the third trimester of pregnancy a sugar called erythritol is produced by the placenta (Keppie et al. 1965), which promotes the growth of *Brucella* bacteria in those tissues. Bacteria preferentially colonize the reproductive tract during this period causing inflammation to the uterus and placenta. Often this inflammation leads to a spontaneous loss of the calf, 50% or more in elk, especially during the pregnancy immediately after infection (Payne 1959). Subsequent pregnancies have lower probabilities of abortion, but the host can still have bacteria present in birth fluids.

There are few microscopic lesions observed in tissues of infected adult elk and aborted fetuses other than nonspecific inflammation. Immunohistochemistry may reveal the presence of *Brucella* in affected tissues. More detailed specific histologic changes may be found in Thorne (1982).

Diagnosis

Before death, the initial diagnosis of brucellosis is made by evaluation of the blood for the presence of antibodies to the bacteria. Mature animals of both sexes should be sampled in order to obtain an accurate picture of the prevalence in the herd. There are multiple laboratory tests to evaluate the immune response, but all suffer from imperfect diagnostic accuracy (Thorne et al 1978*b*). If a blood sample is drawn from an animal too soon after they are infected, they might not have had enough time to build up antibodies and may test negative. Alternatively, chronic infections can also test negative, because the antibody levels may decrease over time (Benavides et al. 2017). Another major challenge with serologic testing is that numerous other bacteria share the identical sugar structure to *Brucella*, and their antibodies are indistinguishable. Because of these difficulties with antemortem testing, the definitive diagnosis for brucellosis is made by postmortem evaluation of the suspect animal and culturing relevant lymph nodes and reproductive tissues. This is also challenging as *Brucella spp.* are slow growing and require special growth conditions. To date, despite its own deficiencies, bacterial culture is the reference standard for the diagnosis of brucellosis.

Treatment and Outcomes

There is no accepted treatment for animals infected with bovine brucellosis. Brucellosis is zoonotic and therefore capable of causing illness in humans. A six-week treatment regimen of multiple antibiotics (commonly doxycycline and either streptomycin or rifampicin) is the accepted therapy for human infections. Due to the difficulty in regularly handling wild ungulates and domestic livestock, culling seropositive animals from a livestock herd is the only accepted course of action. In wild populations, this is controversial as the seropositive animals have often already aborted and are less likely to abort subsequent pregnancies. Therefore, these animals may provide some herd immunity to the rest of the population, making their removal questionable (Cotterill et al. 2019). General prevalence estimates in wild populations seem to indicate that elk may be more resistant to brucellosis than cattle or bison (Thorne 1982).

Bovine brucellosis is not a major source of mortality in wild elk and bison. In the absence of concurrent infections, such as bovine tuberculosis, there is not a meaningful effect on pregnancy probability nor survival of bison (Joly and Messier 2005). There is a meaningful decrease in the pregnancy rates of elk however (Cotterill et al. 2018b; Yang et al. 2019). Regardless of any population-level effects, wildlife management agencies spend a significant amount of time and money dealing with bovine brucellosis mainly due to the political and financial implications for and impacts on domestic livestock, which will be discussed in some detail in the subsequent sections.

Management and Control Strategies

Livestock Management

Appreciating the management constraints for bovine brucellosis in wildlife requires an understanding of the way in which the disease is managed in domestic livestock. In livestock, bovine brucellosis has historically been managed using a stamp-out approach to disease control, whereby infected livestock premises were completely depopulated and restocked. Prior to 2010, two or more affected premises in a two-year period of time would cause an entire state to lose its brucellosis-free status, forcing additional testing of animals prior to interstate movement or change of ownership. Currently, bovine brucellosis is only endemic in the GYA, making it inefficient to treat the entire livestock inventories of Idaho, Montana, and Wyoming the same regardless of whether the livestock spend time in the GYA or not. Instead, the US Department of Agriculture Animal and Plant Health Inspection Service has been operating under an interim rule to regionalize brucellosis control, whereby cases within the bounds of the GYA do not cause the entire state to lose its free status. Instead, each GYA state maintains a Designated Surveillance Area (DSA), the boundaries of which being determined based on the risk of brucellosis transmission from wildlife to livestock. Livestock that spend time within the DSA are required to undergo additional testing. Herds testing positive within the DSA boundaries are not routinely depopulated unless the herd prevalence is extremely high. Instead, the property in question is quarantined and animals may only be sold to slaughter or to an approved feedlot until the herd undergoes three consecutive whole-herd tests for the disease, separated by at least 30 days, all of which must be completely free from additional cases. A final post-calving assurance test must be conducted after all pregnant animals in the herd have given birth in order to catch any late-pregnancy seroconversion.

While the overall number of livestock premises affected by bovine brucellosis and resultant impacts to the US livestock industry have fallen dramatically over the last century, the cost to an individual producer can be devastating. While under quarantine, no animals can be sold for breeding purposes. Bred cows, which are much more valuable than many other categories of cattle, can only be sold to slaughter, which brings much less compensation to the owner. Additionally, while producers typically graze their animals on public lands in the summer, they may be unable to release their animals from quarantine for public land grazing and may be forced to provide hay for them throughout the summer, dramatically increasing the costs of their operation. If the prevalence of the disease is very high, they may also choose, or be asked to, depopulate their herd, losing valuable genetics that have been built up over generations. Although all cattle in the GYA are required to be vaccinated with *B. abortus* strain RB51, vaccination does not necessarily prevent infection but instead reduces abortions (Olsen 2000).

Wildlife Management

Even though bovine brucellosis is not a major source of mortality or reproductive failure in wild elk and bison, wildlife management agencies face immense pressures to control the disease in wild populations to protect ranching interests in the GYA. The control methods commonly employed by federal and state regulators for livestock are unfeasible and ineffective for wildlife. Therefore, management strategies to control the spillover of brucellosis from wild elk and bison populations to neighboring cattle populations focus on spatiotemporal separation of wild and domestic animals during the months leading up to elk and bison calving seasons.

In winter months, bison living in Yellowstone National Park naturally migrate north and west off of federal lands (and away from federal protections) (Plumb et al. 2009). The number of animals leaving YNP is correlated with the severity of winter snowpack and the size of the bison population occupying YNP. Bison are not well tolerated outside of the boundaries of YNP. Damage to fencing, competition for pasture, and brucellosis have all been used as arguments for bison to be aggressively managed outside of YNP boundaries. Hazing (forced movement) of bison back into YNP regularly occurs in the spring.

Outside of YNP boundaries in Montana, wild bison are classified as livestock and managed by the Montana Department of Livestock. Routinely this means most bison leaving YNP near Gardner, Montana, are shot during special local hunts.

The Interagency Bison Management Plan was formed in 2000 through a court-mediated settlement with the goals of (1) maintaining a wild free-ranging bison population; (2) reducing the risk of brucellosis transmission from bison to cattle, primarily through preventing spatiotemporal overlap between cattle and bison; (3) managing bison that leave YNP and enter Montana; and (4) maintaining Montana's brucellosis-free status for domestic livestock (IBMP 2016). There are now few cattle operations left in the areas where bison leave YNP, which reduces the opportunity for commingling between wild bison and domestic livestock and the associated risk of *B. abortus* transmission (National Academies of Sciences, Engineering, and Medicine 2017).

To help control the bison population, a variable number of animals are trapped as they leave YNP and sent to slaughter. This is a hotly contested practice, and alternatives have been sought. One such alternative is a bison quarantine program, whereby captured animals are tested repeatedly over a long period of time and then sent to various groups who are interested in acquiring the valuable "pure" bison genetics of the YNP population into their outside herds (Halbert et al. 2012; Herman et al. 2014).

Elk Feedgrounds

Winter elk feeding in the southern GYA in Wyoming was started in 1910 to mitigate animal starvation during harsh winters (Smith 2001). Artificial feeding has also been used to give these animals another source of food and reduce their incentive to visit winter cattle haystacks, feedlines, and stack yards. The unintended consequence of this practice is that artificial feeding of elk has increased their densities during the brucellosis abortion window, increasing the prevalence of brucellosis in fed populations of elk. Other diseases have occurred on feedgrounds, such as elk hoof rot (Couch et al. 2021). More recently, the threat posed by the slow but incessant expansion of chronic wasting disease (CWD) toward fed elk populations has increased pressure on Wyoming to reduce supplemental feeding practices.

Any reduction in artificial feeding would likely reduce elk population numbers. Discussions of feedground closures have faced stark opposition from hunting groups, outfitters, and the ranching community. Artificial feedgrounds allow the Wyoming Game and Fish Department to manage elk populations in those herd units well above their natural carrying capacity. Hunters enjoy having a supply of readily available elk to harvest each hunting season. Outfitters benefit financially from this practice as well, since nonresident hunters of big or trophy game in National Forest wilderness areas in Wyoming must be accompanied by a licensed guide or resident companion. The ranching community has argued that the immediate closure of feedgrounds would lead to elk populations dispersing to ranching operations on private land, leading to an immediate increase in brucellosis cases in livestock in the near term.

In 2006, a pilot program to test elk on feedgrounds for brucellosis and send animals with *Brucella* antibodies to slaughter was initiated in the Pinedale Elk Herd Unit in Wyoming (Scurlock and Edwards 2010). The sample prevalence was greatly reduced on the treated Pinedale feedgrounds during this time in comparison to the control feedground. However, the program was extremely costly. It was estimated that each removed animal would cost an average of over $9,000 (Boroff et al. 2016). Once the pilot project was discontinued, the prevalence rebounded to pretreatment levels within four years (Cotterill et al. 2018*a*; National Academies of Sciences, Engineering, and Medicine 2017). It is unlikely that such a practice would be tolerable long term without a defined policy end goal, such as feedground closure, once prevalence reached a nadir.

Without the political support to decrease the number of active feedgrounds, wildlife managers in

Wyoming have focused on strategies to decrease animal densities on feedgrounds throughout the winter and reduce the number of elk remaining on the feedgrounds during the early spring months. Distributing small piles of supplemental feedstuff over a larger area on feedgrounds in order to lower animal densities has been estimated to reduce total contacts with an aborted fetus by over 70% (Creech et al. 2012) (Figure 10.5). Managers have also attempted to stop feeding as soon as early spring conditions allow in an effort to further reduce the probability of brucellosis transmission (Cross et al. 2007). When

Elk Feeding Strategies

Low-density feeding enables elk to eat where they would like, more spread out. This decreases spread of diseases like brucellosis and CWD.

High-density feeding enables easy over winter feeding, but with less freedom to spread out. This enables brucellosis, CWD, and other disease spread.

Fig. 10.5. Differences in artificial feeding practices alter the risks of disease transmission in Wyoming elk. Illustration by Laura Donohue.

elk are found commingling with livestock operations, producers can call the local wildlife management agency who will haze animals off those properties. This is variably successful, and elk routinely will return to the same properties even after being hazed multiple times in a season if feed is still available.

Vaccination

Two main *B. abortus* vaccines are available for use in livestock. Since 1941, Strain 19 (S19) was used to great effect in reducing the number of affected livestock herds in the United States. The challenge with S19 is a transiently elevated antibody response that will cross-react with blood tests for diagnosis of exposure to naturally acquired *B. abortus* field strain. In 1996, use of S19 was largely discontinued in favor of Rough Brucella 51 (RB51), which does not have the sugar side chain that cross-reacts on the diagnostic tests (Ragan 2002). RB51 is used preferentially in the United States today; in Wyoming, S19 use requires prior authorization from the state veterinarian. While RB51 greatly reduces abortions and subsequent intraherd transmission, it does not prevent infection with *Brucella*.

While calfhood vaccination is required for all cattle and domestic bison grazing in the DSAs and adult vaccination with RB51 is coming into favor to help boost immunity levels, neither S19 nor RB51 have proven to be a good tool in preventing infections in elk. Elk calves vaccinated with RB51 at multiple doses in a captive research facility and challenged with *B. abortus* did not have any meaningful protection against abortion induced by *B.abortus* compared to unvaccinated calves (Cook et al. 2002; Kreeger et al. 2002). A captive study of elk calves hand-injected with S19 and challenged with field strain *B. abortus* showed minimal efficacy at preventing infection and only modest efficacy at preventing abortion (Roffe et al. 2004). This was confirmed when an S19 ballistic vaccination program that had been initiated on elk feedgrounds in Wyoming in 1985 was terminated after 30 years (Figure 10.6). The program was discontinued since there was little evidence that the vaccine was having a meaningful effect on *B. abortus* prevalence nor the number of reproductive failures (Maichak et al. 2017). The RB51 vaccine has been used in bison in YNP; however, it is extremely inefficient to deliver enough doses to meaningfully lower the disease prevalence in the entire bison population inhabiting YNP, and a sustained effort is required in order to keep the prevalence at a low level (Treanor et al. 2010). The recent National Academies of Sciences (NAS) review noted that development of effective vaccines for elk or more effective vaccines for cattle that might prevent infection may be feasible and could do a great deal to mitigate the ongoing brucellosis problem in the GYA (National Academies of Sciences, Engineering, and Medicine 2017).

Financial, Legal, Political and Social Factors

All *Brucella* spp. are able to infect humans and cause intermittent fevers (called undulant fever) along with other nonspecific flu-like symptoms such chills, weakness, headache, inappetence, body aches, and weight loss (Thorne 1982). There have been two confirmed cases of brucellosis acquired by hunters in the Greater Yellowstone Area, specifically in Madison County in southwestern Montana (Greater Yellowstone Interagency Brucellosis Committee 2010). Human brucellosis in the developed world is largely regulated by intake pasteurization systems; however, the raw milk movement has led to RB51 infections in humans from consuming unpasteurized milk from dairy cattle persistently infected with the modified-live vaccine strain, which is also resistant to rifampin and penicillin (Sfeir 2018, Negrón, et al. 2019). *Brucella abortus* is also recognized as the most common laboratory-acquired infection in the world (Weinstein and Singh 2009).

It is easier to prevent a disease becoming endemic in a wildlife host than to control or eradicate a disease once endemic (Wobeser 1994). The preponder-

Fig. 10.6. (A) Brucella abortus Strain 19 biobullet used in ballistic vaccination of elk on Wyoming feedgrounds. (B) Wyoming Game and Fish technician Grant Gertsch vaccinates elk on a Wyoming feedground. Photographs by Mark Gocke, Wyoming Game and Fish Department.

ance of managers (and the NAS review) state that, without other tools at their disposal, especially an easily deliverable and efficacious vaccine, eradication of brucellosis in wildlife is currently beyond our reach. Without aggressive monitoring studies, the effectiveness of any specific strategy to manage brucellosis in wild populations is difficult to determine. In general, lower cost strategies like low-density feeding have shown more "bang per buck" than higher cost strategies like test-and-slaughter (Boroff et al. 2016). From a purely economic perspective, management strategies that reduce elk populations (such as test-and-slaughter and feedground closures) must be considered alongside the downstream effects of decreased hunter success and corresponding demand for hunting licenses and paid outfitting. One study predicted a 10% reduction in elk populations would cause a 3.5% decrease in resident elk hunting applicants and a 0.4–1.4% decrease in nonresident applicants (Kauffman et al. 2012). This would have negative financial consequences for the hunting-based economy of the GYA and, based on the Muddy Creek experience, may do little to reduce disease prevalence in the long run.

As mentioned previously, increased animal numbers in close proximity to one another from management practices such as artificial feeding increase the risk of density-dependent disease transmission. In the face of diseases such as CWD, a systematic evaluation of the best way to maximize elk numbers in the southern GYA indicated that a reduction in feeding was most likely to result in a more robust elk population in the long term (Maloney et al. 2020). The study noted that, while the benefits of a more robust elk population were provided to the public, the costs of associated potential increase in brucellosis cases were born by private ranchers. The transfer of some of the increased benefits born by society to offset the private costs of probable increases in brucellosis cases remains a challenge to disease management in a complex system such as the GYA.

Even after CWD was detected in Grand Teton National Park in 2020, a few miles from the National Elk Refuge, the state of Wyoming has not shown any interest in reducing the number of elk feedgrounds. Instead, in the 2021 legislative session, the Wyoming legislature passed a bill that stripped the Wyoming Game and Fish Department of the authority to close any feedgrounds and instead gave that authority to the governor. Multiple lawsuits have been filed to challenge feeding practices and prevent new feedgrounds from being opened on federal land and, in 2019, the National Elk Refuge near Jackson, Wyoming, finalized an environmental assessment and step-down plan aimed at reducing feeding. Bison management in the northern GYA has also been fought in the courts with environmental groups challenging the slaughter of any bison coming out of YNP.

Wildlife management agencies are given the purview of conserving and managing all nonlisted wildlife species in their respective states. However, too often management resources and attention is allocated to species tied to revenue generation from the sale of hunting tags. In the case of elk in the GYA, the single-species focus on elk has led to the persistence of feedgrounds and management of population numbers far exceeding natural carrying capacities. Livestock and wildlife interests should be natural allies with the ever-increasing price of land and pressures to sell ranches and create more housing subdivisions in the GYA. Too often, however, creative solutions to wildlife and livestock disease management have remained out of reach and the differing sides have chosen to resort to the courts to impose decisions instead of continuing to work toward mutually beneficial compromise positions.

Summary

"Eliminating *B. abortus* transmission within wildlife populations (elk and bison) and from wildlife to cattle and domestic bison in the GYA—and by extension, eliminating it from the United States—is not feasible unless critical knowledge gaps are addressed. . . . Efforts to reduce brucellosis in the GYA will depend on significant cooperation among federal, state, and tribal entities and private stakeholders as they determine priorities and next steps in moving forward." The above quote from the National Academies of Sciences, Engineering, and Medicine report entitled, *Revisiting Brucellosis in the Greater Yellowstone Area* (National Academies of Sciences, Engineering, and Medicine 2017), exemplifies the One Health nature of this complex problem.

The question of public versus private good is a significant factor in the stand-off over brucellosis in the GYA. It remains to be seen whether the good to society of abundant wildlife and open spaces for wild and domestic animals to graze will be permanently impacted by our current inability to find viable solutions to important wildlife disease issues such as brucellosis.

LITERATURE CITED

Baker, M. F., G. J. Dills, and F. A. Hayes. 1962. Further experimental studies on brucellosis in white-tailed deer. Journal of Wildlife Management 26:27–31.

Becton, P. 1977. The national brucellosis program of the United States. Pages 403–411 *in* R. P. Crawford and R. J. Hidalgo, editors. Bovine brucellosis. An international symposium. Texas A&M University Press, College Station, USA.

Benavides, J. A., D. Caillaud, B. M. Scurlock, E. J. Maichak, W. H. Edwards, and P. C. Cross. 2017. Estimating loss of *Brucella abortus* antibodies from age-specific serological data in elk. EcoHealth 14:234–243.

Boroff, K., M. Kauffman, D. Peck, E. Maichak, B. Scurlock, and B. Schumaker. 2016. Risk assessment and management of brucellosis in the southern Greater Yellowstone Area (II): cost–benefit analysis of reducing elk brucellosis prevalence. Preventive Veterinary Medicine 134:39–48.

Brennan, A., P. C. Cross, K. Portacci, B. M. Scurlock, and W. H. Edwards. 2017. Shifting brucellosis risk in livestock coincides with spreading seroprevalence in elk. PLoS One 12:e0178780.

Coe, P., B. K. Johnson, J. W. Kern, S. L. Findholt, J. G. Kie, and M. Wisdom. 2001. Responses of elk and mule deer to cattle in summer. Journal of Range Management 54:205.

Cook, W. E., E. S. Williams, E. T. Thorne, T. J. Kreeger, G. Stout, K. Bardsley, H. Edwards, G. Schurig, L. A. Colby, F. Enright, et al. 2002. *Brucella abortus* strain RB51 vaccination in elk I. Efficacy of reduced dosage. Journal of Wildlife Diseases 38:18–26.

Corner, A. H. and R. Connell. 1958. Brucellosis in bison, elk, and moose in Elk Island National Park, Alberta, Canada. Canadian Journal of Comparative Medicine and Veterinary Science 22:9–21.

Cotterill, G. G., P. C. Cross, E. K. Cole, R. K. Fuda, J. D. Rogerson, B. M. Scurlock, and J. T. du Toit. 2018*a*. Winter feeding of elk in the Greater Yellowstone Ecosystem and its effects on disease dynamics. Philosophical Transactions of the Royal Society B Biological Sciences 373:20170093.

Cotterill, G. G., P. C. Cross, J. A. Merkle, J. D. Rogerson, B. M. Scurlock, and J. T. du Toit. 2019. Parsing the effects of demography, climate, and management on recurrent brucellosis outbreaks in elk. Journal of Applied Ecology 57:379–389.

Cotterill, G. G., P. C. Cross, A. D. Middleton, J. D. Rogerson, B. M. Scurlock, and J. T. du Toit. 2018*b*. Hidden cost of disease in a free-ranging ungulate: brucellosis reduces mid-winter pregnancy in elk. Ecology and Evolution 8:10733–10742.

Couch, C. E., B. L. Wise, B. M. Scurlock, J. D. Rogerson, R. K. Fuda, E. K. Cole, K. E. Szcodronski, A. J. Sepulveda, P. R. Hutchins, and P. C. Cross. 2021. Effects of supplemental feeding on the fecal bacterial communities of Rocky Mountain elk in the Greater Yellowstone Ecosystem. PLoS One 16:e0249521.

Creech, T. G., P. C. Cross, B. M. Scurlock, E. J. Maichak, J. D. Rogerson, J. C. Henningsen, and S. Creel. 2012. Effects of low-density feeding on elk–fetus contact rates on Wyoming feedgrounds. The Journal of Wildlife Management 76:877–886.

Cross, P. C., W. H. Edwards, B. M. Scurlock, E. J. Maichak, and J. D. Rogerson. 2007. Effects of management and climate on elk brucellosis in the Greater Yellowstone Ecosystem. Ecological Applications 17:957–964.

Cuzzocreo, M. 2017. Assessing spatial associations between Rocky Mountain elk and cattle for disease transmission potential. Thesis, University of Wyoming, Laramie, USA.

Davis, D. S., W. J. Boeer, J. P. Mims, F. C. Heck, and L. G. Adams. 1979. *Brucella abortus* in coyotes. I. A. serologic and bacteriologic survey in eastern Texas. Journal of Wildlife Diseases 15:367–372.

Davis, D. S., F. C. Heck, and L. G. Adams. 1984. Experimental infection of captive axis deer with *Brucella abortus*. Journal of Wildlife Diseases 20:177–179.

Fenstermacher, R., and O. W. Olsen. 1942. Further studies of diseases affecting moose III. Cornell Veterinarian 32:241–254.

Greater Yellowstone Interagency Brucellosis Committee. 2010. Wildlife and brucellosis in the Greater Yellowstone Area. An educational guide for hunters. Montana Fish, Wildlife, and Parks, Billings, USA. https://myfwp.mt.gov/getRepositoryFile?objectID=34841. Accessed 1 Oct 2022.

Halbert, N. D., P. J. P. Gogan, P. W. Hedrick, J. M. Wahl, and J. N. Derr. 2012. Genetic population substructure in bison at Yellowstone National Park. Journal of Heredity 103:360–370.

Herman, J. A., A. J. Piaggio, N. D. Halbert, J. C. Rhyan, and M. D. Salman. 2014. Genetic analysis of a *Bison bison* herd derived from the Yellowstone National Park population. Wildlife Biology 20:335–343.

Hobbs, N. T., C. Geremia, J. Treanor, R. Wallen, P. J. White, M. B. Hooten, and J. C. Rhyan. 2015. State-space modeling to support management of brucellosis in the Yellowstone bison population. Ecological Monographs 85:525–556.

Interagency Bison Management Plan (IBMP). 2016. 2016 IBMP Adaptive Management Plan. US Department of the Interior, National Park Service, Yellowstone National Park, West Yellowstone, Wyoming, USA. http://ibmp.info/Library/AdaptiveMgmt/2016_IBMP_Adaptive_Management_Plan_signedFINAL.pdf. Accessed 1 Oct 2022.

Jellison, W. L., C. W. Fishel, and E. L. Cheatum. 1953. Brucellosis in a moose, *Alces americanus*. Journal of Wildlife Management 17:1–22.

Joly, D. O., and F. Messier. 2005. The effect of bovine tuberculosis and brucellosis on reproduction and survival of wood bison in Wood Buffalo National Park. Journal of Animal Ecology 74:543–551.

Kamath, P. L., J. T. Foster, K. P. Drees, G. Luikart, C. Quance, N. J. Anderson, P. R. Clarke, E. K. Cole, M. L. Drew, W. H. Edwards, et al. 2016. Genomics reveals historic and contemporary transmission dynamics of a bacterial disease among wildlife and livestock. Nature Communications 7:11448.

Kauffman, M. E., B. S. Rashford, and D. E. Peck. 2012. Unintended consequences of bovine brucellosis management on demand for elk hunting in northwestern Wyoming. Human-Wildlife Interactions 6:12–29.

Keppie, J., A. E. Williams, K. Witt, and H. Smith. 1965. The role of erythritol in the tissue localization of the *Brucellae*. British Journal of Experimental Pathology 46:104–108.

Kreeger, T. J., W. E. Cook, W. H. Edwards, and T. Cornish. 2004. Brucellosis in captive Rocky Mountain bighorn sheep (*Ovis canadensis*) caused by *Brucella abortus* biovar 4. Journal of Wildlife Diseases 40:311–315.

Kreeger, T. J., W. E. Cook, W. H. Edwards, P. H. Elzer, and S. C. Olsen. 2002. *Brucella abortus* strain RB51 vaccination in elk II. Failure of high dosage to prevent abortion. Journal of Wildlife Diseases 38:27–31.

Maichak, E. J., B. M. Scurlock, P. C. Cross, J. D. Rogerson, W. H. Edwards, B. Wise, S. G. Smith, and T. J. Kreeger. 2017. Assessment of a strain 19 brucellosis vaccination program in elk. Wildlife Society Bulletin 41:70–79.

Maloney, M., J. A. Merkle, D. Aadland, D. Peck, R. D. Horan, K. L. Monteith, T. Winslow, J. Logan, D. Finnoff, C. Sims, et al. 2020. Chronic wasting disease undermines efforts to control the spread of brucellosis in the Greater Yellowstone Ecosystem. Ecological Applications 30:e02129.

Meagher, M., and M. E. Meyer. 1994. On the origin of brucellosis in bison in Yellowstone National Park: a review. Conservation Biology 8:645–653.

Mohler, J. R. 1917. Report of the Chief of the Bureau of Animal Industry, Pathological Division. United States Department of Agriculture, Washington, D.C., USA.

Murie, O. J. 1951. The elk of North America. The Stackpole Co., Harrisburg, PA, USA and Wildlife Management Institute, Washington, D.C., USA.

National Academies of Sciences, Engineering, and Medicine. 2017. Revisiting brucellosis in the greater Yellowstone area. National Academies Press, Washington, D.C., USA.

Negrón, M. E., G. A. Kharod, W. A. Bower, and H. Walker. 2019. Notes from the field: human *Brucella abortus* RB51 infections caused by consumption of unpasteurized domestic dairy products—United States, 2017–2019. Morbidity and Mortality Weekly Report 68:185.

Olsen, S. C. 2000. Immune responses and efficacy after administration of a commercial *Brucella abortus* strain RB51 vaccine to cattle. Veterinary Therapeutics 1:183–191.

Payne, J. M. 1959. The pathogenesis of experimental brucellosis in the pregnant cow. Journal of Pathogenic Bacteriology 78:447–463.

Plumb, G. E., P. J. White, M. B. Coughenour, and R. L. Wallen. 2009. Carrying capacity, migration, and dispersal in Yellowstone bison. Biological Conservation 142:2377–2387.

Ragan, V. E. 2002. The Animal and Plant Health Inspection Service (APHIS) brucellosis eradication program in the United States. Veterinary Microbiology 90:11–18.

Rhyan, J. C., K. Aune, T. Roffe, D. Ewalt, S. Hennager, T. Gidlewski, S. Olsen, and R. Clarke. 2009. Pathogenesis and epidemiology of brucellosis in Yellowstone bison: serologic and culture results from adult females and their progeny. Journal of Wildlife Diseases 45:729–739.

Rhyan, J. C., T. Gidlewski, T. J. Roffe, K. Aune, L. M. Philo, and D. R. Ewalt. 2001. Pathology of brucellosis in bison from Yellowstone National Park. Journal of Wildlife Diseases 37:101–109.

Rhyan, J. C., P. Nol, C. Quance, A. Gertonson, J. Belfrage, L. Harris, K. Straka, and S. Robbe-Auserman. 2013. Transmission of brucellosis from elk to cattle and bison, Greater Yellowstone Area, USA, 2002–2012. Emerging Infectious Diseases 19:1992–1995.

Roffe, T. J., L. C. Jones, K. Coffin, M. L. Drew, S. J. Sweeney, S. D. Hagius, P. H. Elzer, and D. Davis. 2004. Efficacy of single calfhood vaccination of elk with *Brucella abortus* strain 19. Journal of Wildlife Management 68:830–836.

Scurlock, B. M., and W. H. Edwards. 2010. Status of brucellosis in free-ranging elk and bison in Wyoming. Journal of Wildlife Diseases 46:442–449.

Sfeir, M. M. 2018. Raw milk intake: beware of emerging brucellosis. Journal of Medical Microbiology 67:681–682.

Smith, B. L. 2001. Winter feeding of elk in western North America. Journal of Wildlife Management 65:173–190.

Tessaro, S. V., L. B. Forbes, and C. Turcotte. 1990. A survey of brucellosis and tuberculosis in bison in and around Wood Buffalo National Park, Canada. Canadian Veterinary Journal 31:174–180.

Thorne, E. T. 1982. Brucellosis. Pages 54–63 *in* E. T. Thorne, N. Kingston, W. R. Jolley, and R. C. Bergstrom, editors. Diseases of wildlife in Wyoming, Second edition. Wyoming Game and Fish Department, Cheyenne, USA.

Thorne, E. T., J. K. Morton, and F. M. Blunt, and H. A. Dawson. 1978*a*. Brucellosis in elk. II. Clinical effects and means of transmission as determined through artificial infections. Journal of Wildlife Diseases 14:280–291.

Thorne, E. T., J. K. Morton, and W. C. Ray. 1979. Brucellosis, its effect and impact on elk in western Wyoming. Pages 212–220 *in* M. S. Boyce and L. D. Hayden-Wing, editors. North American elk: ecology, behavior, and management. University of Wyoming, Laramie, USA.

Thorne, E. T., J. K. Morton, and G. M. Thomas. 1978*b*. Brucellosis in elk. I. Serologic and bacteriologic survey in Wyoming. Journal of Wildlife Diseases 14:74–81.

Treanor, J. J., J. S. Johnson, R. L. Wallen, S. Gilles, P. H. Crowley, J. J. Cox, D. S. Maehr, P. J. White, and

G. E. Plumb. 2010. Vaccination strategies for managing brucellosis in Yellowstone bison. Vaccine 28(Supplement 5):F64–F72.

Tunnicliff, E. A. and H. Marsh. 1935. Bang's disease in bison and elk in the Yellowstone National Park and on the National Bison Range. Journal of the American Veterinary Medical Association 86:745–752.

Weinstein, R. A. and K. Singh. 2009. Laboratory-acquired infections. Clinical Infectious Diseases 49:142–147.

Wobeser, G. A. 1994. Investigation and management of disease in wild animals. Springer, Boston, Massachusetts, USA.

Yang, A., J. P. Gomez, C. G. Haase, K. M. Proffitt, and J. K. Blackburn. 2019. Effects of brucellosis serologic status on physiology and behavior of Rocky Mountain elk (*Cervus canadensis nelsoni*) in southwestern Montana, USA. Journal of Wildlife Diseases 55:304–315.

zu Dohna, H., D. E. Peck, B. K. Johnson, A. Reeves, and B. A. Schumaker. 2014. Wildlife-livestock interactions in a western rangeland setting: quantifying disease-relevant contacts. Preventive Veterinary Medicine 113:447–456.

11 Bluetongue, Epizootic Hemorrhagic Disease, and Cervid Adenoviral Hemorrhagic Disease

MARK G. RUDER, SONJA A. CHRISTENSEN

Introduction

Bluetongue (BT) and epizootic hemorrhagic disease (EHD) are caused by distinct viruses that are clinically indistinguishable in wild ruminants, particularly white-tailed deer (*Odocoileus virginianus*) and are collectively referred to as hemorrhagic disease (HD). Hemorrhagic disease outbreaks may be explosive mortality events under some situations. As the viruses are transmitted by insect vectors (i.e., *Culicoides* biting midges), the epidemiology of EHD and BT is complex, involving multiple virus serotypes and numerous ruminant hosts interacting with the insect vectors that span diverse habitats and climates. Globally, these patterns are changing rapidly, as reviewed by MacLachlan et al. (2019).

Another important viral disease of wild cervids that results in a hemorrhagic disease syndrome is adenoviral hemorrhagic disease (AHD). This disease impacts multiple native cervid species in western North America, where the virus is enzootic and causes significant mortality events in deer. While the virus causing AHD is transmitted directly between hosts and does not involve an insect vector, the clinical signs and lesions can mimic those of EHD and BT. Therefore, these diseases are often considered together and are discussed in this chapter.

Illustration by Laura Donohue.

Bluetongue and Epizootic Hemorrhagic Disease

History

The history of HD in North America was reviewed by Nettles and Stallknecht (1992). Although EHD and BT were first confirmed in North America in the 1950s–1960s, the first written accounts consistent with EHD and BT (i.e., dead deer being found along waterways) date back to the late 1800s (Shope et al. 1960, Stair et al. 1968). It was also the 1800s when BT virus (BTV) was first described in South Africa when European sheep were imported; however, there is evidence that BTV has circulated in ungulate populations in some parts of the world for over 1000 years (MacLachlan 2011, Alkhamis et al. 2021). The global history of EHD virus (EHDV) has not been as well investigated.

Etiology

CAUSATIVE AGENTS AND GEOGRAPHIC DISTRIBUTION

Both EHDV and BTV are double-stranded RNA viruses in the genus *Orbivirus* (family Reoviridae). Worldwide, seven EHDV serotypes and at least 27 BTV serotypes exist. Historically, only EHDV-1 and -2 and BTV-2, -10, -11, -13, and -17 were known to circulate in North America (Stallknecht et al. 2002). However, since ~2000, multiple virus serotypes have been detected for the first time in North America, including EHDV-6 and 11 different BTV serotypes (Rivera et al. 2021). While many of these viruses have only been detected on one or two occasions, others are likely established (e.g., EHDV-6, BTV-1, and BTV-3) (Ruder et al. 2016, 2017; Schirtzinger et al. 2018). The factors underlying this emergence and their impacts on wildlife are unclear, but complex and often interacting factors—including climate change—may be involved.

These insect-transmitted viruses are distributed throughout temperate and tropical regions around the globe (~40–50°N and 35°S latitude) in climates supportive of the *Culicoides* spp. vectors (e.g., biting midges, no-see-ums, and punkies). In the United States, these viruses are broadly distributed, and EHDV or BTV has been detected in all but a few New England states (Ruder et al. 2015, Stallknecht et al. 2015); however, significant variation in virus distribution, diversity, and frequency of detection exists within this distribution (Ruder et al. 2016). In Canada, sporadic reports of BTV or EHDV have occurred in southern British Columbia, Alberta, and Saskatchewan (Pare et al. 2012, Pybus et al. 2014). Recently, EHDV-2 was confirmed in Ontario (Allen et al. 2020), consistent with the northern expansion of HD observed in the upper midwestern and northeastern United States (Stallknecht et al. 2015).

Demographics and Field Biology

HOST RANGE

Despite the nearly global distribution of EHDV and BTV, severe disease in wildlife has only been consistently reported in North America. Large-scale outbreaks and severe disease are most common in white-tailed deer, although large BT and EHD outbreaks have occurred in the western United States in pronghorn (*Antilocapra americana*), mule deer (*Odocoileus hemionus*), and bighorn sheep (*Ovis canadensis*) (Thorne et al. 1988; Robinson et al. 1967; Howerth et al. 2001; Ruder et al. 2015, 2016). Both EHDV and BTV also infect elk (*Cervus canadensis*), bison (*Bison bison*), and some livestock species, although outbreaks of severe clinical disease are infrequent. Cattle (*Bos taurus*) typically have subclinical infections with EHDV and BTV, although mild disease is occasionally reported in the United States (Baldwin et 1991, Stevens et al. 2015). Domestic sheep (*Ovis aries*) are susceptible to BTV but not EHDV, and BT outbreaks in sheep occasionally occur in North America (Miller et al. 2010).

CULICOIDES VECTORS AND TRANSMISSION

Both EHDV and BTV are transmitted to ruminant hosts by *Culicoides* biting midges (Diptera:

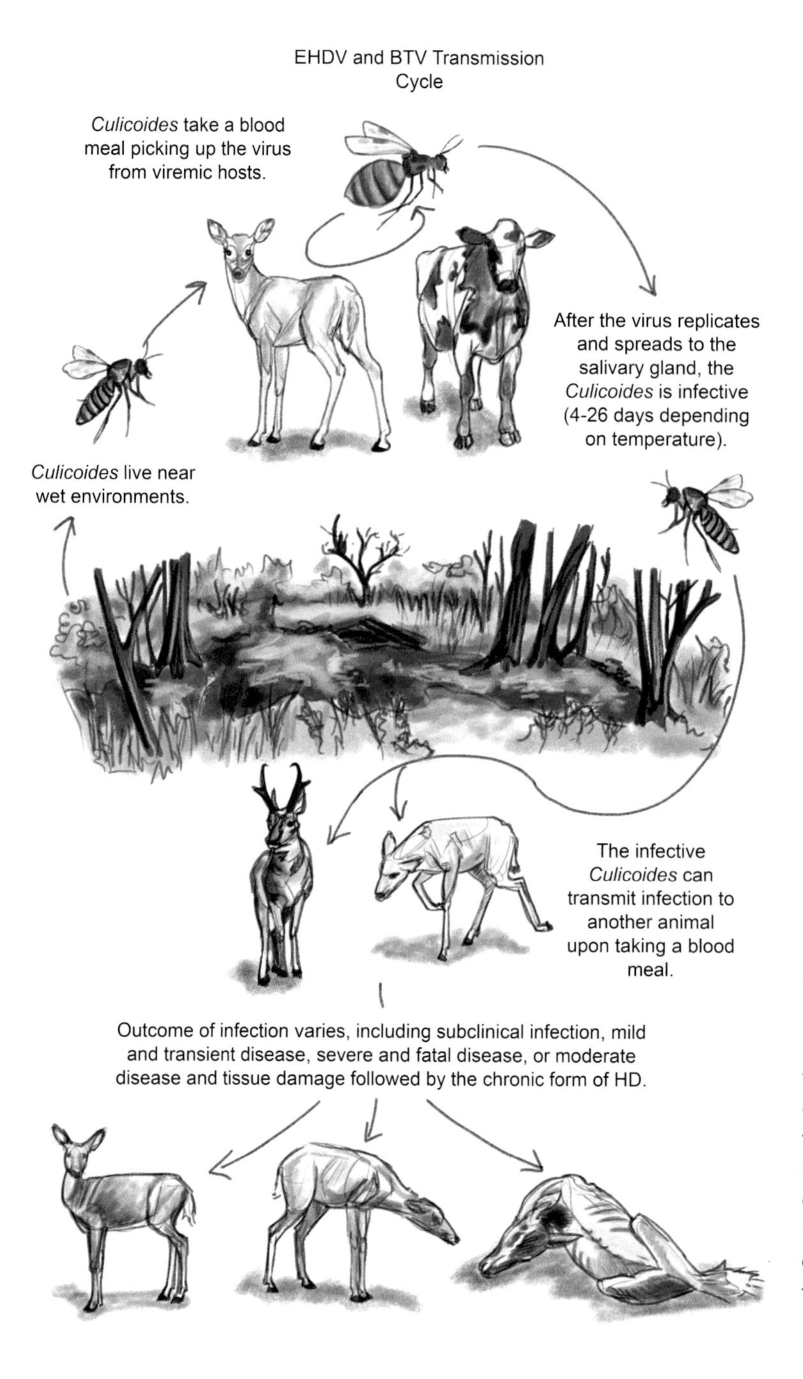

Fig. 11.1. Ruminant host–*Culicoides* vector transmission cycle of epizootic hemorrhagic disease virus (EHDV) and bluetongue virus (BTV). The two viruses are maintained in nature in a ruminant-biting midge vector cycle. This cycle is dependent on adult *Culicoides* biting a viremic host (wild or domestic ruminant) and then transmitting the virus to a susceptible host. The transmission cycle is heavily influenced by temperature because replication of the virus in *Culicoides* biting midges is temperature dependent. Further, many life history characteristics of biting midges are influenced by temperature (e.g., survivorship and larval development). Therefore, virus transmission is favored by hot summertime conditions and may continue until *Culicoides* are no longer infective, or temperatures decrease to 0°C (freezing will kill the adult *Culicoides*). Illustration by Laura Donohue.

Ceratopogonidae), which are small (1–3 mm), primarily blood-feeding flies (Figure 11.1). Approximately 150 *Culicoides* species exist in North America, but only two (*Culicoides sonorensis* and *C. insignis*) are proven vectors of EHDV and BTV (Gibbs and Greiner 1989 Mellor et al. 2000, McGregor et al. 2021). However, other *Culicoides* species (e.g., *C. stellifer*, *C. debilipalpis*, and *C. venustus*) are suspected vectors (Smith et al. 1996, Ruder et al. 2015, McGregor et al. 2019). Individual *Culicoides* species breed in diverse semiaquatic habitats, including tree cavities and mud along streams, ponds, wetlands, and more ephemeral habitats (Figure 11.2), and breeding habitat requirements (e.g., nutrient

Fig. 11.2. *Culicoides* larval habitat. Adult *Culicoides* lay eggs in diverse semiaquatic habitats, ranging from mud along streams, seeps, springs, and ponds, to leaf litter, dung, or tree cavities. Specific locations vary with species. The larval stages then develop in the substrate until pupation and the eventual emergence of winged adults. In temperate locations, *Culicoides* overwinter in the substrate as larvae. The rate of larval development is faster at higher temperatures, which can help fuel large vector populations during outbreaks, further amplifying transmission to ruminant hosts. Some *Culicoides* species, such as *C. sonorensis*, are known to associate with livestock (e.g., cattle) and take advantage of abundant hosts and breeding habitats enriched by animal waste. Illustration by Laura Donohue.

loading and soil and water chemistry) may in part define species distribution. *Culicoides* are not strong fliers and often remain near productive breeding sites (Mellor et al. 2000) but may passively disperse long distances on wind currents, which may facilitate virus spread (Sellers and Maarouf 1991, Kedmi et al. 2010). Aspects of the ruminant host–*Culicoides* vector transmission cycle of EHDV and BTV in North America have been reviewed in greater detail in recent literature (Pfannenstiel et al. 2015, Ruder et al. 2015, Noronha et al. 2021).

EPIDEMIOLOGIC PATTERNS

Hemorrhagic disease outbreaks predictably occur in the late summer to early fall and persist until after the first frost when adult *Culicoides* are killed off (Couvillion et al. 1981). In temperate regions, adult *Culicoides* are not active during the winter and populations overwinter as larvae in substrate. In southern areas, this seasonal transmission may extend throughout much of the year in locations where adult *Culicoides* can survive cooling temperatures for extended durations (Mayo et al. 2014). Numerous environmental factors impacting insect populations also influence EHDV and BTV transmission cycles and disease outbreaks. For instance, some weather variables (e.g., temperature and precipitation) can impact the life history of *Culicoides* spp., and populations can reach high densities under ideal conditions (Mellor et al. 2000, Mullens et al. 2004, Lysyk and Danyk 2007). During seasonal peaks in abundance, *Culicoides* biting rates on ruminant hosts are highest on warm summer nights lacking significant winds. Additionally, virus replication in *Culicoides* is temperature dependent (Mullens et al. 2004). Thus, weather-related factors (e.g., high temperature and low summer precipitation), in addition to many habitat and landscape variables, may be important in creating ideal conditions for large HD outbreaks (Sleeman et al. 2009, Berry et al. 2013, Stallknecht et al. 2015). Severe drought, which is often correlated with high temperature and low summer precipitation, is associated with increased risk for HD outbreaks at northern latitudes of the eastern United States (Christensen et al. 2020), likely related to impacts on biting midge life history, host density around water sources, and other unknown factors.

In North America, EHD and BT exist in enzootic and epizootic cycles, likely driven by interacting climatic, habitat, and landscape factors that in turn impact local and regional host and vector populations (Ruder et al. 2015). Most reported HD outbreaks occur in a broad band extending from the southeastern and mid-Atlantic United States, diagonally through the central and northern Great Plains and Rocky Mountains to California and the Pacific Northwest, extending into southwestern Canada (Figure 11.3). Reports of outbreaks are less common in the highly enzootic regions of the US Southwest but are increasing in epizootic regions of the upper Midwest and Northeast (Howerth et al. 2001, Ruder et al. 2015). Latitudinal gradients are common, and in general, enzootic southern latitudes experience longer transmission seasons, more frequent (e.g., annually or every other year) virus activity, and milder outbreaks compared with northern epizootic regions where HD outbreaks with high mortality typically occur on multiyear cycles (e.g., every 3–5 years). However, these spatial and temporal patterns are general trends, and isolated outbreaks likely occur regularly in both enzootic and epizootic regions. Shifts in these patterns have been recognized in recent years (Stallknecht et al. 2015, MacLachlan et al. 2019).

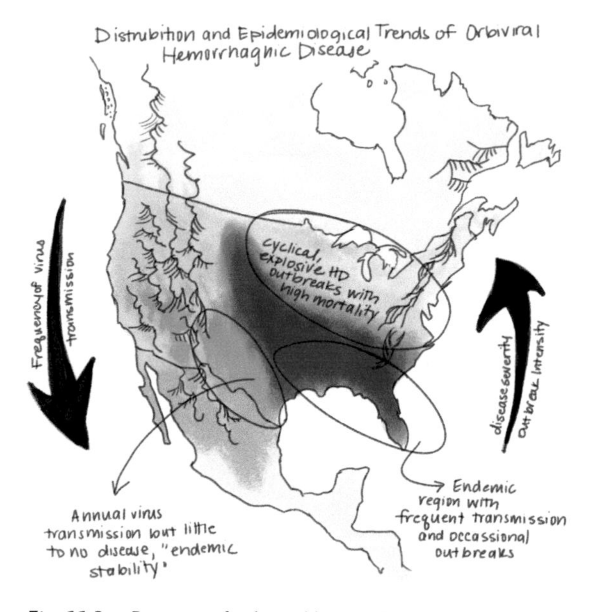

Fig. 11.3. Patterns of orbiviral hemorrhagic disease in North America. The map approximates the geographic distribution and general epidemiological patterns of hemorrhagic disease in North America. Both epizootic hemorrhagic disease virus and bluetongue virus have broad geographic distributions, but disease frequency, severity, and distribution vary markedly across different regions. The distribution of adenoviral hemorrhagic disease overlaps with epizootic hemorrhagic disease and bluetongue in western North America. Illustration by Laura Donohue.

In some enzootic regions of the United States (e.g., parts of Texas, Kansas, and Florida), white-tailed deer coexist with EHDV and BTV in a state of enzootic stability (Stallknecht et al. 1996, Flacke et al. 2004). These regions are characterized by high levels of infection (antibody prevalence may approach 100%) with very little disease. Enzootic stability is the result of acquired immunity (i.e., antibody from a previous infection), high passive immunity in fawns (i.e., maternally derived antibodies), and mechanisms of innate immunity (i.e., genetic resistance) via coevolution of the host and pathogen (Quist et al. 1997; Gaydos et al. 2002*a*, *b*; Stilwell et al. 2021).

Fig. 11.4. Field signs of epizootic hemorrhagic disease. Photograph showing a classic field sign of hemorrhagic disease: sick and dead deer observed along waterways. A male white-tailed deer with respiratory difficulty (note the open-mouth breathing) was observed staggering down a small stream with stagnant and receding shorelines during a severe drought in Kansas. The buck is walking past the carcass of a severely decomposed buck. After death, epizootic hemorrhagic disease virus serotype 2 was isolated from the lung and spleen. Photograph by Shane Hesting, Kansas Department of Wildlife and Parks.

Clinical Signs and Gross Lesions

Detailed descriptions and images of clinical signs together with postmortem lesions have been described elsewhere (Howerth et al. 2001; Ruder et al. 2012, 2016; Nemeth and Yabsley 2021). The signs and postmortem findings caused by EHD and BT are indistinguishable, and the typical incubation period after being bitten by an infectious *Culicoides* is 4–8 days. In deer, HD ranges from inapparent or mild to rapid death without premonitory signs. In many cases, the signs may initially include fever, loss of appetite, rough hair coat, lethargy, salivation, and reddening of mucous membranes and nonhaired skin. Additional signs may include lameness, bleeding from the rectum or oronasal cavity (hence the name hemorrhagic disease), respiratory difficulty, recumbency, and terminal convulsions (Howerth et al. 2001). Others may succumb due to chronic sequelae many weeks after acute disease. Deer behavior makes it difficult to observe clinical signs, and dead or moribund deer are often the most common field sign (Figure 11.4).

Interpretation of necropsy lesions can be complicated by decomposition often encountered in wildlife die-offs. Three forms of HD in white-tailed deer have been described: peracute, acute, and chronic (Howerth et al. 2001). The peracute form may include fluid accumulation in body cavities (effusions), such as the chest cavity and heart sac, or inside tissues (edema) (Figure 11.5). Edema is most common in the lungs, conjunctiva, and under the skin of the head, neck, and ventrum. Gross lesions may be subtle or absent when animals die peracutely (within days of infection). As the disease progresses to the acute form and a bleeding disorder ensues, widespread hemorrhages are common, especially in the gastrointestinal tract and heart. With time, erosions and ulcers may be observed in the oral cavity. Deer surviving acute HD may present with lesions of the chronic form, including healing or infected ulcers in the oral cavity, scarring of the rumen mucosa, and cracked or sloughing hooves in multiple limbs. Given this tissue damage, deer with chronic HD are usually in poor nutritional condition and may present as overwinter mortalities subsequent to malnutrition and secondary problems. The chronic form does not represent continued active infection but, rather, sequelae after recovery from acute HD. In the western

Fig. 11.5. White-tailed deer with epizootic hemorrhagic disease (EHD). Photograph showing classic necropsy lesions from a white-tailed deer that died from EHD in North Carolina. Note the straw-colored fluid accumulating in the chest cavity (pleural effusion), interlobular septae widened by fluid (pulmonary edema), and hemorrhage in the lungs. After death, epizootic hemorrhagic disease virus serotype 2 was isolated from the lung and spleen. Photograph by Jason Allen, North Carolina Wildlife Resources Commission.

United States, EHDV infection has been associated with testicular lesions and antler abnormalities in mule deer (Fox et al. 2015, Roug et al. 2022).

Pathogenesis and Pathology

Virus is introduced to deer through the skin by the bite of infected *Culicoides* midges and disseminates throughout the body in the bloodstream. The high concentration of EHDV or BTV in the blood (viremia) serves as a source of virus for noninfected *Culicoides* that take a blood meal from the infected animal, facilitating spread of infections. Peak viremia typically coincides with a high fever 5–8 days after infection, and a large percentage of feeding *Culicoides* may become infected (Mendiola et al. 2019). Widespread virus replication occurs in endothelial cells (cells lining blood vessels). The ensuing virus-induced damage of endothelium exhausts the body's blood clotting system and results in fluid and blood loss into surrounding tissues. The degree of EHDV- and BTV-induced endothelial damage may play a role in variation in susceptibility to clinical disease between species (MacLachlan et al. 2009). Additionally, many complex immunological factors may be important in disease development and severity, as reviewed elsewhere (MacLachlan et al. 2009, Howerth 2015).

Diagnosis

Syndromic surveillance has proven effective for long-term monitoring of HD (Stallknecht and Howerth 2004, Ruder et al. 2016). This simple approach takes advantage of HD seasonality, field signs, and classic gross necropsy findings, which collectively may lead to a presumptive diagnosis of HD. Throughout temperate regions of North America, epizootic mortality in susceptible species during late summer through early fall should lead to a suspicion of HD. Gross necropsy findings can support this suspicion, histopathology may be incriminating, but a confirmed diagnosis of EHD or BT must be based on virus detection (i.e., reverse transcriptase polymerase chain reaction assay and virus isolation). Field necropsy and diagnostic testing are important steps in not only confirming HD, but also in identifying other causes of deer mortality. Other enzootic (e.g., deer adenovirus, bacterial septicemia, and vesicular stomatitis) and foreign animal diseases (e.g., heartwater and foot-and-mouth disease) can mimic EHD and BT (Howerth et al. 2001).

Preferred samples include fresh (not frozen) spleen and lung, although lymph node and whole blood also are suitable. Since subclinical EHDV or BTV infections occur, positive test results must be supported by consistent field signs or necropsy findings. This consideration is especially true when testing cattle or other less susceptible wildlife species. Multiple factors can interfere with the success of virus detection from field submissions such as stage of infection, postmortem interval, ambient temperature, and sample storage. Success of virus isolation typically is high when a deer dies early in the course of disease (e.g., 5–10 days after infection) and tissues are collected soon after death, promptly refrigerated, and shipped to the laboratory. However, deer dying >2–3 weeks after infection likely have a low level of virus remaining in the tissues, complicating virus detection. Diagnosis of the chronic form of HD is based on consistent field and necropsy findings because the virus typically has been cleared from the body and cannot be isolated at this stage.

As reviewed by others, multiple serological techniques are available for detecting antibodies against EHDV and BTV (Wilson et al. 2015, Rojas et al. 2019). Detecting EHDV or BTV antibodies is not a tool for individual animal diagnosis. However, serology can give an indication of virus activity within a region and is an excellent tool for surveillance and research, provided the interpretation carefully considers both the population and regional epidemiology.

Outcome and Treatment

The outcome of infection in wild and domestic ruminant hosts can vary from recovery to severe disease and death, both within and between species, even in highly susceptible species (e.g., white-tailed deer) (Howerth et al. 2001). Although not completely understood, this variation in clinical outcome may relate to passive and acquired immunity and mechanisms of innate resistance. Previous infection with EHDV or BTV will result in immunity to that particular virus and can also improve the infection outcome for other serotypes (Quist et al. 1997; Gaydos et al. 2002*a*). However, previous infection with EHDV does not protect against subsequent challenge with BTV and vice versa (Quist et al. 1997). In white-tailed deer fawns, maternally derived antibodies against EHDV persist for up to 18 weeks and greatly reduce clinical disease and viremia, helping fawns survive early virus exposure (Stilwell et al. 2021). In addition to these mechanisms of passive and acquired immunity, there is evidence that some southern white-tailed deer subspecies (i.e., *O. virginianus texanus*) have a level of innate resistance to clinical disease caused by EHDV (Gaydos et al. 2002*b*). Collectively, these mechanisms define many of the patterns of infection and disease observed across North America, including enzootic stability in parts of the southern United States and explosive outbreaks of fatal disease in the northeastern states.

There is no treatment for EHD and BT in wild ruminants. Although vaccination of captive cervids is common and is being explored for captive wildlife in zoological institutions (McVey and MacLachlan 2015, Sunwoo et al. 2020), there is no practical or effective application of such vaccines to free-ranging wild ruminant populations.

Management

Hemorrhagic disease was once considered a potential threat to white-tailed deer restoration efforts in the eastern United States, but populations proved resilient and thrived despite periodic HD. However, the recent detections of new viruses, expanding range of HD, and increasing frequency and intensity of outbreaks, are all indicators that changes are occurring (Stallknecht et al. 2015). Despite the potential for significant localized mortality events, few studies have examined the impact of EHD or BT on wild ruminant populations. Estimated white-tailed deer population losses of 6–16% in Missouri, 20% in West Virginia, and 33% in Delaware have been documented during localized EHD outbreaks (Fischer et al. 1995, Gaydos et al. 2004, Haus et al. 2020), and losses in penned white-tailed deer can be much higher. Postoutbreak white-tailed deer population assessment and recovery has been documented in Michigan, although population effects were only significant within 1 km of a riparian corridor (Christensen et al. 2021). Future HD outbreak investigations should aim to document regional and local mortality rates, their short-term and long-term impact, and the effectiveness of reduction in legal deer harvest following large-scale HD outbreaks (essentially the only current management tool for populations experiencing high HD mortality) (Ruder et al. 2016). While disease management options are limited for HD outbreaks, risk communication (and managing expectations) for stakeholders in years with optimal environmental conditions may assist with surveillance, population monitoring, and ensuring appropriate harvest effort. Landowners and outdoor recreationists are critical for detecting and reporting dead or dying deer to state wildlife agencies (Figure 11.6).

Continuing to monitor HD is important to furthering our understanding of HD risk factors and the potential impact on wild ruminant populations,

Fig. 11.6. Hemorrhagic disease outbreaks occur annually in North America but vary greatly in their intensity and geographic scope. Some general trends and scenarios are depicted here. White-tailed deer populations in parts of the northern United States are highly susceptible to epizootic hemorrhagic disease (EHD) virus and bluetongue virus (BT) virus, and outbreaks may become severe under ideal conditions (e.g., hot and dry). In such situations, wildlife managers may be faced with carcass removal and disposal because of exceptionally high mortality encountered during large outbreaks. Illustration by Laura Donohue.

especially white-tailed deer, mule deer, pronghorn, and bighorn sheep, in a changing landscape. While these North American species may have lived with EHDV and BTV for more than a century, if the epidemiological patterns continue to change, so too could the impact on populations. In a changing environment (e.g., climate change, growing human populations, and changing land-use patterns), the future challenges facing wild ruminant populations are difficult to predict. If HD cycles increase in frequency and intensity and expand north in a sustained manner, HD may become a more prominent and frequent nonhunting mortality factor. Further, there is no evidence of building herd immunity among deer populations in areas of northern expansion (Southeastern Cooperative Wildlife Disease Study, unpublished data). The impact these changes may have on species management goals set by agencies or on the health and trajectory of certain populations are not clear. As trends in disease patterns shift, so too must our perception of the potential for HD to impact populations.

Financial, Legal, Political, and Social Factors

The exploitation of one of North America's greatest natural resources, the white-tailed deer, in the captive cervid industry for personal gain has surged in recent years. Within this industry, the natural selection that has defined the vigor and resiliency of free-ranging white-tailed deer is largely being replaced with selective breeding for certain morphological characteristics (e.g., antler mass). In southern enzootic regions of the United States, free-ranging deer have evolved with orbiviruses and, through natural selection, coexist in a state of enzootic stability. Within the captive cervid industry, the artificial practices and genetic manipulation for traits important to humans without regard for overall health or susceptibility to infectious diseases is concerning. Such practices may elevate disease risk, as is suggested by high mortality rates of captive deer in southern locations compared with wild deer on the other side of the fence. The impact of captive cervid operations on wild populations is not well understood but may elevate HD risk to adjacent wild deer populations through the concentration of highly susceptible captive deer, amplification of circulating virus in a localized area, and the impact of the altered land-use on vector species composition and abundance, as some important vector species are known to associate with livestock and potentially other captively held animals.

Hunter perceptions of the magnitude of mortality from an HD outbreak may alter hunter effort in a region following an outbreak. This self-regulation of hunter harvest has been anecdotally described but may have important implications for deer management goals in a given state or jurisdiction. Some wildlife agencies have closed hunting seasons, refunded hunting licenses, or lowered the number of permits issued following severe EHD outbreaks because of the perceived impact on the local or regional population (A. Norton, South Dakota Game, Fish, and Parks, personal communication). The financial impact of such management actions has not been studied. The effect of hunter perceptions of outbreak severity on deer harvest rates has not been documented to our knowledge, although the majority of deer harvest typically occurs after EHD or BT mortality events have occurred, and deer have largely recovered from infection. Neither EHDV nor BTV are infectious to humans; however, deer with the chronic form of HD that develop secondary bacterial septicemia or abscessation may not be suitable for consumption.

Adenoviral Hemorrhagic Disease

History

The first confirmed outbreak of AHD occurred in California in 1993 and resulted in a die-off of >1,000 wild mule and black-tailed deer in a 17-county area (Woods et al. 1996). However, retrospective analysis of archived deer tissues from California suggest

AHD has occurred since at least 1981 (Woods et al. 2018). Since 1993, AHD has been documented most frequently among free-ranging Columbian black-tailed deer (*O. h. columbianus*) in California, as well as a variety of other wild cervid species, primarily in the western United States (Miller et al. 2017).

Etiology

Adenoviral hemorrhagic disease is caused by *Deer atadenovirus A* (OdAdV-1), a nonenveloped double-stranded DNA virus within the *Atadenovirus* genus (Woods et al. 1997, Kauffman et al. 2021). The *Atadenovirus* genus contains seven adenovirus species that infect diverse taxonomic groups, including avian, mammalian, and reptilian species. Despite antigenic similarities, OdAdV-1 is genetically and biologically distinct from other ruminant adenoviruses (e.g., ovine and bovine) (Lapointe et al. 1999; Lehmkuhl et al. 2001; Zakhartchouk et al. 2002; Miller et al. 2017). Phylogenetic analysis of OdAdV-1 obtained from multiple cervid species in the western United States spanning nearly two decades suggests OdAdV-1 is highly conserved between species, and across time and space, but that multiple genotypes exist (Miller et al. 2017).

Demographics and Field Biology

HOSTS AND GEOGRAPHIC DISTRIBUTION

Known hosts for OdAdV-1 include mule deer, white-tailed deer, elk, and moose (*Alces alces*). Outbreaks of severe AHD are most common in certain mule deer subspecies, particularly Columbian black-tailed deer, whereas reports of AHD in white-tailed deer, elk, and moose have typically been sporadic outbreaks involving a small number of animals (Sorden et al. 2000, Shilton et al. 2002, Fox et al. 2017, Woods et al. 2018, Burek-Huntington et al. 2021). Disease has been replicated under experimental conditions in black-tailed deer and white-tailed deer (Woods et al. 1997, 1999, 2001). Experimental infection studies have shown domestic cattle and sheep are not susceptible to infection with OdAdV-1 (Woods et al. 2008, Imus et al. 2019).

The geographic distribution and prevalence of OdAdV-1 are poorly understood and have not been comprehensively investigated. Sporadic and often severe AHD outbreaks are well documented in California and Oregon, but OdAdV-1 has also been documented in nearby states in western North America (e.g., Washington, Idaho, Wyoming, and Colorado) (Miller et al. 2017, Fox et al. 2017) with sporadic detections in more distant locations (e.g., Iowa, Ontario, and Alaska) (Sorden et al. 2000, Shilton et al. 2002, Burek-Huntington et al. 2021).

EPIDEMIOLOGY

The current understanding of AHD epidemiology has been discussed in more detail elsewhere (Woods et al. 2018, Kauffman et al. 2021). While AHD has been documented in wild cervids in many areas of western North America, large-scale mortality events have not been reported outside of California and Oregon (Miller et al. 2017), possibly related to host and virus factors. Although cases can occur throughout the year, findings in California from 1990–2014 suggest most confirmed AHD cases occur May–October, with a peak in July and August (Woods et al. 2018). Juveniles are more susceptible to fatal AHD than yearlings and adults, and fatalities in many outbreaks predominately involve juveniles (Boyce et al. 2000; Woods et al. 1999, 2018). However, all-age die-offs do occur (Woods et al. 2018). In highly enzootic areas (e.g., California), there is evidence of annual AHD activity with intermittent large outbreaks (Woods et al. 2018).

Transmission through direct (cervid-to-cervid) contact is efficient, although the exact routes are unknown, and the potential for indirect (environment-to-cervid) transmission is not well investigated (Woods et al. 2018). Infected animals shed virus from oral and nasal secretions soon after infection, which appears to be an important source of virus for other cervids (Woods et al. 1999). Arthropod vectors are not involved. How the virus is maintained in a population of deer or an ecosystem is not clear, especially during interepizootic periods (Woods et al. 2018, Kauffman et al. 2021). However, intermittent, low-level shedding by cervids having subclinical infections may play a role (Kauffman et al. 2021). Subclinically infected dams may be a source of virus to neonatal offspring (Kauffman et al. 2021).

Clinical Signs and Gross Lesions

Based on experimental infections, the incubation period for AHD is 4–16 days (Woods et al. 1999). Two forms of the disease are recognized: systemic and localized. Common signs of systemic disease may include salivation, foaming at the mouth, labored to open-mouth breathing, lethargy, and diarrhea that may be bloody. Often, affected deer are found dead without observation of clinical signs. Animals with the localized form may have swollen muzzles with ulcers and abscesses in the oral region that may lead to salivation and anorexia. With time, these animals often have chronic loss of condition.

Animals may concurrently develop lesions of both the systemic and localized forms. Most common systemic form lesions are those of systemic vasculitis and include fluid-filled lungs (pulmonary edema), fluid accumulation in the chest cavity (pleural effusion), and dark, blood-filled intestines (Woods et al. 1999). In addition to pulmonary edema, fibrin accumulation in the chest cavity is common (Figure 11.7). With the localized form, vasculitis is associated with necrotic ulcers in the forestomachs and oral cavity (i.e., lips, hard palate, pharynx, and tongue) that may extend into the nasal cavity and surrounding tissues and are typically complicated by secondary bacterial infections. Draining lymph nodes may be enlarged. Histologically, vasculitis involving larger vessels and endothelial intranuclear inclusions are consistent features. Clinical signs and lesions of AHD mimic those of BT and EHD and must be distinguished with diagnostic assays, especially in western North America where the viral ranges overlap. See Woods et al. (1999, 2018) for additional images of lesions.

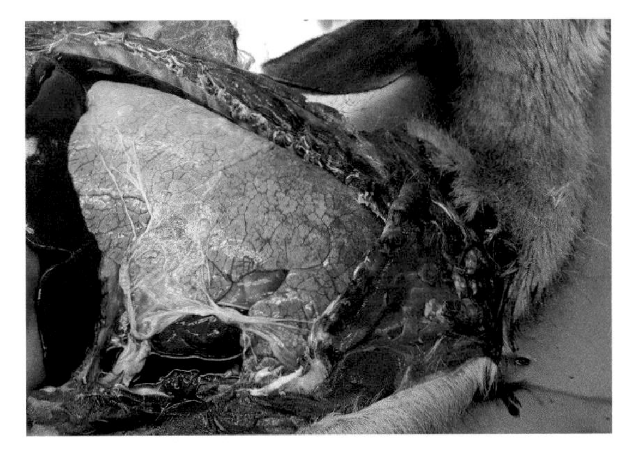

Fig. 11.7. Mule deer fawn with the systemic form of adenoviral hemorrhagic disease (AHD). Photograph showing gross necropsy lesions from a mule deer fawn that died from AHD in Colorado. Note the fibrin sheet accumulating on the surface of the lungs, interlobular septae widened by fluid (pulmonary edema), and the accumulation of red-tinged fluid in the chest cavity (pleural effusion). Photograph by Karen Fox, Colorado Parks and Wildlife.

Pathogenesis and Pathology

Once infected, the endothelial cell lining of medium to large blood vessels is the primary target of the virus (Woods et al. 1999). Comparatively, EHDV and BTV target the endothelial cells lining the microvasculature (i.e., capillaries; Howerth et al. 2001). In the systemic form, vasculitis can occur throughout the body but is most common in vessels of the lungs and alimentary tract. The vascular damage and inflammation in the lungs results in proteinacous fluid accumulating in the lungs and chest cavity, which can ultimately lead to severe respiratory disease and death. In some cases, virus-induced damage to blood vessels is localized to the oral cavity and upper alimentary tract. This localized vascular damage disrupts blood flow to tissues and results in necrosis and ulceration of those tissues, which are then easily invaded by opportunistic bacteria to form larger abscesses (Woods et al. 1999). It is not clear why some animals infected with OdAdV-1 develop the systemic form and others the localized form (Woods et al. 1999). Some animals may develop lesions consistent with both forms (Woods et al. 2018). The vascular damage OdAdV-1, EHDV, and BTV cause is similar enough that the body's clotting factors can be depleted (sometimes referred to as consumptive coagulopathy), resulting in the common gross lesions of hemorrhage and edema that makes it hard to initially determine which virus is the cause.

Diagnosis

Outbreaks of mortality in susceptible cervid species involving predominantly animals <1 year of age should raise suspicion for AHD. A presumptive diagnosis can be made based on observation of the aforementioned clinical signs and postmortem lesions, including intranuclear inclusions in endothelial cells. Confirmation is made by virus isolation or molecular detection of viral DNA from tissues of an affected animal. Immunohistochemistry is also helpful in the diagnosis of AHD. Euthanized or recently dead suspect animals should be submitted to a diagnostic laboratory for necropsy. If a field necropsy is performed, samples of major organs should be fixed in 10% buffered formalin for histopathology. For suspected systemic AHD, fresh, refrigerated lung tissue (and other tissues with lesions) should be submitted for virus isolation. Diagnosing localized infections can be challenging because of bacterial contamination and the presence of virus can be transient (Woods et al. 1999, 2018). Diagnosis depends on documenting gross and histologic lesions, ruling out other diseases, and confirming other cases of AHD in the herd. Oral swabs have been useful diagnostic samples during survey of live animals (Kauffman et al. 2021). A reliable and commercially available serological test would be an extremely beneficial surveillance and research tool when exploring epidemiological questions (e.g., host range, geographic distribution, and prevalence), but cross-reactivity of OdAdV-1 antibodies with those of related ruminant viruses is problematic. Finally, the many similarities in gross lesions between AHD, BT, and EHD make it important to verify suspected field cases with appropriate diagnostic assays.

Outcome and Treatment

There is no vaccine or treatment for AHD. Infection outcome ranges from inapparent disease with recovery to rapid death. Death may occur in as little as 3–5 days with the systemic form of AHD (Boyce et al. 2000, Woods et al. 2018). With the localized form, animals may survive longer and lose nutritional condition as lesions become complicated by bacterial infections, ultimately dying from starvation or sepsis (Woods et al. 1999, 2018). The underlying reasons AHD is more common in neonates and juveniles are not clear but may be related to factors impacting the immune response, such as age-related variation, stress, and nutrition (Miller et al. 2017). The prevalence of subclinical infections or mild cases that deer survive are poorly investigated regardless of age class but could provide a better understanding of AHD epidemiology (Miller et al. 2017). Further, the impact of genetic variation observed with distinct strains of OdAdV-1 on virulence remains an important area of study (Miller et al. 2017).

Management

To minimize risk of OdAdV-1 introduction and spread, wildlife managers often restrict certain activities that may artificially congregate deer (e.g., supplemental water, mineral licks, supplemental feeding, or baiting). Caution and hesitancy should always be exercised when considering the translocation of wild animals. Certainly, AHD should be an important consideration when translocating rehabilitated, captive, or wild cervids, especially in areas with a history of AHD (Boyce et al. 2000). Although AHD has not been documented in free-ranging white-tailed deer in the eastern United States, the experimental susceptibility of white-tailed deer to OdAdV-1, occasional documentation of AHD in this species in enzootic areas, and the documentation of AHD in captive white-tailed deer raises concerns that the disease could emerge in the eastern United States (Sorden et al. 2000, Woods et al. 2001, Miller et al. 2017). Within the captive cervid industry, vigilance for AHD-like disease that includes diagnostic investigation should be expected to help identify and mitigate risk to both captive and wild cervids.

Financial, Legal, Political, and Social Factors

Considering most disease mitigation approaches for AHD center on restricting what people can do (e.g., rehabilitation, providing supplemental resources, and translocating), wildlife managers sometimes face pushback from certain stakeholder groups. It is well understood that baiting and feeding of wild cervids increases risk of direct or indirect exposure to a variety of pathogens, and these practices impact numerous disease systems (e.g., CWD, bovine tuberculosis, and bovine brucellosis). This is also likely true of AHD, as the feeding and baiting of deer by people concentrates cervids and can result in frequent nose-to-nose contact, as well as indirect contact at the point source of food.

In a highly enzootic region of northern California, AHD was most commonly diagnosed (1990–2014) in nonmigratory black-tailed deer that reside in coastal areas with dense human populations (Woods et al. 2018). This may simply be related to the high visibility of deer in this area. However, these suburban and exurban areas also support dense populations of deer, and such cohabitation of both people and deer are likely only to become more common in the future, potentially facilitating deer-to-deer transmission of directly transmitted viruses and increasing overall AHD risk. Other artificial activities, such as feeding and baiting, would further increase disease risk. Public outreach and education are important management endeavors, such as agency efforts to communicate the risk of providing supplemental feed and water to wild deer populations.

Summary

Three different viruses (OdAdV-1, EHDV, and BTV) can cause large-scale die-offs in North American deer (and the latter two may involve other species) that are impossible for the biologist to distinguish from one another without significant veterinary diagnostic support. All three viruses attack the endothelial linings of blood vessels, which in many cases results in a hemorrhagic disease presentation. There is no treatment for these diseases once animals show signs; however, management approaches differ somewhat because BTV and EHDV are vectored by biting midges, and OdAdV-1 is directly transmitted from deer-to-deer.

LITERATURE CITED

Alkhamis, M. A., C. Aguilar-Vega, N. M. Fountain-Jones, K. Lin, A. M. Perez, and J. M. Sanchez-Vizcaino. 2020. Global emergence and evolutionary dynamics of bluetongue virus. Scientific Reports 10:21677.

Allen, S. E., J. L. Rothenburger, C. M. Jardine, A. Ambagala, K. Hooper-McGrevy, N. Colucci, T. Furukawa-Stoffer, S. Vigil, M. Ruder, and N. M. Nemeth. 2019. First detection of epizootic hemorrhagic disease in wild white-tailed deer in Ontario, Canada. Emerging Infectious Diseases 25:832–834.

Baldwin, C. A., D. A. Mosier, S. J. Rogers, and C. R. Bragg. 1991. An outbreak of disease in cattle due to bluetongue virus. Journal of Veterinary Diagnostic Investigation 3:252–255.

Berry, B. S., K. Magori, A. C. Perofsky, D. E. Stallknecht, and A. W. Park. 2013. Wetland cover dynamics drive hemorrhagic disease patterns in white-tailed deer in the United States. Journal of Wildlife Diseases 49:501–509.

Boyce, W. M., L. W. Woods, M. K. Keel, N. J. MacLachlan, C. A. Porter, and H. D. Lehmkuhl. 2000. An epizootic of adenovirus-induced hemorrhagic disease in captive black-tailed deer (*Odocoileus hemionus*). Journal of Zoo and Wildlife Medicine 31:370–373.

Burek-Huntington, K., M. M. Miller, and K. Beckmen. 2021. Adenovirus hemorrhagic disease in moose (*Alces americanus gigas*) in Alaska, USA. Journal of Wildlife Diseases 57:418–422.

Christensen, S. A., M. G. Ruder, D. M. Williams, W. F. Porter, and D. E. Stallknecht. 2020. The role of drought as a determinant of hemorrhagic disease in the eastern United States. Global Change Biology 26:3799–3808.

Christensen, S. A., D. M. Williams, B. A. Rudolph, and W. F. Porter. 2021. Spatial variation of white-tailed deer (*Odocoileus virginianus*) population impacts and recovery from epizootic hemorrhagic disease. Journal of Wildlife Diseases 57:82–93.

Couvillion, C. E., V. F. Nettles, W. R. Davidson, J. E. Pearson, and G. A. Gustafson. 1981. Hemorrhagic disease among white-tailed deer in the Southeast from 1971 through 1980. Proceedings of the United States Animal Health Association 85:522–537.

Fischer, J. R., L. P. Hansen, J. R. Turk, M. A. Miller, W. H. Fales, and H. S. Gosser. 1995. An epizootic of hemorrhagic disease in white-tailed deer (*Odocoileus virginianus*) in Missouri: necropsy findings and population impact. Journal of Wildlife Diseases 31:30–36.

Flacke, G. L., M. J. Yabsley, B. A. Hanson, and D. E. Stallknecht. 2004. Hemorrhagic disease in Kansas: enzootic stability meets epizootic disease. Journal of Wildlife Diseases 40:288–293.

Fox, K. A., L. Atwater, L. Hook-Hanks, and M. Miller. 2017. A mortality event in elk (*Cervus elaphus nelsoni*) calves associated with malnutrition, pasteurellosis, and deer adenovirus in Colorado, USA. Journal of Wildlife Diseases 53:674–676.

Fox, K. A., B. Diamond, F. Sun, A. Clavijo, L. Sneed, D. N. Kitchen, and L. L. Wolfe. 2015. Testicular lesions and antler abnormalities in Colorado, USA mule deer (*Odocoileus hemionus*): a possible role for epizootic hemorrhagic disease virus. Journal of Wildlife Diseases 51:166–176.

Gaydos, J. K., J. M. Crum, W. R. Davidson, S. S. Cross, S. F. Owen, and D. E. Stallknecht. 2004. Epizootiology of an epizootic hemorrhagic disease outbreak in West Virginia. Journal of Wildlife Diseases 40:383–393.

Gaydos, J. K., W. R. Davidson, F. Elvinger, E. W. Howerth, M. Murphy, and D. E. Stallknecht. 2002*a*. Cross-protection between epizootic hemorrhagic disease virus serotypes 1 and 2 in white-tailed deer. Journal of Wildlife Diseases 38:720–728.

Gaydos, J. K., W. R. Davidson, F. Elvinger, D. G. Mead, E. W. Howerth, and D. E. Stallknecht. 2002*b*. Innate resistance to epizootic hemorrhagic disease in white-tailed deer. Journal of Wildlife Diseases 38:713–719.

Gibbs, E. P. J., and E. C. Greiner. 1989. Bluetongue and epizootic hemorrhagic disease. Pages 39–70 *in* T. P. Monath, editor. The arboviruses: epidemiology and ecology. Volume 2. CRC Press, Boca Raton, Florida, USA.

Haus, J. M., J. R. Dion, M. M. Kalb, E. L. Ludwig, J. E. Rogerson, and J. L. Bowman. 2020. Interannual variability in survival rates for adult female white-tailed deer. The Journal of Wildlife Management 84:675–684.

Howerth, E. W. 2015. Cytokine release and endothelial dysfunction: a perfect storm in orbivirus pathogenesis. Veterinaria Italiana 51:275–281.

Howerth, E. W., D. E. Stallknecht, and P. D. Kirkland. 2001. Bluetongue, epizootic hemorrhagic disease, and other orbivirus-related diseases. Pages 77–97 *in* E. S. Williams and I. K. Barker, editors. Infectious diseases of wild mammals. Third edition. Iowa State University Press, Ames, USA.

Imus, J. K., H. D. Lehmkuhl, and L. W. Woods. 2019. Resistance of colostrum-deprived domestic lambs to infection with deer adenovirus. Journal of Veterinary Diagnostic Investigation 31:78–82.

Kauffman, K. M., T. Cornish, K. Monteith, B. Schumaker, T. LaSharr, K. Huggler, and M. Miller. 2021. Detection of *Deer Atadenovirus A* DNA in dam and offspring pairs of Rocky Mountain mule deer (*Odocoileus hemionus hemionus*) and Rocky Mountain elk (*Cervus canadensis nelsoni*). Journal of Wildlife Diseases 57:313–320.

Kedmi, M., Y. Herziger, N. Galon, R. M. Cohen, M. Perel, C. Batten, Y. Braverman, Y. Gottlieb, N. Shpigel, and E. Klement. 2010. The association of winds with the spread of EHDV in dairy cattle in Israel during an outbreak in 2006. Preventive Veterinary Medicine 96:152–160.

Lapointe, J.-M., J. F. Hedges, L. W. Woods, G. H. Reubel, and N. J. MacLachlan. 1999. The adenovirus that causes hemorrhagic disease of black-tailed deer is closely related to bovine adenovirus-3. Archives of Virology 144:393–396.

Lehmkuhl, H. D., L. A. Hobbs, and L. W. Woods. 2001. Characterization of a new adenovirus isolated from black-tailed deer in California. Archives of Virology 146:1187–1196.

Lysyk, T. J., and T. Danyk. 2007. Effect of temperature on life-history parameters of adult *Culicoides sonorensis* (Diptera: Ceratopogonidae) in relation to geographic origin and vectorial capacity for bluetongue virus. Journal of Medical Entomology 44:741–751.

MacLachlan, N. J. 2011. Bluetongue: history, global epidemiology, and pathogenesis. Preventive Veterinary Medicine 102:107–111.

MacLachlan, N. J., C. P. Drew, K. E. Darpel, and G. Worma. 2009. The pathology and pathogensis of bluetongue. Journal of Comparative Pathology 141:1–16.

MacLachlan, N. J., S. Zientara, W. C. Wilson, J. A. Richt, and G. Savini. 2019. Bluetongue and epizootic hemorrhagic disease viruses: recent developments with these globally re-emerging arboviral infections of ruminants. Current Opinion in Virology 34:56–62.

Mayo, C. E., B. A. Mullens, W. K. Reisen, C. J. Osborne, E. P. J. Gibbs, I. A. Gardner, and N. J. MacLachlan. 2014. Seasonal and interseasonal dynamics of bluetongue virus infection of dairy cattle and *Culicoides sonorensis* midges in northern California—implications for virus overwintering in temperate zones. PLoS One 9:e106975.

McGregor, B. L., D. Erram, B. W. Alto, J. A. Lednicky, S. M. Wisely, and N. D. Burkett-Cadena. 2021. Vector competence of Florida *Culicoides insignis* (Diptera: Ceratopogonidae) for epizootic hemorrhagic disease virus serotype 2. Viruses 13:410.

McGregor, B. L., K. E. Sloyer, K. A. Sayler, O. Goodfriend, J. M. Campos Kraur, C. Acevedo, X. Zhang, D. Mathias, S. M. Wisely, and N. D. Burkett-Cadena. 2019. Field data implicating *Culicoides stellifer* and *Culicoides venustus* (Diptera: Ceratopogonidae) as vectors of epizootic hemorrhagic disease virus. Parasites and Vectors 12:258.

McVey, D. S., and N. J. MacLachlan, 2015. Vaccines for prevention of bluetongue and epizootic hemorrhagic disease in livestock: a North American perspective. Vector-Borne and Zoonotic Diseases 15:385–396.

Mellor, P. S., J. Boorman, and M. Baylis. 2000. *Culicoides* biting midges: their role as arbovirus vectors. Annual Review of Entomology 45:307–340.

Mendiola, S. Y., M. K. Mills, E. Maki, B. S. Drolet, W. C. Wilson, R. Berghaus, D. E. Stallknecht, J. Breitenbach, D. S. McVey, and M. G. Ruder. 2019. EHDV-2 infection prevalence varies in *Culicoides sonorensis* after feeding on infected white-tailed deer over the course of viremia. Viruses 11:271:1–15.

Miller, M. M., J. Brown, T. Cornish, G. Johnson, J. O. Mecham, W. K. Reeves, and W. Wilson. 2010. Investigation of a bluetongue disease epizootic caused by bluetongue virus serotype 17 in sheep in Wyoming. Journal of the American Veterinary Medical Association 237:955–959.

Miller, M. M., T. E. Cornish, T. E. Creekmore, K. Fox, W. Laegreid, J. McKenna, M. Vasquez, and L. W. Woods. 2017. Whole-genome sequences of *Odocoileus hemionus* deer adenovirus isolates from deer, moose, and elk are highly conserved and support a new species in the genus *Atadenovirus*. Journal of General Virology 9: 2320–2328.

Mullens, B. A., A. C Gerry, T. J. Lysyk, and E. T. Schmidtmann. 2004. Environmental effects on vector competence and virogensis of bluetongue virus in *Culicoides*; interpreting laboratory data in a field context. Veterinaria Italiana 40:160–166.

Nemeth, N. M., and M. J. Yabsley, editors. 2021. Field manual of wildlife diseases in the southeastern United States. Fourth edition. Southeastern Cooperative Wildlife Disease Study, Athens, Georgia, USA.

Nettles, V. F., and D. E. Stallknecht. 1992. History and progress in the study of hemorrhagic disease of deer. Pages 449–516 *in* R. E. McCabe, editor. Transactions of the 57th North American Wildlife and Natural Resources Conference. Wildlife Management Institute, Washington, D.C., USA.

Noronha, L. E., L. W. Cohnstaedt, J. A. Richt, and W. C. Wilson. 2021. Perspectives on the changing landscape of epizootic hemorrhagic disease virus control. Viruses 13:1–13.

Pare, J., D. W. Geale, M. Koller-Jones, K. Hooper-McGrevy, E. J. Golsteyn-Thomas, and C. A. Power. 2012. Serological status of Canadian cattle for brucellosis, anaplasmosis, and bluetongue in 2007–2008. Canadian Veterinary Journal 53:949–956.

Pfannenstiel, R. S., B. A. Mullens, M. G. Ruder, L. Zurek, L. W. Cohnstaedt, and D. Nayduch. 2015. Management of North American *Culicoides* biting midges: current knowledge and research needs. Vector-Borne and Zoonotic Diseases 15:374–384.

Pybus, M. J., M. Ravi, and C. Pollock. 2014. Epizootic hemorrhagic disease in Alberta, Canada. Journal of Wildlife Diseases 50:720–722.

Quist, C. F., E. W. Howerth, D. E. Stallknecht, J. Brown, T. Pisell, and V. F. Nettles. 1997. Host defense responses associated with experimental hemorrhagic disease in white-tailed deer. Journal of Wildlife Diseases 33:584–599.

Rivera, N. A., C. Varga, M. G. Ruder, S. J. Dorak, A. L. Roca, J. E. Novakofski, and N. E. Mateus-Pinilla. 2021. Bluetongue and epizootic hemorrhagic disease in the United States of America at the wildlife–livestock interface. Pathogens 10:915.

Robinson, R. M., T. L. Hailey, C. W. Livingston, and J. W. Thomas. 1967. Bluetongue in the desert bighorn sheep. Journal of Wildlife Management 31:165–168.

Rojas, J. M., D. Rodriguez-Martin, V. Martin, and N. Sevilla. 2019. Diagnosing bluetongue virus in domestic ruminants: current perspectives. Veterinary Medicine: Research and Reports 10:17–27.

Roug, A., J. Shannon, K. Hersey, W. Heaton, and A. van Wettere. 2022. Investigation into causes of antler deformities in mule deer (*Odocoileus hemionus*) bucks in southern Utah, USA. Journal of Wildlife Diseases 58:222–227.

Ruder, M. G., A. B. Allison, D. E. Stallknecht, D. G. Mead, S. M. McGraw, D. L. Carter, S. V. Kubiski, C. A. Batten, E. Klement, and E. W. Howerth. 2012. Susceptibility of white-tailed deer (*Odocoileus virginianus*) to experimental infection with epizootic hemorrhagic disease virus serotype 7. Journal of Wildlife Diseases 48:676–685.

Ruder, M. G., D. Johnson, E. Ostlund, A. B. Allison, C. Kienzle, J. E. Phillips, R. L. Poulson, and D. E. Stallknecht. 2017. The first 10 years (2006–15) of epizootic hemorrhagic disease virus serotype 6 in the USA. Journal of Wildlife Diseases 53:901–905.

Ruder, M. G., T. J. Lysyk, D. E. Stallknecht, L. D. Foil, D. J. Johnson, C. C. Chase, D. A. Dargatz, and E. P. J. Gibbs. 2015. Transmission and epidemiology of bluetongue and epizootic hemorrhagic disease in North America: current perspectives, research gaps, and future directions. Vector-Borne and Zoonotic Diseases 15:348–363.

Ruder, M. G., D. E. Stallknecht, A. B. Allison, D. G. Mead, D. L. Carter, and E. W. Howerth. 2016. Host and potential vector susceptibility to an emerging orbivirus in the United States, epizootic hemorrhagic disease virus serotype 6. Veterinary Pathology 53:574–584.

Schirtzinger, E. E., D. C. Jasperson, E. N. Ostlund, D. J. Johnson, and W. C. Wilson. 2018. Recent US bluetongue virus serotype 3 isolates found outside of Florida indicate evidence of reassortment with co-circulating endemic serotypes. Journal of General Virology 99:157–168.

Sellers, R. F., and A. R. Maarouf. 1991. Possible introduction of epizootic hemorrhagic disease of deer virus (serotype2) and bluetongue virus (serotype 11) into British Columbia in 1987 and 1988 by infected *Culicoides* carried on the wind. Canadian Journal of Veterinary Reseach 55:367–370.

Shilton, C. M., D. A. Smith, L. W. Woods, G. J. Crawshaw, and H. D. Lehmkuhl. 2002. Adenoviral infection in captive moose (*Alces alces*) in Canada. Journal of Zoo and Wildlife Medicine 33: 73–79.

Shope, R. E., L. G. MacNamara, and R. Mangold. 1960. A virus-induced epizootic hemorrhagic disease of the Virginia white-tailed deer (*Odocoileus virginianus*). The Journal of Experimental Medicine 111:155–170.

Sleeman, J. M., J. E. Howell, W. M. Knox, and P. J. Stenger. 2009. Incidence of hemorrhagic disease in white-tailed deer is associated with winter and summer climatic conditions. EcoHealth 6:11–15.

Smith, K. E., D. E. Stallknecht, and V. F. Nettles. 1996. Experimental infection of *Culicoides lahillei* (Diptera: Ceratopogonidae) with epizootic hemorrhagic disease virus serotype 2 (Orbivirus: Reoviridae). Journal of Medical Entomology 33:117–122.

Sorden, S. D., L. W. Woods, and H. D. Lehmkuhl. 2000. Fatal pulmonary edema in white-tailed deer (*Odocoileus virginianus*) associated with adenovirus infection. Journal of Veterinary Diagnostic Investigation 12:378–380.

Stair, E. L., R. M. Robinson, and L. P. Jones. 1968. Spontaneous bluetongue in Texas white-tailed deer. Pathologia Veterinaria 5:164–173.

Stallknecht, D. E., A. B. Allison, A. W. Park, J. E. Phillips, V. H. Goekjian, V. F. Nettles, and J. R. Fischer. 2015. Apparent increase of reported hemorrhagic disease in the midwestern and northeastern USA. Journal of Wildlife Diseases 51:348–361.

Stallknecht, D. E., and E. W. Howerth. 2004. Epidemiology of bluetongue and epizootic hemorrhagic disease in wildlife: surveillance methods. Veterinaria Italiana 40:203–207.

Stallknecht, D. E., E. W. Howerth, and J. K. Gaydos. 2002. Hemorrhagic disease in white-tailed deer: our current understanding of risk. Pages 75–86 *in* Transactions of the 67th North American Wildlife and Natural Resources Conference. Wildlife Management Institute, Washington, D.C., USA.

Stallknecht, D. E., M. P. Luttrell, K. E. Smith, and V. F. Nettles. 1996. Hemorrhagic disease in white-tailed deer in Texas: a case for enzootic stability. Journal of Wildlife Diseases 32:695–700.

Stevens, G., B. McCluskey, A. King, E. O'Hearn, and G. Mayr. 2015. Review of the 2012 epizootic hemorrhagic disease outbreak in domestic ruminants in the United States. PLoS One 10:e0133359.

Stilwell, N. K., L. L. Clarke, E. W. Howerth, C. Kienzle-Dean, A. Fojtik, L. P. Hollander, D. Carter, D. A. Osborn, G. J. D'Angelo, K. V. Miller, et al. 2021. The effect of maternal antibodies on clinical response to infection with epizootic hemorrhagic disease virus in white-tailed deer fawns. Journal of Wildlife Diseases 37:189–193.

Sunwoo, S. Y., L. E. Noronha, I. Morozov, J. D. Trujillo, I. J. Kim, E. E. Schirtzinger, B. Faburay, B. S. Drolet, K. Urbaniak, D. S. McVey, et al. 2020. Evaluation of a baculovirus-expressed VP2 subunit vaccine for the protection of white-tailed deer (*Odocoileus virginianus*) from epizootic hemorrhagic disease. Vaccines 8:1–15.

Thorne, E. T., E. S. Williams, T. R. Spraker, W. Helms, and T. Segerstrom. 1988. Bluetongue in free-ranging pronghorn antelope (*Antilocapra americana*) in Wyoming: 1976 and 1984. Journal of Wildlife Diseases 24:113–119.

Wilson, W. C., P. Daniels, E. N. Ostlund, D. E. Johnson, R. D. Oberst, T. B. Hairgrove, J. Mediger, and M. T. McIntosh. 2015. Diagnostic tools for bluetongue and epizootic hemorrhagic disease viruses applicable to North American veterinary diagnosticians. Vector-Borne and Zoonotic Diseases 15:364–373.

Woods, L. W., R. S. Hanley, P. H. W. Chiu, M. Burd, R. W. Nordhausen, M. H. Stillian, and P. K. Swift. 1997.

Experimental adenovirus hemorrhagic disease in yearling black-tailed deer. Journal of Wildlife Diseases 33:801–811.

Woods, L. W., R. S. Hanley, R. H. Chiu, H. D. Lehmkuhl, R. W. Nordhausen, M. H. Stillian, and P. K. Swift. 1999. Lesions and transmission of experimental adenovirus hemorrhagic disease in black-tailed deer fawns. Veterinary Pathology 36:100–110.

Woods, L. W., H. D. Lehmkuhl, L. A. Hobbs, J. C. Parker, and M. Manzer. 2008. Evaluation of the pathogenic potential of cervid adenovirus in calves. Journal of Veterinary Diagnostic Investigation 20:33–37.

Woods, L. W., H. D. Lehmkuhl, P. K. Swift, P. H. Chiu, R. S. Hanley, R. W. Nordhausen, M. H. Stillian, and M. L. Drew. 2001. Experimental adenovirus hemorrhagic disease in white-tailed deer fawns. Journal of Wildlife Diseases 37:153–158.

Woods, L. W., B. A. Schumaker, P. A. Pesavento, B. M. Crossley, and P. K. Swift. 2018. Adenoviral hemorrhagic disease in California mule deer, 1990–2014. Journal of Veterinary Diagnostic Investigation 30:530–537.

Woods, L. W., P. K. Swift, B. C. Barr, M. C. Horzinek, R. W. Nordhausen, M. H. Stillian, J. F. Patton, M. N. Oliver, K. R. Jones, and N. J. MacLachlan. 1996. Systemic adenovirus infection associated with high mortality in mule deer (*Odocoileus hemionus*) in California. Veterinary Pathology 33:125–132.

Zakhartchouk, A., A. Bout, L. W. Woods, H. D. Lehmkuhl, and M. J. E. Havenga. 2002. *Odocoileus hemionus* deer adenovirus is related to the members of *Atadenovirus* genus. Archives of Virology 147:841–847.

12 African Swine Fever

Vienna R. Brown, Julianna B. Lenoch, Courtney F. Bowden

Introduction

At the time of writing this chapter, a global pandemic of African swine fever (ASF) is ongoing with the virus having moved from Eastern Europe, Asia, and into the Caribbean—leaving swine production in devastation along the way. Due to the global spread of African swine fever virus (ASFV), the persistence of the virus, and the increasing number of endemic countries, this disease poses an imminent threat of introduction into North America and other countries that are currently ASF free.

Throughout the chapter, we reference Eurasian wild boar (*Sus scrofa*) which are charismatic megafauna that are native to Europe and Asia. Wild boar were introduced into numerous areas in the southeastern United States and California by early settlers and they subsequently augmented and hybridized with established feral domestic swine (*Sus scrofa*) to give rise to contemporary populations of feral swine, a highly invasive species that are present across much of the United States. Feral swine are referred to by various terms, including wild hogs, feral pigs, wild boar, wild swine, razorbacks, and other regional names in North America. African swine fever has never been introduced into the United States; as such, we do not discuss feral swine in specific within

Illustration by Laura Donohue.

the chapter. However, experimental inoculations demonstrate that feral swine are acutely susceptible to ASFV and given its current rapid global movement we anticipate similar patterns of exposure, infection, and risk amongst these populations.

History and Global Distribution

In 1909, ASF was first detected following the importation of European domestic swine into Kenya when nearly 100% of the animals succumbed to a hemorrhagic disease (Gallardo et al. 2015, Sánchez-Cordón et al. 2018). The disease was also detected in central and western Africa but confined to sub-Saharan Africa until it was reported in Portugal in 1957. In 1960, ASF spread to the Iberian Peninsula and other countries in Europe such as France (1964), Italy (1967, 1969, and 1983), Malta (1978), Belgium (1985), and the Netherlands (1986). Various countries in the Americas were also affected by ASF during this period including Cuba (1971, 1980), Brazil (1978), the Dominican Republic (1978), and Haiti (1979). By the 1990s, the disease had been successfully eradicated from these countries, with the exception being the island of Sardinia off the coast of Italy.

Transcontinental spread of ASF occurred for a second time in 2007 with the disease reaching Georgia within the Caucasus region and subsequently spreading further into eastern Europe (Figure 12.1; Revilla et al. 2018, Sánchez-Cordón et al. 2018). Specifically, ASF outbreaks were reported in Armenia, Russia, Belarus, Ukraine, Estonia, Lithuania, Latvia, Romania, Moldova, the Czech Republic, and Poland (Revilla et al. 2018). In August 2018, ASF was reported on a small-scale pig farm in China and has since spread throughout the country as well as through much of Asia and Southeast Asia including Mongolia, Korea, Vietnam, Laos, Cambodia,

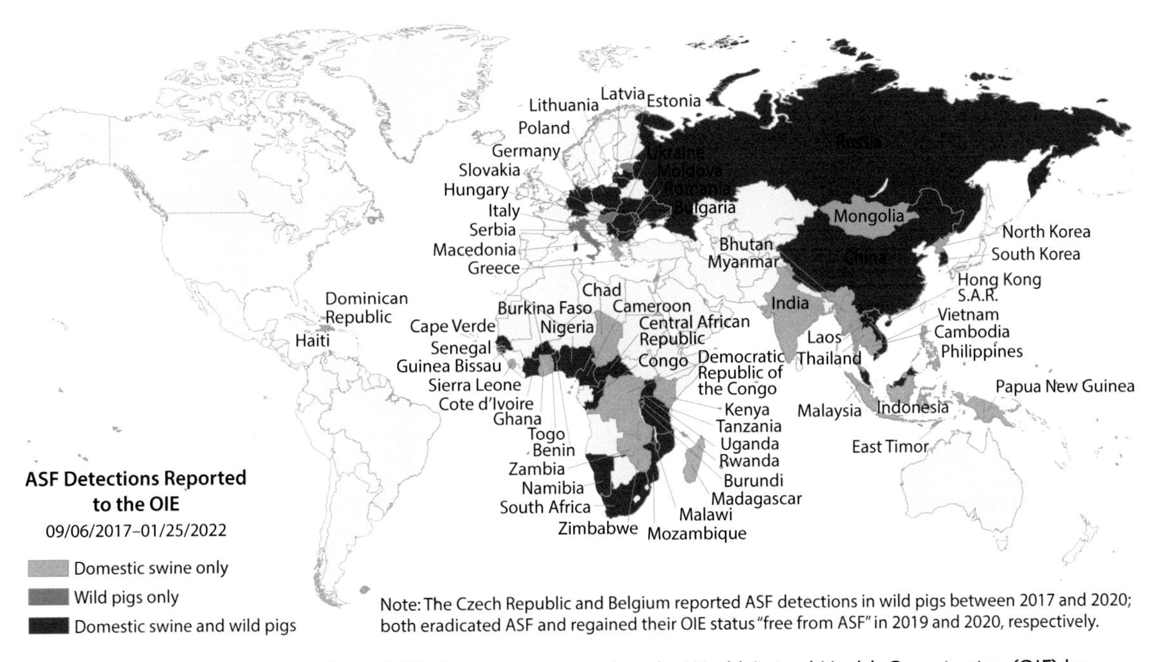

Fig. 12.1. Global African swine fever (ASF) detections reported to the World Animal Health Organization (OIE) by immediate notifications, follow-up reports, and biannual reports occurring in domestic swine and wild pigs from 6 September 2017 to 25 January 2022. Data and map sources: OIE WAHIS, USDA APHIS.

Myanmar, the Philippines, Hong Kong, Indonesia, Timor-Leste, Papua New Guinea, and India (Ge et al. 2018, Zhao et al. 2019, Li et al. 2020, Mighell and Ward 2021). Hungary, Bulgaria, and Belgium reported their first ASF outbreaks in 2018 (Martinez-Aviles et al. 2020), and Germany experienced their first outbreak in 2020 (Sauter-Louis et al. 2021*b*). In the Americas, African swine fever was detected in the Dominican Republic in July 2021 (USDA APHIS 2021*a*) and in Haiti in September 2021 (USDA APHIS 2021*b*).

Etiology

African swine fever is a hemorrhagic disease caused by ASFV, which is a large, linear double-stranded DNA virus (Sanchez-Vizcaino et al. 2019). African swine fever virus is the only member of the family *Asfarviridae* (Alonso et al. 2018) and the only known DNA arbovirus (Gaudreault et al. 2020). Whole-genome sequencing and subtyping has revealed that ASF viruses demonstrate considerable variation in virulence and significant genomic diversity (de Villiers et al. 2010). Wild and domestic members of the Suidae family (see picture on page 198) are the only natural, vertebrate hosts of ASF (Gallardo et al. 2015, Golnar et al. 2019), and soft ticks of the genus *Ornithodoros* are natural, arthropod vectors that play a significant role in pathogen maintenance and transmission.

Epidemiology and Transmission

African swine fever virus transmission dynamics are especially complex: there are three epidemiological cycles, and a fourth has been proposed, that exist independent of one another (Figure 12.2).

Fig. 12.2. Cycles of transmission for African swine fever virus (ASFV). Illustration by Laura Donohue.

Sylvatic Cycle

African swine fever virus evolved in eastern and southern Africa in the burrows of common warthogs (*Phacochoerus africanus*) with the *Ornithodoros moubata* soft ticks that inhabit these dwellings (Penrith et al. 2019). Warthogs are believed to be the original vertebrate host of ASFV, and transmission within the warthog population occurs exclusively between infected ticks and the neonatal warthogs that they feed upon (Figure 12.3; Costard et al. 2013). Newly infected juvenile warthogs exhibit viremia for a short duration, which can serve to infect naïve ticks and allows pathogen maintenance in the absence of vertical and horizontal transmission amongst warthogs. Infected warthogs are asymptomatic carriers, although

Fig. 12.3. Components of the sylvatic epidemiological cycle of African swine fever virus: (A) a mother and juvenile warthog, (B) a warthog burrow, and (C) *Ornithodoros spp.* ticks removed from a warthog burrow. Photograph courtesy of Dr. Charles Masembe and the African swine fever research consortium, Makerere University, Kampala, Uganda.

they often maintain the virus for the duration of their lives in infected lymph nodes (Costard et al. 2013).

Similar to warthogs, bushpigs (*Potamochoerus larvatus*) exhibit moderate viremia when infected with ASFV via the bite of an infected tick but do not develop clinical signs (Anderson et al. 1998, Oura et al. 1998). The contribution of bushpigs to ASFV maintenance and transmission is relatively unknown as they do not inhabit burrows (Netherton et al. 2019); however, bushpigs are thought to play a minor role in ASFV epidemiology as their populations are less dense and they appear to have lower infection rates (Jori and Bastos 2009). Very little is known about red river hogs (*Potamochoerus porcus*) in relation to the epidemiology of ASF; however, ASFV genomic DNA was detected in a single free-ranging red river hog in Nigeria (Luther et al. 2007). Similarly, reports of ASFV infections have been limited for giant forest hogs (*Hylochoerus meinertzhageni*), and the contribution of giant forest hogs to the sylvatic cycle is believed to be negligible.

Tick–Pig Cycle

The soft tick–pig cycle is characterized by ASFV-infected ticks transmitting the pathogen to domestic swine (Costard et al. 2013). Competent tick vectors of the genus *Ornithodoros* (Table 12.1) become infected after feeding on viremic animals and are then capable of transmitting the virus during blood meals (Costard et al. 2013). In the absence of viremic hosts, transstadial, transovarial, and sexual transmission has been documented in *O. moubata* ticks, which contributes to ASFV persistence (Plowright et al. 1970, 1974; Rennie et al. 2001). This cycle is particularly prevalent in sub-Saharan Africa with *Ornithodoros moubata* ticks infecting domestic swine; however, it also played an important role in the outbreak on the Iberian Peninsula with *O. erraticus* ticks transmitting ASFV to domestic swine (Gaudreault *et al.* 2020). The tick–pig epidemiologic cycle is particularly important in outdoor swine production areas that have endemic disease in soft ticks in nearby areas (Sanchez-Vizcaino *et al.* 2012).

Table 12.1. Tick vectors from the genus *Ornithodoros* capable of transmitting African swine fever virus to pigs

Ornithodoros species	Global distribution
O. coriaceus (*O. marocanus*)	North America
O. erraticus	Africa, Asia, Europe
O. moubata complex	Africa, Madagascar
O. puertoricensis	Caribbean, North America
O. savignyi	Africa, Asia
O. turicata	North America

Domestic (Pig–Pig) Cycle

Once ASFV is introduced into domestic pig or Eurasian wild boar populations, virus can be efficiently transmitted between swine hosts and does not require a vector (Costard et al. 2013, Dixon et al. 2020, Blome et al. 2021). ASFV spreads systemically within infected swine, and all secretions and excretions contain virus (Guinat et al. 2016). Direct and indirect contact as well as the consumption of contaminated meat products (e.g., swill feeding practices), allow for the ready transmission of ASFV between pigs (Costard et al. 2013). As an example, interactions between free-ranging domestic swine and wild boar are believed to have contributed to ASFV endemicity in some parts of Europe (Cadenas-Fernandez et al. 2019).

Wild Boar–Habitat

The wild boar–habitat cycle consists of direct transmission through contact with infected wild boar as well as indirect transmission from dead conspecifics or contaminated environments (e.g., soil, drinking water, crops, or feed; Chenais et al. 2018, O'Neill et al. 2020, Sauter-Louis et al. 2021*a*, Viltrop et al. 2021). Direct transmission may occur within or between herds of wild boar (i.e., sounders; O'Neill et al. 2020) and is particularly troublesome in areas with high wild boar density (Guberti et al. 2019). Reports of wild boar scavenging on carcasses are common (Cukor et al. 2019, Probst et al. 2020); in wild boar populations in Europe, carcass-based transmission is the primary route of infection (EFSA 2014, Cukor et al. 2019, Fischer et al. 2020). Carcass-based transmission is estimated to account for 53–66% of transmission events in the ASF outbreak that took place in eastern Poland in 2014 and 2015 (Pepin et al. 2020). Failure to remove carcasses from the landscape provides ample opportunity for ASFV transmission via direct interactions between live and dead conspecifics. Additionally, ASFV is highly stable in a proteinaceous environment, and as such, indirect transmission can occur through environmental contamination during carcass decomposition (Dixon et al. 2020; O'Neill et al. 2020).

Clinical Signs and Gross Lesions

With highly virulent strains of ASFV, death occurs very quickly in nearly 100% of affected domestic swine or wild boar (OIE 2019). Clinical disease associated with peracute and acute cases typically manifests as fever, increased pulse and respiratory rate, reddening of the skin (especially of the extremities; Figure 12.4), anorexia, vomiting, and diarrhea (sometimes bloody). Moderate- and low-virulence strains of ASFV cause weight loss, intermittent fever, and general malaise and result in subacute and chronic forms of the disease, respectively. In swine that succumb to ASFV, gastrohepatic and renal lymph node hemorrhage as well as enlarged, friable spleens can often be observed (Figure 12.4).

Pathogenesis and Pathology

Infection with ASFV in domestic swine and wild boar is associated with severe lymphoid depletion and hemorrhages (Salguero 2020). Once the virus

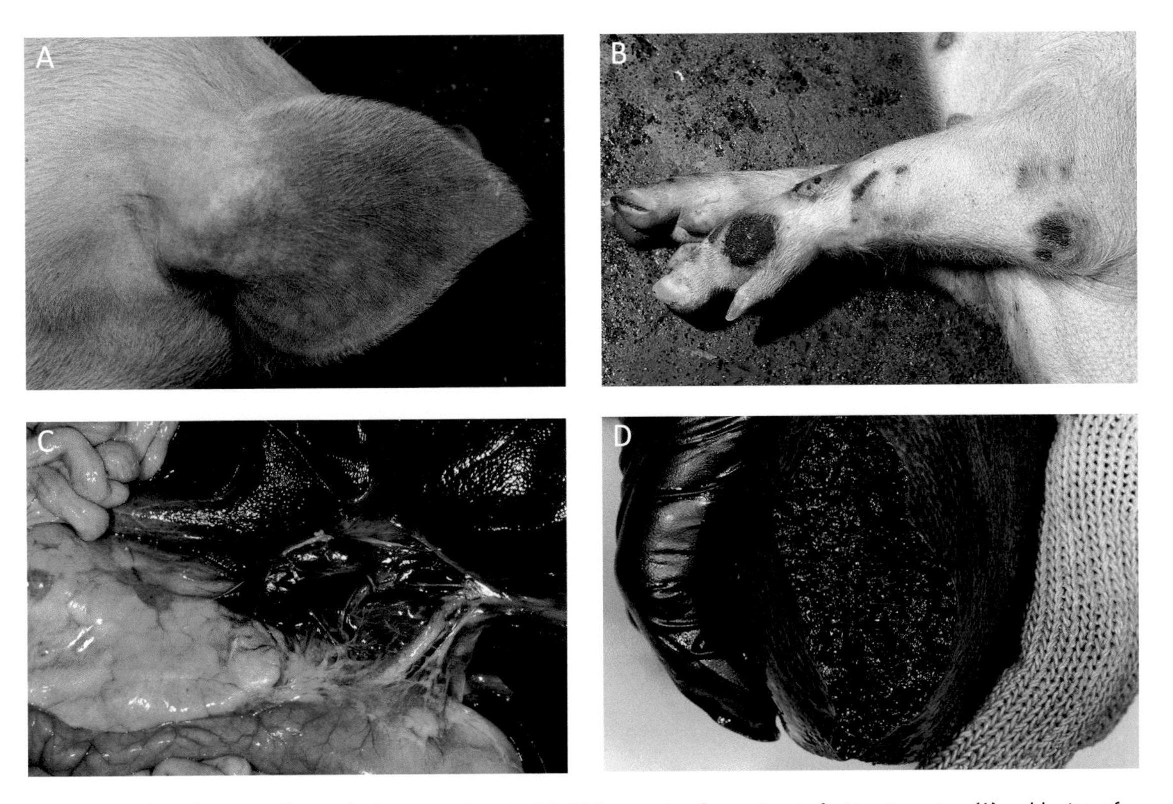

Fig. 12.4. Clinical signs and gross lesions associated with African swine fever virus infection in swine: (A) reddening of the skin on the ear; (B) red–purple skin discoloration of the skin on the leg; (C) hemorrhagic gastrohepatic lymph nodes; and (D) enlarged, friable spleen. Photograph courtesy of Plum Island Animal Disease Center, Plum Island, New York, USA.

enters the body, tonsils are the primary site of replication (Heuschele 1967; Greig 1972) before spreading to local lymph nodes, disseminating to the secondary organs of replication, and ultimately spreading systemically via the lymphatic system. The lymphoid depletion characterized by ASF results from massive destruction of lymphoid organs, tissues, and cell types. Experimental infection of domestic swine with ASFV has demonstrated that tissues abundant in reticuloendothelial cells, including the lymph nodes, spleen, bone marrow, and liver, commonly have the highest virus titers and serve as secondary sites of viral replication (Heuschele 1967).

The pathogenesis of ASFV is largely mediated by the host immune response and the corresponding cytokine storm (Zhu et al. 2019). Monocytes and macrophages are the main target for ASFV (Howey et al. 2013), and virus replication induces an intense upregulation in proinflammatory cytokines that ultimately results in significant apoptosis in both infected cells and uninfected lymphocytes (Howey et al. 2013; Wozniakowski et al. 2016). The vascular changes are those of a disseminated intravascular coagulopathy leading to petechial and ecchymotic hemorrhages in multiple organs, hyperemic splenomegaly, and pulmonary edema (Salguero 2020).

Diagnostics

Given that there is no effective vaccine or treatment, early detection of ASFV is paramount in managing the disease and limiting outbreak size. Validated diagnostic tests are classified into two primary types: antigen and antibody-based assays (OIE 2019). Polymerase chain reaction (PCR) is the gold standard antigen-based test as it is highly sensitive and specific, allows for high throughput, and generates rapid results (Gallardo et al. 2019). Virus isolation can be used to detect ASFV and direct immunofluorescence and antigen-based enzyme-linked immunosorbent assays (ELISA) can detect ASFV antigens (Gallardo et al. 2019, OIE 2019).

Whole blood captured in EDTA tubes is the best suited sample for ASFV detection (Pikalo et al. 2021); however, serological assays can be useful for low to moderately virulent strains of ASF, when clinical signs are not significant enough to trigger an ASF investigation (OIE 2019). It is worth noting that antibody testing is suboptimal for screening apparently healthy animals (Pikalo et al. 2021). A number of diagnostic tests are available for detecting ASFV antibodies, including ELISAs, indirect fluorescent antibody test (IFAT), indirect immunoperoxidase test (IPT), and immunoblotting test (IBT; OIE, 2019). The recommendation for ASF antibody testing is screening via ELISA with nonnegatives being confirmed with IB, IFAT, or IPT (Beltran-Alcrudo et al. 2019).

Treatment and Vaccination

To date, there are no efficacious treatments to reduce the severity of the disease. Infected animals will either die rather quickly, especially with virulent strains of ASFV, or become convalescent (Costard et al. 2009). Natural infection with a low to moderately virulent strain of ASFV confers a high degree of protection against infection with a highly virulent homologous strain of ASFV however (Sang et al. 2020).

The quest to develop a safe, efficacious vaccine for ASFV has proven to be extremely challenging, and there is not a licensed vaccine available to date (Das et al. 2021). This is driven in large part by gaps in knowledge related to ASFV entry and replication within host cells, virus immune evasion and modulation, and the primary antigens responsible for triggering immune activity (Arias et al. 2017, Rock 2017, Sang et al. 2020, Wang et al. 2020). Several vaccine platforms have been evaluated and are showing varying degrees of promise. For example, inactivated ASFV vaccines are often nonprotective as they fail to induce a cellular immune response, even when administered with adjuvant (Teklue et al. 2019; Gavier-Widén et al. 2020). Subunit vaccines, recombinant proteins, and DNA vaccines have demonstrated partial protection against wild-type ASFV (Teklue et al. 2019; Gavier-Widén et al. 2020; Sang et al. 2020). Several live attenuated vaccines have been developed, and this is currently the most promising vaccine candidate (Bosch-Camos et al. 2020; Wu et al. 2020; Muñoz -Pérez et al. 2021), as the host responds to the vaccine as if it were naturally encountering ASFV (Sang et al. 2020).

In live attenuated vaccines, the virus can be modified naturally, through multiple passages in a laboratory setting, or by intentional alterations (Liu et al. 2021). The genes deleted in live attenuated vaccines are typically associated with a reduction in virulence, a decrease in viral replication rate and dissemination, and a dampening of the proinflammatory immune response (Sang et al. 2020). One live attenuated vaccine strain, developed by the United States Department of Agriculture, has been shown to provide protection against homologous virulent strains of ASFV in pigs with varying backgrounds (Tran et al. 2022), although testing is still ongoing. The significant limitation with live attenuated vaccines is that unpredictable, strain-specific virus phenotypes are created following genetic alteration of ASFV (Turlewicz-Podbielska et al. 2021). Some animals can develop clinical disease following vaccination, or vaccinated animals can contribute to virus transmission via shedding patterns (Bosch-Camos et al. 2020, Gavier-Widén et al. 2020). In addition to safety concerns in vaccinated animals, developing these vaccines is challenging due to the need for high-level biocontainment facilities for production, stable and suitable cell lines, and cell culture optimization (Sang et al. 2020). However, efficacious, attenuated live vaccine strains have been cultivated and can be produced (Borca et al. 2021*a*).

An important consideration for vaccine development is the ability to differentiate between infected and vaccinated animals (DIVA). DIVA-compatible vaccines are typically generated by the inclusion of a marker within the vaccine that can induce an immune response in the host that is distinct from that of a natural infection with a wild-type pathogen (Hardham et al. 2020). This differentiation is paramount for understanding pathogen epidemiology on the landscape, for informing management practices and policies, and for demonstrating disease elimination following an outbreak.

Developing vaccines for wild pigs presents significant ecological, logistical, and moral considerations that make it challenging to reach desired vaccine efficacy and herd-level immunity (Maki et al. 2017, Barnett and Civitello 2020, Edwards et al. 2021). Vaccines for wild pigs are especially challenging in that they probably need to be delivered orally necessitating that they are both attractive and palatable to the swine species of interest (Balseiro et al. 2020). Additionally, a sufficient quantity of immunogen must be present and available at the induction site to stimulate a protective response requiring a biocompatible encapsulation process (Cross et al. 2007). A DIVA-compatible, oral vaccine for classical swine fever (CSF) was successfully engineered and delivered to wild boar in Europe (Blome et al. 2011; Rossi et al. 2015), and this achievement could be used as a foundation for the successful implementation of an oral ASF vaccine in wild boar (Teklue et al. 2019).

Given the nature of the global ASFV pandemic, vaccine development for wild boar has been an international research priority, and studies may indicate that oral vaccination is plausible. An experimental ASF vaccine that was originally developed using domestic swine and delivered via intramuscular injection (Tran et al. 2022; Borca et al. 2020, 2021*a*) was found to confer high levels of protection against challenges with a virulent strain of ASFV when the vaccine was administered via the oronasal route (Borca et al. 2021*b*). Additionally, vaccination of wild boar with an attenuated strain of ASFV from Latvia was found to be protective against a virulent strain of ASFV (Barasona et al. 2019), and studying shedding patterns of this vaccination strain demonstrated it to be safe for wild boar and contact animals (Kosowska et al. 2020). Further work is necessary to ensure safety and efficacy for use in free-roaming wild boar.

ASFV Introduction Routes Into Disease-free Countries

Given the lack of a vaccine or treatment for ASF and the devastating consequence of the disease for pork production, preventing viral spread into previously unaffected geographical areas is paramount. Introduction pathways vary by country; however, the primary routes identified for ASFV include legal and illegal importation of live swine, swine products, and swine byproducts for commercial use or personal consumption (Figure 12.5; Brown and Bevins 2018,

Fig. 12.5. Human-mediated routes of introduction of African swine fever (ASF) from infected to disease-free areas. Illustration by Laura Donohue.

Jurado et al. 2019). Another introduction pathway that must be considered is food waste from airlines, trains, or ships that originated in ASF-infected regions. Human food waste may be discarded in landfills that can be accessed by wild boar, improperly disposed of through meat composting or littering, or illegally sold for swill (Figure 12.5; Vergne et al. 2017, Beltran-Alcrudo et al. 2019, Taylor et al. 2020).

Legal importation of live swine poses a risk for ASFV introduction as the virus can go undetected during the incubation period, in which there is an absence of clinical disease (Vergne et al. 2017; Beltran-Alcrudo et al. 2019; Gao et al. 2020; Taylor et al. 2020). This latent stage following infection is problematic for countries that are historically ASF free and do not have any trade barriers in place for live animals or their products. As an example, classical swine fever (CSF) spread to Italy and Spain from the Netherlands in 1997 when infected piglets were shipped prior to a transportation moratorium (Beltran-Alcrudo et al. 2019). In ASF-endemic countries, swine, their products, and their byproducts are often banned for trade and importation; however, at a local level, "emergency sales" of swine can occur during ASF outbreaks to minimize economic impacts. If ASF is undetected in the herd, then these sales are silently contributing to the spread of ASFV into the production system (Costard et al. 2015).

The illegal movement of swine and their products presents a greater risk of ASFV entry than legal activities (Beltran-Alcrudo et al. 2019; Fanelli et al. 2021). Unregulated movement of live swine within or between countries for personal or commercial use creates opportunities for direct transmission via infected animals or swill as well as indirect transmission via fomites. Due to the difficulty of smuggling live animals, this introduction pathway presents a lower risk of ASFV transmission than illegal movement of products or byproducts (Beltran-Alcrudo et al. 2019). Although most countries have systems in place to confiscate and destroy illegally imported animal products, numerous transboundary animal disease outbreaks have been attributed to illegal imports (Costard et al. 2013). Quantifying the magnitude of illegal imports is difficult and reliant on extrapolation from products that have been detected and confiscated, and data are widely variable. Between 2012 and 2016, approximately 68,000 domestic swine products and specimens were confiscated from travelers entering the United States, and many confiscated products originated from countries known to have ASF (Brown and Bevins 2018). Similarly, in August of 2018 custom officials in South Korea identified 4,064 illegal pork products originating from China, and 4 of the 52 submitted for ASFV testing were positive (Kim et al. 2019).

Management Practices

Current management strategies for new ASFV introductions include quarantining infected swine, increased biosecurity practices, epidemiological investigations, restricted swine movement, disposal of infected and exposed swine, and multiple forms of surveillance (i.e., passive, active, and targeted; Sánchez-Cordón et al. 2018; Danzetta et al. 2020). Given that ASFV is highly stable and easily spread, biosecurity is critical for preventing exposure via fomites or a contaminated environment (Nurmoja et al. 2020). Biosecurity, defined as practices to prevent pathogens both coming onto a farm or commercial property or spreading within the premises, is critical. (Mutua and Dione 2021). Producers should minimize the introduction of animals from outside sources, and a quarantine period should be implemented if outsourcing. Access to production areas should be limited for people, birds, and other small mammals, site-specific equipment and supplies need to be maintained, and materials entering and exiting the property need to be thoroughly cleaned and disinfected (Mutua and Dione 2021). Proper disinfection, which includes both mechanical cleaning and the application of an effective disinfectant, is necessary (Juszkiewicz et al. 2019). Swill feeding is widely considered to be a risky practice (EFSA 2014), and the majority of outbreaks in ASFV-free zones have occurred as a result of feeding food waste products from infected pigs to susceptible hosts (Bellini et al. 2016). Although strongly discouraged, swill should be heated to a minimum of 70°C if it is to be fed to swine.

Human-mediated activities are important drivers for the spread of ASFV, and baiting and hunting wild boar present challenges in managing the disease (Guberti et al. 2019). Supplemental feeding increases the carrying capacity of wild boar on the landscape, amplifies the opportunities for direct and indirect contact between animals, and changes movement patterns—all of which contribute to the maintenance and persistence of ASFV in wild boar. Regulations for hunters aimed at minimizing the risk of ASFV spread include safe storage of the carcass until laboratory results are completed, a prohibition of leaving offal piles in the forest, and washing and disinfecting vehicles, boot, knives, and other equipment used during hunting that has been in contact with blood or tissues of excreta (Bellini et al. 2016; Guberti et al. 2019).

Controlling ASFV has proven especially challenging once it is introduced into wild boar populations, and eradication is nearly impossible, although it has been accomplished on a couple of occasions. Early detection, a prompt and coordinated approach to prohibit movement of infected wild boar, and restricted public access have been demonstrated to be effective in stopping the spread of ASFV. Identifying and disposing of carcasses on the landscape, the use of strategic fencing, and culling operations to reduce densities can also help slow or halt the spread of ASFV (Danzetta et al. 2020; Gavier-Widén et al. 2020).

In addition to field activities to manage an ASFV outbreak, disease-dynamic models can be helpful to guide ASFV preparedness efforts or response and management activities (Miller and Pepin 2019). Given ASFV's multiple routes of transmission and stability, transboundary management is extremely challenging. Extensive collaboration and cooperation between governments and industry are paramount (Shi et al. 2021), and science-based policies and approaches must engage stakeholders across many disciplines (Tucker et al. 2021).

Human Dimensions of ASF

The financial, social, and political implications of ASF are potentially tremendous and can simultaneously impact multiple economic sectors. A systematic literature review on the country-level economic impact of ASFV outbreaks found the range from $649,000 (USD 2019) for annual production loss in Nigeria to $94,539,870,064 for the total economic impact of an outbreak in Spain (Brown et al. 2020). This meta-analysis also found that there is a paucity of data that allows estimation of costs and losses from ASF on national or global scales.

In addition to the short-term direct macroeconomic impacts related to outbreak management, production losses and exclusion from international export markets are secondary impacts for associated products. For example, with an ASF outbreak causing greatly reduced pork production, the demand for animal feed would likely decrease drastically in the United States, causing decreased prices and revenue in the feed market (Stancu 2018). The magnitude of a foreign animal disease introduction event ripples through the economy with direct and indirect costs that have short- and long-term impacts.

A significant reduction in live hog prices can be expected in countries where ASF is reported, as the market clears out surplus pork that would otherwise be exported (Carriquiry et al. 2020), as well as increases to pork products with demand exceeding supply in other markets and regions (Mason-D'Croz et al. 2020). Both scenarios are likely to drive-up costs for alternative sources of animal-based protein, including beef and poultry, impacting the most vulnerable communities globally, and negatively impacting food and financial security (Abworo et al. 2017; Plavšić et al. 2019; Matsumoto et al. 2021; Paulino-Ramirez et al. 2021). Pork supply–demand patterns are believed to have played a significant role in ASFV dissemination within China (Yang et al. 2021).

When ASF outbreaks occur, pig trader networks rush to sell their pigs to ensure that their animals do not succumb to infection (Lichoti et al. 2017; Penrith et al. 2019). This has social implications and been linked to rapid viral dissemination. As an interesting aside and a compelling tale on the interconnectedness of human and animal health, it has been postulated that the ASFV outbreak in China reached its worst impacts around December 2019, near the time when SARS-CoV-2 emerged (Xia *et al.* 2021). The authors suggest that the dramatic pork shortage led to an increase in the risk of transmission of a zoonotic disease as human–wildlife interactions were more frequent.

Summary

African swine fever outbreaks in ASF-free countries have become commonplace in the last several years, and the global ASF pandemic is an example of the tremendous implications of diseases shared by livestock and wildlife. In addition to morbidity and mortality events, production losses, and barriers to international trade, an outbreak of ASF has political ramifications. Measures to reduce the risk of ASFV introduction in ASF-free countries should be of the utmost priority as control and eradication is time consuming, costly, and often out of reach.

LITERATURE CITED

Abworo, E. O., C. Onzere, J. O. Amimo, V. Riitho, W. Mwangi, J. Davies, S. Blome, and R. P. Bishop. 2017. Detection of African swine fever virus in the tissues of asymptomatic pigs in smallholder farming systems along the Kenya–Uganda border: implications for transmission in endemic areas and ASF surveillance in East Africa. Journal of General Virology 98:1806–1814.

Alonso, C., M. Borca, L. Dixon, Y. Revilla, F. Rodriquez, J. M. Escribano, and ICTV Report Consortium. 2018. ICTV virus taxonomy profile: *Asfarviridae*. Journal of General Virology 99:613–614.

Anderson, E. C., G. H. Hutchings, N. Mukarati, and P J. Wilkinson. 1998. African swine fever virus infection of the bushpig (*Potamochoerus porcus*) and its significance in the epidemiology of the disease. Veterinary Microbiology. 62:1–15.

Arias, M., A. de la Torre, L. Dixon, C. Gallardo, F. Jori, A. Laddomada, C. Martins, R. M. Parkhouse, Y. Revilla, F. Rodriguez, and J. M. Sánchez-Vizcaíno. 2017. Approaches and perspectives for development of African swine fever virus vaccines. Vaccines 5:35.

Balseiro, A., J. Thomas, C. Gortazar, and M. A. Risalde. 2020. Development and challenges in animal tuberculosis vaccination. Pathogens 9:472.

Barasona J. A., C. Gallardo, E. Cadenas-Fernández, C. Jurado, B. Rivera, A. Rodríguez-Bertos, M. L. Arias, and J. M. Sánchez-Vizcaíno. 2019. First oral vaccination of Eurasian wild boar against African swine fever virus genotype II. Frontiers in Veterinary Science 6:137.

Barnett, K. M., and D. J. Civitello. 2020. Ecological and evolutionary challenges for wildlife vaccination. Trends in Parasitology 36:970–978.

Bellini, S., D. Rutili, and V. Guberti. 2016. Preventive measures aimed at minimizing the risk of African swine fever virus spread in pig farming systems. Acta Veterinaria Scandinavica 58:82.

Beltran-Alcrudo, D., J. R. Falco, E. Raizman, and K. Dietze. 2019. Transboundary spread of pig diseases: The role of international trade and travel. BMC Veterinary Research 15:64.

Blome, S., C. Gabriel, C. Staubach, I. Leifer, G. Strebelow, and M. Beer. 2011. Genetic differentiation of infected from vaccinated animals after implementation of an emergency vaccination strategy against classical swine fever in wild boar. Veterinary Microbiology 153:373–376.

Blome, S., K. Franzke, and M. Beer. 2021. African swine fever—A review of current knowledge. Virus Research 287:198099.

Borca, M. V., A. Rai, E. Ramirez-Medina, E. Silva, L. Valazquez-Salinas, E. Vuono, S. Pruitt, N. Espinoza, and D. P. Gladue. 2021*a*. A cell culture-adapted vaccine virus against the current African swine fever virus pandemic strain. Journal of Virology 95: e0012321.

Borca, M. V., E. Ramirez-Medina, E. Silva, E. Vuono, A. Rai, S. Pruitt, N. Espinoza, L. Velazquez-Salinas, C. G. Gay, and D. P. Gladue. 2021*b*. ASFV-G-ΔI177L as an effective oral nasal vaccine against the Eurasia strain of African swine fever. Viruses 13:765.

Borca, M. V., E. Ramirez-Medina, E. Silva, E. Vuono, A. Rai, S. Pruitt, L. G. Holinka, L. Velazquez-Salinas, J. Zhu, and D. P. Gladue. 2020. Development of a highly effective African swine fever virus vaccine by deletion of the I177L gene results in sterile immunity against the current epidemic Eurasia strain. Journal of Virology 94:e02017–e02019.

Bosch-Camós, L., E. López, and F. Rodriguez. 2020. African swine fever vaccines: a promising work still in progress. Porcine Health Management 6:17.

Brown, V.R., and S. N. Bevins. 2018. A review of African swine fever and the potential for introduction into the United States and the possibility of subsequent establishment in feral swine and native ticks. Frontiers in Veterinary Science 5:11.

Brown, V.R., R. S. Miller, S. C. McKee, K. H. Ernst, N. M. Didero, R. M. Maison, M. J. Grady, and S. A. Shwiff. 2020. Risks of introduction and economic consequences associated with African swine fever, classical swine fever, and foot-and-mouth disease: a review of the literature. Transboundary and Emerging Diseases 68:1910–1965.

Cadenas-Fernández, E., J. M. Sánchez-Vizcaíno, A. Pintore, D. Denurra, M. Cherchi, C. Jurado, J. Vicente, and J. A. Barasona. 2019. Free-ranging pig and wild boar interactions in an endemic area of African swine fever. Frontiers in Veterinary Science 6:376.

Carriquiry, M., A. Elobeid, D. Swenson, and D. Hayes. 2020. Impacts of African swine fever in Iowa and the United States. Iowa State University, Center for Agricultural and Rural Development Working Papers 618, Ames, USA. https://lib.dr.iastate.edu/card_workingpapers/618. Accessed 4 Oct 2022.

Chenais, E., K. Stahl, V. Guberti, and K. Depner. 2018. Identification of wild boar–habitat epidemiologic cycle in African swine fever epizootic. Emerging Infectious Diseases 24:810–812.

Costard, S., L. Mur, J. Lubroth, J. M. Sanchez-Vizcaino, and D. U. Pfeiffer. 2013. Epidemiology of African swine fever virus. Virus Research 173:191–197.

Costard, S., B. Wieland, W. de Glanville, F. Jori, R. Rowlands, W. Vosloo, F. Roger, D. U. Pfeiffer, and L. K. Dixon. 2009. African swine fever: How can global spread be prevented? Philosophical Transactions of the Royal Society B Biological Sciences 364:2683–2696.

Costard, S., F. J. Zagmutt, T. Porphyre, and D. U. Pfeiffer. 2015. Small-scale pig farmers' behavior, silent release of African swine fever virus and consequences for disease spread. Scientific Reports 5:17074.

Cross, M. L., B. M. Buddle, and F. E. Aldwell. 2007. The potential of oral vaccines for disease control in wildlife species. Veterinary Journal 174:472–480.

Cukor, J., R. Linda, P. Vaclavek, K. Mahlerova, P. Stran, and F. Havranek. 2019. Confirmed cannibalism in wild boar and its possible role in African swine fever transmission. Transboundary and Emerging Diseases 67:1068–1073.

Danzetta, M. L., M. L. Marenzoni, S. Iannetti, P. Tizzani, P. Calistri, and F. Feliziani. 2020. African swine fever: lessons to learn from past eradication experiences. A systematic review. Frontiers in Veterinary Science 7:296.

Das, S., P. Deka, P. Deka, K. Kalita, T. Ansari, R. Hazarika, and N. N. Barman. 2021. African swine fever: etiology, epidemiology, control strategies and progress toward vaccine development: a comprehensive review. Journal of Entomology and Zoology Studies 9:919–929.

de Villiers, E. P., C. Gallardo, M. Arias, M. de Silva, C. Upton, R. Martin, and R. P. Bishop. 2010. Phylogenetic analysis of 11 complete African swine fever virus genome sequences. Virology 400:128–136.

Dixon, L. K., K. Stahl, F. Jori, L. Vial, and D. U. Pfeiffer. 2020. African swine fever epidemiology and control. Annual Review of Animal Biosciences 8:221–246.

Dixon, L. K., H. Sun, and H. Roberts. 2019. African swine fever. Antiviral Research 165:34–41.

Edwards, S. J. L., H. J. Chatterjee, and J. M. Santini. 2021. Anthroponosis and risk management: a time for ethical vaccination of wildlife? Lancet Microbe 2:e230–e231.

European Food Safety Authority (EFSA), 2014. Scientific opinion on African swine fever. EFSA Journal 12:3628.

Fanelli, A., O. Muñoz, L. Mantegazza, M. DeNardi, and I. Capua. 2021. Is the COVID-19 pandemic impacting on the risk of African swine fever virus (ASFV) introduction into the United States? A short-term assessment of the risk factors. Transboundary and Emerging Diseases 69:e505–e516.

Fischer, M., J. Huhr, S. Blome, F. J. Conraths, and C. Probst. 2020. Stability of African swine fever virus in carcasses of domestic pigs and wild boar experimentally infected with the ASFV "Estonia 2014" isolate. Viruses 12:1118.

Gallardo, C., A. de la Torre Reoyo, J. Fernández-Pinero, I. Iglesias, J. Muñoz, and M. L. Arias. 2015. African swine fever: a global view of the current challenge. Porcine Health Management 1:21.

Gallardo, C., J. Fernández-Pinero, and M. L. Arias. 2019. African swine fever (ASF) diagnosis, an essential tool in the epidemiological investigation. Virus Research 271:197676.

Gao, X., T. Liu, Y. Liu, J. Xiao, and H. Wang. 2020. Transmission of African swine fever in China through legal trade of live pigs. Transboundary and Emerging Diseases 68:355–360.

Gaudreault, N. N., D. W. Madden, W. C. Wilson, J. D. Trujillo, and J. A. Richt. 2020. African swine fever virus: an emerging DNA arbovirus. Frontiers in Veterinary Science 7:215.

Gavier-Widén, D., K. Ståhl, and L. Dixon. 2020. No hasty solutions for African swine fever. Science 367:622–624.

Ge, S., J. Li, X. Fan, F. Liu, L. Li, Q. Wang, W. Ren, J. Bao, C. Liu, U. Wang, et al. 2018. Molecular characterization of African swine fever virus, China, 2018. Emerging Infectious Diseases 24:2131–2133.

Golnar, A. J., E. Martin, J. D. Wormington, R. C. Kading, P. D. Teel, S. A. Hamer, and G. L. Hamer. 2019. Reviewing the potential vectors and hosts of African swine fever virus transmission in the United States. Vector-Borne and Zoonotic Diseases 19:512–525.

Greig, A. 1972. Pathogenesis of African swine fever in pigs naturally exposed to the disease. Journal of Comparative Pathology 82:73–79.

Guberti V, S. Khomenko, M. Masiulis, and S. Kerba. 2019. African swine fever in wild boar ecology and biosecurity. Food and Agriculture Organization of the United Nations, Office International des Epizooties, and European Commission, FAO Animal Production and Health Manual No. 22, Rome, Italy.

Guinat, C., A. Gogin, S. Blome, G. Keil, R. Pollin, D. U. Pfeiffer, and L. Dixon. 2016. Transmission routes of African swine fever virus to domestic pigs: current knowledge and future research directions. Veterinary Record 178:262–267.

Hardham, J. M., P. Krug, J. M. Pacheco, J. Thompson, P. Dominowski, V. Moulin, C. G. Gay, L. L. Rodriguez, and E. Rieder. 2020. Novel foot-and-mouth disease vaccine platform: formulations for safe and DIVA-compatible FMD vaccines with improved potency. Frontiers in Veterinary Medicine 7:554305.

Heuschele, W. P. 1967. Studies on the pathogenesis of African swine fever. I. Quantitative studies on the sequential development of virus in pig tissues. Archiv für die gesamte Virusforschung 21:349–356.

Howey, E. B., V. O'Donnell, H. C. de Carvalho Ferreira, M. V. Borca, and J. Arzt. 2013. Pathogenesis of highly virulent African swine fever virus in domestic pigs exposed via intraoropharyngeal, intranasopharyngeal, and intramuscular inoculation, and by direct contact with infected pigs. Virus Research 178:328–339.

Jori, F., and A. D. S. Bastos. 2009. Role of wild suids in the epidemiology of African swine fever. Ecohealth 6:296–310.

Jurado, C., L. Mur, M. S. Pérez Aguirreburualde, E. Cadenas-Fernández, B. Martínez-López, J. M. Sánchez-Vizcaíno, and A. Perez. 2019. Risk of African swine fever virus introduction into the United States through smuggling of pork in air passenger luggage. Scientific Reports 9:14423.

Juszkiewicz, M., M. Walczak, and G. Wozniakowski. 2019. Characteristics of selected active substances used in disinfectants and their virucidal activity against ASFV. Journal of Veterinary Research 63:17–25.

Kim, H., M. Lee, S. Lee, D. Kim, S. Seo, H. Kang, and H. Nam. 2019. African swine fever virus in pork brought into South Korea by travelers from China, August 2018. Emerging Infectious Diseases 25:1231–1233.

Kosowska, A., E. Cadenas-Fernández, S. Barrosa, J. M. Sánchez-Vizcaíno, and J. A. Barasona. 2020. Distinct African swine fever virus shedding in wild boar infected with virulent and attenuated isolates. Vaccines 8:767.

Li, Y., M. Salman, C. Shen, H. Yang, Y. Wang, Z. Jiang, J. Edwards, and B. Huang. 2020. African swine fever in a commercial pig farm: outbreak investigation and an approach for identifying the source of infection. Transboundary and Emerging Diseases 67:2564–2578.

Lichoti, J. K., J. Davies, Y. Maru, P. M. Kitala, S. M. Githigia, E. Okoth, S. A. Bukachi, S. Okuthe, and R. P. Bishop. 2017. Pig traders' networks on the Kenya–Uganda border highlight potential for mitigation of African swine fever virus transmission and improved ASF disease risk management. Preventive Veterinary Medicine 140:87–96.

Liu, L., X. Wang, R. Mao, Y. Zhou, J. Yin, Y. Sun, and X. Yin. 2021. Research progress on live attenuated vaccine against African swine fever virus. Microbial Pathogenesis 158:105024.

Luther, N. J., K. A. Majiyagbe, D. Shamaki, L. H. Lombin, J. F. Antiagbong, Y. Bitrus, and O. Owolodun. 2007. Detection of African swine fever virus genomic DNA in a Nigerian red river hog (*Potamochoerus porcus*). Veterinary Record 160:58–59.

Maki, J., A. L. Guiot, M. Aubert, B. Brochier, F. Cliquet, C. A. Hanlon, R. King, E. H. Oertli, C. E. Rupprecht, C. Schumacher, et al. 2017. Oral vaccination of wildlife using a vaccinia-rabies-glycoprotein recombinant virus vaccine (RABORAL V-RB): a global review. Veterinary Research 48:57.

Martínez-Avilés, M., I. Iglesias, and A. de la Torre. 2020. Evolution of the ASF infection stage in wild boar within the EU (2014–2018). Frontiers in Veterinary Science 7:155.

Mason-D'Croz, D., J. R. Bogard, M. Herrero, S. Robinson, T. B. Sulser, K. Wieve, D. Willenbockel, and H. C. J. Godfray. 2020. Modelling the global economic consequences of a major African swine fever outbreak in China. Nature Food 1:221–228.

Matsumoto, N., J. Siengsanan-Lamont, T. Halasa, J. R. Young, M. P. Ward, B. Douangngeun, W. Theppangna, S. Khounsy, J. L. M. L. Toribio, R. D. Bush, et al. 2021. The impact of African swine fever virus on smallholder village pig production: an outbreak investigation in Lao PDR. Transboundary and Emerging Diseases 68:2897–2908.

Mighell, E., and M. P. Ward. 2021. African swine fever spread across Asia, 2018–2019. Transboundary and Emerging Diseases 68:2722–2732.

Miller, R. S., and K. M. Pepin. 2019. Board invited review: prospects for improving management of animal disease introductions using disease-dynamic models. Journal of Animal Sciences 97:2291–2307.

Muñoz-Perez, C., C. Jurado, and J. M. Sánchez-Vizcaíno. 2021. African swine fever vaccine: turning a dream into reality. Transboundary and Emerging Diseases 68:2657–2668.

Mutua, F., and M. Dione. 2021. The context of application of biosecurity for control of African swine fever in smallholder pig systems: current gaps and recommendations. Frontiers in Veterinary Science 8:689811.

Netherton, C. L., S. Connell, C. T. O. Benfield, and L. K. Dixon. 2019. The genetics of life and death: virus–host interactions underpinning resistance to African swine fever, a viral hemorrhagic disease. Frontiers in Genetics 10:402.

Nurmoja, I., K. Mõtus, M. Kristian, T. Niine, K. Schulz, K. Depner, and A. Viltrop. 2020. Epidemiological analysis of the 2015–2017 African swine fever outbreaks in Estonia. Preventive Veterinary Medicine 181:104556.

Office International des Epizooties (OIE) 2019. African swine fever. OIE Technical Disease Cards. World Organization for Animal Health, Paris France.

O'Neill, X., A. White, F. Ruiz-Fons, and C. Gortazar. 2020. Modelling the transmission and persistence of African swine fever in wild boar in contrasting European scenarios. Scientific Reports 10:5895.

Oura, C. A. L., P. P. Powell, E. Anderson, and R. M. E. Parkhouse. 1998. The pathogenesis of African swine fever in the resistant bushpig. Journal of General Virology 79:1439–1443.

Paulino-Ramirez, R., R. Jimenez, and J. Ariel. 2021. Food security and research agenda in African swine fever virus: a new arbovirus threat in the Dominican Republic. Interamerican Journal of Medicine and Health 4:e202101028.

Penrith, M. L., A. D. Bastos, E. M. C. Etter, and D. Beltran-Alcrudo. 2019. Epidemiology of African swine fever in Africa today: sylvatic cycle versus socio-economic imperatives. Transboundary and Emerging Diseases 66:672–686.

Pepin, K. M., A. J. Golnar, Z. Abdo, and T. Podgorski. 2020. Ecological drivers of African swine fever virus persistence in wild boar populations: insight for control. Ecology and Evolution 10:2846–2859.

Pikalo, J., P. Deutschmann, M. Fischer, H. Roszyk, M. Beer, and S. Blome. 2021. African swine fever laboratory diagnosis—lessons learned from recent animal trials. Pathogens 10:177.

Plavšić, B., A. Rozstalnyy, J. Y. Park, V. Guberti, K. Depner, and G. Torres. 2019. Strategic challenges to global control of African swine fever. Proceedings of the 87th General Session of the OIE World Assembly. World Organization for Animal Health, Paris France. https://www.woah.org/fileadmin/Home/eng/Publications_%26_Documentation/docs/pdf/TT/2019_A_87SG_10.pdf, Accessed 4 Oct 2022.

Plowright, W., C. T. Perry, and A. Greig. 1974. Sexual transmission of African swine fever virus in the tick, *Ornithodoros moubata porcinus*, Walton. Research in Veterinary Science 17:106–113.

Plowright, W., C. T. Perry, and M. A. Peirce. 1970. Transovarial infection with African swine fever virus in the asgarid tick, *Ornithodoros moubata porcinus*, Walton. Research in Veterinary Science 11:582–584.

Rennie, L., P. J. Wilkinson, and P. S. Mellor. 2001. Transovarial transmission of African swine fever virus in the argasid tick *Ornithodoros moubata*. Medical and Veterinary Entomology 15:140–146.

Revilla, Y., D. Pérez-Nuñez, and J. A. Richt. 2018. African swine fever virus biology and vaccine approaches. Advances in Virus Research 100:41–74.

Rock, D. L. 2017. Challenges for African swine fever vaccine development—". . . perhaps the end of the beginning." Veterinary Microbiology 206:52–58.

Rossi, S., C. Staubach, S. Blome, V. Guberti, H. H. Thulke, A. Vos, F. Koenen, and M. F. le Potier. 2015. Controlling of CSFV in European wild boar using oral vaccination: a review. Frontiers in Microbiology 6:1141.

Salguero, F. J. 2020. Comparative pathology and pathogenesis of African swine fever infection in swine. Frontiers in Veterinary Science 7:283.

Sánchez-Cordón, P. J., M. Montoya, A. L. Reis, and L. K. Dixon. 2018. African swine fever: a re-emerging viral disease threatening the global pig industry. Veterinary Journal 233:41–48.

Sánchez-Vizcaíno, J. M., A. Laddomada, and M. L. Arias. 2019. African swine fever virus. Pages 443–452 *in* J. J. Zimmerman, L. A. Karriker, A. Ramirez, K. J. Schwartz, G. W. Stevenson, and J. Zhang, editors. Diseases of swine. John Wiley & Sons, Inc., Hoboken, New Jersey, USA.

Sánchez-Vizcaíno, J. M., L. Mur, and B. Martínez-López. 2012. African swine fever: an epidemiological update. Transboundary and Emerging Diseases 59:27–35.

Sang, H., G. Miller, S. Lokhandwala, N. Sangewar, S. D. Waghela, R. P. Bishop, and W. Mwangi. 2020. Progress toward development of effective and safe African swine fever virus vaccines. Frontiers in Veterinary Science 7:84.

Sauter-Louis, C., F. J. Conraths, C. Probst, U. Blohm, K. Schulz, J. Sehl, M. Fischer, J. H. Forth, L. Zani,

K. Depner, et al. 2021*a*. African swine fever in wild boar in Europe—a review. Vaccines 13:1717.

Sauter-Louis, C., J. H. Forth, C. Probst, C. Staubach, A. Hlinak, A. Rudovsky, D. Holland, P. Schlieben, M. Göldner, J. Schatz, et al. 2021*b*. Joining the club: first detection of African swine fever in wild boar in Germany. Transboundary and Emerging Diseases. 68:1744–1752.

Shi, J., L. Wang, and D. S. McVey. 2021. Of pigs and men: the best-laid plans for prevention and control of swine fevers. Animal Frontiers 11:6–13.

Stancu, A. 2018. ASF evolution and its economic impact in Europe over the past decade. The USV Annals of Economics and Public Administration 18:18–27.

Taylor, R. A., R. Condoleo, R. R. L. Simons, P. Gale, L. A. Kelly, and E. L. Snary. 2020. The risk of infection by African swine fever virus in European swine through boar movement and legal trade of pigs and pig meat. Frontiers in Veterinary Science 6:486.

Teklue, T., Y. Sun, M. Abid, Y. Luo, and H. J. Qui. 2019. Current status and evolving approaches to African swine fever vaccine development. Transboundary and Emerging Diseases 67:529–542.

Tran, X. H., T. T. P. Le, Q. H. Nguyen, T. T. Do, V. D. Nguyen, C. G. Gay, M. V. Borca, and D. P. Gladue. 2022. African swine fever virus vaccine candidate ASFV-G-ΔI177L efficiently protects European and native pig breeds against circulating Vietnamese field strain. Transboundary and Emerging Diseases 69:e497–e504.

Tucker, C., A. Fagre, G. Wittemyer, T. Webb, E. O. Abworo, and S. VandeWoude. 2021. Parallel pandemics illustrate the need for One Health solutions. Frontiers in Microbiology 12:718546.

Turlewicz-Podbielska, H., A. Kuriga, R. Niemyjski, G. Tarasiuk, and M. Pomorska-Mól. 2021. African swine fever virus as a difficult opponent in the fight for a vaccine—current data. Viruses 13:1212.

US Department of Agriculture Animal and Plant Health Inspection Service (USDA APHIS). 2021*a*. USDA statement on confirmation of African swine fever in the Dominican Republic. USDA APHIS, Riverdale, Maryland, USA. https://www.aphis.usda.gov/aphis/newsroom/news/sa_by_date/sa-2021/asf-confirm. Accessed 30 Dec 2021.

US Department of Agriculture Animal and Plant Health Inspection Service (USDA APHIS). 2021*b*. USDA statement on confirmation of African swine fever in Haiti. USDA APHIS, Riverdale, Maryland, USA. https://www.aphis.usda.gov/aphis/newsroom/stakeholder-info/sa_by_date/sa-2021/sa-09/asf-haiti. Accessed 30 Dec 2021.

Vergne, T., C. Chen-Fu, S. Li, J. Cappelle, J. Edwards, V. Martin, D. U. Pfeiffer, G. Fusheng, and F. L. Roger. 2017. Pig empire under infectious threat: risk of African swine fever introduction into the People's Republic of China. Veterinary Record 181:117.

Viltrop, A., F. Boinas, K. Depner, F. Jori, D. Kolbasov, A. Laddomada, K. Ståhl, and E. Chenais. 2021. African swine fever epidemiology, surveillance and control. Pages 229–261 *in* L. Iacolina, M. Penrith, S. Bellini, E. Chenais, F. Jori, M. Montoya, K. Ståhl, and D. Gavier-Widén, editors. Understanding and combatting African swine fever. Wageningen Academic Publishers, Wageningen, The Netherlands.

Wang, T., Y. Sun, S. Huang, and H. J. Qiu. 2020. Multifaceted immune responses to African swine fever virus: implications for vaccine development. Veterinary Microbiology 249:108832.

Wozniakowski, G., M. Fraczyk, K. Niemczuk, and Z. Pejsak. 2016. Selected aspects related to epidemiology, pathogenesis, immunity, and control of African swine fever. Journal of Veterinary Research 60:119–125.

Wu, K., J. Liu, L. Wang, S. Fan, Z. Li, Y. Li, L. Yi, H. Ding, M. Zhao, and J. Chen. 2020. Current state of global African swine fever vaccine development under the prevalence and transmission of ASF in China. Vaccines 8:531.

Xia, W., J. Hughes, D. L. Robertson, and X. Jiang. 2021. How one pandemic led to another: ASFV, the disruption contributing to SARS-CoV-2 emergence in Wuhan. Preprints 2021:2021020590.

Yang, J., K. Tang, Z. Cao, D. U. Pfeiffer, K. Zhao, Q. Zhang, and D. D. Zeng. 2021. Demand-driven spreading patterns of African swine fever in China. Chaos 31(6):061102.

Zhao, D., R. Liu, X. Zhang, F. Li, J. Wang, J. Zhang, X. Liu, L. Wang, J. Zhang, X. Wu, et al. 2019. Replication and virulence in pigs of the first African swine fever virus isolated in China. Emerging Microbes and Infections 8:438.

Zhu, J. J., P. Ramanathan, E. A. Bishop, V. O'Donnell, D. P. Gladue, and M. V. Borca. 2019. Mechanisms of African swine fever virus pathogenesis and immune evasion inferred from gene expression changes in infected swine macrophages. PLoS One 14:e0223955.

Wildlife Disease and Health in Carnivores, Rodents, and Bats

13 Plague and Distemper

Threats to Black-footed Ferret Conservation

Tonie E. Rocke

Introduction

In 1981, a dog named Shep that lived on a ranch near Meeteetse, Wyoming, retrieved a carcass its owners did not recognize (Santymire et al. 2014). Curious, they took it to a local taxidermist who identified it and immediately notified state wildlife officials. It was a black-footed ferret (BFF, *Mustela nigripes*), a species thought to have gone extinct several decades earlier (Figure 13.1). Shortly after this discovery, biologists located the remnant population near Meeteetsee, likely the last remaining BFFs in the wild, and worked quickly to protect the species. Ultimately, 18 animals were brought into captivity to prevent losses from disease and to establish a captive breeding and reintroduction program (Thorne and Williams 1988), undoubtedly saving the species from extinction (Santymire et al. 2014).

As the only native ferret species in North America, BFFs once occupied grassland ecosystems from northern Mexico to Canada (Figure 13.2). Their populations declined precipitously through the first half of the 20th century, and the BFF was declared endangered in 1967 and officially listed in 1973 when the Endangered Species Act (ESA) was signed (USFWS 2013). Despite early efforts to establish a captive breeding program in the 1970s, the last known

Yersinia pestis (plague) and canine distemper are major threats to black-footed ferret recovery. Illustration by Laura Donohue.

Fig. 13.1. A wild black-footed ferret. Photograph by Kathy Hargrove (Getty Images).

BFF died in captivity in 1979, and the species was thought to have gone extinct at that time (USFWS 2013).

Numerous factors contributed to the species demise. Black-footed ferrets eat mostly prairie dogs (*Cynomys* spp.; Figure 13.3), and they rely on prairie dog burrows for shelter, breeding, whelping, and raising young, Thus, the fate of the BFF is inexorably linked to that of several species of prairie dogs. Considered vermin by ranchers and other landowners, prairie dog populations were decimated by encroaching urbanization, conversion of grasslands to agriculture, and outright eradication efforts via extensive poisoning and hunting (Thorne and Williams 1988). By the 1960s, prairie dog habitat had plunged from 40 million hectares to 600 thousand hectares (approximately 1.5% of historic range), and numbers of BFFs also plunged because survival of their populations require large prairie dog complexes, thousands of hectares in size (Bevers and Hof 1997).

About the same time (1940s to 1950s), two fatal diseases emerged or were first recognized in their grassland ecosystems that are thought to have played a significant role in the near extinction of BFFs (Carr 1986, Thorne and Williams 1988, Lockhart, et al. 2004): canine distemper (CD), which causes severe disease in ferrets and other carnivores, and sylvatic plague, which afflicts both BFFs and prairie dogs (among other species). Without intervention, BFFs rarely survive exposure to either of these diseases. Despite a highly successful recovery program currently led by the USFWS National Black -Footed Ferret Conservation Center (NBFFCC) and the Black-footed Ferret Reintroduction Implementation Team (BFFRIT), CD and sylvatic plague remain existential threats to BFFs (USFWS 2013, 2019). Managing these disease threats, especially sylvatic plague, is considered fundamental to the BFF's continued recovery and eventual delisting from the ESA (Santymire et al. 2014).

Canine Distemper

Canine distemper is caused by a highly infectious, often fatal virus in the family Paramyxoviridae and genus Morbilliviridae, closely related to human measles and rinderpest (Uhl et al. 2019) (see Chapter 7). First recognized in domestic dogs, the disease now occurs worldwide in a wide variety of carnivores and is considered a threat to the conservation of several endangered species (Martinez-Gutierrez and Ruiz-Saenz 2016), including the BFF.

History

Canine distemper virus (CDV), identified in 1905 (Carré, 1905), is believed to have been the cause of severe disease in domestic dogs for several centuries, originating after infection and subsequent mutation of human measles virus in New World dogs (Figure 13.4; Uhl et al. 2019). Canine distemper was recognized in wild species in the early 1900s, and outbreaks in free-ranging raccoon (*Procyon lotor*), skunk (*Mephitis mephitis*), red fox (*Vulpes vulpes*) and gray fox (*Urocyon cinereoargenteus*) were described about 50 years later (Helmboldt and Jungherr 1955). One of the earliest reports of CDV in the western United States was in an American badger (*Taxidea taxus*) (Armstrong 1942). The disease has since been reported in a wide range of species from five different orders, including two species of nonhuman primates (Martinez-Gutierrez and Ruiz-Saenz, 2016), with varying levels of morbidity and mortality.

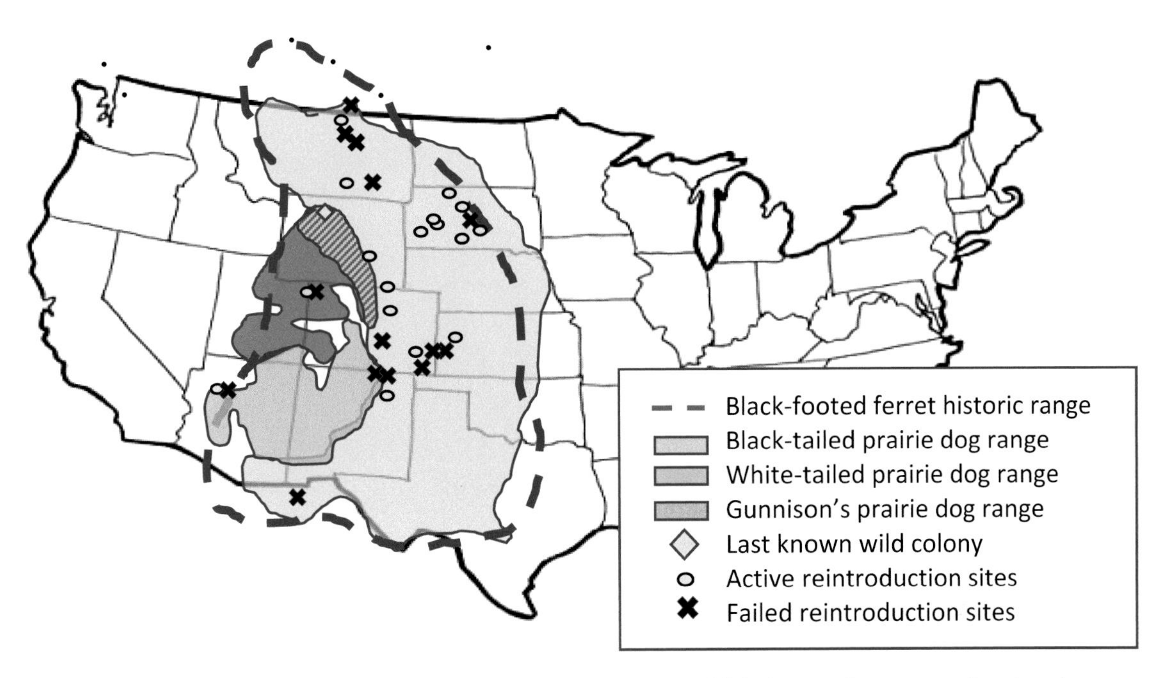

Fig. 13.2. Black-footed ferret historic range, active introduction sites, and failed reintroduction sites. Based on data from the US Fish and Wildlife Service.

Fig. 13.3. Black-tailed prairie dog on a burrow. US Geological Survey photograph by Tonie Rocke.

In the fall of 1985, six BFFs were captured in the only known free-living colony near Meeteetse, Wyoming, and transferred to a Wyoming Game and Fish facility for captive propagation (Williams et al. 1988). Within a few days of capture, one animal became extremely pruritic (itchy), and another became depressed; CD was diagnosed, and, ultimately, all six animals succumbed to the disease. This incident and the subsequent decline in the wild population are what prompted the capture and removal of all remaining ferrets in the colony to prevent further losses and provide animals for the captive breeding program (Williams et al. 1988).

Transmission and Epizootiology

Although CD occurs worldwide in a variety of captive animals (both domestic and wild), little is known about CDV transmission and maintenance in free-ranging populations (Deem et al. 2000; Martinez-Gutierrez and Ruiz-Saenz, 2016), including the BFF. Transmission of CDV between mammals is best known from studies of captive mink (*Mustela vison*). The high susceptibility of ranch-raised foxes, mink, and other mustelidae led to the establishment of a USDA support program in the 1960s that developed several early CDV vaccines (Gorham et al. 1954; Gorham 1966). The virus is fragile, susceptible to ultraviolet light and high temperatures, and thus, does not persist long in outdoor environments and must be

Fig. 13.4. Schematic representation of the possible evolution and spread of canine distemper virus. Illustration by Laura Donohue.

passed from animal to animal to survive. Transmission of CDV primarily occurs via aerosolized oral, respiratory, or ocular exudates (Budd 1981). Although the virus is also shed from skin and in feces and urine, transmission from these sources is less likely to cause disease, except in confined or highly social animals.

Although dense populations are usually required to sustain direct CDV transmission, other factors may play a role in free-living settings, including virus dose, host susceptibility, density and susceptibility of associated species, and inter- and intra-species behavior (Williams 2001). In short-grass prairies, where BFFs reside, the density of carnivores is relatively low. Williams (2001) reported that CD outbreaks occurred in these landscapes every 3–7 years, during all seasons, mostly involving coyotes (*Canis latrans*), but also red fox, swift fox (*Vulpes velox*), badgers, and likely BFFs. Coyotes are more gregarious, and their behavior may facilitate transmission, whereas badgers and ferrets are relatively solitary. Most mustelid species are highly susceptible to CD and likely to die if they contract the virus. Canine distemper is known to be 100% fatal in domestic ferrets (*Mustela putorius furo*; Appel et al. 1981). Given that six of six infected BFFs died shortly after capture (Williams et al. 1998), this species appears equally susceptible to CDV and unlikely to recover if infected. However, the prevalence and frequency of CDV exposure in unvaccinated, free-ranging BFFs is unknown. At present, all released ferrets and any trapped, wild-borne kits are vaccinated against CD as a standard protocol (Wright et al. 2022).

Clinical signs, Pathology and Pathogenesis

Williams et al. (1988) described the clinical signs, gross lesions, and histopathology observed in the six

BFFs with naturally acquired CD. Clinically, the most striking sign was pruritis, causing affected animals to rub and scratch their faces and bodies. Hyperemia was observed, and the skin, ears, nose, eyelids, and footpads became hyperkeratotic as the disease progressed. The animals were depressed and had varying signs of diarrhea, anorexia, and polydipsia. Microscopically, large eosinophilic cytoplasmic and intranuclear inclusion bodies were evident in epithelial cells of many tissues, including urinary bladder, bronchi and bronchioles, bile ducts, pancreatic ducts and islets, tongue, stomach, and skin (Figure 13.5).

In dogs and other animals, early development of antibodies appears to be crucial to prevent symptomatic infection and recovery from the disease (Williams 2001). After aerosolized virus enters the epithelium of the upper respiratory tract, it multiplies in macrophages and spreads to tonsils and regional lymph nodes. Within a week, CDV spreads to systemic lymphoid tissues, with consequent signs of fever and leukopenia. At 1–2 weeks postinfection, resistant animals develop a strong antibody response, and the virus is cleared with no clinical illness. If the immune response is weak, dogs and other animals may show signs of illness but recover within 3 weeks; viral shedding may persist for 2–3 months. If the virus overwhelms the host immune response, severe systemic disease and death occur in 3–4 weeks. In the six BFFs naturally infected with CDV, time to death from onset of clinical signs was 14–48 days. Of four BFFs tested, all were negative for CDV neutralizing antibody at or near the time of death, indicating these animals had no effective humoral immune response.

Diagnosis

Clinical signs and external lesions are suggestive of CD in sick animals but must be differentiated from other diseases (Deem et al. 2000). Immunohistochemistry can be used to detect CDV antigen using cytological smears from epithelial tissues (conjunctival, genital, ocular, or respiratory), blood and buffy coat smears, cerebrospinal fluid, skin, and foot pads. For carcasses, CD diagnosis is presumptive upon observation of typical viral inclusion bodies in epithelium, lymphoid tissues, and brain by histopathologic examination of formalin fixed tissues; immunohistochemistry provides definitive evidence of CDV infection (Deem et al. 2000). Isolation of CDV is also evidence of infection; samples are cultured in pulmonary alveolar macrophages and the presence of CDV is confirmed by specific immunofluorescence. Although serology is often used to assess the

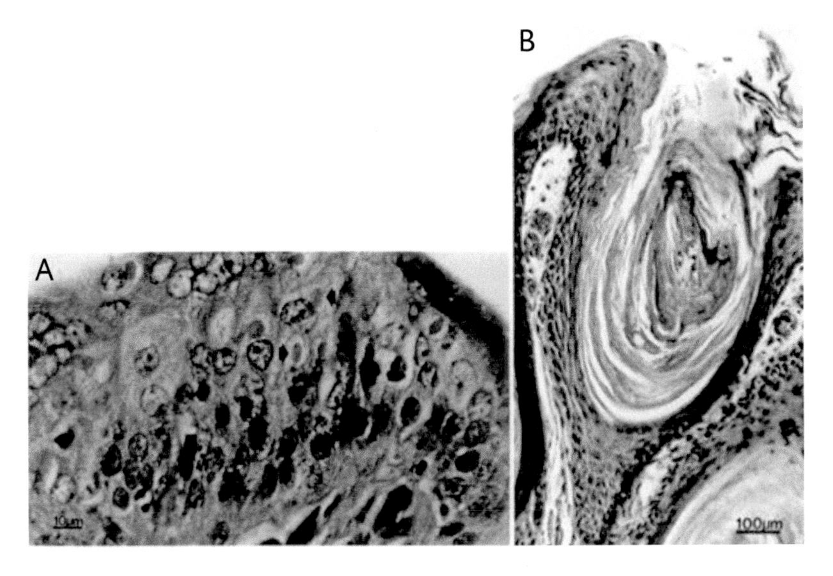

Fig. 13.5. (A) Typical inclusion bodies of canine distemper (arrows) in urinary bladder epithelium of an affected black-footed ferret. (B) Follicular and epidermal hyperkeratosis and dermatitis in a black-footed ferret. Hematoxylin and eosin stain. Reprinted from Williams et al. (1988).

prevalence of CD in wild animal populations, highly susceptible species that experience high mortality rates after exposure, like BFFs, do not develop antibodies against CDV. Therefore, serosurveys of wild ferrets are not useful for assessing prevalence of exposure. Instead, serosurveys of local coyotes are often used to estimate prevalence of disease that might affect BFFs (Schuler et al. 2021) since they occupy similar habitat and are fairly resistant to CDV and other pathogens (e.g., *Yersinia pestis*).

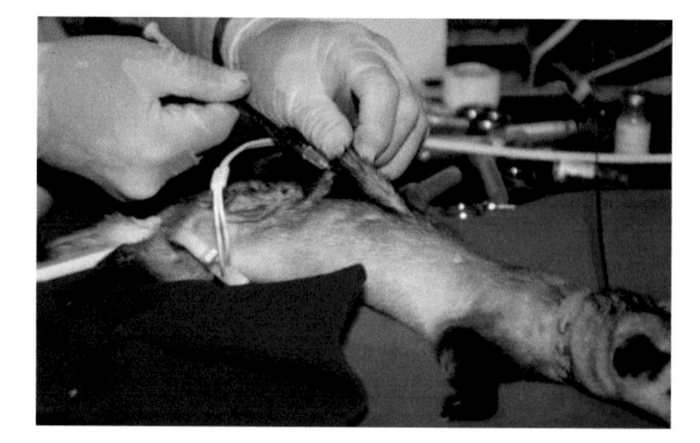

Fig. 13.6. Vaccination of wild black-footed ferret. Photograph by Travis Livieri.

Disease Management

Canine distemper is one of the most contagious and threatening diseases of free-ranging wild carnivores, and thus, it has the potential to significantly affect populations of highly susceptible species (see Chapter 7). The disease is a major threat to wild carnivores in East Africa, particularly in social species like Cape hunting dogs (*Lycaon pictus*) and lions (*Panthera leo*), and for captive carnivores in zoos and other facilities. Animals incubating the virus have been brought into captivity, spreading the disease to others (Williams et al. 1988).

Currently, the primary method for managing distemper in captive carnivores is vaccination. However, modified live vaccines, developed primarily for dogs, have induced CD in grey fox (Halbrooks et al. 1981), BFFs (Carpenter et al. 1976), and domestic ferrets (Budd, 1981). The use of killed vaccines does not provide complete protection (Williams et al. 1996). A canarypox-vectored vaccine (Purevax Ferret Distemper®, Boehringer-Ingelheim/Merial) developed for domestic ferrets has proven safe and effective in preventing disease in a close relative, the Siberian polecat (*M. eversmanni*) (Wimsatt et al. 2003), and was subsequently tested in BFFs (Marinari and Kreeger 2004). Purevax Ferret Distemper® is currently the vaccine of choice to protect both breeding BFFs and those intended for reintroduction. Primary vaccination and boosters are administered at 60 and 90 days of age, respectively (Figure 13.6). Purevax Ferret Distemper® vaccine is also administered to wild-born individuals at most reintroduction sites (e.g., Conata Basin, South Dakota; Rocky Mountain Arsenal, Colorado; and Lower Brule Sioux Tribe, South Dakota), where wild-born free-ranging BFFs are captured on an annual basis for monitoring and other purposes (Wright et al. 2022). Another canarypox-vectored vaccine (Recombitek®, Boehringer-Ingelheim/Merial) used in dogs has also been tested in BFFs, and it induces similar protective antibody levels (Wright et al. 2022).

Given the ubiquity of CDV and the number of carnivores that are susceptible, it is unlikely that other management actions beyond vaccination are feasible. It is unknown how many wild BFFs perish from CD on a regular basis or how effective vaccination is at the population level, but the disease will likely remain a threat to the species until their population numbers are large enough to sustain periodic losses.

Sylvatic Plague

Plague, caused by the bacterium *Yersinia pestis*, is an ancient zoonotic disease, first introduced in North America in the early 1900s. It is primarily a disease of wild rodents that has become endemic in several North American ecosystems, occasionally spilling over into humans and other animals. Because *Y. pestis* was introduced to the continent relatively recently, many North American mammals, for example, the BFF and all five prairie dog species (*Cynomys* spp.), are highly susceptible to the pathogen, not having evolved resistance. Currently, sylvatic plague is considered the biggest hinderance to the recovery of the BFF (USFWS 2013, 2019).

History

Three notable pandemics of plague have been documented in humans. The most notorious, known as the Black Death, was responsible for killing approximately 30–40% of the human population in Europe during the 14th century (Perry and Fetherston 1997). The last, known as the Modern Pandemic, began in China in 1772 and later spread worldwide via steamships and trains, establishing sylvatic cycles in new locations wherever appropriate host and vector assemblages supported its maintenance. Human cases of plague still occur annually where these sylvatic cycles persist (Figure 13.7), mostly in Africa (Barbieri et al. 2020).

In the early 1900s, rats infected with *Y. pestis* jumped ship at several ports in California and elsewhere, introducing the bacterium to North America. Most of these introduction events were short lived. However, the disease established a stronghold in commensal rodents in San Francisco, spreading into ground squirrels (*Spermophilus* spp.) and then eastward into numerous susceptible rodent species before becoming endemic along the Pacific coast and the southwestern United States (Figure 13.8). Prominent wild rodent hosts in endemic North American foci include prairie dogs, ground squirrels, chipmunks (*Neotamias* spp.), wood rats (*Neotoma* spp.), and short-lived rodents, such as deer mice (*Peromyscus* spp.) and voles (*Microtus* spp.) (Abbott and Rocke 2012). Within 40 years of its introduction, the plague bacterium spread approximately 2200 km to the 103rd meridian west (Adjemian et al. 2007) but did not spread beyond that boundary for another 60 years, despite similar rodent and flea assemblages to the east. More recent plague outbreaks in South Dakota (Keuler et al. 2020) and in Saskatchewan, Canada (Antonation et al. 2014), however, indicate the potential for further expansion.

The first documented case of plague in a BFF occurred in 1993 in a captive animal that was accidently exposed to infected captive white-tailed prairie dogs (Williams et al. 1994). In a similar incident in 1995, 27 of 30 captive BFFs died from plague, or went missing in underground burrows and were presumed dead, after they were accidently fed meat from *Y. pestis* infected prairie dogs (Godbey et al. 2006). Prior to these events, BFFs were thought to be resistant to plague like most other carnivores, except felines (both wild and domestic), which are also susceptible. Other mustelids, such as badgers and weasels, rarely show clinical signs of infection (Gage 1998, Gasper and Watson 2001), and in experimental infection studies, domestic ferrets did not develop

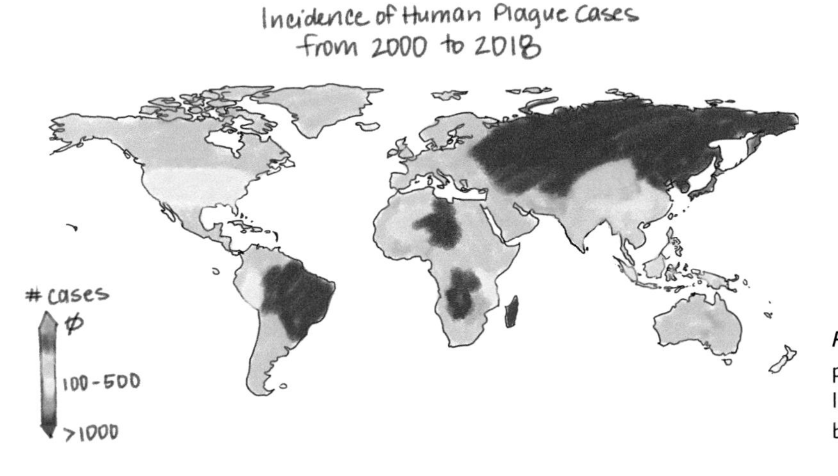

Fig. 13.7. World map of human plague cases from 2000 to 2018. Illustration by Laura Donohue based on Barbieri et al. (2020).

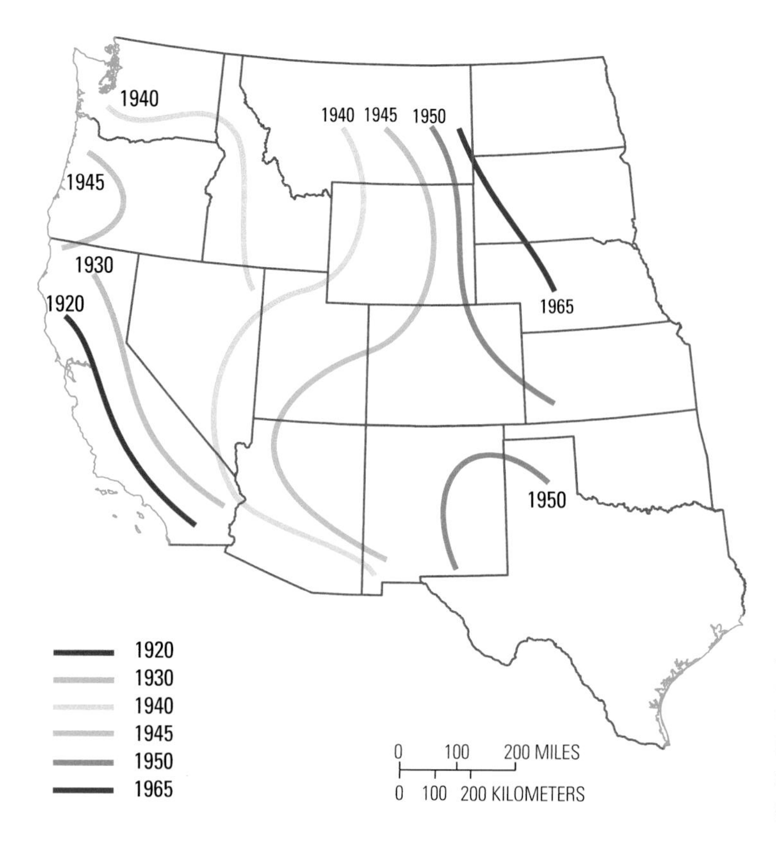

Fig. 13.8. Predicted waves of plague movement from the west coast of the United States eastward into the Great Plains. Based on Adjemian et al. (2007); reprinted from Abbott and Rocke (2012).

clinical plague after subcutaneous injection with >10,000,000 colony-forming units (cfu) (Williams et al. 1991). In contrast, experimentally infected BFFs, included as controls in vaccine efficacy studies with USFWS authorization, succumbed to infection within 3–4 days after injection of 7800 cfu (Rocke et al. 2004, 2006). In another study, BFFs were challenged by feeding them mice infected with *Y. pestis*, and one animal that licked or sniffed the carcass—but refused to eat it—died from plague within four days, demonstrating their extreme susceptibility (Rocke et al. 2006).

Transmission and Epizootiology

The epizootiology and transmission pathways of plague in grassland species and ecosystems are complex, involving numerous wild rodent species of varying susceptibility, numerous flea species (both specialists and generalists with varying transmission efficiencies), and additional reservoirs of the plague bacterium, such as carcasses of animals that died of the disease and plague-resistant carnivores that transport infectious fleas (Figure 13.9). Although transmission of *Y. pestis* occurs mostly via infected flea bites, it can also spread through respiratory droplets, direct contact, or by consumption of infected prey or carcasses, which may be particularly important for BFFs. The primary prey of BFFs are prairie dogs, which suffer sporadic outbreaks of plague that often decimate entire colonies. While infectious fleas (both on and off prairie dogs) are no doubt a source of exposure, consumption of prairie dogs infected with *Y. pestis* may lead to even greater levels of exposure by direct contact and aerosolized bacteria (the most lethal form of plague) as ferrets rip and tear at the meat. Even decomposing carcasses pose a threat of plague infection. Godbey et al. (2006) reported

Fig. 13.9. Transmission of *Yersinia pestis* between prairie dogs, black-footed ferrets, and other animals. Illustration by Laura Donohue.

that *Y. pestis* survived for more than two months in underground carcasses that were subsequently consumed by a captive Siberian polecat that became infected and died of plague. Because most prairie dogs and ferrets succumb to plague underground, and *Y. pestis* is relatively stable in these conditions, the likelihood of encountering infectious carcasses, both during and after an outbreak, is high.

Despite numerous studies, many questions remain about the host, vector, and environmental conditions most important for maintaining *Y. pestis* on the landscape and initiating large-scale plague outbreaks in prairie dogs. Low levels of *Y. pestis* are thought to circulate among short-lived rodents and their fleas and possibly among some prairie dog species. Although the role of short-lived rodents (e.g., deer mice and other small rodents) in maintaining plague on the landscape has long been suspected, strong evidence is lacking. Short-lived rodents are less susceptible to plague than prairie dogs, and plague antibodies have been found in deer mice following outbreaks in prairie dogs (Salkeld and Stapp 2008). Host switching of fleas between short-lived rodents and prairie dogs appears to occur rarely (Russell et al. 2018, Bron 2017), since most are host specific (Figure 13.10). Observations of prairie dog fleas (e.g., *Oropsylla hirsuta*) on deer mice and on northern grasshopper mice (*Onchomys leucogaster*), during and after prairie dog die-offs, indicate short-lived rodents are most often spillover hosts (Stapp et al. 2008; Salkeld and Stapp 2008). However, *Y. pestis* detection in rodent fleas on deer mice and northern grasshopper mice cohabiting with prairie dog colonies, months prior to plague detection and subsequent die-offs in prairie dogs (Bron et al. 2019; Colman et al. 2021), indicate that small rodents may also act as a bridge for *Y. pestis* transmission between other reservoirs that may reside outside prairie dog colonies. Northern grasshopper mice may also facilitate intracolony transmission in regions where they occur (Stapp et al. 2009). In Colorado, the abundance of grasshopper mice appeared to drive outbreaks in prairie dogs by increasing the connectivity between coteries (family groups within a colony), thus permitting more effective "percolation" of the disease through prairie dog populations (Salkeld et al. 2010). It is also possible that short-lived rodents and their fleas may be a source of plague for BFFs, as these rodents make up about 10% of their diet.

Although prairie dog species often suffer high mortality rates from plague (Figure 13.11), susceptibility varies between species and among populations within a species, most likely driven by changes in host genetics resulting from historical exposure to the disease or genetic isolation (Rocke et al. 2012; Russell et al. 2019). In laboratory studies conducted

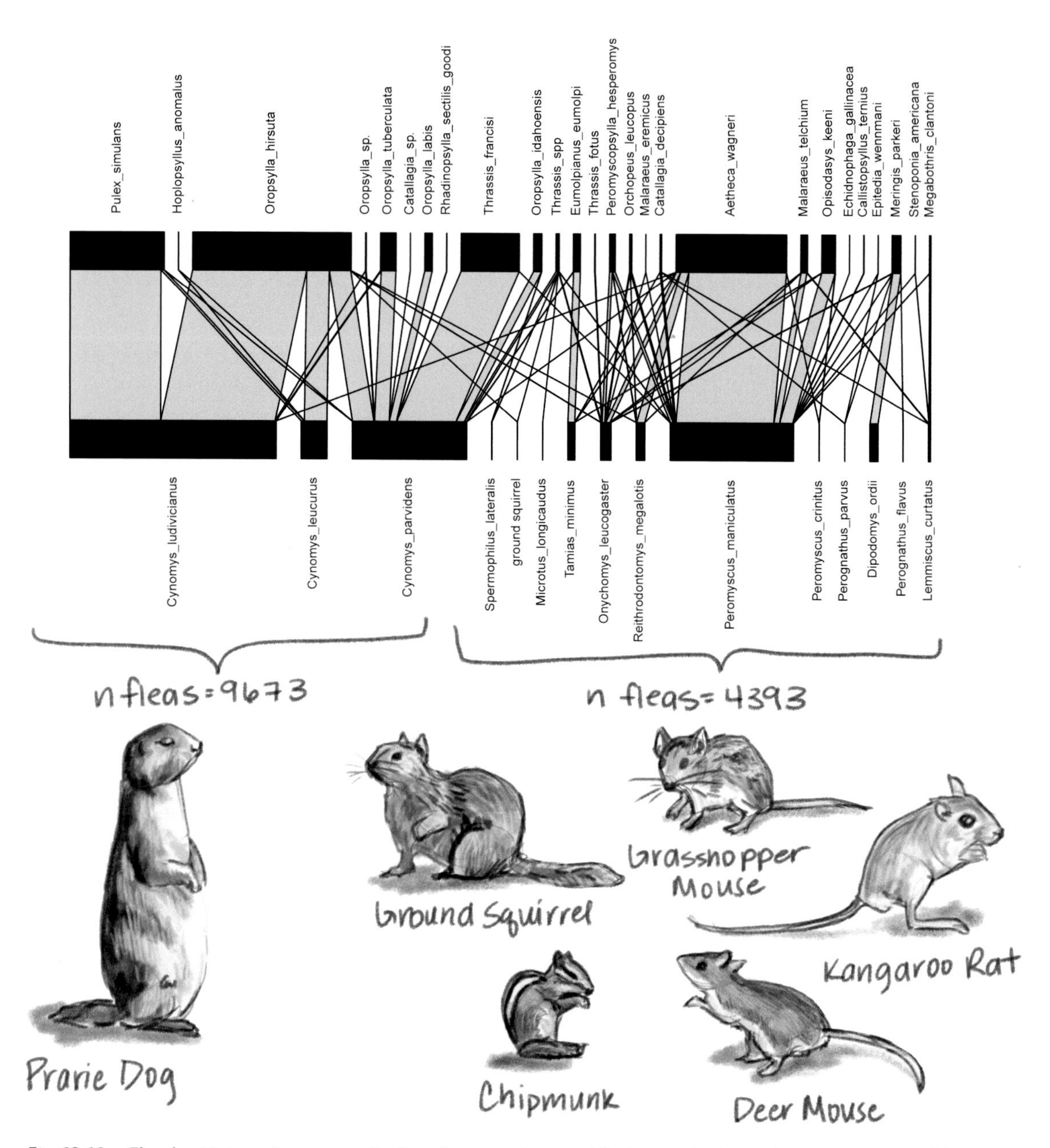

Fig. 13.10. Flea–host interactions network of study areas where prairie dogs and other rodents were captured from 2013–2015. Black boxes (nodes) represent the relative numbers of fleas by species encountered in the upper level and numbers of rodents sampled by species in the lower level. Links (grey and black lines) are the species interactions, and the width of lines represents the frequency of the interaction. Flea switching between prairie dogs and short-lived rodents was rare; 0.05% of fleas on prairie dogs were specific to short-lived rodents, while 0.31% of fleas on short-lived rodents were specific to prairie dogs. Illustration by Laura Donohue based on data from Bron (2017).

at the USGS National Wildlife Health Center using a standard *Y. pestis* challenge dose (Russell et al. 2019), Utah prairie dogs (*Cynomys parvidens*) were least likely to survive (0%), with a mean time to death of 5.7 ± 0.33 days, and white-tailed prairie dogs (*Cynomys leucurus*) from Colorado were most likely to survive (67%), with a mean time to death of 22.1 ± 2.26 days (Table 13.1). Overall, lower susceptibility to plague was observed in populations of black-tailed prairie dogs from Texas and Colorado with a long history of the disease compared to populations from South Dakota with no history of the disease (Rocke et al. 2012, Russell et al. 2019). In other words, in less susceptible hosts, a larger dose of bacteria is required to cause morbidity and mortality, presumably the result of genetically induced changes in susceptibility. Lowered susceptibility to *Y. pestis* could allow the bacterium to circulate within prairie dog colonies at low levels for longer periods of time before a noticeable die-off occurs. Evidence indicates BFFs are so susceptible to plague that even low levels of *Y. pestis* circulation within their home range can affect their survival (Matchett et al. 2010).

Fig. 13.11. Utah prairie dog, confirmed dead from plague. Photograph by Gebbienna Bron.

Whether prairie dogs are involved in enzootic circulation of plague is still questionable (Colman et al. 2021). Genotyping of *Y. pestis* isolates from fleas collected before and during die-offs of prairie dogs in Colorado suggest multiple introductions of the bacterium from unknown reservoirs, with subsequent spread within and between neighboring colonies (Colman et al. 2021). In addition to short-lived rodents, spread between colonies may occur via dispersing prairie dogs (Stapp et al. 2004) or via resistant mesocarnivores (e.g., coyotes and swift fox that harbor common prairie dog fleas such as *O. hirsuta* and *Pulex simulans*) (McGee et al. 2006; Salkeld et al. 2007). Black-footed ferrets are considered spillover hosts and probably contribute very little to the maintenance and circulation of *Y. pestis* in rodents, their primary reservoir hosts. However, fleas infected with *Y. pestis* have been found on BFFs that are vaccinated against plague; therefore, vaccinated animals could potentially transport infected fleas within and between colonies (Mize et al. 2017).

Results of recent model simulations of plague transmission events in prairie dog populations (Russell et al. 2021) revealed that a severe plague outbreak

Table 13.1. Numbers of prairie dogs that survived or died from challenge with *Yersinia pestis* at a challenge dose of 3,500 times the mouse 50% lethal dose (except where noted) and mean day of death of animals that died (based on data from Russell et al. 2019)

Location	Species[a]	Survived	Died	% Survival	Day of death[b]
Colorado	BT	12	12	50.0	8.9 (1.41)
South Dakota	BT	9	95	8.6	6.9 (0.27)
Texas	BT	9	6	60.0	10.3 (1.14)
Arizona	GU	36	24	60.0	10.4 (0.73)
Colorado	GU	11	29	27.5	5.8 (0.35)
Utah	UT	0	21	0.0	5.7 (0.33)
Colorado	WT	10	5	67.0	22.1 (2.26)
Colorado[c]	WT	0	10	0.0	6.4 (0.33)

[a] Abbreviations: BT, black-tailed prairie dog (*Cynomys ludovicianus*); GU, Gunnison's prairie dog (*Cynomys gunnisoni*); UT, Utah prairie dog (*Cynomys parvidens*); WT, white-tailed prairie dog (*Cynomys leucurus*).
[b] Values are means with standard errors given in parenthesis.
[c] Challenged with 10× higher dose.

leading to prairie dog colony depopulation was more likely to occur (95% of simulations) when initiated by direct transmission of *Y. pestis* to a host, either by consuming a carcass or through direct contact with an infected neighbor (Figure 13.12). Depopulation was less likely to occur (10% of simulations) when plague outbreaks were initiated by transmission via infectious fleas that were imported into the colony by short-lived rodents, dispersing prairie dogs from other colonies, or carnivores. This disparity is due to the lowered overall probability of infected fleas finding and infecting a prairie dog host and fleas on that host becoming infected (or staying infected), leaving that host and finding a new host, allowing the disease to spread within the colony. However, these models did not measure the frequency of these events, and flea-initiated transmission of *Y. pestis* is more likely reflective of how most outbreaks occur in prairie dogs. The initiation and resulting outcome of plague outbreaks in prairie dogs may resemble human plague dynamics, where direct contact or pneumonic plague outbreaks are more severe than bubonic plague resulting from flea-borne transmission, but pneumonic outbreaks occur far less frequently (Andrianaivoarimanana et al. 2019). In summary, the epizootiology of plague in grassland ecosystems is complicated. Understanding how the plague bacterium survives in the environment, circulates within and between prairie dog colonies, and ultimately infects BFFs is critical for devising optimal management strategies to prevent or control the disease for both species.

Clinical Signs and Pathology

Signs of plague in BFFs and other animals can include fever, lethargy, anorexia, enlarged lymph nodes (Figure 13.13), dyspnea, nasal discharge, coughing, and sneezing. Gross lesions in BFFs depend on the route of exposure and include acute hemorrhage and necrosis of cervical and mesenteric lymph nodes, subcutaneous hemorrhage, intestinal hemorrhage, and pulmonary edema (Williams et al. 1994). Microscopic lesions include multiple foci of acute hemorrhage and marked perivascular and alveolar edema in the lung.

Diagnosis

A diagnosis of plague is often presumed based on symptoms and exposure history but should be confirmed. In most states, a diagnosis of plague is reportable to public health agencies. Stained smears of peripheral blood, lymph node aspirates, or cultured bacteria (using Gram, Wright, Giemsa or Wayson stain) that reveal bipolar-staining, ovoid, Gram-negative organisms with a "safety pin" appearance are presumptive for plague. Cultured bacteria can be confirmed as *Y. pestis* using direct fluorescent antibody (DFA) or polymerase chain reaction (PCR).

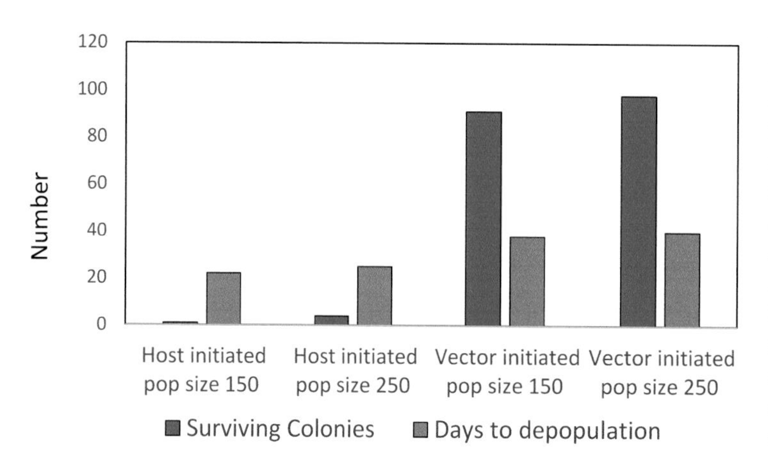

Fig. 13.12. Simulation outcomes of plague outbreaks when initiated by direct transmission to hosts (host initiated) versus those initiated by flea transmission (vector initiated) in starting population (pop) sizes of 150 and 250 prairie dogs. The number of simulations with surviving prairie dog colonies was considerably lower for host-initiated outbreaks (1–4) versus vector-initiated outbreaks (91–98), and the days to depopulation were also shorter: 22–25 versus 38–40. Based on data from Russell et al. (2021).

Fig. 13.13. Wild black-footed ferret with apparent swollen left prescapular lymph node, possibly indicative of plague, although not confirmed. Photograph by Diane Hargreaves.

However, live *Y. pestis* is often difficult to culture from wildlife carcasses as they are rarely found in fresh condition. For these cases, lymphoid, spleen, lung, liver tissue, or bone marrow samples may be tested for evidence of *Y. pestis* by indirect detection methods, such as PCR, although to avoid false positives, targeting of multiple *Y. pestis* genes (e.g., F1 and *pla*) is recommended. Fleas collected from carcasses can also be tested similarly (Engelthaler et al. 1999); however, without corresponding confirmation in animal tissue, positive fleas should not be considered definitive evidence that plague was the cause of the host's death. Plague can be diagnosed retrospectively by serological testing of paired serum samples but is rarely used for wild animals, except perhaps for those in captivity. Because *Y. pestis* is considered a Select Agent (a zoonosis and potential bioweapon), any samples confirmed positive for live *Y. pestis* are required to be reported to the Federal Select Agent Program and then either destroyed or transferred to an appropriate Select Agent-registered laboratory.

Disease Management

Plague is managed effectively in captive BFFs by vaccination and careful screening of prairie dogs that are used to feed them. Purified F1-V fusion protein administered with an adjuvant has been shown to be an effective plague vaccine for BFFs (Rocke et al. 2004, 2006, 2008). Ferret kits, immunized by subcutaneous injection at 60 and 120 days of age, were fully protected against virulent plague challenge with antibody titers persisting for 2 years and likely much longer (Rocke et al. 2008). Currently all BFFs housed, bred, reared, and released for reintroduction by the NBFFCC are vaccinated against plague. As an added precaution, live prairie dogs trapped from healthy, monitored colonies are dusted to remove fleas prior to feeding them to BFFs. Any unhealthy prairie dogs received or caught on the grounds of NBFFCC are euthanized and tested for *Y. pestis* as described above, and prairie dog meat that is processed and frozen is routinely batch tested.

Management of plague in free-ranging BFFs is more complicated and necessarily linked to management of the disease in prairie dogs. Vaccination of free-ranging ferrets with F1-V fusion protein has been shown to improve their survival even without noticeable die-offs in prairie dogs (Matchett et al. 2010), and annual efforts to capture and vaccinate wild-born ferrets have been conducted at several sites; however, this is not a sustainable management strategy in the long term. The current plague vaccine used in ferrets has a limited shelf life, must be prepared in advance, and requires capture and injection of animals twice to provide optimal protection, although even a single dose has been shown to be useful (Travis Livieri, Prairie Wildlife Research, personal communication, 2019). Vaccinated ferrets appear to have life-long immunity against plague, but they are likely to perish if their prey base succumbs to the disease.

For the last several decades, management of plague for both prairie dogs and BFFs relied on dusting prairie dog burrows with 0.05% deltamethrin dust (Seery et al. 2003). In some locations, burrows are dusted proactively to prevent a die-off, especially where BFFs are present, but in most cases, dusting is applied after declines in prairie dogs are noticed. Although successful in abating most outbreaks in prairie dogs, dusting is labor intensive and expensive (approximately \$62/ha), and concerns have been raised about effects on nontarget invertebrates (Cully et al. 2006) and development of insecticide resistance in fleas (Eads et al. 2018). Despite dusting burrows with insecticide, severe outbreaks of plague in prairie dogs have been reported at some locations (e.g., Tripp et al. 2017, Hoogland et al. 2017); thus, dusting is not always effective or alternative modes of transmission are more predominant in those events. More recently, another method of flea control for prairie dogs has been proposed and is currently being tested: the use of grains containing 0.005% fipronil, a broad-spectrum insecticide (Eads et al. 2021). Early studies applying fipronil bait to limited acreages are promising, achieving similar or even better flea control than dusting burrows. However, toxicity studies assessing the effects of long-term bioaccumulation in BFFs and other predatory species are lacking.

Another strategy recently tested for preventing plague in prairie dogs is immunization through bait-delivered oral vaccines. Like the oral rabies vaccine used to prevent rabies in wild fox, raccoon, and other species, a sylvatic plague vaccine (SPV) was developed and registered for experimental field use in prairie dogs utilizing raccoon poxvirus genetically engineered to express two plague antigens (Rocke et al. 2014). A three-year study to assess the effectiveness of bait-delivered SPV (Figure 13.14) was conducted on 29 paired placebo and treatment plots (1–59 ha in size; average 16.9 ha) in seven western states (Rocke et al. 2017). Relative abundance of prairie dogs was higher on SPV treated plots compared to placebo treated plots, and SPV treatment significantly improved prairie dog survival on paired plots where plague occurred, especially among juveniles. A companion study by Tripp and colleagues (2017) at prairie dog colonies in Colorado that also included a pesticide-treated plot demonstrated similar effects. Both SPV and pesticide treatment of burrows similarly enhanced prairie dog survival and abundance, preventing colony collapse if applied proactively prior to disease emergence. Further testing at landscape scales (400 ha or more), where entire prairie dog colonies could be treated, would help determine if SPV application could reduce plague's impact on prairie dogs and BFFs.

Fig. 13.14. Gunnisons's prairie dog consuming bait with sylvatic plague vaccine. US Geological Survey photograph by Tonie Rocke.

For a disease as complex as plague with numerous host and vector species and alternative methods of transmission (e.g., flea-borne, direct contact, and respiratory aerosols) and reservoirs (e.g., small rodents, carnivores, and carcasses), an integrated disease management strategy would be most effective. Enhancing host immunity through vaccination (of both BFFs and prairie dogs) in combination with flea control would be ideal, but such a program would be prohibitively expensive until less costly methods of vaccine production and less laborious methods of application of both vaccine and flea control products become available.

Financial, Legal, Political, and Social Factors

In addition to NBFFCC, BFFs continue to be maintained and bred at five zoos, and captive-bred animals are reintroduced to appropriate sites within their historic range annually. To date, more than 10,000 BFFs have been propagated in captivity, (Figure 13.15A) and approximately 5,500 have been reintroduced across 32 sites (Table 13.2) in eight US states, Mexico, and Canada (USFWS 2019; Pete Gober, USFWS, personal communication, 2021).

Fig. 13.15. (A) Black-footed ferret kits born in captivity at NBFFCC and (B) Elizabeth Ann, the first cloned black-footed ferret. Photographs courtesy of US Fish and Wildlife Service.

Table 13.2. Thirty-two black-footed ferret (*Mustela nigripes*) reintroduction sites (1991–2021) in Wyoming (WY), South Dakota (SD), Montana (MT), Arizona (AZ), Colorado (CO), Utah, (UT), Kansas (KS), New Mexico (NM), Mexico, and Canada

Location	Year	Location	Year
Shirley Basin, WY	1991	Northern Cheyenne Reservation, MT	2008
Badlands National Park, SD	1994	Vermejo Park Ranch, NM	2008
UL Bend NWR,[a] MT	1994	Grasslands National Park, Saskatchewan, Canada	2009
Conata Basin, SD	1996	Vermejo Park Ranch, NM	2012
Aubrey Valley, AZ	1996	Walker Ranch, CO	2013
Fort Belknap Reservation, MT	1997	Soapstone Complex, CO	2014
Coyote Basin, CO and UT	1999	North Holly, CO	2014
Cheyenne River Reservation, SD	2000	Liberty, CO	2014
Wolf Creek, CO	2001	Rocky Mountain Arsenal NWR[a], CO	2015
BLM-40 Complex, MT	2001	Crow Reservation, MT	2015
Janos, Mexico	2001	South Holly, CO	2016
Rosebud Reservation, SD	2004	Meeteetse, WY	2016
Lower Brule Reservation, SD	2006	Bad River Ranch, SD	2017
Wind Cave National Park, SD	2006	Moore Ranch, NM	2018
Espee Ranch, AZ	2007	Standing Rock Reservation, SD	2021
Butte Creek Ranches, KS	2007	May Ranch, CO	2021

[a] NWR, National Wildlife Refuge.

Also, in an effort to increase the genetic diversity and fitness of the species, a program was initiated to clone BFFs from cryopreserved cell lines of wild individuals caught in the 1980s, with the first cloned kit (Figure 13.15B) born in December 2020 (Robyn Bortner, USFWS, personal communication, 2021).

By these metrics, the captive breeding program has been highly successful, yet establishing sustainable wild populations has been challenging, and full recovery of the BFF has yet to be achieved. Of the 32 sites where BFFs have been reintroduced since 1991, 19 sites remain active, and the other 13 have failed (see Figure 13.2) due to sylvatic plague, drought, or a lack of funding. Currently, a minimum of 340 ferrets reside in the wild (perhaps 1,500 at the highest from 1991–2022), but monitoring is difficult and counts vary dramatically between years and sites, subject to precipitation patterns and disease events.

Notably, reproduction has been observed at most sites in most years (Pete Gober, USFWS, personal communication, 2021). Although no routine accounting of the costs incurred by dozens of BFF recovery partners over the past four decades has occurred since remnant wild individuals were rescued from Meeteetse, Wyoming, total captive breeding, reintroduction, and management in the wild, and research effort expenditures have been estimated at \$75–\$100 million (USD) from 1981–2021 (Pete Gober, USFWS, personal communication, 2022).

In the near term, continued captive propagation appears necessary to the survival of the species, but protection and management of adequate prairie dog habitat by public, private, and tribal land managers across their historical range will help ensure the BFF's recovery and long-term survival. Key strategies to achieve that goal include increased public education, landowner incentives to maintain and enhance grassland ecosystems, and disease management, particularly of sylvatic plague. The presence of an ESA listed species, like the BFF, provides justification for protection of public lands on which they live. Protecting larger areas to ensure the BFF's prey base is available contributes to the amount of land set aside for conservation.

Summary

Managing disease to specifically protect an endangered species may be a challenging and costly prospect, but it can have ecosystem-level benefits far beyond a single species. In grassland ecosystems, managing plague in prairie dogs not only protects BFFs, but also directly benefits other threatened and endangered species (e.g., Utah prairie dog) and numerous other species that depend or rely heavily on prairie dog habitat for their survival, such as mountain plovers (*Charadrius montanus*) and burrowing owls (*Athene cunicularia*) and predators that feed on prairie dogs, including the ferruginous hawk (*Buteo regalis*) and swift fox (Kotliar et al. 1999). Protecting prairie dogs and ferrets from plague may also help protect ranchers and their domestic animals. By controlling plague, conservation is served, and human and domestic animal health are protected, another example of the One Health approach in action. To that end, disease management and the BFFRIT recovery goals of 30 successful reintroduction sites and at least 3000 wild BFFs (USFWS, 2013) should be viewed as more broadly representative of ecosystem conservation of grasslands and their biodiversity.

ACKNOWLEDGEMENTS

The author is grateful to Rachel Abbott, Della Garelle, and John Hughes for critical review of the manuscript. Any use of trade, firm, or product names is for descriptive purposes only and does not imply endorsement by the US Government.

LITERATURE CITED

Abbott, R.C., and T. E. Rocke. 2012. Plague. US Geological Survey Circular 1372, Reston, Virginia, USA.

Adjemian, J. Z., P. Foley, K. L. Gage, K. L., and J. E. Foley. 2007. Initiation and spread of traveling waves of plague, *Yersinia pestis*, in the western United States. American Journal of Tropical Medicine and Hygiene 76:365–375.

Andrianaivoarimanana, V., P. Piola, and D. M. Wagner. 2019. Trends of human plague, Madagascar, 1998–2016. Emerging Infectious Diseases 25:220–228.

Antonation, K. S., T. K. Shury, T.K. Bollinger, A. Olson, P. Mabon, G. Van Domselaar, and C. R. Corbett. 2014. Sylvatic plague in a Canadian black-tailed prairie dog (*Cynomys ludovicianus*). Journal of Wildlife Diseases 50:699–702.

Appel., M. J. C., E. P. J. Gibbs, S. J. Martin, V. Ter Meulen, B. K. Rima, J. R. Stephenson, and W. P. Taylor. 1981. Morbillivirus diseases of animals and man. Pages 259–273 *in* E. Kurstak and C. Kurstak, editors. Comparative diagnosis of viral diseases. Volume IV. Vertebrate animal and related viruses. Part B. RNA viruses. Academic Press, New York, New York, USA.

Armstrong W. H. 1942, Canine distemper in the American badger. Cornell Veterinarian 32:447.

Barbieri, R., M. Signoli, D. Chevé, C. Costedoat, S. Tzortzis G. Aboudharam, D. Raoult, and M. Drancourt. 2020. *Yersinia pestis*: the natural history of plague. Clinical Microbiology Reviews 34:e00044-19.

Bevers, M., and J. Hof. 1997. Spatial optimization of prairie dog colonies for black-footed ferret recovery. Operations Research 45:495–507.

Bron, G. M. 2017. The role of short-lived rodent in plague ecology on prairie dog colonies, Ph.D. Thesis, University of Wisconsin, Madison, USA.

Bron, G. M., C. M. Malavé, J. T. Boulerice, J. E. Osorio, and T. E. Rocke. 2019. Plague-positive mouse fleas on mice before plague induced die-offs in black-tailed and white-tailed prairie dogs. Vector Borne Zoonotic Diseases. 19:486–493.

Budd, J. 1981. Distemper. Pages 31–44 *in* J. W. Davis, L. H. Karstad and D. O. Trainer (eds.). Infectious Diseases of Wild Mammals. Iowa State University Press, Ames, USA.

Carr, A. 1986. Introduction. Great Basin Naturalist Memoirs 9:1–7.

Carré, H. 1905. Sur la maladie des jeunes chiens. Comptes Rendus de l'Académie des Sciences 140:689–690,1489–1491.

Carpenter, J. W., M. J. Appel, R. C. Erickson, and M. N. Novilla. 1976. Fatal vaccine-induced canine distemper virus infection in black-footed ferrets. Journal of the American Veterinary Medical Association 169:961–964.

Colman, R. E., R. J. Brinkerhoff, J. D. Busch, C. Ray, A. Doyle, J. W. Sahl, P. Keim, S. K. Collinge, and D. M. Wagner. 2021. No evidence for enzootic plague within black-tailed prairie dog (*Cynomys ludovicianus*) populations. Integrative Zoology 16:834–851.

Cully, J. F., D. E. Biggins and D. B Seery. 2006. Conservation of prairie dogs in areas with plague. Pages 157–168 *in* J. L. Hoogland (editor). Conservation of the black-tailed prairie dog, Island Press, Washington, D.C., USA.

Deem, S. L., L. H. Spelman, R. A. Yates, and R. J. Montali. 2000. Canine distemper in terrestrial carnivores: a review. Journal of Zoo and Wildlife Medicine 31:441–451.

Eads, D. A., D. E. Biggins, J. Bowser J. C. McAllister, R. L. Griebel, E. Childers, T. M. Livieri, C. Painter, L. S. Krank, and K. Bly. 2018. Resistance to deltamethrin in prairie dog (*Cynomys ludovicianus*) fleas in the field and in the laboratory. Journal of Wildlife Diseases 54:745–754.

Eads D. A., T. M. Livieri, P. Dobesh, E. Childers, L. E. Noble, M. C. Vasquez and D. E. Biggins. 2021. Fipronil pellets reduce flea abundance on black-tailed prairie dogs: potential tool for plague management and black-footed ferret conservation. Journal of Wildlife Diseases 57:434–438.

Engelthaler, D. M., K. L. Gage, J. A. Montenieri, M. Chu, and L. G. Carter. 1999. PCR detection of *Yersinia pestis* in fleas: comparison with mouse inoculation. Journal of Clinical Microbiology 37:1980–1984.

Gage, K. L. 1998. Plague. Pages 885–906 *in* L. Collier, A, Balows, and M. Sussman, editors. Topley and Wilson's microbiology and microbial infections. Ninth edition. Oxford University Press, Inc. New York, New York, USA.

Gasper, P., and R. Watson. 2001. Plague and yersiniosis. Pages 312–329 *in* E. S. Williams and I. K. Barker, editors. Infectious diseases of wild mammals. Iowa State University Press, Ames, USA.

Godbey, J. L., D. E. Biggins, and D. Garelle. 2006. Exposure of captive black-footed ferrets to plague and implications for species recovery. Pages 233–237 *in* J. Roelle, B. Miller, J. Godbey, and D. Biggins, editors. Recovery of the black-footed ferret: progress and continuing challenges. US Geological Survey Scientific Investigations Report 2005-5293, Reston, Virginia, USA.

Gorham, J. R. 1966. The epizootiology of distemper. Journal of the American Veterinary Medical Association 149:610–622.

Gorham, J. R., R. Leader, and J. C. Gutierrez. 1954. Distemper immunization of mink by air-borne infection with egg-adapted virus. Journal of the American Veterinary Medical Association 125:134–136.

Halbrooks, R. D., L. J. Swango, P. R. Schnurrenberger, F. E. Mitchell, and E. P. Hill. 1981. Response of gray foxes to modified live-virus canine distemper vaccines. Journal American Veterinary Medical Association 179:1170–1174.

Helmboldt, C. F., and E. L. Jungherr. 1955. Distemper complex in wild carnivores simulating rabies. American Journal of Veterinary Research 16:463–469.

Hoogland, J. L, D. E. Biggins, N. Blackford, D. A. Eads, D. Long, L. M. Ross, S. Tobey, and E. M. White. 2017. Plague in a colony of Gunnison's prairie dogs (*Cynomys gunnisoni*) despite three years of infusions of burrows with 0.05% deltamethrin to kill fleas. Journal of Wildlife Diseases 54:347–351.

Keuler, K. M., G. M. Bron, R. Griebel, and K. L. D. Richgels. 2020. An invasive disease, sylvatic plague, increases fragmentation of black-tailed prairie dog (*Cynomys ludovicianus*) colonies. PLoS One 15:e0235907.

Kotliar, N. B., B. W. Baker, and A. D. Whicker. 1999. A critical review of assumptions about the prairie dog as a keystone species. Environmental Management 24:177–192.

Lockhart, J. M., T. E. Thorne, and D. R. Gober. 2004. A historical perspective on recovery of the black-footed ferret and the biological and political challenges affecting its future. Pages 6–19 *in* J. Roelle, B. Miller, J. Godbey, and D. Biggins, editors. Recovery of the black-footed ferret: progress and continuing challenges. US Geological Survey Scientific Investigations Report 2005-5293, Reston, Virginia, USA.

Marinari, P. E., and J. S. Kreeger. 2004. An adaptive management approach for black-footed ferrets in captivity. Pages 23–27 *in* J. Roelle, B. Miller, J. Godbey, and D. Biggins, editors. Recovery of the black-footed ferret: progress and continuing challenges. US Geological Survey Scientific Investigations Report 2005-5293, Reston, Virginia, USA.

Martinez-Gutierrez, M., and J. Ruiz-Saenz. 2016. Diversity of susceptible hosts in canine distemper virus infection: a systematic review and data synthesis. BMC Veterinary Research 12:78–88.

Matchett, M. R. D. E. Biggins, V. Carlson, B. Powell, and T. Rocke. 2010. Enzootic plague reduces black-footed ferret (*Mustela nigripes*) survival in Montana. Vector-Borne and Zoonotic Diseases 10:17–35.

McGee, B. K., M. J. Butler, D. B. Pence, J. L. Alexander, J. B. Nissen, W. B. Ballard, and K. L. Nicholson. 2006. Possible vector dissemination by swift foxes following a plague epizootic in black-tailed prairie dogs in northwestern Texas. Journal of Wildlife Diseases 42:415–420.

Mize, E. L, S. M. Grassel, and H. B. Britten. 2017. Fleas of black-footed ferrets (*Mustela nigripes*) and their potential role in the movement of plague. Journal of Wildlife Diseases 53:521–531.

Perry, R., and J. Fetherston. 1997. *Yersinia pestis*—etiologic agent of plague. Clinical Microbiology Reviews 10:35–66.

Rocke, T. E., B. Kingstad-Bakke, W. Berlier, and J. E. Osorio. 2014. A recombinant raccoon poxvirus vaccine expressing both *Yersinia pestis* F1 and truncated V antigens protects animals against lethal plague. Vaccines 2:772–784.

Rocke, T. E., J. Mencher, S. R. Smith, A. M. Friedlander, G. P. Andrews, and L. A. Baeten. 2004. Recombinant F1-V fusion protein protects black-footed ferrets (*Mustela nigripes*) against virulent *Yersinia pestis* infection. Journal of Zoo and Wildlife Medicine 35:142–146.

Rocke, T. E., P. Nol, P. Marinari, J. Kreeger, S. Smith, G. P. Andrews, and A. M. Friedlander. 2006. Vaccination as a potential means to prevent plague in black-footed ferrets. Pages 243–247 *in* J. Roelle, B. Miller, J. Godbey, and D. Biggins, editors. Recovery of the black-footed ferret: progress and continuing challenges. US Geological Survey Scientific Investigations Report 2005-5293, Reston, Virginia, USA.

Rocke, T. E., S. Smith, P. Marinari, J. Kreeger, J. T. Enama, and B.S. Powell. 2008. Vaccination with F1-V fusion protein protects black-footed ferrets (*Mustela nigripes*) against plague upon oral challenge with *Yersinia pestis*. Journal of Wildlife Diseases 44:1–7.

Rocke, T. E., D. W. Tripp, R. E. Russell, R. C. Abbott, K. Richgels, M. R. Matchett, D. E. Biggins, R. Griebel, G. Schroeder, S. M. Grassel, et al. 2017. Sylvatic plague vaccine partially protects prairie dogs (*Cynomys* spp.) in field trials. EcoHealth 14:438–450.

Rocke, T. E., J. Williamson, K. R. Cobble, J. D. Busch, M. F. Antolin, and D. M. Wagner. 2012. Resistance to plague among black-tailed prairie dog populations. Vector-Borne and Zoonotic Diseases 12:111–116.

Russell, R. E., R. A. Abbott, D. W. Tripp, and T.E. Rocke. 2018. Local factors predict on-host flea distributions and plague occurrence on prairie dog colonies. Ecology and Evolution 8:8951–8972.

Russell, R. E., D. W. Tripp, and T. E. Rocke. 2019. Differential plague susceptibility in species and populations of prairie dogs. Ecology and Evolution 9:11962–11971.

Russell, R. E., D. P. Walsh, M. D. Samuel, M. D. Grunnill, and T.E. Rocke. 2021. Space matters; host spatial structure and the dynamics of plague transmission. Ecological Modelling 443:109450.

Salkeld, D. J., R. J. Eisen, P. Stapp, A. P. Wilder, J. Lowell, D. W. Tripp, D. Albertson, and M. F. Antolin. 2007. The potential role of swift foxes (*Vulpes velox*) and their fleas in plague outbreaks in prairie dogs. Journal of Wildlife Diseases 43:425–431.

Salkeld, D.J., M. Salathé, P. Stapp, and J. H. Jones. 2010. Plague outbreaks in prairie dog populations explained by percolation thresholds of alternate host abundance. Proceedings of the National Academy of Sciences USA 107:14247–14250.

Salkeld, D. J., and P. Stapp. 2008. No evidence of deer mouse involvement in plague (*Yersinia pestis*) epizootics in prairie dogs. Vector Borne Zoonotic Diseases. 8:331–337.

Santymire, R., H. Branvold-Faber, and P. E. Marinari. 2014. Recovery of the black-footed ferret. Pages 219–231 *in* J.G. Fox and R. P. Marini, editors. Biology and diseases of the ferret. Third edition. John Wiley and Sons, Hoboken, New Jersey, USA.

Schuler, K., M. Claymore, H. Schnitzler, E. Dubovi, T. Rocke, M. J. Perry, D. Bowman, and R. C. Abbott. 2021. Sentinel coyote pathogen survey to assess declining black-footed ferret (*Mustela nigripes*) population in South Dakota, USA. Journal of Wildlife Diseases 57:264–272.

Seery, D. B., D. E. Biggins, J. A. Montenieri, R. E. Enscore, D. T. Tanda, and K. L. Gage. 2003. Treatment of black-tailed prairie dog burrows with deltamethrin to control fleas (Insecta: Siphonaptera) and plague. Journal of Medical Entomology 40:718–722.

Stapp, P., M. F. Antolin, and M. Ball. 2004. Patterns of extinction in prairie-dog metapopulations: plague outbreaks follow El Niño events. Frontiers in Ecology and the Environment 2:235–240.

Stapp, P., D. J. Salkeld, R. J. Eisen, R. Pappert, J. Young, L. G. Carter, K. L. Gage, D. W. Tripp, and M. F. Antolin. 2008. Exposure of small rodents to plague during epizootics in black-tailed prairie dogs. Journal of Wildlife Diseases 44:724–730.

Stapp, P., D. J. Salkeld, H. A. Franklin, J. P. Kraft, D. W. Tripp, M. F. Antolin, and K. L. Gage. 2009. Evidence for the involvement of an alternate rodent host in the dynamics of introduced plague in prairie dogs. Journal of Animal Ecology 78:807–817.

Thorne, E. T., and E. S. Williams. 1988. Disease and endangered species: the black-footed ferret as a recent example. Conservation Biology 2:66–73.

Tripp, D. W., T. E. Rocke, J. P. Runge, R. C. Abbott, and M. W. Miller. 2017. Burrow dusting or oral vaccination prevents plague-associated prairie dog colony collapse. EcoHealth 14:451–462.

Uhl, E. W., C. Kelderhouse, J. Buikstra, J. P. Blick, B,\. Bolon, and R. J. Hogan. 2019. New World origin of canine distemper: interdisciplinary insights. International Journal of Paleopathology 24:266–278.

US Fish and Wildlife Service (USFWS). 2013. Recovery plan for the black-footed ferret (*Mustela nigripes*). Second revision. US Fish and Wildlife Service, Denver, Colorado, USA.

US Fish and Wildlife Service (USFWS). 2019. Species status assessment report for the black-footed ferret (*Mustela nigripes*). US Fish and Wildlife Service, Denver, Colorado, USA.

Williams, E. S. 2001. Canine distemper. Pages 50–59 *in* E. S. Williams and I. K. Barker, editors. Infectious diseases of wild mammals. Iowa State University Press, Ames, USA.

Williams, E. S., S. L. Anderson, J. Cavender, C. Lynn, K. List, C. Hearn, C., and M. J. Appel. 1996. Vaccination of black-footed ferret (*Mustela nigripes*) × Siberian polecat (*M. eversmanni*) hybrids and domestic ferrets (*M. putorius furo*) against canine distemper. Journal of Wildlife Diseases 32:417–423.

Williams, E. S., K. Mills, D. R. Kwiatkowski, E. T. Thorne, and A. Boerger-Fields. 1994. Plague in a black-footed ferret (*Mustela nigripes*). Journal of Wildlife Diseases 30:581–585.

Williams, E. S., E. T. Thorne, M. J. Appel, and D. W. Belitsky. 1988. Canine distemper in black-footed ferrets (*Mustela nigripes*) from Wyoming. Journal of Wildlife Diseases 24:385–398.

Williams, E. S., E. T. Thorne, T. J. Quan, and S. L. Anderson. 1991. Experimental infection of domestic ferrets (*Mustela putorius furo*) and Siberian polecats (*Mustela eversmanni*) with *Yersinia pestis*. Journal of Wildlife Diseases 27: 441–445.

Wimsatt, J., D. E. Biggins, K. Innes, B. Taylor, and D. Garrell. 2003. Evaluation of oral and subcutaneous delivery of an experimental canarypox recombinant canine distemper vaccine in the Siberian polecat (*Mustela eversmanni*). Journal of Zoo and Wildlife Medicine 34:25–35.

Wright, M.L., T.M. Livieri, and R.M. Santymire. 2022. Recombitek canine distemper vaccine as an alternative for Purevax distemper vaccine in endangered black-footed ferrets (*Mustela nigripes*). Journal of Zoo and Wildlife Medicine 53:194–199.

14 Rabies and Wildlife Conservation

THOMAS MÜLLER, CHARLES E. RUPPRECHT

Introduction

Rabies is an acute, progressive encephalitis (Jackson 2018). Warm-blooded vertebrates are susceptible (Gilbert 2018), although cases among birds are rare. Reservoirs, those taxa responsible for sylvatic cycles (i.e., related to disease perpetuation in wildlife), include bats and mesocarnivores (Table 14.1). Globally, most cases in humans are from dogs (*Canis familiaris*; Layan et al. 2021).

The causative agents are lyssaviruses, which are neurotropic RNA viruses (Kuhn et al. 2020). Within reservoir species populations, lyssaviruses undergo purifying selection, by the removal of deleterious mutations that arise during replication and demonstrate adaptation to certain specific mammalian hosts, although broad cross species transmission occurs (Troupin et al. 2016). Viral transmission is usually via bites (Figure 14.1). Less commonly, infection follows aerosol or mucosal contamination (Johnson et al. 2006, Zhao et al. 2019*a*). Virions pass to peripheral nerves, replicate within the central nervous system (CNS), progress to organs (e.g., salivary glands), and are excreted in saliva (Fooks et al. 2017).

Susceptibility varies by species, age, immune status, infectious dose, route, and exposure severity (Rohde and Rupprecht 2020). Incubation periods are 4–6

Examples of global rabies occurrence. Illustration by Laura Donohue.

Table 14.1. Distribution of selected commonly reported rabid wildlife by continent

Location	Example	Reference
Africa	Bats (e.g., *Eidolon*, *Miniopterus*, and *Rousettus* spp.); African wild dog (*Lycaon pictus*); bat-eared fox (*Otocyon megalotis*); civet (*Civettictis civetta*); Ethiopian wolf (*Canis simensis*); jackals (*Lupulella* spp.); kudu (*Tragelaphus* spp.); meerkat (*Suricata suricatta*); mongoose (*Cynictis penicillata*)	Swanepoel et al. 1993, Zulu et al. 2009, van Zyl et al. 2010, Marino et al. 2017, Stuchin et al. 2018, Canning et al. 2019, Hikufe et al. 2019, Markotter et al. 2020, Sabeta et al. 2020, Koeppel et al. 2021, McMahon et al. 2021
Asia	Bats (e.g., *Myotis*, *Murina*, and *Pteropus* spp.); ferret badgers (*Melogale* spp.); foxes (e.g., *Alopex* and *Vulpes* spp.); jackal (*Canis mesomelas*); mongoose (*Urva* spp.); raccoon dog (*Nyctereutes procyonoides*)	WHO 1990, Kuzmin et al. 2008*a*, Seimenis 2008, Shao et al. 2011, Zhao et al. 2019*b*, Seidlova et al. 2020, Yakovchits et al. 2021
Australia	Bats (e.g., *Pteropus* and *Saccolaimus* spp.)	Field 2018, Merritt et al. 2018, Prada et al. 2019, Barrett et al. 2020, May et al. 2020
Europe	Bats (e.g., *Eptesicus* and *Myotis* spp.); foxes (e.g., *Alopex* and *Vulpes* spp.); raccoon dog (*Nyctereutes procyonoides*)	Freuling et al. 2012, Schatz et al. 2013, Marston et al. 2017, Robardet et al. 2019
North America	Bats (e.g., *Eptesicus*, *Lasiurus*, and *Tadarida* spp.); coyote (*Canis latrans*); foxes (e.g., *Alopex*, *Urocyon*, and *Vulpes* spp.); mongoose (*Herpestes auropunctatus*); raccoon (*Procyon lotor*); skunks (e.g., *Mephitis* and *Spilogale* spp.)	Seetahal et al. 2018, Nadin-Davis and Fehlner-Gardiner 2019, Jaramillo-Reyna et al. 2020, Ma et al. 2022
South America	Bats (e.g., *Artibeus*, *Desmodus*, and *Tadarida*, spp.); foxes (*Cerdocyon* and *Lycalopex*, spp.); nonhuman primates (e.g., *Callithrix* spp.), procyonids (e.g., *Nasua* and *Potos* spp.)	Kirandjiski et al. 2012, Marcelo Azevedo de Paula Antunes, J. et al. 2017, Botto Nuñez et al. 2019, Kotait et al. 2019, Zucca et al. 2020, Meske et al. 2021

weeks but may be longer. Signs manifest as behavioral alterations. Coma and death occur days after onset. Rabies has the highest case fatality of any infectious disease. There is no proven treatment. Historically, control involved population reduction. Today, culling is supplanted by vaccination. Human rabies is preventable by minimizing exposures, preexposure vaccination (PrEP), and postexposure prophylaxis (PEP).

Rabies is found on all continents, except Antarctica (Rupprecht et al. 2018*b*). Wildlife are reservoirs, vectors, and victims in different regions (Table 14.1). Some localities (e.g., Hawaii) have no history of rabies due to "splendid isolation." Others achieved "rabies-free" status after elimination (e.g., Japan) and institution of measures (e.g., quarantine) to prevent reestablishment.

The burden of rabies is incalculable. Laboratory-based surveillance is less than ideal. Likely, hundreds of millions of rabid animals succumb (e.g., assuming a prevalence of ~ 0.1 % in a single maternity colony in Texas inhabited by ~10–15 million free-tailed bats, *Tadarida brasiliensis*). Estimates suggest >12 million human exposures and tens of thousands of deaths annually, predominantly in Africa and Asia (Hampson et al. 2015). Considering impacts and management, some infectious diseases are rather restricted in scope by geography or host range and are more amenable to prevention, control, local elimination, or—albeit rarely—directed, operational extinction. Unlike smallpox and rinderpest, rabies is not a candidate for eradication because of its broad global distribution, extensive adaptations among wild mammalian reservoirs, and perpetuation of genetically diverse viral species (Rupprecht et al. 2008, 2020*a*). Use of PrEP for mass animal immunization and PEP in bitten persons are the major tools that are hoped will eliminate human rabies from rabid dogs by 2030 (WHO 2018). If successful, enhanced wildlife rabies surveillance is anticipated (Baker et al. 2019).

History

Rabies is an ancient infectious disease (Tarantola 2017). The notion of a "virus" (i.e., a "slimy liquid") was

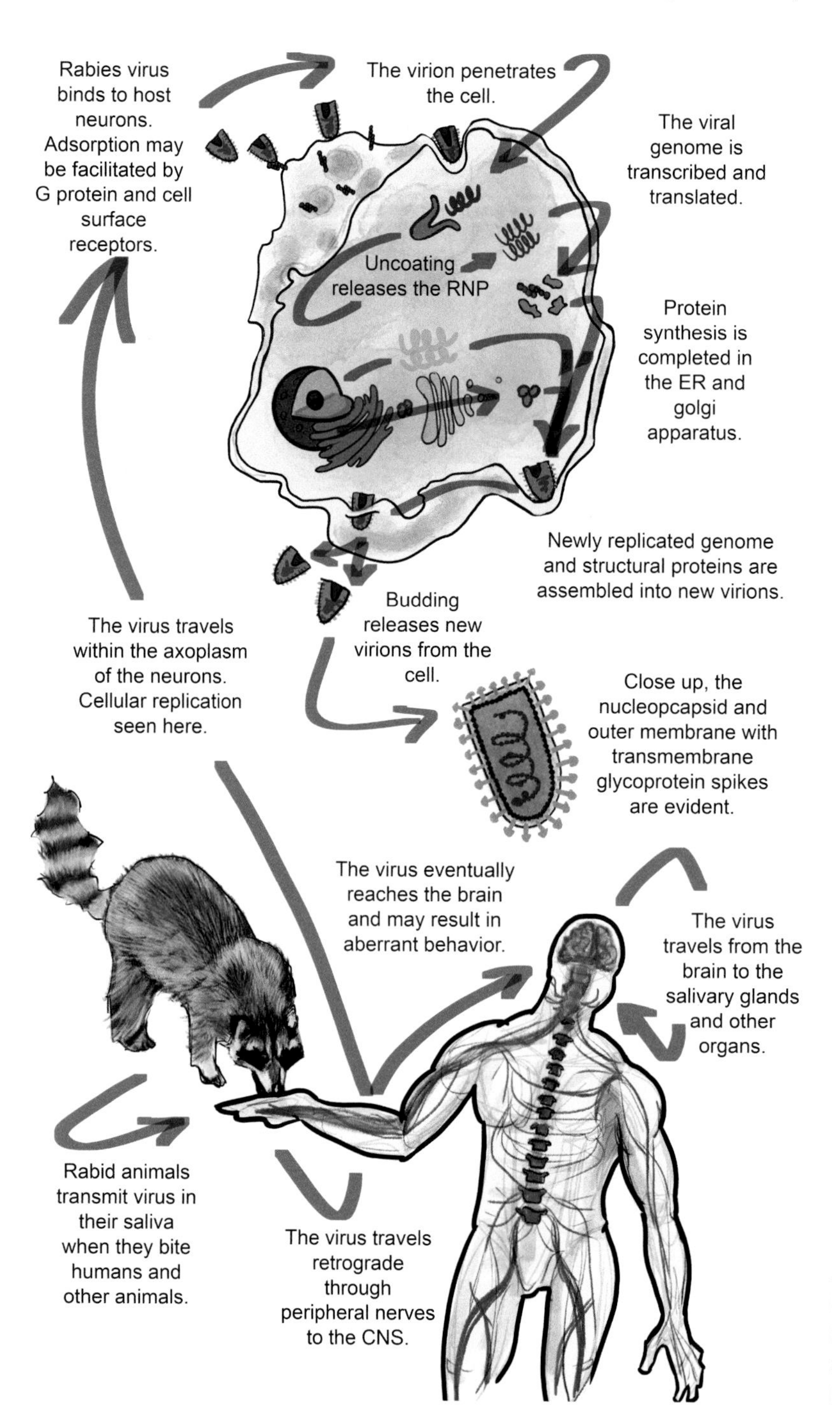

Fig. 14.1. Generalized model of lyssavirus pathobiology in an individual, susceptible (i.e., nonimmune) host. Illustration by Laura Donohue.

an apt term for the viscous saliva of rabid animals. Physicians in several ancient cultures observed a fatal syndrome associated with animal bites (Rupprecht et al. 2020*b*). Phylogenetic studies project that ancestral lyssaviruses arose thousands of years ago, whereas other investigators suggest an older genealogy (Hayman et al. 2016). Until recently, the role of wildlife was poorly appreciated. Nevertheless, rabies is the first described zoonosis in bats and likely predates colonization of the Americas. During the 20th century, as canine rabies was controlled, investigators examined wildlife more thoroughly. Globally, researchers progressed in laboratory-based surveillance, epizootiology, pathobiology, and—perhaps most importantly—the concept and application of wildlife vaccination (Table 14.2).

Etiology

The infectious nature of rabies (i.e., associated with the saliva of rabid dogs) was realized since the 19th century (Tarantola 2017). Knowledge on rabies

Table 14.2. Selective examples of prominent 20th-century researchers on wildlife rabies: in memoriam

Researcher	Contribution	Reference
G. M. Baer	The "father" of oral wildlife vaccination	Baer et al. 1971
W. J. Bigler	Raccoon rabies epizootiology	Bigler et al. 1973
J. Blancou	Insights into dual laboratory and field techniques	Blancou and Wandeler 1989
K. Bögel	Support for wildlife rabies control in Europe	Bögel et al. 1981
K. M. Charlton	Experimental pathobiology of infection and immunity	Charlton et al. 1997
K. A. Clark	Coyote rabies prevention and control along the Mexico–United States border	Clark et al. 1994
D. G. Constantine	Wellspring to bat rabies investigations in the Americas	Constantine 1962
J. G. Debbie	Focus upon diagnostics and control in the United States	Debbie 1974
B. Dietzschold	Pathobiology and vaccinology	Dietzschold et al. 2008
C. O. Everard	Study of mongoose rabies in the Caribbean region	Everard et al. 1981
M. Fekadu	Diagnostics, pathobiology, and immunization	Fekadu et al. 1991
E. H. Follmann	Arctic fox rabies epizootiology and control	Follmann et al. 1996
A. M. Greenhall	Vampire bat control techniques in Latin America	Greenhall 1971
S. D. S. Greval	Mongoose rabies in Asia	Greval 1932
A. N. Hamir	Pathobiological studies of experimental and natural infections, especially in raccoons	Hamir et al. 1998
O. J. Hübschle	Kudu rabies in southern Africa	Hübschle 1988
R. Kantorovich	Arctic and red fox rabies	Kantorovich et al. 1963
A. A. King	Characterization of wildlife lyssaviruses, particularly in Africa	King et al. 1994
I. Kotait	Diagnostics, vaccinology, and identification of nonhuman primates as a reservoir in South America	Kotait et al. 2019
K. Kulonen	Diagnostics, control, and epizootiology of wildlife rabies in Europe	Kulonen and Boldina 1996
K. Lawson	Dedication to the use of viral vaccines for oral immunization in Canada	Lawson et al. 1997
S. B. Linhart	Biology of rabies control among mammalian carnivores	Linhart et al. 1997
R. D. Lord	Ecology and control of rabies in vampire bat populations, particularly in Venezuela	Lord 1992
C. D. MacInnes	Prevention and control of wildlife in Canada	MacInnes et al. 2001
G. Malkov	Epizootiology of wildlife rabies in the former USSR	Malkov and Gribanova 1980
C. D. Meredith	Virological insights into wildlife rabies of southern Africa	Meredith and Standing 1981
R. L. Parker	Wildlife rabies epizootiology	Parker 1969
P. P. Pastoret	Fox rabies immunization in Belgium with recombinant vaccine	Pastoret et al. 1988
J. L. Pawan	Link of cross-species transmission of rabies virus from vampire bats during major Trinidadian outbreak	Pawan 1936
L. G. Schneider	Initiation of oral vaccination program of foxes in Germany and initiation of machine-made baits	Schneider 1989
F. Steck	First oral vaccine studies in Switzerland	Steck et al. 1982*a*
J. H. Steele	Global history of wildlife rabies	Steele 1988
G. Wachendörfer	Red fox rabies epizootiology and control	Wachendörfer et al. 1978
W. G. Winkler	Epizootiology of wildlife rabies within North America	Winkler 1988

virus (RABV), its morphology and biochemistry emerged during the 1960s (Rupprecht et al. 2020*b*). By enhanced surveillance and virus "hunting," RABV and "rabies-related viruses" were detected throughout the world. While not differing morphologically, lyssaviruses vary antigenically and genetically (Dietzschold et al. 1988, Badrane et al. 2001, Troupin et al. 2016, Coertse et al. 2020). The original assumption of a single viral species is obsolete—all lyssaviruses cause rabies.

Lyssaviruses are nonsegmented, negative-sense, single-stranded, bullet-shaped RNA viruses, classified in the genus Lyssavirus, family Rhabdoviridae, order Mononegavirales. The genus comprises at least 17 viral species, according to sequences (Walker et al. 2020). In contrast to RABV, commercially available biologics do not provide adequate efficacy against genetically divergent lyssaviruses (Fooks et al. 2021).

Demographics and Field Biology

Although rabies is a mammalian disease, not all species are reservoirs, and none are "carriers" (Hayman et al. 2016). For millennia, virus–host coevolution resulted in adaptations to specific hosts (Table 14.1). Stochastic events during infectious interactions determine host–viral population dynamics. Genetically distinct variants exist as ensembles, circulating in complex ecological communities (Marston et al. 2017). As a result, "adapted" variants perpetuate within populations but are not sustained outside of reservoirs (i.e., a rabid skunk infects a raccoon, but the raccoon is unlikely to infect other raccoons nor maintain infectious chains with this variant; Figure 14.2). Individually, a case can occur as a single, dead-end event, or variants may be maintained for centuries (Kuzmina et al. 2013). Outcomes (e.g., death or survival) are

Fig. 14.2. Simplified sylvatic scheme of epizootiological cycles of lyssaviruses, illustrating the concepts of intraspecific perpetuation within reservoir species (e.g., fox to fox), suggested cross-species transmission (CST) pathways (e.g., bat to cat), and incidental dead-end, spillover infections (e.g., any case of rabies in humans). Note that, in so-called urban rabies, rabid dogs maintain intraspecific cycles, with opportunities for CST to other domestic animals, wildlife, and humans. Illustration by Laura Donohue.

predicated by viral type, dose, route, host, life-history stage, and immune status.

Epidemiologically, host populations exceeding a minimum size achieve disease sustainability (i.e., critical community sizes, CCS). These maintain pathogens independently, while populations smaller than CCS cannot. Together with certain genetic constraints, this may explain why only certain taxa are reservoirs (WHO 2018). Host vagility and community engagement also offer ecological opportunities. For example, rabid domestic dogs are keystone reservoirs throughout Africa and Asia. Other canids, such as jackals, are infected by rabid dogs (Troupin et al. 2016). Occasionally, outbreaks are maintained jackal-to-jackal for short periods before disappearance. In addition, surveillance may be inadequate to define a longer term epizootiology (Hikufe et al. 2019).

Predator–prey interactions provide unpredictable cross-species transmission episodes. For example, on Prince Edward Island, bat RABV was detected in a local fox population but without long-term maintenance (Daoust et al. 1996). Alternatively, for unknown reasons, recurring transmissions from bats to mesocarnivores have been detected in Arizona from 2000 to date, the so-called 'Flagstaff phenomenon' (Leslie et al. 2006, Kuzmin et al. 2012). Why certain events abate, while others perpetuate, is not understood (Badrane and Tordo 2001, Wallace et al. 2014, Marston et al. 2018, Mollentze et al. 2020).

Spatial distribution and viral dynamics are delineated by the distinctive ecology of their reservoirs. The native, but also introduced, range of species may predetermine viral geographic occurrence. For example, circumpolar circulation of Arctic RABV lineages is identical to Arctic fox (*Vulpes lagopus*) distribution (Mansfield et al. 2006, Kuzmin et al. 2008*b*). In contrast, raccoons (*Procyon lotor*) only maintain RABV in parts of their native North American distribution but not in introduced ranges of Europe and the Caucases (Jenkins et al. 1988, Müller et al. 2015*a*). Similarly, red (*Vulpes vulpes*) and corsac foxes (*Vulpes corsac*) are RABV reservoirs in Eurasia (Kuzmin et al. 2001) but not where red foxes are aliens, such as Australia and New Zealand (WHO 2018, Müller and Freuling 2020*a*). However, raccoon dogs (*Nyctereutes procyonoides*) in native (Far East) and nonnative (central and eastern Europe) ranges are RABV reserviors (Botvinkin et al. 1981, Kim et al. 2006, Müller et al. 2015*a*). While only one mesocarnivore may serve as a reservoir in some locales (e.g., mongooses in the Caribbean (Everard and Everard 1992) and ferret badgers in parts of China (Zhao et al. 2019*b*)), elsewhere rabies is a multispecies problem (e.g., North America), in coati, coyotes, foxes, raccoons, skunks, and bats (Slate et al. 2009, Willoughby et al. 2017, Ma et al. 2022). Complex ecoepidemic settings facilitate occasional spillover or even sustained transmission (Wallace et al. 2014). Unpredictably, spillover rates of bat RABV variants are higher compared to other bat lyssaviruses (Messenger et al. 2002, Marston et al. 2018).

Topographical features representing permanent or temporal barriers for disease spread may determine areas with viral circulation (Wandeler et al. 1988). Natural barriers are seas, mountain ranges, large rivers, lakes, and valleys, while anthropogenic features include highways, dams, and less favorable habitats (e.g., pine plantations). For example, the Appalachian Mountains present a barrier to westward spread of raccoon rabies. Alternatively, environmental factors and climate change may enhance disease spread. Urbanization and suburban sprawl create opportunities for peridomestic species adaptations. Red foxes, raccoons, skunks, and Arctic foxes are opportunistic, proliferating alongside human expansions (Deplazes et al. 2004, Elmhagen et al. 2014). Similar observations extend to house bats (O'Shea et al. 2014). Such urban areas pose challenges to interventions (Slate et al. 2009, Vos et al. 2012). Global climate models imply alterations upon mesocarnivore distributions (Mulatti et al. 2012) and population growth of vampire bats (*Desmodus rotundus*) in the Americas (Lee et al. 2012) with likely expansion of sylvatic rabies cycles to other bats and mesocarnivores (Meske et al. 2021).

Population densities, contact frequencies, and spatial structures influence transmission dynamics and disease prevalence. Lyssavirus transmission is assumed to be density dependent, with incidence increasing directly with host abundance (Morters et al. 2013). Perceptions that rabies perpetuates in high-density but, less frequently, in lower density populations does not always apply. For example, rabies is maintained in low-density populations of foxes and jackals on the Arab peninsula, Namibia, and Turkey (Vos et al. 2009, Memish et al. 2015, Hikufe et al. 2019).

When rabies enters new areas, an infection "front" spreads as a band, typically 40–100 km (27–67 miles) wide (Holmala and Kauhala 2006). These epizootics show distinct seasonal and cyclical patterns in spread and intensity. Periodic flare-ups occur as reservoir populations rebound (i.e., after an epizootic, 80% of resident reservoirs may succumb). High mortality causes populations to fall below the CCS, and disease spread is hampered. During this silent phase, rabies perpetuates at a low level. When population densities increase, viral transmissions result in characteristic waves (Blancou 1988, Holmala and Kauhala 2006). Disease cycles occur in 3- to 8-year periods, with significant synchrony across regions (Hampson et al. 2007). Differences are marginal by reservoir: 3–6 years (e.g., dogs), 4–5 or 6–8 years (e.g., skunks), 5 years (e.g., raccoons), and 3–5 years (e.g., red and Arctic foxes) (Steck and Wandeler 1980, Gremillion-Smith and Woolf 1988, Pybus 1988, Hampson et al. 2007, Recuenco et al. 2007, Kim et al. 2014).

Interspecific contact networks, mating, birth pulses, and dispersals predict seasonal peaks in prevalence due to decreased or increased viral susceptibility. Seasonal patterns, with peaks in spring and autumn, have been described for raccoons, skunks, and foxes (Aubert 1992, Oertli et al. 2009, Hirsch et al. 2016). Reports are the "tip of the iceberg" as less than 10% of cases are recorded (Holmala and Kauhala 2006).

Host–viral ecology determine bat rabies seasonality and maintenance (George et al. 2011). Bats are among the few true hibernators and select hibernacula carefully. Swarming sites offer mating opportunities and early hibernacula selection, where activity increases in late summer to early autumn, accelerating viral transmission prehibernation (Davis et al. 2016). Significant physiological changes occur at this time, including decreases in body temperature, metabolism, and immunity (Bouma et al. 2010). As metabolism slows, a concurrent reduction in viral replication occurs (Davis et al. 2016). Whether such seasonality occurs in nonhibernating, tropical bats remains elusive.

Signs and Gross Lesions

Rabies is a neurological disease, with a propensity to present in uncharacteristic ways. Much is known about manifestations in wildlife (Jenkins et al. 1988, Everard and Everard 1992), but disease cannot be confirmed by signs alone. Other infections, toxicosis, trauma, neoplasia, metabolic conditions, and behavioral habituations can all mimic rabies.

During incubation, the animal is not infectious, and rabies may not be suspected unless lesions that are compatible with a bite wound are observed (Canning et al. 2019). A rabid animal may seem normal and die acutely without premonitory signs, or signs may progress over days to weeks (Gough and Niemeyer 1975). The prodromal phase is nonspecific, with fever, inappetence, restlessness, altered activity patterns, and general depression (Cliquet et al. 2009). Rabies is best known for the acute neurological phase, a state ranging between mania or "furious rabies" and paralytic or "dumb rabies" (Wang et al. 2011). Animals become aggressive without provocation or in response to auditory or visual stimuli, lose their avoidance of humans (Walroth et al. 1996), and attack conspecifics or other species (Rosatte et al. 2007), often with dramatic consequences (Figure 14.3). Some mesocarnivores chase vehicles and bite inanimate objects. Pica and self-inflicted injuries occur. Activity patterns shift (Andral et al. 1982). Bats lose the ability to fly and become grounded, paralyzed, or aggressive (Kuzmin

Fig. 14.3. A rabid raccoon (*Procyon lotor*) presenting with embedded porcupine (*Erethizon dorsatum*) quills. Courtesy US Department of Agriculture, Wildlife Services.

et al. 2008*a*, Davis et al. 2012). Cranial nerve deficits may produce anisocoria, a prolapsed nictitans (i.e., "third eyelid"), head tilts, hypersalivation, and partial facial paralysis or lolling tongues. Vocalization changes may be evident. Alternatively, individuals may appear abnormally shy, and unusually inquisitive with domestic animals, in periurban and rural settings (Oertli et al. 2009). Increasing disorientation, incoordination, hyperexcitability, tremors, paresis, paralysis, and convulsions appear, followed rapidly by coma and death from respiratory, cardiac, and renal failure associated with generalized encephalitis.

Upon necropsy, all findings may fall within normal limits. In contrast, carcasses may appear emaciated, with evidence of self-mutilation or broken teeth. Alopecia may be scant or profound, fur may be matted with fecal and urine staining, and pads or hooves may be worn (Franco-Molina et al. 2021). Mucosal membranes can appear slightly congested. Stomach contents may be empty or contain soil, sticks, and stones. Ingesta may occur in the airways from aspiration, and lung lobes may seem consolidated (Stoltenow et al. 2000). The brain may appear normal or slightly congested. No finding is pathognomonic (i.e., specifically indicative for rabies). Constellations indicative of a neurological condition heighten suspicion.

Pathogenesis and Pathology

Lyssaviruses transmit through transdermal or mucosal exposures (Jackson 2018, Zhao et al. 2019*a*; Figure 14.1). Bites, scratches, and licks can produce infection. Other routes include airborne transmission in large bat nursery colonies (Constantine 1962), fomites in kudu groups browsing upon thorny *Acacia* (Hübschle 1988), and organ transplantation (Ross et al. 2015).

After receptor-mediated entry and local replication, lyssaviruses travel by retrograde transport via axoplasmic microtubules (Lafon 2005). Transsynaptic spread ensues to the CNS. Affected brain areas depend upon peripheral nerves infected at entry (Fooks et al. 2017, Jackson 2018). In the CNS, replication occurs in neurons and astrocytes (Potratz et al. 2020*a*). Cellular structural integrity is largely preserved. Illness is likely concomitant with neuronal dysfunction, but mechanisms remain unclear (Fooks et al. 2017). Despite viral virulence, histological lesions are minimal. Punctate inflammation, mononuclear infiltrates, and intracytoplasmic inclusion (Negri) bodies (Figure 14.4) are distinctive changes. Afterwards, virions undergo centrifugal anterograde spread to peripheral organs (Jackson et al.

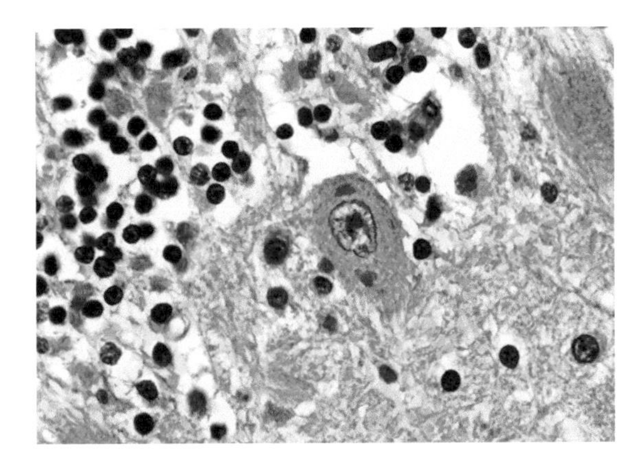

Fig. 14.4. Photomicrograph of encephalitic changes in the brain, including mononuclear infiltrates and intracytoplasmic inclusions, known as Negri bodies, within a rabies virus-infected neuron, as visualized by hematoxylin and eosin staining. Courtesy US Public Health Image Library.

1999). In salivary glands, mucogenic acinar cell infection results in high but variable salivary shedding (Dierks et al. 1969, Fekadu et al. 1982). Ocular tissues (e.g., cornea and retina) and organs innervated by sympathetic, parasympathetic, and sensory nerves (e.g., heart, kidneys, liver, and tactile hairs) are also infected (Potratz et al. 2020*b*). Detailed descriptions on lyssavirus pathophysiology and immune evasion and manipulation are available (Brzozka et al. 2006, Lafon 2011, Begeman et al. 2017).

Diagnosis

Due to unique lyssavirus pathogenesis, there are no tools to diagnose animal rabies prior to onset. Testing of healthy animals is unmeaningful (Thulke et al. 2009), but antemortem testing can be confirmatory in humans displaying compatible symptoms (e.g., hydrophobia or priapism). For an unequivocal diagnosis, postmortem laboratory testing of animals is paramount (WHO 2018, OIE 2021).

Brain tissue is preferred for postmortem diagnosis. Several tests identify specific viral components. The direct fluorescent antibody test (DFA) is the gold standard (Figure 14.5). The direct rapid immunohistochemical test (dRIT) and classical immunohistochemistry (IHC) provide sensitive and specific means to detect viral antigens in the CNS (i.e., with impressions or in formalin-fixed tissues, respectively) using light microscopy (Lembo et al. 2006). In national reference laboratories, sensitive reverse-transcriptase polymerase chain reactions (RT-PCR), both conventional and real-time tests, detect viral nucleic acids (WHO 2018). Electron microscopy visualizes intact virions or replicating virus. For isolation or diagnostic confirmation, animal inoculation has been replaced by cell culture (Robardet et al. 2011). Although point-of-care tests are available, diligence is necessary under field conditions. Rapid immunochromatographic tests are not yet recommended due to a lack of sensitivity, lot-to-lot variation, and inadequate validation (Eggerbauer et al. 2016, WHO 2018, Klein et al. 2020).

Fig. 14.5. Photomicrograph of specific viral fluorescence (apple-green inclusions), representing rabies virus antigens in brain impressions (medulla oblongata and cerebellum) of a naturally infected red fox (*Vulpes vulpes*), indicative of a positive diagnostic result, as visualized by the direct fluorescent antibody test.

Serological assays measure RABV-specific antibodies in response to vaccination (or exposure to viral antigens). Methods include antigen-binding assays or virus neutralization tests (Moore 2021, Fooks et al. 2017). Detailed protocols for routine diagnostic tests are obtainable from OIE and WHO manuals (Rupprecht et al. 2018*a*, WHO 2018, OIE 2021).

Treatment

Most rabies cases are fatal. There is no proven treatment. Individual animal management is not practiced, even by authorized wildlife rehabilitators. When suspected, animals are euthanized. For humans and domestic animals at risk, PrEp with a licensed pure, potent, safe, and efficacious vaccine, administered parenterally, is highly effective before exposure. Valuable mammals in zoological exhibitions may be vaccinated (National Association of State Public Health Veterinarians 2016, WHO 2018). After human exposure, PEP is highly efficacious, if provided appropriately in a timely manner. Culling or vaccination are the major prevention and control strategies used in certain wildlife populations.

Management

Traditional aspects included education (i.e., to minimize bites), legislation (i.e., on animal translocations), and environmental alterations (i.e., habitat modifications or exclusions). Tactics specific to the animal source were once considered impossible. Research during the past century created new paradigms. Given the impact on human health, dog rabies control was the priority. By the late 19th century, several countries eliminated canine rabies by strict sanitary measures, including muzzling, quarantine, movement restrictions, and culling (Müller et al. 2012*b*, Müller and Freuling 2018). Post-Pasteur, development of inactivated rabies vaccines achieved large-scale breakthroughs for mass dog vaccination and canine rabies elimination, applied in Europe and North America (Vigilato et al. 2018, Müller and Freuling 2018, WHO 2018, González-Roldán et al. 2021, Ma et al. 2022). Latin American countries are approaching selective regional elimination of canine rabies, but true eradication is not achievable, due to diverse reservoirs among wild mesocarnivores and bats.

Wildlife control is complex and once involved extensive culling. Population reduction was born with recognition that RABV transmission was density dependent (Morters et al. 2013). Culling, including trapping, poisoning, or gassing, was widespread in Europe and North America (Aubert 1992, Winkler and Bögel 1992). Success occurred in a few cases, such as in Alberta, Corsica, Denmark, or small areas of France (Westergaard 1982, Aubert 1994). Typically, culling could not suppress densities below a threshold of $R < 1$ to depress viral transmission (Aubert 1992). Population reduction outweighed advantages, as it was costly, labor intensive, largely ineffective, and counterproductive—maintaining populations in flux—resulting in more aggression over new or emptied niches, which exacerbated disease (Rupprecht et al. 2001). Cruelty was problematic, as control based on wholesale slaughter became both ethically and ecologically unacceptable (Winkler and Bögel 1992, Aubert 1994, Rupprecht et al. 2001). Vampire bat population reduction using anticoagulants may be viewed similarly (Mayen 2003).

Recognizing that parenteral dog vaccination was effective (Lawson et al. 1982), CDC researchers devised ways to entice wildlife to self-vaccinate during the late 1960s, refashioning equipment to poison predators into alternative devices (Winkler and Bögel 1992). The "humane coyote getter" delivered vaccine instead of cyanide, hoping to elicit a mucosal immune response. Another contraption was the "Vac-Trap," a self-injector using vaccine-laden syringes that injected an animal when a plate was triggered. Both ideas did not work well, were harmful for target and nontargets, and could not economically vaccinate populations (Winkler and Bögel 1992). Nevertheless, parenteral vaccination became an integral part of labor-intensive tactics, known as trap–vaccinate–release (TVR) and point infection control (PIC). Both TVR and PIC, developed and used in Canada as an emergency response to raccoon rabies introduction into Ontario, relied on parenteral vaccination in combination with oral vaccination and population reduction in surrounding areas to prevent further spread (Rosatte et al. 2001). Despite local success, such tactics were the most expensive per unit area and impractical for large-scale use (Sterner et al. 2009).

Today, the almost "magical" focus for management is oral rabies vaccination (ORV), involving strategic, large-scale environmental release of vaccine-laden baits. Application of ORV is based on an ingenious idea from US researchers and enthusiastic commitment of European partners (Baer et al. 1971; Steck et al. 1982*a*, 1982*b*). This concept was based on three components: effective vaccines; attractive baits; and appropriate distribution (Freuling et al. 2019). Over the past five decades, numerous ORV constructs, based on replication-competent viruses (i.e., inactivated vaccines were not efficacious per os), were developed, both attenuated and recombinant, with different safety profiles (Müller and Freuling 2020*b*).

Efficacious vaccines are only useful if baits are highly attractive. Therefore, bait development requires careful preparation and must address species-specific concerns. Machine-made baits replaced older experimental products made from offal (Schneider et al. 1987, Hanlon et al. 1989, Rosatte et al. 2008, Rosatte et al. 2009), such as chicken heads and intestines, or sausages (Winkler and Baer 1976, Steck et al. 1982*a*, Schneider and Cox 1983, Pastoret et al. 1988). Requirements for ORV biologics and baits are in the OIE Terrestrial Manual (OIE 2021).

An effective distribution system is equally important. Optimal large-scale strategies regarding bait densities, timing, and distribution require an understanding of species biology: foraging behaviors, community dynamics, bait uptake, and knowledge of bait competitors (Slate et al. 2009). Aerial bait distribution using fixed-wing aircrafts or helicopters is preferred for rural habitats (Johnston et al. 1988, Rosatte et al. 2009, Müller et al. 2012*a*). In urban and suburban areas, TVR, hand baiting, and bait stations are applied depending on the reservoir (European Commission 2002, Selhorst et al. 2006, Bjorklund et al. 2017, Haley et al. 2017). To achieve maximum success, vaccination strategies should be flexible (Figure 14.6).

Since the late 1970s, ORV has been used to control wildlife rabies, particularly in the Northern Hemisphere, where campaigns and programs were implemented in 30 European countries, Canada, the United States, Israel, South Korea, and Turkey (Linhart et al. 1997, Slate et al. 2009, Ün et al. 2012, Freuling et al. 2013, Fehlner-Gardiner 2018, Müller and Freuling 2018, Cho et al. 2020). More than 1.3 billion vaccines baits have been distributed (Müller et al. 2015*b*, Müller and Freuling 2020*b*). Use of ORV has eliminated red fox rabies in parts of Canada and western Europe, gray fox rabies in Texas, and coyote rabies along the Mexico–US border (MacInnes et al. 2001, Sidwa et al. 2005, Freuling et al. 2013, Maki et al. 2017, Fehlner-Gardiner 2018, Müller and Freuling 2018, Robardet et al. 2019, Ma et al. 2022). In the United States, large-scale ORV programs

Fig. 14.6. Management of wild mesocarnivore populations by vaccination against rabies may include aerial vaccine bait distribution, bait stations, and trapping and vaccination. Illustration by Laura Donohue.

prevented further spread of raccoon rabies (Slate et al. 2009, Sterner et al. 2009, Ma et al. 2022). In Canada, raccoon rabies incursions from the United States were controlled by targeted ORV campaigns (Sterner et al. 2009, Stevenson et al. 2016, Fehlner-Gardiner 2018). Due to ongoing strife and repeated introductions from eastern Europe, the European Union has not yet become rabies free (Robardet et al. 2019).

Despite successes, ORV faces challenges, including costs in vast endemic areas like Canada, Europe, Russia, and the United States; lack of effective biologics, attractive baits, and vaccination strategies for certain reservoirs; inappropriate baiting densities; vaccine refractoriness in certain reservoirs (e.g., skunks and mongoose); risks of sustained spillover from bat-associated lyssaviruses; increased mesocarnivore urbanization; and rabies in highly developed suburban settings (Rupprecht et al. 2008, Slate et al. 2009, Müller et al. 2015*a*, Te Kamp et al. 2020). Often, authorities succumb to just copying a successful strategy without a thorough understanding of target-species ecology. Even when a strategy follows sound recommendations, setbacks occur that quickly destroy gains. Reasons for failure are myriad (Stöhr and Meslin 1998, Selhorst et al. 2006, Rupprecht et al. 2008). Authorities should take lessons learned at the earliest planning stages to minimize errors.

Besides mass parenteral vaccination, ORV has a role as a complementary tool in dog-mediated rabies, experiencing a renaissance in dog rabies management (Wallace et al. 2020, Chanachai et al. 2021). Could similar strategies be used for bats, building upon herd immunity? Bats respond well to experimental vaccination (Setien et al. 1998, Aguilar-Setien et al. 2002, Stading et al. 2017). Thus, ORV might seem applicable. Yet, the size of some local populations (tens of millions of individuals), diversity of >1,200 bat species, antigenic complexity of bat-associated lyssaviruses (Fooks et al. 2021), conservation status, and inaccessibility make this a utopian dream.

Financial, Legal, Political, and Social Factors

Health Economics

Rabies exacts a tremendous economic toll, more than USD $120 billion annually (Anderson and Shwiff 2015). Most costs are associated with canine rabies, prevention of premature human death, domestic animal vaccination, and surveillance (Hampson et al. 2015). In a One Health context, death reduction in multiple species by dog vaccination and human prophylaxis result in substantial savings (Shwiff et al. 2013). In one 1990–2015 Mexican analysis, dog rabies vaccination prevented ~13,000 human cases at an incremental cost of USD $300 million (González-Roldán et al. 2021). Investigators estimated an average cost of USD $23,000 per human death averted, USD $410 per additional year-of-life, and USD $190 per dog rabies death averted. Mexico is now free of canine rabies (Ma et al. 2022). Similar trends were reported in Asia and Africa (Hatch et al. 2017, Shwiff et al. 2018). Besides canine rabies, vampire bats create a substantial burden due to livestock losses and human outbreaks (Anderson et al. 2014, Johnson and Montano Hirose 2018). For mesocarnivores, ORV can be highly cost beneficial, leading to long-term prevention and selective elimination (Freuling et al. 2008, Shwiff et al. 2008, 2009, Freuling et al. 2013, Muller et al. 2015*b*, Elser et al. 2016, Bedeković et al. 2019).

Legal Issues

Due to the impacts of rabies on public health, agriculture, and conservation biology, multiple entities have developed regional, national, or local legislative frameworks (European Commission 2002, National Asscociation of State Public Health Veterinarians 2016). Jurisdictional topics include animal import and export; biosafety; culling, hunting, trapping, and toxicants for population reduction; border controls; domestic and exotic pet ownership and sales; leash laws; outbreak responses; quarantine; regulatory oversight on biologics; reporting, notification, and surveillance; sanitation for handling of food and game animals; stray animal control; trade; translocation restrictions; vaccination; waste management; and wildlife rehabilitation. Some laws last for centuries with little alteration, such as quarantine imposition in "rabies-free" areas, whereas others are reactionary (Cliquet and Wasniewski 2018, Wright et al. 2021). Even shorter actions carry broader repercussions. For example, the 2021 US ban on canine importations was intended as temporary, yet larger implications are still being felt, with recent extension to Canada (Pieracci et al. 2021).

Political Factors

Global institutions (e.g., Food and Agriculture Organization of the United Nation, World Organisation for Animal Health, and World Health Organization), in collaboration with partners and stakeholders, established recommendations for elimination of human rabies caused from dogs by 2030 (Abela-Ridder et al. 2018). Stable political institutions are expected to be responsible for enacting and enforcing laws on disease detection, prevention, and control. Such national, regional, or local applications improve intersectoral cooperation and responsibilities in codified detail. Unified precepts of health, justice, and law have been explored in kind to minimize zoonotic impacts (Zucca et al. 2020). Unfortunately, neither public will nor local assignment of practical responsibility exist in many developing countries. Recognizing limitations, the Pan American Health Organization (PAHO) strove towards a progressive regional rabies program in the Americas in 1983 by uniting agricultural and medical sectors (Vigilato et al. 2018). Although the PAHO focus is upon canine rabies and secondly upon bat rabies, other international (e.g., World Organisation for Animal Health), regional (e.g., European Union) and national (e.g., US Department of Agriculture) agencies also focused on wildlife rabies management (Robardet et al. 2019).

Social Concerns

Rabies is greatly feared in all societies, providing motivation within societies for its prevention and control. Given public needs, wants, and expectations, health professionals and wildlife managers can harness such energy and engage with citizens. Applications, such as "willingness to pay" or community "knowledge, attitudes, and practices" surveys, are key to decision making and best practices including prescriptive resource allocations, aid in identification of related gaps, removal of real and perceived obstacles to progress, focus upon translational research, and prioritization areas for education (Lankau et al. 2015, Straily and Trevino-Garrison 2017, Neevel et al. 2018). Besides academia, government, and industry in such endeavors, other stakeholders include international nongovernmental organizations. For example, the need for health communications, public awareness and education, political advocacy, action, and outreach led to the creation of World Rabies Day, the Global Alliance for Rabies Control, Mission Rabies, the North American Rabies Management Plan, the Rabies in the Americas Conference, among others (Gamble et al. 2019, Rupprecht et al. 2020*a*, 2020*b*).

Summary

Rabies has a highly evolved lifecycle in a diverse array of species, producing fear and human mortality for millennia—this zoonosis remains so in many parts of the world today.

Multifaceted, transdisciplinary global activities, heightened during the 21st century, provide evidence that human rabies is preventable, canine rabies is eliminable, and wildlife rabies is controllable. Such progressive models illustrate the current challenges at hand in dealing with "diseases of nature" yet, more importantly, exemplify theoretical possibilities that evolve and drive real-world practicalities for other wildlife health priorities, even during a pandemic (Nadal et al. 2021).

ACKNOWLEDGEMENTS

We are fortunate to have engaged Laura Donohue, recognizing her creativity and excellent illustrations for our chapter. We thank the US Department of Health and Human Services Public Health Image Library for the Negri body micrograph, USDA Wildlife Services for the rabid raccoon photograph, Friedrich Loeffler Institute librarian Viola Damrau for her wonderful assistance with this rather complicated reference format and apologize in advance for those important but omitted historical citations, due to space constraints! Finally, we recognize all the diverse professionals around the world, living and deceased, that have engaged passionately to set the rich foundation for the ever-emerging field of wildlife rabies and invite the next generation of investigators to join us to improve its detection, prevention, control, and selective elimination in a One Health context.

LITERATURE CITED

Abela-Ridder, B., K. Balogh de, J. A. Kessels, I. Dieuzy-Labaye, and G. Torres. 2018. Global rabies control: the role of international organisations and the Global Strategic Plan to eliminate dog-mediated rabies. Revue Scientifique et Technique 37:741–749.

Aguilar-Setien, A., Y. L. Campos, E. T. Cruz, R. Kretschmer, B. Brochier, and P. P. Pastoret. 2002. Vaccination of vampire bats using recombinant vaccinia–rabies virus. Journal of Wildlife Diseases 38:539–544.

Anderson, A., S. Shwiff, K. Gebhardt, A. J. Ramirez, D. Kohler, and L. Lecuona. 2014. Economic evaluation of vampire bat rabies prevention in Mexico. Transboundary and Emerging Diseases 61:140–146.

Anderson, A., and S. A. Shwiff. 2015. The cost of canine rabies on four continents. Transboundary and Emerging Diseases 62:446–452.

Andral, L., M. Artois, M. Aubert, and J. Blancou. 1982. Radio-tracking of rabid foxes. Comparative Immunology, Microbiology and Infectious Diseases 5:285–291. (In French.)

Aubert, M. 1992. Epidemiology of fox rabies. Pages 9–18 *in* K. Bögel, F.-X. Meslin, and M. Kaplan, editors. Wildlife rabies control. Wells Medical Ltd., Kent, England.

Aubert, M. 1994. Control of rabies in foxes: what are the appropriate measures? Veterinary Record 134:55–59.

Badrane, H., C. Bahloul, P. Perrin, and N. Tordo. 2001. Evidence of two lyssavirus phylogroups with distinct pathogenicity and immunogenicity. Journal of Virology 75:3268–3276.

Badrane, H., and N. Tordo. 2001. Host switching in lyssavirus history from the Chiroptera to the Carnivora orders. Journal of Virology 75:8096–8104.

Baer, G. M., M. K. Abelseth, and J. G. Debbie. 1971. Oral vaccination of foxes against rabies. American Journal of Epidemiology 93:487–490.

Baker, L., J. Matthiopoulos, T. Müller, C. Freuling, and K. Hampson. 2019. Optimizing spatial and seasonal deployment of vaccination campaigns to eliminate wildlife rabies. Philosophical Transactions of the Royal Society of London B Biological Sciences 374:20180280.

Barrett, J., A. Höger, K. Agnihotri, J. Oakey, L. F. Skerratt, H. E. Field, J. Meers, and C. Smith. 2020. An unprecedented cluster of Australian bat lyssavirus in *Pteropus conspicillatus* indicates pre-flight flying fox pups are at risk of mass infection. Zoonoses and Public Health 67:435–442.

Bedeković, T., I. Šimić, N. Krešić, and I. Lojkić. 2019. Influence of different factors on the costs and benefits of oral vaccination of foxes against rabies. Zoonoses and Public Health 66:526–532.

Begeman, L., C. GeurtsvanKessel, S. Finke, C. M. Freuling, M. Koopmans, T. Müller, T. J. H. Ruigrok, and T. Kuiken. 2017. Comparative pathogenesis of rabies in bats and carnivores, and implications for spillover to humans. Lancet Infectious Diseases 18:e147–e159.

Bigler, W. J., R. G. McLean, and H. A. Trevino. 1973. Epizootiologic aspects of raccoon rabies in Florida. American Journal of Epidemiology 98:326–335.

Bjorklund, B., B. Haley, R. Bevilacqua, M. Chandler, A. Duffiney, K. von Hone, D. Slate, R. Chipman, A. Martin, and T. Algeo. 2017. Progress towards bait station integration into oral rabies vaccination programs in the United States: field trials in Massachusetts and Florida. Tropical Medicine and Infectious Disease 2:40.

Blancou, J. 1988. Ecology and epidemiology of fox rabies. Reviews of Infectious Diseases 10:606–609.

Blancou, J., and A. I. Wandeler. 1989. Rabies virus and its vectors in Europe. Revue Scientifique et Technique 8:927–929.

Bögel, K., H. Moegle, F. Steck, W. Krocza, and L. Andral. 1981. Assessment of fox control in areas of wildlife rabies. Bulletin of the World Health Organization 59:269–279.

Botto Nuñez, G., D. J. Becker, and R. K. Plowright. 2019. The emergence of vampire bat rabies in Uruguay within a historical context. Epidemiology and Infection 147:e180.

Botvinkin, A. D., V. P. Savitskii, G. N. Sidorovand, and V. G. Judin. 1981. Importance of the racoon dog in the epidemiology and epizootiology of rabies in the Far East. Zhurnal mikrobiologii, epidemiologii, i immunobiologii. 12:79–82. (In Russian.)

Bouma, H. R., H. V. Carey, and F. G. M. Kroese. 2010. Hibernation: the immune system at rest? Journal of Leukocyte Biology 88:619–624.

Brzozka, K., S. Finke, and K. K. Conzelmann. 2006. Inhibition of interferon signaling by rabies virus phosphoprotein P: activation-dependent binding of STAT1 and STAT2. Journal of Virology 80:2675–2683.

Canning, G., H. Camphor, and B. Schroder. 2019. Rabies outbreak in African wild dogs (*Lycaon pictus*) in the Tuli region, Botswana: interventions and management mitigation recommendations. Journal for Nature Conservation 48:71–76.

Chanachai, K., V. Wongphruksasoong, A. Vos, K. Leelahapongsathon, R. Tangwangvivat, O. Sagarasaeranee, P. Lekcharoen, P. Trinuson, and S. Kasemsuwan. 2021. Feasibility and effectiveness studies with oral vaccination of free-roaming dogs against rabies in Thailand. Viruses 13:571.

Charlton, K. M., S. Nadin-Davis, G. A. Casey, and A. I. Wandeler. 1997. The long incubation period in rabies: delayed progression of infection in muscle at the site of exposure. Acta Neuropathologica 94:73–77.

Cho, H.-K., Y.-J. Shin, N.-S. Shin, and J.-S. Chae. 2020. Efficient distribution of oral vaccines examined by infrared triggered camera for advancing the control of raccoon dog rabies in South Korea. Journal of Veterinary Medical Science 82:1685–1692.

Clark, K. A., S. U. Neill, J. S. Smith, P. J. Wilson, V. W. Whadford, and G. W. McKirahan. 1994. Epizootic canine rabies transmitted by coyotes in south Texas. Journal of the American Veterinary Medical Association 204:536–540.

Cleaveland, S., and C. Dye. 1995. Maintenance of a microparasite infecting several host species: rabies in the Serengeti. Parasitology 111(Supplement S1):S33–S47.

Cliquet, F., E. Picard-Meyer, J. Barrat, S. M. Brookes, D. M. Healy, M. Wasniewski, E. Litaize, M. Biarnais, L. Johnson, and A. R. Fooks. 2009. Experimental infection of foxes with European bat lyssaviruses type-1 and 2. BMC Veterinary Research 5:19.

Cliquet, F., and M. Wasniewski. 2018. Maintenance of rabies-free areas. Revue Scientifique et Technique 37:691–702.

Coertse, J., C. Grobler, C. Sabeta, E. C. Seamark, T. Kearney, J. Paweska, and W. Markotter. 2020. Lyssaviruses in insectivorous bats, South Africa, 2003–2018. Emerging Infectious Diseases 26:3056–3060.

Constantine, D. G. 1962. Rabies transmission by nonbite route. Public Health Reports 77:287–289.

Daoust, P. Y., A. I. Wandeler, and G. A. Casey. 1996. Cluster of rabies cases of probable bat origin among red foxes in Prince Edward Island, Canada. Journal of Wildlife Diseases 32:403–406.

Davis, A., P. Gordy, R. Rudd, J. A. Jarvis, and R. A. Bowen. 2012. Naturally acquired rabies virus infections in wild-caught bats. Vector Borne and Zoonotic Diseases 12:55–60.

Davis, A. D., S. M. Morgan, M. Dupuis, C. E. Poulliott, J. A. Jarvis, R. Franchini, A. Clobridge, and R. J. Rudd. 2016. Overwintering of rabies virus in silver haired bats. PLoS One 11:e0155542.

Debbie, J. G. 1974. Use of inoculated eggs as a vehicle for the oral rabies vaccination of red foxes. Infection and Immunity 9:681–683.

Deplazes, P., D. Hegglin, S. Gloor, and T. Romig. 2004. Wilderness in the city: the urbanization of *Echinococcus multilocularis*. Trends in Parasitology 20:77–84.

Dierks, R. E., F. A. Murphy, and A. K. Harrison. 1969. Extraneural rabies virus infection. Virus development in fox salivary gland. American Journal of Pathology 54:251–273.

Dietzschold B., J. Li, M. Faber, and M. Schnell. 2008. Concepts in the pathogenesis of rabies. Future Virology 3:481–490.

Dietzschold, B., C. E. Rupprecht, M. Tollis, M. Lafon, J. Mattei, T. J. Wiktor, and H. Koprowski. 1988. Antigenic diversity of the glycoprotein and nucleocapsid proteins of rabies and rabies-related viruses: implications for epidemiology and control of rabies. Reviews of Infectious Diseases 10:785–797.

Eggerbauer, E., P. de Benedictis, B. Hoffmann, T. C. Mettenleiter, K. Schlottau, E. C. Ngoepe, C. T. Sabeta, C. M. Freuling, and T. Müller. 2016. Evaluation of six commercially available rapid immunochromatographic tests for the diagnosis of rabies in brain material. PLoS Neglected Tropical Diseases 10:e0004776.

Elmhagen, B., P. Hersteinsson, K. Noren, E. R. Unnsteinsdottir, and A. Angerbjorn. 2014. From breeding pairs to fox towns: the social organisation of arctic fox populations with stable and fluctuating availability of food. Polar Biology 37:111–122.

Elser, J. L., L. L. Bigler, A. M. Anderson, J. L. Maki, D. H. Lein, and S. A. Shwiff. 2016. The economics of a successful raccoon rabies elimination program on Long Island, New York. PLoS Neglected Tropical Diseases 10:e0005062.

European Commission. 2002. The oral vaccination of foxes against rabies. Report of the Scientific Committee on Animal Health and Animal Welfare. European Commission, Brussels, Belgium.

Everard, C., G. M. Baer, M. E. Alls, and S. A. Moore. 1981. Rabies serum neutralizing antibody in mongooses from Grenada. Transactions of the Royal Society of Tropical Medicine and Hygiene 75:654–666.

Everard, C. O., and J. D. Everard. 1992. Mongoose rabies in the Caribbean. Annals of the New York Academy of Sciences 653:356–366.

Fehlner-Gardiner, C. 2018. Rabies control in North America—past, present and future. Revue Scientifique et Technique 37:421–437.

Fekadu, M., J. H. Shaddock, and G. M. Baer. 1982. Excretion of rabies virus in the saliva of dogs. Journal of Infectious Diseases 145:715–719.

Fekadu, M., J. H. Shaddock, J. W. Sumner, D. W. Sanderlin, J. C. Knight, J. J. Esposito, and G. M. Baer. 1991. Oral vaccination of skunks with raccoon poxvirus recombinants expressing the rabies glycoprotein or the nucleoprotein. Journal of Wildlife Diseases 27:681–684.

Field, H. E. 2018. Evidence of Australian bat lyssavirus infection in diverse Australian bat taxa. Zoonoses and Public Health 65:742–748.

Follmann, E. H., D. G. Ritter, and G. M. Baer. 1996. Evaluation of the safety of two attenuated oral rabies vaccines, SAG1 and SAG2, in six Arctic mammals. Vaccine 14:270–273.

Fooks, A. R., F. Cliquet, S. Finke, C. Freuling, T. Hemachudha, R. S. Mani, T. Müller, S. Nadin-Davis, E. Picard-Meyer, H. Wild, et al. 2017. Rabies. Nature Reviews Disease Primers 3:17091.

Fooks, A. R., R. Shipley, W. Markotter, N. Tordo, C. M. Freuling, T. Müller, L. M. McElhinney, A. C. Banyard, and C. E. Rupprecht. 2021. Renewed public health threat from emerging lyssaviruses. Viruses 13:1769.

Franco-Molina, M. A., S. E. Santana-Krímskaya, B. Cortés-García, J. A. Sánchez-Aldana-Pérez, O. García-Jiménez, and J. Kawas. 2021. Fatal case of rabies in a captive white-tailed deer: a case report from Chiapas, Mexico. Tropical Medicine and Infectious Disease 6:135.

Freuling, C. M., K. Hampson, T. Selhorst, R. Schroder, F. X. Meslin, T. C. Mettenleiter, and T. Müller. 2013. The elimination of fox rabies from Europe: determinants of success and lessons for the future. Philosophical Transactions of the Royal Society of London B Biological Sciences 368:20120142.

Freuling, C., D. Kloess, A. Kliemt, R. Schröder, and T. Müller. 2012. The WHO Rabies Bulletin Europe: a key source of information on rabies and a pivotal tool for surveillance and epidemiology. Revue Scientifique et Technique 31:799–807.

Freuling, C. M., T. Selhorst, H. J. Batza, and T. Müller. 2008. The financial challenge of keeping a large region rabies-free—the EU example. Developments in Biological Standardization 131:273–282.

Freuling, C. M., A. Vos, and T. F. Müller. 2019. Controlling rabies in foxes—unprecedented success in Europe. International Animal Health Journal 6:56–61.

Gamble, L., A. Gibson, S. Mazeri, B. M. de C Bronsvoort, I. Handel, and R. J. Mellanby. 2019. Development of non-governmental organisation–academic partnership to tackle rabies in Africa and Asia. Journal of Small Animal Practice 60:18–20.

George, D. B., C. T. Webb, M. L. Farnsworth, T. J. O'Shea, R. A. Bowen, D. L. Smith, T. R. Stanley, L. E. Ellison, and C. E. Rupprecht. 2011. Host and viral ecology determine bat rabies seasonality and maintenance. Proceedings of the National Academy of Sciences USA 108:10208–10213.

Gilbert, A. T. 2018. Rabies virus vectors and reservoir species. Revue Scientifique et Technique 37:371–384.

González-Roldán, J. F., E. A. Undurraga, M. I. Meltzer, C. Atkins, F. Vargas-Pino, V. Gutiérrez-Cedillo, and J. R. Hernández-Pérez. 2021. Cost-effectiveness of the national dog rabies prevention and control program in Mexico, 1990–2015. PLoS Neglected Tropical Diseases 15:e0009130.

Gough, P. M., and C. Niemeyer. 1975. A rabies epidemic in recently captured skunks. Journal of Wildlife Diseases 11:170–176.

Greenhall, A. M. 1971. Control of vampire bats. Study and program project for Latin America. Boletin de la Oficina Sanitaria Panamericana 73:231–245. (In Spanish.)

Gremillion-Smith, C., and A. Woolf. 1988. Epizootiology of skunk rabies in North America. Journal of Wildlife Diseases 24:620–626.

Greval, S. D. S. 1932. Rabies in the mongoose. The Indian Medical Gazette 67:451–453.

Haley, B., T. P. Algeo, B. Bjorklund, A. G. Duffiney, R. E. Hartin, A. Martin, K. M. Nelson, R. B. Chipman, and D. Slate. 2017. Evaluation of bait station density for oral rabies vaccination of raccoons in urban and rural habitats in Florida. Tropical Medicine and Infectious Disease 2:41.

Hamir, A. N., B. A. Summers, and C. E. Rupprecht. 1998. Concurrent rabies and canine distemper encephalitis in a raccoon. Journal of Veterinary Diagnostic Investigation 10:194–196.

Hampson, K., L. Coudeville, T. Lembo, M. Sambo, A. Kieffer, M. Attlan, J. Barrat, J. D. Blanton, D. J. Briggs, S. Cleaveland, et al. 2015. Estimating the global burden of endemic canine rabies. PLoS Neglected Tropical Diseases 9:e0003709.

Hampson, K., J. Dushoff, J. Bingham, G. Brückner, Y. H. Ali, and A. Dobson. 2007. Synchronous cycles of domestic dog rabies in sub-Saharan Africa and the impact of control efforts. Proceedings of the National Academy of Sciences USA 104:7717–7722.

Hanlon, C. A., D. E. Hayes, A. N. Hamir, D. E. Snyder, S. R. Jenkins, C. P. Hable, and C. E. Rupprecht. 1989. Proposed field evaluation of a rabies recombinant vaccine for raccoons: site selection, target species characteristics, and placebo baiting trials. Journal of Wildlife Diseases 25:555–567.

Hatch, B., A. Anderson, M. Sambo, M. Maziku, G. McHau, E. Mbunda, Z. Mtema, C. E. Rupprecht, S. A. Shwiff, and L. Nel. 2017. Towards canine rabies elimination in south-eastern Tanzania: assessment of health economic data. Transboundary and Emerging Diseases 64:957–958.

Hayman, D. T., A. R. Fooks, D. A. Marston, and R. J. Garcia. 2016. The global phylogeography of lyssaviruses—challenging the 'out of Africa' hypothesis. PloS Neglected Tropical Diseases 10:e0005266.

Hikufe, E. H., C. M. Freuling, R. Athingo, A. Shilongo, E.-E. Ndevaetela, M. Helao, M. Shiindi, R. Hassel, A. Bishi, S. Khaiseb, et al. 2019. Ecology and epidemiology of rabies in humans, domestic animals and wildlife in Namibia, 2011–2017. PloS Neglected Tropical Diseases 13:e0007355.

Hirsch, B. T., J. Reynolds, S. D. Gehrt, and M. E. Craft. 2016. Which mechanisms drive seasonal rabies outbreaks in raccoons? A test using dynamic social network models. Journal of Applied Ecology 53:804–813.

Holmala, K., and K. Kauhala. 2006. Ecology of wildlife rabies in Europe. Mammal Review 36:17–36.

Hübschle, O. J. 1988. Rabies in the kudu antelope. Reviews of Infectious Diseases 10:629–633.

Jackson, A. C. 2018. Rabies: a medical perspective. Revue Scientifique et Technique 37:569–580.

Jackson, A. C., H. Ye, C. C. Phelan, C. Ridaura-Sanz, Q. Zheng, Z. Li, X. Wan, and E. Lopez-Corella. 1999. Extraneural organ involvement in human rabies. Laboratory Investigation 79:945–951.

Jaramillo-Reyna, E., C. Almazán-Marín, M. E. de la O-Cavazos, R. Valdéz-Leal, A. H. Bañuelos-Álvarez, M. A. Zúñiga-Ramos, M. Melo-Munguía, M. Gómez-Sierra, A. Sandoval-Borja, S. Chávez-López, et al. 2020. Rabies virus variants identified in Nuevo Leon State, Mexico, from 2008 to 2015. Journal of the American Veterinary Medical Association 256:438–443.

Jenkins, S. R., B. D. Perry, and W. G. Winkler. 1988. Ecology and epidemiology of raccoon rabies. Reviews of Infectious Diseases 10:620–625.

Johnston, D. H., D. R. Voigt, C. D. MacInnes, P. Bachmann, K. F. Lawson, and C. E. Rupprecht. 1988. An aerial baiting system for the distribution of attenuated or recombinant rabies vaccines for foxes, raccoons, and skunks. Reviews of Infectious Diseases 10(Supplement 4):S660–S664.

Johnson, N., and J. A. Montano Hirose. 2018. The impact of paralytic bovine rabies transmitted by vampire bats in Latin America and the Caribbean. Revue Scientifique et Technique 37:451–459.

Johnson, N., R. Phillpotts, and A. R. Fooks. 2006. Airborne transmission of lyssaviruses. Journal of Medical Microbiology 55:785–790.

Kantorovich, R. A., V. P. Riutova, I. A. Buzinov, and G. V. Konovalo. 1963. Experimental investigations into rage and rabies in polar foxes, natural hosts of infection. Acta Virologica 7:554–560.

Kim, B. I., J. D. Blanton, A. Gilbert, L. Castrodale, K. Hueffer, D. Slate, and C. E. Rupprecht. 2014. A conceptual model for the impact of climate change on fox rabies in Alaska, 1980–2010. Zoonoses and Public Health 61:72–80.

Kim, C. H., C. G. Lee, H. C. Yoon, H. M. Nam, C. K. Park, J. C. Lee, M. I. Kang, and S. H. Wee. 2006. Rabies, an emerging disease in Korea. Journal of Veterinary Medicine. 53:111–115.

King, A. A., C. D. Meredith, and G. R. Thomson. 1994. The biology of southern African lyssavirus variants. Current Topics in Microbiology and Immunology 187:267–295.

Kirandjiski, T., S. Mrenoski, I. Celms, D. Mitrov, I. Dzadzovski, I. Cvetkovikj, K. Krstevski, E. Picard-Meyer, P. Viviani, D. Malinovski, et al. 2012. First reported cases of rabies in the Republic of Macedonia. Veterinary Record 170:312.

Klein, A., A. Fahrion, S. Finke, M. Eyngor, S. Novak, B. Yakobson, E. Ngoepe, B. Phahladira, C. Sabeta, P. De Benedicti, et al. 2020. Further evidence of inadequate quality in lateral flow devices commercially offered for the diagnosis of rabies. Tropical Medicine and Infectious Disease 5:13.

Koeppel, K. N., O. L. van Schalkwyk, and P. N. Thompson. 2021. Patterns of rabies cases in South Africa between 1993 and 2019, including the role of wildlife. Transboundary and Emerging Diseases 69:836–848.

Kotait, I., R. N. Oliveira, M. L. Carrieri, J. G. Castilho, C. I. Macedo, P. M. C. Pereira, V. Boere, L. Montebello, and C. E. Rupprecht. 2019. Non-human primates as a reservoir for rabies virus in Brazil. Zoonoses and Public Health 66:47–59.

Kuhn, J. H., S. Adkins, D. Alioto, S. V Alkhovsky, G. K Amarasinghe, S. J Anthony, T. Avšič-Županc, M. A Ayllón, J. Bahl, A. Balkema-Buschmann, et al. 2020. 2020 taxonomic update for phylum Negarnaviricota, including the large orders Bunyavirales and Mononegavirales. Archives of Virology 165:3023–3072.

Kulonen, K., and I. Boldina. 1996. No rabies detected in voles and field mice in a rabies-endemic area. Zentralblatt fur Veterinarmedizin 43:445–447.

Kuzmin, I. V., R. Franka, and C. E. Rupprecht. 2008*a*. Experimental infection of big brown bats with West Caucasian bat virus. Developments in Biological Standardization 131:327–337.

Kuzmin, I. V., G. J. Hughes, A. D. Botvinkin, S. G. Gribencha, and C. E. Rupprecht. 2008*b*. Arctic and Arctic-like rabies viruses: distribution, phylogeny and evolutionary history. Epidemiology and Infection 136:509–519.

Kuzmin, I. V., M. Shi, L. A. Orciari, P. A. Yager, A. Velasco-Villa, N. A. Kuzmina, D. G. Streicker, D. L. Bergman, and C. E. Rupprecht. 2012. Molecular inferences suggest multiple host shifts of rabies viruses from bats to mesocarnivores in Arizona during 2001–2009. PLoS Pathogens 8:e1002786.

Kuzmin, I. V., G. N. Sidorov, A. D. Botvinov, and E. I. Rekhov. 2001. Epizootic situation and prospectives of rabies control among wild animals in the south of western Siberia. Zhurnal mikrobiologii, epidemiologii, i immunobiologii 2001(3):28–35. (In Russian.)

Kuzmina, N. A., P. Lemey, I. V. Kuzmin, B. C. Mayes, J. A. Ellison, L. A. Orciari, D. Hightower, S. T. Taylor, and C. E. Rupprecht. 2013. The phylogeography and spatiotemporal spread of south-central skunk rabies virus. PLoS One 8:e82348.

Lafon, M. 2005. Rabies virus receptors. Journal of Neurovirology 11:82–87.

Lafon, M. 2011. Evasive strategies in rabies virus infection. Advances in Virus Research 79:33–53.

Lankau, E. W., S. W. Cox, S. C. Ferguson, J. D. Blanton, D. M. Tack, B. W. Petersen, and C. E. Rupprecht. 2015. Community survey of rabies knowledge and exposure to bats in homes—Sumter County, South Carolina, USA. Zoonoses and Public Health 62:190–198.

Lawson, K. F., H. Chiu, S. J. Crosgrey, M. Matson, G. A. Casey, and J. B. Campbell. 1997. Duration of immunity in foxes vaccinated orally with ERA vaccine in a bait. Canadian Journal of Veterinary Research 61:39–42.

Lawson, K. F., D. H. Johnston, and J. M. Patterson. 1982. Immunization of foxes by the oral and intramuscular route with inactivated rabies vaccines. Canadian Journal of Comparative Medicine 46:382–385.

Layan, M., S. Dellicour, G. Baele, S. Cauchemez, and H. Bourhy. 2021. Mathematical modelling and phylodynamics for the study of dog rabies dynamics and control: a scoping review. PLoS Neglected Tropical Diseases 15:e0009449.

Lee, D. N., M. Papeş, and R. A. Van den Bussche. 2012. Present and potential future distribution of common vampire bats in the Americas and the associated risk to cattle. PLoS One 7:e42466.

Lembo, T., M. Niezgoda, A. Velasco-Villa, S. Cleaveland, E. Ernest, and C. E. Rupprecht. 2006. Evaluation of a direct, rapid immunohistochemical test for rabies diagnosis. Emerging Infectious Diseases 12:310–313.

Leslie, M. J., S. Messenger, R. E. Rohde, J. Smith, R. Cheshier, C. Hanlon, and C. E. Rupprecht. 2006. Bat-associated rabies virus in skunks. Emerging Infectious Diseases 12:1274–1277.

Linhart, S. B., R. King, S. Zamir, U. Naveh, M. Davidson, and S. Perl. 1997. Oral rabies vaccination of red foxes and golden jackals in Israel: preliminary bait evaluation. Revue Scientifique et Technique 16:874–880.

Lord, R. D. 1992. Seasonal reproduction of vampire bats and its relation to seasonality of bovine rabies. Journal of Wildlife Diseases 28:292–294.

Ma, X., S. Bonaparte, M. Toro, L. A. Orciari, C. M. Gigante, J. D. Kirby, R. B. Chipman, C. Fehlner-Gardiner, V. Gutiérrez Cedillo, N. Aréchiga-Ceballos, et al. 2022. Rabies surveillance in the United States during 2020. Journal of the American Veterinary Medical Association 260:1157–65.

MacInnes, C. D., S. M. Smith, R. R. Tinline, N. R. Ayers, P. Bachmann, D. G. Ball, L. A. Calder, S. J. Crosgrey, C. Fielding, P. Hauschildt, et al. 2001. Elimination of rabies from red foxes in eastern Ontario. Journal of Wildlife Diseases 37:119–132.

Maki, J., A.-L. Guiot, M. Aubert, B. Brochier, F. Cliquet, C. A. Hanlon, R. King, E. H. Oertli, C. E. Rupprecht, C. Schumacher, et al. 2017. Oral vaccination of wildlife using a vaccinia–rabies–glycoprotein recombinant virus vaccine: a global review. Veterinary Research 48:57.

Malkov, G. B., and L. YA. Gribanova. 1980. Main results of the wildlife rabies study in Siberia and the Far East. Pages 52–61 *in* Modern methods in study of natural foci infections. Institute for Natural Foci Infections, Omsk, Russia. (In Russian.)

Mansfield, K. L., V. Racloz, L. M. McElhinney, D. A. Marston, N. Johnson, L. Rønsholt, L. S. Christensen, E. Neuvonen, A. D. Botvinkin, C. E. Rupprecht, et al. 2006. Molecular epidemiological study of Arctic rabies virus isolates from Greenland and comparison with isolates from throughout the Arctic and Baltic regions. Virus Research 116:1–10.

Marcelo Azevedo de Paula Antunes, J., L. de Castro Demoner, Morosini de Andrade Cruvinel, T., A. Paula Kataoka, L. Fatima Alves Martorelli, G. Puglia Machado, and J. Megid. 2017. Rabies virus exposure of Brazilian free-ranging wildlife from municipalities without clinical cases in humans or in terrestrial wildlife. Journal of Wildlife Diseases 53:662–666.

Marino, J., C. Sillero-Zubiri, A. Deressa, E. Bedin, A. Bitewa, F. Lema, G. Rskay, A. Banyard, and A. R. Fooks. 2017. Rabies and distemper outbreaks in smallest Ethiopian wolf population. Emerging Infectious Diseases 23:2102–2104.

Markotter, W., J. Coertse, L. de Vries, M. Geldenhuys, and M. Mortlock. 2020. Bat-borne viruses in Africa: a critical review. Journal of Zoology 311:77–98.

Marston, D. A., D. L. Horton, J. Nunez, R. J. Ellis, R. J. Orton, N. Johnson, A. C. Banyard, L. M. McElhinney, C. M. Freuling, M. Fırat, et al. 2017. Genetic analysis of a rabies virus host shift event reveals within-host viral dynamics in a new host. Virus Evolution 3:vex038.

Marston, D. A., A. C. Banyard, L. M. McElhinney, C. M. Freuling, S. Finke, X. de Lamballerie, T. Müller, and A. R. Fooks. 2018. The lyssavirus host-specificity conundrum—rabies virus—the exception not the rule. Current Opinion in Virology 28:68–73.

May, F., K. Mann, D. Francis, and M. Young. 2020. Identification of focus areas for Australian bat lyssavirus potential exposure prevention in the Metro North Hospital and Health Service region. Zoonoses and Public Health 67:732–741.

Mayen, F. 2003. Haematophagous bats in Brazil, their role in rabies transmission, impact on public health, livestock industry and alternatives to an indiscriminate reduction of bat population. Journal of Veterinary Medicine. 50:469–472.

McMahon, W. C., J. Coertse, T. Kearney, M. Keith, L. H. Swanepoel, and W. Markotter. 2021. Surveillance of the rabies-related lyssavirus, Mokola in non-volant small mammals in South Africa. Onderstepoort Journal of Veterinary Research 88:e1–e13.

Memish, Z. A., A. M. Assiri, and P. Gautret. 2015. Rabies in Saudi Arabia: a need for epidemiological data. International Journal of Infectious Diseases 34:99–101.

Meredith, C. D., and E. Standing. 1981. Lagos bat virus in South Africa. Lancet 317:832–833.

Merritt, T., K. Taylor, K. Cox-Witton, H. Field, K. Wingett, D. Mendez, M. Power, and D. Durrheim. 2018. Australian bat lyssavirus. Australian Journal of General Practice 47:93–96.

Meske, M., A. Fanelli, F. Rocha, L. Awada, P. C. Soto, N. Mapitse, and P. Tizzani. 2021. Evolution of rabies in South America and inter-species dynamics (2009–2018). Tropical Medicine and Infectious Disease 6:98.

Messenger, S. L., J. S. Smith, and C. E. Rupprecht. 2002. Emerging epidemiology of bat-associated cryptic cases of rabies in humans in the United States. Clinical Infectious Diseases 35:738–747.

Mollentze, N., D. G. Streicker, P. R. Murcia, K. Hampson, and R. Biek. 2020. Virulence mismatches in index hosts shape the outcomes of cross-species transmission. Proceedings of the National Academy of Sciences 117:28859–28866.

Moore, S. M. 2021. Challenges of rabies serology: defining context of interpretation. Viruses 13:1516.

Morters, M. K., O. Restif, K. Hampson, S. Cleaveland, J. L. Wood, and A. J. Conlan. 2013. Evidence-based control of canine rabies: a critical review of population density reduction. Journal of Animal Ecology 82:6–14.

Mulatti, P., T. Müller, L. Bonfanti, and S. Marangon. 2012. Emergency oral rabies vaccination of foxes in Italy in 2009–2010: identification of residual rabies foci at higher altitudes in the Alps. Epidemiology and Infection 140:591–598.

Müller, T., P. Demetriou, J. Moynagh, F. Cliquet, A. R. Fooks, F. J. Conraths, T. C. Mettenleiter, and C. M. Freuling. 2012*b*. Rabies elimination in Europe—a success story. Pages 31–44 *in* A. R. Fooks and T. Müller, editors. Rabies control—towards sustainable prevention at the source, Compendium of the OIE Global Conference on Rabies Control. World Organisation for Animal Health, Paris, France.

Müller, F. T., and C. M. Freuling. 2018. Rabies control in Europe: an overview of past, current and future strategies. Revue Scientifique et Technique 37:409–419.

Müller, T., and C. M. Freuling. 2020*a*. Rabies in terrestrial animals. Pages 195–230 *in* A. R. Fooks and A. C. Jackson, editors. Rabies: scientific basis of the disease and its management. Fourth edition. Academic Press, Cambridge, Massachusetts, USA.

Müller, T., and C. M. Freuling. 2020*b*. Rabies vaccines for wildlife. Pages 45–70 *in* H. C. J. Ertl, editor. Rabies and rabies vaccines. Springer Nature, Cham, Switzerland.

Müller, T., C. M. Freuling, P. Gschwendner, E. Holzhofer, H. Mürke, H. Rüdiger, P. Schuster, D. Klöss, C. Staubach, K. Teske, et al. 2012*a*. SURVIS: a fully-automated aerial baiting system for the distribution of vaccine baits for wildlife. Berliner und Münchener Tierärztliche Wochenschrift 125:197–202.

Müller, T., C. M. Freuling, P. Wysocki, M. Roumiantzeff, J. Freney, T. C. Mettenleiter, and A. Vos. 2015*a*. Terrestrial rabies control in the European Union: historical achievements and challenges ahead. Veterinary Journal 203:10–17.

Müller, T., R. Schröder, P. Wysocki, T. C. Mettenleiter, and C. M. Freuling. 2015*b*. Spatio-temporal use of oral rabies vaccines in fox rabies elimination programmes in Europe. PLoS Neglected Tropical Diseases 9:e0003953.

Nadal, D., S. Beeching, S. Cleaveland, K. Cronin, K. Hampson, R. Steenson, and B. Abela-Ridder. 2021. Rabies and the pandemic: lessons for One Health. Transactions of the Royal Society of Tropical Medicine and Hygiene 116:197–200.

Nadin-Davis, S. A., and C. Fehlner-Gardiner. 2019. Origins of the Arctic fox variant rabies viruses responsible for recent cases of the disease in southern Ontario. PLoS Neglected Tropical Diseases 13:e0007699.

National Association of State Public Health Veterinarians. 2016. Rabies compendium. http://www.nasphv.org/documentsCompendiaRabies.html. Accessed 21 June 2022.

Neevel, A. M. G., T. Hemrika, E. Claassen, and L. H. M. van de Burgwal. 2018. A research agenda to reinforce rabies control: a qualitative and quantitative prioritization. PLoS Neglected Tropical Diseases 12:e0006387.

Oertli, E. H., P. J. Wilson, P. R. Hunt, T. J. Sidwa, and R. E. Rohde. 2009. Epidemiology of rabies in skunks in Texas. Journal of the American Veterinary Medical Association 234:616–620.

Office International des Epizooties (OIE). 2021. Manual of diagnostic tests and vaccines for terrestrial animals. World Organisation for Animal Health, Paris, France.

O'Shea, T. J., R. A. Bowen, T. R. Stanley, V. Shankar, and C. E. Rupprecht. 2014. Variability in seroprevalence of rabies virus neutralizing antibodies and associated factors in a Colorado population of big brown bats. PLoS One 9:e86261.

Parker, R. L. 1969. Epidemiology of rabies. Archives of Environmental Health 19:857–861.

Pastoret, P. P., B. Brochier, B. Languet, I. Thomas, A. Paquot, B. Bauduin, M. P. Kieny, J. P. Lecocq, J. De Bruyn, F. Costy, et al. 1988. First field trial of fox vaccination against rabies using a vaccinia–rabies recombinant virus. Veterinary Record 123:481–484.

Pawan, J. L. 1936. The transmission of paralytic rabies in Trinidad by the vampire bat. Annals of Tropical Medicine and Parasitology 30:101–130.

Pieracci, E. G., C. E. Williams, R. M. Wallace, C. R. Kalapura, and C. M. Brown. 2021. US dog importations during the COVID-19 pandemic: do we have an erupting problem? PLoS One 16:e0254287.

Potratz, M., L. Zaeck, M. Christen, V. Te Kamp, A. Klein, T. Nolden, C. M. Freuling, T. Müller, and S. Finke. 2020*a*. Astrocyte infection during rabies encephalitis depends on the virus strain and infection route as demonstrated by novel quantitative 3d analysis of cell tropism. Cells 9:412.

Potratz, M., L. M. Zaeck, C. Weigel, A. Klein, C. M. Freuling, T. Müller, and S. Finke. 2020*b*. Neuroglia infection by rabies virus after anterograde virus spread in peripheral neurons. Acta Neuropathologica Communications 8:199.

Prada, D., V. Boyd, M. Baker, B. Jackson, and M. O'Dea. 2019. Insights into Australian bat lyssavirus in insectivorous bats of Western Australia. Tropical Medicine and Infectious Disease 4:46.

Pybus, M. J. 1988. Rabies and rabies control in striped skunks in three prairie regions of western North America. Journal of Wildlife Diseases 24:434–449.

Recuenco, S., M. Eidson, M. Kulldorff, G. Johnson, and B. Cherry. 2007. Spatial and temporal patterns of enzootic raccoon rabies adjusted for multiple covariates. International Journal of Health Geographics 6:14.

Robardet, E., D. Bosnjak, L. Englund, P. Demetriou, P. R. Martín, and F. Cliquet. 2019. Zero endemic cases of wildlife rabies (classical rabies virus, RABV) in the European Union by 2020: an achievable goal. Tropical Medicine and Infectious Disease 4:124.

Robardet, E., E. Picard-Meyer, S. Andrieu, A. Servat, and F. Cliquet. 2011. International interlaboratory trials on rabies diagnosis: an overview of results and variation in reference diagnosis techniques (fluorescent antibody test, rabies tissue culture infection test, mouse inoculation test) and molecular biology techniques. Journal of Virological Methods 177:15–25.

Rohde, R. E., and C. E. Rupprecht. 2020. Update on lyssaviruses and rabies: will past progress play as prologue in the near term towards future elimination? Faculty Reviews 9:9.

Rosatte, R., M. Allan, P. Bachmann, K. Sobey, D. Donovan, J. C. Davies, A. Silver, K. Bennett, L. Brown, B. Stevenson, et al. 2008. Prevalence of tetracycline and rabies virus antibody in raccoons, skunks, and foxes following aerial distribution of V-RG baits to control raccoon rabies in Ontario, Canada. Journal of Wildlife Diseases 44:946–964.

Rosatte, R., D. Donovan, M. Allan, L. A. Howes, A. Silver, K. Bennett, C. MacInnes, C. Davies, A. Wandeler, and B. Radford. 2001. Emergency response to raccoon rabies introduction into Ontario. Journal of Wildlife Diseases 37:265–279.

Rosatte, R. C., D. Donovan, J. C. Davies, M. Allan, P. Bachmann, B. Stevenson, K. Sobey, L Brown, A. Silver, K. Bennett, et al. 2009. Aerial distribution of ONRAB baits as a tactic to control rabies in raccoons and striped skunks in Ontario, Canada. Journal of Wildlife Diseases 45:363–374.

Rosatte, R., A. Wandeler, F. Muldoon, and D. Campbell. 2007. Porcupine quills in raccoons as an indicator of rabies, distemper, or both diseases: disease management implications. Canadian Veterinary Journal 48:299–300.

Ross, R. S., B. Wolters, B. Hoffmann, L. Geue, S. Viazov, N. Grüner, M. Roggendorf, and T. Müller. 2015. Instructive even after a decade: complete results of initial virological diagnostics and re-evaluation of molecular data in the German rabies virus "outbreak" caused by transplantations. International Journal of Medical Microbiology 305:636–643.

Rupprecht, C. E., B. Abela-Ridder, R. Abila, A. Charinna Amparo, A. Banyard, J. Blanton, K. Chanachai, K. Dallmeier, K. de Balogh, V. Del Rio Vilas, et al. 2020*a*. Towards rabies elimination in the Asia-Pacific region: from theory to practice. Biologicals 64:83–95.

Rupprecht, C. E., H. Bannazadeh Baghi, V. J. Del Rio Vilas, A. D. Gibson, F. Lohr, F. X. Meslin, J. F. R. Seetahal, K. Shervell, and L. Gamble. 2018*b*. Historical, current and expected future occurrence of rabies in enzootic regions. Revue Scientifique et Technique 37:729–739.

Rupprecht, C. E., J. Barrett, D. Briggs, F. Cliquet, A. R. Fooks, B. Lumlertdacha, F. X. Meslin, T. Müller, L. H. Nel, C. Schneider, et al. 2008. Can rabies be eradicated? Developments in Biological Standardization 131:95–121.

Rupprecht, C. E., A. Fooks, and B. Abela-Ridder. 2018*a*. Laboratory techniques in rabies. Fifth edition. World Health Organization, Geneva, Switzerland.

Rupprecht, C. E., C. M. Freuling, R. S. Mani, C. Palacios, C. T. Sabeta, and M. Ward. 2020*b*. A history of rabies—the foundation for global canine rabies elimination. Pages 1–42 *in* A. R. Fooks and A. C. Jackson, editors. Rabies—scientific basis of the disease and its management. Fourth edition. Academic Press, Cambridge, Massachusetts, USA.

Rupprecht, C. E., K. Stöhr, and C. Meredith. 2001. Rabies. Pages 3–36 *in* E. S. Williams and K. Barker, editors. Infectious diseases of wild mammals. Iowa State University Press, Ames, USA.

Sabeta, C. T., D. A. Marston, L. M. McElhinney, D. L. Horton, B. M. N. Phahladira, and A. R. Fooks. 2020. Rabies in the African civet: an incidental host for lyssaviruses? Viruses 12:368.

Schatz, J., A. R. Fooks, L. McElhinney, D. Horton, J. Echevarria, S. Vázquez-Moron, E. A. Kooi, T. B. Rasmussen, T. Müller, and C. M. Freuling. 2013. Bat rabies surveillance in Europe. Zoonoses and Public Health 60:22–34.

Schneider, L. G. 1989. Rabies control by oral vaccination of wildlife. Revue Scientifique et Technique 8:923–924.

Schneider, L. G., and J. H. Cox. 1983. Ein Feldversuch zur oralen Immunisierung von Füchsen gegen die Tollwut in der Bundesrepublik Deutschland: Unschädlichkeit, Wirksamkeit und Stabilität der Vakzine SAD B 19. Tierärztliche Umschau 38:315–324. (In German.)

Schneider, L. G., J. H. Cox, W. W. Müller, and K. P. Hohnsbeen. 1987. Der Feldversuch zur oralen Immunisierung von Füchsen gegen die Tollwut in der Bundesrepublik Deutschland - Eine Zwischenbilanz. Tierärztliche Umschau 3:184–198. (In German.)

Seetahal, J., A. Vokaty, M. A. N. Vigilato, C. V. F. Carrington, J. Pradel, B. Louison, A. Van Sauers, R. Roopnarine, J. C. González Arrebato, M. F. Millien, et al. 2018. Rabies in the Caribbean: a situational analysis and historic review. Tropical Medicine and Infectious Disease 3:89.

Seidlova, V., J. Zukal, J. Brichta, N. Anisimov, G. Apoznański, H. Bandouchova, T. Bartonička, H. Berková, A. D. Botvinkin, T. Heger, et al. 2020. Active surveillance for antibodies confirms circulation of lyssaviruses in Palearctic bats. BMC Veterinary Research 16:482.

Seimenis, A. 2008. The rabies situation in the Middle East. Developments in Biological Standardization 131:43–53.

Selhorst, T., T. Müller, and H. J. Bätza. 2006. Epidemiological analysis of setbacks in oral vaccination in the final stage of fox rabies elimination in densely populated areas in Germany. Developments in Biological Standardization 125:127–132.

Setien, A. A., B. Brochier, N. Tordo, O. de Paz, P. Desmettre, D. Peharpre, and P. P. Pastoret. 1998. Experimental rabies infection and oral vaccination in vampire bats. Vaccine 16:1122–1126.

Shao, X. Q., X. J. Yan, G. L. Luo, H. L. Zhang, X. L. Chai, F. X. Wang, J. K. Wang, J. J. Zhao, W. Wu, S. P. Cheng, et al. 2011. Genetic evidence for domestic raccoon dog rabies caused by Arctic-like rabies virus in Inner Mongolia, China. Epidemiology and Infection 139:629–635.

Shwiff, S. A., J. L. Elser, K. H. Ernst, S. S. Shwiff, and A. M. Anderson. 2018. Cost–benefit analysis of controlling rabies: placing economics at the heart of rabies control to focus political will. Revue Scientifique et Technique 37:681–689.

Shwiff, S., K. Hampson, and A. Anderson. 2013. Potential economic benefits of eliminating canine rabies. Antiviral Research 98:352–356.

Shwiff, S. A., K. N. Kirkpatrick, and R. T. Sterner. 2008. Economic evaluation of an oral rabies vaccination program for control of a domestic dog–coyote rabies epizootic: 1995–2006. Journal of the American Veterinary Medical Association 233:1736–1741.

Shwiff, S. A., R. T. Sterner, R. Hale, M. T. Jay, B. Sun, and D. Slate. 2009. Benefit cost scenarios of potential oral rabies vaccination for skunks in California. Journal of Wildlife Diseases 45:227–233.

Sidwa, T. J., P. Wilson, G. M. Moore, E. H. Oertli, B. N. Hicks, R. E. Rohde, and D. H. Johnston. 2005. Evaluation of oral rabies vaccination programs for control of rabies epizootics in coyotes and gray foxes: 1995–2003. Journal of the American Veterinary Medical Association 227:785–792.

Slate, D., T. P. Algeo, K. M. Nelson, R. B. Chipman, D. Donovan, J. D. Blanton, M. Niezgoda, and C. E. Rupprecht. 2009. Oral rabies vaccination in North America: opportunities, complexities, and challenges: opportunities, complexities, and challenges. PLoS Neglected Tropical Diseases 3:e549.

Stading, B., J. A. Ellison, W. C. Carson, P. S. Satheshkumar, T. E. Rocke, and J. E. Osorio. 2017. Protection of bats against rabies following topical or oronasal exposure to a recombinant raccoon poxvirus vaccine. PLoS Neglected Tropical Diseases 11:e0005958.

Steck, F., and A. Wandeler. 1980. The epidemiology of fox rabies in Europe. Epidemiologic Reviews 2:71–96.

Steck, F., A. Wandeler, P. Bichsel, S. Capt, U. Hafliger, and L. Schneider. 1982*a*. Oral immunization of foxes against rabies. Laboratory and field studies. Comparative Immunology Microbiology and Infectious Diseases 5:165–171.

Steck, F., A. I. Wandeler, P. Bichsel, S. Capt, and L. G. Schneider. 1982*b*. Oral immunisation of foxes against rabies. Zentralblatt Veterinarmedizin Reihe B 29:372–396.

Steele, J. H. 1988. Rabies in the Americas and remarks on global aspects. Reviews of Infectious Diseases 10:585–597.

Sterner, R. T., M. I. Meltzer, S. A. Shwiff, and D. Slate. 2009. Tactics and economics of wildlife oral rabies vaccination, Canada and the United States. Emerging Infectious Diseases 15:1176–1184.

Stevenson, B., J. Goltz, and A. Massé. 2016. Preparing for and responding to recent incursions of raccoon rabies variant into Canada. Canada Communicable Disease Report 42:125–129.

Stöhr, K., and F. X. Meslin. 1998. Rabies progress and setbacks in the oral immunisation of foxes against rabies in Europe. WHO Mediterranean Zoonosis Control Centre 45:8–10.

Stoltenow, C. L., K. Solemsass, M. Niezgoda, P. Yager, and C. E. Rupprecht. 2000. Rabies in an American bison from North Dakota. Journal of Wildlife Diseases 36:169–171.

Straily, A., and I. Trevino-Garrison. 2017. Knowledge, attitudes and practices of law enforcement officers on rabies and animal control issues in Kansas. Zoonoses and Public Health 64:111–117.

Stuchin, M., C. M. Machalaba, K. J. Olival, M. Artois, R. G. Bengis, P. Caceres, F. Diaz, E. Erlacher-Vindel, S. Forcella, F. A. Leighton, et al. 2018. Rabies as a threat to wildlife. Revue Scientifique et Technique 37:341–357.

Swanepoel, R., B. J. Barnard, C. D. Meredith, G. C. Bishop, G. K. Bruckner, C. M. Foggin, and O. J. Hubschle. 1993. Rabies in southern Africa. Onderstepoort Journal of Veterinary Research 60:325–346.

Tarantola, A. 2017. Four thousand years of concepts relating to rabies in animals and humans, its prevention and its cure. Tropical Medicine and Infectious Disease 2:5.

Te Kamp, V., C. M. Freuling, A. Vos, P. Schuster, C. Kaiser, S. Ortmann, A. Kretzschmar, S. Nemitz, E. Eggerbauer, R. Ulrich, et al. 2020. Responsiveness of various reservoir species to oral rabies vaccination correlates with differences in vaccine uptake of mucosa associated lymphoid tissues. Scientific Reports 10:2919.

Thulke, H., D. Eisinger, C. Freuling, A. Frohlich, A. Globig, V. Grimm, T. Müller, T. Selhorst, C. Staubach, and S. Zips. 2009. Situation-based surveillance: adapting investigations to actual epidemic situations. Journal of Wildlife Diseases 45:1089–1103.

Troupin, C., L. Dacheux, M. Tanguy, C. Sabeta, H. Blanc, C. Bouchier, M. Vignuzzi, S. Duchene, E. C. Holmes, and H. Bourhy. 2016. Large-scale phylogenomic analysis reveals the complex evolutionary history of rabies virus in multiple carnivore hosts. PLoS Pathogens 12:e1006041.

Ün, H., S. Eskiizmirliler, N. Ünal, C. Freuling, N. Johnson, A. R. Fooks, T. Müller, A. Vos, and O. Aylan. 2012. Oral vaccination of foxes against rabies in Turkey between 2008 and 2010. Berliner und Münchener Tierärztliche Wochenschrift 125:203–208.

van Zyl, N., W. Markotter, and L. H. Nel. 2010. Evolutionary history of African mongoose rabies. Virus Research 150:93–102.

Vigilato, M. A. N., B. Molina-Flores, V. J. Del Rio Vilas, J. C. Pompei, and O. Cosivi. 2018. Canine rabies elimination: governance principles. Revue Scientifique et Technique 37:703–709.

Vos, A., C. Freuling, S. Eskiizmirliler, H. Un, O. Aylan, N. Johnson, S. Gürbüz, W. Müller, N. Akkoca, T. Müller, et al. 2009. Rabies in foxes, Aegean region, Turkey. Emerging Infectious Diseases 15:1620–1622.

Vos, A., S. Ortmann, A. S. Kretzschmar, B. Köhnemann, and F. Michler. 2012. The raccoon as potential rabies reservoir species in Germany: a risk assessment. Berliner und Münchener Tierärztliche Wochenschrift 125:228–235.

Wachendörfer, G., R. Farrenkopf, W. Lohrbach, U. Förster, J. W. Frost, and W. A. Valder. 1978. Passageversuche mit einer Variante des Tollwut-Impfstammes ERA bei wildlebenden Spezies - Ein Beitrag zur oralen Immunisierung von Füchsen gegen Tollwut. Deutsche tierärztliche Wochenschrift 85:279–285. (In German.)

Walker, P. J., S. G. Siddell, E. J. Lefkowitz, A. R. Mushegian, E. M. Adriaenssens, D. M. Dempsey, B. E. Dutilh, B. Harrach, R. L. Harrison, R. C. Hendrickson, et al. 2020. Changes to virus taxonomy and the statutes ratified by the International Committee on Taxonomy of Viruses (2020). Archives of Virology 165:2737–2748.

Wallace, R. M., F. Cliquet, C. Fehlner-Gardiner, A. R. Fooks, C. T. Sabeta, A. Aguilar Setién, C. Tu, V. Vuta, B. Yakobson, D.-K. Yang, et al. 2020. Role of oral rabies vaccines in the elimination of dog-mediated human rabies deaths. Emerging Infectious Diseases 26(12).

Wallace, R. M., A. Gilbert, D. Slate, R. Chipman, A. Singh, W. Cassie, and J. D. Blanton. 2014. Right place, wrong species: a 20-year review of rabies virus cross species transmission among terrestrial mammals in the United States. PLoS One 9:e107539.

Walroth, R., N. Brown, A. Wandeler, A. Casey, and C. MacInnes. 1996. Rabid black bears in Ontario. Canadian Veterinary Journal 37:492.

Wandeler, A. I., S. Capt, H. Gerber, A. Kappeler, and R. Kipfer. 1988. Rabies epidemiology, natural barriers and fox vaccination. Parassitologia 30:53–57.

Wang, X., B. G. Werner, S. Smole, V. Pani, and L. L. Han. 2011. Signs observed among animal species infected with raccoon rabies variant virus, Massachusetts, USA, 1992–2010. Animals (Basel) 1:396–401.

Westergaard, J. M. 1982. Measures applied in Denmark to control the rabies epizootic in 1977–1980. Comparative Immunology, Microbiology and Infectious Diseases 5:383–387.

Willoughby, A., K. Phelps, and K. Olival. 2017. A comparative analysis of viral richness and viral sharing in cave-roosting bats. Diversity 9:35.

Winkler, W. G. 1988. Wildlife rabies: overview of ecology and epidemiology. Reviews of Infectious Diseases 10:604–605.

Winkler, W. G., and G. M. Baer. 1976. Rabies immunization of red foxes with vaccine in sausage baits. American Journal of Epidemiology 103:408–415.

Winkler, W. G., and K. Bögel. 1992. Control of rabies in wildlife. Scientific American 266:86–92.

World Health Organization (WHO). 1990. Report of A WHO/NVI Workshop on Arctic rabies, Uppsala, Sweden, 24–27 April 1990. World Health Organization, Geneva Switzerland

World Health Organization (WHO). 2018. WHO expert consultation on rabies, third report. World Health Organization Technical Report Series 1012:195.

Wright, N., D. Subedi, S. Pantha, K. P. Acharya, and L. H. Nel. 2021. The role of waste management in control of rabies: a neglected issue. Viruses 13:225.

Yakovchits, N. V., R. V. Adelshin, I. D. Zarva, S. A. Chupin, O. V. Melnikova, E. I. Andaev, M. I. Shulpin, A. E. Metlin, and A. D. Botvinkin. 2021. Fox rabies outbreaks in the Republic of Buryatia: connections with neighbouring areas of Russia, Mongolia and China. Transboundary and Emerging Diseases 68:427–434.

Zhao, H., Zhao, H., J. Zhang, C. Cheng, and Y.-H. Zhou. 2019*a*. Rabies acquired through mucosal exposure, China, 2013. Emerging Infectious Diseases 25:1028–1029.

Zhao, J.-H., L.-F. Zhao, F. Liu, H.-Y. Jiang, and J.-L. Yang. 2019*b*. Ferret badger rabies in Zhejiang, Jiangxi and Taiwan, China. Archives of Virology 164:579–584.

Zucca, P., M.-C. Rossmann, J. E. Osorio, K. Karem, P. De Benedictis, J. Haißl, P. De Franceschi, E. Calligaris, M. Kohlweiß, G. Meddi, et al. 2020. The "bio-crime model" of cross-border cooperation among veterinary public health, justice, law enforcements, and customs to tackle the illegal animal trade/bio-terrorism and to prevent the spread of zoonotic diseases among human population. Frontiers of Veterinary Science 7:593683.

Zulu, G. C., C. T. Sabeta, and L. H. Nel. 2009. Molecular epidemiology of rabies: focus on domestic dogs (*Canis familiaris*) and black-backed jackals (*Canis mesomelas*) from northern South Africa. Virus Research 140:71–78.

15 Rabbit Hemorrhagic Disease

J. JEFFREY ROOT, THOMAS GIDLEWSKI

Introduction

The etiologic agents of rabbit hemorrhagic disease (RHD) are rabbit hemorrhagic disease viruses (RHDVs). These highly contagious viruses are members of the viral family Caliciviridae, genus *Lagovirus* (Schoch et al. 2020) and negatively affect various lagomorph species, often causing high mortality rates in select taxa. While other lagomorph-associated caliciviruses exist, we focus on RHDV (also known as classical RHDV including subtype RHDVa) and RHDV2 (also know as subtype RHDVb and RHDV serotype 2) within this chapter, with a major emphasis on recent outbreaks of RHDV2 in wildlife in North America. Although the bulk of the research associated with environmental persistence, viral shedding, and transmission has been directed towards classical RHDV, RHDV2 is thought to have similar traits (USDA 2020*a*). As will be evident in this chapter, the environmental stability of RHDVs can be so long that the epidemiology of these viruses in wild populations can be quite complicated. This disease is a classic example of the dangers associated with the introduction and establishment of a foreign animal pathogen transmitted from domestic animals into wildlife (Tom Gidlewski, personal observation).

Rabbits, hares, and pika represent the major lagomorph taxa found throughout much of the world. Illustration by Laura Donohue.

Emergence of RHDV and RHDV2

An outbreak of a viral-induced disease in adult rabbits imported from Germany was reported in Jiangsu province of the People's Republic of China during early 1984 (Liu et al. 1984, Chasey 1994). This observation presumably represents the first documented case cluster of RHD in farmed rabbits. Likely based on the importation mentioned above, it has been suggested that RHDV emerged in Europe during the 1970s or 1980s (CFSPH 2020). The European rabbit (*Oryctolagus cuniculus*) is the source of dozens of breeds of commercial and pet rabbits, and RHDV initially caused disease in only this species. During the 1980s, the virus spread to other parts of Asia, Europe, North Africa, and to the New World (Abrantes et al. 2012).

While it did not become established at all locations where it was introduced, RHDV eventually affected countries in a nearly worldwide distribution (Abrantes et al. 2012) (Figure 15.1). Prior to the emergence of RHDV2, RHDV was thought to have been responsible for the mortality of nearly 250 million free-range European rabbits and domestic rabbits (McIntosh et al. 2007). This estimate from 2007 is now undoubtedly low. Other studies have indicated that between 1987 and 1990, RHDV was responsible for the death of millions of rabbits in Italy alone (Capucci and Lavazza 1998).

During the summer of 2010, a new serotype of RHDV emerged in France (Le Gall-Reculé et al. 2011), now known as RHDV2. Following the initial detection, the virus spread throughout Europe and reached Australia by 2015 (Mahar et al. 2018). The first detection of RHDV2 in North America occurred in Quebec, Canada, during 2016 in domestic rabbits associated with small farms (USDA 2019, Ambagala et al. 2021). In 2018, RHDV2 was confirmed in feral domestic rabbits in British Columbia, Canada (USDA 2018, Ambagala et al. 2021). That same year, a virus closely related to the 2018 Canadian virus was detected in a pet rabbit in Ohio, USA (USDA 2019). The following year, RHDV2 was detected in domestic and feral rabbits on Orcas Island, Washington,

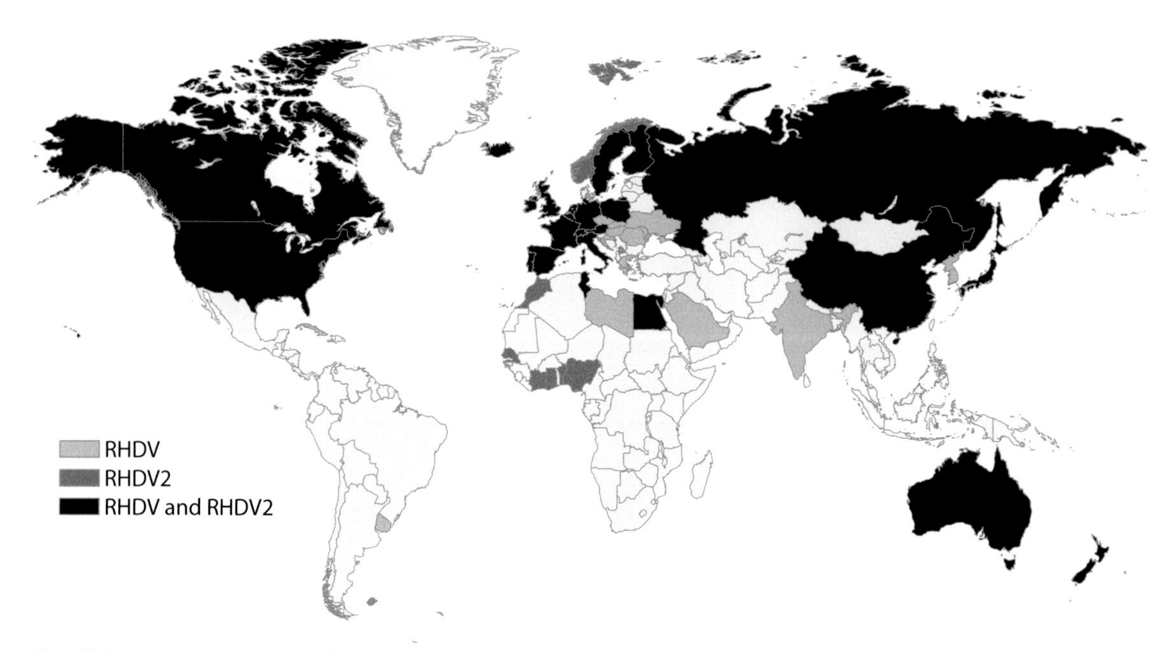

Fig. 15.1. Approximate global distribution of rabbit hemorrhagic disease virus (RHDV) and RHDV2 in the world as of April 2021.

USA (USDA 2019). During early 2020, RHDV2 was confirmed in a domestic rabbit in New York, USA. Subsequently, a major outbreak of RHDV2 in domestic rabbits and native North American lagomorphs, primarily including jackrabbits (*Lepus* spp.) and cottontails (*Sylvilagus* spp.), was reported from the southwestern United States and Mexico during the spring of 2020 (Figure 15.2). The initial detections in the United States were reported from New Mexico, USA. At the time of this writing, RHDV2 has been confirmed in more than 10 US states, as far east as Florida and as far north as Montana. Thus, it appears RHDV2 may still be expanding its distribution throughout North America (Duff et al. 2020), as susceptible wild and domestic hosts are widely distributed. RHDV2 may have a competitive advantage over RHDV in areas where RHDV has been previously established because RHDV2 has a broader host range, the capacity to infect young animals, and the ability to infect and cause mortality in rabbits that have antibodies to classical RHDV (Peacock et al. 2017, Taggart et al. 2022).

Rabbit Hemorrhagic Disease Virus Introductions

Precisely how RHDV and RHDV2 were introduced into new countries located significant distances from endemic countries is poorly understood, including the recent introduction of RHDV2 into North America. However, importation of rabbit meat from the People's Republic of China is believed to be one source (Gregg et al. 1991, McIntosh et al. 2007). Some introductions of RHDV have been accidental, such as from field trials on an island spilling over to the mainland (i.e., Australia); others have been illegal introductions (i.e., New Zealand) (O'Hara 2006, Effler 2015). It has been suggested that the rapid spread of RHDV2 to many countries would not have been possible without human assistance (Rouco et al. 2019). As outlined

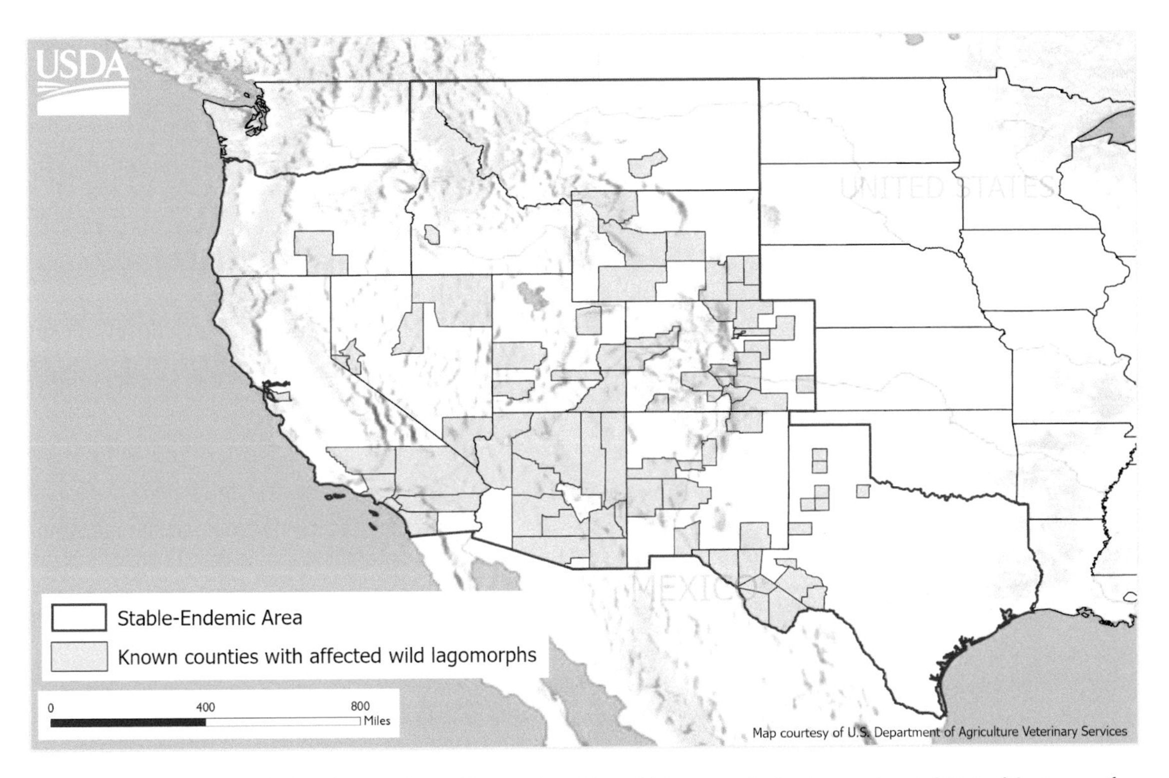

Fig. 15.2. Distribution of rabbit hemorrhagic disease virus 2 in wild lagomorphs in the continental United States as of 8 July 2021.

below, there are many anthropogenic and natural factors that could result in the dispersal of RHD viruses.

Etiology

While RHDV infections in wild lagomorph species are largely limited (e.g., primarily wild European rabbits), RHDV2 has a broader host range. For example, while RHDV did not cause disease in select experimentally infected North American lagomorphs (Lavazza et al. 2015, Mohamed et al. 2022), RHDV2 does cause mortality in many naturally infected North American lagomorph species including cottontails (*Sylvilagus* spp.) and jackrabbits (*Lepus* spp.). An additional major distinction between the two viral serotypes is that young rabbits (<6–8 weeks) tend to be resistant or subclinical in regards to RHDV but are susceptible to RHDV2 (2–3 weeks and older) (OIE 2019).

While multiple rabbit and hare species are susceptible to RHDV2, we are unaware of any experimental or field evaluations of this virus in pika (*Ochotona* spp.; see picture on page 259). Considering that some pika species (e.g., American pika; *O. princeps*) are thought to be negatively affected by climate change (Erb et al. 2011), and others are considered rare and endangered lagomorphs (e.g., Kozlov's pika, *O. koslowi*) (Lin et al. 2010), this virus could pose a serious threat to their fragile populations if they are susceptible to RHDV2. The rapid range expansion of RHDV2 throughout the United States suggests that the virus may have the capacity to reach alpine sites, although anthropogenic-associated transmission to these habitats could be lower than urban and suburban areas.

Demographics and Field Biology

Affected Species

While RHDV primarily affects wild and domestic European rabbits (note that a small number of cases have been reported in Iberian hares, *Lepus granatensis*; Lopes et al. 2014), RHDV2 has a broader host range and has had negative impacts on rabbits and hares in both the Old and New Worlds. Some examples of species affected by RHDV2 include domestic and feral European rabbits, desert cottontails (*Sylvilagus audubonii*), mountain cottontails (*S. nuttallii*), eastern cottontails (*S. floridanus*), brush rabbits (*S. bachmani*), riparian brush rabbits (*S. bachmani riparius*), black-tailed jackrabbits (*Lepus californicus*), antelope jackrabbits (*L. alleni*), and pygmy rabbits (*Brachylagus idahoensis*) in the New World (USDA 2020*b*, Mohamed et al. 2022) and wild, feral, and domestic European rabbits, brown hares (*Lepus europaeus*), Italian hares (*L. corsicanus*), Iberian hares, and cape hares (*L. capensis*) in the Old World (Puggioni et al. 2013, Camarda et al. 2014, Hall et al. 2017, Velarde et al. 2021) (Table 15.1). The fact that the viral host changed over time demonstrates the need to be ever vigilant to the adaptable nature of viruses (Tom Gidlewski, personal observation).

There have been reports of RHD viruses in species outside of the order Lagomorpha (e.g., mammalian orders Artiodactyla, Carnivora, Eulipotyphla, and Rodentia) (Calvete et al. 2019, Bao et al. 2020, Abade dos Santos et al. 2022). In some of these cases, it was possible to demonstrate RHD virus or RHD nucleic acid associated with lesions; however, cause and effect has not been experimentally demonstrated.

Climate

Based on field studies in Australia, RHDV has had a greater impact on rabbits occupying arid regions as compared to cooler, more humid regions; therefore, climate and temperature may play a role in regional differences in infection rates (Cooke 2002). It is unknown whether RHDV2 will show similar epidemiologic patterns in New World lagomorphs.

Seasonal Patterns

In parts of Europe, RHDV2 is more prevalent in late spring, while previous observations of RHDV cases (prior to its apparent replacement by RHDV2) were more common during the fall (Duff et al.

Table 15.1. Examples of lagomorph species that have been negatively affected by rabbit hemorrhagic disease virus 2 as of July 2021 (see page 262 for more examples.)

Common name	Scientific name	Status	Reference
European rabbit	*Oryctolagus cuniculus*	Wild, domestic, feral	Camarda et al. 2014, USDA 2020*b*
Desert cottontail	*Sylvilagus audubonii*	Wild	USDA 2020*b*
Mountain cottontail	*Sylvilagus nuttallii*	Wild	USDA 2020*b*
Eastern cottontail	*Sylvilagus floridanus*	Wild, experimental	USDA 2020*b*, Mohamed et al. 2022
Western brush rabbit	*Sylvilagus bachmani*	Wild	USDA 2021
Black-tailed jackrabbit	*Lepus californicus*	Wild	USDA 2020*b*
Antelope jackrabbit	*Lepus alleni*	Wild	USDA 2020*b*
Brown hare	*Lepus europaeus*	Wild	Hall et al. 2017, Velarde et al. 2017
Italian hare	*Lepus corsicanus*	Wild	Camarda et al. 2014
Cape hare	*Lepus capensis*	Wild	Puggioni et al. 2013
Iberian hare	*Lepus granatensis*	Wild	Velarde et al. 2021

2020). Surveillance of wildlife for RHDV2 infections in the United States has been based largely upon mortality events during 2020–2021. Currently, unless a new wild lagomorph species presents with a probable RHDV2 infection, cases in wildlife are generally not confirmed in counties that have already had positive wild lagomorphs or domestic rabbits. As a result, identifying temporal trends in cases in wildlife populations in the United States is difficult at present. However, domestic rabbit mortality events in the United States from February 2020 to June 2021 peaked in the spring, similar to what has been reported in Europe (see above; Figure 15.3).

Outbreak Signs in Field Settings

Outbreaks of RHD viruses may go unnoticed in rural environments. However, when enhanced surveillance is warranted, some simple techniques can be used to aid in the detection of rabbit carcasses. First, rabbit carcasses, especially in large numbers, will release volatile organic compounds indicative of "the smell of death." Thus, investigating the cause of the odors of decaying flesh can prove useful in finding rabbit carcasses. Second, the behavior of scavenging birds can also aid in finding lagomorph mortalities, especially when animals have recently expired. Similarly, observations of increased numbers of mammalian scavengers could aid in the detection of rabbit carcasses, but they may not be as visible as scavenging birds. Depending on how quickly carcasses are consumed, scavenging animals can also hinder searches for deceased lagomorphs in field settings.

Clinical Signs

The incubation period for RHDV2 has been estimated to range from 3–5 days (CFSPH 2020), although shorter incubation periods have been noted in experimental studies (Mohamed et al. 2022). Clinical signs of lagomorphs infected with RHD viruses can consist of ataxia, inappetence, dullness, and respiratory and neurological signs (Duff et al. 2020). Highly susceptible rabbits infected with RHD viruses may show few, if any, clinical signs prior to death. The rapid course of RHD typically results in carcasses in good physical condition (OIE 2019). In experimental settings, animals can demonstrate no clinical signs until they are discovered moribund or dead, and some rabbits die with fresh grass in their mouths, suggesting that they had eaten a short time prior to death (Cooke 2002, Mohamed et al. 2022). Bloody nares have been noted in experimentally infected cottontails that have succumbed to RHDV2 infection (Mohamed et al. 2022) and have been suggested as an indicator of RHDV in various species (Cooke 2002). Because of this striking sign, bloody discharge from the nose or mouth may be

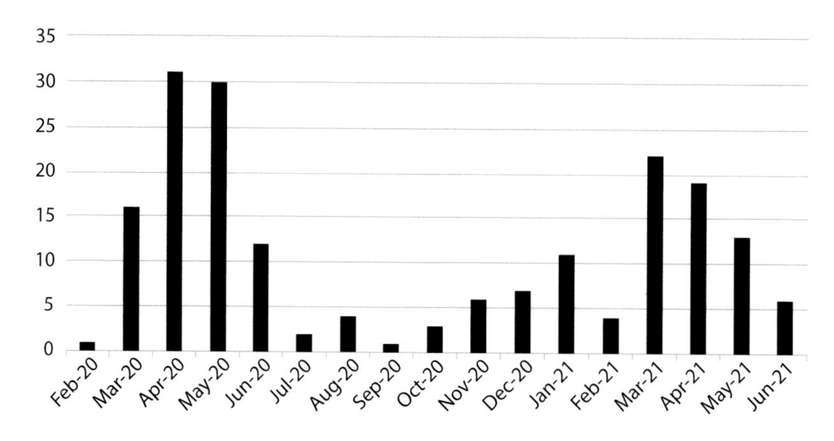

Fig. 15.3. Approximate monthly rabbit hemorrhagic disease virus 2 (RHDV2) domestic rabbit mortality events reported in the United States from February 2020 to June 2021. This timeframe represents the initial detections and the expansion of the RHDV2 outbreak that was first documented in the southwestern United States. This figure is based upon the mortality start dates of mortality events reported to the US Department of Agriculture, Animal and Plant Health Inspection Service.

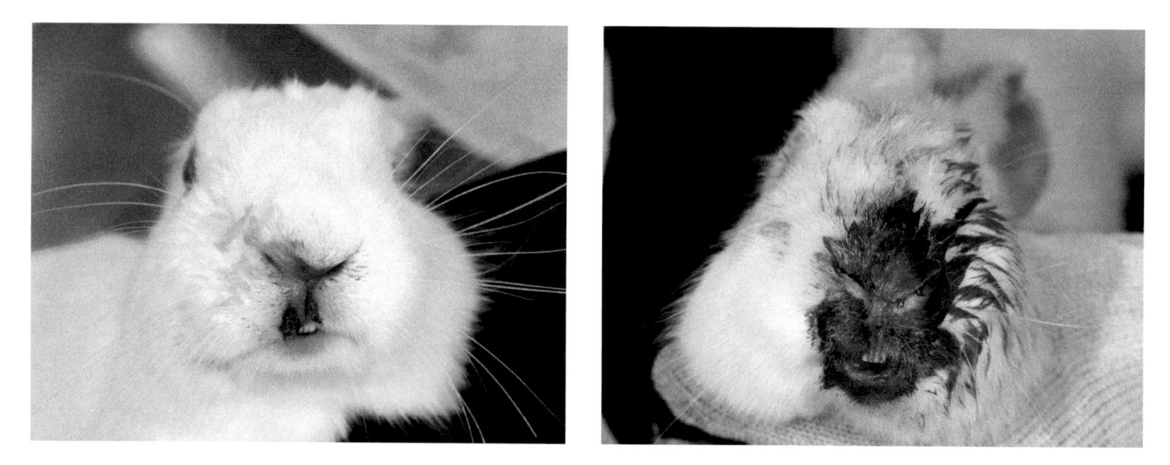

Fig. 15.4. Bloody discharge (i.e., epistaxis) associated with the nares of rabbits that succumbed to rabbit hemorrhagic disease. Images by Fawzi Mohamed.

a useful field indicator when no apparent signs of trauma are noted (Figure 15.4).

Pathogenesis and Pathology

The primary RHD lesions in rabbits are acute necrotizing hepatitis (Abrantes et al. 2012), splenic enlargement, and hemorrhages caused by disseminated intravascular coagulation (Henning et al. 2005, OIE 2019). Concurrent experimental infections of New Zealand white rabbits and eastern cottontails with RHDV2 demonstrated very similar lesions in both species (Mohamed et al. 2022). The onset of clinical signs and death appeared 24–36 hours sooner in New Zealand white rabbits inoculated with RHDV2 compared to eastern cottontail rabbits (Mohamed et al. 2022).

Diagnosis

The definitive diagnostic tests for RHD infection in US lagomorphs are the antigen enzyme-linked immunosorbent assay (AG-ELISA) and reverse transcription-polymerase chain reaction (RT-PCR).

Some specific AG-ELISAs can distinguish which RHD virus may be present. The RT-PCR assay distinguishes between RHDV and RHDV2 (OIE 2021*b*). Electron microscopy, hemagglutination testing, immunostaining, western blot, and rabbit inoculation can also be used to identify virus. Diagnostic assays are limited, as neither RHDV nor RHDV2 can be grown in cell culture (CFSPH 2020); the virus can only be propagated by animal inoculation.

The highest titers of virus are found in the livers of infected animals during acute or peracute disease. Virus can also be detected in spleen, serum, and various bodily excretions (OIE 2019). Liver is usually considered the best diagnostic tissue for collection from intact rabbit and hare carcasses to diagnose RHD.

Outcome and Treatment

There is no practical treatment for RHD at this time; however, the potential use of small-molecule inhibitors as a treatment has been reported (Perera et al. 2022). Nonetheless, not all animals infected with these viruses succumb to their infections. For example, survival in both field and experimental infection studies have been reported for RHDV (Cooke 2002). In addition, 2 of 5 (40%) eastern cottontails (*Sylvilagus floridanus*) experimentally infected with RHDV2 survived infection (Mohamed et al. 2022). Among animals that survive, it is unknown how long those animals maintain infectious virus (OIE 2019). Based upon a highly sensitive PCR assay, however, viral RNA persistence (which does not always indicate the presence of live virus) has been documented for up to two months for rabbits that have recovered from RHD (OIE 2019). It would be of interest to know if New World lagomorphs that recover from RHDV2 infections have similar survival probabilities as compared to naïve animals (Figure 15.5). If New World lagomorph survival following RHDV-2 infection is confirmed in multiple species, serosurveys coupled with survival analyses could be used to help address the long-term impacts of this virus on wild lagomorph populations.

Management

The management of RHD viruses in wild lagomorph populations poses many challenges. The populations of certain lagomorphs (e.g., rabbits and hares) can be very large, animals within these populations can have overlapping home ranges, and the perimeter of a population is not readily definable unlike other wild mammalian disease systems (e.g., prairie dogs, *Cynomys* spp., and *Yersinia pestis*). The management of RHD viruses is complicated by the stability of the virus in rabbit carcasses and meat. This suggests mechanical vectoring by humans (e.g., movement of the virus on

Fig. 15.5. A pair of desert cottontails (*Sylvilagus audubonii*) foraging within close proximity to each other. As rabbit hemorrhagic disease virus 2 (RHDV2) is highly contagious and environmentally stable for long periods of time, close contact and/or environmental contamination may facilitate transmission. Experimental infections suggest that some *Sylvilagus* spp. exposed to RHDV2 may survive their infections (Mohamed et al. 2022).

clothes, shoes, and equipment) and by predators and scavengers, as well as mechanical transmission from insects may all contribute to virus spread and new outbreaks (Chasey 1994, Asgari et al. 1998, Cooke 2002, McColl et al. 2002, OIE 2019) (Figure 15.6). In many ways, foreign animal disease introduction and establishment in wildlife is a worst-case scenario. Our management options are severely limited, and we very well may have an established reservoir forever (Thomas Gidlewski, personal observation).

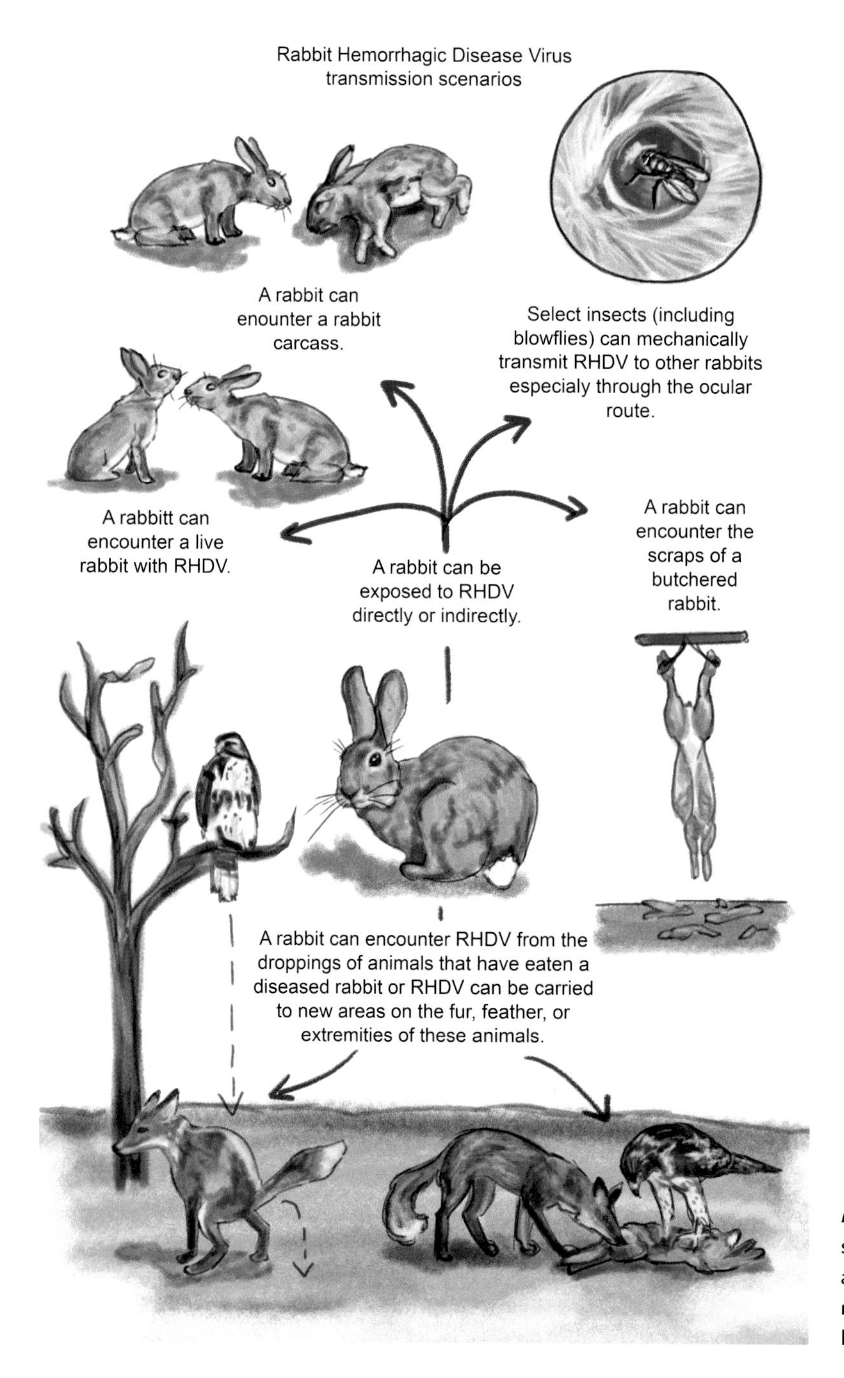

Fig. 15.6. Several transmission scenarios have been proposed and are possible for rabbit hemorrhagic disease viruses (RHDVs). Illustration by Laura Donohue.

Environmental Stability

The environmental stability of RHD viruses presents a major challenge to management. Homogenized liver supernatant from samples collected from rabbit carcasses at 20 days post-mortem contained sufficient live virus to kill naïve rabbits (McColl et al. 2002). Laboratory experiments conducted at cool temperatures (4°C) have indicated that the virus can survive for at least 225 days in organ suspensions; survival times of at least 2 days were noted for organ suspensions and virus placed in a dry state when temperatures were increased to 60°C (Šmíd et al. 1991). An additional study reported that RHDV can remain viable in animal tissues for at least three months in field settings, but dried excreted virus remained viable for a shorter period (e.g., <1 month) (Henning et al. 2005). Overall, these studies suggest that the long-term survival of RHDV in carcasses and tissues may help to maintain RHDV in lagomorph populations, even in the absence of active infections (Henning et al. 2005). Indeed, assuming optimal conditions for virus survival, it is conceivable that a wild population could become reinfected months after an epizootic has ended.

Virus Dispersion by Non-lagomorph Species

While they are not thought to have the capacity to replicate the virus (but see recent information on this topic in the affected species subsection above), predators and scavengers can excrete RHDV in fecal material following the consumption of infected rabbits (OIE 2019). It has been suggested that RHDV may survive on feet, claws, and regurgitated materials of scavenging birds (Chasey 1994). Thus, it is reasonable to assume that the same scenario may be possible for RHDV2. Depending on the mobility of the animal species in question, this represents a possible mechanism for the moderate to reasonably long-distance movement of the virus. While the chance of this type of transmission event may be small (Chasey 1994), raptors predating or scavenging on RHD infected rabbits or carcasses and subsequently moving the virus significant distances could help explain the "jumps" in the distribution of known RHDV2 cases in the United States during early 2020 to 2021. While natural transmission pathways have certainly expanded the distribution of RHDV2, anthropogenic pathways likely played a significant role in spread in the United States.

Mechanical Transmission from Insects

Insects (e.g., mechanical vectors), as well as the microclimates and vegetative characteristics that influence their density, have been suggested to be involved in small-scale patterns of spread of RHDV (Henning and Davies 2005). For example, various insects were suggested as potential mechanical vectors for the movement of RHDV out of a quarantine enclosure on Wardang Island, Australia (Cooke 2002). Furthermore, oral and anal excretions from flies have been suggested as a potential transmission mechanism to rabbits via the oral or conjunctival routes (Asgari et al. 1998). Overall, many insect species and taxonomic groups can be contaminated with RHDV (Cooke 2002).

Citizen Science

Considering that moribund rabbits and rabbit carcasses can be quickly removed from the landscape by predators and scavengers and that many rabbits may die underground (Duff et al. 2020), outbreaks of RHD could go unnoticed in some situations. This provides an excellent opportunity for citizen science to play a significant role in wildlife management, as few agencies have the staff needed to survey the vast habitat of native rabbit species. Outreach to encourage the general public to report dead and moribund rabbits could be a high priority for any region with lagomorphs that may become infected. Citizen science in the form of organized hikers, hunters, and other types of outdoor enthusiasts with a prearranged on-line reporting system could be extremely valuable for the early detection of case clusters. Notably, citizen science efforts from rabbit hunting events in New Zealand proved to be a useful and inexpensive method to obtain large quantities of data from a vast geographic region regarding potential RHDV effects in rabbit populations (Rouco et al. 2014).

Field Biosecurity

When working with wild lagomorphs, it is important to consider the potential for people and their equipment to spread RHD viruses. At a minimum, equipment should be decontaminated, and clothing should be sanitized whenever the location of study sites changes. RHDV can be inactivated with a household bleach solution diluted with water with appropriate contact time (USDA 2020*a*). See additional guidance on this topic (e.g., disinfectants, dilutions, contact time, and personal protective equipment) from the US Department of Agriculture (USDA 2020*a*).

During the US RHDV2 outbreak, field necropsies of rabbit carcasses have been discouraged to avoid the unintentional spread of the virus; rather, it has been recommended that carcasses be contained in at least two layers and transferred to a qualified laboratory with appropriate biocontainment for necropsy (USGS 2020). Considering that not every carcass discovered during a large mortality event will be submitted for diagnostic evaluation as well as the fact that RHD viruses can survive in carcasses for long periods (Henning et al. 2005), managing a large number of carcasses in an outbreak situation may become an issue. Some suggestions for dealing with carcasses include incineration or burial sufficient in depth to inhibit access by scavengers (USGS 2020). Caution should be exercised in dealing with carcass removal, packaging, or disposal, as rabbits are frequently infected with other pathogens, some of which are zoonotic (e.g., *Francisella tularensis* and other pathogens). Rabbit ectoparasites that can spread pathogens will leave to find a new host when a carcass begins to cool.

Alternative Surveillance Systems

There may be value in utilizing carnivore sera as a means to assess the previous presence of RHD in rabbit populations following a rabbit population decline or when rabbits are not available to sample. Red fox (*Vulpes vulpes*) given oral doses of homogenized liver from rabbits that succumbed to RHD developed antibody responses that lasted for at least six months in some individuals (Leighton et al. 1995). Thus, assuming availability of appropriate serological assays, this surveillance method could be used to assess previous RHD activity at locations where lagomorph populations have experienced die-offs for unknown reasons. Serological pathogen surveillance programs in wild canids (e.g., coyotes, *Canis latrans*) have previously been used in the United States to estimate the prevalence of *Yersinia pestis* in various ecosystems (USDA 2010). Assuming the continued surveillance of coyotes for multiple pathogens and that they develop an antibody response, adding RHDV2 to the list of pathogens monitored could be a cost-effective means for large-scale surveillance.

Vaccination

No licensed RHDV vaccines currently exist in the United States, but two killed vaccines (Filavac VHD K C+V and Eravac RHDV-2) from the European Union may be imported for use under special permit (USDA 2020*c*). However, as of the fall of 2021, a US vaccine manufacturer has received emergency use authorization for an experimental RHDV2 vaccine from the USDA. Injectable vaccines may have utility for protecting species or populations of concern to a limited extent (e.g., populations in enclosures, islands, or small populations). Strategic use of one of the European vaccines to protect listed and sensitive cottontail populations like riparian brush rabbits

(*Sylvilagus bachmani riparius*) began in 2021. In the absence of vaccines adapted for oral vaccination through a baiting program, the widespread use of vaccination to control RHD viruses in wildlife populations will remain limited.

Financial, Legal, and Political Factors

Financial

In Europe, RHDV2 has been responsible for significant economic losses to commercial rabbit production and to geographic regions where rabbit hunting has a large influence on the local economy (Rouco et al. 2019). While the rabbit industry in the United States is smaller in size than in some other countries, it is estimated to be worth over $2 billion (USD), the bulk of which is associated with pet rabbits and associated supplies (USDA 2020*b*) (Figure 15.7).

Legal

At present, rabbit hemorrhagic disease is listed as a notifiable terrestrial animal disease by the World Organization for Animal Health (OIE 2021*a*). Within the United States, state and federal agricultural authorities should be informed of RHDV cases without delay (CFSPH 2020). Because RHDV2 affects both domestic and wild lagomorphs, state, tribal, and federal wildlife officials should be informed of any suspected or confirmed cases immediately.

In addition to the health of lagomorph populations, decimated populations of these species could negatively affect other trophic levels, especially predators that rely on rabbits and hares as a major food source. In Mediterranean Spain, the reduction in European rabbit populations (for which RHD viruses were thought to be at least partially responsible) has been implicated as a major threat to endangered predators that specialize on rabbits as a food source in this region (Moreno et al. 2004). Similarly, drastic declines in lagomorph populations in key regions in the United States could negatively affect select predators, potentially leading to changes in legal status under state and federal laws. For example, a preferred food of lynx (*Lynx canadensis*) is often snowshoe hares (*Lepus americanus*) (O'Donoghue et al. 1998).

Fig. 15.7. Domestic rabbits are popular pets in many countries. In situations where domestic rabbits are housed outdoors, wild and domestic rabbits could interact and spread pathogens to each other. Illustration by Laura Donohue.

Political

Political factors associated with RHD viruses are more prevalent in countries or regions where the virus has been accidently, intentionally, or illegally introduced. Prior to 2020, RHDV2 was considered a foreign animal disease in the United States. However, now that it has become established in wildlife populations, RHDV2 is considered to be endemic and stable in certain US locations. Without the intervention of large-scale oral vaccination or other control programs, native lagomorph populations will continue to be at risk. Importantly, the United States has multiple lagomorph species that are of conservation concern. If RHDV2 were to severely affect these fragile populations, they could be at risk of extinction, a situation likely to provoke political action. A few potential impacts on policy and public resource use could involve restriction of public access and changes in hunting seasons and bag limits.

Summary

With RHDV2 now established in the United States and spreading among and within wild rabbit and hare populations, population and ecosystem level impacts (including conservation issues involving rare and sensitive subspecies) may be realized. Currently, the only practical management tools available are vaccines, and as yet there is no broad scale ability to utilize them, because capture–vaccinate–release of individual animals is the only option at this time. It remains to be seen whether surviving lagomorph populations will develop tolerance or natural immunity.

ACKNOWLEDGMENTS

We are indebted to J. Ellis (USDA National Wildlife Research Center) and J. Ringenberg (USDA National Wildlife Disease Program) for assistance with literature searches; J. Ellis, S. Maroney (USDA Veterinary Services), and M.J. McCool (USDA Veterinary Services) for assistance with maps; and F. Mohamed (USDA Veterinary Services) and the USDA RHDV2 Coordination cell for their review of a previous version of this chapter. The salaries for the authors of this chapter were provided by the US Department of Agriculture, Animal and Plant Health Inspection Service.

LITERATURE CITED

Abade dos Santos, F. A., A. Pinto, T. Burgoyne, K. P. Dalton, C. L. Carvalho, D. W. Ramilo, C. Carneiro, T. Carvalho, M. C. Peleteiro, F. Parra, et al. 2022. Spillover events of rabbit haemorrhagic disease virus 2 (recombinant GI.4P–GI.2) from Lagomorpha to Eurasian badger. Transboundary and Emerging Diseases 69:1030–1045.

Abrantes, J., W. van der Loo, J. Le Pendu, and P. J. Esteves. 2012. Rabbit haemorrhagic disease (RHD) and rabbit haemorrhagic disease virus (RHDV): a review. Veterinary Research 43:12.

Ambagala, A., H. Schwantje, S. Laurendeau, H. Snyman, T. Joseph, B. Pickering, K. Hooper-McGrevy, S. Babiuk, E. Moffat, L. Lamboo, et al. 2021. Incursions of rabbit haemorrhagic disease virus 2 in Canada—clinical, molecular and epidemiological investigation. Transboundary and Emerging Diseases 68:1711–1720.

Asgari, S., J. R. E. Hardy, R. G. Sinclair, and B. D. Cooke. 1998. Field evidence for mechanical transmission of rabbit haemorrhagic disease virus (RHDV) by flies (Diptera: Calliphoridae) among wild rabbits in Australia. Virus Research 54:123–132.

Bao, S., K. An, C. Liu, X. Xing, X. Fu, H. Xue, F. Wen, X. He, and J. Wang. 2020. Rabbit hemorrhagic disease virus isolated from diseased alpine musk deer (*Moschus sifanicus*). Viruses 12:897.

Calvete, C., M. Mendoza, M. P. Sarto, M. P. J. de Bagues, L. Lujan, J. Molın, A. J. Calvo, F. Monroy, and J. H. Calvo. 2019. Detection of rabbit hemorrhagic disease virus Gi.2/RHDV2/b in the Mediterranean pine vole (*Microtus duodecimcostatus*) and white-toothed shrew (*Crocidura russula*). Journal of Wildlife Diseases 55:467–472.

Camarda, A., N. Pugliese, P. Cavadini, E. Circella, L. Capucci, A. Caroli, M. Legretto, E. Mallia, and A. Lavazza. 2014. Detection of the new emerging rabbit haemorrhagic disease type 2 virus (RHDV2) in Sicily from rabbit (*Oryctolagus cuniculus*) and Italian hare (*Lepus corsicanus*). Research in Veterinary Science 97:642–645.

Capucci, L., and A. Lavazza. 1998. A brief update on rabbit hemorrhagic disease virus. Emerging Infectious Diseases 4:343–344.

Center for Food Security and Public Health (CFSPH). 2020. Rabbit hemorrhagic disease and other lagoviruses. The

Center for Food Security and Public Health, Iowa State University, Ames, USA. https://www.cfsph.iastate.edu/Factsheets/pdfs/rabbit_hemorrhagic_disease.pdf. Accessed 15 Jan 2021.

Chasey, D. 1994. Possible origin of rabbit haemorrhagic disease in the United Kingdom. Veterinary Record 135:496–499.

Cooke, B. D. 2002. Rabbit haemorrhagic disease: field epidemiology and the management of wild rabbit populations. OIE Revue Scientifique et Technique 21:347–358.

Duff, P., C. Fenemore, P. Holmes, B. Hopkins, J. Jones, M. He, D. Everest, and M. Rocchi. 2020. Rabbit haemorrhagic disease: a re-emerging threat to lagomorphs. Veterinary Record 187:106–107.

Effler, P. 2015. Australia's war against rabbits: the story of rabbit haemorrhagic disease. Emerging Infectious Diseases 21:735.

Erb, L. P., C. Ray, and R. Guralnick. 2011. On the generality of a climate-mediated shift in the distribution of the American pika (*Ochotona princeps*). Ecology 92:1730–1735.

Gregg, D. A., C. House, R. Meyer, and M. Berninger. 1991. Viral haemorrhagic disease of rabbits in Mexico: epidemiology and viral characterization. OIE Revue Scientifique et Technique 10:435–451.

Hall, R., D. Peacock, J. Kovaliski, J. Mahar, R. Mourant, M. Piper, and T. Strive. 2017. Detection of RHDV2 in European brown hares (*Lepus europaeus*) in Australia. Veterinary Record 180:121.

Henning, J., and P. Davies. 2005. A serial cross-sectional study of the prevalence of rabbit haemorrhagic disease on three farms in the lower North Island of New Zealand. New Zealand Veterinary Journal 53:149–153.

Henning, J., J. Meers, P. R. Davies, and R. S. Morris. 2005. Survival of rabbit haemorrhagic disease virus (RHDV) in the environment. Epidemiology and Infection 133:719–730.

Lavazza, A., P. Cavadini, I. Barbieri, P. Tizzani, A. Pinheiro, J. Abrantes, P. J. Esteves, G. Grilli, E. Gioia, M. Zanoni, et al. 2015. Field and experimental data indicate that the eastern cottontail (*Sylvilagus floridanus*) is susceptible to infection with European brown hare syndrome (EBHS) virus and not with rabbit haemorrhagic disease (RHD) virus. Veterinary Research 46:13.

Le Gall-Reculé, G., F. Zwingelstein, S. Boucher, B. Le Normand, G. Plassiart, Y. Portejoie, A. Decors, S. Bertagnoli, J. L. Guérin, and S. Marchandeau. 2011. Virology: detection of a new variant of rabbit haemorrhagic disease virus in France. Veterinary Record 168:137–138.

Leighton, F. A., M. Artois, L. Capucci, D. Gavier-Widén, and J. P. Morisse. 1995. Antibody response to rabbit viral hemorrhagic disease virus in red foxes (*Vulpes vulpes*) consuming livers of infected rabbits (*Oryctolagus cuniculus*). Journal of Wildlife Diseases 31:541–544.

Lin, G., H. Ci, S. J. Thirgood, T. Zhang, and J. Su. 2010. Genetic variation and molecular evolution of endangered Kozlov's pika (*Ochotona koslowi* Büchner) based on mitochondrial cytochrome B gene. Polish Journal of Ecology 58:563–568.

Liu, S. J., H. P. Xue, B. Q. Pu, and N. H. Qian. 1984. A new viral disease in rabbits. Animal Husbandry and Veterinary Medicine 16:253–255.

Lopes, A. M., S. Marques, E. Silva, M. J. Magalhães, A. Pinheiro, P. C. Alves, J. Le Pendu, P. J. Esteves, G. Thompson, and J. Abrantes. 2014. Detection of RHDV strains in the Iberian hare (*Lepus granatensis*): earliest evidence of rabbit lagovirus cross-species infection. Veterinary Research 45:94.

Mahar, J. E., R. N. Hall, D. Peacock, J. Kovaliski, M. Piper, R. Mourant, N. Huang, S. Campbell, X. Gu, A. Read, et al. 2018. Rabbit hemorrhagic disease virus 2 (RHDV2; GI.2) is replacing endemic strains of RHDV in the Australian landscape within 18 months of its arrival. Journal of Virology 92:e01374-17.

McColl, K. A., C. J. Morrissy, B. J. Collins, and H. A. Westbury. 2002. Persistence of rabbit haemorrhagic disease virus in decomposing rabbit carcases. Australian Veterinary Journal 80:298–299.

McIntosh, M. T., S. C. Behan, F. M. Mohamed, Z. Lu, K. E. Moran, T. G. Burrage, J. G. Neilan, G. B. Ward, G. Botti, L. Capucci, et al. 2007. A pandemic strain of calicivirus threatens rabbit industries in the Americas. Virology Journal 4:96.

Mohamed, F. M., T. Gidlewski, M. L. Berninger, H. M. Petrowski, A. J. Bracht, C. Bravo de Rueda, M. Grady, E. S. O'Hearn, C. E. Lewis, K. E. Moran, et al. 2022. Comparative susceptibility of eastern cottontails and New Zealand white rabbits to classical rabbit hemorrhagic disease virus (RHDV) and RHDV2. Transboundary and Emerging Diseases 69:e968–e978.

Moreno, S., R. Villafuerte, S. Cabezas, and L. Lombardi. 2004. Wild rabbit restocking for predator conservation in Spain. Biological Conservation 118:183–193.

O'Donoghue, M., S. Boutin, C. J. Krebs, D. L. Murray, and E. J. Hofer. 1998. Behavioural responses of coyotes and lynx to the snowshoe hare cycle. Oikos 82:169–183.

Office International des Epizooties (OIE). 2019. Rabbit haemorrhagic disease. OIE Technical Disease Cards. World Organisation for Animal Health, Paris, France.

Office International des Epizooties (OIE). 2021a. OIE-listed diseases, infections, and infestations in force in 2021. World Organisation for Animal Health, Paris, France.

Office International des Epizooties (OIE). 2021b.. Manual of diagnostic tests and vaccines for terrestrial animals 2021. World Organisation for Animal Health, Paris, France.

O'Hara, P. 2006. The illegal introduction of rabbit haemorrhagic disease virus in New Zealand. OIE Revue Scientifique et Technique 25:119–123.

Peacock, D., J. Kovaliski, R. Sinclair, G. Mutze, A. Iannella, and L. Capucci. 2017. RHDV2 overcoming RHDV immunity in wild rabbits (*Oryctolagus cuniculus*) in Australia. Veterinary Record 180:280.

Perera, K.D., D. Johnson, S. Lovell, W. C. Groutas, K.-O. Chang, and Y. Kim. 2022. Potent protease inhibitors of highly pathogenic lagoviruses: rabbit hemorrhagic disease virus and European brown hare syndrome virus. Microbiology Spectrum 10:e00142-22

Puggioni, G., P. Cavadini, C. Maestrale, R. Scivoli, G. Botti, C. Ligios, G. Le Gall-Reculé, A. Lavazza, and L. Capucci. 2013. The new French 2010 *Rabbit Hemorrhagic Disease Virus* causes an RHD-like disease in the Sardinian Cape hare (*Lepus capensis mediterraneus*). Veterinary Research 44:96.

Rouco, C., J. A. Aguayo-Adán, S. Santoro, J. Abrantes, and M. Delibes-Mateos. 2019. Worldwide rapid spread of the novel rabbit haemorrhagic disease virus (GI. 2/ RHDV2/b). Transboundary and Emerging Diseases 66:1762–1764.

Rouco, C., G. Norbury, and D. Ramsay. 2014. Kill rates by rabbit hunters before and 16 years after introduction of rabbit haemorrhagic disease in the southern South Island, New Zealand. Wildlife Research 41:136–140.

Schoch, C. L., S. Ciufo, M. Domrachev, C. L. Hotton, S. Kannan, R. Khovanskaya, D. Leipe, R. McVeigh, K. O'Neill, B. Robbertse, et al. 2020. NCBI taxonomy: a comprehensive update on curation, resources and tools. Database (Oxford) 2020:baaa062.

Šmíd, B., L. Valíček, L. Rodák, J. Štěpánek, and E. Jurák. 1991. Rabbit haemorrhagic disease: an investigation of some properties of the virus and evaluation of an inactivated vaccine. Veterinary Microbiology 26:77–85.

Taggart, P. L., R. N. Hall, T. E. Cox, J. Kovaliski, S. R. McLeod, and T. Strive. 2022. Changes in virus transmission dynamics following the emergence of RHDV2 shed light on its competitive advantage over previously circulating variants. Transboundary and Emerging Diseases 69:1118–1130.

US Department of Agriculture (USDA). 2010. Plague surveillance update. USDA Animal and Plant Health Inspection Service, Riverdale, Maryland, USA.

US Department of Agriculture (USDA). 2018. Emerging risk notice—animal health: rabbit hemorrhagic disease in British Columbia, Canada. USDA Animal Plant and Health Inspection Service, Riverdale, Maryland, USA.

US Department of Agriculture (USDA). 2019. Emerging risk notice: Rabbit hemorrhagic disease virus, serotype 2. USDA Animal and Plant Health Inspection Service, Riverdale, Maryland, USA.

US Department of Agriculture (USDA). 2020*a*. General guidance for cleaning and disinfection of rabbit hemorrhagic disease virus (RHDV) contaminated premises. USDA Animal and Plant Health Inspection Service, Riverdale, Maryland, USA.

US Department of Agriculture (USDA). 2020*b*. Rabbit hemorrhagic disease virus, type 2. USDA Animal and Plant Health Inspection Service, Riverdale, Maryland, USA.

US Department of Agriculture (USDA). 2020*c*. RHDV2 vaccine frequently asked questions (FAQs). USDA Animal and Plant Health Inspection Service, Riverdale, Maryland, USA.

US Department of Agriculture (USDA). 2021. Situation Repot # 32: Mutlitstate outbreak of rabbit hemorrhagic disease virus 2 (RHDV2). USDA Animal and Plant Health Inspection Service, Riverdale, Maryland, USA.

US Geological Survey (USGS). 2020. Continued expansion of rabbit hemorrhagic disease virus 2 in North America and additional instructions regarding mortality event investigations. USGS, National Wildlife Health Center, Wildlife Health Bulletin 2020-05, Madison, Wisconsin, USA.

Velarde, R., J. Abrantes, A. M. Lopes, J. Estruch, J. V. Côrte-Real, P. J. Esteves, I. García-Bocanegra, J. Ruiz-Olmo, and C. Rouco. 2021. Spillover event of recombinant *Lagovirus europaeus*/GI.2 into the Iberian hare (*Lepus granatensis*) in Spain. Transboundary and Emerging Diseases 68:3187–3193.

Velarde, R., P. Cavadini, A. Neimanis, O. Cabezón, M. Chiari, A. Gaffuri, S. Lavín, G. Grilli, D. Gavier-Widén, A. Lavazza, et al. 2017. Spillover events of infection of brown hares (*Lepus europaeus*) with rabbit haemorrhagic disease type 2 virus (RHDV2) caused sporadic cases of an European brown hare syndrome-like disease in Italy and Spain. Transboundary and Emerging Diseases 64:1750–1761.

16

White-nose Syndrome in Bats

Conservation, Management, and Context-dependent Decision Making

MICHELLE L. VERANT, RILEY F. BERNARD

Introduction

White-nose syndrome (WNS) is a disease that affects hibernating bats and is caused by the fungal pathogen *Pseudogymnoascus destructans* (*Pd*; Lorch et al. 2011, Minnis and Lindner 2013). This cold-loving fungus infects unfurred skin of bats during hibernation, appearing as white fuzzy material on the wings, ears, and muzzle. The disease was first discovered in New York in 2006–2007 following mass mortality of bats observed during winter hibernacula surveys (Blehert et al. 2009). There is evidence the fungus was present along the east coast prior to 2005, with significant geographic spread occurring before the first dead bats were observed (Thapa et al. 2021). *Pseudogymnoascus destructans* is presumed to have been introduced to North America from Eurasia via human activities, where it is widespread on bats and within caves, yet does not cause substantial bat mortality or population declines in European or Asian species (Hoyt et al. 2020).

Since its introduction to North America, *Pd* has continued to spread across the continent, infecting over a dozen species of bats (USFWS 2021*b*). Seasonal migrations and within-season movements of bats are primarily responsible for transmission of *Pd* through contact between bats or with environmental

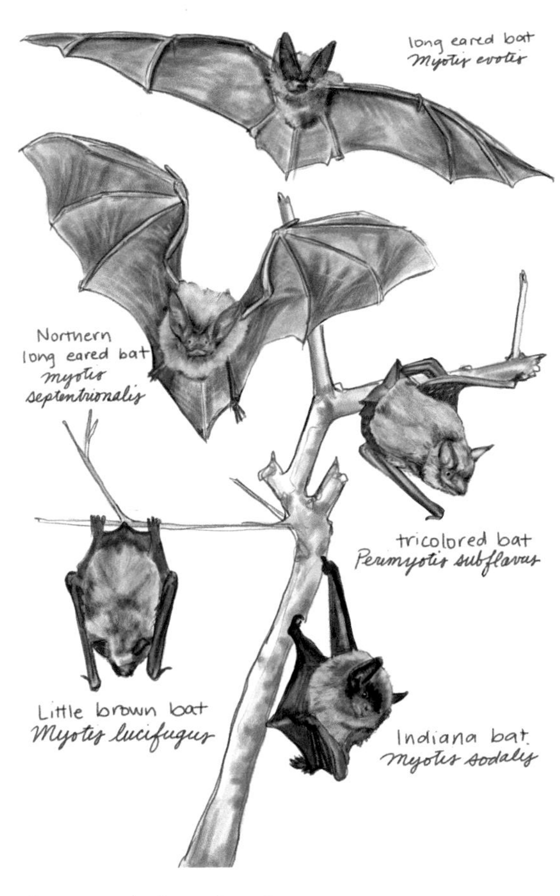

Illustration by Laura Donohue.

reservoirs, such as hibernacula (Miller-Butterworth et al. 2014, Langwig et al. 2021). However, long-distance spread events, such as the appearance of WNS in Washington in 2016, approximately 2,100 km from the nearest known occurrence of the disease (Lorch et al. 2016), are presumably human mediated. *Pseudogymnoascus destructans* has been detected on clothing and gear (e.g., mist nets) after contact with WNS-affected bats or environmental surfaces within hibernacula (e.g., caves and mines) indicating potential for transport of fungal spores between sites (Carpenter et al. 2016, Ballmann et al. 2017). Additionally, bats may hitchhike long distances on ships or other means of transit to novel locations (Constantine 2003).

Population-level declines for some species have been greater than 90% (Cheng et al. 2021) and have significantly altered the composition of bat communities within eastern North America (Frick et al. 2015). Differences in susceptibility to infection and severity of disease have driven these disparate trends across species. In North America, Europe, and Asia, *Pd* has been detected on over 40 species of bats, with diagnostic signs of WNS confirmed in most infected species (USFWS 2021*b*). High mortality rates caused by WNS in multiple North American bat species have elevated conservation concerns and protections for these species (see picture on page 273. For example, the northern long-eared bat (*Myotis septentrionalis*) was given protection under the United States Endangered Species Act (USFWS 2016), and additional species are under review for listing due to severe population declines caused by WNS. Canada has listed three species, little brown bat (*Myotis lucifugus*), northern long-eared bat, and tricolored bat (*Perimyotis subflavus*), on their List of Wildlife Species at Risk (COSEWIC 2014). In contrast, specific populations of some species, such as little brown bats, have demonstrated resilience to WNS with stabilization or even slight increases in colony size following precipitous declines upon first arrival of the disease (Frick et al. 2017, Dobony and Johnson 2018, Turner et al. 2022). It has been hypothesized that populations demonstrating persistence in the face of this emerging disease may exhibit eco-evolutionary rescue, which is the concept that evolution may occur quickly, minimize population decline, and allow for population recovery prior to species extinction (DiRenzo et al. 2018, Gignoux-Wolfsohn et al. 2021). Susceptibility to WNS has been predicted and documented in bat species in the western United States; however, transmission dynamics, disease severity, and population-level impacts of WNS remain uncertain (Haase et al. 2020, Hranac et al. 2021, Neubaum and Siemers 2021). Despite these uncertainties, WNS presents the greatest conservation threat to North American bats, since over 50% of the 47 species rely on hibernation for overwinter survival.

Pathogenesis and Pathology

Pseudogymnoascus destructans is a primary pathogen that grows on the skin of bats when they enter hibernation, reducing their body temperature and suppressing their immune system to conserve energy over winter. Fungal hyphae invade the epidermis and underlying connective tissue causing erosion and ulceration of wing membranes (Meteyer et al. 2009). This destruction of wing tissue inhibits vital physiologic functions, resulting in disruption of hibernation physiology and behavior including more frequent arousals from torpor, increased energy use, electrolyte imbalances, dehydration, starvation, and death prior to spring emergence (Reeder et al. 2012, Warnecke et al. 2013, Verant et al. 2014). Some species of bats have apparent resistance or tolerance to infection with *Pd* (Frick et al 2017) and either do not develop skin lesions (i.e., do not develop WNS; see WNSNRT 2023 for a current list) or experience less severe clinical disease and lower mortality rates (e.g., big brown bats, *Eptesicus fuscus*; Frank et al. 2014, Cheng et al. 2021). Mechanisms for this resistance or tolerance within and between species of bats have continued to be a focus of investigation and have been attributed to differences in hibernation temperatures (Langwig et al. 2016, Turner et al.

2022), fat reserves in the fall (Cheng et al. 2019), skin components and microbiomes (Frank et al. 2016, Vanderwolf et al. 2021), genetics (Gignoux-Wolfsohn et al. 2021), and immune responses (Moore et al. 2013).

Diagnosis of WNS

Diagnosing WNS requires identification of the pathogen in conjunction with skin pathology. Whole bat carcasses provide the best diagnostic specimen and allow for detection of *Pd* via culture or molecular methods and histologic examination of wing membranes to confirm disease manifestation. Histologically, the presence of "cupping erosions" within the epidermis of the wings, ears and muzzle in combination with periodic acid–Schiff stained hyphae and curved conidia of *Pd* is characteristic of WNS (Meteyer et al. 2009). Long-wave ultraviolet light (385 nm) can be used to screen bats for WNS and has become a useful, nonlethal tool for disease surveillance in the field (Turner et al. 2014). When *Pd* invades the epidermis, a characteristic orange-yellow fluorescence is visible under ultraviolet light (Figure 16.1). Other patterns and colors of fluorescence can be seen on the skin of bats, so training or experience with this technique is helpful for recognizing the characteristic appearance of WNS. If fluorescence is observed, a biopsy sample can be collected from the affected area for histopathology in combination with a wing swab sample to confirm *Pd* infection. Histopathology remains the gold standard for diagnosis of WNS (Meteyer et al. 2009). The presence of histologic lesions in conjunction with detection of *Pd* is required to confirm WNS in a bat (WNSNRT 2019*a*).

Fig. 16.1. Illumination of a bat wing membrane under long-wave ultraviolet light shows an orange-yellow fluorescence characteristic of white-nose syndrome skin lesions.

Disease Surveillance

When WNS first emerged, disease surveillance relied primarily on detection of dead bats and signs of disease during winter hibernacula surveys. As WNS has spread into the western United States, winter hibernacula surveys have become less feasible since most bats do not hibernate in large aggregations within caves or mines, as is common in the eastern United States (Weller et al. 2018). Observations of dead or sick bats continue to be an effective means of passive surveillance, with many state agencies requesting reports from the public or wildlife rehabilitators. The discovery of WNS in Washington in 2016 came from a sick bat found by a hiker and brought to a wildlife rehabilitator who recognized potential signs of WNS and submitted the bat for testing (Lorch et al. 2016). Bats submitted to public health laboratories for rabies testing during the winter have also been a useful surveillance method in some states (Griggs et al. 2012). Field signs of WNS include unusual mortality of bats or substantial population declines within hibernacula; sick or dead bats found on the landscape during winter; visible white fungal growth on the muzzle, ears, or wings of hibernating bats; abnormal activity during winter, such as day-time flights; or atypical shifts in roosting locations to near the entrance of a hibernaculum (Turner et al. 2011, Carr et al. 2014, USGS National Wildlife Health Center 2021). However, winter bat activity outside of hibernacula may be normal for some species and locations (Lausen and Barclay 2006*a*, Schwab and

Mabee 2014, Bernard and McCracken 2017, Langwig et al. 2021, Whiting et al. 2021). Bats may normally have moderate to severe wing damage (Reichard and Kunz 2009, Fuller et al. 2011) or poor body condition in early spring when they emerge from hibernation (Storm and Boyles 2010). These signs are suggestive of WNS, but diagnostic testing is required to confirm disease.

The development of a highly sensitive and specific real-time polymerase chain reaction (PCR) for *Pd* (Muller et al. 2013) has enabled more active and noninvasive surveillance methods for early detection of the fungus in bat populations, often prior to the onset of visible signs of WNS and mortality (Janicki et al. 2015). Swab samples of the wing membranes and muzzle can be collected from live bats either hand-captured or sampled in place within a hibernaculum or from bats captured in early spring near hibernacula, within or near summer roosts, or on the landscape. Capturing and sampling bats during the summer is an option (Dobony et al. 2011, Carpenter et al. 2016, Ballmann et al. 2017), but the probability of detecting *Pd* is lower. Fresh guano produced by captured bats can be an effective opportunistic sample to test via PCR in conjunction with wing swabs because bats can ingest *Pd* during grooming and then concentrate and pass the fungal material in detectable amounts through their gastrointestinal tract (Ballmann et al. 2017). If bats are not accessible for direct sampling or if physical capture of bats is not desired, environmental samples collected within a hibernaculum (i.e., sediment or swabs of substrate surfaces) can be tested for *Pd* by PCR. Pooled guano samples collected from under summer roost sites, such as maternity colonies or day roosts in the spring, after bats have been present at the site for a few weeks has demonstrated utility in detecting *Pd* by PCR and doesn't require handling of bats (Niaqually Valley News 2018, Montana Fish, Wildlife and Parks 2020). Guano or environmental substrates can be cultured for *Pd*, but culturing is time intensive and requires use of specialized media (Vanderwolf et al. 2016). For details on each of these techniques, please see the US Geological Survey (USGS) National Wildlife Health Center Bat Submission Guidelines in literature cited (USGS National Wildlife Health Center 2021).

The recommended time of year for conducting *Pd* or WNS surveillance is during the winter hibernation period or within a few weeks postemergence from hibernation (Figure 16.2). This timing can vary by location, elevation, or even annual weather patterns but is generally November through May. This time period coincides with the annual cycle of in-

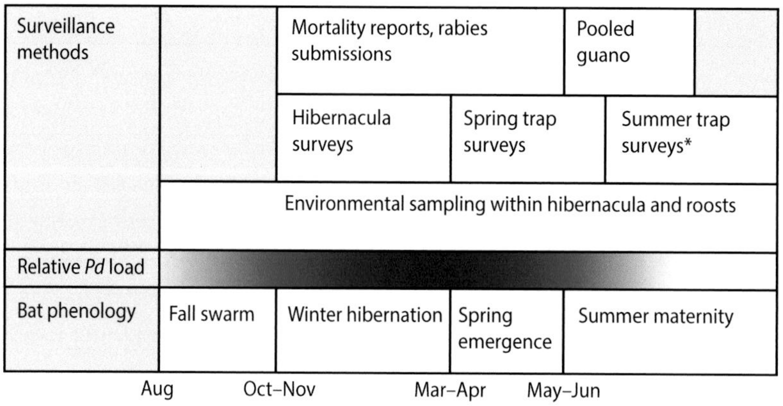

Surveillance methods		Mortality reports, rabies submissions		Pooled guano	
		Hibernacula surveys	Spring trap surveys	Summer trap surveys*	
	Environmental sampling within hibernacula and roosts				
Relative *Pd* load					
Bat phenology	Fall swarm	Winter hibernation	Spring emergence	Summer maternity	
	Aug	Oct–Nov	Mar–Apr	May–Jun	

Fig. 16.2. Timing of surveillance methods based on the phenology of bats and annual cycle of *Pseudogymnoascus destructans* (*Pd*) infection and progression of white-nose syndrome. The amount of *Pd* on bats (*Pd* load or infection intensity) increases during winter hibernation (darker shading) and declines rapidly in spring to near-undetectable levels in summer (light shading). The asterisk indicates a lower probability of detecting *Pd* on bats.

fection and progression of the disease (Lorch et al. 2011, Langwig et al. 2015). But, bats can become infected with *Pd* in the fall when they contact infected bats or environmental reservoirs within hibernacula during swarming (Neubaum and Siemers 2021).

During winter, the fungus proliferates on the cold, torpid bat with peak fungal loads and skin lesions typically occurring in middle to late winter (January–March). There is evidence that infected bats who survive until spring are capable of mounting an immune response and may fully heal the damaged wing membranes within several weeks once they become active and start feeding (Dobony et al. 2011, Meteyer et al. 2011). Thus, diagnostic skin lesions disappear within a couple weeks, and the amount of fungus on the bat declines precipitously after it emerges from hibernation, typically decreasing to near-undetectable levels in the summer (Langwig et al. 2015). In contrast, environmental samples from a hibernaculum can be collected at any time of the year because *Pd* is known to persist in caves and mines year-round (Lorch et al 2013). *Pseudogymnoascus destructans* has also been detected within rock crevices and bridges used by bats during winter, in transitory roosts in spring, or in maternity sites during summer (Niaqually Valley News 2018, Montana Fish, Wildlife and Parks 2020), but persistence of the fungus in these environments has not been investigated.

An active national surveillance strategy for *Pd* was developed in 2019 to identify priority sampling regions for early detection of the fungus in new geographic areas within the United States (USGS 2021). The strategy is based on a dynamic model-based approach (Grider et al. 2021) using cumulative surveillance data and employs active sampling of bats or bat roosts for presence of *Pd* using multiple methods. Species known to be susceptible to WNS, such as *Myotis* sp. and tricolored bats, are targeted for sampling to maximize the probability of detection. The USGS National Wildlife Health Center is leading this project in partnership with states, federal agencies, and other land managers. Secondary priorities for surveillance include detection of *Pd* or WNS in new species of bats or in populations at sites with critical biological or management significance. This strategy also prioritizes sampling at locations where results were deemed inconclusive in previous years.

Finding Bats

Active surveillance for *Pd* or WNS requires locating bats and bat roosts. This can be challenging given the cryptic nature of bats and the diversity of roost types used by different species throughout the year (Figure 16.3). Seasonal migration and within-season movements of bats between roosts can also complicate

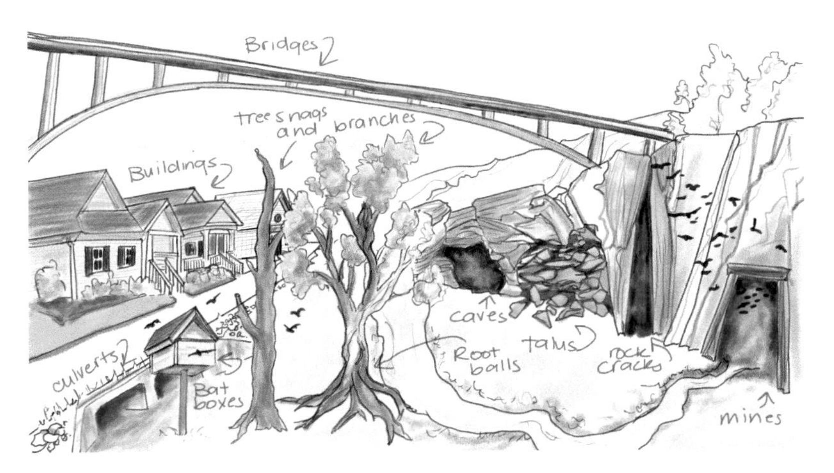

Fig. 16.3. Bats in North America use a variety of habitats for roosting, many of which are shown. The types of preferred roosts can vary by species, geographic location, and time of year based on a bat's physiologic and behavioral requirements. Illustration by Laura Donohue.

monitoring populations over time (Bernard et al. 2020). Some species are more gregarious and can be easier to find when aggregated at winter or summer roosts in caves, mines, buildings, or bat boxes. Relatively high site fidelity (Gumbert et al. 2002) allows for annual monitoring during winter hibernacula surveys. This has been the most common method used to track the spread and impacts of WNS in the eastern United States, primarily with little brown bats, big brown bats, northern long-eared bats, Indiana bats (*Myotis sodalis*), and tricolored bats. Recently, use of alternative hibernaculum types have been observed among remnant populations of these species post-WNS, such as big brown bats hibernating in talus and rock crevices in Virginia (Moosman et al. 2017); northern-long eared bats overwintering in rock crevices in Nebraska (White et al. 2020) and anthropogenic structures on Long Island (Grider et al. 2016); and tricolored bats roosting in tree cavities, bridges, and foliage during winter within coastal areas of the mid-Atlantic (Newman 2020). Tricolored bats have also been found hibernating in culverts across the southeastern United States and Texas (Meierhofer et al. 2019). These observations are likely due to more broad and concerted efforts to find bats in areas affected by WNS but may also represent changes in roosting behavior in response to the disease or other factors.

Species that tend to roost singly or in smaller groups dispersed across a landscape, such as eastern small-footed bats (*Myotis leibii*) commonly found in talus slopes (Moosman et al. 2017), are more difficult to find and require more time-intensive and creative approaches. The use of rock roosts (e.g., talus, cliffs, cracks, crevices, and rock outcrops) by bats has been increasingly reported and studied (Lausen and Barclay 2006*b*, Klüg-Baerwald et al. 2017, Neubaum 2018, White et al. 2020), particularly as WNS has expanded westward where most bat species, including little brown bats and big brown bats, do not often hibernate in caves or mines (Weller et al. 2018). Winter bat activity has been demonstrated near these geologic features, suggesting bats are using these locations for hibernation (Lemen et al. 2017, Hammesfahr and Ohms 2018). Finally, little brown bats in Alaska have even been documented hibernating under root balls (Blejwas et al. 2021). The rare use of caves and mines or aggregations of bats in the west provides fewer opportunities for winter hibernacula surveys or *Pd* or WNS surveillance. Thus, active surveillance in these areas has relied predominantly on spring trapping of bats or collection of pooled guano from under summer roosts. As an example, *Pd* was detected for the first time in Montana in 2020 in samples collected underneath bridges used by bats (Montana Fish, Wildlife and Parks 2020).

Responding to WNS: Challenges for Managing an Emerging Wildlife Disease

Wildlife diseases are challenging to manage often due to limited resources, lack of knowledge or feasible management options, competing objectives, and multijurisdictional responsibilities (Bernard and Grant 2019, Canessa et al. 2019). Although the One Health concept emphasizes connections between the health of animals, people, and the environment, diseases that solely affect wildlife (e.g., WNS, amphibian chytridiomycosis, and Tasmanian devil facial tumor disease) are often a lower priority for resources, because they lack a direct human nexus (e.g., zoonotic diseases such as West Nile virus, highly pathogenic avian influenza in avifauna, or diseases that infect game species such as chronic wasting disease in cervids). Despite this gap in funding opportunities, there continues to be a concerted effort to make WNS a priority for funding and conservation action by demonstrating the value of bats to the ecosystem, the economy, and human well-being (Kunz et al. 2011; also see Summary).

Emerging diseases that affect cryptic species, like WNS, often go undetected longer or are more difficult to detect and understand, because we typically do not have established monitoring programs, disease surveillance systems, or validated diagnostic tests for these species and pathogens (Russell et al.

2020). Prior to WNS, management priorities for bats were focused on species listed as threatened or endangered under the Endangered Species Act or as state Species of Greatest Conservation Need. As WNS spread throughout eastern North America, it became apparent that researchers and natural resource managers were lacking general baseline knowledge on historically common species that were now rapidly disappearing (i.e., tricolored and northern long-eared bats). These recognized data gaps became the focus of intensive research and coordinated national responses within the United States and Canada and coordination with Mexico through a trilateral agreement focused on the conservation of North American bat species. The urgency of WNS prompted development of the North American Bat Monitoring Program in 2015 (Loeb et al. 2015), which is an extensive network of partners using standardized protocols to collect and collate data from bat surveys. The data are then used to assess species status and trends at local, regional, and range-wide scales to improve the collective understanding of bat populations, identify effects of WNS and other stressors, and provide essential data to inform management decisions.

Impacts of emerging wildlife diseases can be difficult to predict, because pathogens often affect populations and species differently. Impacts of WNS are variable across individuals, specific colonies or populations, species, and ecosystems (Langwig et al. 2016), making it difficult to predict the spread, population-level effects and to determine when, where, and how to intervene to reduce anticipated impacts (Bernard et al. 2020). Even once common and widely distributed species, like the little brown bat, may exhibit differential susceptibility to *Pd* and WNS across its range, suggesting there is no "one size fits all" solution. Thus, proactive surveillance and established population monitoring programs are important for identifying disease effects among all potentially susceptible hosts. This initial data can be used as the basis of focused research to understand species-specific differences and other uncertainties in order to determine appropriate management responses. As an example, continued annual counts of hibernating bats within known hibernacula in the eastern United States helped to identify sites that supported persistent colonies of bats following precipitous populations declines caused by WNS. Futher investigation of these sites demonstrated the important role of temperature in moderating mortality from WNS (Langwig et al. 2012) and eventually lead to purposeful manipulation of some hibernacula to achieve colder temperatures to provide more suitable habitat for bats hibernating with WNS (Turner et al. 2022).

While there is a general consensus that the best available scientific information should inform wildlife and disease management decisions, a disconnect between researchers and managers still remains (Merkle et al. 2019). Even when effective interventions are available, these control strategies may have secondary or nontarget effects on other natural or cultural resources, livestock, or human health. Consequently, certain strategies, such as year-round cave closures, may conflict with other values and interests of stakeholders and the public as has been demonstrated with WNS and other diseases of wildlife (Sells et al. 2016, Bernard et al. 2020, Bernard and Grant 2021). A recent study by Bernard and Grant (2019), focused on fungal diseases of wildlife, found 52% of managers surveyed stated that their most pressing concern was the loss or decline of species. However, these managers were not only concerned about the focal species directly, since they were required by agency mandates and long-term strategic plans to consider all threatened and endangered species and habitats within their jurisdiction, as well as public access and recreation, and multiuse management such as timber harvest or extraction. Thus, natural resource managers must contend with competing management objectives (Box 16.1), often leaving them with a limited set of actions that fulfill the needs of their agency mandates, stakeholders, and wildlife goals (Sells et al. 2016, Mitchell et al. 2018, Bernard and Grant 2019, Bernard et al. 2019, Wright et al. 2020).

Management responsibilities for wildlife and their diseases often fall under multiple or overlapping jurisdictions and authorities. For example, bats can migrate long distances seasonally, meaning summer and winter roost sites for a population are likely to be managed by separate entities. Furthermore, when bats come into contact with people or make use of anthropogenic structures, public health departments may also get involved due to the potential risk of rabies exposure for people. This can contribute to inconsistent and variable implementation of mitigation measures (Schwab et al. 2018) due to differences in priorities and resources among agencies. The absence of collaboration around shared goals, objectives, and methods can lead to a patchwork of disease-management strategies for contiguous populations and a potentially less effective response for slowing disease spread or preserving affected species.

WNS Treatment and Management

Minimizing the spread of *Pd* and maximizing the survival of North America's diverse bat species are common goals shared by bat biologists, natural resource managers, and conservation organizations (Bernard et al. 2020). Although there is no silver bullet in combating the disease, there are a number of actions that bat advocates, researchers, and managers can take to reach these goals (Figure 16.4). The actions listed below can be implemented proactively, that is, before *Pd* and WNS have been identified in a region, which may prevent the introduction of the pathogen to an area or minimize population declines (Table 16.1). Some actions may be implemented reactively, such as trying to keep fungal loads low, minimizing transmission of *Pd* via human or bat movement, or minimizing population loss. Other actions can be implemented in a "state-independent" method, meaning it can be implemented at any time with the intent of improvements in habitat (Cheng et al. 2019, Newman 2020, Bernard et al. 2021), public education and engagement (Shapiro et al. 2021), or conservation-minded risk communication about bats (Lu et al. 2017, 2020; Siemer et al. 2021). State-independent actions can have long-lasting effects on the persistence of local bat populations and public support for bat conservation.

Hibernating bats with WNS can be individually brought into a captive setting and treated with supportive care (e.g., warmth and food). During the healing process, bats develop a systemic inflammatory immune response that helps to clear the fungal infection (Meteyer et al. 2012). If the systemic disease and energetic requirements during early recovery are not too severe, bats can completely heal wing membranes and recover within about five weeks (Meteyer et al. 2011, Fuller et al. 2020). The bats can then be released in spring. However, bats that recover from WNS remain susceptible and can be infected with *Pd* and develop WNS again. Therefore, this treatment is not a long-term or scaleable solution. Preventative treatments, such as vaccines and probiotics, have demonstrated efficacy in reducing mortality from WNS and are current priorities for research (Palmer et al. 2018, Hoyt et al. 2019, Rocke et al. 2019). These methods may be applied to specific populations (e.g., at a hibernaculum or maternity roost), but a sufficient proportion of the population must still be effectively treated to have a population-level effect (Fletcher et al. 2020). This requires access to sufficient numbers of bats and locations where bats are known to aggregate, which may not be feasible for all species or populations.

Other Impacts on Bats

White-nose syndrome is one of many conservation challenges for North American bat populations. Additional threats include climate change, predation by domestic free-ranging cats (Oedin et al. 2021, Salinas-Ramos et al. 2021), habitat alteration, and wind energy development (Frick et al. 2019). Challenges such as climate change are predicted to become more extreme and may lead to a reduction in preferred habitat (Mammola et al. 2019) and prey

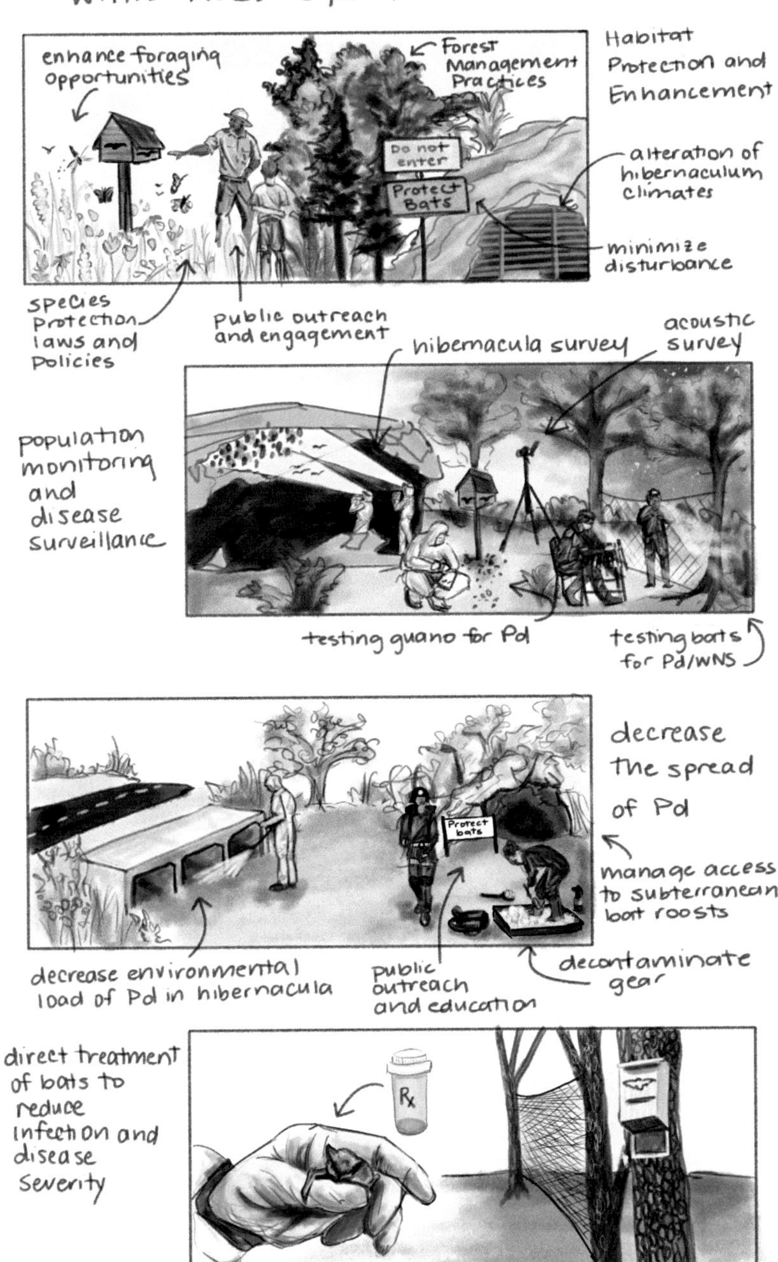

Fig. 16.4. Managing white-nose syndrome (WNS) in bats can be accomplished using a multilayered approach that (1) protects and enhances habitat to foster resilient populations and assist with species recovery, (2) monitors populations and surveys for disease to assess impacts on susceptible species and better understand disease dynamics, (3) decreases the spread of *Pseudogymnoascus destructans* (*Pd*) to bats and new locations, and (4) treats bats to reduce infection and disease severity. Illustration by Laura Donohue.

availability (Hallmann et al. 2017). While there are plans to combat the effects of climate change, the conversion of land for the development of wind farms also has a toll. Wind energy development has been found to be one of the primary contributors in the decline of migratory species such as silver-haired bats (*Lasionycteris noctivagans*) and tricolored bats (Rodhouse et al. 2019), leading to additive mortality for some species affected by WNS. Direct persecution of bats by humans due to fear and misguided perceptions of risk can also have local impacts on vulnerable species and erode public support for bat

Table 16.1. White-nose syndrome (WNS) and *Pseudogymnoascus destructans* (*Pd*) management actions

Action	Description	Timing	References
Habitat enhancement and protection			
Restrict access to roosts	Seasonal or partial closures of roost sites (i.e., hibernacula, maternity, or other sensitive locations); bat-friendly cave gates, permits, and organized tours	SI	USFWS 2016, Fagan et al. 2018
Public outreach and engagement	Educating the public on bat ecology and benefits to human systems and health; community science projects	SI	Lu et al. 2017, MacFarlane and Rocha 2020, Shapiro et al. 2021, Siemer et al. 2021
Species and habitat protection laws, policies, and guidance	Provide legal support for regulations to preserve or recover bat species; management practices for bat-control activities in structures	P and R	Federal and state listing documents, Daniel 2020, WNSNRT 2019b
Forest management practices	Enhance or protect roosting habitat; enhance foraging areas; maintain riparian zones	SI	Daniel 2006, Johnson and King 2018, Bernard et al. 2021
Improved foraging conditions	Attract or release prey at fall swarming sites or hibernacula and maternity sites in spring	R	Cheng et al. 2019, Bernard et al. 2021
Population monitoring and disease surveillance			
Hibernacula surveys	*Pd* or WNS surveillance via visual surveys or diagnostic samples; population counts	SI	NABat 2021, USGS 2021
Acoustic surveys	Species occupancy; seasonal or site-specific bat activity; population estimation	SI	NABat 2021, USFWS 2019
Test bats for *Pd* or WNS	Capture of bats to collect diagnostic samples for *Pd* or WNS; can be used for early detection or determining species affected, spatial distribution, or prevalence	P and R	USGS 2021
Test guano for *Pd*	Collection of pooled guano at bat roosts; can be used for early detection or determining spatial distribution	P and R	Mulec et al. 2013, Swift et al. 2018, USGS 2021
Decrease the spread of *Pd*			
Reduce environmental load of *Pd*	Clean or decontaminate artificial roosts (bat boxes, culverts, mines); clean natural roosts when and where appropriate	R	Gabriel et al. 2018, Palmer et al. 2018, Bernard et al. 2019, Sewall et al. 2023
Reduce anthropogenic spread of *Pd*	Manage access to subterranean bat roosts; clean and decontaminate gear that contacts bats or environments; use site-dedicated gear	P and R	USFWS 2016, NPS 2017, Bernard et al. 2020, USFWS et al. 2022
Public outreach and engagement	Education about WNS and measures to prevent human-assisted spread of *Pd*	P and R	WDFW 2016, Fagan et al. 2018, WNSNRT 2019b
Direct treatment of bats			
Vaccine	Induction of antifungal immune response	P	Rocke et al. 2019
Ultraviolet light	Implement during infection to minimize fungal load	R	Palmer et al. 2018
Anti-*Pd* microbes or volatile compounds	Apply during infection to minimize WNS mortality	R	Cornelison et al. 2014, Cheng et al. 2017, Gabriel et al. 2018, Hoyt et al. 2019
Rehabilitation of infected bats	Supportive care during winter, release in spring	R	Meteyer et al. 2011

Timings of actions are state independent (SI), proactive (P), and reactive (R).

conservation (Buttke et al. 2015, O'Shea et al. 2016). O'Shea et al. (2016) found that the intentional killing of bats by humans accounted for 39% of mass mortality events. Emerging zoonotic diseases, such as the COVID-19 pandemic and associations with bats as reservoirs of these diseases, have amplified these fears and animosity towards bats, an additional hurdle for bat conservation (Sasse and Gramza 2021). Addressing misinformation about bats, moderating risk perceptions of bat-associated diseases, and highlighting bat-friendly social norms through psychological approaches to science communication (MacFarlane and Rocha 2020) and balanced conservation messaging (Lu et al. 2017) are essential for

Box 16.1. Weighing the alternatives with context-dependent decisions

Management of a multihost pathogen with a continental distribution is complex and involves numerous objectives, considerations, and uncertainties. Specific management actions to achieve a goal may not always be implemented, at least not immediately or consistently. Here we provide a set of questions that can be used when weighing available alternatives. These have helped guide management decisions in various areas of the United States (Bernard et al. 2019, 2020).

What is the problem and what ecological and biological (i.e., host or environment) information is imperative for making a management decision?

The first step is the creation of a problem statement where decision makers, stakeholders, and biological experts cogenerate a statement comprised of goals and the actions that need to be taken. This statement includes goals to be met; legal considerations; decision maker(s); scope, frequency, and timing of the decision; available knowledge; and the uncertainties related to the problem (Runge et al. 2013). The information required to determine the best course of action for white-nose syndrome (WNS) is context dependent, driven by the problem statement and varies by host species, roost site, and management entity. While the goal for any decision is to identify the optimal alternative for the objectives identified (e.g., maximize bat persistence, minimize cost and undesirable nontarget effects, and maintain multiuse management of the cave site), the information needed to choose between alternative management actions will depend on the decision context identified prior to coming up with alternatives (Canessa et al. 2018).

Which species can or should be targeted?

Priorities for management can be based on a list of factors, including but not limited to their Endangered Species Act status, state listing status, susceptibility to *Pseudogymnoascus destructans* infection and observed disease severity, access to and size of hibernating colonies, knowledge of roosting locations, habitat preferences, and migration ecology.

Which sites should be targeted?

For some species, such as the little brown bat (*Myotis lucifugus*) in the eastern portion of its range, there is only a single large colony (>10,000 bats) remaining in the area affected by WNS (Cheng et al. 2021). These sites can improve our understanding as to why and how bats are persisting within WNS endemic locations (Maslo et al. 2015, 2017). For northern long-eared (*Myotis septentrionalis*) and tricolored bats (*Perimyotis subflavus*), most sites still occupied by bats in winter contain fewer than 10 individuals (Cheng et al. 2021). There is growing evidence that northern long-eared bats are overwintering in coastal habitats and appear to be surviving and reproducing despite drastic declines from WNS in other areas (Grider et al. 2016). Identifying, protecting, or enhancing these refugia may be an effective conservation measure. Many of these coastal overwintering sites are anthropogenic structures, so public education and partnerships with private landowners are important for survival of remnant populations (Dowling and O'Dell 2018).

Large aggregations of bats are rare in the western United States providing limited opportunities for targeted treatments at a particular site (Weller et al. 2018). While management actions are often site or species specific and implemented by one jurisdiction, achieving population or species-level impacts likely requires collaboration and collective investments across jurisdictions.

When is the best time to treat individuals or sites?

Timing will depend on the delivery method and mechanism of action for the treatment (Grider et al. 2022). Direct treatment of bats may be most successful when bats aggregate, either in hibernacula or at summer roosts. Some treatments, like vaccines, are likely to be most effective prior to infection and early in the course of the epidemic. In contrast, actions to reduce fungal loads within a hibernaculum can be applied in the summer when bats are absent and may be most impactful in years following first detection of WNS to limit environmental accumulation of the fungus. Other

(continued)

considerations include frequency of treatment (e.g., multiple months or years or until a desired outcome is reached) or number of sites treated (e.g., population-wide impact versus location specific). Again, the optimal management parameters will be directly tied to the specified actions and objectives identified by the natural resource agency and stakeholder groups (Bernard and Grant 2019, Bernard et al. 2019).

How do I choose among management actions?

With every decision, there are trade-offs to consider. For example, (1) how many times should an individual or site be treated; (2) how much will an action cost (including dollars for the treatment, personnel time, and necessary equipment); (3) what is the safety and efficacy; (4) how many individuals or sites should be treated to have a site- or population-level impact (Fletcher et al. 2020); (5) what are the nontarget effects; and (6) what is the extent of disturbance to bats and other nontarget species? The answers and uncertainties for each of these questions are factored into decisions about how well each potential action will achieve the desired management objectives (see Bernard et al. 2020). If an action fails to meet higher ranking objectives, it is less likely to be chosen as the optimal action (see Sells et al. 2016, Bernard et al. 2019 for examples of disease decision problems).

maintaining public support for management actions to conserve bats.

Summary

WNS differentially affects multiple bat species across diverse landscapes, which requires incorporating context-dependent information and decision making in management actions and conservation plans. The potential risk of extinction or regional extirpation for some species due to WNS has resulted in petitions for Endangered Species Act protection for multiple species (USFWS et al. 2022). The northern long-eared bat was reclassified as endangered and the tricolored bat has been proposed for listing due to continued declines across their range. These protections, while important for helping to recover imperiled species, can create additional obstacles for private and public sectors. Necessary actions may include biological assessments (i.e., US Fish and Wildlife Service section 7 consultation) and stipulations on timing or mitigation measures for large projects (USFWS 2016, 2021*a*). The drastic population declines and significant alterations in bat community composition observed across North America due to WNS also have important implications for human health and economic well-being. Bats consume large numbers of insects, including biting insects like mosquitos (Wray et al. 2018)—many of which are disease vectors (Maslo et al. 2022)—and pests that damage crops and forestry products (Brown et al. 2015, Maine and Boyles 2015). These pest-control services have been valued at about $3 billion (USD) annually for the US agricultural industry (Boyles et al. 2011). Thus, a substantial loss of insectivorous bats may result in economic losses and increased pesticide use.

Responding to this urgent threat to North American bats over the past decade, researchers and natural resource managers have filled numerous gaps in knowledge regarding bat biology and behavior, as well as host-pathogen ecology for this novel fungal disease. The collaborative response framework and lessons learned from the bat–WNS system can be applied to other taxonomic groups facing similar challenges, such as emerging disease and anthropogenic activities that threaten species conservation and global biodiversity.

LITERATURE CITED

Ballmann, A. E., M. R. Torkelson, E. A. Bohuski, R. E. Russell, and D. S. Blehert. 2017. Dispersal hazards of *Pseudogymnoascus destructans* by bats and human activity at hibernacula in summer. Journal of Wildlife Diseases 53:725–735.

Bernard, R. F., J. Evans, N. W. Fuller, J. D. Reichard, J. T. H. Coleman, C. J. Kocer, E. H. C. Grant, and E. H. Campbell Grant. 2019. Different management strategies are optimal for combating disease in East Texas cave versus culvert hibernating bat populations. Conservation Science and Practice 1:e106.

Bernard, R. F., and E. H. C. Grant. 2019. Identifying common decision problem elements for the management of emerging fungal diseases of wildlife. Society and Natural Resources 32:1040–1055.

Bernard, R. F., and E. H. C. Grant. 2021. Rapid assessment indicates context-dependent mitigation for amphibian disease risk. Wildlife Society Bulletin 45:290–299.

Bernard, R. F., and G. F. McCracken. 2017. Winter behavior of bats and the progression of white-nose syndrome in the southeastern United States. Ecology and Evolution 7:1487–1496.

Bernard, R. F., J. D. Reichard, J. T. H. Coleman, J. C. Blackwood, M. L. Verant, J. L. Segers, J. M. Lorch, J. White, M. S. Moore, A. L. Russell, et al. 2020. Identifying research needs to inform white-nose syndrome management decisions. Conservation Science and Practice 2:e220.

Bernard, R. F., E. V. Willcox, R. T. Jackson, V. A. Brown, and G. F. McCracken. 2021. Feasting, not fasting: winter diets of cave hibernating bats in the United States. Frontiers in Zoology 18:48.

Blehert, D. S., A. C. Hicks, M. J. Behr, C. U. Meteyer, B. M. Berlowski-Zier, E. L. Buckles, J. T. H. Coleman, S. R. Darling, A. Gargas, R. Niver, et al. 2009. Bat white-nose syndrome: an emerging fugal pathogen? Science 323:227.

Blejwas, K. M., G. W. Pendleton, M. L. Kohan, and L. O. Beard. 2021. The Milieu Souterrain Superficiel as hibernation habitat for bats: implications for white-nose syndrome. Journal of Mammalogy 102:1110–1127.

Boyles, J. G., P. M. Cryan, G. F. McCracken, and T. H. Kunz. 2011. Economic importance of bats in agriculture. Science 332:41–42.

Brown, V. A., E. Braun de Torrez, and G. F. McCracken. 2015. Crop pests eaten by bats in organic pecan orchards. Crop Protection 67:66–71.

Buttke, D. E., D. J. Decker, and M. A. Wild. 2015. The role of One Health in wildlife conservation: a challenge and opportunity. Journal of Wildlife Diseases 51:1–8.

Canessa, S., C. Bozzuto, E. H. Campbell Grant, S. S. Cruickshank, M. C. Fisher, J. C. Koella, S. Lötters, A. Martel, F. Pasmans, B. C. Scheele, et al. 2018. Decision-making for mitigating wildlife diseases: from theory to practice for an emerging fungal pathogen of amphibians. Journal of Applied Ecology 55:1987–1996.

Canessa, S., C. Bozzuto, F. Pasmans, and A. Martel. 2019. Quantifying the burden of managing wildlife diseases in multiple host species. Conservation Biology 33:1131–1140.

Carpenter, G. M., E. V. Willcox, R. F. Bernard, and W. H. Stiver. 2016. Detection of *Pseudogymnoascus destructans* on free-flying male bats captured during summer in the southeastern USA. Journal of Wildlife Diseases 52:922–926.

Carr, J. A., R. F. Bernard, and W. H. Stiver. 2014. Unusual bat behavior during winter in Great Smoky Mountains National Park. Southeastern Naturalist 13:N18–N21.

Cheng, T. L., H. Mayberry, L. P. McGuire, J. R. Hoyt, K. E. Langwig, H. Nguyen, K. L. Parise, J. T. Foster, C. K. R. Willis, A. M. Kilpatrick, et al. 2017. Efficacy of a probiotic bacterium to treat bats affected by the disease white-nose syndrome. Journal of Applied Ecology 45:701–708.

Cheng, T. L., A. Gerson, M. S. Moore, J. D. Reichard, J. DeSimone, C. K. R. Willis, W. F. Frick, and A. M. Kilpatrick. 2019. Higher fat stores contribute to persistence of little brown bat populations with white-nose syndrome. Journal of Animal Ecology 88:591–600.

Cheng, T. L., J. D. Reichard, J. T. H. Coleman, T. J. Weller, W. E. Thogmartin, B. E. Reichert, A. B. Bennett, H. G. Broders, J. Campbell, K. Etchison, et al. 2021. The scope and severity of white-nose syndrome on hibernating bats in North America. Conservation Biology 35:1586–1597.

Committee on the Status of Endangered Wildlife in Canada (COSEWIC). 2014. COSEWIC Annual Report 2013–2014. COSEWIC, Gatineau, Quebec, Canada. https://species-registry.canada.ca/index-en.html#/documents/1346. Accessed 27 Oct 2021.

Constantine, D. G. 2003. Geographic translocation of bats: known and potential problems. Emerging Infectious Diseases 9:17–21.

Cornelison, C. T., M. K. Keel, K. T. Gabriel, C. K. Balament, T. A. Tucker, G. E. Pierce, and S. A. Crow Jr. 2014. A preliminary report on the contact-independent antagonism of *Pseudogymnoascus destructans* by *Rhodococcus rhodochorous* strain DAP96253. BMC Microbiology 14:246.

DiRenzo, G. V., E. F. Zipkin, E. H. C. Grant, J. A. Royle, A. V. Longo, K. R. Zamudio, and K. R. Lips. 2018. Eco-evolutionary rescue promotes host–pathogen coexistence. Ecological Applications 28:1948–1962.

Dobony, C. A., A. C. Hicks, K. E. Langwig, R. I. von Linden, J. C. Okoniewski, and R. E. Rainbolt. 2011. Little brown myotis persist despite exposure to white-nose syndrome. Journal of Fish and Wildlife Management 2:190–195.

Dobony, C. A., and J. B. Johnson. 2018. Observed resiliency of little brown myotis to long-term white-nose syndrome exposure. Journal of Fish and Wildlife Management 9:168–179.

Dowling, Z. R., and D. I. O'Dell. 2018. Bat use of an island off the coast of Massachusetts. Northeastern Naturalist 25:362–382.

Fagan, K. E., E. V. Willcox, L. T. Tran, R. F. Bernard, and W. H. Stiver. 2018. Roost selection by bats in buildings, Great Smoky Mountains National Park. Journal of Wildlife Management 82:424–434.

Fletcher, Q. E., Q. M. R. Webber, and C. K. R. Willis. 2020. Modelling the potential efficacy of treatments for white-nose syndrome in bats. Journal of Applied Ecology 57:1283–1291.

Frank, C. L., M. R. Ingala, R. E. Ravenelle, K. Dougherty-Howard, S. O. Wicks, C. Herzog, and R. J. Rudd. 2016. The effects of cutaneous fatty acids on the growth of *Pseudogymnoascus destructans*, the etiological agent of white-nose syndrome (WNS). PLoS One 11:e0153535.

Frank, C. L., A. Michalski, A. A. Mcdonough, and M. Rahimian. 2014. The resistance of a North American bat species (*Eptesicus fuscus*) to white-nose syndrome (WNS). PLoS One 9:e113958.

Frick, W. F., T. L. Cheng, K. E. Langwig, J. R. Hoyt, A. F. Janicki, K. L. Parise, J. T. Foster, A. M. Kilpatrick, and A. Marm Kilpatrick. 2017. Pathogen dynamics during invasion and establishment of white-nose syndrome explain mechanisms of host persistence. Ecology 98:624–631.

Frick, W. F., T. Kingston, and J. Flanders. 2019. A review of the major threats and challenges to global bat conservation. Annals of the New York Academy of Sciences 1469:5–25.

Frick, W. F., S. J. Puechmaille, J. R. Hoyt, B. A. Nickel, K. E. Langwig, J. T. Foster, K. E. Barlow, T. Bartonička, D. Feller, A.-J. J. Haarsma, et al. 2015. Disease alters macroecological patterns of North American bats. Global Ecology and Biogeography 24:741–749.

Fuller, N. W., L. P. McGuire, E. L. Pannkuk, T. Blute, C. G. Haase, H. W. Mayberry, T. S. Risch, and C. K. R. Willis. 2020. Disease recovery in bats affected by white-nose syndrome. Journal of Experimental Biology 223:eb211912.

Fuller, N. W., J. D. Reichard, M. L. Nabhan, S. R. Fellows, L. C. Pepin, and T. H. Kunz. 2011. Free-ranging little brown myotis (*Myotis lucifugus*) heal from wing damage associated with white-nose syndrome. EcoHealth 8:154–162.

Gabriel, K. T., L. Kartforosh, S. A. Crow Jr., and C. T. Cornelison. 2018. Antimicrobial activity of essential oils against the fungal pathogens *Ascosphaera apis* and *Pseudogymnoascus destructans*. Mycopathologia 183:921–934.

Gignoux-Wolfsohn, S. A., M. L. Pinsky, K. Kerwin, C. Herzog, M. Hall, A. B. Bennett, N. H. Fefferman, and B. Maslo. 2021. Genomic signatures of selection in bats surviving white-nose syndrome. Molecular Ecology 30:5643–5657.

Grider, J. F., A. L. Larsen, J. A. Homyack, and M. C. Kalcounis-Rueppell. 2016. Winter activity of coastal plain populations of bat species affected by white-nose syndrome and wind energy facilities. PLoS One 11:e0166512.

Grider, J. F., R. E. Russell, A. E. Ballmann, and T. J. Hefley. 2021. Long-term *Pseudogymnoascus destructans* surveillance data reveal factors contributing to pathogen presence. Ecosphere 12:e03808.

Grider, J., W. E. Thogmartin, E. H. C. Grant, R. F. Bernard, and R. E. Russell. 2022. Early treatment of white-nose syndrome is necessary to stop population decline. Journal of Applied Ecology 59:2531–2541.

Griggs, A., M. K. Keel, K. Castle, and D. Wong. 2012. Enhanced surveillance for white-nose syndrome in bats. Emerging Infectious Diseases 18:530–532.

Gumbert, M. W., J. M. O'Keefe, and J. R. MacGregor. 2002. Roost fidelity in Kentucky. Pages 143–152 *in* A. Kurta and J. Kennedy, editors. The Indiana bat: biology and management of an endangered species. Bat Conservation International, Austin, Texas, USA.

Haase, C. G., N. W. Fuller, Y. A. Dzal, C. R. Hranac, D. T. S. Hayman, C. L. Lausen, K. A. Silas, S. H. Olson, and R. K. Plowright. 2020. Body mass and hibernation microclimate may predict bat susceptibility to white-nose syndrome. Ecology and Evolution 11:506–515.

Hallmann, C. A., M. Sorg, E. Jongejans, H. Siepel, N. Hofland, H. Schwan, W. Stenmans, A. Müller, H. Sumser, T. Hörren, et al. 2017. More than 75 percent decline over 27 years in total flying insect biomass in protected areas. PLoS One 12:e0185809.

Hammesfahr, A. M., and R. E. Ohms. 2018. Winter bat activity in a landscape without traditional hibernacula. US National Park Service Publications and Papers 232, Washington, D.C., USA. https://digitalcommons.unl.edu/cgi/viewcontent.cgi?article=1232&context=natlpark. Accessed 17 Oct 2021.

Hoyt, J. R., K. E. Langwig, K. Sun, K. L. Parise, A. Li, Y. Wang, X. Huang, L. Worledge, H. Miller, J. P. White, et al. 2020. Environmental reservoir dynamics predict global infection patterns and population impacts for the fungal disease white-nose syndrome. Proceedings of the National Academy of Sciences USA 117:7255–7262.

Hoyt, J. R., K. E. Langwig, J. P. White, H. M. Kaarakka, J. A. Redell, K. L. Parise, Wi. F. Frick, J. T. Foster, A. Marm Kilpatrick, and A. M. Kilpatrick. 2019. Field trial of a probiotic bacteria and a chemical, chitosan, to protect bats from white-nose syndrome. Scientific Reports 9:9158.

Hranac, C., C. Haase, N. Fuller, M. Mcclure, J. Marshall, C. Lausen, L. Mcguire, S. Olson, and D. Hayman. 2021. What is winter? Modelling spatial variation in bat host traits and hibernation and their implications for overwintering energetics. Ecology and Evolution 11:11604–11614.

Janicki, A. F., W. F. Frick, A. M. Kilpatrick, K. L. Parise, J. T. Foster, and G. F. McCracken. 2015. Efficacy of visual surveys for white-nose syndrome at bat hibernacula. PLoS One 10:e0133390.

Johnson, C., and R. King. 2018. Beneficial forest management practices for WNS-affected bats: voluntary guidance for land managers and woodland owners in the eastern United States. https://www2.illinois.gov/sites/naturalheritage/speciesconservation/EndangeredandThreatenedSpeciesProgram/Documents/BeneficialForestMgmtWNSAffectedBats.pdf. Accessed 15 Dec 2021.

Klüg-Baerwald, B. J., C. L. Lausen, C. K. R. Willis, and R. M. Brigham. 2017. Home is where you hang your bat: winter roost selection by prairie-living big brown bats. Journal of Mammalogy 98:752–760.

Kunz, T. H., E. Braun de Torrez, D. Bauer, T. Lobova, and T. H. Fleming. 2011. Ecosystem services provided by bats. Annals of the New York Academy of Sciences 1223:1–38.

Langwig, K. E., W. F. Frick, J. T. Bried, A. C. Hicks, T. H. Kunz, and A. M. Kilpatrick. 2012. Sociality, density-dependence and microclimates determine the persistence of populations suffering from a novel fungal disease, white-nose syndrome. Ecology Letters 15:1050–1057.

Langwig, K. E., W. F. Frick, J. R. Hoyt, K. L. Parise, K. P. Drees, T. H. Kunz, J. T. Foster, and A. M. Kilpatrick. 2016. Drivers of variation in species impacts for a multi-host fungal disease of bats. Philosophical Transactions of the Royal Society B Biological Sciences 371:20150456.

Langwig, K. E., W. F. Frick, R. Reynolds, K. L. Parise, K. P. Drees, J. R. Hoyt, T. L. Cheng, T. H. Kunz, J. T. Foster, and A. M. Kilpatrick. 2015. Host and pathogen ecology drive the seasonal dynamics of a fungal disease, white-nose syndrome. Proceedings of the Royal Society B Biological Sciences 282:10–12.

Langwig, K. E., J. P. White, K. L. Parise, H. M. Kaarakka, J. A. Redell, J. E. DePue, W. H. Scullon, J. T. Foster, A. M. Kilpatrick, and J. R. Hoyt. 2021. Mobility and infectiousness in the spatial spread of an emerging fungal pathogen. Journal of Animal Ecology 90:1134–1141.

Lausen, C. L., and R. M. R. Barclay. 2006*a*. Winter bat activity in the Canadian prairies. Canadian Journal of Zoology 84:1079–1086.

Lausen, C. L., and R. M. R. Barclay. 2006*b*. Benefits of living in a building: big brown bats (*Eptesicus fuscus*) in rocks versus buildings. Journal of Mammalogy 87:362–370.

Lemen, C. A., P. W. Freeman, and J. A. White. 2017. Acoustic evidence of bats using rock crevices in winter: a call for more research on winter roosts in North America. Transactions of the Nebraska Academy of Sciences and Affiliated Societies 36: 9–13. http://digitalcommons.unl.edu/tnas/506. Accessed 1 Oct 2021.

Loeb, S. C., T. J. Rodhouse, L. E. Ellison, C. L. Lausen, J. D. Reichard, K. M. Irvine, T. E. Ingersoll, J. T. H. Coleman, W. E. Thogmartin, J. R. Sauer, et al. 2015. A plan for the North American Bat Monitoring Program (NABat). USDA Forest Service, Southern Research Station General Technical Report SRS-208, Asheville, North Carolina, USA.

Lorch, J. M., C. U. Meteyer, M. J. Behr, J. G. Boyles, P. M. Cryan, A. C. Hicks, A. E. Ballmann, J. T. H. Coleman, D. N. Redell, D. M. Reeder, et al. 2011. Experimental infection of bats with *Geomyces destructans* causes white-nose syndrome. Nature 480:376–378.

Lorch, J. M., L. K. Muller, R. E. Russell, M. O'Connor, D. L. Lindner, and D. S. Blehert. 2013. Distribution and environmental persistence of the causative agent of white-nose syndrome, *Geomyces destructans*, in bat hibernacula of the eastern United States. Applied and Environmental Microbiology 79:1293–1301.

Lorch, J. M., J. M. Palmer, D. L. Lindner, A. E. Ballmann, K. G. George, K. Griffin, S. Knowles, J. R. Huckabee, K. H. Haman, C. D. Anderson, et al. 2016. First detection of bat white-nose syndrome in western North America. mSphere 1:e00148-16.

Lu, H., K. A. McComas, D. E. Buttke, S. Roh, M. A. Wild, and D. J. Decker. 2017. One Health messaging about bats and rabies: how framing of risks, benefits and attributions can support public health and wildlife conservation goals. Wildlife Research 44:200–206.

Lu, H., K. McComas, H. Kretser, and T. B. Lauber. 2020. Scared yet compassionate? Exploring the order effects of threat versus suffering messages on attitude toward scary victims. Science Communication 42:3–30.

MacFarlane, D., and R. Rocha. 2020. Guidelines for communicating about bats to prevent persecution in the time of COVID-19. Biological Conservation 248:e108650.

Maine, J. J., and J. G. Boyles. 2015. Bats initiate vital agroecological interactions in corn. Proceedings of the National Academy of Sciences USA 112:12438–12443.

Mammola, S., E. Piano, P. Cardoso, P. Vernon, D. Domínguez-Villar, D. C. Culver, T. Pipan, and M. Isaia. 2019. Climate change going deep: the effects of global climatic alterations on cave ecosystems. Anthropocene Review 6:98–116.

Maslo, B., R. L. Mau, K. Kerwin, R. McDonough, E. McHale, and J. T. Foster. 2022. Bats provide a critical ecosystem service by consuming a large diversity of agricultural pest insects. Agriculture, Ecosystems and Environment 324:107722.

Maslo, B., O. C. Stringham, A. J. Bevan, A. Brumbaugh, C. Sanders, M. Hall, and N. H. Fefferman. 2017. High annual survival in infected wildlife populations may veil a persistent extinction risk from disease. Ecosphere 8:e02001.

Maslo, B., M. Valent, J. F. Gumbs, and W. F. Frick. 2015. Conservation implications of ameliorating survival of little brown bats with white-nose syndrome. Ecological Applications 25:1832–1840.

Meierhofer, M. B., J. S. Johnson, S. J. Leivers, B. L. Pierce, J. E. Evans, and M. L. Morrison. 2019. Winter habitats of bats in Texas. PLoS One 14:e0220839.

Merkle, J. A., N. J. Anderson, D. L. Baxley, M. Chopp, L. C. Gigliotti, J. A. Gude, T. M. Harms, H. E. Johnson, E. H. Merrill, M. S. Mitchell, et al. 2019. A collaborative approach to bridging the gap between wildlife managers and researchers. Journal of Wildlife Management 83:1644–1651.

Meteyer, C. U., D. Barber, and J. N. Mandl. 2012. Pathology in euthermic bats with white-nose syndrome suggests a natural manifestation of immune reconstitution inflammatory syndrome. Virulence 3:583–588.

Meteyer, C. U., E. L. Buckles, D. S. Blehert, A. C. Hicks, D. E. Green, V. Shearn-Bochsler, N. J. Thomas, A. Gargas, and M. J. Behr. 2009. Histopathologic criteria to confirm white-nose syndrome in bats. Journal of Veterinary Diagnostic Investigation 21:411–414.

Meteyer, C. U., M. Valent, J. Kashmer, E. L. Buckles, J. M. Lorch, D. S. Blehert, A. Lollar, D. Berndt, E. Wheeler, C. L. White, et al. 2011. Recovery of little brown bats (*Myotis lucifugus*) from natural infection with *Geomyces destructans*, white-nose syndrome. Journal of Wildlife Diseases 47:618–626.

Miller-Butterworth, C. M., M. J. Vonhof, J. Rosenstern, G. G. Turner, and A. L. Russell. 2014. Genetic structure of little brown bats (*Myotis lucifugus*) corresponds with spread of white-nose syndrome among hibernacula. Journal of Heredity 105:354–364.

Minnis, A. M., and D. L. Lindner. 2013. Phylogenetic evaluation of *Geomyces* and allies reveals no close relatives of *Pseudogymnoascus destructans*, comb. nov., in bat hibernacula of eastern North America. Fungal Biology 117:638–649.

Mitchell, M. S., H. Cooley, J. A. Gude, J. Kolbe, J. J. Nowak, K. M. Proffitt, S. N. Sells, and M. Thompson. 2018. Distinguishing values from science in decision making: setting harvest quotas for mountain lions in Montana. Wildlife Society Bulletin 42:13–21.

Montana Fish, Wildlife, and Parks. 2020. White-nose fungus found in Montana. Montana Fish, Wildlife, and Parks, Helena, USA. https://wildlife.org/white-nose-fungus-found-in-montana/. Accessed 20 Jan 2021.

Moore, M. S., J. D. Reichard, T. D. Murtha, M. L. Nabhan, R. E. Pian, J. S. Ferreira, and T. H. Kunz. 2013. Hibernating little brown myotis (*Myotis lucifugus*) show variable immunological responses to white-nose syndrome. PLoS One 8:e58976.

Moosman, P. R., P. R. Anderson, and M. G. Frasier. 2017. Use of rock-crevices in winter by big brown bats and eastern small-footed bats in the Appalachian Ridge and Valley of Virginia. Banisteria 48:9–13.

Mulec, J., E. Covington, and J. Walochnik. 2013. Is bat guano a reservoir of *Geomyces destructans*? Open Journal of Veterinary Medicine 3:161–167.

Muller, L. K., J. M. Lorch, D. L. Lindner, M. O'Connor, A. Gargas, and D. S. Blehert. 2013. Bat white-nose syndrome: a real-time TaqMan polymerase chain reaction test targeting the intergenic spacer region of *Geomyces destructans*. Mycologia 105:253–259.

Neubaum, D. J. 2018. Unsuspected retreats: autumn transitional roosts and presumed winter hibernacula of little brown myotis in Colorado. Journal of Mammalogy 99:1294–1306.

Neubaum, D. J., and J. L. Siemers. 2021. Bat swarming behavior among sites and its potential for spreading white-nose syndrome. Ecology 102:e03325.

Newman, B. A. 2020. Winter torpor and roosting ecology of tri-colored bats (*Perimyotis subflavus*) in trees and bridges. Thesis, Department of Forestry and Environmental Conservation, Clemson University, Clemson, South Carolina, USA.

Niaqually Valley News. 2018. Deadly bat fungus found at Mount Rainier National Park. https://www.yelmonline.com/stories/deadly-bat-fungus-found-at-mount-rainier-national-park,104626. Accessed 29 Mar 2018.

North American Bat Monitoring Program (NABat). 2021. https://www.nabatmonitoring.org/about-1. Accessed 10 Oct 2021.

Oedin, M., F. Brescia, A. Millon, B. P. Murphy, P. Palmas, J. C. Z. Woinarski, and E. Vidal. 2021. Cats *Felis catus* as a threat to bats worldwide: a review of the evidence. Mammal Review 51:323–337.

O'Shea, T. J., P. M. Cryan, D. T. S. Hayman, R. K. Plowright, and D. G. Streicker. 2016. Multiple mortality events in bats: a global review. Mammal Review 46:175–190.

Palmer, J. M., K. P. Drees, J. T. Foster, and D. L. Lindner. 2018. Extreme sensitivity to ultraviolet light in the fungal pathogen causing white-nose syndrome of bats. Nature Communications 9:35.

Reeder, D. M., C. L. Frank, G. G. Turner, C. U. Meteyer, A. Kurta, E. R. Britzke, M. E. Vodzak, S. R. Darling, C. W. Stihler, A. C. Hicks, et al. 2012. Frequent arousal from hibernation linked to severity of infection and mortality in bats with white-nose syndrome. PLoS One 7:e38920.

Reichard, J. D., and T. H. Kunz. 2009. White-nose syndrome inflicts lasting injuries to the wings of little brown myotis (*Myotis lucifugus*). Acta Chiropterologica 11:457–464.

Rocke, T. E., B. Kingstad-Bakke, M. Wüthrich, B. Stading, R. C. Abbott, M. Isidoro-Ayza, H. E. Dobson, L. dos Santos Dias, K. Galles, J. S. Lankton, et al. 2019. Virally-vectored vaccine candidates against white-nose syndrome induce anti-fungal immune response in little brown bats (*Myotis lucifugus*). Scientific Reports 9:6788.

Rodhouse, T. J., R. M. Rodriguez, K. M. Banner, P. C. Ormsbee, J. Barnett, and K. M. Irvine. 2019. Evidence of region-wide bat population decline from long-term monitoring and Bayesian occupancy models with empirically informed priors. Ecology and Evolution 9:11078–11088.

Runge, M. C., J. B. Grand, and M. S. Mitchell. 2013. Structured decision making. Pages 51–72 *in* P. R. Krausman and J. W. Cain, editors. Wildlife management and conservation: contemporary principles and practice. Johns Hopkins University Press, Baltimore, Maryland, USA.

Russell, R. E., G. V. DiRenzo, J. A. Szymanski, K. E. Alger, and E. H. C. Grant. 2020. Principles and mechanisms of wildlife population persistence in the face of disease. Frontiers in Ecology and Evolution 8: 569016.

Salinas-Ramos, V. B., E. Mori, L. Bosso, L. Ancillotto, and D. Russo. 2021. Zoonotic risk: one more good reason why cats should be kept away from bats. Pathogens 10:304.

Sasse, D. B., and A. R. Gramza. 2021. Influence of the COVID-19 pandemic on public attitudes toward bats in Arkansas and implications for bat management. Human Dimensions of Wildlife 26:90–93.

Schwab, N. A., and T. J. Mabee. 2014. Winter acoustic activity of bats in Montana. Northwestern Naturalist 95:13–27.

Schwab, S. R., C. M. Stone, D. M. Fonseca, and N. H. Fefferman. 2018. The importance of being urgent: the impact of surveillance target and scale on mosquito-borne disease control. Epidemics 23:55–63.

Sells, S. N., M. S. Mitchell, V. L. Edwards, J. A. Gude, and N. J. Anderson. 2016. Structured decision making for managing pneumonia epizootics in bighorn sheep. Journal of Wildlife Management 80:957–969.

Sewall, B. J., G. G. Turner, M. R. Scafini, M. F. Gagnon, J. S. Johnson, M. K. Keel, E. Anis, T. M. Lilley, J. P. White, C. L. Hauer, and B. E. Overton. 2023. Environmental control reduces white-nose syndrome infection in hibernating bats. Animal Conservation. https://doi.org/10.1111/acv.12852. Accessed 10 April 2023.

Shapiro, H. G., A. S. Willcox, E. V. Willcox, and M. L. Verant. 2021. US National Park visitor perceptions of bats and white-nose syndrome. Biological Conservation 261:109248.

Siemer, W. F., T. B. Lauber, H. E. Kretser, K. L. Schuler, M. Verant, C. J. Herzog, and K. A. McComas. 2021. Predictors of intentions to conserve bats among New York property owners. Human Dimensions of Wildlife 26:275–292.

Storm, J. J., and J. G. Boyles. 2010. Body temperature and body mass of hibernating little brown bats *Myotis lucifugus* in hibernacula affected by white-nose syndrome. Acta Theriologica 56:123–127.

Swift, J. F., R. F. Lance, X. Guan, E. R. Britzke, D. L. Lindsay, and C. E. Edwards. 2018. Multifaceted DNA metabarcoding: Validation of a noninvasive next-generation approach to studying bat populations. Evolutionary Applications 11:1120–1138.

Taylor, D. A. R., R. W. Perry, D. A. Miller, and W. M. Ford. 2020. Forest management and bats. https://www.srs.fs.usda.gov/pubs/misc/misc_2020_perry_001.pdf. Accessed on 27 October 2021.

Thapa, V., G. G. Turner, and M. J. Roossinckid. 2021. Phylogeographic analysis of *Pseudogymnoascus destructans* partitivirus-pa explains the spread dynamics of white-nose syndrome in North America. PLoS Pathogens 17: e1009236.

Turner, G. G., C. U. Meteyer, H. A. Barton, J. F. Gumbs, D. M. Reeder, B. Overton, H. Bandouchova, T. Bartonička, N. Martínková, J. Pikula, et al. 2014. Nonlethal screening of bat-wing skin with the use of ultraviolet fluorescence to detect lesions indicative of white-nose syndrome. Journal of Wildlife Diseases 50:566–573.

Turner, G. G., D. M. Reeder, and J. T. H. Coleman. 2011. A five-year assessment of mortality and geographic spread of white-nose syndrome in North American bats and a look to the future. Bat Research News 52:13–27.

Turner, G. G., B. J. Sewall, M. R. Scafini, T. M. Lilley, D. Bitz, and J. S. Johnson. 2022. Cooling of bat hibernacula to mitigate white-nose syndrome. Conservation Biology 36:e13803.

US Fish and Wildlife Service (USFWS). 2016. 4(d) rule for the northern long-eared bat. Federal Register 81:1900–1922.

US Fish and Wildlife Service (USFWS). 2021*a*. Endangered Species Act: Section 7(a)(2). Section 7 Consultation. https://www.fws.gov/midwest/endangered/section7/index.html. Accessed 16 Dec 2021.

US Fish and Wildlife Service (USFWS). 2021*b*. Bats affected by WNS. White-nose syndrome response team. https://www.whitenosesyndrome.org/static-page/bats-affected-by-wns. Accessed 27 Oct 2021.

US Fish and Wildlife Service (USFWS), US Geological Survey (USGS), and Bat Conservation International (BCI). 2022. Analytical assessments in support of the US Fish and Wildlife Service 3-bat species status assessment. https://usgs-cru-individual-data.s3.amazonaws.com/wford/tech_publications/NABat-Analyses-SSA-MYLU-PESU-MYSE-20220108-1.pdf. Accessed 5 Sept 2022.

US Geological Survey (USGS). 2021. White-nose syndrome surveillance. https://www.usgs.gov/centers/nwhc/science/white-nose-syndrome-surveillance?qt-science_center_objects=0#qt-science_center_objects. Accessed 27 Oct 2021.

USGS National Wildlife Health Center. 2021. National Wildlife Health Center bat white-nose syndrome (WNS)/ Pd surveillance submission guidelines winter 2020–2021. https://www.usgs.gov/media/files/bat-white-nose-syndromepd-surveillance-submission-guidelines. Accessed 27 Oct 2021.

Vanderwolf, K. J., L. J. Campbell, T. L. Goldberg, D. S. Blehert, and J. M. Lorch. 2021. Skin fungal assemblages of bats vary based on susceptibility to white-nose syndrome. ISME Journal 15:909–920.

Vanderwolf, K. J., D. Malloch, and D. F. McAlpine. 2016. Detecting viable *Pseudogymnoascus destructans* (Ascomycota: Pseudeurotiaceae) from walls of bat hibernacula: effect of culture media. Journal of Cave and Karst Studies 78:158–162.

Verant, M. L., C. U. Meteyer, J. R. Speakman, P. M. Cryan, J. M. Lorch, and D. S. Blehert. 2014. White-nose syndrome initiates a cascade of physiologic disturbances in the hibernating bat host. BMC Physiology 14:10.

Warnecke, L., J. M. Turner, T. K. Bollinger, V. Misra, P. M. Cryan, D. S. Blehert, G. Wibbelt, and C. K. R. Willis. 2013. Pathophysiology of white-nose syndrome in bats: a mechanistic model linking wing damage to mortality. Biology Letters 9:20130177.

Weller, T. J., T. J. Rodhouse, D. J. Neubaum, P. C. Ormsbee, R. D. Dixon, D. L. Popp, J. A. Williams, S. D. Osborn, B. W. Rogers, L. O. Beard, et al. 2018. A review of bat hibernacula across the western United States: implications for white-nose syndrome surveillance and management. PLoS ONE 13:e0205647.

Washington Department of Fish and Wildlife (WDFW). 2016. Bats astray. http://happyvalleybats.org/wp-content/uploads/2018/05/bats_astray_Washington.pdf. Accessed 27 Oct 2021.

White, J. A., P. W. Freeman, H. W. Otto, and C. A. Lemen. 2020. Winter use of a rock crevice by northern long-eared myotis (*Myotis septentrionalis*) in Nebraska. Western North American Naturalist 80:114–119.

White-Nose Syndrome National Response Team (WNSNRT). 2019*a*. WNS case definitions. US Fish and Wildlife Service, White-Nose Syndrome Response Team 3, Hadley, Massachusetts, USA. https://s3.us-west-2.amazonaws.com/prod-is-cms-assets/wns/prod/de91e7d0-9c0e-11e9-ad22-19882a049409-WNS-Case-Definitions_v5162019_FINAL-clean-logo.pdf. Accessed 27 Oct 2021.

White-Nose Syndrome National Response Team (WNSNRT). 2019*b*. White-nose syndrome show cave guidance: recommended practices to reduce risks of people spreading the fungus *Pseudogymnoascus destructans*. https://s3.us-west-2.amazonaws.com/prod-is-cms-assets/wns/prod/920cf500-0c6e-11ea-a154-67ca1cde5e5d-FINAL-WNS-Show_cave_guidance_11122019.pdf. Accessed 27 Oct 2021.

White-Nose Syndrome National Response Team (WNSNRT). 2023. Bats affected by white-nose syndrome. US Fish and Wildlife Service, White-Nose Syndrome Response Team 3, Hadley, Massachusetts, USA. https://www.whitenosesyndrome.org/static-page/bats-affected-by-wns. Accessed 10 April 2023.

Whiting, J. C., B. Doering, K. Aho, and J. Rich. 2021. Long-term patterns of cave-exiting activity of hibernating bats in western North America. Scientific Reports 11:8175.

Wray, A. K., M. A. Jusino, M. T. Banik, J. M. Palmer, H. Kaarakka, J. P. White, D. L. Lindner, C. Gratton, M. Z. Peery, J. P. White, et al. 2018. Incidence and taxonomic richness of mosquitoes in the diets of little brown and big brown bats. Journal of Mammalogy 99:668–674.

Wright, A. D., R. F. Bernard, B. A. Mosher, K. M. O'Donnell, T. Braunagel, G. V. DiRenzo, J. Fleming, C. Shafer, A. B. Brand, E. F. Zipkin, et al. 2020. Moving from decision to action in conservation science. Biological Conservation 249:108698.

IV | Wildlife Disease and Health in Avian Species

17 Waterfowl Diseases

A Catalyst for Wildlife Health Professions in North America

David A. Jessup

Introduction

It can be argued that repeated major mortality events in wild waterfowl (specifically ducks and geese) were a significant factor in North American state, provincial, and federal wildlife management agencies recognizing the importance of wildlife diseases in the latter half of the 20th century. Type C botulism die-offs, primarily a warm-weather phenomenon known as "western duck sickness," had been described for over 150 years (Rosen 1971*a*) but were considered essentially a "natural phenomenon" until links with water management and farming practices became apparent (Hunter and Clark 1971). Beginning in the early 20th century, factors converged to increase recognition of waterfowl diseases: waterfowl harvesting, which had been a source of food for pioneers and early settlers, was beginning to be regulated with the establishment of government wildlife agencies; knowledge about waterfowl numbers and seasonal migration culminated in flyway delineation and establishment of refuges; and international treaties provided transboundary protection. The draining and conversion of vast wetlands, prairie potholes, swamps, and marshes to farmland greatly reduced waterfowl habitat (by some estimates 90% or more). This concentrated waterfowl, likely exacerbating

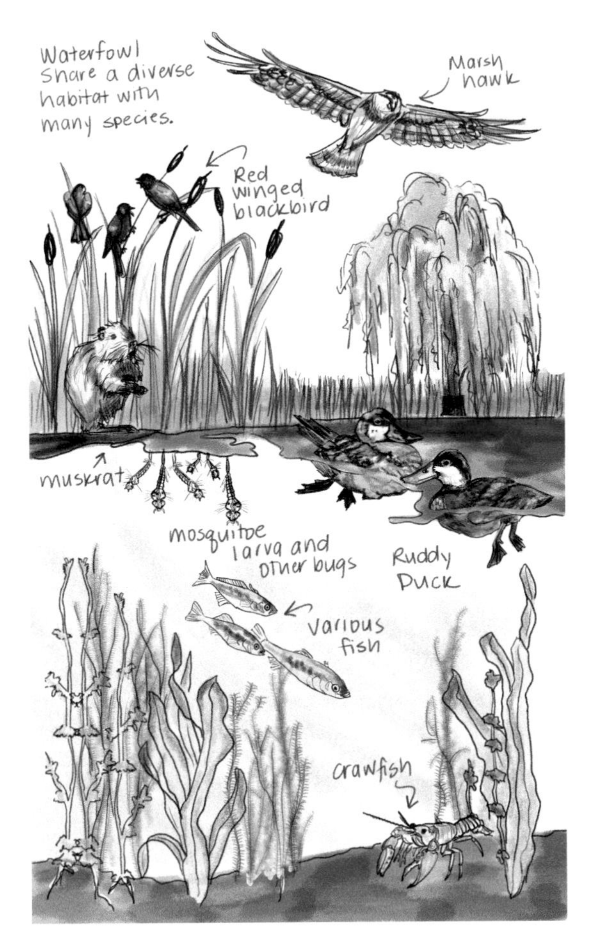

Illustration by Laura Donohue.

disease mortality and making die-offs more visible. Economic prosperity and American traditions promoted waterfowl hunting, which had been primarily a subsistence activity, into a popular form of recreation. The sciences that would become wildlife management were coalescing, making way for studies on waterfowl and the wetlands on which they depended (Leopold 1933). United States federal (US Fish and Wildlife Service [USFWS]) and state wildlife agencies established waterfowl refuge systems. During and after the Great Depression, popular organizations like Ducks Unlimited preserved and enhanced waterfowl habitat on private lands in North America. In short, waterfowl became a more valued, valuable, and quantifiable natural resource and were no longer viewed as an unlimited commodity. Waterfowl acquired increased biological recognition, legal protection, and social and political constituencies.

Therefore, it's little wonder that piles of thousands—even tens of thousands—of dead ducks and geese stacking up regularly became a cause for concern to duck hunters, with state wildlife agencies, the USFWS, and biologists in academia investigating large-scale waterfowl mortality events (Figure 17.1). Around 1950, a new phenomenon—massive winter mortality events caused by *Pasteurella multocida*—were described in both California and the Rainwater Basin of Nebraska (Friend 2006). Although "avian cholera" (*P. multocida* septicemia) may have existed in North America, it is unlikely that winter epidemics killing tens of thousands of waterfowl—having such recognizable gross lesions—would have been missed (Jessup 1986). A virulent strain of bacteria (now believed to have come from domestic turkeys) emerged to become a second recurring scourge of increasingly valued waterfowl, with die-offs accounting for hundreds of thousands of birds annually. During the formative years (1950–1960) of the Wildlife Disease Association, waterfowl diseases were one of the most important areas of research, and the primary charge of the federal government wildlife disease investigation unit at Patuxent, Maryland, USA

Fig. 17.1. Burning duck carcasses, part of a pile of approximately 10,000 dead waterfowl resulting from a botulism die-off in California's Central Valley in 1974. Such scenes inspired political, social, and financial support for wildlife health research and management efforts to reduce massive waterfowl die-offs. California Department of Fish and Game file photograph.

(USFWS), as well as laboratories in California, Michigan, and other states.

Lead poisoning due to ingestion of lead shotgun pellets had been noted for over 100 years, and although it did not cause spectacular focal mortality events, it did kill many birds year after year and was clearly not a "natural phenomenon." Frank Bellrose's work was vital in helping document and quantify losses, and in galvanizing serious discussion about the need for non-toxic shot (Sanderson and Bellrose 1986). Federal legal efforts to curtail lead poisoning from various sources were being considered by the 1970s. Lead shot for waterfowl hunting was phased out in the United States beginning in 1987–1988 and was banned by 1991.

A herpes virus of European origin (Netherlands) caused extensive mortality in the commercial duck-raising operations of Long Island, New York, USA, in the late 1960s, and the potential for it to infect wild waterfowl was a great concern (Leibovitz 1971). Small outbreaks in urban ponds and wild waterfowl occurred sporadically, but an epornitic at Lake Andes, South Dakota, involving over 140,000 birds in 1973 (Friend 2006, Wobeser 1981) seemed to confirm the worst fears: duck viral enteritis (DVE; also called duck plague) was becoming endemic and could cause major mortality events under certain conditions. Although outbreaks clustered around the onset of breeding season (April to May in many areas), if or how the virus might be maintained in wild populations was unknown.

These four waterfowl diseases (botulism, avian cholera, DVE, and lead poisoning) caused repeated significant mortality during the latter half of the 20th century. Biologists worried about the sustainability of waterfowl populations in North America—high mortality forced changes in some waterfowl hunting regulations and seasons and appeared to threaten the viability of the now-extensive and very popular public and private management efforts in North America. In 1976, these diseases became a primary focus of the new USFWS National Wildlife Health Research Center in Madison, Wisconsin, USA. Avian influenza in wild waterfowl did not become a high priority until the beginning of the 21st century (see Chapter 18). This chapter summarizes common diseases causing major waterfowl mortality in North America and others with similar signs (Table 17.1). These case examples are well studied and helped to catalyze political, social, financial, and legal actions toward improving waterfowl management and promoting wildlife health research.

Table 17.1. Gross lesions of the most common waterfowl diseases seen in North America

Gross lesion	Diseases associated with gross finding
No obvious gross lesions	Usually indicates acute death, often due to botulism; can also be due to hypothermia and electrocution
Hatchet breast	Commonly due to emaciation or malnutrition secondary to aspergillosis, lead poisoning, crippling injuries from hunting or predators, or parasitism; not seen in acute death due to botulism, avian cholera, duck viral enteritis, and avian influenza
Stained vent area	Green staining commonly seen due to lead poisoning and occasionally avian cholera; blood staining may be caused by duck viral enteritis
Prolapsed penis	Rare lesion; seen due to duck viral enteritis and organophosphate poisoning
Petechial or ecchymotic hemorrhages on heart and pinpoint pale liver infarcts	Seen commonly due to avian cholera and duck viral enteritis and occasionally avian influenza

Under field conditions, these can be used to reach a presumptive diagnosis, but submission of a representative sample of fresh bird carcasses to a diagnostic laboratory is strongly recommended. See also Figure 17.4.

Type C Botulism (Alkali Disease, Limberneck, Western Duck Sickness)

Etiology

Clostridium botulinum is a gram-positive bacterium whose spores are ubiquitous and persistent in alkaline soils in North America. When spores germinate in the presence of a common bacteriophage, growth of vegetative forms result in production of very potent toxins that can cause paralysis in a wide variety of species. Type C botulinum toxin is the cause of waterfowl botulism, although type E may be found in piscivorous birds like loons and gulls (Friend 2006, Wobeser 1981). Warmer temperatures (around 21–27°C) and neutral to alkaline conditions (above pH 5.7) optimize clostridial growth and toxin production. Growth of the vegetative form requires decomposing proteinaceous matter (often small dead animals or invertebrates) under anaerobic conditions (often under water) (Wobeser 1981, Jensen and Allen 1960). Drowned burrowing animals, birds that strike powerlines, even aquatic insect mortality can provide the decomposing protein source (Rosen 1971*a*, Jensen and Allen 1960). The maggots produced by flies feeding on the decomposing animals concentrate the botulinum toxin (as few as two maggots may be a lethal dose for 50% of the population) and serve as an irresistible lure for birds (Hunter and Clark 1971, Jessup 1986). Each bird that dies provides additional decomposing protein in warm, anaerobic conditions and, in just a day or two, an outbreak can undergo exponential growth. This is the recipe for "botulism soup," and the ingredients for this soup are widely available during the warmer months in much of the western United States and Canada.

Although the conditions under which waterfowl botulism may occur are common and mostly natural, modern agricultural practices can exacerbate the problem (Hunter and Clark 1971, Jessup 1986). Summer and early fall flooding of recently cultivated alkaline soils containing dead insects and small animals and fluctuating water levels provide extensive shallow water and muddy soil that tends to stay warm. Sewage or waters high in agricultural wastes can cause nutrient blooms that reduce dissolved oxygen and kill fish and invertebrates, another recipe for botulism soup.

Fig. 17.2. A northern pintail duck (*Anas acuta*) in stage 1 type C botulism intoxication with flaccid paralysis of the muscles of the legs and wings: because of their inability to fly, they are often referred to as "floppers." California Department of Fish and Game file photograph.

Demographics and Field Biology

Although waterfowl botulism is distinctly seasonal (warm half of the year), the conditions for vegetative *C. botulinum* growth can occasionally occur within a carcass during colder months, resulting in small focal off-season die-offs. As noted, there are distinct geographic, temperature and weather conditions but no waterfowl species, age, or sex predilection. Carcasses are almost always in excellent body condition (rapid death) with essentially no lesions. Type C toxin is fairly specific for birds. Retrieving dogs (and other mammals) are essentially immune and can be used to gather live and dead birds. Under optimal conditions outbreaks may be massive, as many as 500,000 birds, with outbreaks involving 50,000 or more birds being fairly common (Rosen 1971*a*, Jessup 1986).

Signs and Gross Lesions

The term "limberneck" describes the partially paralyzed condition of ducks and geese that have ingested a sublethal dose of toxin or are early in progression of intoxication. The toxin blocks neuromuscular transmission resulting in a distinct flaccid paralysis. The muscles of the legs are affected first followed by the wings, then the back and neck, but birds may still breath and pupils may be responsive (Figure 17.2). If birds survive, muscle control returns in the opposite order. Essentially, a rapidly expanding waterfowl die-off in summer or fall in western North America, where birds are in excellent body condition and exhibit no lesions, is likely to be botulism until proven otherwise, though representative sampling and postmortem examination is recommended.

Pathogenesis and Pathology

Botulism intoxication leaves no observable gross lesions and no specific microscopic lesions in waterfowl. Ingestion of the preformed toxin causes a progressive curariform flaccid paralysis that can be classified by stages, and affected birds may haul out, drown, or die of asphyxiation, predation, or other causes. Stage 1 paralysis begins with the muscles of the wings and legs (floppers), stage 2 extends to the neck muscles (limberneck), and few birds survive stage 3 paralysis involving the back and chest (waterfowl have no muscular diaphragm), but striated muscles remain functional (heart beating, pupils responsive) for a while. Banding studies suggest that birds that survive Stage 1 or 2 paralysis have some resistance to subsequent exposure to botulism (Jessup 1986). Scavenging species such as vultures have several orders of magnitude greater resistance to the toxin.

Diagnosis

Characteristic field conditions and birds showing neurologic behavior as well as the lack of gross lesions in carcasses is presumptive of botulism, but demonstration of botulinum toxin is the only way to get a definitive diagnosis. The mouse protection assay (MPA), where serum from affected waterfowl is inoculated into laboratory mice, half of which are given type C specific antitoxin, remains the "gold standard" (Wobeser 1981). An enzyme-linked immunosorbent assay test has also been developed, but it is not as sensitive as the MPA (Friend 1987).

Outcome and Treatment

Individual birds can be treated with type C botulism antitoxin, but it is expensive and not easily obtained in quantity. Birds in stage 1 paralysis (floppers) have a significant chance of recovery, particularly if gavaged with water to flush the alimentary tract and provided supportive care, shade, and protection for 24–48 hours (D. A. Jessup, unpublished). However, triage may be necessary, saving the antitoxin and intensive care for rare or sensitive species when large flocks of free-flying waterfowl are involved. Management and preventive measures may save more birds than treatment during an outbreak.

Management

Bird-scaring devices like propane cannons, fireworks, and airboat herding may help keep birds away from smaller outbreak foci but are difficult to implement over very large areas. The pickup of dead and affected live birds, often by airboat crews, can reduce sources of decomposing protein, maggot production, and botulinum toxin. Lowering water levels to dry out the land or flooding deeply with cold water have been attempted (Hunter and Clark 1971). Constructing water barriers (or checks) having steeper sides may help. Conversion away from flood irrigation to less wasteful water management practices or to agricultural crops that don't require summer or fall flooding can help eliminate botulism soup (Hunter and Clark 1971). Waterfowl managers of state, federal, and private lands can adapt to the biology of botulism by altering infrastructure design, flooding regimes, timing and planning of land use, and cover crop management practices (Figure 17.3).

The carcasses of birds dying of botulism contain massive amounts of bacteria and toxin, and the maggots of flies concentrate it in a form attractive to birds, expanding and prolonging outbreaks. In many locations, the water table may not allow deep burial, so on-site incineration of carcasses may be the most cost-effective disposal option. Complete incineration essentially desiccates and destroys toxic tissues and maggots, but it is not easy to accomplish when dealing with truckloads of water-saturated dead waterfowl. A more complete discussion of waterfowl carcass incineration (also used with avian cholera and DVE) including pictures showing several effective methods can be found in Friend (1987).

Financial, Legal, Political, and Social Factors

Massive waterfowl die-offs are concerning to the public and may be alarming when words like "die-off," "massive mortality event," "toxin," and "botulism" appear in press reports. Public outreach, including preprepared media releases and background information are recommended. Although die-offs may occur on waterfowl refuges, they often occur on, or spread to, private lands. Despite suspicion of government in general, many ranchers and farmers have conservation interests and may be willing to help with botulism response. They may also appreciate help with cleanup and reduced financial and legal concerns over air and water quality. Disease response efforts can be an opportunity to elicit common ground, cooperation, and property access that may be beneficial in the future and presents an

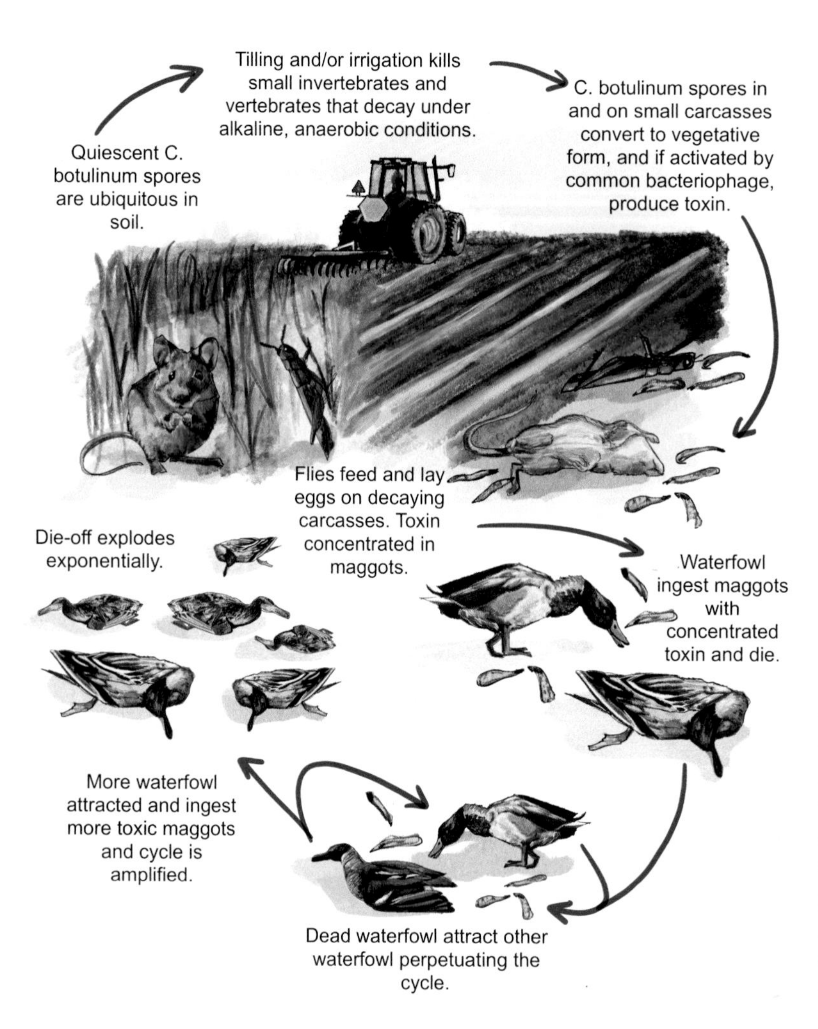

Fig. 17.3. Waterfowl botulism die-offs involve a cycle of events that include a source of protein decomposing under warm anaerobic, alkaline conditions, where *Clostridium botulinum* spores can convert to the vegetative state and bacteriophages that activate toxin production are all present. This is the recipe for botulism soup. Illustration by Laura Donohue.

opportunity to better inform farmers about agricultural practices that help preclude outbreaks and the mass mortality of botulism (Figure 17.6*a*).

Avian Cholera (Waterfowl Cholera or Avian Pasteurellosis)

History

Fowl cholera caused by *P. multocida* in waterfowl appears to have originated in domestic turkeys (Samuel et al. 2007), where it had been a recognized cause of mortality since at least the 1930s. Wintering waterfowl have always congregated in flocks, but the loss of marshlands (90–95%) led to a greater concentration of ducks and geese. Heavy rains, cold weather, and storms have been posited as potential stressing factors, although waterfowl cholera occurs in mild and harsh winter weather. The first reports of avian cholera mortality in North American waterfowl coincides with some of the most widespread post-World War II use of dichlorodiphenyltrichloroethane (DDT) and other chlorinated hydrocarbon insecticides for control of mosquito-borne diseases and crop pests. Reduced immune function might explain increased susceptibility to a bacterial pathogen—mallard ducks (*Anas platyrhynchos*) dosed with petrochemicals are more susceptible to fatal infection with *P. multocida* (Friend 2006)—but little evidence exists to support chemical-induced immune suppression as a significant or ongoing factor. Some evidence suggests that temperature and water-quality conditions can prolong

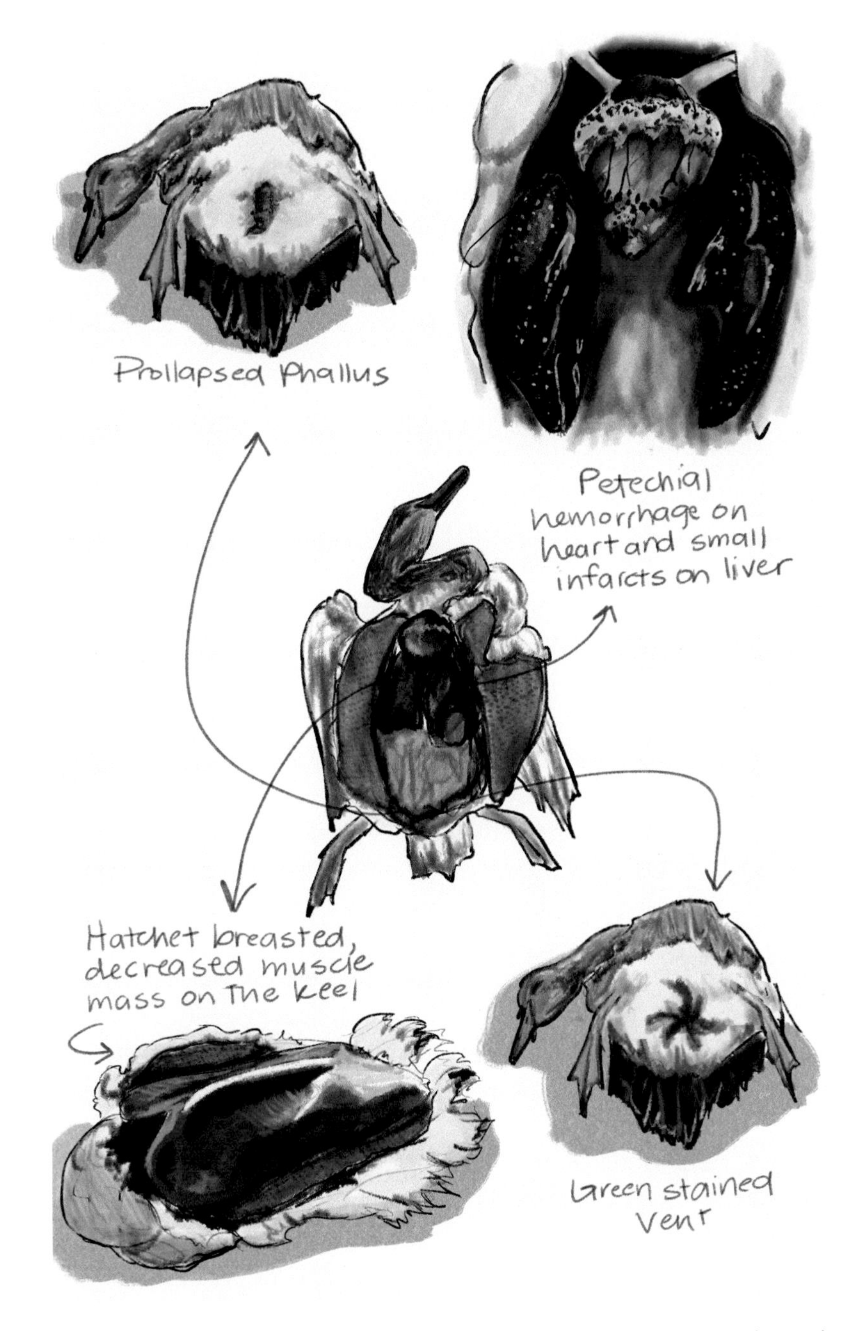

Fig. 17.4. Although the gross lesions of avian cholera (multiple small petechial to ecchymotic hemorrhages on the surface of the heart, great vessels, and other organs and multiple small white spots—septic infarcts—in the liver) are striking, they are not pathognomonic. Similar lesions may be seen with duck viral enteritis and avian influenza. Characteristic gross lesions and field signs of the more common waterfowl diseases are shown. Illustration by Laura Donohue.

bacterial survival outside the host, but more recent studies suggest this is not a significant factor in temporal persistence and recurrence at the same locations (often refuges) year after year (Blanchong et al. 2006). The most convincing evidence as to year-to-year maintenance of fowl cholera comes from the work of Samuel et al. (2005) implicating lesser snow geese (*Chen caerulescens*) and Ross's geese (*Chen rossii*) as carriers of *P. multocida*. The geese can maintain the bacteria during the Arctic summer and may even suffer low-level sporadic mortality. This observation supports anecdotal reports by California biologists that "white geese" are the first to die (Figure 17.6*b*), with their carcasses often observed from the air at the center of early spreading outbreaks (D. A. Jessup, unpublished data). Although some questions remain as to the exact role of biological, environmental, and anthropogenic factors in the epidemiology of avian cholera, fundamental changes occurred in the host–pathogen–environment relationship between waterfowl, *P. multocida*, and North American waterfowl habitat around 1950, and this resulted in the emergence of an apparently new disease in the mid-20th century.

Etiology

The gram-negative coccobacillus *Pasteurella multocida* causes avian (fowl or waterfowl) cholera (Rosen 1971*b*). It is a nonmotile, penicillin-sensitive organism. Many strains exist, which are currently classified into five serogroups based on capsular composition and 16 somatic serovars (Wobeser 1981). Different strains of *P. multocida* are a cause of disease in a wide variety of wild and domestic animals, with strain 1 primarily in migratory waterfowl. The microbiology of this organism was first described by Pasteur (1880) and was reviewed by Rosen (1971*b*). High concentrations of *P. multocida* are present in mucus and feces of dead and dying birds, and bird to bird transmission occurs via ingestion and inhalation. As with many similar bacterial diseases, contact rate between individuals and flocks, spatial concentration of populations (crowding or "reduced social distancing"), and attraction of unexposed birds to carcasses (decoy effect) at foci of infection exacerbate severity and spread of disease.

Demographics and Field Biology

Infected waterfowl die very quickly (6–12 hours after exposure), sometimes literally falling from the sky (Friend 1987). More subacute and less spectacular infections are also observed. Raptors and scavenging birds feeding on carcasses are relatively resistant and have a longer course of disease.

Signs and Gross Lesions

Ducks, geese, and swans with avian cholera die acutely and peracutely and are almost always in good to excellent body condition (Jessup 1986). A thousand or more birds a day may be found dead, as is true of botulism die-offs. Outbreaks are most often seen in middle to late fall and early winter as waterfowl complete migration to central and northern California, southern Oregon, the Texas Panhandle, and the Rainwater Basin of Nebraska (the Pacific and Central flyways) (Friend 1987). No consistent pattern of annual disease emergence is observed in the Mississippi and Atlantic flyways, although repeated avian cholera die-offs of eider ducks (*Somateria mollissima*) off the Maine coast are reported (Friend 1987).

Although acute death with no premonitory signs is the most common presentation, waterfowl with pasteurellosis may appear sleepy or convulse throwing their heads back against their wings (Friend 1987), behaviors also seen with DVE and some forms of pesticide poisoning (particularly organophosphates) (D. A. Jessup, unpublished data). Characteristic gross lesions consist of multiple small (petechial to ecchymotic) hemorrhages across the surface of the heart and on great vessels and multiple small white spots (septic infarcts) in the liver (Rosen 1971b, Friend 1987). The liver and spleen may be swollen, friable, and copper colored in some cases. The surface of the gizzard and major vessels of the abdominal organs may also have petechial and ecchymotic hemorrhages (Friend 1987). Birds die so quickly that there is often undigested food in the crop. Lower portions of the digestive tract commonly contain a thick yellow fluid rich in *Pastuerella* bacteria, and the vent may be pasted with feces (Figure 17.4) (Friend 1987).

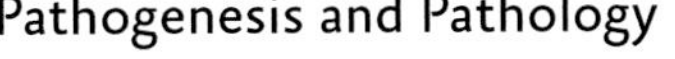

Pathogenesis and Pathology

Inhaled or ingested *P. multocida* colonize the upper alimentary or respiratory tract and invade the bloodstream; in a few hours, the bacteria reproduce in large numbers, which causes septicemia that spreads to all body organs (Jessup 1986). Clots of bacteria and inflammatory products plug blood vessels, and the toxins produced by dying bacteria cause tissue damage and septic shock. Microscopic lesions are those of acute bacterial septicemia (septic thrombi), and most organs are affected to some degree (Wobeser 1981).

Diagnosis

A simple blood smear stained with Wright's or Giemsa can reveal large numbers of bipolar staining, rod-shaped bacteria in the blood of infected birds (Jessup 1986). A presumptive diagnosis of fowl cholera is warranted for waterfowl in good body condition

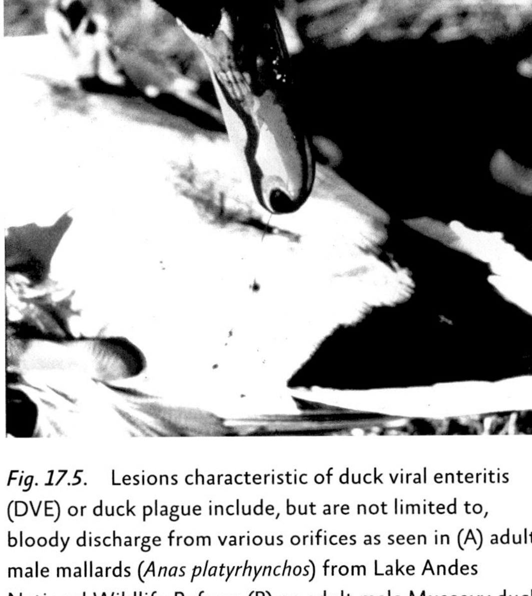

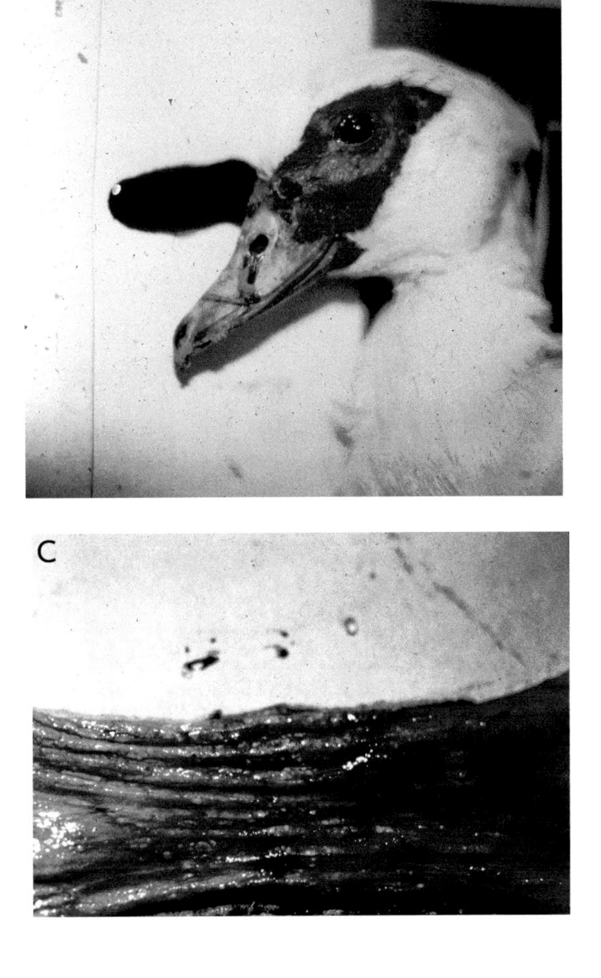

Fig. 17.5. Lesions characteristic of duck viral enteritis (DVE) or duck plague include, but are not limited to, bloody discharge from various orifices as seen in (A) adult male mallards (*Anas platyrhynchos*) from Lake Andes National Wildlife Refuge; (B) an adult male Muscovy duck (*Cairina moschata*) from a Sacramento park. Muscovy are often affected early when outbreaks occur in mixed-species flocks in park ponds, not because Muscovy ducks are carriers—as first assumed—but because they are highly susceptible. The enanthematous linear plaques in the esophagus of (C) the adult male Muscovy is a pathognomonic, although not a common, lesion in DVE infected waterfowl. (A) US Geological Survey file photograph; (B and C) Photographs by the author.

with characteristic gross lesions and with positive blood smear stains (Davidson 2006). Although it is beneficial to do a full postmortem examination, bacterial culture, and perhaps histopathology, these tests may be impractical for all but a subset of birds when hundreds to thousands are dying daily. Sterile collection of the whole heart and liver (separately) in Whirl-Pak collection bags followed by rapid refrigeration and shipment to a laboratory will allow for diagnosis by culture and staining. *Pastuerella multocida* survives reasonably well, for as long as a month in the bone marrow of the wings in some species (Friend 1987), when die-offs are late to be discovered. One problem with taking shortcuts when diagnosing avian cholera is that two significant viral diseases (i.e., avian influenza and duck virus enteritis) can cause similar, often indistinguishable gross lesions (Jessup 1986) and can occur under the same temporal and environmental circumstances (Figure 17.4). A complete postmortem examination of a subset of fresh dead birds conducted at a state, federal, or university diagnostic laboratory is recommended to verify the cause(s) of winter waterfowl die-offs.

Outcome and Treatment

Once in the bloodstream, *P. multocida* septicemia is invariably fatal to waterfowl. There is no treatment.

Management

Avian cholera often occurs on public and private waterfowl areas during the hunting season and is a relatively easy disease to diagnose; however, management has proven frustrating. Efforts to manage avian cholera are similar to those for botulism and largely consist of picking up and incinerating carcasses to prevent seeding the environment with more bacteria (as opposed to toxin and maggots) and so that dead birds do not act as decoys attracting live birds (Wobeser 1994). State, provincial, and federal agency personnel, often assisted by volunteers and waterfowl advocacy groups, can coordinate carcass removal (Figure 17.6c), which reduces the "decoy effect," although it is unclear if carcass collection reduces overall mortality (Wobeser 1994). Pyrotechnics, noise makers, and hazing with airboats have also been used but not in organized ways that foster scientific analysis of their effectiveness. Currently, waterfowl constituencies prefer action over inaction when large outbreaks occur. Regionally coordinated efforts to keep waterfowl populations dispersed across a flyway or a series of refuges and private wetlands may be of some benefit.

Financial, Legal, Political, and Social Factors

Although fowl cholera could theoretically spread to poultry, there is little historical evidence for this occurring. However, there may be social and political as well as biological motives and support for carcass cleanup. Waterfowl hunting is a popular and an increasingly expensive recreational activity that attracts wealthy and politically powerful people. Although there is no legal mandate to report or manage avian cholera outbreaks, decreasing mortality serves conservation interests and may reduce loss of income for supporting businesses, although economic studies to quantify this are not available. As noted in the introduction, social appreciation for waterfowl for both nonconsumptive recreation (birdwatching) and hunting as early as the 1950s combined to make the investigation and management of avian cholera a priority for government wildlife agencies. When appropriate, this public support may be translated into political support to improve waterfowl health and management.

Duck Viral Enteritis (DVE, Dutch Duck Plague)

History

Duck virus enteritis was described in duck-raising facilities in the Netherlands in the late 1950s. The first outbreak in the United States in 1967 occurred on commercial Pekin duck farms of Long Island,

New York, USA, and mortality was noted in black ducks (*Anas rubripes*) in nearby Flanders Bay (Leibovitz 1971). Over the next 5–6 years, a number of small outbreaks were reported in various locations across the United States, almost exclusively in semitame birds or park pond situations. In January 1973 at the Lake Andes National Wildlife Refuge in South Dakota, a massive mortality event killed approximately 100,000 mallards and 40,000 Canada geese (*Branta canadensis*) (Friend 1987). The US Department of Agriculture (USDA) subsequently dropped the "foreign animal disease" designation for DVE, although the USFWS expressed doubt that it had yet become endemic in wild waterfowl. This was due in part to difficulties in isolating the virus as a means of verification of infection. Small to moderate-sized outbreaks have continued to the present day, and it is now generally accepted that DVE is likely widespread, particularly in dense feral mixed flocks of waterfowl in community ponds.

Etiology

Anatid alphaherpesvirus 1 of the family Herpesviridae causes DVE. It is highly infectious, somewhat persistent in the environment, and can become latent in individual birds that may become lifelong virus shedders.

Demographics and Field Biology

About 75% of DVE outbreaks cluster around the months of April and May, and some connection to reproduction is posited, possibly hormone-induced immunosuppression that allows expression and shedding of latent virus. As noted, contact with domestic type, mixed breed (funny ducks), and park pond situations is a common, but not exclusive, finding.

Signs and Gross Lesions

Sick birds are seldom observed in the field possibly because they seek cover due to sensitivity to light. Affected birds may be weak, show CNS signs, or have convulsions, and most die rapidly (Friend 1987). Few other waterfowl diseases cause a bloody discharge from the vent or bill that is characteristic of DVE (Figures 17.5A and 17.5B). Males often have a prolapsed penis (penile prolapse and convulsions are also seen with organophosphate poisoning) (D. A. Jessup unpublished). Another classic gross lesion is a small area of necrosis just inside the rim of the upper bill (Friend 1987).

Pathogenesis and Pathology

The anatid herpesvirus attacks the hematopoetic tissues, particularly the lymphoid tissues of the digestive tract, resulting in band-shaped areas of necrosis in duck intestines, necrotic discs (lymphoid patches) in goose intestines, and necrosis in the cloaca of both (Jessup 1986). These lesions and extensive bleeding into the digestive tract should trigger strong suspicion of DVE. One fairly uncommon, but nearly pathognomonic lesion, is linear enanthematous plaques in the esophagus (Figure 17.5C; Jessup 1986). The extensive tissue damage and viremia can cause septic shock and lesions similar (and grossly indistinguishable from) those of avian cholera, vis-à-vis petechial and ecchymotic hemorrhages on the heart and great vessels and septic infarcts (small white spots) on the liver (Figure 17.4). For a more comprehensive description of microscopic lesions of DVE, see Leibovitz (1971) and Wobeser (1981).

Diagnosis

Although the lesions of DVE on full postmortem examination of fresh specimens are usually adequate for a presumptive diagnosis, definitive diagnosis is based on virus isolation.

Outcome and Treatment

Duck viral enteritis mortality events tend to be somewhat self-limiting. Many infected birds will die

Fig. 17.6. Waterfowl mortality events can be sudden and alarming in size. (A) Early irrigation of San Joaquin Valley alkaline soils during warm weather can cause large avian botulism die-offs, in this case over 5,000 waterfowl carcasses (which are a potent source of toxin) were gathered from adjacent fields for disposal. (B) Early season avian cholera epidemics in California often occur in Ross and snow geese (*Chen rossi* and *Chen caerulescens*, respectively) and can be spotted from the air as clusters of large, dead, white birds. This historically engendered rapid efforts to pick up carcasses to "get ahead of the outbreak." The realization that Ross and snow geese may be carriers of *P. multocida* has altered this thinking. (C) The removal of waterfowl carcasses from die-off sites has been one of the primary management responses to the more common waterfowl diseases as a means to reduce sources of toxins, pathogenic bacteria, or viruses from the environment and to prevent the presence of carcasses from attracting more potentially susceptible birds. Although its overall efficacy may be questionable, it is politically and socially popular. California Department of Fish and Game file photographs.

quickly, but others may survive to become carriers. There is no treatment. A commercial duck vaccine is available but does not show efficacy in many species of wild ducks, geese and swans.

Management

Elimination or reduction of contact between feral domestic ducks and wild waterfowl, to the extent this can be done, is to be encouraged. Keeping feral domestic-type ducks is not allowed on most federal and state waterfowl refuges. The DVE virus can remain viable for long periods of time, notably for 60 days in water from Lake Andes held at 4°C (Friend 2006). For many years, the USFWS recommended destruction of infected flocks, but this is no longer favored by most wildlife agencies. Carcass pick-up and incineration, or deep burial, area cleaning, and disinfection are the most accepted management responses (Friend 1987). Response equipment and vehicles should be decontaminated to reduce the potential of fomite transmission.

Financial, Legal, Political and Social Factors

Although DVE is no longer a USDA reportable disease, with no legal mandate for response or elimination, response to outbreaks is usually at least as vigorous as for other waterfowl diseases. In part, this is due to its relative rarity, the alarming signs (bleeding from nares and vent and large amounts of bloody feces), the name "duck plague," and the reduction of potential spread of DVE to wild and domestic waterfowl is seen as being in the public interest. The relative rarity of DVE outbreaks compared to those of botulism or avian cholera make the financial consequences of the disease pale by comparison. Efforts to round up and euthanize feral ducks from affected parks and private lakes have become controversial and have led to protests and lawsuits (both threatened and filed).

Lead Poisoning (Plumbism)

History and Etiology

In water birds, lead poisoning usually results from ingestion of lead pellets from shotgun shells. Some wetlands have been hunted using lead shot for 100–200 years fostering its accumulation. Fishing weights, and occasionally mining or smelter slag, are ingested by loons and swans. Annual mortality estimates reached as high as 1–2 million bird deaths before lead shot for waterfowl hunting was banned nationwide in 1991. See Chapter 20 for discussion of lead poisoning in raptors and scavenging birds.

Demographics and Field Biology

All bird species are susceptible to lead poisoning, but the condition is less common in ducks and geese having piscivorous diets. An exception to this is that loons often ingest lead fishing weights. Lead poisoning has been reported in all areas of North America and much of Europe as low-level mortality, not as large-scale die-offs, most often in areas that have been heavily hunted. Cases in ducks, geese, and swans are more commonly observed in early spring or at the onset of migration and may be mistaken for cripples as the birds are awkward or reluctant to fly when approached and may have an unsteady gait when attempting to escape into cover. A good retrieving dog can be very helpful in catching these birds. In many lead-poisoned birds, the vent will be pasted with bright green feces.

Signs and Gross Lesions

Most lead-poisoned waterfowl are emaciated (hatchet breasted) and devoid of normal fat reserves (Friend 1987). The crop and esophagus may be impacted with ingesta as the lead paralyzes the smooth muscles of the digestive tract. The gizzard lining is often stained bright green and the gall bladder distended with bile (Friend 1987). Lead pellets may or may not be found in the ingesta, but X-ray or fluoroscopy examination may reveal them. Lead shotgun pellets embedded in tissues are inert and do not cause lead poisoning unless the bird (and pellets) are ingested by another animal. See chapter 20 for discussion of pathogenesis and pathology.

Diagnosis

In waterfowl, a depressed and emaciated bird with an enlarged gall bladder, bile-stained gizzard lining, and green feces may be presumptively diagnosed as having lead poisoning. Finding lead fragments in the gizzard or intestinal tract or high blood or tissue lead levels can confirm the diagnosis. For discussion of treatment and outcomes in raptors and other bird species, see Chapter 20.

Management

After a phase-out period beginning in 1987–1988, the use of lead shot for hunting waterfowl was banned nationwide in 1991. It is also illegal to possess lead ammunition when hunting on state and federal refuges. Some states also ban the use of lead shot for upland game bird hunting. Recent changes in laws regarding lead rifle ammunition are covered in Chapter 20.

Financial, Legal, Political, and Social Factors

The toxicity of lead to a wide variety of species has been known for over 150 years (Sanderson and Bellrose 1986). By the 1970s, a strong body of evidence was accruing that spent lead shot accumulating in wetlands was a significant cause of waterfowl mortality. Calls for banning lead shot for duck and goose hunting were met by opposition from many hunting groups who were concerned about being regulated, the higher cost and potential negative effects of steel and other hard metal ammunition on expensive legacy shotguns, the killing versus crippling capacity of nonlead ammunition, and whether other wetland management practices might significantly mitigate lead poisoning (D. A. Jessup, unpublished data). The resulting controversy also took on a states-rights aura. Essentially all of the opposing arguments were fairly well resolved by the mid-1980s; a strong hunter education effort was made to help support the transition and eventual end of the legal use of lead shot for waterfowl hunting in the United States. Thirty years later, there remains solid social, political, and legal support for reducing and, as much as possible eliminating, lead ammunition. The financial impacts on hunters are low.

Avian Influenza

Avian influenza is a serious and complicated contemporary waterfowl disease problem. It is covered in Chapter 18.

Less Common North American Waterfowl Diseases

Avian Tuberculosis

Avian tuberculosis in waterfowl is caused by *Mycobacterium avium*, although other mycobacterial species have occasionally been identified (Friend 1987). Although the bacteria are fairly ubiquitous, disease is relatively rare in wild waterfowl, mostly being picked up during large-scale waterfowl necropsy surveys, but it is more common in captive and exhibit waterfowl. Due to the chronic nature of the disease, infected birds are often in poor body condition. The lesions are similar to those caused by other mycobacteria with yellow to white caseated abscesses (Friend 1987). They may vary in size and are usually crumbly or soft, although older ones may be partially mineralized. The larger abscesses may have the many layered "onion skin" appearance on cut cross sections. Nodular abscesses are commonly found in the liver, spleen, or scattered through the cavities (in some cases looking like many corn kernels). Characteristic lesions are adequate for a presumptive diagnosis, but more subtle forms of disease are also seen. Mycobacteria are difficult to culture, and confirmation may be best accomplished by acid-fast staining or molecular-diagnostic methods.

Aspergillosis

Aspergillosis occurs when waterfowl inhale spores of fungi, most often *Aspergillus fumigatus* (Friend 1987). The fungus is ubiquitous, so disease generally occurs only when a concentrated source of fungus is present, like moldy hay, grain, or other vegetable matter left to rot in the field, and when birds that are nutritionally deprived or injured are foraging on rotten vegetation. It is most often seen in late winter or early spring. Infected birds may be weak and show respiratory distress and are often emaciated (hatchet breasted). Aspergillosis is a serious disease problem in seabirds, particularly under rehabilitation conditions (see Chapter 5). The most common gross lesion is multiple yellow or white cottony or cheesy plaques in the air sacs, lungs and pleural spaces. In some cases, the lungs are heavy, wet, and red and peppered with small roundish nodules. As the infection advances and more oxygen gets to the growing vegetative fungus, mats of white, grey, and blue mold can be seen elaborating spores. This "bread mold" appearance is fairly diagnostic (Figure 17.7), although other fungi including other forms of *Aspergillus* can have a similar appearance. It is not very communicable from bird to bird in the wild (until badly infected birds start shedding spores) and cleaning up, plowing under, or burning large sources of moldy feed can prevent continued infections. In captive and confined situations, aerosolized spores can result in rapid spread of disease between birds and may pose a human health hazard.

Pesticide Poisonings

Organophosphate and carbamate insecticides are acutely toxic to waterfowl and ingestion of very small amounts (six granules of carbofuran, for example) is lethal (D. A. Jessup, unpublished). Sublethal amounts result in weakness or spastic paralysis. Die-offs of dozens to hundreds of birds are sudden with affected birds usually in excellent body condition. Dead birds may superficially appear similar to avian cholera, DVE, and botulism. One characteristic sign or gross lesion commonly seen in males is prolapse of the penis, also seen with DVE, but not with cholera or botulism. No other gross lesions are seen with organophosphate or carbamate poisoning (as with botulism), but as noted, characteristic lesions are seen with DVE and cholera. Pesticide use falls under state agriculture laws, but incidental mortality of waterfowl can be prosecuted under the Migratory Bird Treaty Act (MBTA) and other laws.

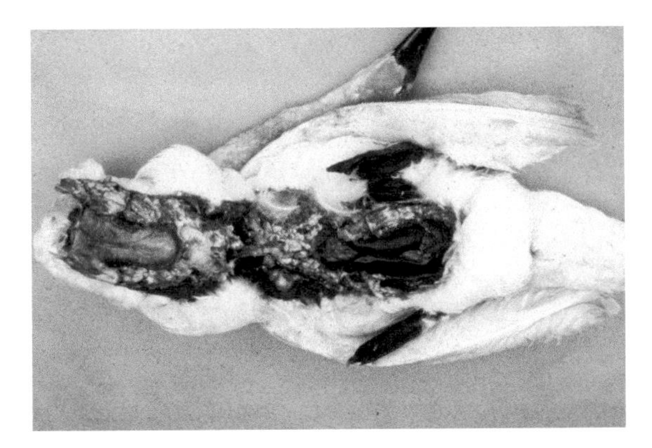

Fig. 17.7. A tundra swan (*Cygnus columbianus*) carcass opened in the field by reflecting the breast plate forward reveals an advanced case of aspergillosis affecting the lungs and thoracic and abdominal air sacs. In the presence of oxygen, fungal fruiting bodies form, which look much like bread mold, and are easy to recognize. Photograph by the author.

Summary

Most people value wild waterfowl, and relatively few people have negative attitudes toward them and their management, due in large part to the efforts of waterfowl conservation organizations like Ducks Unlimited, California Waterfowl Association, and Audubon Societies. Recently, however, with the rise of fear over avian influenza and other pandemic zoonotic infections, some people may see waterfowl as a potential source of disease. Waterfowl strikes can damage airplanes and have caused crashes. They can damage crops and be a nuisance in public places like parks and golf courses due aggression or fecal accumulation. Therefore, it is wise for waterfowl management agencies to be proactive in providing both the public and agriculturalists with positive and cooperative ideas for mitigating such problems. It can also be helpful for duck and goose hunters to increase their knowledge of disease and parasite identification and about waterfowl public and environmental health issues. State wildlife agencies can teach hunters basic disease-monitoring skills that will help augment data from field research with citizen science information. At a minimum, hunters should understand protocols for handling and processing waterfowl and the appropriate agency to contact in the event an infectious disease outbreak is observed in hunter-killed birds.

Under the MBTA it is illegal to kill waterfowl except as prescribed by law. "Incidental take" of birds, as a result of things like pesticide applications, oil spills or toxic settlement ponds, may be subject to legal action and financial penalty under MBTA and state laws. It is notable that the Trump administration attempted to reinterpret and exempt incidental take, but this was reversed by the Biden administration. Although one could perhaps argue that farming practices that cause botulism die-offs or aspergillosis deaths are forms of incidental take, no legal precedents exist for using the MBTA in such cases, and government agencies tend to see small scale, nonrepeated "takes" as "teachable moments."

As noted, waterfowl diseases in North America played an important role in the overall recognition of wildlife health and disease as an area of professional concern to wildlife management and, therefore, worthy of investment and support. These diseases, and the pioneering scientists who studied their ecology, helped pave the way for major improvements in waterfowl management and wildlife health in the succeeding decades.

LITERATURE CITED

Blanchong, J. A., M. D Samuel, D. R.Goldberg, D. J. Shadduck, and M. A. Lehr. 2006. Persistence of *Pasteurella multocida* in wetlands following avian cholera outbreaks. Journal of Wildlife Diseases 42:33–39.

Davidson, W. 2006. Field manual of wildlife diseases in the southeastern United States. Third edition. Southeastern Cooperative Wildlife Disease Study, Athens, Georgia, USA.

Friend, M. 1987. Field guide to wildlife diseases: general field procedures and diseases of migratory birds. US Geological Survey, Reston, Virginia, USA. https://pubs.er.usgs.gov/publication/2001165. Accessed 17 Nov 2020.

Friend, M. 2006. Disease emergence and resurgence: the wildlife human connection. US Geological Survey, Circular 1285, Reston, Virginia, USA. http://www.nwhc.usgs.gov/publications/disease_emergence/index.jsp. Accessed 16 Nov 2020.

Hunter, B. F., and W. E. Clark. 1971. Avian botulism. Resources Agency of California, Department of Fish and Game, Wildlife Management Leaflet 14, Sacramento, USA.

Jensen, W. I., and J. P. Allen. 1960. A possible relationship between aquatic invertebrates and avian botulism. Transactions North American Wildlife Natural Resources Conference 25:171–180.

Jessup, D. 1986. Anseriformes: avian cholera, waterfowl botulism, duck virus enteritis. Pages 342–353 *in* M. E. Fowler, editor. Zoo and wild animal medicine. Second edition. Iowa State University Press, Ames, USA.

Leibovitz, L. 1971. Gross and histopathologic changes of duck plague (duck virus enteritis). American Journal of Veterinary Research 32:275–290.

Leopold, A. 1933. Game management. C. Scribner's Sons, New York, New York, USA, and London, England.

Pasteur, L. 1880. The attenuation of the causal agent of fowl cholera. Comptes rendus de l'Academie des sciences, 91:673–680. (In French.)

Rosen, M.N. 1971*a*. Botulism. Pages 100–117 *in* J. W. Davis, R. C. Anderson, L. Karstad, and D. O. Trainer, editors. Infectious and parasitic diseases of wild birds. Iowa State University Press, Ames, USA.

Rosen, M.N. 1971*b*. Avian cholera. Pages 59–74 *in* J.W. Davis, R.C. Anderson, L. Karstad, and D.O. Trainer, editors. Infectious and parasitic diseases of wild birds. Iowa State University Press, Ames, USA.

Samuel, M. D., R. G. Botzler, and G. A. Wobeser. 2007. Avian cholera. Pages 239–269 *in* N.J. Thomas, D.B. Hunter, and C.T. Atkinson, editors. Infectious diseases of wild birds. Wiley-Blackwell, Hoboken, New Jersey, USA.

Samuel, M. D., D. J. Shadduck, D. R.Goldberg, and W. P. Johnson. 2005. Avian cholera in waterfowl: the role of lesser snow geese and Ross's geese carriers in the Playa Lakes region. Journal of Wildlife Diseases 41:48–57.

Sanderson, G. C., and F. C. Bellrose. 1986. A review of the problem of lead poisoning in waterfowl. Illinois Natural History Survey 172., Special Publication 4, Champaign, USA.

Wobeser, G. A., 1981. Diseases of wild waterfowl. Plenum Press, New York, New York, USA.

Wobeser, G.A. 1994. Investigation and management of diseases in wild animals. Plenum Press, New York, New York, USA.

18 Avian Influenza in Wild Birds

ANDREW M. RAMEY

Introduction

Avian influenza has been recognized as an important poultry disease since the 1870s (Lupiani and Reddy 2009). It wasn't until the 1960s and 1970s that wild birds of the orders Anseriformes and Charadriiformes (particularly waterfowl, gulls, and shorebirds) were identified as reservoir hosts for influenza A viruses (IAVs) that could periodically spill over into domestic and commercial birds and various mammalian hosts (Easterday et al. 1968, Dasen and Laver 1970, Slepuškin et al. 1972, Zakstel'skaja et al. 1972, Slemons et al. 1974; Figure 18.1). The significance of spillover as a mechanism by which IAVs from wild birds lead to the emergence of poultry disease was supported through research conducted throughout the 1980s and 1990s (Halvorson et al. 1983, Deibel et al. 1985, Röhm et al. 1995), though documented evidence for avian influenza as an important disease in wild birds was limited prior to 2002, with the single exception of an outbreak in terns in South Africa in 1961 (Becker 1966).

In 1996, a highly pathogenic (HP) H5N1 subtype IAV emerged in domestic geese in Guangdong, China (Xu et al. 1999), which went on to cause significant poultry losses and periodic cases of serious, and sometimes fatal, human disease throughout East

Illustration by Laura Donohue.

and Southeast Asia (Wan 2012). In 2002, genetically related HP H5N1 IAVs were reported to cause mortality events among captive and free-ranging aquatic birds in Hong Kong (Ellis et al. 2004). In 2005, a genetically similar HP H5N1 IAV of the same goose Guangdong (Gs/GD) lineage was identified as the cause of an outbreak of avian influenza among more than 6,000 wild birds at Qinghai Lake, China (Chen et al. 2006), which resulted in the mortality of large numbers of wild bar-headed geese (*Anser indicus*), brown-headed gulls (*Chroicocephalus brunnicephalus*), and great black-headed gulls (*Larus ichthyaetus*). Since 2005, HP viral descendants of this Gs/GD lineage have continued to cause periodic outbreaks among domestic poultry as well as mortality among wild aquatic birds in Asia, Europe, Africa, and North America (Sakoda et al. 2010, Ip et al. 2015, Bevins et al. 2016, Kleyheeg et al. 2017, Li et al. 2017, Pohlmann et al. 2017, Abolnik et al. 2019). While the role of wild birds in maintaining these HP Gs/GD IAVs independently of domestic poultry is not entirely clear, there is considerable evidence that wild birds have contributed to the dispersal of these viruses throughout Eurasia and to North America (Lee et al. 2015, Pasick et al. 2015, Global Consortium for H5N8 and Related Influenza Viruses 2016, Ramey et al. 2016, Lycett et al. 2020, Caliendo et al., unpublished data*). Thus, HP avian influenza appears to have become an emergent disease in wildlife facilitated by direct or indirect transmission between wild and domestic birds (Verhagen et al. 2021*b*, Ramey et al. 2022).

Causative Agent and Disease in Birds

Etiology

Avian influenza is a viral disease affecting wild and domestic birds resulting from infection with one or more IAVs. The IAVs causing avian influenza are assigned to the same taxonomic species within the family *Orthomyxoviridae* as the viral agents causing swine influenza and human seasonal influenza (Suarez 2008; Figure 18.1). For example, both avian and human origin IAVs may, in cases, infect swine. Coinfection of hosts, such as swine, with multiple IAVs may lead to novel, and potentially even pandemic, reassortant viruses (Zhou et al. 1999, Smith et al. 2009).

Influenza A viruses have a negative sense, eight segment RNA genome coding for at least 11 proteins. The hemagglutinin (HA) gene segment is the primary viral antigen of an IAV and facilitates binding to host cells. The neuraminidase (NA) gene segment is the other surface glycoprotein and enables new virions to be released from host cells following replication. IAVs are often referred to by their combined HA and NA subtypes (e.g., H5N1 or H7N9) which corresponds to the general viral serotype. To date, 16 HA and 9 NA subtypes have been identified in avian hosts (Olsen et al. 2006). Given the lack of RNA polymerase proofreading mechanisms and segmented genome of IAVs, these viruses may evolve rapidly through both antigenic drift (i.e., mutation) and antigenic shift (i.e., reassortment). Antigenic shift may be particularly common among IAVs maintained in wild birds (Dugan et al. 2008, Lebarbenchon et al. 2012, Wille et al. 2013). Antigenic drift and antigenic shift lead to viral mutations conferring selective advantages, such as immune system evasion and host adaptation. Furthermore, these two evolutionary mechanisms may play important roles in the emergence of disease among diverse avian, swine, and human hosts caused by IAVs.

Signs and Gross Lesions

Infection of birds with IAVs may result in a broad spectrum of signs and outcomes ranging from no evidence of clinical disease to 100% mortality. Clinical signs of avian influenza in wild and domestic birds are highly variable among individuals and species and may include coughing; sneezing; decreased activity; decreased egg production; inflammation,

* Caliendo, V., N. S. Lewis, A. Pohlmann, S. R. Baillie, A. C. Banyard, M. Beer, I. H. Brown, R. A. M. Fouchier, R. D. E. Hansen, T. K. Lameris, et al. 2022. Transatlantic spread of highly pathogenic avian influenza H5N1 by wild birds from Europe to North America in 2021. bioRxiv https://doi.org/10.1101/2022.01.13.476155. Accessed 27 Aug 2021.

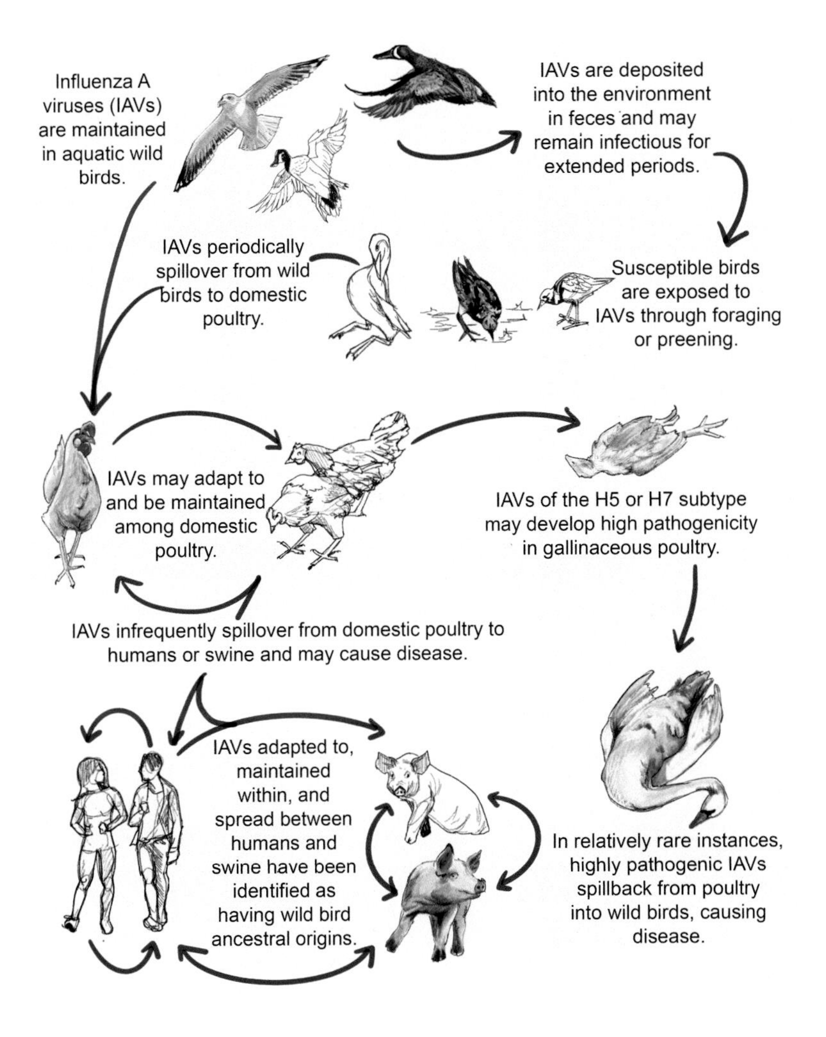

Fig. 18.1. Generalized ecology of influenza A viruses in wild birds and spillover hosts. Illustration by Laura Donohue.

hemorrhaging, and necrosis within various tissues and organs; neurologic signs; paralysis; and death (Swayne and Pantin-Jackwood 2008; Figure 18.2).

Pathogenesis and Pathology

Route of infection, exposure dose, and prior exposure history may all contribute to clinical outcomes among affected birds; however, genotypic and phenotypic characteristics of IAVs are important predictors of viral pathogenicity in avian hosts. More specifically, IAVs of two HA subtypes, H5 and H7, may have a polybasic cleavage site conferring an HP phenotype, particularly among gallinaceous domestic poultry (Wood et al. 1993, Swayne and Suarez 2000). This polybasic cleavage site facilitates viral replication in a broad spectrum of host cells. Highly pathogenic H5 and H7 viruses typically produce paralysis and death among intravenously inoculated six-week-old specific pathogen free chickens as measured through a standardized pathogenicity index (Allan et al. 1977, Swayne and Suarez 2000). Viruses lacking a polybasic cleavage site and causing limited paralysis and death through a standardized pathogenicity index are considered to be of low pathogenicity (LP). To date, evidence for the evolution of HP H5 and H7 IAVs is generally limited to commercial poultry production systems following the spillover of LP precursor viruses from wild bird hosts (Figure 18.1). There is currently very little evidence for evolution of

HP IAV genotypes or phenotypes in wild birds; however, such viruses can clearly spillback from domestic poultry into wild birds. The apparent frequency of occurrence of HP IAVs in wild birds, particularly those of the Gs/GD lineage, appears to have dramatically increased since 2002 (Ramey et al. 2022).

Morbidity or mortality of wild birds following HP IAV infection may be the result of systemic viral disease. Lesions associated with the presence of HP IAV may be observed in the brain, heart, kidneys, liver, lungs, pancreas, and respiratory tract (Tanimura et al. 2006, Bröjer et al. 2009, Kim et al. 2015, Shearn-Bochsler et al. 2019, Caliendo et al. 2020). Common histopathological lesions previously reported among HP IAV-affected wild birds include inflammation and necrosis of brain tissues; hemorrhages and necrosis of heart muscle; congestion, edema, hemorrhages, and necrosis within the respiratory tract; and hepatic and pancreatic necrosis (Tanimura et al. 2006; Bröjer et al. 2009, 2012; Kim et al. 2015; Krone et al. 2018; Shearn-Bochsler et al. 2019; Caliendo et al. 2020; Figure 18.2).

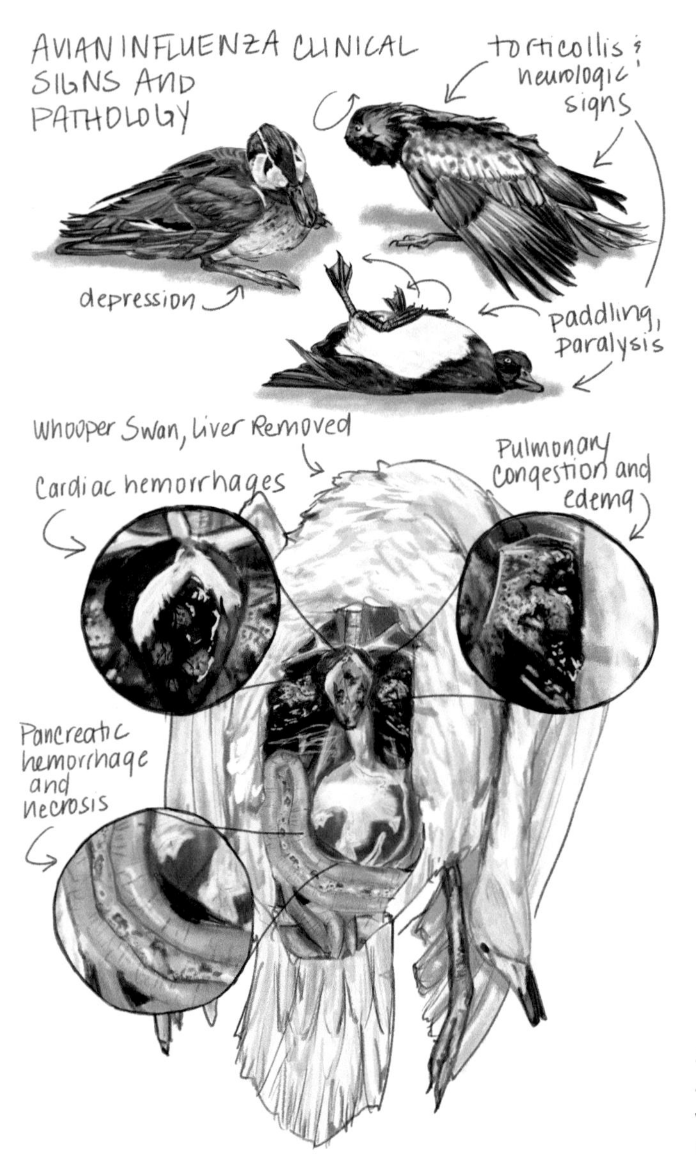

Fig. 18.2. Common clinical signs and lesions associated with highly pathogenic influenza A virus infection in wild birds. Illustration by Laura Donohue.

Diagnosis of IAV Infection

Active infection of wild birds with IAVs may be diagnosed through numerous methods. Since at least the 1960s, diagnosis of IAVs in wild birds has been accomplished through collection of oropharyngeal swabs, cloacal swabs, and other tissues and subsequent viral culture in embryonating chicken eggs (Becker 1966, Slemons et al. 1974, Spackman et al. 2008). More recently, molecular diagnostics, such as real-time reverse-transcriptase polymerase chain reaction assay, have been applied to RNA extractions from oropharyngeal or cloacal swabs collected from apparently healthy live-captured or hunter-harvested wild birds (Spackman et al. 2008). Molecular diagnostics and immunohistochemical staining have been used to identify IAVs in tissues recovered from wild bird carcasses (Shearn-Bochsler et al. 2019). A variety of methods may also be used to infer prior exposure of wild birds to IAVs. Antibodies to IAVs may be detected in wild bird sera using agar gel immunodiffusion, hemagglutinin inhibition, virus microneutralization, and enzyme-linked immunosorbent assays (Spackman et al. 2008, Ramey et al. 2019).

Demographics and Field Biology

Reservoir Hosts of LP IAVs

Wild aquatic water birds of the orders Anseriformes and Charadriiformes are considered the natural reservoir of the greatest genetic and antigenic diversity of IAVs, particularly those of LP (Figures 18.1 and 18.3; see also Figure 18.4). Influenza A viruses are presumably spread among wild aquatic reservoir species primarily through a fecal–oral route, in which infected birds shed virus into the wetlands or coastal environments where virions may remain infectious for extended periods (Lebarbenchon et al. 2011, Ramey et al. 2020). The most common wild bird hosts of these LP IAVs include ducks, geese, swans, gulls, shorebirds, and seabirds (Olsen et al. 2006, Munster et al. 2007, Ip et al. 2008, Lang et al. 2016).

The mallard (*Anas platyrhynchos*), a common species of dabbling duck, is the most widely recognized reservoir host of IAVs, a function of the ubiquity of this species across the Holarctic, widespread sampling of this taxon, and the susceptibility of this species to infection with antigenically diverse viruses (Latorre-Margalef et al. 2014, Hollander et al. 2019, Verhagen et al. 2021*a*). IAVs have also commonly been isolated from other dabbling ducks inhabiting Asia, Europe, and North America including American green-winged teal (*Anas carolinensis*), blue-winged teal (*Spatula discors*), common teal (*Anas crecca*), Eurasian wigeon (*Mareca penelope*), spot-billed duck (*Anas zonorhyncha*), northern pintail (*Anas acuta*), and northern shoveler (*Spatula clypeata*) (Munster et al. 2007, Kim et al. 2010, Hill et al. 2012, Reeves et al. 2018, Carter et al. 2019). The prevalence of IAVs appears to reach an annual epidemiologic peak among dabbling ducks each autumn within the northern hemisphere, because of the congregation of immunologically naïve hatch year birds at staging and stopover areas en route to

Fig. 18.3. Capture of blue-winged teal (*Spatula discors*) and sympatric waterbirds in Louisiana using a rocket net as part of research on and surveillance for influenza A viruses in wild birds inhabiting North America. Photograph by Paul Link, Louisiana Department of Wildlife and Fisheries.

wintering areas (Hinshaw et al. 1980, Van Dijk et al. 2014, Papp et al. 2017).

In the southern hemisphere, IAVs have also been isolated from numerous species of dabbling ducks including cinnamon teal (*Anas cyanoptera*), silver teal (*Anas versicolor*), yellow-billed pintail (*Anas georgica*), yellow-billed teal (*Anas flavirostris*), and white-cheeked pintail (*Anas bahamensis*) in South America (Spackman et al. 2006, Ghersi et al. 2009, Jiménez-Bluhm et al. 2018, Rimondi et al. 2018, Bravo-Vasquez et al. 2020); garganey (*Anas querquedula*) and red-billed teal (*Anas erythrorhyncha*) in Africa (Gaidet et al. 2007, Cumming et al. 2011); and grey teal (*Anas gracilis*) and Pacific black duck (*Anas superciliosa*) in Australia (Haynes et al. 2009). Peak prevalence of IAVs among dabbling ducks in the southern hemisphere may be related to concentration of birds (Gaidet et al. 2012*a*), precipitation (Ferenczi et al. 2016) or other factors (Ruiz et al. 2021). Given the frequent detection of antigenically diverse viruses in dabbling ducks sampled on six continents, this taxonomic group is posited to be among the most important for global maintenance of IAVs (Olsen et al. 2006).

Though less well-sampled than dabbling ducks, diverse IAVs have also been isolated from numerous species of diving ducks, sea ducks, tree ducks, and shelducks sampled throughout the northern and southern hemispheres (Gaidet et al. 2007, Spackman et al. 2009, Karlsson et al. 2013, Reeves et al. 2013, Rimondi et al. 2018, Carter et al. 2021, McBride et al. 2021). While these birds have generally been found to exhibit lower IAV prevalence as compared to dabbling ducks (Olsen et al. 2006, Munster et al. 2007, Ip et al. 2008, Gaidet et al. 2012*a*, Bevins et al. 2014, Carter et al. 2021), they may still play a role in the maintenance of LP viruses that ultimately lead to the emergence of economically costly HP poultry disease (Xu et al. 2017, McBride et al. 2021). Timing of peaks of infection among diving ducks, sea ducks, tree ducks, and shelducks are unclear.

Geese and swans are frequently infected with diverse IAVs, though generally at lower prevalence than dabbling ducks (Olsen et al. 2006, Munster et al. 2007, Ip et al. 2008, Bevins et al. 2014). Antibody seroprevalence data corroborates the premise that geese and swans may be commonly exposed to IAVs, with seropositivity reported to be higher among adult as compared to juvenile birds (Hoye et al. 2011, Ely et al. 2013, Wilson et al. 2013, Curran et al. 2015, Samuel et al. 2015, Lambrecht et al. 2016). Influenza A viruses have been detected in geese and swans throughout the annual cycle, though annual seasonal peaks in prevalence remain unclear (Hoye et al. 2011, Ely et al. 2013, Lambrecht et al. 2016, Ramey et al. 2019). Although geese and swans may play a less prominent role in the maintenance and dispersal of IAVs at a global scale as compared to dabbling ducks, they may be important hosts at more regional scales. For example, emperor geese (*Anser canagicus*), which are endemic to western Alaska and northeastern Russia, are frequently exposed to antigenically diverse IAVs, may be infected with IAVs throughout the annual cycle, and likely play an important role in the dispersal of viruses across the Bering Strait between East Asia and North America through intercontinental migrations (Hupp et al. 2007; Ely et al. 2013; Ramey et al. 2015, 2019).

Gulls appear to be other globally important hosts of LP IAVs, particularly those of the H13 and H16 HA subtypes (Haynes et al. 2009, Pereda et al. 2008, Wille et al. 2011, Sharshov et al. 2014, Mathieu et al. 2015). Investigations in Europe and North America have reported elevated IAV prevalence among gulls during summer and in autumn upon the recruitment of immunologically naïve hatch-year birds into the population (Lewis et al. 2013, Tønnessen et al. 2013, Huang et al. 2014, Verhagen et al. 2014, Froberg et al. 2019). Genomic data for IAVs isolated from gulls suggest that these birds may also play an important role in the dispersal of viruses between continents (Wille et al. 2011, Dusek et al. 2014, Huang et al. 2014, Mathieu et al. 2015, Verhagen et al. 2020).

Low-pathogenicity IAVs have been sporadically detected from samples collected from diverse species of shorebirds inhabiting Africa, Asia, Australia,

Europe, North America, and South America, though prevalence has generally been low (Hurt et al. 2006, Iverson et al. 2008, Gaidet et al. 2012*b*, Hall et al. 2014, Nelson et al. 2016, Jiménez-Bluhm et al. 2018, Kakogawa et al. 2020). An important exception is a well-described annual epidemic occurring each spring at Delaware Bay, in which ruddy turnstones (*Arenaria interpres*) and other sympatric shorebirds and gulls may exhibit relative high prevalence of infection by LP IAVs, coincident with the congregation of birds feeding on horseshoe crab (*Limulus polyphemus*) spawn during northward spring migration (Krauss et al. 2010). Ruddy turnstones have also been reported to be hosts of IAVs elsewhere within the United States, South America, and Australia and, therefore, may facilitate the intra- and inter-continental dispersal of viruses through long-distance migratory movements (De Araujo et al. 2014, Hurtado et al. 2015, Nelson et al. 2016, Jiménez-Bluhm et al. 2018, Poulson et al. 2020, Hoye et al. 2021).

At least 98 different species of seabirds have been tested for IAVs (Lang et al. 2016). Prevalence has been reported to be very low among most taxa sampled, with the possible exceptions of cormorants, murres, noddies, terns, tubenoses, and penguins. Of these birds, the greatest number and diversity of viruses have been isolated from common (*Uria aalge*) and thick-billed (*Uria lomvia*) murres (Lang et al. 2016). Adélie penguin (*Pygoscelis adeliae*), chinstrap penguin (*Pygoscelis antarcticus*), and snowy sheathbill (*Chionis albus*) are the only avian species inhabiting Antarctica from which LP IAVs have been isolated (Barriga et al. 2016; Hurt et al. 2014, 2016). It is unclear what role, if any, seabirds may play in the maintenance and dispersal of IAVs leading to disease outbreaks in poultry.

Low-pathogenicity IAVs have been sporadically isolated from numerous other taxa of wild birds including egrets, grebes, pelicans, skimmers, tinamous, and songbirds (Borovská et al. 2011; Gronesova et al. 2008; Simulundu et al. 2009; Lebarbenchon et al. 2010, 2015; Slusher et al. 2014; Nelson et al. 2016). Antibodies reactive to LP IAVs have also been detected in a relatively high proportion of ibis and loons, suggesting that these taxa could also be competent wild bird hosts (Uher-Koch et al. 2019, Bahnson et al. 2020). It remains unclear, however, if any of these taxa contribute appreciably to the maintenance of LP IAVs in the wild bird reservoir or may instead be spillover hosts.

Wild Birds as Hosts of HP IAVs

The detection of HP IAVs in wild birds had occurred only once (Becker 1966) prior to the emergence of HP H5 IAVs of the Gs/GD lineage. However, since 2002, HP IAVs, predominately of the Gs/GD lineage, have been reported to affect thousands of wild birds in Asia, Europe, Africa, and North America including diverse taxa such as ducks, geese, swans, gulls, shorebirds, ibis, terns, grebes, pelicans, raptors, pheasants, pigeons, coots, cranes, storks, bitterns, egrets, herons, spoonbills, corvids, and songbirds (Chen et al. 2006; Gaidet et al. 2008; Sakoda et al. 2010; Ip et al. 2015; Verhagen et al. 2015, 2021*b*: Bevins et al. 2016; Kleyheeg et al. 2017; Li et al. 2017; Pohlmann et al. 2017; Abolnik et al. 2019; Navarro-López et al., unpublished data*). Wild birds testing positive for HP IAVs have displayed a wide range of clinical presentations ranging from apparent asymptomatic infection to severe neurological signs and death. The relative influences of species susceptibility, population immunity, and other exposure factors (e.g., route and dose) on wild bird health outcomes are not entirely clear, though there is some evidence that mortality may be higher in juvenile birds (Krone et al. 2018, Hill et al. 2019) and that some taxa (e.g., diving ducks, geese, swans, raptors, and grebes) may be particularly susceptible to clinical disease (Nagy et al. 2007, Hesterberg et al. 2009, Kleyheeg et al.

* Navarro-López, R., M. Solís-Hernández, M. A. Márquez-Ruiz, A. Rosas-Téllez, C. A. Guichard-Romero, G. D. J. Cartas-Heredia, R. Morales-Espinosa, H. E. Valdez-Gómez, and C. L. Afonso. 2020. Epizootic of highly pathogenic H7N3 avian influenza in an ecologic reserve in Mexico. bioRxiv. https://www.biorxiv.org/content/10.1101/2020.03.05.978502v1. Accessed 27 Aug 2021.

2017, Shearn-Bochsler et al. 2019). Given the frequency and magnitude of Gs/GD lineage IAV outbreaks in wild birds inhabiting regions of Europe and Asia, HP IAVs may now be a relevant consideration to the conservation and management of populations of susceptible birds, particularly threatened, endangered, or rare species of wetland birds within enzootic regions. Though some dabbling ducks have also been reported to display clinical signs of disease from Gs/GD HP IAV infection, infected individuals have also repeatedly been characterized as apparently asymptomatic leading to speculation that such birds may play an important role in the maintenance and dispersal of viruses through time and space (Jeong et al. 2014, Verhagen et al. 2021*b*). It is still unclear, however, if wild birds can maintain IAVs exhibiting an HP phenotype in the absence of episodic spillback from poultry.

Nonpoultry Spillover Hosts of Wild Bird–Origin IAVs

It has been posited that all IAVs infecting mammals share genetic ancestry with viruses maintained in wild aquatic birds; however, domestic poultry and swine likely play important roles in facilitating the transmission of IAVs across species (Webby and Webster 2001; Figure 18.1). Nonetheless, there is evidence, albeit limited, for the spillover of wild-bird-origin IAVs directly to other hosts including marine mammals, other domestic animals, and humans. For example, there have been periodic outbreaks of phocine influenza affecting wild seals in North America and Europe since 1980 for which the best evidence suggests wild bird origin (Hinshaw et al. 1984, Anthony et al. 2012, Zohari et al. 2014). The isolation of H13N2 and H13N9 IAVs from a long-finned pilot whale (*Globicephala melas*) stranded in Massachusetts, USA, in 1984 was attributed to probable spillover from gulls (Hinshaw et al. 1986). An outbreak of H10N4 influenza in farmed mink in southern Sweden in 1984 was also purported to be the result of the introduction of a gull-origin IAV (Klingeborn et al. 1985, Berg et al. 1990). Influenza A viruses of the H1N1, H3N8, H5N1, and H7N2 subtypes detected in swine, horses, marten, and pikas, respectively, have been characterized as sharing antigenic and genetic similarity with IAVs in wild birds, though it is unclear if these viruses represent direct spillover from wild birds or if other unidentified intermediate hosts were involved (Pensaert et al. 1981, Klopfleisch et al. 2007, Su et al. 2016, Bravo-Vasquez et al. 2020). Several reports provide evidence for human infection with IAVs resulting from exposure to wild birds, though such infections appear to be relatively rare and associated with the handling, harvesting, or processing wild birds (Gill et al. 2006, Gilsdorf et al. 2006, Gray et al. 2011, Shafir et al. 2012, Thornton et al. 2019).

Management of IAV Outbreaks in Wild Birds

To date, the detection and management of IAVs in wild birds have been directed largely towards the protection of poultry holdings from the introduction of HP IAVs or LP H5 and H7 subtype IAVs with the potential to develop an HP phenotype. For example, large-scale active surveillance for Gs/GD lineage HP IAVs in wild birds has been conducted within countries and across regions to provide early detection of HP IAVs in nonendemic areas (Haynes et al. 2009; Hesterberg et al. 2009; Bevins et al. 2014, 2016; Figures 18.3 and 18.4). Findings have been useful for elevating situational awareness and emphasizing the need for rigorous biosecurity among poultry-production facilities in areas where Gs/GD lineage HP IAVs have been detected. Passive surveillance has also been identified as a useful tool for identifying and characterizing HP IAV outbreaks in wild birds. Passive surveillance efforts have helped to provide information on species affected and the geographic extent of HP IAV outbreaks (Ip et al. 2016, Kleyheeg et al. 2017, Li et al. 2017). Other possible manage-

Fig. 18.4. Lindsay Carlson swabbing hunter-harvested ducks in western Alaska as part of research on and surveillance for influenza A viruses in wild birds inhabiting North America. Photograph by Andrew Reeves, US Geological Survey.

ment actions that have been proposed to help mitigate detrimental effects of HP IAV outbreaks in wild birds, for which there is some scientific basis for implementation, include

- the suspension of wild bird banding and capture efforts during outbreaks, particularly those that use bait and promote the concentration of wildlife, to reduce transmission among wild birds and minimize possible human exposure;
- the implementation of access restrictions to wetlands affected during outbreaks to limit possible human exposure to viruses and potential spread via fomites;
- the development and implementation of disinfection protocols for vehicles, boats, and other gear used at potentially affected sites;
- the suspension of hunting in outbreak-affected areas to limit possible human exposure and spread via contaminated bird carcasses or fomites; and
- the manipulation of water levels of small managed wetlands affected during outbreaks to facilitate more rapid inactivation of viruses maintained in surface water (e.g., through increasing penetration of ultraviolet radiation or raising the water temperature).

Social, Financial, Political, and Legal Considerations of Avian Influenza in Wild Birds

Avian influenza in wild birds may engender numerous social, financial, political, and legal considerations depending upon the nature of causative agent identified. For example, the detection of HP IAVs in wild birds are required to be reported to the World Organization for Animal Health as detailed through the Terrestrial Animal Health Code, whereas the detection of LP IAVs in wild birds are not (OIE 2021). Thus, the report of HP IAV in wild birds may trigger government-mandated biosecurity measures and lead to restrictions in the movement or trade of commercial poultry within an affected region. Furthermore, spillover of Gs/GD lineage HP IAVs from wild birds to domestic birds, or even LP H5 or H7 IAVs that ultimately develop high pathogenicity in poultry, may trigger government-mandated control strategies (e.g., depopulation of poultry) resulting in significant economic and food-security consequences.

For example, the 2014–2015 outbreak of HP avian influenza in North America was the result of the introduction of a Gs/GD lineage HP IAV from East Asia to North America via migratory birds and subsequent spillover to domestic poultry (Lee et al. 2015, Pasick et al. 2015, Ramey et al. 2016). Following introduction, Gs/GD IAVs quickly spread among domestic poultry in western and central Canada and the United States (Xu et al. 2016, Lee et al. 2018) resulting in partial or complete bans on poultry exports (Greene 2015). This outbreak ultimately resulted in the loss of >50 million domestic birds in the United States through direct mortality of diseased birds or the depopulation of potentially exposed flocks (Ramos et al. 2017). The estimated cost of mitigation measures in the United States (indemnity payments and response activities) was estimated to be $879 million (USD) (Hagerman and Marsh 2016), while the total economic burden (including lost revenue) was estimated to exceed $3 billion (USD) (Greene 2015).

The detection of IAVs in wild birds that have previously been associated with human disease (i.e., those viruses that have caused disease in persons handling infected poultry) may result in additional or different sociopolitical considerations. For example, the detection of IAVs in wild birds (or poultry) that have previously been associated with human disease may prompt public health agencies to recommend or institute (1) surveillance activities for IAVs of concern, (2) social measures and personal protection equipment guidelines to prevent or mitigate disease, (3) chemoprophylaxis for potentially exposed persons, or (4) investment in vaccine development (WHO 2006, Schünemann et al. 2007, CDC 2021). Information on societal and financial costs associated with detections of IAVs in wild birds that represent a potential threat to human health is currently limited and represents an important data gap.

Finally, as HP IAVs of the Gs/GD lineage become increasingly common in wild birds, it is plausible that such viruses may raise legal considerations regarding the conservation of threatened, endangered, or rare wild bird species. That is, in countries and regions where the protection of threatened, endangered, or rare wild birds is mandated by law or international treaty and where outbreaks of Gs/GD lineage HP IAVs occur, management agencies may be prompted to develop or adopt strategies to mitigate effects of disease to potentially vulnerable populations of birds.

Summary

Influenza A viruses are maintained among wild waterfowl, gulls, shorebirds, and some seabirds and typically do not cause disease. Some avian-origin IAVs that have previously spilled over into domestic poultry have developed an HP phenotype and have become important pathogens affecting both wild and domestic birds, specifically those of the Gs/GD lineage. These Gs/GD lineage HP IAVs have important implications for the production and trade of domestic birds, for conservation and management of wild birds, and for zoonotic spillover risk to humans handling infected birds.

LITERATURE CITED

Abolnik, C., R. Pieterse, B M. Peyrot, P. Choma, T. P. Phiri, K. Ebersohn, C. V. Heerden, A. A. Vorster, G. V. Zel, P. J. Geertsma, et al. 2019. The incursion and spread of highly pathogenic avian influenza H5N8 clade 2.3.4.4 within South Africa. Avian Diseases 63:149–156.

Allan, W. H., D. J. Alexander, B. S. Pomeroy, and G. Parsons. 1977. Use of virulence index tests for avian influenza viruses. Avian Diseases 21:359–363.

Anthony, S. J., J. S. Leger, K. Pugliares, H. S. Ip, J. M. Chan, Z. W. Carpenter, Navarrete-I. Macias, M. Sanchez-Leon, J. T. Saliki, J. Pedersen, et al. 2012. Emergence of fatal avian influenza in New England harbor seals. MBio 3:4.

Bahnson, C. S., S. M. Hernandez, R. L. Poulson, R. E. Cooper, S. E. Curry, T. J. Ellison, H. C. Adams, C. N. Welch, and D. E. Stallknecht. 2020. Experimental infections and serology indicate that American white ibis (*Eudociumus albus*) are competent reservoirs for type A influenza virus. Journal of Wildlife Diseases 56:530–537.

Barriga, G. P., D. Boric-Bargetto, M. C. San Martin, V. Neira, H. van Bakel, M. Thompsom, R. Tapia, D. Toro-Ascuy, L. Moreno, Y. Vasquez, et al. 2016. Avian influenza virus H5 strain with North American and Eurasian lineage genes in an Antarctic penguin. Emerging Infectious Diseases 22:2221–2223.

Becker, W. B. 1966. The isolation and classification of tern virus: influenza virus A/tern/South Africa/1961. Epidemiology & Infection 64:309–320.

Berg, M., L. Englund, I. A. Abusugra, B. Klingeborn, and T. Linne. 1990. Close relationship between mink influenza (H10N4) and concomitantly circulating avian influenza viruses. Archives of Virology 113:61–71.

Bevins, S. N., R. J. Dusek, C. L. White, T. Gidlewski, B. Bodenstein, K. G. Mansfield, P. DeBruyn, D. Kraege, E. Rowan, C. Gillin, et al. 2016. Widespread detection of highly pathogenic H5 influenza viruses in wild birds from the Pacific Flyway of the United States. Scientific Reports 6:28980.

Bevins, S. N., K. Pedersen, M. W. Lutman, J. A. Baroch, B. S. Schmit, D. Kohler, T. Gidlewski, D. L. Nolte, S. R. Swafford, and T. J. DeLiberto. 2014. Large-scale avian influenza surveillance in wild birds throughout the United States. PLoS One 9:e104360.

Borovská, P., P. Kabát, M. Ficová, A. Trnka, D. Svetlíková, and T. Betáková. 2011. Prevalence of avian influenza viruses, *Mycobacterium avium*, and *Mycobacterium avium*, subsp. *paratuberculosis* in marsh-dwelling passerines in Slovakia, 2008. Biologia 66:282–287.

Bravo-Vasquez, N., J. Yao, P. Jimenez-Bluhm, V. Meliopoulos, P. Freiden, B. Sharp, L. Estrada, A. Davis, S. Cherry, B. Livingston, et al. 2020. Equine-like H3 avian influenza viruses in wild birds, Chile. Emerging Infectious Diseases 26:2887–2898.

Bröjer, C., E. O. Ågren, H. Uhlhorn, K. Bernodt, D. S. Jansson, and D. Gavier-Widén. 2012. Characterization of encephalitis in wild birds naturally infected by highly pathogenic avian influenza H5N1. Avian Diseases 56: 144–152.

Bröjer, C., E. O. Ågren, H. Uhlhorn, K. Bernodt, T. Mörner, D. S. Jansson, R. Mattsson, S. Zohari, P. Thorén, et al. 2009. Pathology of natural highly pathogenic avian influenza H5N1 infection in wild tufted ducks (*Aythya fuligula*). Journal of Veterinary Diagnostic Investigation 21: 579–587.

Caliendo, V., L. Leijten, L. Begeman M. J., Poen, R. A. Fouchier, N. Beerens, and T. Kuiken. 2020. Enterotropism of highly pathogenic avian influenza virus H5N8 from the 2016/2017 epidemic in some wild bird species. Veterinary Research 51:1–10.

Carter, D. L., P. Link, G. Tan, D. E. Stallknecht, and R. I. Poulson. 2021. Influenza A viruses in whistling ducks (subfamily Dendrocygninae). Viruses 13:192.

Carter, D., P. Link, P. Walther, A. Ramey, D. Stallknecht, and R. Poulson. 2019. Influenza A prevalence and subtype diversity in migrating teal sampled along the United States Gulf Coast. Avian Diseases 63:165–171.

Centers for Disease Prevention and Control (CDC). 2021. Avian influenza: information for health professionals and laboratorians. https://www.cdc.gov/flu/avianflu/healthprofessionals.htm. Accessed 2 Mar 2021.

Chen, H., Y. Li, Z. Li, J. Shi, K. Shinya, G. Deng, Q. Qi, G. Tian, S. Fan, H. Zhao, et al. 2006. Properties and dissemination of H5N1 viruses isolated during an influenza outbreak in migratory waterfowl in western China. Journal of Virology 80:5976–5983.

Cumming, G. S., A. Caron, C. Abolnik, G. Cattoli, L. W. Bruinzeel, C. E. Burger, K. Cecchettin, N. Chiweshe, B. Mochotlhoane, G. L. Mutumi, et al. 2011. The ecology of influenza A viruses in wild birds in southern Africa. EcoHealth 8:4–13.

Curran, J. M., T. M. Ellis, and I. D. Robertson. 2015. Serological surveillance of wild waterfowl in northern Australia for avian influenza virus shows variations in prevalence and a cyclical periodicity of infection. Avian Diseases 59:492–497.

Dasen, C. A., and W. G. Laver. 1970. Antibodies to influenza viruses (including the human A2/Asian/57 strain) in sera from Australian shearwaters (*Puffinus pacificus*). Bulletin of the World Health Organization 42:885–889.

De Araujo, J., S. M. de Azevedo Júnior, N. Gaidet, R. F. Hurtado, D. Walker, L. M. Thomazelli, T. Ometto, M. M. Seixas, R. Rodrigues, D. B. Galindo, et al. 2014. Avian influenza virus (H11N9) in migratory shorebirds wintering in the Amazon Region, Brazil. PLoS One 9:e110141.

Deibel, R., D. E. Emord, W. Dukelow, V. S. Hinshaw, and J. M. Wood. 1985. Influenza viruses and paramyxoviruses in ducks in the Atlantic Flyway, 1977–1983, including an H5N2 isolate related to the virulent chicken virus. Avian Diseases 29:970–985.

Dugan, V.G., R. Chen, D. J. Spiro, N. Sengamalay, J. Zaborsky, E. Ghedin, J. Nolting, D.E. Swayne, J. A. Runstadler, G. M. Happ, et al. 2008. The evolutionary genetics and emergence of avian influenza viruses in wild birds. PLoS Pathogens 4:e1000076.

Dusek, R. J., G. T. Hallgrimsson, H. S. Ip, J. E. Jónsson, S. Sreevatsan, S. W. Nashold, J. L. TeSlaa, S. Enomoto, R. A. Halpin, X. Lin, et al. 2014. North Atlantic migratory bird flyways provide routes for intercontinental movement of avian influenza viruses. PLoS One 9:e92075.

Easterday, B. C., D. O. Trainer, B. Tůmová, and H. G. Pereira. 1968. Evidence of infection with influenza viruses in migratory waterfowl. Nature 219:523–524.

Ellis, T. M., R. B. Bousfield, L. A. Bissett, K. C. Dyrting, G. S. Luk, S. T. Tsim, K. Sturm- Ramirez, R. G. Webster, Y. Guan, and J. S. Malik Pieris. 2004. Investigation of outbreaks of highly pathogenic H5N1 avian influenza in waterfowl and wild birds in Hong Kong in late 2002. Avian Pathology 33:492–505.

Ely, C. R., J. S. Hall, J. A. Schmutz, J. M. Pearce, J. Terenzi, J. S. Sedinger, and H. S. Ip. 2013. Evidence that life history characteristics of wild birds influence infection and exposure to influenza A viruses. PLoS One 8:e57614.

Ferenczi, M., C. Beckmann, S. Warner, R. Loyn, K. O'Riley, X. Wang, and M. Klaassen. 2016. Avian influenza infection dynamics under variable climatic conditions, viral prevalence is rainfall driven in waterfowl from temperate, south-east Australia. Veterinary Research 47:23.

Froberg, T., F. Cuthbert, C. S. Jennelle, C. Cardona, and M. Culhane. 2019. Avian influenza prevalence and viral shedding routes in Minnesota ring-billed gulls (*Larus delawarensis*). Avian Diseases 63:120–125.

Gaidet, N., A. Caron, J. Cappelle, G. S. Cumming, G. Balança, S. Hammoumi, G. Cattoli, C. Abolnik, R. Servan de Almeida, P. Gil, et al. 2012*a*. Understanding the ecological drivers of avian influenza virus infection in wildfowl: a continental-scale study across Africa. Proceedings of the Royal Society B Biological Sciences 279:1131–1141.

Gaidet, N., G. Cattoli, S. Hammoumi, S. H. Newman, W. Hagemeijer, J. Y. Takekawa, J. Cappelle, T. Dodman, T. Joannis, P. Gil, et al. 2008. Evidence of infection by H5N2 highly pathogenic avian influenza viruses in healthy wild waterfowl. PLoS Pathogens 4:e1000127.

Gaidet, N., T. Dodman, A. Caron, G. Balança, S. Desvaux, F. Goutard, G. Cattoli, F. Lamarque, W. Hagemeijer, and F. Monicat. 2007. Avian influenza viruses in water birds, Africa. Emerging Infectious Diseases 13:626–629.

Gaidet, N., A. B. El Mamy, J. Cappelle, A. Caron, A., G. S. Cumming, V. Grosbois, P. Gil, S. Hammoumi, R. S. De Almeida, S. R. Fereidouni, et al. 2012*b*. Investigating avian influenza infection hotspots in old-world shorebirds. PLoS One 7:e46049.

Ghersi, B. M., D. L. Blazes, E. Icochea, R. I. Gonzalez, T. Kochel, Y. Tinoco, M. M. Sovero, S. Lindstrom, B. Shu, A. Klimov, et al. 2009. Avian influenza in wild birds, central coast of Peru. Emerging Infectious Diseases 15:935–938.

Gill, J. S., R. Webby, M. J. Gilchrist, and G. C. Gray. 2006. Avian influenza among waterfowl hunters and wildlife professionals. Emerging Infectious Diseases 12:1284–1286.

Gilsdorf, A., N. Boxall, V. Gasimov, I. Agayev, F. Mammadzade, P. Ursu, E. Gasimov, C. Brown, S. Mardel, D. Jankovic, et al. 2006. Two clusters of human infection with influenza A/H5N1 virus in the Republic of Azerbaijan, February–March 2006. Eurosurveillance 11:3–4.

Global Consortium for H5N8 and Related Influenza Viruses. 2016. Role for migratory wild birds in the global spread of avian influenza H5N8. Science 354:213–217.

Gray, G. C., D. D. Ferguson, P. E. Lowther, G. L. Heil, and J. A. Friary. 2011. A national study of US bird banders for evidence of avian influenza virus infections. Journal of Clinical Virology 51:132–135.

Greene, J. L. 2015. Update on the highly-pathogenic avian influenza outbreak of 2014–2015. https://nationalaglawcenter.org/wp-content/uploads/assets/crs/R44114.p Accessed 2 Mar 2021.

Gronesova, P., M. Ficova, A. Mizakova, P. Kabat, A. Trnka, and T. Betakova. 2008. Prevalence of avian influenza viruses, *Borrelia garinii*, *Mycobacterium avium*, and *Mycobacterium avium* subsp. *paratuberculosis* in waterfowl and terrestrial birds in Slovakia, 2006. Avian Pathology 37:537–543.

Hagerman, A. D., and T. L. Marsh. 2016. Theme overview: economic consequences of highly pathogenic avian influenza. Choices 31:1–2.

Hall, J. S., G. T. Hallgrimsson, K. Suwannanarn, S. Sreevatsen, H. S. Ip, E. Magnusdottir, J. L. TeSlaa, S. W. Nashold, and R. J. Dusek, 2014. Avian influenza virus ecology in Iceland shorebirds: intercontinental reassortment and movement. Infection, Genetics and Evolution 28:130–136.

Halvorson, D., D. Karunakaran, D. Senne, C. Kelleher, C. Bailey, A. Abraham, V. Hinshaw, and J. Newman. 1983. Epizootiology of avian influenza: simultaneous monitoring of sentinel ducks and turkeys in Minnesota. Avian Diseases 27:77–85.

Haynes, L., E. Arzey, C. Bell, N. Buchanan, G. Burgess, V. Cronan, C. Dickason, H. Field, S. Gibbs, P. M. Hansbro, et al. 2009. Australian surveillance for avian influenza viruses in wild birds between July 2005 and June 2007. Australian Veterinary Journal 87:266–272.

Hesterberg, U., K. Harris, D. Stroud, V. Guberti, L. Busani, M. Pittman, V. Piazza, A. Cook, and I. Brown. 2009. Avian influenza surveillance in wild birds in the European Union in 2006. Influenza and Other Respiratory Viruses 3:1–14.

Hill, N. J., J. Y. Takekawa, C. J. Cardona, B. W. Meixell, J. T. Ackerman, J. A. Runstadler, and W. M. Boyce, 2012. Cross-seasonal patterns of avian influenza virus in breeding and wintering migratory birds: a flyway perspective. Vector-Borne and Zoonotic Diseases 12:243–253.

Hill, S. C., R. Hansen, S. Watson, V. Coward, C. Russell, J. Cooper, S. Essen, H. Everest, K. V. Parag, S. Fiddaman, et al. 2019. Comparative micro-epidemiology of pathogenic avian influenza virus outbreaks in a wild bird population. Philosophical Transactions of the Royal Society B Biological Sciences 374:20180259.

Hinshaw, V. S., W. J. Bean, J. Geraci, P. Fiorelli, G. Early, and R. G. Webster. 1986. Characterization of two influenza A viruses from a pilot whale. Journal of Virology 58:655–656.

Hinshaw, V. S., W. J. Bean, R. G. Webster, J. E. Rehg, P. Fiorelli, G. Early, J. R. Geraci, and D. J. St. Aubin. 1984. Are seals frequently infected with avian influenza viruses? Journal of Virology 51:863–865.

Hinshaw, V. S., R. G. Webster, and B. Turner. 1980. The perpetuation of orthomyxoviruses and paramyxoviruses in Canadian waterfowl. Canadian Journal of Microbiology 26:622–629.

Hollander, L. P., A. Fojtik, C. Kienzle-Dean, N. Davis-Fields, R. L. Poulson, B. Davis, C. Mowry, and D. E. Stallknecht. 2019. Prevalence of influenza A viruses in ducks sampled in northwestern Minnesota and evidence for predominance of H3N8 and H4N6 subtypes in mallards, 2007–2016. Avian Diseases 63:126–130.

Hoye, B. J., C. M. Donato, S. Lisovski, Y. M. Deng, S. Warner, A. C. Hurt, M. Klaassen, and D. Vijaykrishna. 2021. Reassortment and persistence of influenza A viruses from diverse geographic origins within Australian wild birds: evidence from a small, isolated population of ruddy turnstones. Journal of Virology 95:5.

Hoye, B. J., V. J. Munster, H. Nishiura, R. A. M. Fouchier, J. Madsen, and M. Klaassen. 2011. Reconstructing an annual cycle of interaction: natural infection and antibody dynamics to avian influenza along a migratory flyway. Oikos 120:748–755.

Huang, Y., M. Wille, J. Benkaroun, H. Munro, A. L. Bond, D. A. Fifield, G. J. Robertson, D. Ojkic, H. Whitney, H., and A. S. Lang, 2014. Perpetuation and reassortment of gull influenza A viruses in Atlantic North America. Virology 456:353–363.

Hupp, J. W., J. A. Schmutz, C. R. Ely, E. E. Syroechkovski, Jr., A. V. Kondratyev, W. D. Eldridge, and E. Lappo. 2007. Moult migration of emperor geese *Chen canagica* between Alaska and Russia. Journal of Avian Biology 38:462–470.

Hurt, A. C., P. M. Hansbro, P. Selleck, B. Olsen, C. Minton, A. W. Hampson, and I. G. Barr. 2006. Isolation of avian influenza viruses from two different transhemispheric migratory shorebird species in Australia. Archives of Virology 151:2301.

Hurt, A. C., Y. C. Su, M. Aban, H. Peck, H. Lau, C. Baas, Y. M. Deng, N. Spirason, P. Ellström, J. Hernandez, et al. 2016. Evidence for the introduction, reassortment, and persistence of diverse influenza A viruses in Antarctica. Journal of Virology 90:9674–9682.

Hurt, A. C., D. Vijaykrishna, J. Butler, C. Baas, S. Maurer-Stroh, M. C. Silva-de-la-Fuente, G. Medina-Vogel, B. Olsen, A. Kelso, A., Barr, et al. 2014. Detection of evolutionarily distinct avian influenza A viruses in Antarctica. MBio 5:3.

Hurtado, R., T. Fabrizio, Vanstreels, R. E. S. Krauss, R. J. Webby, R. G. Webster, and E. L. Durigon, 2015. Molecular characterization of subtype H11N9 avian influenza virus isolated from shorebirds in Brazil. PLoS One 10:e0145627.

Ip, H. S., R. J. Dusek, B, Bodenstein, M. K. Torchetti, P. DeBruyn, K. G. Mansfield, T. DeLiberto, and J. M. Sleeman. 2016. High rates of detection of clade 2.3.4.4 highly pathogenic avian influenza H5 viruses in wild birds in the Pacific Northwest during the winter of 2014–15. Avian Diseases 60:354–358.

Ip, H. S., P. L. Flint, J. C. Franson, R. J. Dusek, D. V. Derksen, R. E. Gill, C. R. Ely, J. M. Pearce, R. B. Lanctot, S. M. Matsuoka, et al. 2008. Prevalence of influenza A viruses in wild migratory birds in Alaska: patterns of variation in detection at a crossroads of intercontinental flyways. Virology Journal 5:71.

Ip, H. S., M. K. Torchetti, R. Crespo, P. Kohrs, P. DeBruyn, Mansfield, K. G., T. Baszler, L. Badcoe, B. Bodenstein, V. Shearn-Bochsler, et al. 2015. Novel Eurasian highly pathogenic avian influenza A H5 viruses in wild birds, Washington, USA, 2014. Emerging Infectious Diseases 21:886–890.

Iverson, S. A., J. Y. Takekawa, S. Schwarzbach, C. J. Cardona, N. Warnock, M. A. Bishop, G. A. Schirato, S. Paroulek, J. T. Ackerman, H. Ip, et al. 2008. Low prevalence of avian influenza virus in shorebirds on the Pacific Coast of North America. Waterbirds 31:602–610.

Jeong, J., H.-M. Kang, E.-K. Lee, B.-M. Song, Y.-K. Kwon, H.-R. Kim, K.-S. Choi, J.-Y. Kim, H.-J. Lee, O.-K. Moon, et al. 2014. Highly pathogenic avian influenza virus (H5N8) in domestic poultry and its relationship with migratory birds in South Korea during 2014. Veterinary Microbiology 173:249–257.

Jiménez-Bluhm, P., E. A. Karlsson, P. Freiden, B. Sharp, F. Di Pillo, F., J. E. Osorio, C. Hamilton-West, and S. Schultz-Cherry. 2018. Wild birds in Chile harbor diverse avian influenza A viruses. Emerging Microbes and Infections 7:44.

Kakogawa, M., M. Onuma, K. Saito, Y. Watanabe, K. Goka, and M. Asakawa. 2020. Epidemiologic survey of avian influenza virus infection in shorebirds captured in Hokkaido, Japan. Journal of Wildlife Diseases 56: 651–657.

Karlsson, E. A., K. Ciuoderis, P. J. Freiden, B. Seufzer, J. C. Jones, J. Johnson, R. Parra, A. Gongora, D. Cardenas, D. Barajas, et al. 2013. Prevalence and characterization of influenza viruses in diverse species in Los Llanos, Colombia. Emerging Microbes and Infections 2:1–10.

Kim, H.-R., Y.-K. Kwon, I. Jang, Y.-J. Lee, H.-M. Kang, E.-K. Lee, B.-M. Song, H.-S. Lee, Y.-S. Joo, K.-H. Lee, et al. 2015. Pathologic changes in wild birds infected with highly pathogenic avian influenza A (H5N8) viruses, South Korea, 2014. Emerging Infectious Diseases 21: 775–780.

Kim, H.-R., Y.-J. Lee, K.-K. Lee, J.-K. Oem, S.-H. Kim, M.-H. Lee, O.-S. Lee, and C.-K. Park. 2010. Genetic relatedness of H6 subtype avian influenza viruses isolated from wild birds and domestic ducks in Korea and their pathogenicity in animals. Journal of General Virology 91:208–219.

Kleyheeg, E., R. Slaterus, R. Bodewes, J. M. Rijks, M. A. H. Spierenburg, N. Beerens, L. Kelder, M. J. Poen, J. A. Stegeman, R. A. M. Fouchier, et al. 2017. Deaths among wild birds during highly pathogenic avian influenza A (H5N8) virus outbreak, the Netherlands. Emerging Infectious Diseases 23:2050–2054.

Klingeborn, B., L. Englund, R. Rott, N. Juntti, and G. Rockborn. 1985. An avian influenza A virus killing a mammalian species—the mink. Archives of Virology 86:347–351.

Klopfleisch, R., P. U. Wolf, C. Wolf, T. Harder, E.Starick, M. Niebuhr, M., T. C. Mettenleiter, and J. P. Teifke. 2007. Encephalitis in a stone marten (*Martes foina*) after natural infection with highly pathogenic avian influenza virus subtype H5N1. Journal of Comparative Pathology 137:155–159.

Krauss, S., D. E. Stallknecht, N. J. Negovetich, L. J. Niles, R. J. Webby, and R. G. Webster. 2010. Coincident ruddy turnstone migration and horseshoe crab spawning creates an ecological 'hot spot' for influenza viruses. Proceedings of the Royal Society B Biological Sciences 277:3373–3379.

Krone, O., A. Globig, R. Ulrich, T. Harder, J. Schinköthe, C. Herrmann, S. Gerst, F. J. Conraths, and M. Beer. 2018. White-tailed sea eagle (*Haliaeetus albicilla*) die-off due to infection with highly pathogenic avian influenza virus, subtype H5N8, in Germany. Viruses 10:478.

Lambrecht, B., S. Marché, P. Houdart, T. van den berg, and D. Vangeluwe. 2016. Impact of age, season, and flowing vs. stagnant water habitat on avian influenza prevalence in mute swan (*Cygnus olor*) in Belgium. Avian Diseases 60:322–328.

Lang, A. S., C. Lebarbenchon, A. M. Ramey, G. J. Robertson, J. Waldenström, and M. Wille. 2016. Assessing the role of seabirds in the ecology of influenza A viruses. Avian Diseases 60:378–386.

Latorre-Margalef, N., C. Tolf, V. Grosbois, A. Avril, D. Bengtsson, M. Wille, A. D. Osterhaus, R. A. Fouchier, B. Olsen, and J. Waldenström. 2014. Long-term variation in influenza A virus prevalence and subtype diversity in migratory mallards in northern Europe. Proceedings of the Royal Society B Biological Sciences 281:20140098.

Lebarbenchon, C., S. Sreevatsan, T. Lefèvre, M. Yang, M. A. Ramakrishnan, J. D. Brown, and D. E. Stallknecht. 2012. Reassortant influenza A viruses in wild duck populations: effects on viral shedding and persistence in water. Proceedings of the Royal Society B Biological Sciences 279:3967–3975.

Lebarbenchon, C., S. Sreevatsan, M. A. Ramakrishnan, R. Poulson, V. Goekjian, J. J. Di Matteo, B. Wilcox, and D. E. Stallknecht. 2010. Influenza A viruses in American white pelican (*Pelecanus erythrorhynchos*). Journal of Wildlife Diseases 46:1284–1289.

Lebarbenchon, C., B. R. Wilcox, R. L. Poulson, M. J. Slusher, N. B. Fedorova, D. A. Katzel, C. J. Cardona, G. A. Knutsen, D. E. Wentworth, and D. E. Stallknecht, 2015. Isolation of type A influenza viruses from red-necked grebes (*Podiceps grisegena*). Journal of Wildlife Diseases 51:290–293.

Lebarbenchon, C., M. Yang, S. P. Keeler, M. A. Ramakrishnan, J. D. Brown, D. E. Stallknecht, and S. Sreevatsan. 2011. Viral replication, persistence in water and genetic characterization of two influenza A viruses isolated from surface lake water. PLoS One 6:e26566.

Lee, D. H., M. K. Torchetti, J. Hicks, M. L. Killian, J. Bahl, M. Pantin-Jackwood, and D. E. Swayne. 2018. Transmission dynamics of highly pathogenic avian influenza virus A (H5Nx) clade 2.3.4.4, North America, 2014–2015. Emerging Infectious Diseases 24:1840–1848.

Lee, D. H., M. K. Torchetti, K. Winker, H. S. Ip, C. S. Song, and D. E. Swayne. 2015. Intercontinental spread of Asian-origin H5N8 to North America through Beringia by migratory birds. Journal of Virology 89:6521–6524.

Lewis, N. S., Z. Javakhishvili C. A., Russell, A. Machablishvili, P. Lexmond, J. H. Verhagen, O. Vuong, T. Onashvili, M. Donduashvili, D. J. Smith, et al. 2013. Avian influenza virus surveillance in wild birds in Georgia: 2009–2011. PLoS One 8:e58534.

Li, M., H. Liu, Y. Bi, J. Sun, G. Wong, D. Liu, L. Li, J. Liu, Q. Chen, H. Wang, et al. 2017. Highly pathogenic avian influenza A (H5N8) virus in wild migratory birds, Qinghai Lake, China. Emerging Infectious Diseases 23:637–641.

Lupiani, B., and S. M. Reddy. 2009. The history of avian influenza. Comparative Immunology, Microbiology and Infectious Diseases 32:311–323.

Lycett, S. J., A, Pohlmann, C. Staubach, V. Caliendo, M. Woolhouse, M. Beer, T. Kuiken, and Global Consortium for H5N8 and Related Influenza Viruses. 2020. Genesis and spread of multiple reassortants during the 2016/2017 H5 avian influenza epidemic in Eurasia. Proceedings of the National Academy of Sciences USA 117:20814–20825.

Mathieu, C., V. Moreno, J. Pedersen, J. Jeria, M. Agredo, C. Gutiérrez, A. García, M. Vásquez, P. Avalos, and P. Retamal. 2015. Avian Influenza in wild birds from Chile, 2007–2009. Virus Research 199:42–45.

McBride, D. S., S. E. Lauterbach, Y.-T. Li, G. J. D. Smith, M. L. Killian, J. M. Nolting, Y. C. Su, and A. S. Bowman. 2021. Genomic evidence for sequestration of influenza A virus lineages in sea duck host species. Viruses 13:172.

Munster, V. J., C. Baas, P. Lexmond, J. Waldenström, A. Wallensten, T. Fransson, G. F. Rimmelzwaan, W. E. Beyer, M. Schutten, B. Olsen, et al. 2007. Spatial, temporal, and species variation in prevalence of influenza A viruses in wild migratory birds. PLOS Pathogens 3:e61.

Nagy, A., J. Machova, J. Hornickova, M. Tomci, I. Nagl, B. Horyna, and I. Holko. 2007. Highly pathogenic avian influenza virus subtype H5N1 in mute swans in the Czech Republic. Veterinary Microbiology 120:9–16.

Nelson, M. I., S. Pollett, B. Ghersi, M. Silva, M. P. Simons, E. Icochea, A. E., Gonzalez, K. Segovia, M. R. Kasper, J. M. Montgomery, et al. 2016. The genetic diversity of

influenza A viruses in wild birds in Peru. PLoS One 11:e0146059.
Office International des Epizooties (OIE). 2021. Avian influenza portal. World Organsation for Animal Health, Paris, France. https://www.oie.int/en/animal-health-in-the-world/web-portal-on-avian-influenza/ Accessed 14 Jan 2021.
Olsen, B., V. J. Munster, A. Wallensten, J. Waldenström, A. D. Osterhaus, and R. A. Fouchier. 2006. Global patterns of influenza A virus in wild birds. Science 312:384–388.
Papp, Z., R. G. Clark, E. J. Parmley, F. A. Leighton, C. Waldner, and C. Soos. 2017. The ecology of avian influenza viruses in wild dabbling ducks (*Anas* spp.) in Canada. PLoS One 12:e0176297.
Pasick, J., Y. Berhane, T. Joseph, V. Bowes, T. Hisanaga, K. Handel, and S. Alexandersen. 2015. Reassortant highly pathogenic influenza A H5N2 virus containing gene segments related to Eurasian H5N8 in British Columbia, Canada, 2014. Scientific Reports 5:9484.
Pensaert, M., K. Ottis, J. Vandeputte, M. M. Kaplan, and P. A. Bachmann. 1981. Evidence for the natural transmission of influenza A virus from wild ducks to swine and its potential importance for man. Bulletin of the World Health Organization 59:75–78.
Pereda, A. J., M. Uhart, A. A. Perez, M. E. Zaccagnini, L. La Sala, J. Decarre, A. Goijman, L. Solari, R. Suarez, M. I. Craig, et al. 2008. Avian influenza virus isolated in wild waterfowl in Argentina: evidence of a potentially unique phylogenetic lineage in South America. Virology 378:363–370.
Pohlmann, A., E. Starick, T. Harder, C. Grund, D. Höper, A. Globig, C. Staubach, K. Dietze, G. Strebelow, R. G. Ulrich, et al. 2017. Outbreaks among wild birds and domestic poultry caused by reassorted influenza A (H5N8) clade 2.3.4.4 viruses, Germany, 2016. Emerging Infectious Diseases 23:633–636.
Poulson, R., D. Carter, S. Beville, L. Niles, A. Dey, C. Minton, P. McKenzie, S. Krauss, R. Webby, R. Webster, et al. 2020. Influenza A viruses in ruddy turnstones (*Arenaria interpres*): connecting wintering and migratory sites with an ecological hotspot at Delaware Bay. Viruses 12:1205.
Ramey, A. M., N. J. Hill, T. J. DeLiberto, S. E. J. Gibbs, M. C. Hopkins, A. S. Lang, R. L. Poulson, D. J. Prosser, J. M. Sleeman, D. E. Stallknecht, et al. 2022. Highly pathogenic avian influenza is an emerging disease threat to wild birds in North America. Journal of Wildlife Management 86:e22171.
Ramey, A. M., A. B. Reeves, J. Z. Drexler, J. T. Ackerman, S. De La Cruz, A. S. Lang, C. Leyson, P. Link, D. J. Prosser, G. J. Robertson, et al. 2020. Influenza A viruses remain infectious for more than seven months in northern wetlands of North America. Proceedings of the Royal Society B Biological Sciences 287:20201680.
Ramey, A. M., A. B. Reeves, S. A. Sonsthagen, J. L. TeSlaa, S. Nashold, T. Donnelly, B. Casler, and J. S. Hall, 2015. Dispersal of H9N2 influenza A viruses between East Asia and North America by wild birds. Virology 482:79–83.
Ramey, A. M., A. B. Reeves, J. L. TeSlaa, S. Nashold, T. Donnelly, J. Bahl, and J. S. Hall, 2016. Evidence for common ancestry among viruses isolated from wild birds in Beringia and highly pathogenic intercontinental reassortant H5N1 and H5N2 influenza A viruses. Infection, Genetics and Evolution 40:176–185.
Ramey, A. M., B. D. Uher-Koch, A. B. Reeves, J. A. Schmutz, R. L. Poulson, and D. E. Stallknecht. 2019. Emperor geese (*Anser canagicus*) are exposed to a diversity of influenza A viruses, are infected during the non-breeding period and contribute to intercontinental viral dispersal. Transboundary and Emerging Diseases 66:1958–1970.
Ramos, S., M. MacLachlan, and A. Melton. 2017. Impacts of the 2014–2015 highly pathogenic avian influenza outbreak on the US poultry sector. USDA Economic Research Service LDPM-282-02, Washington, D.C., USA.
Reeves, A. B. J. M. Pearce, A. M. Ramey, C. R. Ely, J. A. Schmutz, P. L. Flint, D. V. Derksen, H. S. Ip, and K. A. Trust. 2013. Genomic analysis of avian influenza viruses from waterfowl in western Alaska, USA. Journal of Wildlife Diseases 49:600–610.
Reeves, A. B., J. S. Hall, R. L. Poulson, T. Donnelly, D. E. Stallknecht, and A. M. Ramey. 2018. Influenza A virus recovery, diversity, and intercontinental exchange: a multi-year assessment of wild bird sampling at Izembek National Wildlife Refuge, Alaska. PLoS One 13:e0195327.
Rimondi, A., A. S. Gonzalez-Reiche, V. S. Olivera, J. Decarre, G. J. Castresana, M. Romano, M. I. Nelson, H. van Bakel, A. J., Pereda, L. Ferreri, et al. 2018. Evidence of a fixed internal gene constellation in influenza A viruses isolated from wild birds in Argentina (2006–2016). Emerging Microbes & Infections 7:194.
Röhm, C., T. Horimoto, Y. Kawaoka, J. Süss, and R. G. Webster. 1995. Do hemagglutinin genes of highly pathogenic avian influenza viruses constitute unique phylogenetic lineages? Virology 209:664–670.
Ruiz, S., P. Jimenez-Bluhm, F. Di Pillo, C. Baumberger, P. Galdames, V. Marambio, C. Salazar, C. Mattar, J. Sanhueza, S. Schultz-Cherry, et al. 2021. Temporal dynamics and the influence of environmental variables on the prevalence of avian influenza virus in main

wetlands in central Chile. Transboundary and Emerging Diseases 68:1601–1614.

Sakoda, Y., S. Sugar, D. Batchluun, T.-O. Erdene-Ochir, M. Okamatsu, N. Isoda, K. Soda, H. Takakuwa, Y. Tsuda, N. Yamamoto, et al. 2010. Characterization of H5N1 highly pathogenic avian influenza virus strains isolated from migratory waterfowl in Mongolia on the way back from the southern Asia to their northern territory. Virology 406:88–94.

Samuel, M. D., J. S. Hall, J. D. Brown, D. R. Goldberg, H. Ip, and V. V. Baranyuk. 2015. The dynamics of avian influenza in lesser snow geese: implications for annual and migratory infection patterns. Ecological Applications 25:1851–1859.

Schünemann, H. J., S. R. Hill, M. Kakad, R. Bellamy, T. M. Uyeki, F. G. Hayden, Y. Yazdanpanah, J. Beigel, T. Chotpitayasunondh, C. Del Mar, et al. 2007. WHO rapid advice guidelines for pharmacological management of sporadic human infection with avian influenza A (H5N1) virus. Lancet Infectious Diseases 7:21–31.

Shafir, S. C., T. Fuller, T. B. Smith, and A. W. Rimoin, 2012. A national study of individuals who handle migratory birds for evidence of avian and swine-origin influenza virus infections. Journal of Clinical Virology 54:364–367.

Sharshov, K., M. Sivay, D. Liu, M. Pantin-Jackwood, V. Marchenko, A. Durymanov, A. Alekseev, T. Damdindorj, G. F. Gao, D. E. Swayne, et al. 2014. Molecular characterization and phylogenetics of a reassortant H13N8 influenza virus isolated from gulls in Mongolia. Virus Genes 49:237–249.

Shearn-Bochsler, V. I., S. Knowles, and H. Ip. 2019. Lethal infection of wild raptors with highly pathogenic avian influenza H5N8 and H5N2 viruses in the USA, 2014–15. Journal of Wildlife Diseases 55:164–168.

Simulundu, E., A. S. Mweene, D. Tomabechi, B. M. Hang'ombe, A. Ishii, Y. Suzuki, I. Nakamura, H. Sawa, C. Sugimoto, K. Ito, et al. 2009. Characterization of H3N6 avian influenza virus isolated from a wild white pelican in Zambia. Archives of Virology 154:1517–1522.

Slemons, R. D., D. C. Johnson, J. S. Osborn, and F. Hayes. 1974. Type-A influenza viruses isolated from wild free-flying ducks in California. Avian Diseases 18:119–124.

Slepuškin, A. N., T. V. Pysina, F. K. Gonsovsky, A. A. Sazonov, V. A. Isačenko, N. N. Sokolova, V. M. Polivanov, D. K. Lvov, and L. J. Zakstel'skaja. 1972. Haemagglutination-inhibiting activity to type A influenza viruses in the sera of wild birds from the far east of the USSR. Bulletin of the World Health Organization 47:527–530.

Slusher, M. J., B. R. Wilcox, M. P. Lutrell, R. L. Poulson, J. D. Brown, M. J. Yabsley, and D. E. Stallknecht. 2014. Are passerine birds reservoirs for influenza A viruses? Journal of Wildlife Diseases 50:792–809.

Smith, G.J., D. Vijaykrishna, J. Bahl, S. J. Lycett, M. Worobey, O. G. Pybus, S. K. Ma, C. L. Cheung, J. Raghwani, S. Bhatt, et al. 2009. Origins and evolutionary genomics of the 2009 swine-origin H1N1 influenza A epidemic. Nature 459:1122–1125.

Spackman, E., K. G. McCracken, K. Winker, and D. E. Swayne. 2006. H7N3 avian influenza virus found in a South American wild duck is related to the Chilean 2002 poultry outbreak, contains genes from equine and North American wild bird lineages, and is adapted to domestic turkeys. Journal of Virology 80:7760–7764.

Spackman, E., D. L. Suarez, and D. E. Swayne. 2008. Avian influenza diagnostics and surveillace methods. Pages 299–308 *in* D. E. Swayne, editor. Avian influenza. Blackwell Publishing, Ames, Iowa, USA.

Spackman, E., D. E. Swayne, M. Gilbert, D. O. Joly, W. B. Karesh, D. L. Suarez, R. Sodnomdarjaa, P. Dulam, and C. Cardona. 2009. Characterization of low pathogenicity avian influenza viruses isolated from wild birds in Mongolia 2005 through 2007. Virology Journal. 6:190.

Su, S., G. Xing, J. Wang, Z. Li, J. Gu, L. Yan, J. Lei, S. Ji, B. Hu, G. C. Gray, et al. 2016. Characterization of H7N2 avian influenza virus in wild birds and pikas in Qinghai-Tibet Plateau area. Scientific Reports 6:30974.

Suarez, D. L. 2008. Influenza A virus. Pages 3–22 *in* D. E. Swayne, editor. Avian influenza. Blackwell Publishing, Ames, Iowa, USA.

Swayne, D. E., and M. Pantin-Jackwood. 2008. Pathobiology of avian influenza virus infections in birds and mammals. Pages 87–122 *in* D. E. Swayne, editor. Avian influenza. Blackwell Publishing, Ames, Iowa, USA.

Swayne, D. E., and D. L. Suarez. 2000. Highly pathogenic avian influenza. OIE Revue Scientifique et Technique 19:463–482.

Tanimura, N., K. Tsukamoto, M. Okamatsu, M. Mase, T. Imada, K. Nakamura, M. Kubo, S. Yamaguchi, W. Irishio, M. Hayashi, et al. 2006. Pathology of fatal highly pathogenic H5N1 avian influenza virus infection in large-billed crows (*Corvus macrorhynchos*) during the 2004 outbreak in Japan. Veterinary Pathology 43:500–509.

Thornton, A. C., F. Parry-Ford, E. Tessier, N. Oppilamany, H. Zhao, J. Dunning, R. Pebody, and G. Dabrera. 2019. Human exposures to H5N6 avian influenza, England, 2018. The Journal of Infectious Diseases 220:20–22.

Tønnessen, R., A. B. Kristoffersen, C. M. Jonassen, M. J. Hjortaas, E. F. Hansen, Rimstad, and A. G. Hauge. 2013. Molecular and epidemiological characterization of avian influenza viruses from gulls and dabbling ducks in Norway. Virology Journal 10:112.

Uher-Koch, B. D., T. J. Spivey, C. R. Van Hemert, J. A. Schmutz, K. Jiang, X. F. Wan, and A. M. Ramey. 2019. Serologic evidence for influenza A virus exposure in three loon species breeding in Alaska, USA. Journal of Wildlife Diseases 55:862–867.

Van Dijk, J. G., B. J. Hoye, J. H. Verhagen, B. A. Nolet, R. A. Fouchier, and M. Klaassen. 2014. Juveniles and migrants as drivers for seasonal epizootics of avian influenza virus. Journal of Animal Ecology 83:266–275.

Verhagen, J. H., P. Eriksson, L. Leijten, O. Blixt, B. Olsen, J. Waldenström, P. Ellström, and T. Kuiken. 2021*a*. Host range of influenza A virus H1 to H16 in Eurasian ducks based on tissue and receptor binding studies. Journal of Virology 95:e01873-20.

Verhagen, J. H., R. A. Fouchier, and N. Lewis. 2021*b*. Highly pathogenic avian influenza viruses at the wild–domestic bird interface in Europe: future directions for research and surveillance. Viruses 13:212.

Verhagen, J. H., F. Majoor, P. Lexmond, O. Vuong, G. Kasemir, D. Lutterop, A. D. Osterhaus, R. A. Fouchier, and T. Kuiken. 2014. Epidemiology of influenza A virus among black-headed gulls, the Netherlands, 2006–2010. Emerging Infectious Diseases 20:138–141.

Verhagen, J. H., M. Poen, D. E. Stallknecht, S. van der Vliet, P. Lexmond, S. Sreevatsan, R. L. Poulson, R. A.Fouchier, and C. Lebarbenchon. 2020. Phylogeography and antigenic diversity of low-pathogenic avian influenza H13 and H16 viruses. Journal of Virology 94:e00537-20.

Verhagen, J. H., H. P. van der Jeugd, B. A. Nolet, R. Slaterus, S. P. Kharitonov, P. P. de Vries, O. Vuong, F. Majoor, T Kuiken, and R. A. Fouchier, 2015. Wild bird surveillance around outbreaks of highly pathogenic avian influenza A (H5N8) virus in the Netherlands, 2014, within the context of global flyways. Eurosurveillance 20:21069.

Wan, X. F. 2012. Lessons from emergence of A/Goose/Guangdong/1996-like H5N1 highly pathogenic avian influenza viruses and recent influenza surveillance efforts in southern China. Zoonoses and Public Health 59:32–42.

Webby, R. J., and R. G. Webster. 2001. Emergence of influenza A viruses. Philosophical Transactions of the Royal Society of London B Biological Sciences 356:1817–1828.

Wille M., G. J. Robertson, H. Whitney, M. A. Bishop, J. A. Runstadler, A. S. Lang. 2011. Extensive geographic mosaicism in avian influenza viruses from gulls in the northern hemisphere. PLoS One 6:e20664.

Wille, M., C. Tolf, A. Avril, N. Latorre-Margalef, S. Wallerström, B. Olsen, and J. Waldenström. 2013. Frequency and patterns of reassortment in natural influenza A virus infection in a reservoir host. Virology 443:150–160.

Wilson, H. M., J. S. Hall, P. L. Flint, J. C. Franson, C. R. Ely, J. A. Schmutz, M. D. Samuel. 2013. High seroprevalence of antibodies to avian influenza viruses among wild waterfowl in Alaska: implications for surveillance. PLoS One 8:e58308.

Wood, G. W., J. W. McCauley, J. B. Bashiruddin, and D. J. Alexander. 1993. Deduced amino acid sequences at the haemagglutinin cleavage site of avian influenza A viruses of H5 and H7 subtypes. Archives of Virology 130:209–217.

World Health Organization (WHO). 2006. Rapid advice guidelines on pharmacological management of humans infected with avian influenza A (H5N1) virus. https://apps.who.int/iris/bitstream/handle/10665/69416/WHO_CDS_EPR_GIP_2006_4r1.pdf. Accessed 2 Mar 2021.

Xu, W., Y. Berhane, C. Dubé, B. Liang, J. Pasick, G. VanDomselaar, and S. Alexandersen. 2016. Epidemiological and evolutionary inference of the transmission network of the 2014 highly pathogenic avian influenza H5N2 outbreak in British Columbia, Canada. Scientific Reports 6:30858.

Xu, Y., A. M. Ramey, A. S. Bowman, T. J. DeLiberto, M. L. Killian, S. Krauss, J. M. Nolting, M. K. Torchetti, A. B. Reeves, R. J. Webby, et al. 2017. Low-pathogenic influenza A viruses in North American diving ducks contribute to the emergence of a novel highly pathogenic influenza A (H7N8) virus. Journal of Virology 91:9.

Xu, X., K. Subbarao, N. J. Cox, and Y. Guo. 1999. Genetic characterization of the pathogenic influenza A/Goose/Guangdong/1/96 (H5N1) virus: similarity of its hemagglutinin gene to those of H5N1 viruses from the 1997 outbreaks in Hong Kong. Virology 261:15–19.

Zakstel'skaja, L. J., V. A. Isačenko, N. G. Osidze, C. C. Timofeeva, A. N. Slepuškin, and N. N. Sokolova. 1972. Some observations on the circulation of influenza viruses in domestic and wild birds. Bulletin of the World Health Organization 47:497–501.

Zhou, N. N., D. A. Senne, J. S. Landgraf, S. L. Swenson, G. Erickson, K. Rossow, L. Liu, K. J. Yoon, S. Krauss, and R. G. Webster 1999. Genetic reassortment of avian, swine, and human influenza A viruses in American pigs. Journal of Virology 73:8851–8856.

Zohari, S., A. Neimanis., T. Härkönen, C. Moraeus, and J. F. Valarcher. 2014. Avian influenza A (H10N7) virus involvement in mass mortality of harbour seals (*Phoca vitulina*) in Sweden, March through October 2014. Eurosurveillance 19:20967.

19 Avian Malaria and the Extinction of Hawaiian Forest Birds

Carter T. Atkinson

Introduction

Isolated island ecosystems and their endemic and indigenous fauna and flora are highly vulnerable to the introduction of new diseases and disease vectors. In the Pacific Basin, introduced avian pox virus (*Avipoxvirus* spp.), malaria (*Plasmodium* spp.), and mosquito vectors (*Culex* spp.) have had impacts on the avifauna of the Galapagos Islands (Parker et al. 2011), New Zealand (Schoener et al. 2014, Alley et al. 2010), and Hawai'i (Warner 1968), and these archipelagoes face continual threats from the introduction of new diseases such as West Nile Virus (LaPointe et al. 2009, Harvey-Samuel et al. 2021). In the Hawaiian Islands, introduced diseases; changing climate; continued habitat loss; stochastic environmental events; and invasive plants, invertebrates, and vertebrates have degraded native ecosystems and played a significant role in the decline and extinction of multiple species and whole families of native forest birds (van Riper and Scott 2001). Recently, invasive fungal pathogens of the keystone forest tree, 'ōhi'a lehua (*Metrosideros polymorpha*) have created new threats for long-term restoration of native forest bird communities (Mortensen et al. 2016, Camp et al. 2019). Introduced *Aedes* mosqui-

Illustration by Laura Donohue.

toes and the viruses (Zika and dengue) they transmit are also threats to the health of human populations.

In the past 15 years, the ʻIʻiwi (*Drepanis coccinea*), the ʻAkikiki (*Oreomystis bairdi*) and the ʻAkekeʻe (*Loxops caeruleiostris*) have been listed as endangered (USFWS 2010, 2017). Native forest bird populations on Kauaʻi are collapsing (Paxton et al. 2016), and the introduced mosquito vector of avian malaria (*Culex quinquefasciatus*) is becoming more abundant in high-elevation forests that were once refuges from disease transmission (Atkinson et al. 2014, Fortini et al. 2015). This review examines the history and dynamics of introduced *Plasmodium relictum* and *Culex quinquefasciatus*, their impacts on native forest birds in the Hawaiian Islands, multiple factors that affect transmission of this disease, and options for management. While introduced avian pox virus (*Avipoxvirus* spp.) is also a significant factor in decline of native birds (Jarvi et al. 2008, Samuel et al. 2018), the focus will be on malaria, since both diseases are transmitted by *Culex* and most management strategies are the same. Long-term conservation of native species may ultimately depend on how effectively a One Health approach to landscape-scale vector control can be adopted to maintain biodiversity of native ecosystems.

A Paradise Lost

The endemic Hawaiian honeycreepers (Drepanidinae) are rivaled only by the Galapagos finches in terms of divergent evolution of bill shapes and life-history characteristics (Pratt 2009). While there were once over 100 native forest bird species on the Hawaiian archipelago, including native thrushes, honeyeaters, monarch flycatchers, crows, and more than 50 species of honeycreepers, only 21 remain, and 12 survive as small, fragmented populations limited to high elevations on Kauaʻi, Maui, and Hawaiʻi Islands (Paxton et al. 2018).

The Hawaiian archipelago was never colonized naturally by biting arthropods that are vectors of protozoan, helminth, viral, and bacterial pathogens. The only known indigenous haemosporidian parasite to have reached the islands without human assistance is *Haemoproteus iwa*, which arrived in great frigatebirds (*Fregata minor*). The frigatebird carried their own host-specific, ectoparasitic hipposcid fly vector, but neither the fly nor the parasite made a host switch to forest birds (Work and Raymeyer 1996, Levin et al. 2011). Unlike the Galapagos Islands where indigenous mosquitoes are present and capable of bridging transmission of vector-borne parasites between migrants and endemic birds (Levin et al. 2013), the few parasites that may have reached Hawaiʻi in shorebirds, waterfowl, and other rare migrants encountered a dead-end with no transmission routes available to new hosts (Warner 1968).

In the early 19th century, a whaling ship unintentionally released larvae of *Culex quinquefasciatus*, the southern house mosquito, while filling water casks in a stream on Maui (van Dine 1904). *Culex quinquefasciatus*, a peridomestic mosquito with a preference for feeding on avian as well as mammalian hosts, most likely spread rapidly across the islands, making humans, domestic animals, and forest birds vulnerable to introduction of mosquito-transmitted diseases. Avian pox virus (*Avipoxvirus*) was the first widespread vector-borne disease to be reported in native birds, with evidence that it was having population-level impacts by the late 19th century (Warner 1968). As colorful native birds began to disappear from lowland habitats where mosquitoes were most prevalent, there were campaigns to introduce nonnative birds for aesthetic reasons, to control agricultural pests, and to provide new hunting opportunities (Foster 2009). Releases reached a peak in the first half of the 20th century, with the introduction of at least 140 species from six continents (Moulton et al. 2001). These releases were not done with the benefit of health screenings or quarantines, and this may be how a lineage of *Plasmodium relictum* that is common in Asian and African passerines became established in Hawaiʻi (Beadell et al. 2006; Hellgren et al. 2015).

The first report of *P. relictum* was in nonnative species on Hawai'i Island (Baldwin 1941), but it is not clear where the parasite was first introduced. By the early 1950s however, it was present at lower elevations on the islands of Kaua'i and O'ahu (Warner 1968). By 1973 when the Endangered Species Act was passed and many large-scale conservation efforts to save native forest birds began in earnest, introduced malaria, avian pox, and *Culex* mosquitoes were well established on Kaua'i, O'ahu, Maui, and Hawai'i Islands (Warner 1968, van Riper et al. 1986).

Etiology and Life Cycle

The avian plasmodia comprise a large group of over 100 species of pigmented intraerythrocytic apicomplexan parasites with asexual stages of reproduction in circulating red blood cells and sexual and asexual reproduction in mosquitoes (Figure 19.1) (Atkinson 2008).

Transmission between hosts occurs when mosquitoes seek a blood meal from a suitable avian host. Mosquitoes become infected by ingesting gametocytes in circulating blood cells that subsequently undergo gametogenesis and fertilization in the midgut bloodmeal. A motile zygote called the ookinete forms after fertilization and penetrates the midgut wall. Here, it rounds up into a sphere to become a growing oocyst and eventually undergoes asexual reproduction (sporogony) to produce sporozoites that rupture from the oocyst and invade salivary glands. New infections are initiated when mosquitoes take a second blood meal and release infective sporozoites into the tissues of a new host. Here, they undergo asexual reproduction (merogony) and form merozoites that invade circulating erythrocytes, undergo additional cycles of merogony and eventually form gametocytes (gametogony) that are infective to new mosquito hosts.

While many species of avian *Plasmodium* have limited host or geographic ranges, a few are more generalist and cosmopolitan parasites. *Plasmodium relictum* is probably the most widely distributed species of avian *Plasmodium*, with reports from more 200 avian species from all continents except Antarctica (Valkiūnas et al. 2018). *Plasmodium relictum* is composed of at least five mitochondrial cytochrome b lineages that are morphologically identical but with differences in geographic distribution, host range, and pathology (Valkiūnas et al. 2018). The lineage present in Hawai'i has been identified as GRW4 based on a 478-bp region of the cytochrome b gene (Beadell et al. 2006, Valkiūnas et al. 2018).

Malaria is primarily a disease of the blood and reticuloendothelial system (Atkinson 2008). After a short prepatent period when the parasite cannot be found in the bloodstream, parasitemia in circulating erythrocytes increases until a peak is reached approximately 6–12 days after infection. During this time, rapidly reproducing asexual meronts destroy red blood cells through rupture and removal from circulation by macrophages in the liver and spleen. Most pathology is associated with severe anemia that accompanies erythrocyte destruction, although development of tissue meronts in capillary endothelial cells in some species of avian *Plasmodium* can obstruct blood flow in the brain and lead to death. Signs of acute infection include declines in food consumption, weight loss, and reduced activity levels associated with severe anemia. In surviving individuals, parasitemia rapidly falls as the humoral and cellular immune response begins to bring the infection under control (van Riper et al. 1994). This is followed by a lengthy chronic, low-intensity infection where birds have immunity to reinfection with the same strains of the parasite (Atkinson et al. 2001a). Birds are generally believed to remain infected for life, although experimental data to confirm this is sparse (Atkinson 2008). Fitness costs associated with chronic infections are more difficult to quantify but may result in long-term effects on lifetime reproductive success, mate choice, and life span (LaPointe et al. 2012, Asghar et al. 2016).

The very high susceptibility of Hawaiian honeycreepers to malaria is almost unparalleled among species of birds. Captive penguins are particularly susceptible as are many other species of birds that

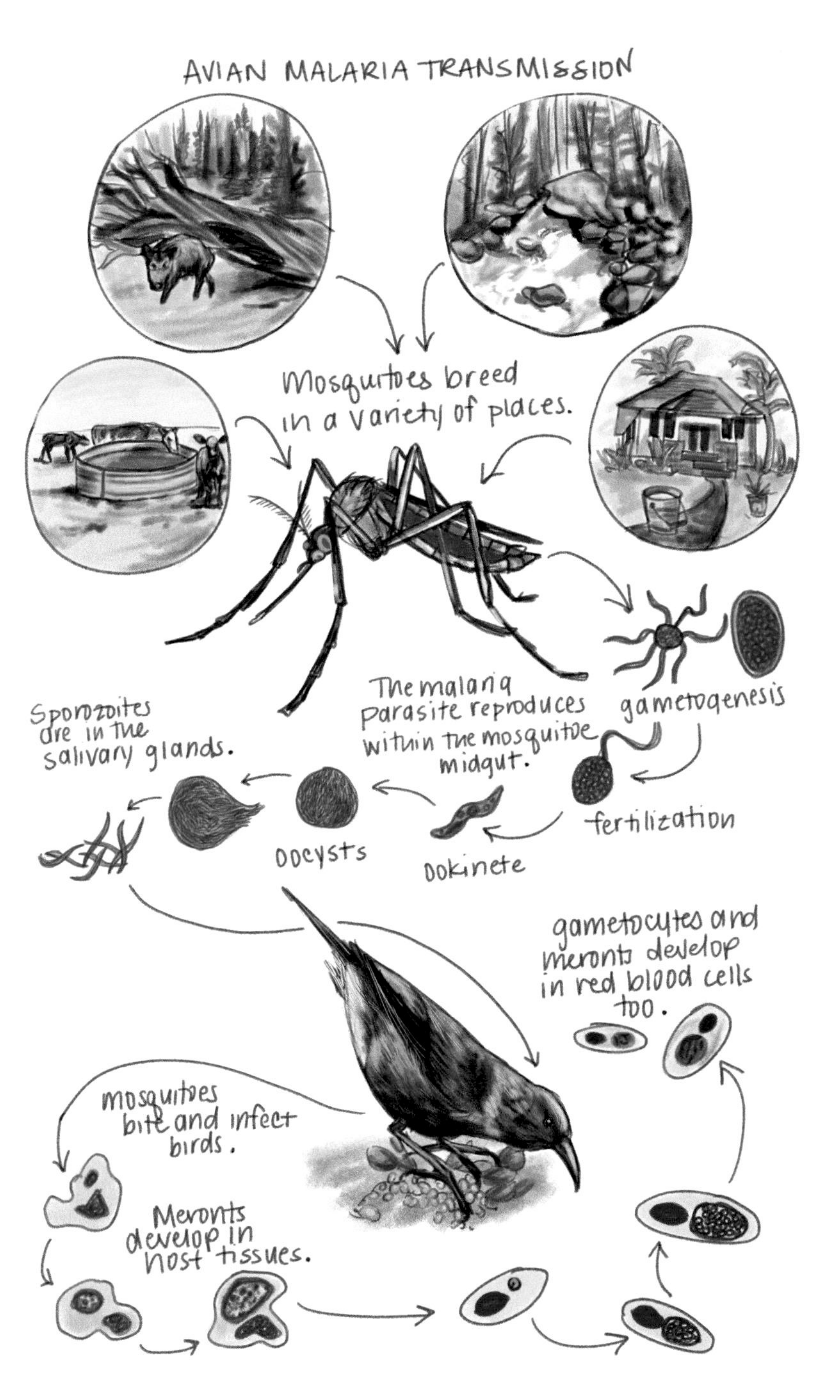

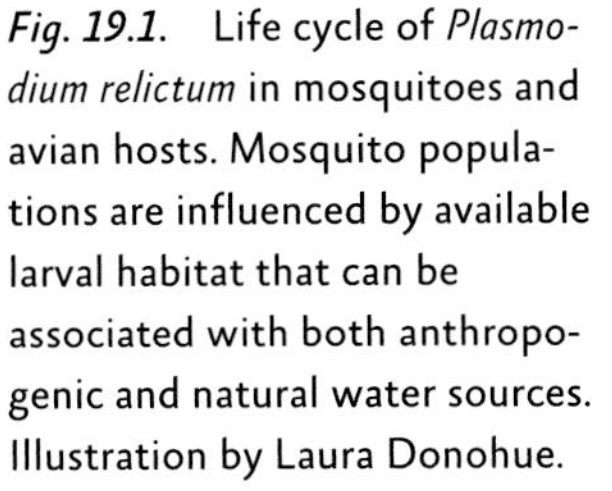

Fig. 19.1. Life cycle of *Plasmodium relictum* in mosquitoes and avian hosts. Mosquito populations are influenced by available larval habitat that can be associated with both anthropogenic and natural water sources. Illustration by Laura Donohue.

have been moved outside of their natural range and exposed to novel mosquito vectors in outdoor zoological collections (Atkinson 2008, LaPointe et al. 2012). The physiological and immunological mechanisms responsible for high susceptibility of honeycreepers are unknown, but recent work on the transcriptomics of both the host and parasite has identified expression of multiple immune system genes that may help to shed some light on this question and make it possible to identify host genotypes that are more tolerant or resistant to disease (Cassin-Sackett et al. 2019; Videval et al. 2015, 2021).

Early Research

Studies by Richard Warner in 1950s and early 1960s using sentinel studies on both Kaua'i and O'ahu established that native honeycreeper species are extremely susceptible to avian malaria (Warner 1968). He exposed caged honeycreepers at low elevations on both islands and observed fulminating infections with both pox and malaria. These classic early experiments led to the hypothesis that populations of highly susceptible native forest birds were restricted to elevations higher than roughly 600 m where mosquito numbers and transmission of both diseases were limited by cooler temperatures.

It was not until the 1970s that the first extensive field surveys in Hawai'i were conducted to document differences in prevalence and intensity of infection in wild populations of native and nonnative species. Van Riper and colleagues very effectively combined field investigations, surveys for mosquito vectors and larval habitat, and experimental studies to demonstrate differential susceptibility and mortality in native and nonnative birds and apparent "immunogenetic resistance" in some native species from lower elevation (Goff and van Riper 1980, van Riper et al. 1986). Their work established that only a single species of *Plasmodium*, *P. relictum*, is present in Hawai'i (Laird and van Riper 1981), showed how landscape features may increase interactions between birds and malaria (Goff and van Riper 1980), and offered a model describing how prevalence, intensity, and vector abundance varied across large landscapes on Hawai'i Island (van Riper et al. 1986) (Figure 19.2).

These studies overlapped with the Hawai'i Forest Bird Survey, the first state-wide assessment of remote forest habitats to describe remaining forest bird communities, estimate population sizes, determine island by island geographic and altitudinal ranges for remaining species, and identify anomalies in distribution that were potentially related to disease transmission (Scott et al. 1986). Together, these investigations helped to establish that introduced malaria and pox were important limiting factors for native forest bird distribution and survival. This early research laid the groundwork for preservation and management of key high-elevation forest bird habitats on Maui, Moloka'i, and Hawai'i Islands (Scott et al. 2001) and formed the foundation for more detailed studies on the epizootiology of the disease.

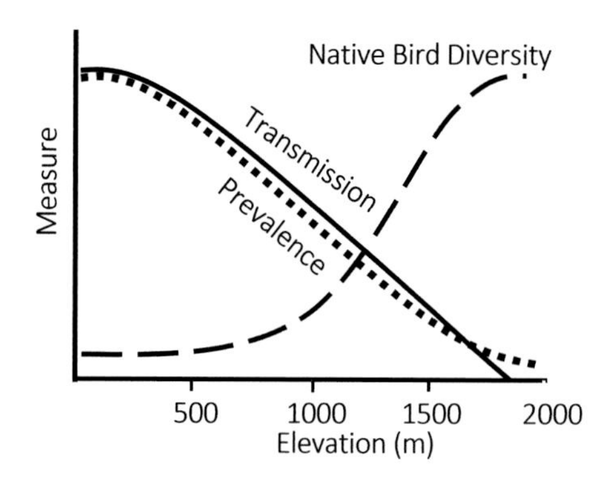

Fig. 19.2. Revised van Riper et al. (1986) transmission model for avian malaria based on studies conducted on the eastern slopes of Mauna Loa and Kīlauea volcanoes. The revised model incorporates recently emergent low-elevation populations of Hawai'i 'Amakihi (*Chlorodrepanis virens*) that help to support high rates of transmission and high prevalence of infection at low elevations. Mosquito populations, extrinsic development of *Plasmodium*, and disease transmission decrease at cooler, high elevations, allowing higher diversity of susceptible native forest birds.

Characteristics Of Malaria Transmission in Hawai'i

Host Susceptibility

Native Hawaiian honeycreepers are extremely susceptible to infection with *P. relictum*. Mortality from acute infections ranges from almost 100% in the most susceptible species (Warner 1968, van Riper et al. 1986, Atkinson et al. 1995) to as low as 17% in some populations of low-elevation Hawai'i 'Amakihi (*Chlorodrepanis virens*) that have evolved tolerance to the disease (Atkinson et al. 2013). After a short prepatent period, birds develop high peripheral parasitemias that can exceed 80% of circulating erythrocytes in the most susceptible species (van Riper et al. 1986, Atkinson et al. 1995). They also show dramatic declines in activity that may make them more susceptible to predation (Yorinks and Atkinson 2000). Birds that recover from acute infections remain chronically infected for years (Atkinson et al. 2001*a*). These characteristics make native honeycreepers ideal reservoir hosts for maintaining transmission of the parasites. This is particularly true in some low-elevation habitats on Hawai'i Island, where transmission is year-round and prevalence can exceed 80% in malaria-tolerant Hawai'i 'Amakihi (Woodworth et al. 2005, Samuel et al. 2011).

Native monarch flycatchers including Kaua'i 'Elepaio (*Chasiempis sclateri*), O'ahu 'Elepaio (*Chasiempis ibidis*), and Hawai'i 'Elepaio (*Chasiempis sandwichensis*) and extant species of native thrushes including Puaiohi (*Myadestes palmeri*) and 'Ōma'o (*Myadestes obscurus*) also appear to have tolerance to infection and are efficient reservoir hosts. They have a high prevalence of infection in the wild and persist in some low- and mid-elevation habitats with high rates of disease transmission (Atkinson et al. 2001*b*, VanderWerf et al. 2006).

By contrast, the importance of nonnative species as reservoir hosts is less clear. While they were instrumental in the introduction of *Plasmodium* to the islands, they are more naturally resistant to malaria as a result of long coevolutionary associations with the parasite (van Riper et al. 1986, Atkinson et al. 2005, McClure et al. 2020). While nonnative species are capable of supporting malaria transmission in lowland locations without the presence of native reservoir hosts, they are less infective to mosquitoes and support a lower prevalence of infection in the vector population than what is found at midelevations where native species overlap (McClure et al. 2020).

Vector Competency

Culex quinquefasciatus is an exceptionally competent vector for *P. relictum*, with over 90% of engorged females developing midgut oocysts and salivary gland infections under ideal temperatures ranging from 28–30°C (LaPointe et al. 2005). *Culex quinquefasciatus* will take blood meals from a wide range of vertebrates but prefers to feed on birds (Tempelis et al. 1970). This mosquito is active at dusk and at night when roosting birds are most vulnerable to blood-feeding insects (van Riper et al. 1986). Among other introduced mosquitoes that are common in some forest habitats, only 7% of engorged *Aedes albopictus* and *Wyeomyia mitchellii* develop oocysts and sporozoites after infective blood meals (LaPointe et al. 2005). The diurnal host-seeking patterns of *Aedes* and *Wyeomyia* mosquitoes and their preference for mammalian rather than avian blood meals make it unlikely that they play important roles in transmission of avian malaria in Hawai'i (LaPointe et al. 2005, van Riper et al. 1986).

Host Community

High rates of malaria transmission are supported in areas where native reservoir hosts overlap seasonally and altitudinally with *Culex* populations, particularly in midelevation habitats (Samuel et al. 2011) and low-elevation habitats that support tolerant native species such as Hawai'i 'Amakihi (Woodworth et al. 2005, McClure et al. 2020). The dynamics of malaria transmission in low-elevation forest bird communities may facilitate selection for disease tolerance in native forest birds (Woodworth et al. 2005, Foster et al. 2007, Atkinson et al. 2013). By themselves, highly susceptible native hosts would rapidly succumb to epidemic malaria transmission in the presence of mosquito populations that remain high throughout the year. Integration into a larger community of less susceptible nonnative forest birds, however, may help to facilitate selection for disease tolerance by dampening transmission (McClure et al. 2020). Other factors that may be essential for this to happen include the presence of suitable lowland habitat to support native forest birds, sufficient isolation from human and agricultural development where mosquito populations and predators may reach their greatest density, host genetic diversity, and host gene flow (McClure et al. 2020, Atkinson et al. 2013, Foster et al. 2007, Samuel et al. 2015, Cassin-Sackett et al. 2019).

Landscape Heterogeneity, Rainfall, and Larval Habitat

Culex quinquefasciatus is attracted to larval habitats that are heavily polluted with organic matter (Hughes et al. 2010). In forested habitats, they will use a variety of water sources including ground pools, rock holes along stream margins, puddles in roads, heavily vegetated margins of slow flowing streams, and tree cavities (LaPointe 2000). Larval abundance and density may be greatest around human and agricultural development where water catchment tanks, discarded containers, roads, tarps, and roofing material create abundant standing water (Reiter and LaPointe 2007, 2009; McClure et al. 2018). Ranching infrastructure including stock tanks and man-made reservoirs with liners weighted by old tires can become excellent larval habitat at higher elevations, particularly when contaminated with fecal matter and warmed by sunshine (Goff and van Riper 1980). Rainfall plays a key role by providing sufficient moisture to maintain these habitats for the duration of larval development. Droughts can significantly reduce mosquito populations by allowing larval habitat to dry out. Periods of excessively high rainfall can flood larval habitats and cause high adult mosquito mortality (Ahumada et al. 2004, Aruch et al. 2007).

Standing water tends to be ephemeral on the porous volcanic soils found on Kīlauea and Mauna Loa Volcanoes on Hawai‘i Island and preferred larval habitats are often associated with fallen tree ferns (LaPointe 2000). The ferns have starchy cores that are consumed by feral pigs, creating open troughs that catch and hold rainwater (Figure 19.3).

Residual starch in the ferns helps to create a rich organic broth that is ideal habitat for larval *Culex*. Both the abundance and altitudinal distribution of tree ferns and the overlap of tree ferns with feral pig populations are key determinants of mosquito populations in these habitats (Samuel et al. 2011).

On Kohala and Mauna Kea volcanoes on Hawai‘i Island and the older islands of Maui, Moloka‘i, O‘ahu, Lāna‘i, and Kaua‘i that are deeply dissected by streams on their wet windward slopes, mosquito habitat may be associated with a combination of tree ferns damaged by pigs, perched rock pools along the margins of stream beds, and occasional ground pools (Aruch et al. 2007; Atkinson et al. 2014; LaPointe et al. 2016, 2021). The relationship between stream flow and rainfall may be particularly important in determining abundance and persistence of mosquito habitat in these riparian systems. During seasonal periods of heavy rainfall, stream flow may be high enough to scour stream beds and flush larvae that may be present in rock pools along stream margins to the ocean. During extended periods of drought, isolated stream pools may persist long enough to sustain multiple generations of mosquito larvae, thus leading to a somewhat counterintuitive association between periods of low rainfall and high mosquito populations (Aruch et al. 2007, Reiter and LaPointe 2009, Atkinson et al. 2014).

Temperature and Elevation

The extrinsic asexual sporogonic cycle of *Plasmodium* takes place on the midgut wall of mosquitoes after an infective blood meal. The cycle is temperature dependent, typically taking approximately 6 days at optimum temperatures to produce infective sporozoites that can invade salivary glands and transmit the parasite at the next blood meal (LaPointe et al. 2010). Hawaiian isolates of *P. relictum* will develop at temperatures between 13° C and 30° C, but there is a significant reduction in prevalence and intensity of oocysts in experimentally infected mosquitoes at temperatures ranging from 13–19°C; oocysts are unable to complete sporogony at temperatures lower than 17° C within the expected life span of the mosquito. Based on the adiabatic lapse rate of temperatures across altitudinal gradients in Hawai‘i (i.e., de-

Fig. 19.3. (A) Tree fern (*Cibotium* sp.) in a midelevation Hawaiian rainforest. (B) Fallen trunk of a tree fern that has been hollowed out by a feral pig. The hollowed trough catches rainwater and provides larval habitat for *Culex quinquefasciatus.* Density of these cavities can be high across some midelevation forests with large feral pig populations.

clines of 6.5°C for every increase in 1000 m of elevation; Giambelluca and Schroeder 1998), ambient temperatures begin to limit growth and survivorship of the parasite in the vector in midelevation forests higher than 900 m (LaPointe et al. 2010), with forest bird habitat above 1500 m generally too cold to allow completion of sporogony. Vector abundance in forests between 900 and 1800 m is seasonal and highly variable based on availability of larval habitat and thermal constraints on larval development (LaPointe 2000, Ahumada et al. 2004). The combination of reduced vector populations, cooler temperatures that prevent completion of extrinsic parasite development, and availability of suitable forest bird habitat define the limits of high-elevation forest bird refugia from disease transmission. These refugia form a narrow band of forest between 1500 m and 2000 m on the high islands of Hawai'i and Maui with the upper limit defined by the trade wind inversion layer and the limits it places on rainfall and forest growth (Loope and Giambelluca 1998, Cao et al. 2007) (Figure 19.4).

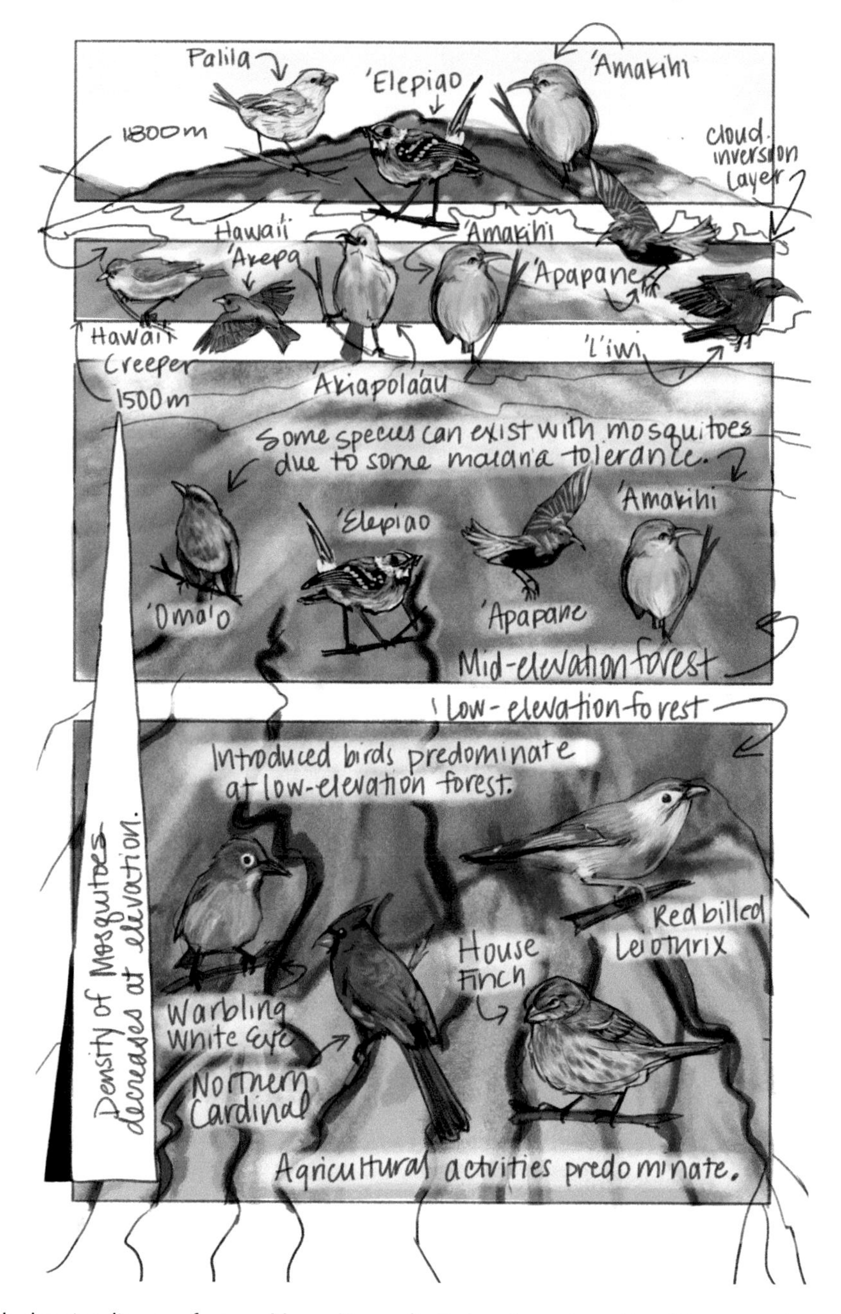

Fig. 19.4. High-elevation disease refugia on Mauna Kea are limited on the eastern side of the mountain to a narrow band of forest above 1,500 m that is capped by both the inversion layer and loss of forest cover by ranching. A few native species including Palila (*Loxioides bailleui*), ʻElepaio (*Chasiempis sandwichensis*), and Hawaiʻi ʻAmakihi (*Chlorodrepanis virens*) are also adapted to living in dry woodland habitats above the inversion layer. High-elevation refugia provide a haven from disease transmission for threatened and endangered species, allowing the highest diversity of native species, including endangered Hawaiʻi ʻAkepa (*Loxops coccineus*), ʻAkiapolaʻau (*Hemignathus wilsoni*), and Hawaii creeper or ʻAlawī (*Loxops mana*). These refugia also provide a source of uninfected juvenile and adult honeycreepers that help to maintain seasonal epidemics of malaria and pox at elevations below 1,500 m, particularly highly mobile species like ʻApapane (*Himatione sanguinea*) and ʻIʻiwi (*Drepanis coccinea*) that track nectar resources across a wide range of elevations. In a warming climate, high-elevation populations of Hawaiʻi's most endangered birds may be threatened by expanding disease transmission, particularly if upward expansion of critical forest habitat is capped by a stable inversion layer. Illustration by Laura Donohue.

Seasonal and Altitudinal Dynamics

The high-elevation volcanoes on Hawai'i Island (>2500 m) create a range of transmission seasons that are linked to their corresponding temperature gradients and seasonal fluctuations in mosquito populations. These range from continuous transmission at lower elevations in areas that support year-round populations of *C. quinquefasciatus* to seasonal transmission during the warmest months of the year between August and December at middle elevations. At elevations above 1500 m, transmission may be sporadic and limited to unusual periods of warm weather (Samuel et al. 2011, Fortini et al. 2020). Rather than being limited to distinct zones, transmission moves up and down the slopes of higher volcanoes in sync with seasonal fluctuations of temperature and associated availability of mosquito vectors (Samuel et al. 2011).

Peak breeding seasons for forest birds are from January–June (Woodworth and Pratt 2009), thus generating large numbers of highly susceptible juvenile birds that disperse from high-elevation habitats into middle-elevation habitats as conditions for malaria transmission reach a seasonal peak in the warmer fall months. Yearly epizootics are common in the middle elevations of Kīlauea Volcano, where resident native species serve as reservoir hosts for initiating mosquito infections that lead to transmission to highly susceptible juveniles that originate locally or from dispersal from adjacent high-elevation habitats. During peak years of malaria transmission, mortality in juvenile and adult 'Apapane (*Himatione sanguinea*) can exceed 50% and 25%, respectively (Atkinson and Samuel 2010, Samuel et al. 2015).

Both movement of *Culex* mosquitoes and avian hosts play a role in the dynamics of this system (LaPointe 2008). In locations like Kīlauea Volcano on Hawai'i Island where a patchwork of native forest, residential communities, and ranching and agricultural land surrounds Hawai'i Volcanoes National Park, movement of mosquitoes into native forests and movement of native birds into more anthropogenic habitats can help to support high rates of disease transmission. This is particularly true for species such as 'Apapane and 'I'iwi that track seasonal nectar resources across the landscape and move to low-elevation habitats where they are more exposed to disease transmission (Ralph and Fancy 1995, Hart et al. 2011, Guillaumet et al. 2017). These lower elevation native forests with seasonally abundant food resources can act as ecological traps, particularly if birds spend the night and risk exposure to infective mosquito bites. Similarly, movement of infected mosquitoes from lowland areas into higher elevation habitats may help to explain periodic disease outbreaks that do not occur annually (Freed and Cann 2013). While *C. quinquesfaciatus* can disperse at least 3 km from a central release point, it is likely that they may be able to move considerably farther if adult mosquitoes are carried by wind above the forest canopy or move along natural and man-made corridors such as lava flows, fence lines, and roads (LaPointe 2008).

A detailed epidemiological model for disease transmission across the altitudinal gradient on the eastern slope of Mauna Loa and Kīlauea Volcanoes was developed by Samuel et al. (2011) by linking forest bird and mosquito demographic models (Ahumada et al. 2004) to estimate daily changes in susceptible, infected, and recovered birds and susceptible, exposed, and infectious mosquitoes. The model successfully reproduced several key attributes of the disease system including the seasonal and altitudinal variation in transmission patterns linked to temperature, rainfall, and availability of larval habitat and made it possible to evaluate impacts of specific conservation actions on dynamics of the system. An overarching conclusion from the model was that malaria recovery and mortality rates were the most important factors affecting native bird abundance because of their high susceptibility to the disease, outweighing demographic parameters related to habitat, fecundity, and proportion of breeding adults. Consequently, conservation programs that reduce malaria transmission or malaria mortality through mosquito control or improvements in host disease tolerance would be more likely to have larger impacts on bird abundance than controlling or reducing impacts from predation and other limiting factors (Samuel et al. 2011, Hobbelen et al. 2012).

Modeling the Effects of Climate Change on Malaria Transmission

Presence of environmentally stable, high-elevation refugia on the higher islands that were cool enough to limit mosquito populations, extrinsic development of *Plasmodium*, and disease transmission probably played a key role in preventing widespread extinction of highly susceptible honeycreepers and allowed persistence of the rarest species up through the end of the 20th century (Atkinson and LaPointe 2009, Samuel et al. 2011). While long-term maintenance of these refugia through habitat restoration and control of predators and invasive species might have once been sufficient to stabilize population sizes and restore rare species, it became evident in the early 2000s (see Case Studies 1 and 2; Boxes 19.1 and 19.2) that changing climatic conditions were beginning to destabilize the system and allow disease transmission to move to higher elevations. Accelerating increases in mean air temperatures at high elevations have enabled temperature fluctuations that support extrinsic development of the parasite and its resultant creep to higher elevations (Giambelluca et al. 2008). Declines in rainfall, frequency of high rainfall events, and stream base flow (Chu and Chen 2005, Oki 2004, Bassiouni and Oki 2012, Elison Timm et al. 2011) and increases in the duration of dry periods (Chu et al. 2010) between flooding events have led to increases in the persistence of available mosquito habitat along stream margins in riparian systems (Aruch et al. 2007, Atkinson et al. 2014; Case Study 1).

There is also evidence that the trade wind inversion layer (loosely defined as the boundary between warmer, dry convective air masses formed from equatorial solar heating and cooler, moist subtropical easterly winds) has been increasing in frequency (Cao et al. 2007). This inversion layer limits clouds and atmospheric mixing and is decreasing in elevation which may result in reduced cloud thickness, rainfall, and overall size of the rain belt on the high mountains of Hawai'i and Maui (Cao et al. 2007, Diaz et al. 2011). This shift could reduce the extent of wet and mesic forests at their upper elevations and limit their upward expansion by capping precipitation at current elevations. This would lead to steady erosion of high-elevation refugia as mosquito numbers and disease transmission move higher on the mountain, effectively squeezing remaining susceptible forest birds against the upper limits of remaining high-elevation montane forests (Figure 19.3) (Benning et al 2002).

Wei et al. (2017) used spatial transmission models developed for Kīlauea Volcano on Hawai'i Island (Samuel et al. 2011) to evaluate effectiveness of different management strategies for controlling malaria transmission under different conditions of climate change. Options that were evaluated included factors that improve host demographics, such as evolved tolerance to malaria and predator removal, and factors that reduce vector abundance, such as controlling feral pigs to reduce larval habitats, competition with other introduced mosquitoes for larval habitat, release of transgenic refractory mosquitoes, and release of male mosquitoes that produce infertile offspring through either irradiation or cytoplasmic incompatibility (Table 19.1).

Under conditions modeled for the most severe climate warming, evolved malaria tolerance in native birds, release of male mosquitoes that produce infertile offspring, or release of transgenic refractory mosquitoes were the most likely scenarios to lead to long-term persistence of highly susceptible native birds because of their positive effects on host demography (tolerance) or reductions on malaria transmission (refractory transgenic mosquitoes and sterile or incompatible male mosquitoes). Combinations of other methods, for example, feral pig control and

Box 19.1. Case Study 1: Changing climate and the demise of Kaua'i's endemic avifauna

The Alaka'i Plateau on Kaua'i was a stronghold for native forest birds in Hawai'i through the 1970s, with an extraordinary avifauna that included the last remaining species of 'Ō'ō (*Moho braccatus*), two species of native thrush, an island-endemic species of native flycatcher, and eight species of honeycreepers that included both island-endemic and archipelago-wide species (Conant et al. 1998). The plateau is deeply dissected by stream valleys and deep canyons but is flat and boggy, sloping from elevations of 1200 m at its western end to 1569 m at Mount Wai'ale'ale, where total yearly rainfall averages 11.4 m. Unlike higher elevation refugia on Maui and Hawai'i islands, temperatures on the Alaka'i Plateau have historically been able to support both mosquitoes and seasonal malaria transmission at the lower end of the plateau (Fortini et al. 2020). While *C. quinquefasciatus* was locally abundant at a few locations on the plateau (Hermann and Snetsinger 1997), prevalence of infection in native forest birds between 1994–1997 was below 10% and endangered birds such as the 'I'iwi, 'Akikiki, and 'Akeke'e were relatively common in forests near Kōke'e State Park (Atkinson et al. 2014).

Range contractions of the 'Akikiki and 'I'iwi from the lower elevations of the plateau first became evident during bird surveys that were conducted in 2000 (Foster et al. 2004), but other native species appeared to be holding their own. When forest bird populations were resampled between 2007 and 2013 to measure prevalence of *Plasmodium*, infection had increased roughly threefold in native species from 11.4% to 32.6% (Atkinson et al. 2014). By 2012, honeycreeper populations at the lower, western elevations of the plateau had declined by more than 89% and retreated to the highest elevations where they currently face high risk of extinction (Paxton et al. 2016).

Analysis of stream flow data from several watersheds on the plateau suggests that a combination of warmer temperatures and hydrologic changes related to long-term drying trends in the islands may have contributed to increases in available mosquito habitat along stream margins as number of days between high rainfall and stream scouring events increased (Atkinson et al. 2014). Increasing numbers of *Culex* mosquitoes moving up stream canyons and valleys from lower elevations may have also played a role, but the importance of this phenomenon is still unknown.

Treatment of rock pools and standing water with biopesticides like Bti (*Bacillus thuringiensis israelensis*) and Ls (*Lysinibacillus sphaericus*) has been attempted for localized control of mosquito populations near home ranges of critically endangered birds. While these efforts were effective against the target larvae, there was little or no observable impact on adult mosquito populations (LaPointe et al. 2021). Long-term sustainability of forest bird biodiversity in this remote, rugged habitat may ultimately depend on development of landscape-scale mosquito control. Remaining options to prevent impending extinction include moving birds into captive propagation facilities or translocations to other islands with suitable, disease-free, high-elevation habitat.

predator removal, were not as effective but might, under some circumstances, help to buy time for implementation of more effective approaches (Wei et al. 2017) (Table 19.1).

Malaria Control in the 21st Century

Larval habitat reduction through both feral pig control and small-scale applications of biopesticides such as *Bacillus thuringensis israeliensis* (BTI) in standing water may provide adequate mosquito control for suppression of local transmission (LaPointe et al. 2009) but is often impractical at larger scales, given the movement of birds and mosquitoes and the difficulty of targeting small and widely dispersed larval habitat (LaPointe et al. 2021). Recent advances in genetic and incompatible insect technologies that can be applied over large landscapes for mosquito control

Box 19.2. Case study 2: Lessons learned from the 2019 Kiwikiu translocation on Maui

The Kiwikiu (*Pseudonestor xanthophyrs*) is a highly endangered honeycreeper endemic to the island of Maui. Surveys conducted in 2017 estimated that only 157 ± 67 individuals remain in high-elevation forests on the northeastern slopes of Haleakalā Volcano (Judge et al. 2019). Kiwikiu are particularly threatened by unusual mortality events because of their low reproductive rate (one chick per one or two breeding seasons) and low population density. Their current restriction to the highest remaining forest habitat on Haleakalā and recent range contractions from lower elevation forests suggested that they are highly susceptible to malaria (Judge et al. 2019).

The US Fish and Wildlife Service Recovery Plan for the Kiwikiu (USFWS 2006) recommended establishing a second population within its historical range to protect the species from catastrophic loss from severe weather events and frequent rainfall that have been shown to reduce reproductive success. The Kahikinui region of Maui on the leeward (southern) slope of Haleakalā was selected for establishing a new population of Kiwikiu. Nakula Natural Area Reserve (NAR) was chosen as the first release site because of reduced storm frequency and rainfall relative to eastern windward slopes, high elevation and presumed low disease risk, and presence of adequate food resources (Warren et al. 2019, 2021).

Baseline work was conducted in 2015 and 2016 to determine abundance and altitudinal distribution of *Culex* mosquitoes, location of larval habitats and prevalence of *P. relictum* in resident forest birds to assess disease risk for the upcoming translocation and to develop a management plan for treating larval habitats to reduce mosquito numbers (Warren et al. 2019). *Culex* larvae were found in intermittent pools in stream gulches that bisected the release site and adult mosquitoes were captured throughout the year with a malarial prevalence up to 10% based on polymerase chain reaction diagnostics. The malaria prevalence in birds also approximated 10%, equivalent to similar elevations in other parts of Hawai'i. Unlike other locations in Hawai'i, peak numbers of mosquitoes were captured in the spring rather than the fall in 2016, with lower captures in 2015 between June and December. A determination was made that malaria risk was present, but that it could be mitigated by targeted larval mosquito control using VectoMax™ (a combined formulation of Bti and Ls) in stream drainages and release of birds in September to avoid the springtime peak that was observed in adult mosquito numbers in 2016.

After years of preparation that included habitat restoration, building release aviaries at the release site, predator control, and mosquito control in adjacent intermittent stream drainages, 14 birds were transferred to the site in mid-October 2019 for release. The cohort consisted of seven wild birds that had been captured in high-elevation forests at Hanawi NAR and seven captive birds from a captive propagation facility managed by San Diego Zoo Global. All birds were monitored by radio telemetry after release. Within 6 weeks after release, 93% (13 of 14) had either died or disappeared. Necropsy of 10 birds that were recovered near the release site revealed gross and microscopic lesions from acute malaria. The timing of the deaths from 6 to 20 days after release of the birds suggested that they became infected with *P. relictum* soon after being moved to the release site (Warren et al. 2021).

Important lessons from the release will help to guide future translocations. An assumption was made based on mosquito captures from 2016 that the release in September would be done when mosquito numbers were at a seasonal low. Unfortunately, unusually warm weather in 2019 led to a peak in mosquito densities in the fall that coincided with the release. In fact, ongoing monitoring of stream drainages at the release site detected a large spike in the number of pools with mosquito larvae in September, but the translocation proceeded without real-time information about abundance of adult mosquitoes at the release site (Warren et al. 2021). The tragic consequences illustrate how quickly mosquito populations and disease transmission can respond to changes in temperature and how quickly high-elevation refugia may be disappearing on Maui.

Illustration by Laura Donohue.

Illustration by Laura Donohue.

Table 19.1. Likelihood that ʻIʻiwi (*Drepanis coccinea*) will persist in mid- and high-elevation habitats under a carbon dioxide emissions trajectory with an additional 8.5 Watts/m^2 solar radiative forcing (RCP8.5), which is the worst-case climate projection with the most aggressive assumed fossil-fuel use

	Likelihood that ʻIʻiwi will persist under RCP8.5 climate projection[a]	
Management action	High elevation	Mid-elevation
Improvement in host demographics		
Malaria tolerance	Low (2, 3)	High (1, 2)
Predator removal	Low (4)	Low (5)
Reduction of vector abundance		
Controlling feral pigs	Low (6)	Low (7)
Competition with other mosquitoes for larval habitat	Low (8)	Low (8)
Release of sterile or incompatible males	High (9)	High (9)
Release of refractory transgenic mosquitoes	High (10, 11)	High (10, 11)
Combined mitigation strategies		
Feral pig control and predator removal	Low (12)	Low (12)
Feral pig control and sterile or incompatible males	High (13)	High (13)
Malaria tolerance and predator control	High (14)	High (14)
Malaria tolerance and feral pig control	Low (15)	High (16)
Malaria tolerance and refractory mosquitoes	High (17)	High (18)
Malaria tolerance and sterile or incompatible males	High (19)	High (20)

This scenario predicts warming of more than 4° C by 2100.

[a] Explanations of likelihood estimates are (1) high rates of transmission are needed to select and maintain tolerance in the population; (2) evolved tolerance must reduce mortality by 50–75% to be effective; (3) low selection pressure at high elevations is unlikely to maintain tolerance in the population; (4) ʻIʻiwi populations declined by >50% even with substantial improvement in survival after predators are removed; (5) midelevation ʻIʻiwi continue to decline and vanish, as improvements in survival are outweighed by high malaria mortality; (6) reduced mosquito populations from elimination of larval habitat was unable to prevent decline in ʻIʻiwi because of high mortality from malaria but could postpone population declines by several decades; (7) availability of natural cavities and favorable climate for mosquitoes would not reduce decline in ʻIʻiwi because of high malaria mortality; (8) relatively small reduction in larval carrying capacity for *Culex* will be completely overwhelmed by expected increases in mosquito populations; (9) continual releases of sterile or incompatible males reduced vector abundance and malaria transmission; (10) at least 90% of the wild mosquito population needs to be replaced with refractory mosquitoes to be successful; (11) maintaining a lower proportion of refractory mosquitoes (80%) delayed declines in ʻIʻiwi by several decades; (12) combined effects were not successful unless predator removal achieved unlikely demographic improvements of 50% reduction in mortality and 110–150% increase in fecundity; (13) additive effects of reduced mosquito numbers from feral pig control and release of sterile or incompatible males will allow ʻIʻiwi to recover; (14) additive effects of tolerance and predator control were complementary and improved host demographics and survivorship; (15) feral pig control reduced selection pressure to develop and maintain malaria tolerance; (16) selection pressure remained high enough to maintain tolerance; (17) high if >90% of mosquitoes are refractory but lower if the percentage of refractory mosquitoes is below 90%, and selection pressure is unable to maintain tolerance; (18) presence of refractory mosquitoes provides only a slight advantage for selection and maintenance of tolerance; (19) if malaria mortality among tolerant birds is >25%, release of sterile or incompatible males reduced wild mosquito abundance and reduced mortality in susceptible birds; and (20) release of sterile or incompatible males reduced transmission but did not help to maintain tolerance in the population because of reduced selective pressure.

Results are summarized from Wei et al. (2017) based on a model of disease transmission for Kīlauea and Mauna Loa volcanoes on Hawaiʻi Island.

(Harvey-Samuel et al. 2021) have piqued the interest of both the conservation and public health communities (Revive and Restore 2017). This interest has gained significant momentum in the past several years because of the imminent extinction crisis facing Hawaiian birds. Use of this technology to suppress populations of *C. quinquefasciatus* would also significantly benefit human and domestic animal health in Hawaiʻi by reducing nuisance biting, transmission of dog heartworm disease, and threats from importation of arboviruses such as West Nile virus and Japanese encephalitis that are transmitted by this mosquito and not currently established in the islands (LaPointe 2007). Application of these techniques to *Aedes aegypti* and *Aedes albopictus* would also reduce the risks of future outbreaks of dengue virus and potential outbreaks of Zika virus and chikungunya virus (Johnston et al. 2020).

A workshop convened in 2016 considered traditional sterile insect techniques that rely on irradiation, incompatible insect techniques (IIT) based on the naturally occurring bacterial endosymbiont *Wolbachia pipientis*, and approaches based on genetic engineering for suppressing or eliminating mosquito populations (Revive and Restore 2017). Unlike irradiation, incompatible insect techniques rely on sperm–egg incompatibility that occurs when *Wolbachia*-infected males mate with uninfected females or females infected with an incompatible *Wolbachia* strain. This incompatibility can result between naturally occurring strains of the bacteria or through artificial inoculation of *Wolbachia* from unrelated species of mosquitoes (Atyame et al. 2016). Both approaches rely on mass rearing of mosquitoes, separation of males from females, and release of either irradiated males or males containing the incompatible strain of *Wolbachia* into the wild. Since males do not take blood meals, there is no threat from increasing disease transmission. When a released male mates with a wild female, infertile eggs are produced (Harvey-Samuel et al. 2021). If enough sterile irradiated or *Wolbachia*-incompatible males are released and mosquito populations are sufficiently suppressed, malaria transmission in high-elevation habitats can be interrupted (Wei et al. 2017). The effectiveness of the IIT approach has been demonstrated in urban and suburban habitats with *Aedes* mosquitoes, but the technology is still new and has yet to be applied for control of wildlife diseases in natural habitats (Mains et al. 2016, 2019; Crawford et al. 2020).

A consensus was reached that *Wolbachia*-based IIT is more likely to receive enough near-term public support to be acceptable (Revive and Restore 2017, McClure 2020). Incompatible insect techniques have been used successfully to suppress mosquito populations in Burma using naturally infected incompatible strains of *Culex quinquefasciatus* (Laven 1967). A biopesticide registered with the US Environmental Protection Agency that uses male *Aedes albopictus* transinfected with a novel *Wolbachia* is currently being used in the United States to suppress wild populations of this mosquito (Mains et al. 2016) and could be used in Hawai‘i to suppress *Aedes* vectors of dengue, Zika, and chikungunya. This approach is more likely to be acceptable in Hawai‘i where there is strong public resistance to genetically modified organisms (Gupta 2018). While a landscape-scale mosquito-suppression program is a recurring investment of time and resources, it may help to buy enough time to evaluate the safety and public acceptance of technological solutions that depend on gene drives and synthetic biology to either eliminate or permanently replace wild populations with modified mosquitoes that do not transmit targeted pathogens.

Current IIT efforts in Hawai‘i are taking two approaches: importation and release of *C. quinquefasciatus* infected with a naturally occurring incompatible strain of *Wolbachia* (Atkinson et al. 2016) and release of Hawaiian strains of *C. quinquefasciatus* that have been transinfected with a strain of *Wolbachia* from *Aedes albopictus* (Ant et al. 2019). The former approach has been used successfully in small-scale trials to control *C. quinquefasciatus* on La Réunion Island and might be successful if incompatible types in Hawaiian *Culex* are widespread in critical forest bird habitat (Atyame et al. 2015, Atkinson et al. 2016). The latter approach has some important advantages because released mosquitoes are more likely to be fully incompatible with all known wild populations of *C. quinquefasciatus* (Ant et al. 2019).

While promising in theory, many obstacles still need to be overcome before landscape-level mosquito control can move forward. Sufficient public support is essential; substantial stable funding sources are a requirement; and new methods for release and monitoring of *Wolbachia*-infected incompatible mosquitoes in remote, topographically challenging environments need to be developed (Paxton et al. 2018). An approach that applies this technology simultaneously to solve both wildlife and human health problems (One Health) may be the most effective in the long run because of the importance of public engagement and support (Revive and Restore 2017).

Summary

Since its introduction roughly 100 years ago, avian malaria has had a profound impact on geographic and altitudinal distribution of native forest birds and has acted in concert with other risk factors to limit bird populations in midelevation native forests where suitable habitat remains. While modeled predictions about expansion of disease transmission into high elevations suggested that we might have 20–30 years to develop and apply measures for mitigating climate-driven loss of high-elevation refugia (Wei et al. 2017, Fortini et al. 2015), the reality is that declines and extinctions of native honeycreepers have already begun on the Alaka'i Plateau of Kaua'i (Paxton et al. 2016), and mosquitoes and malaria transmission are currently expanding into high-elevation habitats on eastern Maui (Case Study 2; Warren et al. 2019, 2021). As of 2022, fenced and intensively managed high-elevation native forests at Hakalau Forest National Wildlife Refuge on the eastern slope of Mauna Loa Volcano may be the only habitats left in the archipelago where native forest birds are stable or increasing and malaria prevalence and mosquito populations remain low (Camp et al. 2010, 2016; LaPointe et al. 2016).

Large-scale application of incompatible insect techniques, development and release of genetically modified refractory mosquitoes, and development of evolved tolerance to the disease in native birds can all mitigate population impacts of disease based on transmission models (Wei et al. 2017, Hobbelen et al. 2012). Incompatible insect techniques are the most promising because of rapid advancement of this technology in the past 5 years (Harvey-Samuel et al. 2021). Public resistance to release of genetically engineered sterile or refractory mosquitoes (Bloss et al. 2017, Gupta 2018) and incomplete knowledge about immunological mechanisms that control disease tolerance (Cassin-Sackett et al. 2019) are unlikely to make either approach feasible in the short term, but synthetic biology (Harvey-Samuel et al. 2021, Samuel et al. 2020) or natural evolution of disease tolerance may ultimately be the only long-term solutions.

Landscape-scale mosquito control will likely require significant investments in infrastructure to rear and release incompatible male mosquitoes and an ongoing commitment among conservation organizations and local, state, and federal government agencies for necessary monitoring programs to measure success of the releases and make adaptive adjustments to maximize their effectiveness. These efforts will not be cheap and may compete for funding for other actions to benefit native forest birds such as protection or restoration of essential habitat, fencing natural areas to control feral pigs and ungulates, and control of predators and invasive plants. The most recent estimates for costs for these recovery actions could be as high as $2.5 billion (USD) spread over 30 years (Leonard 2008, 2009). In today's dollars, this comes close to $3.2 billion or more than $100 million per year. Social and cultural aspects of the recovery process are also at play. While native forest birds are of immense cultural significance within the native Hawaiian community (Amante-Helweg and Conant 2009), the general public has little awareness of their plight because most remaining species persist in remote, high-elevation areas with little access and few opportunities for public interaction and appreciation (Pratt et al. 2009). Conflicts within local communities have also arisen among groups that favor hunting over forest bird conservation and advocate against fencing of native habitats to remove feral pigs (Leonard 2008, 2009).

While ongoing public outreach efforts by both government agencies and private organizations are raising awareness of native forest birds and may be an important component of the recovery process (USFWS 2006), additional support for landscape-scale mosquito control might be possible by expanding the program to *Aedes* mosquitoes and *C. quinquefasciatus* in urban, suburban, and agricultural areas. A broad application of the technology to both anthropogenic and natural habitats may help to reduce nuisance mosquitoes that affect the tourist industry, reduce threats from human and animal pathogens such as dengue virus and dog heartworm (*Dirofilaria immitis*), and reduce the vulnerability of the islands to new imported diseases such as West Nile virus, Japanese encephalitis, Zika, and chikungunya (LaPointe 2007). Long-term sustainability of native ecosystems in Hawai'i is deeply intertwined with the economic future and quality of life in the islands (HSTF, 2008). A One Health approach that links human, wildlife, and environmental health through improved biosecurity and control of mosquitoes and other invasive species may help improve quality of life, prevent further declines and extinctions among native forest birds, and help to sustain biodiversity in native ecosystems.

ACKNOWLEDGEMENTS

I would like to thank colleagues, staff, and volunteers at the US Geological Survey Pacific Island Ecosystems Research Center and the US Geological Survey National Wildlife Health Center. Support from collaborators and partners at the US Fish and Wildlife Service, National Park Service, State of Hawai'i Department of Forestry and Wildlife, Nature Conservancy, the University of Hawai'i, and the Smithsonian Institution have directly and indirectly made much of the research reported in this chapter possible. The US Geological Survey Ecosystems Mission Area, the National Park Service Natural Resource Protection Program, the US Fish and Wildlife Service Science Support Partnership Program, and National Science Foundation Grant DEB 0083944 provided important financial support for research reported here. Any use of trade, firm, or product names is for descriptive purposes only and does not imply endorsement by the US Government.

LITERATURE CITED

Ahumada, J. A., D. LaPointe, and M. D. Samuel. 2004. Modeling the population dynamics of *Culex quinquefasciatus* (Diptera: Culicidae), along and elevational gradient in Hawaii. Journal of Medical Entomology 41:1157–1170.

Alley, M. R., K. A. Hale, W. Cash, H. J. Ha, and L. Howe. 2010. Concurrent avian malaria and avipox virus infection in translocated South Island saddlebacks (*Philesturnus carunculatus carunculatus*). New Zealand Veterinary Journal 58:218–223.

Amante-Helweg, V. L. U., and S. Conant. 2009. Hawaiian culture and forest birds. Pages 59–79 *in* T. K. Pratt, C. T. Atkinson, P.C. Banko, J. D. Jacobi, and B. L. Woodworth, editors. Conservation biology of Hawaiian forest birds: implications for island avifauna. Yale University Press, New Haven, Connecticut, USA.

Ant, T. H., C. Herd, F. Louis, A. B. Failloux, and S. P. Sinkins. 2019. *Wolbachia* transinfections in *Culex quinquefasciatus* generate cytoplasmic incompatibility. Insect Molecular Biology 29:1–8.

Aruch, S., C. T. Atkinson, A. F. Savage, and D. A. LaPointe. 2007. Prevalence and distribution of pox-like lesions, avian malaria and mosquito vectors in Kipahulu Valley, Haleakala National Park, Hawaii. Journal of Wildlife Diseases 43:567–575.

Asghar, M., V. Palinauskas, N. Zaghdoudi-Allen, G. Valkiūnas, A. Mukhin, E. Platonova, A. Färnert, S. Bensch, and D. Hasselquist. 2016. Parallel telomere shortening in multiple body tissues owing to malaria infection. Proceedings of the Royal Society B Biological Sciences 283:20161184.

Atkinson, C. T. 2008. Avian malaria. Pages 35–53 *in* C. T. Atkinson, N. J. Thomas, D. B. Hunter, editors. Parasitic diseases of wild birds. Wiley-Blackwell. Ames, Iowa, USA.

Atkinson, C.T., R. J. Dusek, and J. K. Lease. 2001*a*. Serological responses and immunity to superinfection with avian malaria in experimentally-infected Hawaii Amakihi. Journal of Wildlife Diseases 37:20–27.

Atkinson, C. T., R. J. Dusek, J. K. Lease, and M. D. Samuel. 2005. Prevalence of pox-like lesions and malaria in forest bird communities on leeward Mauna Loa Volcano, Hawaii. Condor 107:537–546.

Atkinson, C. T., and D. A. LaPointe. 2009. Introduced avian diseases, climate change, and the future of Hawaiian honeycreepers. Journal of Avian Medicine and Surgery 23:53–63.

Atkinson, C. T., J. K. Lease, B. M. Drake, and N. P. Shema. 2001*b*. Pathogenicity, serological responses and diagnosis of experimental and natural malarial infections in native Hawaiian thrushes. Condor 103:209–218.

Atkinson, C. T., K. S. Saili, R. B. Utzurrum, and S. I. Jarvi. 2013. Experimental evidence for evolved tolerance to avian malaria in a wild population of low elevation Hawai'i 'Amakihi (*Hemignathus virens*). EcoHealth 10:366–375.

Atkinson, C. T., and M. D. Samuel. 2010. Avian malaria (*Plasmodium relictum*) in native Hawaiian forest birds: epizootiology and demographic impacts on 'Apapane (*Himatione sanguinea*). Journal of Avian Biology 41:357–366.

Atkinson, C. T., R. B. Utzurrum, D. A. LaPointe, R. J. Camp, L. H. Crampton, J. T. Foster, and T. W. Giambelluca. 2014. Changing climate and the altitudinal range of avian malaria in the Hawaiian Islands—an ongoing conservation crisis on the island of Kaua'i. Global Change Biology 20:2426–2436.

Atkinson, C. T., W. Watcher-Weatherwax, and D. A. LaPointe. 2016. Genetic diversity of *Wolbachia* endosymbionts in *Culex quinquefasciatus* from Hawaii, Midway Atoll and American Samoa. University of Hawai'i, Hawai'i Cooperative Studies Unit Technical Report HCSU-074, Hilo, USA.

Atkinson, C. T., K. L. Woods, R. J. Dusek, L. S. Sileo, and W. M. Iko. 1995. Wildlife disease and conservation in Hawaii: pathogenicity of avian malaria (*Plasmodium relictum*) in experimentally infected Iiwi (*Vestiaria coccinea*). Parasitology 111(Supplement S1):S59–S69.

Atyame, C. M., J. Cattel, C. Lebon, O. Flores, J.-S. Dehecq, M. Weill, L. C. Gouagna, and P. Tortosa. 2015. *Wolbachia*-based population control strategy targeting *Culex quinquefasciatus* mosquitoes proves efficient under semi-field conditions. PLoS One 10:e0119288.

Atyame, C. M., P. Labbé, C. Lebon, M. Weill, R. Moretti, F. Marini, L. C. Gouagna, M. Calvitti, and P. Tortosa. 2016. Comparison of irradiation and *Wolbachia* based approaches for sterile-male strategies targeting *Aedes albopictus*. PLoS One 11:e0146834.

Baldwin, P. H. 1941. Checklist of the birds of the Hawaii National Park, Kilauea-Mauna Loa Section, with remarks on their present status and a field key for their identification. Hawai'i National Park, Natural History Bulletin Number 7, Hawai'i National Park, USA.

Bassiouni, M., and D. S. Oki. 2012. Trends and shifts in streamflow in Hawai'i, 1913–2008. Hydrological Processes 27:1484–1500.

Beadell, J.S., F. Ishtiaq, R. Covas, M. Melo, B. H. Warren, C. T. Atkinson, S. Bensch, G. R. Graves, Y. V. Jhala, M. A. Peirce, et al. 2006. Global phylogeographic limits of Hawaii's avian malaria. Proceedings of the Royal Society B Biological Sciences 273:2935–2944.

Benning, T. L., D. A. LaPointe, C. T. Atkinson, and P. M. Vitousek. 2002. Interactions of climate change with land use and biological invasions in the Hawaiian Islands: modeling the fate of endemic birds using GIS. Proceedings of the National Academy of Sciences USA 99:14246–14249.

Bloss, C. S., J. Stoler, and K. C. Brouwer. 2017. Public response to a proposed field trial of genetically engineered mosquitoes in the United States. Journal of the American Medical Association 318:662–664.

Camp, R. J., K. W. Brink, P. M. Gorresen, and E. H. Paxton. 2016. Evaluating abundance and trends in a Hawaiian avian community using state-space analysis. Bird Conservation International 26:225–242.

Camp, R. J., D. A. LaPointe, P. J. Hart, D. E. Sedgwick, and L. K. Canale. 2019. Large-scale tree mortality from rapid ohia death negatively influences avifauna in Lower Puna, Hawaii Island, USA. Condor 121:duz007.

Camp, R. J., T. K. Pratt, P. M. Gorresen, J. J. Jeffrey, and B. L. Woodworth. 2010. Population trends of forest birds at Hakalau Forest National Wildlife Refuge, Hawai'i. Condor 112:196–212.

Cao, G., T. W. Giambelluca, D. E. Stevens, and T. A. Schroeder. 2007. Inversion variability in the Hawaiian trade wind regime. Journal of Climate 20:1145–1160.

Cassin-Sackett, L., T. E. Callicrate, and R. C. Fleischer. 2019. Parallel evolution of gene classes, but not genes: evidence from Hawai'ian honeycreeper populations exposed to avian malaria. Molecular Ecology 28:568–683.

Chu, P.-S., and H. Chen. 2005. Interannual and interdecadal rainfall variations in the Hawaiian Islands. Journal of Climate 18:4796–4813.

Chu, P.-S., Y. R. Chen, and T. A. Schroeder. 2010. Changes in precipitation extremes in the Hawaiian Islands in a warming climate. Journal of Climate 23:4881–4900.

Conant, S., H. D. Pratt, and R. J. Shallenberger. 1998. Reflections on a 1975 expedition to the lost world of the Alaka'i and other notes on the natural history, systematics, and conservation of Kaua'i birds. Wilson Bulletin 110:1–22.

Crawford, J. E., D. W. Clark, V. Criswell, M. Desnoyer, D. Cornel, B. Deegan, K. Gong, K. C. Hopkins, P. Howell, J. S. Hyde, et al. 2020. Efficient production of male *Wolbachia*-infected *Aedes aegypti* mosquitoes enables large-scale suppression of wild populations. Nature Biotechnology 38:1000.

Diaz, H. F., T. W. Giambelluca, and J. K. Eischeid. 2011. Changes in the vertical profiles of mean temperature and humidity in the Hawaiian Islands. Global and Planetary Change 77:21–25.

Elison Timm, O., H. F. Diaz, T. W. Giambelluca, and M. Takahashi. 2011. Projection of changes in the frequency of heavy rain events over Hawaii based on leading Pacific climate modes. Journal of Geophysical Research Atmospheres 116:D04109.

Fortini, L. B., L. R. Kaiser, and D. A. LaPointe. 2020. Fostering real-time climate adaptation: analyzing past, current, and forecast temperature to understand the dynamic risk to Hawaiian honeycreepers from avian malaria. Global Ecology and Conservation 23:e01069.

Fortini, L. B., A. E. Vorsino, F. A. Amidon, E. H. Paxton, and J. D. Jacobi. 2015. Large-scale range collapse of Hawaiian forest birds under climate change and the need for 21st century conservation options. PLoS One 10:e0140389.

Foster, J. T. 2009. The history and impact of introduced birds. Pages 312–330 *in* T. K. Pratt, C. T. Atkinson, P.C. Banko, J. D. Jacobi, and B. L. Woodworth, editors. Conservation biology of Hawaiian forest birds: implications for island avifauna. Yale University Press, New Haven, Conecticut, USA.

Foster, J. T., E. J. Tweed, R. J. Camp, B. L. Woodworth, C. D. Adler, and T. Telfer. 2004. Long-term population changes of native and introduced birds in the Alaka'i swamp, Kaua'i. Conservation Biology 18:716–725.

Foster, J. T., B. L. Woodworth, L. E. Eggert, P. J. Hart, D. Palmer, D. C. Duffy, and R. C. Fleischer. 2007. Genetic structure and evolved malaria resistance in Hawaiian honeycreepers. Molecular Ecology 16:4738–4746.

Freed, L. A., and R. L. Cann. 2013. Vector movement underlies avian malaria at upper elevation in Hawaii: implications for transmission of human malaria. Parasitology Research 112:3887–3895.

Giambelluca, T. W., H. F. Diaz, and M. S. A. Luke. 2008. Secular temperature in Hawaii. Geophysical Research Letters 35:L12702.

Giambelluca, T. W., and T. A. Schrodeder. 1998. Climate. Pages 49–59 *in* S. P. Juvik and J. O. Juvik, editors. Atlas of Hawai'i. University of Hawai'i Press, Honolulu, USA.

Goff, M. L., and C. van Riper III. 1980. Distribution of mosquitos (Diptera, Culicidae) on the east flank of Mauna-Loa volcano. Hawaii Pacific Insects 22:178–188.

Guillaumet, A., W. A. Kuntz, M. D. Samuel, and E. H. Paxton 2017. Altitudinal migration and the future of an iconic Hawaiian honeycreeper in response to climate change and management. Ecological Monographs 87:410–428.

Gupta, C. 2018. Contested fields: an analysis of anti-GMO politics on Hawai'i Island. Agriculture and Human Values 35:181–192.

Hart, P. J., B. L. Woodworth, R. J. Camp, K. Turner, K. McClure, K. Goodall, C. Henneman, C. Speigel, J. Lebrun, E. Tweed, and M. Samuel. 2011. Temporal variation in bird and resource abundance across an elevational gradient in Hawaii. Auk 128:113–126.

Harvey-Samuel, T., T. Ant, J. Sutton, C. N. Niebuhr, S. Asigau, P. Parker, S. Sinkins, and L. Alphey. 2021. *Culex quinquefasciatus*: status as a threat to island avifauna and options for genetic control. CABI Agricultural and Bioscience 2:9.

Hawai'i 2050 Sustainability Task Force (HSTF).2008. Hawai'i 2050 sustainability plan, charting a course for Hawai'i's sustainable future. State of Hawai'i. Office of the Auditor, Hawai'i 2050 Sustainability Task Force, Honolulu, USA.

Hellgren, O., C.T. Atkinson, S. Bensch, T. Albayrak, D. Dimitrov, J.G. Ewen, K.S. Kim, M.R. Lima, L. Martin, V. Palinauskas, et al. 2015. Global phylogeography of the avian malaria pathogen *Plasmodium relictum* based on MSP1 allelic diversity. Ecography 38:842–850.

Hermann, C. M., and T. J. Snetsinger. 1997. Pox-like lesions on endangered Puaiohi (*Myadestes palmeri*) and occurrence of mosquito (*Culex quinquefasciatus*) populations near Koaie Stream. 'Elepaio 57:73–75.

Hobbelen, P. H. F., M. D. Samuel, D. A. LaPointe, and C. T. Atkinson. 2012. Modeling future conservation of Hawaiian honeycreepers by mosquito management and translocation of disease-tolerant Amakihi. PLoS One 7:e49594.

Hughes, D. T., J. Pelletier, C. W. Luetge, and W. S. Leal. 2010. Odorant receptor from the southern house mosquito narrowly tuned to the oviposition attractant skatole. Journal of Chemical Ecology 36:797–800.

Jarvi, S.I., D. Triglia, A. Giannoulis, M. Farias, K. Bianchi, and C.T. Atkinson. 2008. Diversity and virulence of *Avipoxvirus* in Hawaiian forest birds. Conservation Genetics, 9:339–348.

Johnston, D. I., M. A. Viray J. M. Ushiroda, H. He, A. C. Whelen, R. H. Sciulli, G. Y. Kunimoto, Y. S. Park. 2020. Investigation and response to an outbreak of dengue: Island of Hawai'i, 2015–2016. Public Health Reports 135:230–237.

Judge, S.W., R.J. Camp, C.C. Warren, L.K. Berthold, H.L. Mounce, P.J. Hart, and R.J. Monello. 2019. Pacific island landbird monitoring annual report, Haleakalā National Park and East Maui Island, 2017. US National Park Service Natural Resource Report NPS/PACN/NRR—2019/1949, Fort Collins, Colorado, USA.

Laird, M. and C. van Riper III. 1981. Questionable reports of *Plasmodium* from birds in Hawaii, with the recognition of *P. relictum* ssp. *capistranoae* (Russell, 1932) as the avian malaria parasite there. Pages 159 - 165 *in* E. V. Canning, editor. Parasitological topics. A presentation volume to P. C. C. Garnham on the occasion of his 80th Birthday 1981. Society of Protozoologists, Special Publication No. 1, Lawrence, Kansas, USA.

LaPointe, D.A. 2000. Avian malaria in Hawai'i: the distribution, ecology and vector potential of forest-dwelling mosquitoes. Ph.D. dissertation, University of Hawai'i at Manoa, Honolulu, USA.

LaPointe, D. A. 2007. Current and potential impacts of mosquitoes and the pathogens they vector in the Pacific region. Proceedings of the Hawaiian Entomological Society 39:75–81.

LaPointe, D. A. 2008. Dispersal of *Culex quinquefasciatus* (Diptera: Culicidae) in a Hawaiian rain forest. Journal of Medical Entomology 45:600–609.

LaPointe, D. A., C. T. Atkinson, and M. D. Samuel. 2012. Ecology and conservation biology of avian malaria. Annals of the New York Academy of Sciences 1249:211–216.

LaPointe, D. A., T. V. Black, M. Riney, K. W. Brinck, L. H. Crampton, and J. Hite. 2021. Field trials to test new trap technologies for monitoring *Culex* populations and the efficacy of the biopesticide formulation VectoMax® FG for control of larval *Culex quinquefasciatus* in the Alaka'i Plateau, Kaua'i, Hawaii. University of Hawai'i, Hawai'i Cooperative Studies Unit Technical Report HCSU-096, Hilo, USA http://dspace.lib.hawaii.edu/handle/10790/5384. Accesssed 10 Oct 2022.

LaPointe, D. A., J. M. Gaudioso-Levita, C. T. Atkinson, A. Egan, and K. Hayes. 2016. Changes in the prevalence of avian disease and mosquito vectors at Hakalau Forest National Wildlife Refuge: a 14-year perspective and assessment of future risk. University of Hawai'i, Hawai'i Cooperative Studies Unit Technical Report HCSU-073, Hilo, USA.

LaPointe, D.A., M.L. Goff, and C.T. Atkinson. 2005. Comparative susceptibility of introduced forest-dwelling mosquitoes in Hawaii to avian malaria, *Plasmodium relictum*. Journal of Parasitology 91:843–849.

LaPointe, D. A., M. L. Goff, and C. T. Atkinson. 2010. Thermal constraints to the sporogonic development and altitudinal distribution of avian malaria *Plasmodium relictum* in Hawaii. Journal of Parasitology 96:318–324.

LaPointe, D. A., E. K. Hofmeister, C. T. Atkinson, R. E. Porter, and R. J. Dusek. 2009. Experimental infection of Hawaii amakihi (*Hemignathus virens*) with West Nile virus and competence of a co-occurring vector, *Culex quinquefasciatus*: potential impacts on endemic Hawaiian avifauna. Journal of Wildlife Diseases 45:257–271.

Laven, H. 1967. Eradication of *Culex pipiens fatigans* through cytoplasmic incompatibility. Nature. 216:383–390.

Leonard, D. L., Jr. 2008. Recovery expenditures for birds listed under the US Endangered Species Act: the disparity between mainland and Hawaiian taxa. Biological Conservation 141:2054–2061.

Leonard, D. L., Jr. 2009. Social and political obstacles to saving Hawaiian birds: realities and remedies. Pages 533–551 *in* T. K. Pratt, C. T. Atkinson, P.C. Banko, J. D. Jacobi, and B. L. Woodworth, editors. Conservation biology of Hawaiian forest birds: implications for island avifauna. Yale University Press, New Haven, Connecticut, USA.

Levin, I. I., G. Valkiūnas, D. Santiago-Alarcon, L. L. Cruz, T. A. Iezhova, S. L. O'Brien, F. Hailer, D. Dearborn, E. A. Schreiber, R. C. Fleischer, et al. 2011. Hippoboscid-transmitted *Haemoproteus* parasites (Haemosporida) infect Galapagos pelecaniform birds: evidence from molecular and morphological studies, with a description of *Haemoproteus iwa*. International Journal for Parasitology 41:1019–1027.

Levin I. I., P. Zwiers, S. L. Deem, E. A. Geest, J. M. Higashiguchi, T. A. Iezhova, G. Jiménez-Uzcátegui, D. H. Kim, J. P. Morton, N. G. Perlut, et al. 2013. Multiple lineages of avian malaria parasites (*Plasmodium*) in the Galapagos Islands and evidence for arrival via migratory birds. Conservation Biology 27:1366–1377.

Loope, L. L., and T. W. Giambelluca. 1998. Vulnerability of island tropical montaine cloud forests to climate change, with special reference to east Maui, Hawaii. Climatic Change 39:503–517.

Mains, J. W., C. L. Brelsfoard, R. I. Rose, and S. L. Dobson. 2016. Female adult *Aedes albopictus* suppression by *Wolbachia*-infected male mosquitoes. Scientific Reports 6:33846.

Mains, J. W., P. H. Kelly, K. L. Dobson, W. D. Petrie, and S. L. Dobson. 2019. Localized control of *Aedes aegypti* (Diptera: Culicidae) in Miami, FL, via inundative releases of *Wolbachia*-infected male mosquitoes. Journal of Medical Entomology 56:1296–1303.

McClure, K. M. 2020. Landscape-level mosquito suppression to protect Hawai'i's rapidly vanishing avifauna. Cornell University Wildlife Health Center Blog, Ithaca, New York, USA. https://wildlife.cornell.edu/blog/landscape-level-mosquito-suppression-protect-hawaiis-rapidly-vanishing-avifauna. Accessed 10 Oct 2022.

McClure, K. M., R. C. Fleischer, and A. M. Kilpatrick. 2020. The role of native and introduced birds in the transmission of avian malaria in Hawaii. Ecology 101:e03038.

McClure, K. M., C. Lawrence, and A. M. Kilpatrick. 2018. Land use and larval habitat increase *Aedes albopictus* (Diptera: Culicidae) and *Culex quinquefasciatus* (Diptera: Culicidae) abundance in lowland Hawaii. Journal of Medical Entomology 55:1509–1516.

Mortenson, L.A., R. Flint Hughes, J. B. Friday, L. M. Keith, J. M. Barbosa, N. J. Friday, Z. Liu, and T. G. Sowards.

2016. Assessing spatial distribution, stand impacts and rate of *Ceratocystis fimbriata* induced ohia (*Metrosideros polymorpha*) mortality in a tropical wet forest, Hawaii Island, USA. Forest Ecology and Management 377:83–92.

Moulton, M. P., K. E. Miller, and E. A. Tillman. 2001. Patterns of success among introduced birds in the Hawaiian Islands. Studies in Avian Biology 22:31–46.

Oki, D. S. 2004. Trends in streamflow characteristics at long-term gaging stations, Hawaii. US Geological Survey, US Department of Interior, US Geological Survey, Scientific Investigations Report 2004-5080, Reston, Virginia, USA.

Parker, P. G., E. L. Buckles, H. Farrington, K. Petren, and N. K. Whiteman. 2011. 110 years of Avipoxvirus in the Galapagos Islands. PLoS One 6:e15989.

Paxton, E. H., R. J. Camp, P. M. Gorresen, L. H. Crampton, D. L. Leonard Jr., and E. A. VanderWerf. 2016. Collapsing avian community on a Hawaiian Island. Science Advances 2:e1600029.

Paxton, E. H., M. Laut, J. P. Vetter, and S. J. Kendall. 2018. Research and management priorities for Hawaiian forest birds. Ornithological Applications 120:557–565.

Pratt, T. K. 2009. Origins and evolution. Pages 3–24 *in* T. K. Pratt, C. T. Atkinson, P.C. Banko, J. D. Jacobi, and B. L. Woodworth, editors. Conservation biology of Hawaiian forest birds: implications for island avifauna. Yale University Press, New Haven, Connecticut, USA.

Pratt, T. K., C. T. Atkinson, P. C. Banko, J. J. Jacobi, B. L. Woodworth, and L. A. Mehrhoff. 2009. Can Hawaiian forest birds be saved? Pages 552–580 *in* T. K. Pratt, C. T. Atkinson, P.C. Banko, J. D. Jacobi, and B. L. Woodworth, editors. Conservation biology of Hawaiian forest birds: implications for island avifauna. Yale University Press, New Haven, Connecticut, USA.

Ralph, C. J., and S. G. Fancy. 1995. Demography and movements of Iiwi and Apapane in Hawaii. Condor 97:729–742.

Reiter, M. E., and D. A. LaPointe. 2007. Landscape factors influencing the spatial distribution and abundance of mosquito vector *Culex quinquefasciatus* (Diptera: Culicidae) in a mixed residential–agricultural community in Hawai'i. Journal of Medical Entomology. 44:861–868.

Reiter, M. E., and D. A. LaPointe. 2009. Larval habitat for the avian malaria vector *Culex quinquefasciatus* (Diptera: Culicidae) in altered mid-elevation mesic–dry forests in Hawaii. Journal of Vector Ecology 31:208–216.

Revive and Restore. 2017. To restore a mosquito-free Hawai'i. Summary report of the workshop to formulate strategic solutions for a "Mosquito-Free Hawai'i." http://www.cpc-foundation.org/uploads/7/6/2/6/76260637/report_on_mosquito_free_workshop.pdf. Accessed 10 Mar 2023.

Samuel, M. D., P. H. F. Hobbelen, F. DeCastro, J. A. Ahumada, D. A. LaPointe, C. T. Atkinson, B. L. Woodworth, P. J. Hart, and D. C. Duffy. 2011 The dynamics, transmission and population impact of avian malaria in native Hawaiian birds—an epidemiological modeling approach. Ecological Applications 21:2960–2973.

Samuel, M. D., W. Liao, C. T. Atkinson, and D. LaPointe. 2020. Facilitated adaptation for conservation—can gene editing save Hawaii's endangered birds from climate driven avian malaria? Biological Conservation 241:108390.

Samuel, M. D., B. L. Woodworth, C. T. Atkinson, P. J. Hart, and D. A. LaPointe. 2015. Avian malaria in Hawaiian forest birds: infection and population impacts across species and elevations. Ecosphere 6:104.

Samuel, M. D., B. L. Woodworth, C. T. Atkinson, P. J. Hart, and D. A. LaPointe. 2018. The epidemiology of avian pox and interaction with avian malaria in Hawaiian forest birds. Ecological Monographs 88:621–637.

Schoener, E. R., M. Banda, L. Howe, I. C. Castro, and M. R. Alley. 2014. Avian malaria in New Zealand. New Zealand Veterinary Journal 62:189–198.

Scott, J. M., S. Mountainspring, F. L. Ramsey, and C. B. Kepler 1986. Forest bird communities of the Hawaiian Islands: their dynamics, ecology, and conservation. University of California, Department of Biology, Cooper Ornithological Society Studies in Avian Biology 9, Los Angeles, USA.

Scott, J. M., S. Conant, and C. van Riper III. 1981. Introduction. Pages 1–12 *in* J. M. Scott, S. Conant, and C. van Riper III, editors. Evolution, ecology, conservation, and management of Hawaiian birds: a vanishing avifauna. Cooper Ornithological Society, Studies in Avian Biology 22, Lawrence, Kansas, USA.

Tempelis, C. H., R. O. Hayes, A. D. Hess, and W. C. Reeves. 1970. Blood-feeding habits of four species of mosquito found in Hawai'i. American Journal of Tropical Medicine and Hygiene 19:335–341.

US Fish and Wildlife Service (USFWS). 2006. Revised recovery plan for the Hawaiian forest birds Region 1. US Fish and Wildlife Service, Portland, Oregon, USA.

US Fish and Wildlife Service (USFWS). 2010. Endangered and threatened wildlife and plants; determination of endangered status for 48 species on Kauai and designation of critical habitat; final rule. Federal Register 75(70):18960. https://www.federalregister.gov/documents/2010/04/13/2010-1904/endangered-and-threatened-wildlife-and-plants-determination-of-endangered-status-for-48-species-on. Accessed 10 Oct 2022.

US Fish and Wildlife Service (USFWS). 2017. Endangered and threatened wildlife and plants; threatened species status for the Iiwi (*Drepanis coccinea*). Federal Register 82 (181):43873. https://www.federalregister.gov/documents/2017/09/20/2017-20074/endangered-and-threatened-wildlife-and-plants-threatened-species-status-for-the-iiwi-drepanis. Accessed 10 Oct 2022.

Valkiūnas, G., M. Ilgūnas, D. Bukauskaitė, H. Weissenböck, K. Fragner, H. Weissenböck, C. T. Atkinson, and T. A. Iezhova. 2018. Characterization of *Plasmodium relictum*, a cosmopolitan agent of avian malaria. Malarial Journal 17:184.

VanderWerf, E. A., M. D. Burt, J. L. Rohrer, and S. M. Mosher. 2006. Distribution and prevalence of mosquito-borne diseases in O'ahu 'Elepaio. Condor 108:770–777.

Van Dine, D. L. 1904. Mosquitos in Hawaii. University of Hawai'i at at Mānoa, Hawai'i Agricultural Experiment Station Bulletin No 6, Honolulu, USA.

van Riper, C. III, C.T. Atkinson, and T. Seed. 1994. Plasmodia of birds. Pages 73–140 *in* J. P. Kreier, editor. Parasitic protozoa. Volume 7. Academic Press, New York, New York, USA.

van Riper, C. III, and J. M. Scott. 2001. Limiting factors affecting Hawaiian native birds. Pages 221–233 *in* J. M. Scott, S. Conant, and C. van Riper III, editors. Evolution, ecology, conservation, and management of Hawaiian birds: a vanishing avifauna. Cooper Ornithological Society, Studies in Avian Biology 22, Lawrence, Kansas, USA.

van Riper, C. III, S.G. van Riper, M.L. Goff, and M. Laird. 1986. The epizootiology and ecological significance of malaria in Hawaiian land birds. Ecological Monographs 56:327–344.

Videvall, E., C. K. Cornwallis, V. Palinauskas, G. Valkiūnas, and O. Hellgren. 2015. The avian transcriptome response to malaria infection. Molecular Biology and Evolution 32:1255–1267.

Videvall, E., K. L. Paxton, M. G. Campana, L. Cassin-Sackett, C. T. Atkinson, and R. C. Fleischer. 2021. Transcriptome assembly and differential gene expression of the invasive avian malaria parasite *Plasmodium relictum* in Hawai'i. Ecology and Evolution 11:4935–4944.

Warner, R. E. 1968. The role of introduced diseases in the extinction of the endemic Hawaiian avifauna. Condor 70:101–120.

Warren, C. C., L. K. Berthold, H. L. Mounce, J. T. Foster, and L. C. Sackett. 2019. Evaluating the risk of avian disease in reintroducing the endangered Kiwikiu (Maui Parrotbill: *Pseudonestor xanthophrys*) to Nakula NAR, Maui, Hawai'i. University of Hawai'i at Mānoa, Pacific Cooperative Studies Unit Technical Report 201, Honolulu, USA. https://scholarspace.manoa.hawaii.edu/items/4fd7af2e-b56c-4d24-94c6-67a354258547. Accessed 10 Oct 2022.

Warren, C. C., L. K. Berthold, H. L. Mounce, P. Luscomb, B. Masuda, and L. Berry. 2021. 2019 Kiwikiu conservation translocation report. University of Hawai'i at Mānoa, Pacific Cooperative Studies Unit Technical Report 203, Honolulu, USA. https://scholarspace.manoa.hawaii.edu/items/147e866a-7af7-48cb-bd60-39d359032e03. Accessed 10 Oct 2022.

Wei, L., C. T. Atkinson, D. A. LaPointe, and M. D. Samuel. 2017. Mitigating future avian malaria threats to Hawaiian forest birds from climate change. PLoS One 12:e0168880.

Woodworth, B.L., C.T. Atkinson, D.A. LaPointe, P. J. Hart, C. S. Spiegel, E. J. Tweed, C. Henneman, J. LeBrun, T. Denette, R. DeMots, et al. 2005. Host population persistence in the face of introduced vector-borne diseases: Hawaii Amakihi and avian malaria. Proceedings of the National Academy of Sciences USA 102:1531–1536.

Woodworth, B. L., and T. K. Pratt. 2009. Life history and demography. Pages 194–233 *in* T. K. Pratt, C. T. Atkinson, P.C. Banko, J. D. Jacobi, and B. L. Woodworth, editors. Conservation biology of Hawaiian forest birds: implications for island avifauna. Yale University Press, New Haven, Connecticut, USA.

Work, T. M., and R. A. Raymeyer. 1996. *Haemoproteus iwa* n. sp. in great frigatebirds (*Fregata minor* [Gmelin]) from Hawaii: parasite morphology and prevalence. Journal of Parasitology 82:489–491.

Yorinks, N., and C. T. Atkinson. 2000. Effects of malaria (*Plasmodium relictum*) on activity budgets of experimentally infected juvenile Apapane (*Himatione sanquinea*). Auk, 117:731–738.

20 Lead Intoxication in Raptors and Scavenging Birds

MICHELLE G. HAWKINS, KRYSTAL M. T. WOO, KATHERINE E. CARR

Introduction

Lead is one of the most ubiquitous metals on the planet, with documents recording its use as early as the fifth millennium BCE (Nriagu 1983, Krone 2018). More recently, it was used for plumbing, for ammunition, and as an additive in paint and gasoline. A ban on lead in paint and gasoline in North America and Europe in the 1970s was successful in decreasing human exposure (Krone 2018, US Energy Information Administration 2020, CDC 2021). Despite reduction in lead use, remnants of these products still pose a threat to humans and wildlife. Lead takes decades to degrade in the environment and remains bioavailable to animals. At medieval mining sites in the Cevennes and Morvan National Parks in France, lead is still bioavailable centuries after operations ceased (Camizuli et al. 2018). Some estimate lead ammunition takes 100–300 years to degrade in the environment (De Francisco et al. 2003).

Lead ammunition is the most widespread source of lead exposure and intoxication to wild bird species (West et al. 2017, Krone 2018, Pain et al. 2019, Plaza and Lambertucci 2019). Comparison of lead isotopes between ammunition and tissue samples from lead-poisoned birds along with spatiotemporal associations between big game hunting seasons and lead

Illustration by Laura Donohue.

poisoning in avian predators and scavengers show lead ammunition is an important source of lead poisoning. (West et al. 2017, Krone 2018, Plaza and Lambertucci 2019). Lead ammunition sources include shot from animals not retrieved by the hunters or in offal left behind (i.e., entrails, heart, lungs, and liver) after field dressing (Figure 20.1) (Herring et al. 2016, Mctee et al. 2019). Most scavenging birds are social feeders; one contaminated carcass can poison many individuals (Figure 20.2) (Sheppard et al. 2013).

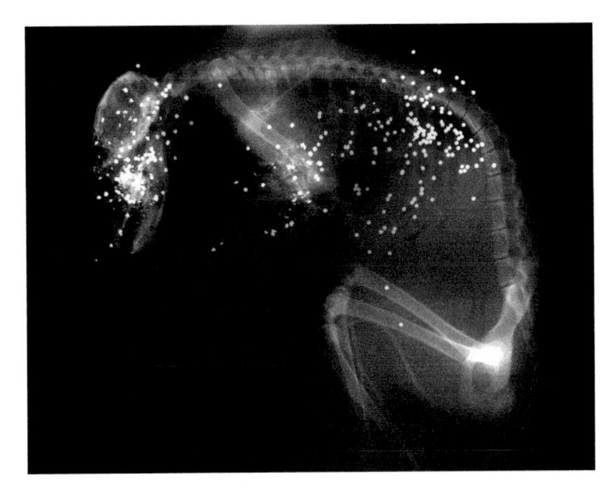

Fig. 20.1. Lateral radiograph of a raccoon shot once in the head and once in the leg with lead ammunition. Note how the shot has fragmented and expanded into tissues distant from the original wound. Pest species such as the raccoon are shot in some areas and left in the field, and scavengers eat from the carcass and become exposed to lead. Photograph courtesy of M. Kevin Keel.

Even with consensus in the peer-reviewed literature that lead ingestion from spent ammunition is the primary source of lead exposure in wildlife today (Bellinger et al. 2013, Haig et al. 2014), this view is not universally shared (Harradine 2004, Bronson 2014, Arnemo et al. 2016). Ingestible lead can reach wildlife from sources other than ammunition. Lead mining and smelting waste in waterways poisoned tundra swans (*Cygnus columbianus*) that consumed contaminated water and vegetation in Idaho, USA (Blus et al. 1999). Upland game birds, waterfowl, predators, and scavengers can still ingest lead from lost fishing weights, in addition to spent ammunition (Friend 1999, Finkelstein et al. 2003, Pokras and Kneeland 2009, Haig et al. 2014). Birds also ingest lead paint used from the 1970s or before when it flakes or is scraped off (Finkelstein et al. 2003, Finkelstein et al. 2012); osprey (*Pandion haliaetus*) and peregrine falcons (*Falco peregrinus*) nest on bridges that have old, exposed lead paint (Katzner et al. 2018). Intranest lead concentration variations in ferruginous hawk (*Buteo regalis*) nestlings, osprey, and—to a lesser extent—golden eagles (*Aquila chrysaetos*) suggest episodic ingestion as would occur with parental feeding of lead-contaminated prey, providing empirical and population-level evidence of the lead source (Katzner et al. 2018).

Not unlike other pollutants, inhalable lead can travel long distances in the atmosphere and can be

Fig. 20.2. California condors (*Gymnogyps californianus*) and some other scavenging birds are social eaters, so one carcass exposed to lead shot can poison multiple animals. Photograph courtesy of Evan McWreath, Ventana Wildlife Society.

present far from its source. This can come from a variety of industrial emissions including coal-fired power plants and lead smelters (Kålås et al. 2000, Pokorny et al. 2009, Gómez-Ramírez et al. 2011). A study after the 2018 Camp Fire in northern California, USA, showed significant lead from structural fires spreads long distances (US Energy Information Administration 2020, California Air Resources Board 2021). Lead concentrations were 50 times higher than average in a town located 15 miles away from the fire; lower lead concentrations traveled in the smoke as far as 150 miles away (California Air Resources Board 2021). From a One Health perspective, if lead is available as particulate matter in smoke it can be a health risk to humans, wildlife, and other animals.

Raptor and Scavenging Species of Concern Affected by Lead Exposure

Despite a large body of scientific literature on the toxic effects of lead exposure on birds, controversy still exists regarding its impacts at the population level (Haig et al. 2014). Raptors, long-lived apex predators at the top of food webs, are an excellent example (Haig et al. 2014). Population-level effects have been described for species with low recruitment rates or small population sizes, such as the California condor (*Gymnogyps californianus*), bald eagles (*Haliaeetus leucocephalus*), Steller's sea eagles (*Haliaeetus pelagicus*), white-tailed eagles (*Haliaeetus albicilla*), Spanish imperial eagles (*Aquila adalberti*), red kites (*Milvus milvus*), griffon vultures (*Gyps fulvus*), and Egyptian vultures (*Neophron percnopterus*) (Pattee and Hennes 1983, García-Fernández et al. 2005, Pain et al. 2005, Cade 2007, Mateo et al. 2007, Gangoso et al. 2009, Krone et al. 2009, Saito 2009). Recent research has documented lead ammunition exposure in new species, taxa, and locations revealing potential impacts on any raptor, scavenging bird, or species consuming gunshot prey (Pain et al. 2019, Descalzo et al. 2021, Johnson et al. 2013).

Lead exposure is a worldwide conservation issue in vulture and condor populations, adding to deliberate poisonings associated with human–wildlife conflict (Naidoo et al. 2017, Plaza and Lambertucci 2019, Safford et al. 2019) and has been identified on all continents where vultures and condors exist (Table 20.1) (Fisher et al. 2006, Hernández and Margalida 2009, Nam and Lee 2009, Berny et al. 2015, Carneiro 2016, Mateo-Tomás et al. 2016, Ganz et al. 2018, Garbett et al. 2018, Plaza and Lambertucci 2019, Plaza et al. 2018, Krüger and Amar 2018, Pain et al. 2019, Descalzo et al. 2021). California condors are long-lived birds and generally hatch their first chick at 6 to 7 years of age and only have one chick every other year. Almost any cause of early mortalities can cause populations to decline.

Clinical Effects and Health Impacts of Lead on Raptors and Scavenging Birds

Knowledge of the long-term sublethal effects of lead on human health continues to grow, but knowledge about intoxication in wildlife lags (Pokras and Kneeland 2009). Measures mitigating environmental lead require monitoring in the environment and in animal tissues. Using raptors as biomonitors or sentinels for humans is being explored by countries as mitigation programs are developed (Gómez-Ramírez et al. 2014, Espín et al. 2016, García-Fernández et al. 2020, Espín et al. 2021). These programs include testing blood and tissue, plucked feathers, failed eggs, regurgitated pellets, excrement, preen oil from living birds, molted feathers, and tissues from carcasses (Espín et al. 2021).

In most adult animals, lead absorption can occur through both the digestive and respiratory systems. Lead ingestion is more likely to cause intoxication than inhalation (Sanderson et al. 1998, Katzner et al. 2018). The health impacts of lead depend on the concentrations reached in blood and tissues. Generally, blood lead concentrations of 20–50 μg/dL are considered subclinical; 50–100 μg/dL, clinically toxic; and >100 μg/dL, severe clinical intoxication, but these breakpoints are not universally accepted (Pain et al. 2019). An individual's susceptibility to *ble 20.1.* Raptor and scavenging species of concern affected by lead exposure

cies	Location	IUCN conservation status[a]	Evidence	References
tures and condors				
ifornia condor *Gymnogyps alifornianus*)	California	Critically endangered, increasing population	From 1992–2009, lead intoxication was the most important mortality factor for immature and adult free-flying condors (26%–67% of adults died from it); from 1997–2010, 48% of condors had blood lead concentrations that required treatment with some individuals intoxicated more than once; 17%–22% probability of free-flying birds having lead concentrations necessitating treatment	Walters et al. 2010, Finkelstein et al. 2012, Rideout et al. 2012, McWreath 2019, IUCN 2020
ffon vulture *Gyps fulvus*)	Europe	Least concern, increasing population	Blood samples showed 45% with lead exposure with 5% exhibiting toxic concentrations; lead isotope ratios were consistent with lead ammunition; highest lead concentrations occurred during game hunting season and in hunted areas	Gangoso et al. 2009, Kelly et al. 2011, Mateo-Tomás et al. 2016, IUCN 2021
ican white-backed vulture (*Gyps africanus*)	Botswana	Critically endangered, decreasing population	30% of birds showed lead exposure; 2.3% had toxic concentrations, which declined steeply between hunting and nonhunting seasons	Garbett et al. 2018, IUCN 2021
	South Africa	Critically endangered, decreasing population	66% of birds (including chicks in the nest) had blood lead concentrations; nonscavenging birds did not have lead concentrations	van den Heever et al. 2019, IUCN 2021
pe vulture (*Gyps coprotheres*)	South Africa	Vulnerable, decreasing population	80% of birds had blood lead concentrations; nonscavenging birds did not	van den Heever et al. 2019, IUCN 2021
arded vulture (*Gypaetus barbatus*)	South Africa	Regionally is critically endangered, near threatened, decreasing population	Bone lead concentrations indicated exposure and accumulation over time, suggesting lead contributed to their deaths	Krüger and Amar 2018, IUCN 2021
	Northern Spain and southern France	Near threatened, decreasing population	Very low lead concentrations in samples from the Pyrenees between 1990 and 2009, with concentrations slightly higher during the hunting season	Hernández and Margalida 2009, IUCN 2021
	Swiss Alps	Near threatened, decreasing population	Bone lead concentrations in clinically toxic range and burdens compatible with acute lead intoxication; higher than elsewhere in Europe or North America	Ganz et al. 2018, IUCN 2021
rkey vulture (*Cathartes aura*)	Virginia	Least concern, stable population	Chronic lead exposure, with potential sublethal, chronic lead burdens	Behmke et al. 2015, IUCN 2018
	Pacific Northwest, USA	Least concern, stable population	9% had lead concentrations consistent with subclinical poisoning, similar exposures in coastal and interior regions, suggesting lead exposure widely distributed	Herring et al. 2018*b*, IUCN 2018
ack vulture (*Coragyps atratus*)	Virginia	Least concern, increasing population	Evidence of chronic lead exposure; insight into the sublethal, chronic lead burdens	Behmke et al. 2015, IUCN 2016
dean condor (*Vultur gryphus*)	Argentina	Vulnerable, decreasing population	Evidence of recent and previous exposures found throughout range in Argentina, with approximately 60% of continental population	Saggese et al. 2009, Wiemeyer et al. 2017, IUCN 2020
	Chile, Ecuador, Peru	Vulnerable, decreasing population	Elevated blood lead concentrations and poisonings after ingestion of ammunition from countries with no previous published data	Wiemeyer et al. 2017, IUCN 2020

(continued)

Table 20.1. (continued)

Species	Location	IUCN conservation status[a]	Evidence	References
Eagles				
Bald eagle (*Haliaeetus leucocephalus*)	Montana, USA	Least concern, increasing population	High lead concentrations during hunting season but not during nonhunting seasons	Bedrosian et al. 2012 IUCN 2016
	Pacific Northwest, USA	Least concern, increasing population	61% had detectable lead; toxic blood concentrations in 48% from 1991–2008	Stauber et al. 2010, IUCN 2016
	Iowa, USA	Least concern, increasing population	Eagles arrive during hunting season; 99% from 4 rehabilitation centers had detectable lead concentrations in blood or tissues	COMS 2014, Yaw et 2017, IUCN 2016
Steller's sea eagle (*Haliaeetus pelagicus*)	Japan	Vulnerable, decreasing population	Significant lead intoxications from lead ammunition-based on isotope analysis 15 years after the total ban	Ishii et al. 2020, IUC 2021
White-tailed eagle (*Haliaeetus albicilla*)	Japan, Poland, Finland, Ireland	Least concern, increasing population	All had detectable lead concentrations with isotope analysis correlated with lead ammunition	Ishii et al. 2017, Kitowski et al. 201 O'Donoghue 2017, Isomursu et al. 20 IUCN 2021
	Japan	Least concern, increasing population	Lethal lead intoxications identified from ammunition, based on isotope analysis 15 years after the ban on lead ammunition	Ishii et al. 2020, IUC 2021
Golden eagle (*Aquila chrysaetos*)	European Alps	Least concern, stable population	Low background concentrations	Kenntner et al. 2007, Smits and Naidoo 2018, IUCN 2021
	Montana, USA	Least concern, stable population	Blood lead concentrations fell in late winter after hunting season	Domenech et al. 202 IUCN 2021
	Pacific Northwest, USA	Least concern, stable population	Lead exposure in up to 82%; toxic blood concentrations in 62% between 1991 and 2008	Stauber et al. 2010, IUCN 2021
	California, Oregon, Idaho, Wyoming, USA	Least concern, stable population	Nestlings near fields with supplemental food had high lead concentrations, which declined exponentially with distance to the food source	Herring et al. 2020, IUCN 2021
	Western USA (Rocky Mountain Migratory Flyway)	Least concern, stable population	Prevalence of lead exposure determined in 1427 migratory birds between 1975 and 2013 in the western USA was 60%; only 10% had clinically toxic concentrations; 4% lethal concentrations	Langner et al. 2015, IUCN 2021
	Swiss Alps	Least concern, stable population	Concentrations compatible with acute lead intoxication found; higher than elsewhere in Europe or North America	Ganz et al. 2018, IUC 2021
Argentine solitary crowned eagle (*Harpyhaliaetus coronatus*)	Central and northern Argentina	Endangered, decreasing population (IUCN 2016)	Lead was identified in the blood and bones	Saggese et al. 2009, IUCN 2016
Wedge-tailed eagle (*Aquila audax*)	Australia and Tasmania	Least concern, increasing population (IUCN 2016)	Lead detected in 100% of femur and liver tissues and in 96% of blood samples from nestlings; 73% of livers had lead $^{207/206}$ ratios compatible with Tasmanian ammunition	Pay et al. 2021, IUCN 2016
Other				
Common raven (*Corvus corax*)	Pacific Northwest, USA	Least concern, increasing population (IUCN 2016)	29% of had concentrations consistent with subclinical poisoning; similar findings in coastal and interior birds; suggests lead exposure is widely distributed	Herring et al. 2018*b*, IUCN 2016
	Northern California, USA	Least concern, increasing population (IUCN 2016)	Lead concentrations increased with distance from the coast; median blood lead concentrations hunting season 6 times higher than birds in nonhunting season	West et al. 2017, IUC 2016

lead intoxication can be influenced by many environmental and biologic factors and sensitivity varies between species. Indirect immunosuppressive effects (increased susceptibility to infectious diseases and parasite infestations) and increased susceptibility to collisions with power lines occur (Ecke et al. 2017).

Ingested lead is dissolved by the acidic pH of the avian ventriculus or "gizzard," the muscular portion of avian stomachs. Absorption varies with gut transit time and type of lead ingested (Franson and Pain 2011). Lead can be ground in the ventriculus of waterfowl and songbirds by "grit"—tiny pebbles, rocks, etc.—or it may pass through the gastrointestinal tract with little absorption (Custer et al. 1984). The raptor ventriculus does not contain grit; raptors instead egest "pellets" of bones, hair, and hard tissues of prey, which may have some protective effects against lead absorption (Duke et al. 1976). This is dependent on the time from ingestion to cast: owls egest pellets after each meal, whereas hawks digest bones more thoroughly and may not egest a pellet until 2–3 meals later (Duke et al. 1976). Lead was identified in egested pellets from the endangered Bonelli's eagle (*Aquila fasciata*), coinciding with feather concentrations (Gil-Sánchez et al. 2018).

Once absorbed, lead mimics calcium in blood and other fluids in all body systems (Pokras and Kneeland 2009, Franson and Pain 2011). It binds to sulfur groups in proteins and enzymes, causes increased production of zinc protoporphyrin, and decreased delta-aminolevulinic acid dehydratase activity, thereby decreasing the overall heme production necessary for red blood cells (Redig et al. 1991, Burger and Gochfeld 2004, Espín et al. 2015). Blood lead concentrations rise sharply after acute absorption and can remain in the blood for several weeks (Franson and Pain 2011); however, the half-life of blood lead concentrations in California condors is estimated to only be 17 days, shorter than other raptors (Green et al. 2008).

With increasing blood concentrations, lead travels to soft tissues, especially kidney and liver. It concentrates in the renal cortex and causes nephrosis and proximal tubular necrosis (Franson and Russell 2014, Manning et al. 2019, Torimoto et al. 2021). Lead is diffusely distributed in the liver causing atrophy, necrosis, and gallbladder enlargement (Figure 20.3) (Franson and Russell 2014, Torimoto et al. 2021).

Bone lead concentration is generally considered the best indicator of exposure over the lifetime of the bird, thus it is an excellent sample for biomonitoring (Franson and Russell 2014). In most species, highest concentrations of lead are found in bone, then kidney, liver, and feathers (Custer et al. 1984, Pattee et al. 2006, Franson and Pain 2011). The opposite

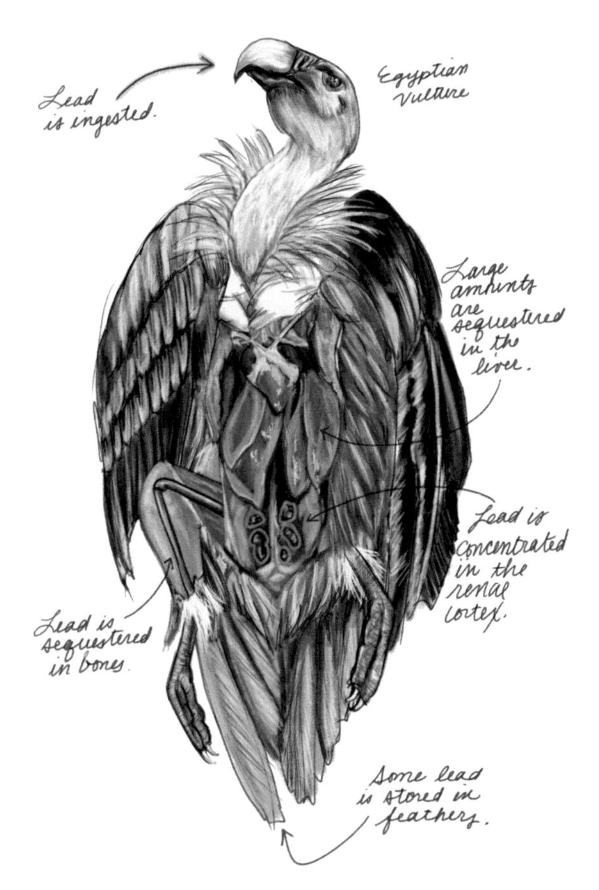

Fig. 20.3. Once absorbed from the gastrointestinal tract into the bloodstream, lead substitutes for calcium and accumulates in liver, kidney, and other soft tissues, as well as in bone and feathers. Illustration by Laura Donohue.

order has been found in some raptors (Martin et al. 2008, Castro et al. 2011).

Failed eggs and feathers are accessible samples for biomonitoring (Espín et al. 2021). Lead is deposited in feathers during growth and bone during exposure events (Finkelstein et al. 2010, Fritsch et al. 2019, Gorski et al. 2021) and is stable in both tissues. Lead can affect reproduction by altering feather iridescence (a visual cue) or can be mobilized from bone during eggshell production (Chatelain et al. 2017, Fritsch et al. 2019). Lead concentrations in the feathers of juvenile birds are a reliable indicator of dietary exposure and can be correlated with lead in kidney and bone (Golden et al. 2003, Pain et al. 2019). However, adult feathers can be problematic because they are subject to atmospheric lead deposition and must be cleaned thoroughly before analysis. Even after cleaning, lead concentrations in Spanish imperial eagle feather specimens from museums were positively correlated with specimen age, suggesting externally deposited lead is difficult to remove (Pain et al. 2005). Analysis of feathers along their length indicates lead circulating in feather pulp during growth in California condors. This suggests ratios of local concentrations to feather growth rates reflect exposure times (Fry and Maurer 2003, Fry 2004). Little lead appears to be transferred from the female to eggs (Scheuhammer 1987, Furness 1993, Lohr et al. 2020), so failed eggs may not be useful for biomonitoring programs.

Effects of lead on heart and gastrointestinal tract are likely species specific. Gizzard changes in eagles and crop stasis in a California condor have been reported (Wynne and Stringfield 2007, Manning et al. 2019). In bald eagles, the most common gross lesions were degeneration, fibrosis, and necrosis in the heart muscle identified on histopathology (Manning et al. 2019). Lead accumulates in arterial walls in red kites. Neither myocardial nor vasculature lesions were identified in Andean condors (*Vultur gryphus*), indicating species-specific tissue tropism (Pattee et al. 2006, Franson and Russell 2014, Torimoto et al. 2021). There are no reports of cardiac lesions in California condors, and since individual birds may be exposed multiple times, it seems likely cardiac lesions are milder than those seen in eagles.

Skeletal muscles do not appear to be affected by increasing blood lead, but ingestion of embedded lead from prey muscle tissue increases blood concentrations (Custer et al. 1984, Glucs 2018). Lead embedded in pectoral muscles appears to be inert, possibly due to scar tissue preventing absorption (Sanderson et al. 1998, Behmke et al. 2017, Berny et al. 2017, Gorski et al. 2021). Lead fragments in joints may cause arthritis due to physical damage and the acidic environment caused by inflammation (Gameiro et al. 2013, Berny et al. 2017).

Histologically, brain appears less affected than other tissues, absorption possibly limited by the blood–brain barrier (Custer et al. 1984, Burger and Gochfeld 2004, Aloupi et al. 2017, Ecke et al. 2017, Bassi 2021). However, lead has been detected in the cerebrum, cerebellar cortex, and midbrain in red kites (Torimoto et al. 2021). Nervous system signs of lead intoxication in wildlife includes learning deficits, behavior changes such as abnormal migratory behavior, blindness, incoordination, head tilt, and seizures (Burger and Gochfeld 2004, 2005; Fallon et al. 2017; Krone 2018; Torimoto et al. 2021) usually resulting in death.

Diagnosis of Lead Poisoning in Raptor and Scavenger Populations

The diagnosis of lead poisoning in wildlife has improved, becoming faster, more economical, and more accessible over time. This facilitates diagnosing lead intoxication or exposure in animals that are brought to wildlife rehabilitation centers, allowing earlier initiation of treatment and making biomonitoring programs feasible.

Scavenging species and eagles are routinely screened for lead on admission to wildlife rehabilitation centers (Figure 20.4). In some centers, pocket gophers, squirrels, and opossums used for food are

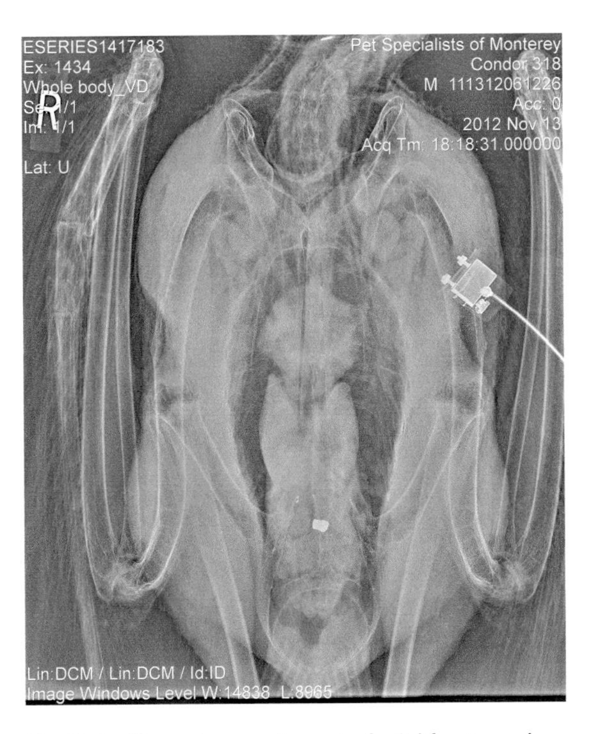

Fig. 20.4. Ventrodorsal radiograph of a California condor (*Gymnogyps californianus*) that ingested an intact 0.22-caliber bullet. One large and four small metallic objects are identified most likely in the proventriculus or ventriculus. Photograph courtesy of Ventana Wildlife Society.

tested if they are from areas with historical sources, such as smelting facilities (Reynolds et al. 2006). Other species are not routinely tested unless they exhibit clinical signs or have metal objects identified in their gastrointestinal tract on radiography. Testing can be performed on tissues and blood of live birds in rehabilitation or dead birds in the field.

Graphite-furnace atomic absorption spectrometry (GFAAS) and inductively coupled plasma mass spectrometry (ICP-MS) are available through most state veterinary diagnostic laboratories and some commercial laboratories to measure blood lead concentrations as well as those in tissues (kidney, liver, and other bodily fluids) (Slabe 2019). Bone samples for biomonitoring are primarily evaluated using ICP-MS (Jenni et al. 2015, Ganz et al. 2018, Gorski et al. 2021). The two methods show linear correlation (Zhang et al. 1997).

A third method, anodic stripping voltammetry, has become popular and is available in a point-of-care whole blood diagnostic device.* Requiring only a drop of blood, these analyzers are easy to use, fast (3 minutes) and a fraction of the cost of laboratory testing. They also eliminate sample transit time and reporting of results. Validation of the device has been performed in many species, from humans to raptor sentinels for California condors (Green et al. 2008, Herring et al. 2018*a*). All studies found the device underestimates blood lead concentrations, but values were consistently low and corrected using species-specific equations and factors (Green et al. 2008, Herring et al. 2018).

Another diagnostic modality is portable x-ray fluorescence (XRF), which can also be used for biomonitoring (Specht et al. 2018, Hampton et al. 2021). Good correlation was reported for XRF device measurements and bone measurements using ICP-MS in California condors and wedge-tailed eagles (*Aquila audax*) (Figure 20.5) (Specht et al. 2018, Hampton et al. 2021).

Treatment

Unfortunately, it is still necessary to treat individuals, particularly threatened and endangered species. Lead in the gastrointestinal tract may be removed via lavage, laxatives, cathartics, or endoscopic or surgical intervention, which should be followed by radiographs to confirm removal of all fragments (Samour and Naldo 2002, Redig and Arent 2008, Fallon et al. 2017). Lead in or around joints may be absorbed by the body and, therefore, should also be removed (Redig and Arent 2008). After removal of lead, chelation therapy is standard-of-care. Edetate calcium disodium (CaEDTA) and dimercaptosuccinic acid (DMSA) are the most commonly used and are relatively safe compared to other chelators (Samour and

* Leadcare® system. Meridian Bioscience, Inc, Cincinatti, OH 45244.

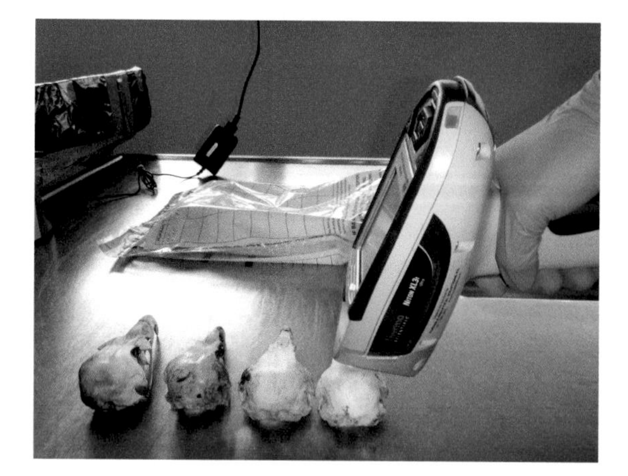

Fig. 20.5. Portable x-ray fluorescence is a rapid, cost-effective method for determining bone lead concentrations. This method shows great promise for biomonitoring programs assessing chronic lead exposure in threatened or endangered species. Photograph courtesy of Jordan Hampton, University of Melbourne, Melbourne, Australia.

Naldo 2002, Hoogesteijn et al. 2003, Wynne and Stringfield 2007, Redig and Arent 2008, Fallon et al. 2017).

The oral bioavailability of CaEDTA is poor, but it can be administered intravenously, intramuscularly, and subcutaneously (Samour and Naldo 2002, Redig and Arent 2008, Fallon et al. 2017). The CaEDTA mobilizes lead from bone and kidney and redistributes it to other tissues (Hammond et al. 1967, Cory-Slechta et al. 1987). It binds to other metals like zinc and iron and is excreted via the kidney, so a washout period is often necessary to allow time for replenishment of the metallic ions as well as to prevent nephrosis (Anderson et al. 2000, Redig and Arent 2008). Some studies report no adverse effects from treatment (Samour and Naldo 2002, Redig and Arent 2008). DMSA has good oral bioavailability, so can be sprinkled on food or given as a suspension (Denver et al. 2000, Fallon et al. 2017). It binds to lead in soft tissues, especially brain and kidneys, but doesn't have any effects on lead in bone (Cory-Slechta 1988). There is a narrow therapeutic index with fewer adverse effects at lower doses. Regurgitation occurs at higher dosages in cockatiels (Denver et al. 2000, Hoogesteijn et al. 2003, Redig and Arent 2008, Fallon et al. 2017). Due to the different actions of CaEDTA and DMSA, they may be used together for treatment with few side effects (Cory-Slechta 1988, Hoogesteijn et al. 2003, Wynne and Stringfield 2007). Unfortunately, these medications must be compounded, which may be expensive or difficult to obtain depending on state laws.

Supportive care is often given if signs of toxicity or other health problems indicate it, including benzodiazepines for seizure control, fluid therapy to correct dehydration, nutritional support, and vitamin supplementation (Redig and Arent 2008, Fallon et al. 2017). Treatment of lead poisoning often requires intensive and prolonged care. Because lead can accumulate in bone and cause chronic health effects, the best treatment is always prevention.

Strategies to Mitigate Lead Intoxication

As was made clear in Flint, Michigan, USA, lead poisoning via water and food is very much a One Health issue. Around the world, legislative actions have been taken to try to reduce lead exposure via ammunition in wildlife and subsequently in humans (Arnemo et al. 2016, Plaza et al. 2018). While this approach has had some success, it has proven difficult to persuade large interest groups to support necessary restrictions and biomonitoring programs to measure progress (Gómez-Ramírez et al. 2014, Arnemo et al. 2016).

Bans on the sale and use of lead ammunition and fishing gear are commonly used strategies on the regional and national level to attenuate lead exposure to wildlife. After years of study of lead intoxication in waterfowl populations of North America and lead poisoning in eagles in the United States, a mandate for use of non-toxic shot was implemented, first in heavily affected wetlands in 1986, followed by a 1991 nationwide ban prohibiting the use of lead shot for hunting waterfowl (Tranel and Kimmel 2009, Johnson et al. 2013).

Regulatory and advisory bodies are implementing global plans to reduce environmental lead for more wide-reaching protection for birds. In 2014, parties from 130 countries at the Convention on the Conservation of Migratory Species of Wild Animals (COMS) advised phase out of all lead ammunition used in terrestrial and wetland environments by 2017 (COMS 2014, Arnemo et al. 2016). A Lead Task Group was created to address this process; however, recent COMS meetings report little progress on that deadline (COMS 2020). Help from other groups has been sought including the Vulture Working Group, Intergovernmental Task Force on Illegal Killing, and Taking and Trade of Migratory Birds in the Mediterranean that already promote welfare and protection of these species (COMS 2020). Despite being a One Health issue, there is still a long way to go globally.

In 2008, California, USA, banned the use of lead ammunition for hunting big game and nongame species in the range of the California condor due to its negative effects on recovery (Kelly et al. 2013). A significant reduction in lead intoxication was reported in turkey vultures (*Cathartes aura*) and golden eagles, sentinel species for California condors within its range within one year, suggesting early success (Kelly et al. 2011). Yet, like eagles after the 1991 lead shot ban, lead intoxication of California condors has continued. In 2013, California passed a statewide, phased lead ammunition ban for hunting any wildlife (Kelly et al. 2013) and, in 2019, the statewide ban was complete (Figure 20.6). The problems associated with lead ammunition, including its spread through big game carcasses along bullet tracks and its toxicity to humans and wildlife are effectively demonstrated with radiographs and ballistic gels in state-mandated hunter-education courses. Data continue to be collected from condor, eagle, and scavenging bird populations to document the effect of these bans.

A series of partial bans on lead ammunition for hunting on the island of Hokkaido, Japan reduced lead poisonings in Steller's sea eagles and white-tailed eagles, but issues with lead intoxication there continue (Pain et al. 2019, Ishii et al. 2020). Regional bans appear to have limited effects on species with migratory behavior (Krone 2018). Partial bans on se-

Fig. 20.6. Use of lead shot for waterfowl hunting was banned nationwide in the United States in 1980. Use of lead rifle bullets was banned in California in the range of the California condor in 2011 and statewide in 2019. Illustration by Laura Donohue.

lect types of lead ammunition or for specific game species have been shown little power to reduce mortality in raptors and scavengers (Franson and Russell 2014, Warner et al. 2014, Yaw et al. 2017, Pain et al. 2019, Ishii et al. 2020).

More geographically expansive regulations for total bans on lead ammunition may be necessary to reduce lead poisoning in raptors and scavenging birds (Krone 2018). Twenty-three European countries have legally restricted the use of lead shot, with Denmark and the Netherlands having total bans (Mateo and Kanstrup 2019). Through the European Chemicals Agency (ECHA), the European Union placed an EU wide ban on lead ammunition for shooting in wetlands effective February 2023. The ECHA is preparing a proposal for a similar ban on lead ammunition in terrestrial habitats and on lead fishing tackle (ECHA 2017). This ban could provide consistency among member states across which the at-risk animals migrate. In member states with more stringent regulations, the stricter provisions will supersede (European Commission 2021). There are few regulations on the use of lead ammunition in South America, except for some provinces in Argentina (Plaza et al. 2018).

Summary

From a One Health perspective, both animal and human health are clearly affected by the pervasiveness of lead in the environment, and global regulations are changing this little by little. Removing and disposing (e.g., via burial or burning) of lead-contaminated offal of sport-hunted animals is one positive step hunters can take. Another is eliminating lead ammunition (Bedrosian et al. 2012, Pain et al. 2019). Lead-free bullets for hunting purposes are available and have similar ballistic properties (Gremse et al. 2014, Pain et al. 2019), and the cost is only nominally higher. Federal and state agencies, the hunting community, and the public must work together toward a transition from lead to nonlead ammunition. A failure to do this would reflect poorly on hunting as a socially supported activity and violates ethical standards of many conservation organizations.

LITERATURE CITED

Aloupi, M., A. Karagianni, S. Kazantzidis, and T. Akriotis. 2017. Heavy metals in liver and brain of waterfowl from the Evros Delta, Greece. Archives of Environmental Contamination and Toxicology 72:215–234.

Anderson, W. L., S. P. Havera and B. W. Zercher. 2000. Ingestion of lead and nontoxic shotgun pellets by ducks in the Mississippi flyway. Journal of Wildlife Management 64:848–857.

Arnemo, J.M., O. Andersen, S. Stokke, V. G. Thomas, O. Krone, D. J. Pain, and R. Mateo. 2016. Health and environmental risks from lead-based ammunition: science versus socio-politics. EcoHealth 13:618–622.

Bassi, E. 2021. Lead contamination in tissues of large avian scavengers in south-central Europe. Science of the Total Environment 778:146130.

Bedrosian, B, D. Craighead, and R. Crandall. 2012. Lead exposure in bald eagles from big game hunting, the continental implications and successful mitigation efforts. PLoS One 7:e51978.

Behmke, S., J. Fallon, A. E. Duerr, A. Lehner, J. Buchweitz, and T. Katzner. 2015. Chronic lead exposure is epidemic in obligate scavenger populations in eastern North America. Environment International 79:51–55.

Behmke, S., P. Mazik, and T. E. Katzner. 2017. Assessing multi-tissue lead burdens in free-flying obligate scavengers in eastern North America. Environmental Monitoring and Assessment 189:139.

Bellinger, D. C., J. Burger, T. J. Cade, D. A. Cory-Slechta, M. Finkelstein, H. Hu, M. Kosnett, P. J. Landrigan, B. Lanphear, and M. A. Pokras. 2013. Health risks from lead-based ammunition in the environment. Environmental Health Perspectives 121:A178–A179.

Berny, P., L. Vilagines, J.-M.Cugnasse, O. Mastain, J.-Y. Chollet, G. Joncour, and M. Razin. 2015. Vigilance poison: illegal poisoning and lead intoxication are the main factors affecting avian scavenger survival in the Pyrenees (France). Ecotoxicology and Environmental Safety 118:71–82.

Berny, P. J., E. Mas, and D. Vey. 2017. Embedded lead shots in birds of prey: the hidden threat. European Journal of Wildlife 63:101.

Blus, L. J., C. J. Henny, D. J. Hoffman, L. Sileo, and D. J. Audet. 1999. Persistence of high lead concentrations and associated effects in tundra swans captured near a mining and smelting complex in northern Idaho. Ecotoxicology 8:125–132.

Bronson, R. 2014. Not so fast everybody. Pages 10–11 *in* Conference program: 2014 annual meeting of the Minnesota Chapter of the Wildlife Society. https://wildlife.org/wp-content/uploads/2016/02/2014-MTWS-AM-Program-Final.pdf. Accesssed 20 Sept 2022.

Burger, J., and M. Gochfeld. 2004. Effects of lead and exercise on endurance and learning in young herring gulls. Ecotoxicology and Environmental Safety 57:136–144.

Burger, J., and M. Gochfeld. 2005. Effects of lead on learning in herring gulls: an avian wildlife model for neurobehavioral deficits. Neurotoxicology 26:4: 615–624.

Cade, T. J. 2007. Exposure of California condors to lead from spent ammunition. Journal of Wildlife Management 71:2125–2133.

California Air Resources Board. 2021. Camp Fire air quality data analysis. https://ww2.arb.ca.gov/sites/default/files/2021-07/Camp_Fire_report_July2021.pdf. Accessed 21 Dec 2021.

Camizuli, E., R. Scheifler, S. Garnier, F. Monna, R. Losno, C. Gourault G. Hamm C. Lachiche G. Delivet, C. Chateau, et al. 2018. Trace metals from historical mining sites and past metallurgical activity remain bioavailable to wildlife today. Scientific Reports 8:3436.

Carneiro, M. A. G. 2016. Biomonitoring of arsenic, cadmium, lead and mercury with raptors from the Iberian Peninsula. Ph.D.Dissertation, Universidade de Tras-os-Montes e Alto Douro, Vila Real, Portugal.

Castro, I., J. Aboal, J. Fernández, and A. Carballeira. 2011. Use of raptors for biomonitoring of heavy metals: gender, age and tissue selection. Bulletin of Environmental Contamination and Toxicology 86:347–351.

Centers for Disease Control and Prevention (CDC). 2021. Lead in paint. https://www.cdc.gov/nceh/lead/prevention/sources/paint.htm. Accessed 19 Nov 2021.

Chatelain, M., A. Pessato, A. Frantz, J. Gasparini, and S. Leclaire. 2017. Do trace metals influence visual signals? Effects of trace metals on iridescent and melanic feather colouration in the feral pigeon. Oikos 126:1542–1553.

Convention on Migratory Species (COMS). 2014. Preventing poisoning of migratory birds. https://www.cms.int/sites/default/files/document/mos2_inf11_cms_res_11_15_e_0.pdf. Accessed 14 Dec 2021.

Convention on Migratory Species (COMS). 2020. Preventing poisoning of migratory birds. https://www.cms.int/en/workinggroup/preventing-poisoning-migratory-birds. Accessed 14 Dec 2021.

Cory-Slechta, D. A. 1988. Mobilization of lead over the course of DMSA chelation therapy and long-term efficacy. Journal of Pharmacology and Experimental Therapeutics 246:84–91.

Cory-Slechta, D. A., B. Weiss, and C. Cox. 1987. Mobilization and redistribution of lead over the course of calcium disodium ethylenediamine tetraacetate chelation therapy. Journal of Pharmacology and Experimental Therapeutics 243:804–813.

Custer, T.W, J. C. Franson, and O. H. Pattee. 1984. Tissue lead distribution and hematologic effects in American kestrels (*Falco sparverius* L.) fed biologically incorporated lead. Journal of Wildlife Diseases 20:39–43.

De Francisco, N., J. D. Ruiz Troya, and E. I. Agüera. 2003. Lead and lead toxicity in domestic and free living birds. Avian Pathology 32:3.

Denver, M. C., L. A. Tell, F. D. Galey, J. G. Trupkiewicz, and P. H Kass. 2000. Comparison of two heavy metal chelators for treatment of lead toxicosis in cockatiels. American Journal of Veterinary Research 6:935–940.

Descalzo, E., P. R. Camarero, I. S. Sánchez-Barbudo, M. Martinez-Haro, M. E. Ortiz-Santaliestra, R. Moreno-Opo, and R. Mateo. 2021. Integrating active and passive monitoring to assess sublethal effects and mortality from lead poisoning in birds of prey. Science of the Total Environment 750:142260.

Domenech, R., A. Shreading, P. Ramsey, and M. McTee. 2021. Widespread lead exposure in golden eagles captured in Montana. Journal of Wildlife Management 85:195–201.

Duke, G., O. Evanson, P. Redig, and D. Rhoades. 1976. Mechanism of pellet egestion in great-horned owls (*Bubo virginianus*). American Journal of Physiology, Legacy Content 231:1824–1829.

Ecke, F., N. J. Singh, J. M. Arnemo, A. Bignert, B. Helander, A. M. M. Berglund, H. Borg, C. Brojer, K. Holm, M. Lanzone, et al. 2017. Sublethal lead exposure alters movement behavior in free-ranging golden eagles. Environmental Science and Technology 51:5729–5736.

Espín, S., J. Andevski, G. Duke, I. Eulaers, P. Gómez-Ramírez, G. T. Hallgrimsson, B. Helander, D. Herzke, V. L. B. Jaspers, O. Krone, et al. 2021. A schematic sampling protocol for contaminant monitoring in raptors. Ambio 50:95–100.

Espín, S., A. J. García-Fernández, D. Herzke, R. F. Shore, B. Van Hattum, E. Martínez-López, M. Coeurdassier, I. Eulaers, C. Fritsch, P. Gómez-Ramírez, et al. 2016. Tracking pan-continental trends in environmental contamination using sentinel raptors—what types of samples should we use? Ecotoxicology 25:777–801.

Espín, S., E. Martínez-López, P. Jiménez, P. María-Mojica, and A. J. García-Fernández. 2015. Delta-aminolevulinic acid dehydratase (δALAD) activity in four free-living bird species exposed to different levels of lead under natural conditions. Environmental Research 137:185–198.

European Chemicals Agency (ECHA). 2017. European Chemicals Agency annex XV restriction report—lead in shot. https://echa.europa.eu/documents/10162/6ef877d5-94b7-a8f8-1c49-8c07c894fff7. Accessed 21 Dec 2021.

European Commission. 2021. Commission Regulation (EU) 2021/57 of 25 January 2021 amending Annex XVII to Regulation (EC) No 1907/2006 of the European Parliament and of the Council concerning the registration, evaluation, authorisation and restriction of chemicals (REACH) as regards lead in gunshot in or around wetlands. https://eur-lex.europa.eu/eli/reg/2021/57/oj. Accessed 21 Dec 2021.

Fallon, J., P. Redig, T. Miller, M. Lanzone, and T. Katzner. 2017. Guidelines for evaluation and treatment of lead poisoning of wild raptors. Wildlife Society Bulletin 41:205–211.

Finkelstein, M. E., D. F. Doak, D. George, J. Burnett, J. Brandt, M. Church, J. Grantham, and D. R. Smith. 2012. Lead poisoning and the deceptive recovery of the critically endangered California condor. Proceedings of the National Academy of Sciences USA 109:11449–11454.

Finkelstein M. E., D. George, S. Scherbinski, R. Gwiazda, M. Johnson, J. Burnett, J. Brandt, S. Lawrey, A. P. Pessier, M. Clark, et al. 2010. Feather lead concentrations and 207/206 Pb ratios reveal lead exposure history of California condors (*Gymnogyps californianus*). Environmental Science and Technology 44:2639–2647.

Finkelstein, M. E., R. H. Gwiazda, and D. R. Smith. 2003. Lead poisoning of seabirds: environmental risks from leaded paint at a decommissioned military base. Environmental Science and Technology 37:3256–3260.

Fisher, I. J., D. J. Pain, and V. G. Thomas. 2006. A review of lead poisoning from ammunition sources in terrestrial birds. Biololgical Conservation 131:421–432.

Franson, J. C., and D. J. Pain. 2011. Lead in birds. Pages 563–593 *in* W. N. Bey and J. P. Meador, editors. Environmental contaminants in biota: interpreting tissue concentrations. Second edition. CRC Press, Boca Raton, Florida, USA.

Franson, J. C., and R. E. Russell. 2014. Lead and eagles: demographic and pathological characteristics of poisoning, and exposure levels associated with other causes of mortality. Ecotoxicology 23:1722–1731.

Friend, M. 1999. Lead. Pages 317–334 *in* M. Friend and J. C. Franson, technical editors. Fireld manual of wildlife diseases: general field procedures and diseases of birds. US Geological Survey, Biological Resources Division Information and Technology Report 1999-001, Madison, Wisconsin, USA.

Fritsch, C., Ł. Jankowiak, and D. Wysocki. 2019. Exposure to Pb impairs breeding success and is associated with longer lifespan in urban European blackbirds. Science Reports 9:486.

Fry, D. M., and J. R. Maurer. 2003. Assessment of lead contamination sources exposing California condors. California Department of Fish and Game, Species Conservation and Recovery Program Report, Sacramento, USA.

Fry, M. 2004. Final report addendum: analysis of lead in California condor feathers: determination of exposure and depuration during feather growth. California Department of Fish and Game, Species Conservation and Recovery Program Report, Sacramento, USA.

Furness, R. W. 1993. Birds as monitors of pollutants. Pages 86–143 *in* R. W. Furness and J. J. D. Greenwood, editors. Birds as monitors of environmental change. Springer, Dordrecht, the Netherlands.

Gameiro, V. S., G. C. S. de Araújo, and F. M. M. Bruno. 2013. Lead intoxication and knee osteoarthritis after a gunshot: long-term follow-up case report. Case Reports 2013:bcr2013009404.

Gangoso, L., P. Alvarez-Lloret, A. A. Rodriguez-Navarro, R. Mateo, F. Hiraldo, and J. A. Donazar. 2009. Long-term effects of lead poisoning on bone mineralization in vultures exposed to ammunition sources. Environmental Pollution 157:569–574.

Ganz, K., J. Lukas, M. M. Madry, T. Kraemer, J. Hannes, and D. Jenny. 2018. Acute and chronic lead exposure in four avian scavenger species in Switzerland. Archives of Environmental Contamination and Toxicology 75:566–575.

Garbett, R., G. Maude, P. Hancock, D. Kenny, R. Reading, and A. Amar. 2018. Association between hunting and elevated blood lead levels in the critically endangered African white-backed vulture *Gyps africanus*. Science of the Total Environment 630:1654–1665.

García-Fernández, A.J, S. Espín, P. Gómez-Ramírez, E. Martínez-López, and I. Navas. 2020. Wildlife sentinels for human and environmental health hazards in ecotoxicological risk assessment. Page 77–94 *in* K. Roy, editor. Ecotoxicological QSARs. Humana Press, New York, New York, USA.

García-Fernández, A., E. Martinez-Lopez, D. Romero, P. Maria-Mojica, A. Godino and P. Jimenez. 2005. High levels of blood lead in griffon vultures (*Gyps fulvus*) from Cazorla Natural Park (southern Spain). Environmental Toxicology 20:459–463.

Gil-Sánchez, J. M., S. Molleda, J. A. Sánchez-Zapata, J. Bautista, I. Navas, R. Godinho, A. J. García-Fernández, and M. Moleón. 2018. From sport hunting to breeding success: patterns of lead ammunition ingestion and its effects on an endangered raptor. Science of the Total Environment 613:483–491.

Glucs, Z. E.. 2018. Lead exposure and hormonal stress response in California condors. PhD Dissertation, University of California, Santa Cruz, USA.

Golden, N. H., B. A. Rattner, J. B. Cohen, D. J. Hoffman, E. Russek-Cohen, and M. A. Ottinger. 2003. Lead accumulation in feathers of nestling black-crowned night herons (*Nycticorax nycticorax*) experimentally treated in the field. Environmental Toxicology and Chemistry 22:1517–1524.

Gómez-Ramírez, P., E. Martínez-López, P. María-Mojica, M. León-Ortega, and A. García-Fernández. 2011. Blood lead levels and δ-ALAD inhibition in nestlings of Eurasian eagle owl (*Bubo bubo*) to assess lead exposure associated to an abandoned mining area. Ecotoxicology 20:131–138.

Gómez-Ramirez, P., R. Shore, N. Van Den Brink, B. Van Hattum, J. Bustnes, G. Duke, C. Fritsch, A. J. García-Fernández, B. O. Helander, V. Jaspers, et al. 2014. An overview of existing raptor contaminant monitoring activities in Europe. Environment International 67:12–21.

Gorski, P. R., S. R. Scott, and E. M. Lemley. 2021. Application of stable isotopic ratio analysis to identify the cause of acute versus chronic lead poisoning of a tundra swan

(*Cygnus columbianus*): a case study. Bulletin of Environmental Contamination and Toxicology 106:250–256.

Green, R. E., W. G. Hunt, C. N. Parish and I. Newton. 2008. Effectiveness of action to reduce exposure of free-ranging California condors in Arizona and Utah to lead from spent ammunition. PLoS One 3:e4022.

Gremse, F., O. Krone, M. Thamm, F. Kiessling, R. H. Tolba, S. Rieger, and C. Gremse. 2014. Performance of lead-free versus lead-based hunting ammunition in ballistic soap. PLoS ONE 9:e102015.

Haig, S. M., J. D'Elia, C. Eagles-Smith, J. M. Fair, J. Gervais, G. Herring, J. W. Rivers, and J. H. Schulz. 2014. The persistent problem of lead poisoning in birds from ammunition and fishing tackle. Condor 116:408–428.

Hammond, P., A. Aronson, and W. Olson. 1967. The mechanism of mobilization of lead by ethylenediaminetetraacetate. Journal of Pharmacology and Experimental Therapeutics 157:196–206.

Hampton, J. O., A. J. Specht, J. M. Pay, M. A. Pokras, and A. J. Bengsen. 2021. Portable X-ray fluorescence for bone lead measurements of Australian eagles. Science of the Total Environment 789:147998.

Harradine, J. 2004. Spent lead shot and wildlife exposure and risks. Pages 119–130 *in* Sport Shooting and the Environment: Sustainable Use of Lead Ammunition. Proceedings of the World Symposium on Lead Ammunition. World Forum on the Future of Sport Shooting Activities, Rome, Italy.

Hernández, M., and A. Margalida. 2009. Assessing the risk of lead exposure for the conservation of the endangered Pyrenean bearded vulture (*Gypaetus barbatus*) population. Environmental Research 109:837–842.

Herring, G., C. A. Eagles-Smith, B. Bedrosian, D, Craighead, R. Domenech, H. W. Langner, C. N. Parish, A. Shreading, A. Welch, and R. Wolstenholme. 2018*a*. Critically assessing the utility of portable lead analyzers for wildlife conservation. Wildlife Society Bulletin 42:284–294.

Herring, G., C. A. Eagles-Smith, J. A. Buck, A. E. Shiel, C. R. Vennum, C. Emery, B. Johnson, D. Leal, J. A. Heath, B. M. Dudek, C. R. Preston and B. Woodbridge. 2020. The lead (Pb) lining of agriculture-related subsidies: enhanced golden eagle growth rates tempered by Pb exposure. Ecosphere 11:e03006.

Herring, G., C. A. Eagles-Smith, and D. E Varland. 2018*b*. Mercury and lead exposure in avian scavengers from the Pacific Northwest suggest risks to California condors: implications for reintroduction and recovery. Environmental Pollution 243:610–619.

Herring, G., C. A. Eagles-Smith, and M. T. Wagner. 2016. Ground squirrel shooting and potential lead exposure in breeding avian scavengers. PLoS One 11:e0167926.

Hoogesteijn, A. L., B. L. Raphael, P. Calle, R. Cook, and G. Kollias. 2003. Oral treatment of avian lead intoxication with meso-2, 3-dimercaptosuccinic acid. Journal of Zoo and Wildlife Medicine 34:82–87.

Ishii, C., Y. Ikenaka, S. M. M. Nakayama, T. Kuritani, M. Nakagawa, K. Saito, Y. Watanabe, K. Ogasawara, M. Onuma, A. Haga, et al. 2020. Current situation regarding lead exposure in birds in Japan (2015–2018): lead exposure is still occurring. Journal of Veterinary Medical Science 82:1118–1123.

Ishii, C., S. M. M. Nakayama, Y. Ikenaka, H. Nakata, K. Saito, Y. Watanabe, H. Mizukawa, S. Tanabe, K. Nomiyama, T. Hayashi, et al. 2017. Lead exposure in raptors from Japan and source identification using Pb stable isotope ratios. Chemosphere 186:367–373.

Isomursu, M., J. Koivusaari, T. Stjernberg, V. Hirvelä-Koski, and E.-R. Venäläinen. 2018. Lead poisoning and other human-related factors cause significant mortality in white-tailed eagles. Ambio 47:858–868.

International Union for the Conservation of Nature (IUCN). 2016. The IUCN Red List of Threatened Species. Version 2016-1. https://www.iucnredlist.org. Accessed on 20 Sept 2022.

International Union for the Conservation of Nature (IUCN). 2018. The IUCN Red List of Threatened Species. Version 2018-1. https://www.iucnredlist.org. Accessed on 20 Sept 2022.

International Union for the Conservation of Nature (IUCN). 2020. The IUCN Red List of Threatened Species. Version 2020-1. https://www.iucnredlist.org. Accessed on 20 Sept 2022.

International Union for the Conservation of Nature (IUCN). 2021. The IUCN Red List of Threatened Species. Version 2021-1. https://www.iucnredlist.org. Accessed on 20 Sept 2022.

Jenni, L., M. M. Madry, T. Kraemer, J. Kupper, H. Naegeli, H. Jenny, and D. Jenny. 2015. The frequency distribution of lead concentration in feathers, blood, bone, kidney and liver of golden eagles *Aquila chrysaetos*: insights into the modes of uptake. Journal of Ornithology 156:1095–1103.

Johnson, C. K., T. R. Kelly, and B. A. Rideout. 2013. Lead in ammunition: a persistent threat to health and conservation. EcoHealth 10:455–464.

Kålås, J., E. Steinnes, and S. Lierhagen. 2000. Lead exposure of small herbivorous vertebrates from atmospheric pollution. Environmental Pollution 107:21–29.

Kanstrup, N., J. Swift. D. A. Stroud, and M. Lewis. 2018. Hunting with lead ammunition is not sustainable: European perspectives. Ambio 47:846–857.

Katzner, T. E., M. J. Stuber, V. A. Slabe, J. T. Anderson, J. L. Cooper, L. L. Rhea, and B. A. Millsap. 2018. Origins of lead in populations of raptors. Animal Conservation 21:232–240.

Kelly, T. R., P. H. Bloom, S. G. Torres, Y. Z. Hernandez, R. H. Poppenga, W. M. Boyce, and C. K. Johnson. 2011. Impact of the California lead ammunition ban on reducing lead exposure in golden eagles and turkey vultures. PLoS One 6:e17656.

Kelly, T. R., R. H. Poppenga, L. A. Woods, Y. Z. Hernandez, W. M. Boyce, F. J. Samaniego, S. G. Torres, and C. K. Johnson. 2013. Causes of mortality and unintentional poisoning in predatory and scavenging birds in California. Veterinary Record Open 1:e000028.

Kenntner, N., Y. Crettenand, H.-J. Fünfstück, M. Janovsky, and F. Tataruch. 2007. Lead poisoning and heavy metal exposure of golden eagles (*Aquila chrysaetos*) from the European Alps. Journal of Ornithology 148:173–177.

Kitowski, I., D. Jakubas, D. Wiącek, and A. Sujak. 2017. Concentrations of lead and other elements in the liver of the white-tailed eagle (*Haliaeetus albicilla*), a European flagship species, wintering in eastern Poland. Ambio 46:825–841.

Krone, O. 2018. Lead poisoning in birds of prey. Pages 251–272 *in* J. H. Sarasola, J. M. Grande, and J. J. Negro, editors. Birds of prey: biology and conservation in the XXI century. Springer International Publishing, Cham, Switzerland.

Krone, O., N. Kenntner, A. Trinogga, M. Nadjafzadeh, F. Scholz, J. Sulawa, K. Totschek, P. Schuck-Wersig, and R. Zieschank. 2009. Lead poisoning in white-tailed sea eagles: causes and approaches to solutions in Germany. Ingestion of lead from spent ammunition: implications for wildlife and humans. The Peregrine Fund, Boise, Idaho, USA.

Krüger, S. C., and A. Amar. 2018. Lead exposure in the critically endangered bearded vulture (*Gypaetus barbatus*) population in southern Africa. Journal of Raptor Research 52:491–499.

Langner, H. W., R. Domenech, V. A. Slabe, and S. P. Sullivan. 2015. Lead and mercury in fall migrant golden eagles from western North America. Archives of Environmental Contamination Toxicology 69:54–61.

Lohr, M. T., J. O. Hampton, S. Cherriman, F. Busetti, and C. Lohr. 2020. Completing a worldwide picture: preliminary evidence of lead exposure in a scavenging bird from mainland Australia. Science of the Total Environment 715:135913.

Manning, L. K., A. Wünschmann, A. G. Armién, M. Willette, K. MacAulay, J. B. Bender, J. P. Buchweitz, and P. T. Redig. 2019. Lead intoxication in free-ranging bald eagles (*Haliaeetus leucocephalus*). Veterinary Pathology 56:289–299.

Martin, P. A., D. Campbell, K. Hughes, and T. McDaniel. 2008. Lead in the tissues of terrestrial raptors in southern Ontario, Canada, 1995–2001. Science of the Total Environment 391:96–103.

Mateo, R., A. J. Green, H. Lefranc, R. Baos, and J. Figuerola. 2007. Lead poisoning in wild birds from southern Spain: a comparative study of wetland areas and species affected, and trends over time. Ecotoxicology and Environmental Safety 66:119–126.

Mateo, R., and N. Kanstrup. 2019. Regulations on lead ammunition adopted in Europe and evidence of compliance. Ambio 48:989–998.

Mateo-Tomás, P., PM. Jiménez-Moreno, P. R. Camarero, I. S. Sánchez-Barbudo, R. C. Rodríguez Martín-Doimeadios, and R. Mateo. 2016. Mapping the spatio-temporal risk of lead exposure in apex species for more effective mitigation. Proceedings of the Royal Society B Biological Sciences 283:20160662.

Mctee, M., B. Hiller, and P. Ramsey. 2019. Free lunch, may contain lead: scavenging shot small mammals. Journal of Wildlife Management 83:1466–1473.

McWreath, E. M. 2019. Ground ingredients: analysis of lead exposure in the California condor's (*Gymnogyps californianus*) ground foraging habitat, Thesis, West Virginia University, Morgantown, USA.

Naidoo, V., K. Wolter, and C. J. Botha. 2017. Lead ingestion as a potential contributing factor to the decline in vulture populations in southern Africa. Environmental Research 152:150–156.

Nam, D. H., and D. P. Lee. 2009. Abnormal lead exposure in globally threatened cinereous vultures (*Aegypius monachus*) wintering in South Korea. Ecotoxicology 18:225–229.

Nriagu, J. O. 1983. Occupational exposure to lead in ancient times. Science of the Total Environment 31:105–116.

O'Donoghue, B. 2017. R.A.P.T.O.R. Recording and addressing persecution and threats to our raptors. Irish National Parks and Wildlife Service, Dublin, Ireland. https://www.npws.ie/sites/default/files/publications/pdf/2017-raptor-report.pdf. Accessed 21 Dec 2021.

Pain, D. J, A. A. Meharg, M. Ferrer, M. Taggart, and V. Penteriani. 2005. Lead concentrations in bones and feathers of the globally threatened Spanish imperial eagle. Biological Conservation 121:603–610.

Pain, D. J., R. Mateo, and R. E. Green. 2019. Effects of lead from ammunition on birds and other wildlife: a review and update. Ambio 48:935–953.

Pattee, O. H., J. W. Carpenter, S. H. Fritts, B. A. Rattner, S. N. Wiemeyer, J. A. Royle, and M. R. Smith. 2006. Lead poisoning in captive Andean condors (*Vultur gryphus*). Journal of Wildlife Diseases 42:772–779.

Pattee, O. H., and S. K. Hennes. 1983. Bald eagles and waterfowl: the lead shot connection. Transactions of the North American Wildlife and Natural Resources Conference 48:230–237.

Pay, J. M., T. E. Katzner, C. E. Hawkins, A. J. Koch, J. M. Wiersma, W. E. Brown, N. J. Mooney, and E. Z. Cameron. 2021. High frequency of lead exposure in the population of an endangered Australian top predator, the Tasmanian wedge-tailed eagle (*Aquila audax fleayi*). Environmental Toxicology and Chemistry 40:219–230.

Plaza, P. I., M. Uhart, A. Caselli, G. Wiemeyer, and S. A. Lambertucci. 2018. A review of lead contamination in South American birds: the need for more research and policy changes. Perspectives in Ecology and Conservation 16:201–207.

Plaza, P. I., and S. A. Lambertucci. 2019. What do we know about lead contamination in wild vultures and condors? A review of decades of research. Science of the Total Environmental 654:409–417.

Pokorny, B., I. Jelenko, U. Kierdorf, and H. Kierdorf. 2009. Roe deer antlers as historical bioindicators of lead pollution in the vicinity of a lead smelter, Slovenia. Water, Air and Pollution 203:317–324.

Pokras, M. A., and M. R. Kneeland. 2009. Understanding lead uptake and effects across species lines: a conservation medicine based approach. Pages 7–22 *in* R. T. Watson, M. Fuller, M. A. Pokras, and W. G. Hunt, editors. Proceedings, ingestion of lead from spent ammunition: implications for wildlife and humans. The Peregrine Fund, Boise, Idaho, USA.

Redig, P. T., and L. R. Arent. 2008. Raptor toxicology. Veterinary Clinics of North America: Exotic Animal Practice 11:261–282.

Redig, P. T., E. M. Lawler, S. Schwartz, J. L. Dunnette, B. Stephenson, and G. E. Duke. 1991. Effects of chronic exposure to sublethal concentrations of lead acetate on heme synthesis and immune function in red-tailed hawks. Archives of Environmental Contamination and Toxicology 21:72–77.

Reynolds, K. D., M. S. Schwarz, C. A. McFarland, T. McBride, B. Adair, R. E. Strauss, G. P. Cobb, M. J. Hooper, and S. T. McMurry. 2006. Northern pocket gophers (*Thomomys talpoides*) as biomonitors of environmental metal contamination. Environmental Toxicology and Chemistry 25:458–469.

Rideout, B. A., I. Stalis, R. Papendick, A. P. Pessier, B. Puschner, M. E. Finkelstein, D. R. Smith, M. Johnson, M. Mace, R. Stroud, et al. 2012. Patterns of mortality in free-ranging California condors (*Gymnogyps californianus*). Journal of Wildlife Diseases 48:95–112.

Safford, R., J. Andevski, A. Botha, C. G. R. Bowden, N. Crockford, R. Garbett, A. Margalida, I. Ramírez, M. Shobrak, J. Tavares, et al. 2019. Vulture conservation: the case for urgent action. Bird Conservation International 29:1–9.

Saggese, M. D., A. Quaglia, S. A. Lambertucci, M. S. Bo, J. H. Sarasola, R. Pereyra-Lobos, and J. J. Maceda. 2009. Survey of lead toxicosis in free-ranging raptors from central Argentina. Pages 223–231 *in* RT Watson, M. Fuller, M. Pokras, and W. G. Hunt, editors. Proceedings, ingestion of lead from spent ammunition: implications for wildlife and humans. The Peregrine Fund, Boise, Idaho, USA.

Saito, K. 2009. Lead poisoning of Steller's sea-eagle (*Haliaeetus pelagicus*) and whitetailed eagle (*Haliaeetus albicilla*) caused by the ingestion of lead bullets and slugs. Pages 302–309 *in* RT Watson, M. Fuller, M. Pokras, and W. G. Hunt, editors. Proceedings, Ingestion of lead from spent ammunition: implications for wildlife and humans. The Peregrine Fund, Boise, Idaho, USA.

Samour, J. H., and J. Naldo. 2002. Diagnosis and therapeutic management of lead toxicosis in falcons in Saudi Arabia. Journal of Avian Medicine and Surgery 16:16–20.

Sanderson, G. C., W. L. Anderson, G. L. Foley, S. P. Havera, L. M. Skowron, J. W. Brawn, G. D. Taylor, and J. W. Seets. 1998. Effects of lead, iron, and bismuth alloy shot embedded in the breast muscles of game-farm mallards. Journal of Wildlife Diseases 34:688–697.

Scheuhammer, A. 1987. The chronic toxicity of aluminium, cadmium, mercury, and lead in birds: a review. Environmental Pollution 46:263–295.

Sheppard, J. K., M. Walenski, M. P. Wallace, J. J. V. Velazco, C. Porras, and R. R. Swaisgood. 2013. Hierarchical dominance structure in reintroduced California condors: correlates, consequences, and dynamics. Behavior, Ecology and Sociobiology 67:1227–1238.

Slabe, V. A.. 2019. Lead exposure of North American raptors. Ph.D. Dissertation, West Virginia University, Morgantown, USA.

Smits, J., and V. Naidoo. 2018. Toxicology of birds of prey. Pages 229–250 *in* J. Sarasola J. Grande, and J. Negro, editors. Birds of prey. Springer, Cham, Switzerland.

Specht, A. J., C. N. Parish, E. K. Wallens, R. T. Watson, L. H. Nie, and M. G. Weisskopf. 2018. Feasibility of a portable X-ray fluorescence device for bone lead measurements of condor bones. Science of the Total Environment 615:398–403.

Stauber E, N. Finch, P. A. Talcott, and J. M. Gay. 2010. Lead poisoning of bald (*Haliaeetus leucocephalus*) and golden (*Aquila chrysaetos*) eagles in the US Inland Pacific

Northwest Region—an 18-year retrospective study: 1991–2008. Journal of Avian Medicine and Surgery 24:279–287.

Torimoto, R., C. Ishii, H. Sato, K. Saito, Y. Watanabe, K. Ogasawara, A. Kubota, T. Matsukawa, K. Yokoyama, A. Kobayashi, et al. 2021. Analysis of lead distribution in avian organs by LA-ICP-MS: study of experimentally lead-exposed ducks and kites. Environmental Pollution 283:117086.

Tranel, M. A., and R. O. Kimmel. 2009. Impacts of lead ammunition on wildlife, the environment, and human health—a literature review and implications for Minnesota. Pages 318–337 *in* R. T. Watson, M. Fuller, M. A. Pokras, and W. G. Hunt, editors. Proceedings, ingestion of lead from spent ammunition: implications for wildlife and humans. The Peregrine Fund, Boise, Idaho, USA.

US Energy Information Administration. 2020. Gasoline and the environment: leaded gasoline. US Energy Information Administration, https://www.eia.gov/energyexplained/gasoline/gasoline-and-the-environment-leaded-gasoline.php. Accessed 19 Nov 2021.

van den Heever L., H. Smit-Robinson, V. Naidoo, and A. E. McKechnie. 2019. Blood and bone lead levels in South Africa's *Gyps* vultures: risk to nest-bound chicks and comparison with other avian taxa. Science of the Total Environment 669:471–480.

Walters, J. R., S. R. Derrickson, D. M. Fry, S. M. Haig, J. M. Marzluff, and J. M. Wunderle Jr. 2010. Status of the California condor (*Gymnogyps californianus*) and efforts to achieve its recovery. The Auk 127:969–1001.

Warner, S. E., E. E. Britton, D. N. Becker, and M. J. Coffey. 2014. Bald eagle lead exposure in the Upper Midwest. Journal of Fish Wildlife Management 5:208–216.

West, C. J., J. D. Wolfe, A. Wiegardt, and T. Williams-Claussen. 2017. Feasibility of California condor recovery in northern California, USA: contaminants in surrogate turkey vultures and common ravens. Condor 119:720–731.

Wiemeyer, G. M., M. A. Pérez, L. Torres Bianchini, I. Sampietro, G. F. Bravo, N. L. Jácome, V. Astore, and S. A. Lambertucci. 2017. Repeated conservation threats across the Americas: high levels of blood and bone lead in the Andean condor widen the problem to a continental scale. Environmental Pollution 220:672–679.

Wynne, J., and C. Stringfield. 2007. Treatment of lead toxicity and crop stasis in a California condor (*Gymnogyps californianus*). Journal of Zoo and Wildlife Medicine 38:588–590.

Yaw, T., K. Neumann, L. Bernard, J. Cancilla, T. Evans, A. Martin-Schwarze, and B. Zaffarano. 2017. Lead poisoning in bald eagles admitted to wildlife rehabilitation facilities in Iowa, 2004–2014. Journal of Fish and Wildlife Management 8:465–473.

Zhang, Z. W., S. Shimbo, N. Ochi, M. Eguchi, T. Watanabe, C.-S. Moon, and M. Ikeda. 1997. Determination of lead and cadmium in food and blood by inductively coupled plasma mass spectrometry: a comparison with graphite furnace atomic absorption spectrometry. Science of the Total Environment 205:179–187.

21 Gyrfalcons as Sentinels for Changing Disease Ecology in the Arctic

ROBIN W. RADCLIFFE, MICHAEL T. HENDERSON

Introduction

The Arctic and Rapid Changes in the Far North

The Arctic tundra supports some of the most charismatic and emblematic taxa on earth, many of which are threatened by the accelerated climate change occurring at the northernmost latitudes. Arctic temperatures are warming two to three times faster than the global average, a trend that is likely to accelerate with time (Bekryaev et al. 2010, IPCC 2021). Climate change is the top conservation concern in Arctic ecosystems (Thomas et al. 2004, Bradley et al. 2005, Gilg et al. 2012). Warming temperatures have precipitated the melting of permafrost and sea ice (Stroeve et al. 2007), shifted species ranges northward (Callaghan et al. 2004, Wheeler et al. 2018), altered predator–prey dynamics (Ims and Fuglei 2005, Gilg et al. 2009), changed precipitation profiles (Min et al. 2011), and reconfigured landscapes (Sturm et al. 2001; Tape et al. 2006, 2016).

These climate change driven phenomena are altering Arctic disease ecology, exposing wildlife to novel diseases and disrupting host–pathogen dynamics (Bradley et al. 2005, Jenkins et al. 2006, Hoberg and Brooks 2015). Pathogens and disease are pivotal

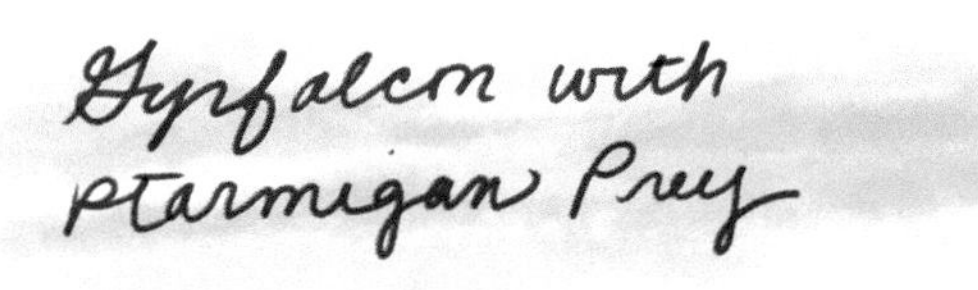

Illustration by Laura Donohue.

components of ecology and can act as ecosystem engineers by affecting nutrient cycles, biodiversity, host behavior, predation risk, energy budgets, and competition (Hatcher et al. 2012). Diseases can benefit ecosystem function by regulating population growth and decreasing interspecific competition for resources. Effects of diseases on individuals are typically either neutral (e.g., asymptomatic carriers) or, more frequently, negative, particularly when hosts are naïve to a pathogen (LaPointe et al. 2012). Naïve hosts, or those that lack appropriate immune and behavioral responses to a pathogen, are particularly susceptible to population declines, extirpation, and extinction (Atkinson and LaPointe 2009). The naïveté of Arctic wildlife to diseases is driven by a lack of historical exposure and, thus, selective pressure to develop adaptive behaviors or immunity to counter a pathogen. This may be compounded by low genetic diversity, both in general and at specific loci responsible for an immune response (Callaghan et al. 2004, Weber et al. 2013). Genetic variability, or lack thereof, is critical in generating functional responses to emergent diseases and changes to host–agent–environment dynamics occurring in the Arctic (O'Brien and Evermann 1988, Van Hemert et al. 2014). In general, warmer and wetter weather leads to greater pathogen diversity and abundance (Epstein 2002), facilitating the northward expansion of pathogens as new areas conform to their thermal niche (e.g., Aleuy and Kutz 2020). More specifically, Arctic climate change alters disease transmission by (1) affecting development rates, mortality, and reproduction of pathogens and (2) affecting survival, reproduction, and distributions of vectors and hosts (both intermediate and definitive; Dobson et al. 2015).

Empirical examples of changing disease ecology in the Arctic are sparse, but several well-studied systems suggest apparent trends. One example is the movement of the protostrongylid nematode, *Umingmakstrongylus pallikuukensis*, a muskox lungworm recently discovered on Canada's Victoria Island. As adjacent areas conformed to the thermal niche of the lungworm, the extent of the parasite's range expanded to fill that space (Kutz et al. 2013, Kafle et al. 2020). This range expansion appears to have been facilitated by longer and warmer summers allowing a complete life cycle in a single year, in contrast to the historical pattern of a two-year cycle (Kutz et al. 2002, 2005). Historically, overwinter mortality of intermdeiate hosts and developing larvae reduced parasite abundance for the subsequent season; thus, the more rapid development has increased infection pressure, a pattern that is likely to continue (Pickles et al. 2013). The diversity and abundance of pathogens are also changing for Arctic wildlife, like polar bears (*Ursus maritimus*), which are facing increased pressure from *Francisella tularensis*, *Bordetella bronchiseptica*, and *Toxoplasma gondii* (Pilfold et al. 2021). Altered host distributions have substantially impacted disease ecology, such as the intentional introduction of elk (*Cervus elaphus*) into the south-central Yukon, which provided an important reservoir host for *Dermacentor albipictus* and led to parasitic pressure on native species, particularly moose (*Alces alces*) and caribou (*Rangifer tarandus*; Kutz et al. 2009). Although trends can be deduced from these examples, many aspects of Arctic disease ecology of critical concern to species conservation remain poorly understood.

Gyrfalcons as Sentinels of Arctic Health

Sentinel, or indicator, species are often used as proxies for ecosystem function and to delineate changes within an ecosystem because of their unique ecology (Bal et al. 2018, Siddig et al. 2016). Researching disease ecology is logistically difficult, time consuming, and expensive (particularly in remote areas like the Arctic), because biological samples (e.g., serum or fecal) frequently require live-animal captures and specialized skills and equipment to collect and analyze samples. It is implausible to collect biological samples from all taxa within the Arctic system; thus, sentinel species are an efficient tool to understand the changing disease ecology (Schuler et al. 2021).

Focusing research on sentinel species can reduce the scale, intricacy, and expense of investigations. Effective sentinel species in the Arctic are frequently habitat specialists with a focused niche, apex predators that rely on (and may reflect the health of) lower trophic levels, and year-round residents that are linked to the system across annual cycles (Sergio et al. 2008; Christensen et al. 2013; Figure 21.1).

The gyrfalcon (*Falco rusticolus*) has been identified as an effective sentinel species, because it is an

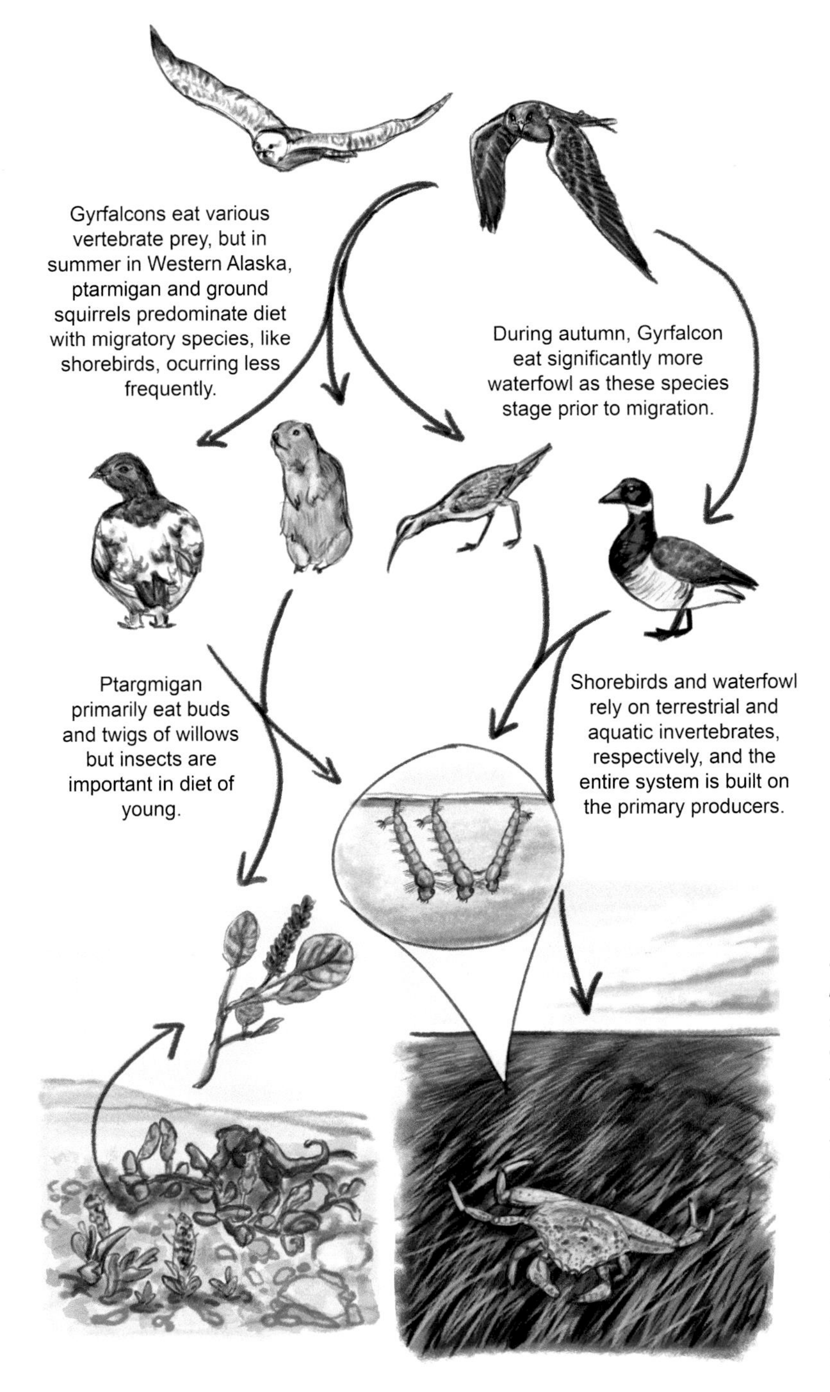

Fig. 21.1. Gyrfalcon (*Falco rusticolus*) is an apex predator and sentinel species of the Arctic ecosystem. The gyrfalcon sits on the top of the Arctic food web, preying on birds and mammals that, in turn, rely on invertebrates and primary producers (plants). Since a disruption anywhere in the food web will impact the top predator (the gyrfalcon), they make a fitting sentinel species for Arctic ecosystem health. Illustration by Laura Donohue.

Arctic specialist, top avian predator, and year-round resident (White and Springer 1965, Christensen et al. 2013). Further, the gyrfalcon is among the bird species most threatened by climate change and is notoriously susceptible to novel diseases (Liebezeit et al. 2012, e.g., USGS 2020). This may be driven in part by their low genetic diversity (Johnson et al. 2007). In this chapter, we focus on gyrfalcon health and how changes in pathogen abundance and emergence patterns may indicate changes in Arctic disease ecology. By surveying gyrfalcon health, we can shed light on emerging disease threats for gyrfalcons, their competitors, and prey.

Gyrfalcons have a circumpolar distribution and populations are strongly linked to rock and willow ptarmigan (*Lagopus muta* and *L. lagopus*, respectively) abundance (Nielsen and Cade 2017), but their dietary niche expands to include shorebirds, seabirds, waterfowl, and small mammals during summer and fall (Nielsen and Cade 1990, Robinson et al. 2019). Disease threats to gyrfalcons are likely shared with other predators, because there is substaintial overlap in the dietary niche of the Arctic raptor guild (Johnson et al. 2020) and many mammalian predators (e.g., Cotter et al. 1992). The parasitic fauna of gyrfalcons has only been described in Iceland, and little is known for other parts of their distribution, including North America. Veterinary literature from captive raptors held for falconry can be particularly helpful, because it can shed light on novel disease pathology and offer context for how wild populations are likely to fare if exposed to the same pathogen (Redig 1992, Redig et al. 1993; Table 21.1).

Alaskan Gyrfalcon Health and Disease Monitoring Program

A collaborative effort of the Cornell University College of Veterinarian Medicine, the Alaska Department of Fish and Game, and The Peregrine Fund is using gyrfalcons as a sentinel species to understand changing disease ecology over time by comparing new data with two decades of archival samples (Henderson et al. 2021). A long-term monitoring program for gyrfalcons on Alaska's Seward Peninsula revealed that falcons at this site are among the few declining populations in the world (Franke et al. 2020); increased knowledge of health and disease could direct further research on apex predator declines in the Arctic. Given that dietary habits can influence transmission of pathogens, a second field site was established in the Izembek National Wildlife Refuge on the remote Alaska Peninsula (a key waterfowl flyway), where gyrfalcons are exposed to novel pathogens from feeding on diverse migratory prey (Figure 21.2).

The Alaska Gyrfalcon Health Program collects information in four key areas: (1) traditional biological and ecological data on individual birds; (2) health data through physical examination and targeted collection of biological samples; (3) spatial geographical data by GPS telemetry and collection of information across widely dispersed field sites; and (4) temporal data through archival sampling across time and demographic comparisons of health between nestling and adult birds (Table 21.2). Our selection of diseases and parasites was based on knowledge gleaned from falconry medicine and parasite surveys from the retrieval of dead gyrfalcons in Iceland (Ó. K. Nielsen, Icelandic Institute of Natural History, personal communication). Preliminary findings reveal widespread coccidia in nestling falcons, mites at eyrie nest sites, high levels of exposure to *Salmonella* in both young and adult birds, oral plaques from parasites more prominent in the Seward Peninsula population, and identification of a novel host for the malaria parasite, *Plasmodium unalis*, in a gyrfalcon nestling from the Seward Peninsula (R. W. Radcliffe, et al., Cornell University, unpublished data).

Disease Transmission and Climate Change

Disease transmission can be categorized as vector borne—transmitted by blood-feeding arthropods (e.g., mosquitos, ticks, and fleas)—or by direct

Table 21.1. Overview of probable disease risks for gyrfalcons (*Falco rusticolus*; gyrs) from a review of the literature, including reports from captive falcons used in falconry and surveys of wild populations

Disease	Agent	Health implications	Wild (W) or captive (C)	Reference
Avian malaria	*Plasmodium relictum*	Acute death or recovery with treatment	W, C	Kingston et al. 1976
Frounce	*Trichomonas gallinae*	Primary carrier is domestic pigeons	W, C	Stabler 1956; Stabler and Hamilton 1954
Capillariasis	*Eucoleus contorta*	Pathology survey of Icelandic gyrs; 36% infected with *Capillaria*	W (Iceland)	Trainer et al. 1968; Clausen and Gundmundsson 1981
Endoparasites and ectoparasites	Multiple	Pathology survey of 46 gyrs in Iceland: *Capillaria contorta* 76% (Frounce-like lesions 43%, correlated with poor body condition but not age); *Mesocestoides* 27%; *Cryptocotyle* trematodes 4–8%; *Mallophagons* (90%); Astigmatan feather mite (47%); *Ixodes caledonicus* tick 20%	W (Iceland)	Christensen et al. 2015
		I. howelli nymphal ticks on gyr chicks Significance: *I. howelli* closely related to Russian tick (*I. berlesi*) suggesting communication between Alaskan and Russian gyr populations	W (Alaska)	White and Springer 1965
Serratospiculiasis	*Serratospiculum* sp.	Incidental finding in gyrs	W (Iceland)	Nielsen, personal communication
		Pneumonia and airsacculitis in falcons	C (Middle East)	Samour and Naldo 2001
			C (USA)	Ward and Fairchild 1972
Avian influenza	Highly pathogenic avian influenza virus	Adult gyr mortality from eating wild duck	C (falconry)	Ip et al. 2015
		Avian influenza movement with birds from Asia to Alaska	W	Ramey et al. 2015
Falcon herpes	Falconid herpesvirus-1	Falcons infected by consuming pigeons	C (falconry)	Raghav and Samour 2019
West Nile virus (WNV)	WNV	Infection by mosquito vector or eating prey	W, C	Nemeth et al. 2006
		WNV antibodies passed egg to chick		Stout et al. 2005
Avian cholera	*Pasteurella multocida*	Gyrfalcon found dead after feeding on infected waterfowl	C (falconry)	Williams et al. 1987
Bacterial flora	Multiple	Normal oral and cloacal flora;	W (Greenland)	Cooper et al. 1980
		Pasteurella anatipestifer in peregrine falcons; *Salmonella* and *Chlamydia* in gyrs	C (UAE)	Wernery et al. 1998
		Salmonella typhimuriam seropositive gyrs	W (Alaska)	Henderson et al. 2021
Fungal pneumonia	*Aspergillus fumigatus*	Historic sensitivity in northern raptors	C (falconry)	Rahim et al. 2013; Hamilton and Stabler 1953

contact with an infected individual (e.g., parent to offspring, in flocks, or from prey to predator). Both vector-borne diseases and those directly transmitted are evolving with climate change, primarily driven by temperature changes and altered ecosystem dynamics. New pathogens, vectors, and novel hosts may be introduced into the Arctic, some via increased shipping traffic facilitated by reduced sea ice (Mudryk

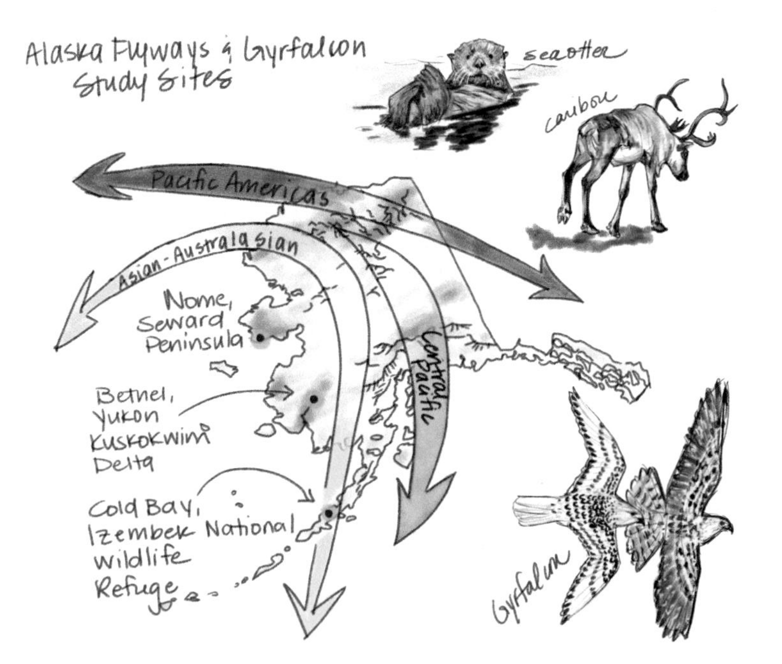

Fig. 21.2. Map of western Alaska delineating Gyrfalcon Health Program study areas. Historical samples were collected on the Yukon–Kuskokwim Delta (61.349186, −160.971789). Historical and contemporary samples are collected from the Seward Peninsula (Nome; 64.729649, −165.352799) near the current northern limit of avian malaria and at the intersection of several migratory pathways. Contemporary samples are collected from the Izembek National Wildlife Refuge (Cold Bay; 55.291630, −162.690932), where gyrfalcons (*Falco rusticolus*) feed predominantly on migratory waterfowl. These study sites leverage existing research infrastructure that allow sampling of gyrfalcons with unique diets across a significant latitudinal gradient within western Alaska. Illustration by Laura Donohue.

Table 21.2. Collaborative Alaska Gyrfalcon Health Program: overview of tests, methods, and implications

Disease or pathogen (sample type)	Analysis method[a]	Implications or importance
Parasites		
Endoparasites		
Coccidia, nematodes (mute/fecal sample)	Parasitology (float and PCR)	Delineate temporal and spatial patterns of endoparasite exposure
Trichomoniasis (oral exam for plaques)	Microscopy and PCR	Oral plaques indicate an emergent disease for Arctic raptors
Capillariasis (oral exam for plaques)	Microscopy and PCR	
Hemoparasites (blood smears/whole blood)		
Avian malaria, *Haemoproteus*, *Leucocytozoon*	Microscopy and PCR	Indicate trends for important vector-borne pathogens
Ectoparasites (collection from bird or isolation in eyrie mutes)	Microscopy and PCR	Describe external parasite fauna and possible vectors
Bacteria	Serology	
Salmonellosis (*S. typhimurium*; plasma)	Tube agglutination	Exposure to pathogens from domestic poultry or wild birds
Viruses		
Avian influenza (plasma)	AGID	Indicates transmission from migratory prey to residents
West Nile virus (plasma)	PRNT	Emerging disease in North America
Avian paramyxovirus (plasma)	HI	Exposure to pathogens from domestic poultry or wild birds
Falcon adenovirus-1 (plasma)	AGID	Emerging disease in falcons

Data span both time (temporal) and geographic (spatial) scales by comparison of archival and contemporary samples across geographic field sites.
[a] Analysis method abbreviations are polymerase chain reaction (PCR), agar gel immunodiffusion assay (AGID), plaque reduction neutralization test (PRNT), and hemagglutination-inhibition test (HI).

et al. 2021). Below we examine vector-borne diseases and those that spread through direct contact, highlight specific examples, and discuss how they relate to climate change. We also examine how gyrfalcon research can help elucidate disease emergence for the Arctic system and highlight what we know about gyrfalcon health, with a primary reliance on knowledge gained from Icelandic studies and the falconry medicine literature.

Part 1: Vector-borne Disease and Arctic Climate Change

Vector-borne disease agents may affect the health of Arctic species, including the gyrfalcon (e.g., haemosporidian blood parasites—avian malaria, *Plasmodium* spp., and leucocytozoons), and their distributions and abundance are likely to expand with increasing temperatures in Arctic and subarctic regions. Warmer temperatures favor vectors, such as *Ixodes scapularis*, the primary tick vector for the pathogen *Borrelia burgdorferi* that causes Lyme disease, which is expanding northward at an estimated 46 km each year (Ogden et al. 2008, Leighton et al. 2012). Although temperature is the primary factor that limits tick populations, the long-distance movement of ticks is facilitated by spring movement of migratory birds, some of which have destinations beyond the current suitable range for *I. scapularis*, suggesting expansion is likely to continue (Ogden et al. 2015). Migratory birds from around the globe breed in the Arctic, potentially carrying vectors that could transmit West Nile or other novel viruses, avian blood parasites, and other pathogens.

Hematozoans: Avian Malaria

Avian malaria is caused by protozoan parasites of the *Plasmodium* genus that infect and destroy red blood cells of birds, resulting in anemia (see Chapter 19). Gross lesions include a dark, swollen liver and spleen, and thin, watery blood. The clinical signs of avian malaria include lethargy, weakness, and weight loss. The primary prevention methods are mosquito control (e.g., eliminating larval stages in pooling water, mosquito trapping, sterile-insect release, and reproduction control). Confirmation of avian malaria can be accomplished inexpensively by direct observation of parasites inside nucleated red blood cells, but up to 67% of chronic infections can be missed on blood smears. Molecular methods are more sensitive and provide accurate species identification. Serological analyses can also detect *Plasmodium* antibodies via enzyme-linked immunosorbent assay and Western blot (Atkinson et al. 2001).

Significant effects on wild populations can result from increasing exposure to *Plasmodium* organisms, particularly when species are naïve to the disease (LaPointe et al. 2012). *Plasmodium relictum* is the most virulent, especially in passerines. *Plasmodium* spp. are typically vectored by *Culex* spp. mosquitos, which are common in continental areas in both tropical and temperate climates. The devastating effect of avian malaria is evidenced by the extinction of 75% of Hawaiian honeycreeper species (subfamily Drepanidinae), following the introduction of *Culex quinquefasciatus* into the Hawaiian archipelago (Atkinson et al. 1995, Atkinson and LaPointe 2009).

Like the honeycreepers, many Arctic wildlife species are naïve to avian malaria because *Culex* and *Plasmodium* hemoparasites have historically been largely absent from the Arctic. Recent research in Alaska revealed two mosquito genera (*Culiseta* and *Aedes*) capable of transmitting *P. circumflexum* with *Culiseta* spp. being the better, though less abundant, vector throughout the state (Smith et al. 2019). *Plasmodium circumflexum* is highly generalized (able to infect individuals across families and even orders of birds) and cold adapted. *Plasmodium circumflexum* is now able to complete its life cycle up to 64°N (Loiseau et al. 2012, Salakij et al. 2012) and infect passerines, waterfowl, shorebirds, and raptors. Avian malaria can be exacerbated by comorbidities, such as avian keratin disorder, an emergent disease characterized by beak deformities (Wilkinson et al. 2016). Expansion of avian malaria into the Arctic represents a substantial threat to the millions of birds that breed there and likely an even greater threat to resident species.

Gyrfalcons are exceptionally vulnerable to avian malaria (Remple 2004), making them an ideal sentinel for early detection of *Plasmodium* spp. Gyrfalcons are primarily year-round residents; so avian malaria infections would represent local transmission and not lingering infections originating on a southern wintering ground. Gyrfalcon nests attract mosquitos (M. T. Henderson, The Peregrine Fund, personal observation); thus, if *Plasmodium* and a competent vector are present, nestlings are likely to contract the disease as they are unable to escape. Kingston et al. (1976) discovered a gyrfalcon in remote Alaska that died from fulminating malaria (*P. relictum*) of unknown origin, but few other examples exist for North American populations. A captive gyrfalcon and two peregrine falcons (*Falco peregrinus*) were concurrently exposed to *P. relictum* and significantly higher parasitemia of red-blood cells was observed in the gyrfalcon compared to the peregrines (16% and 0.01%, respectively), although all falcons recovered after treatment with chloroquine (Kingston et al. 1976). Captive gyrfalcons infected with avian malaria exhibit a loss of stamina and listlessness that can progress to dyspnea and sudden death when restrained for blood sampling (Kingston et al. 1976, Remple 2004). Successful treatment typically consists of chloroquine and primaquine to treat the blood and tissue phases respectively, neither of which are applicable to wild populations. Loiseau et al. (2012) demonstrated that transmission of *Plasmodium* spp. is possible within the gyrfalcon's breeding range, and thus understanding the current and future dynamics of avian malaria is critical for gyrfalcons and possibly other Arctic bird species.

Hematozoans: *Leucozytozoon* and *Haemoproteus*

Leucocytozoon and *Haemoproteus* are relatively benign hemosporidian parasites of wild and domestic birds; black flies of the genus *Simulium* are vectors for *Leucocytozoon*, and hippoboscid flies (*Ornithomyia*) or biting midges (*Culicoides*) vector *Haemoproteus*. The parasites are found globally wherever the vectors exist, including the Arctic. Although more than 60 species of leucocytozoons are known, only a single species, *Leucocytozoon toddi*, is known to infect the Falconiformes. Infections with *Leucocytozoon* spp. and *Haemoproteus* spp. have been described in raptors with little evidence of pathogenicity, although infection with one group may predispose to infections with other hematozoans (Meixell et al. 2016). In the same study, ducks concurrently infected with avian influenza virus did not have higher hemoparasite burdens. It is generally accepted that disease occurs primarily in young birds or those under stress as a result of megaloschizonts invading various tissues. Diagnosis is made by observation of the gametocytes in peripheral blood smears with the large parasites distorting both the red and white blood cells (Leucocytozoons were once thought to only infect white blood cells—hence the name).

There are no reports of Leucocytozoons in gyrfalcons, though a recent study of blood parasites in Alaskan grouse found the highest prevalence (96%) of leucozytozoonosis in Seward Peninsula ptarmigan—a critical prey species and breeding area for gyrfalcons in Alaska (Smith et al. 2016). Although Leucocytozoons are family specific with falcons having different parasites than those of their prey, infection in ptarmigan may help inform the route of infection and role of disease in young falcons. *Haemoproteus* infections vary with age and geography and are so common in saker falcons (*Falco cherrug*) in the Middle East that their absence is noteworthy (Remple 2004; Lierz et al. 2008). Naïve hosts are likely to be most affected by novel infections or coinfections. For example, severe anemia occurs in juvenile snowy owls (*Nyctea scandiaca*) and in adults held in captivity outside the Arctic. As host–parasite relationships shift with climate change, normally innocuous hematozoans may find new hosts and cause disease—this "reshuffling" of host–parasite interactions is similar to the disease observed when healthy animals are translocated outside of their historic range and exposed to novel pathogens.

Part 2: Pathogens of Predatory Lifestyle: Shared Diseases with Prey

Infections of solitary predators, such as gyrfalcons, can be acquired from infected prey, and dietary choices can dictate exposure risk to specific diseases (Mørk et al. 2019, Malmberg et al. 2021; Figure 21.3). Buhler et al. (2020) demonstrated prey-to-predator transmission of *Bartonella* spp. (causal agent of cat scratch fever) from migratory geese to Arctic foxes (*Vulpes lagopus*) when they discovered that a nest flea (*Ceratophyllus vagabundus vagabundus*) could transmit the bacteria during predation events. It is noteworthy that this transmission occurred between different taxonomic classes. Tapeworms in the genus *Mesocestoides* are shared among both predator and prey in the Arctic and may represent the only known parasitic infection passed from the ptarmigan to the gyrfalcon (Ó. K. Nielsen, personal communication). Remarkably, gyrfalcon and ptarmigan share very few diseases despite a linked existence for millennia—but these evolutionary adaptive defense mechanisms between predator and prey (and host and pathogen) are likely to be disrupted as landscapes and species guilds change with the climate.

With rapid climate change in the Arctic, the evolutionary processes required for physiological adaptation are likely too slow; thus, species are forced to respond by altering phenotypically plastic traits, such as distributions, phenology, and dietary habits, all of which are already occurring (Moe et al. 2009, Corrigan et al. 2011, Ehrich et al. 2012, Tape et al. 2016). These phenotypic shifts are likely to alter the disease transmission dynamics and expose naïve wildlife to

Fig. 21.3. Gyrfalcon (*Falco rusticolus*) disease risks within the Arctic landscape. Likely avenues for gyrfalcon disease transmission are the ingestion of infected prey or infection via vectors, such as mosquitoes. As the Arctic is transformed by climate change, the distribution of prey species, pathogens, and vectors shifts, changing the disease ecology for gyrfalcons and other wildlife. Illustration by Laura Donohue.

novel diseases. Prey-to-predator transmission may be further amplified, because predation risk can be greater for infected prey as a result of reduced fitness (Hudson et al. 1992). Long-distance migrants, like waterfowl, likely represent a significant exposure risk for Arctic predators, since they can host various pathogens from around the world, congregate in large numbers, and are important reservoirs for ecologically important diseases (Seekings et al. 2021, Tanikawa et al. 2021).

Avian Influenza ("Bird Flu")

Avian influenza (AI) is an enzootic viral disease caused by infection with the highly contagious avian influenza Type A viruses, which exhibit great potential for *in situ* recombination (i.e., highly variable and prone to mutation; see Chapter 18). Avian influenza can be classified as low-pathogenic (LPAI) or highly pathogenic (HPAI) and can present with mild to severe symptoms, including acute death. Notable gross lesions include multifocal areas of necrosis in the heart, spleen, pancreas, and brain (Causey and Edwards 2008). Avian influenza infects various bird species (and occasionally mammals, including humans) with water birds, particularly waterfowl, being important reservoirs. During HPAI epizootics, diagnosis is typically made in dead birds by a positive immunohistochemical staining of influenza A virus antigen in areas of necrosis. Alternate diagnostics include molecular detection or virus isolation of HPAI virus from oropharyngeal or cloacal swabs or tissues. There is no treatment for HPAI other than supportive care. Highly pathogenic AI emerges sporadically in poultry and can spill back into wild bird populations with lineages such as the Eurasian H5 clade 2.3.4.4 becoming more common in wild birds since ca. 2010 (Ramey et al. 2022). Inter- and intracontinental long-distance movements of HPAI can be facilitated by migratory birds, particularly during spring migration. Low-pathogenic AI circulates year-round in wild birds. Many migratory birds breed in the Arctic where avifauna commingle and provide ample opportunities for transmission and recombination of AI. Changes in migratory phenology and patterns are forming novel assemblages from around the world that are likely to determine emergence of future strains of AI (Lehikoinen and Jaatinen 2012, Adde et al. 2020). Warmer temperatures may facilitate virus survival in bodies of water, increasing persistence, pathogenicity, and transmissibility of AI within the Arctic (Morin et al. 2018).

Migratory birds are an important resource for Arctic predators (including humans) that may be susceptible to AI, especially in late summer and fall when juvenile waterfowl and shorebirds are plentiful and LPAI is circulating in birds throughout much of the Arctic (e.g., Uher-Koch et al. 2019, Van Hemert et al. 2019). The coalescing of distant populations, changing system dynamics, and immeasurable uncertainty inherent in the evolutionary pathway of AI make it an important health concern for Arctic wildlife.

Climate change may alter AI dynamics in gyrfalcons via changes in the distribution of prey species. Rock ptarmigan in Iceland host a suite of pathogens including multiple ectoparasites and endoparasites (e.g., the chewing louse, *Amyrsidea lagopiprevalence*; the nematode, *Capillaria caudinflata*; and the skin mite, *Metamicrolichus islandicus*); however, infection by the coccidian protozoan *Eimeria muta* appears to have the greatest effect on ptarmigan survival, breeding success, and population density (Skirnisson et al. 2012, Stenkewitz et al. 2016). Ptarmigan populations are now undergoing significant range shifts (including reduced overlap with the range of gyrfalcons; Booms et al. 2011), population declines, and genetic isolation (Bech et al. 2009; Revermann et al. 2012; Melin et al. 2020, Scridel et al. 2021). Gyrfalcon diet is plastic (Robinson et al. 2019), and they will select alternate prey if ptarmigan are absent or less availabile. Dietary shifts may change gyrfalcon disease exposure, particularly if the alternative prey are migratory waterfowl (Nielsen 2003) harboring AI.

Gyrfalcons, like all raptors, are highly susceptible to HPAI (Bertran et al. 2012), although there is no evidence that LPAI is pathogenic for gyrfalcons (Lierz et al. 2007). The high morbidity of HPAI and remote habitat of gyrfalcons makes studies of AI in wild populations difficult, but experience from captive birds is suggestive. During the 2014–2015 HPAI epizootic in North American poultry (Ip et al. 2015), a captive gyrfalcon (and three other falcons) suddenly died after feeding on an American widgeon (*Anas americanus*) in Washington state (USGS 2020). The gyrfalcon displayed vomiting, lethargy, and decreased appetite prior to death and gross lesions consisted of severe congestion and hemorrhage in the lungs and other tissues. Highly pathogenic AI H5N8 (Eurasian H5 clade 2.3.4.4 lineage, A/gyrfalcon/Washington/41088-6/2014 (H5N8); GenBank taxon no. 1589663) was isolated from the gyrfalcon's brain. In general, wild raptors with HPAI are invariably found dead with gross lesions consistent with multiorgan failure (Shearn-Bochsler et al. 2019). Vaccination of captive birds is effective at preventing mortalities, suggesting that LPAI could potentially "prime" gyrfalcons for HPAI, reducing pathobiological effects, depending on antigenic similarity of viruses, exposure route, and viral load to which a bird is exposed (Lierz et al. 2007, Bertran et al. 2012). It is unknown if LPAI circulates or recombines in gyrfalcons, but a survey effort on the Yukon–Kuskokwim Delta, Alaska, did not detect any AI in gyrfalcon nestlings (T. L. Booms, Alaska Department of Fish and Game, personal communication). Understanding AI transmission, distribution, and pathogenicity has important conservation and economic concerns, since HPAI outbreaks can significantly impact poultry production and food security for indigenous people. Avian influenza serology in gyrfalcons can provide a better understanding of transmission pathways in the Arctic and serve as an early warning indicator for outbreaks of HPAI.

Frounce

Frounce was historically described in falconry literature as early as the 17th century (Latham 1615; Figure 21.4). It is a disease of the mouth and throat caused by the protozoan parasite, *Trichomonas gallinae* (Eberly 1969). More recently, the term Frounce has been applied broadly to describe the clinical presentation of caseous white to yellow plaques in the oral cavity that could be caused by either *T. gallinae* or by a nematode parasite of the Capillarinae family (Trainer et al. 1968).

Trichomoniasis (*T. gallinae*)

The causal agent of trichomoniasis is *T. gallinae*, a small (5–20 μm) single-celled protozoan with four whip-like anterior flagella, an axostyle protruding from the posterior end, and a fin-like undulating membrane (Stabler 1956). Trichomonads have a direct life cycle, reproduce by binary fission, and birds are its only natural host, although mammals have been experimentally infected. Trichomoniasis disease is characterized by both avirulent and virulent strains of *T. gallinae*, with virulent strains causing significant pathology. Both wild and captive gyrfalcons infected with *T. gallinae* develop the classic oral plaques of Frounce (Christensen et al. 2015, Hamilton and Stabler 1953, Stabler and Hamilton 1954). Clinical signs of trichomoniasis vary widely but include emaciation, difficulty closing the mouth, repeated swallowing, and difficulty eating or drinking—driven by esophagus obstruction that can lead to starvation or suffocation. Acute lesions can vary from a creamy white, sticky exudate to pinpoint raised plaques or ulcers in the oral mucosa. Chronic lesions change to a yellow color with caseous masses often invading the beak, eyes, and even the brain with occasional infiltration of the liver and other abdominal organs. Definitive diagnosis is made by identification of *T. gallinae* organisms in saliva, or wet mount from caseous lesions under light microscopy, as translucent small flagellates in a single pear-shape with jerky motions lacking any clear direction. Treatment is oral dosing of an antiprotozoal medication (metronidazole, dimetradizole (1,2-dimethyl-5-nitroimidazole), carnidazole, 2-amino-5-nitrothiazole,

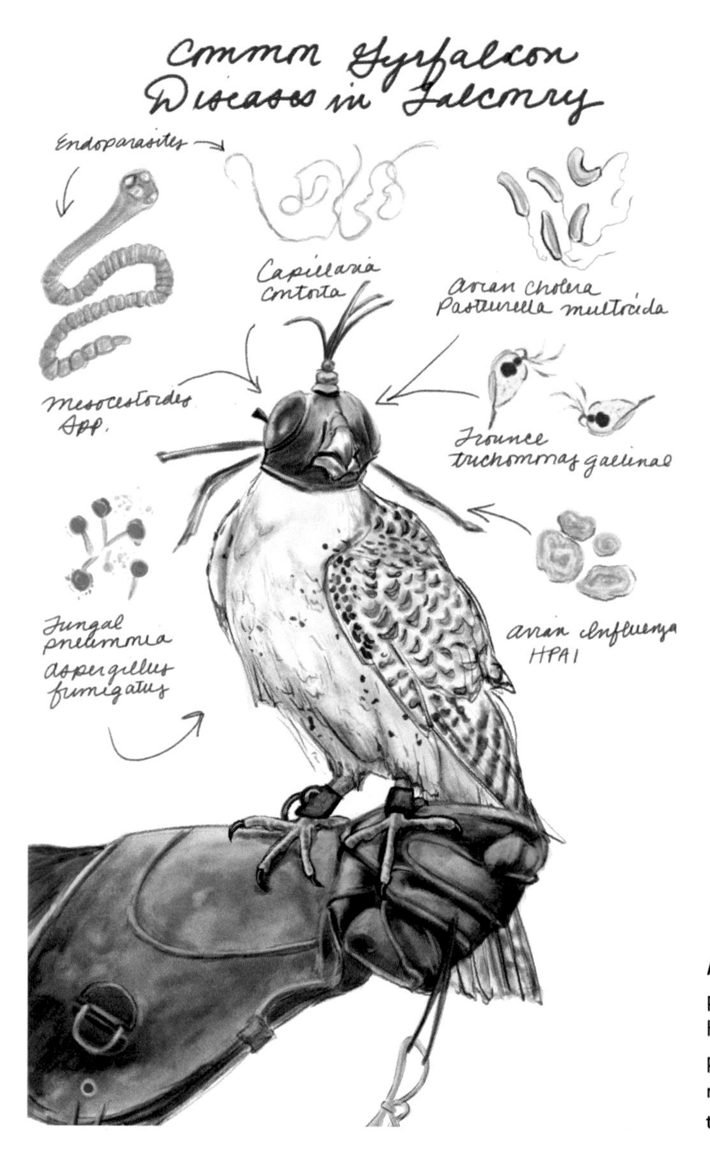

Fig. 21.4. Falconry and falcon medicine provide an in-depth examination of falcon health that cannot be obtained from wild populations, and it offers insights that inform research methodologies and species conservation. Illustration by Laura Donohue.

quaternary ammonia, and copper sulfate), a method that is not applicable for wild raptor populations, except as part of regional surveys of nestling health (Dudek et al. 2018).

The rock pigeon (*Columba livia*) is the primary host and asymptomatic carrier of *T. gallinae* (65% infection rate reported by Stabler 1941), in which the disease is typically referred to as a canker. *Trichomonas gallinae* is transmitted to raptors by ingestion of infected doves and pigeons of the family Columbidae, with disease most notable in nestlings that have lower oral pH than adults (Taylor et al. 2019). Infections during early development can impact population dynamics, such as raptor survival and recruitment (Lindström 1999). Gyrfalcons will readily hunt rock pigeons, and presumably Eurasian collared-doves (*Streptopelia decaocto*) to a lesser extent, when the species cooccur (Lynds et al. 2020). A wild gyrfalcon nestling and hatch-year (birds in juvenile plumage) falcons were recently observed with white plaques characteristic of *T. gallinae* in the mouth, on the Seward Peninsula, only 17 miles from a local pigeon breeder. If *T. gallinae* becomes established in the Arctic, it will be a concern for all Arctic raptor species, and gyrafalcons could serve well to indicate the presence of the pathogen.

Precipitous bird population declines from *T. gallinae* have been proposed. Stabler (1956, 1968) suggested that trichomoniasis may have played a role in the extinction of the passenger pigeon (*Ectopistes migratorizls*) and a contributing factor in the decline of the peregrine falcon. The Arctic has not historically supported any columbids, but warmer temperatures, maintenance of feral populations in cities, accidental introduction of pigeons due to increased shipping traffic and infrastructure, or northward range expansions could facilitate their introduction. Columbids have a propensity for establishing new populations following invasion and rapidly expanding their range to any suitable habitat; in fact, the Eurasian collared-dove is one of the most successfully expanding vertebrates in the world (Bagi et al. 2017). The North American Eurasian collared-dove population has consistently moved northward, with the northern extent of their range reaching British Columbia and southeastern Alaska (Romagosa 2020), whereas rock pigeons breed as far north as Fairbanks, Alaska (Lowther and Johnson 2020). One Eurasian collared-dove was observed for two consecutive days on the Seward Peninsula, an important breeding area for gyrfalcons (M. T. Henderson, personal observation), although it is possible this bird was a vagrant and not representative of any current or future range shifts. Nevertheless, given the ability of columbids to expand their range and their relatively close proximity to the Arctic, *T. gallinae* warrants consideration as a conservation concern.

Capillariasis (*Eucoleus contortus*)

Capillariasis is caused by infection with a nematode from the subfamily Capillariinae, which consists of ca. 300 species that parasitize most vertebrate animals (Okulewicz and Zaleśny 2005). *Eucoleus contortus* is highly prevalent in gyrfalcons but not found in ptarmigan. This nematode can have both a direct life cycle or indirect with an annelid intermediate host. In a direct life cycle, transmission occurs in the nest with nestlings being infected through ingestion of fecal-contaminated food. Indirect transmission could occur in gyrfalcons in Alaska by consumption of Arctic ground squirrels (*Urocitellus parryii*), highly opportunistic feeders that are known to ingest invertebrates. In Iceland, which lies outside the range of the Arctic ground squirrel, the common wood mouse (*Apodemus sylvatica*) is the only terrestrial mammal eaten by the gyrfalcon (Ó. K. Nielsen, personal communication).

Pathology varies greatly by species, but the nematode burrows into the pharyngeal, esophageal, and crop mucosa of raptors causing caseous plaques and masses similar to trichomoniasis with secondary inflammation, necrosis, and stricture (Figure 21.5A and 21.5B). Distinguishing trichomoniasis and capillarisis macroscopically is difficult, but confirmation can be made by observing capillarid nematodes in impression smears of the lesions, scrapings of plaque material, or in histologic sections of tissues sampled by biopsy or on histopathology (Trainer et al. 1968). Raptors infected with *E. contortus* can present with lethargy or emaciation, but finding a carcass is more likely in wild populations (Ó. K. Nielsen, personal communication). Broad-spectrum anthelmintics are effective against *Capillaria* worms, including the benzimidazole drugs albendazole, fenbendazole, flubendazole, mebendazole, and oxfendazole, as well the macrocyclic lactones (e.g., ivermectin), but treatment is impractical for all but captive falconry birds or rehabilitated raptors.

A variety of characteristics make nematodes well suited to exploit the changing Arctic ecosystem (Aleuy and Kutz 2020). A range of factors affect the parasite's ability to capitalize on the warming temperatures including evolutionary history, phenotypic plasticity, and life history (e.g., direct vs. indirect life cycles). Reports of *E. contortus* in gyrfalcons date to at least the 1920s in Iceland, and it may be the most abundant pathogen and greatest cause of mortality for the falcon in that country (Trainer et al. 1968; Christensen et al. 2015). Clausen and Gudmundsson

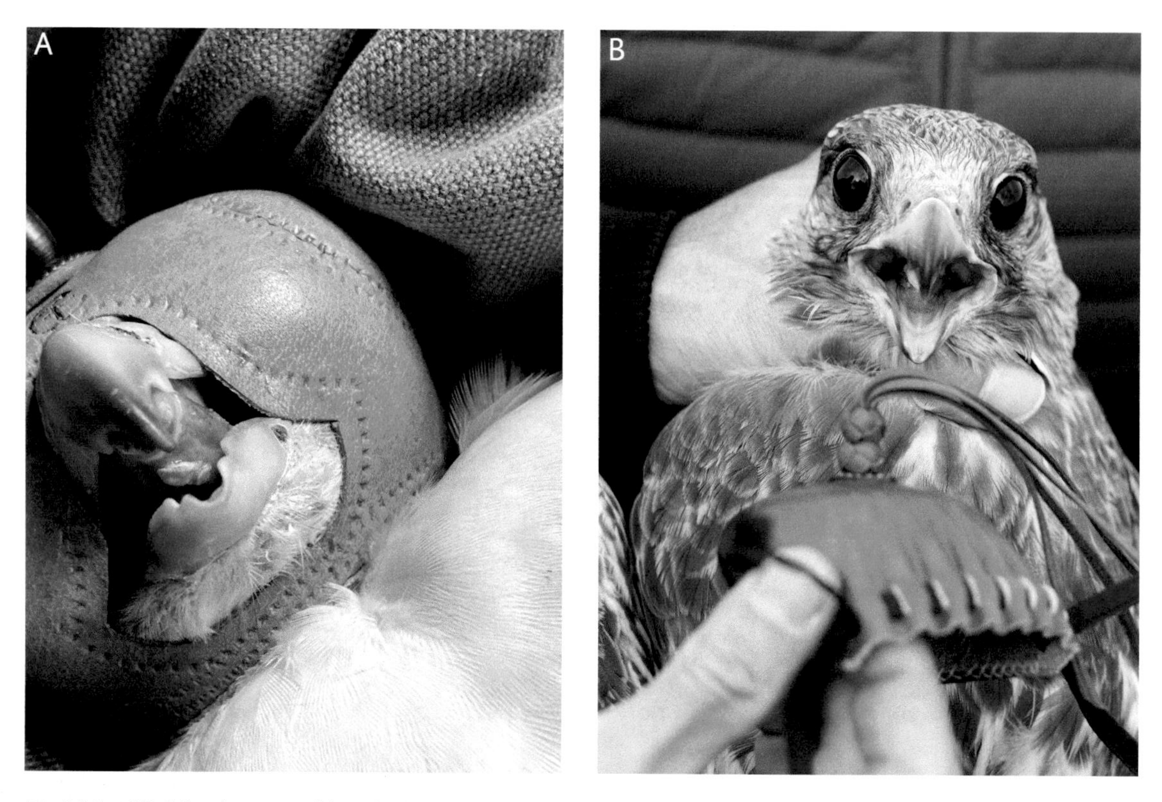

Fig. 21.5. (A) A hatch-year gyrfalcon (*Falco rusticolus*) of the rarer white color phase trapped in September on the Seward Peninsula showing significant plaques in the oral cavity (base and tip of tongue) from infection with capillaria nematodes (*E. contortus*). Note the skin ulcer around the falcon's cere, which is typical of heavy infestations of capillaria in the Iceland gyrfalcon population. (B) A hatch-year gray phase gyrfalcon trapped in autumn in the Izembek National Wildlife Refuge showing mild oral plaques primarily in the commissure of the mouth. Geographic differences in gyrfalcon susceptibility to capillaria may be related to genetics, food choice (higher risk in populations living near humans and domestic pigeons), parasite virulence, or other factors. (A) Photograph by Michael Henderson (The Peregrine Fund) and Chris Barger (Alaska Department of Fish and Game). (B) Photograph by Robin W. Radcliffe (Cornell Conservation Medicine Program).

(1981) identified *E. contortus* as the cause of death for 37% of gyrfalcon carcasses, a trend that was substantiated decades later when Christensen et al. (2015) found 43% (18 out of 46) of carcasses infected with *E. contortus*. Capillariasis in gyrfalcons is poorly understood outside of Iceland, but it is likely that *E. contortus* is a significant cause of mortality for adults and juveniles across their range. In the Arctic, it is unclear whether *E. contortus* has a direct or indirect life cycle, how the infections are transmitted, and what role the gyrfalcon (and other hosts) serve in the life cycle of the pathogen. It will be important to investigate capillarid life history, describe parasite occurrence and abundance, and understand circumpolar distribution of the parasite. Capillariasis in ptarmigan is caused by *Capillaria caudinflata*, rather than *E. contortus*, which has an indirect life cycle with an earthworm intermediate host and causes high juvenile mortality in ptarmigan (Stenkewitz 2016). The evolutionary relationship between capillariasis–gyrfalcon–ptarmigan is a system in balance with unique nematode parasites infecting sympatric Arctic residents—a relationship that is likely to be upset with climate change.

Arctic Health in the Future

Emerging pathogens and altered disease ecology represent a significant threat to Arctic wildlife that are already facing environmental stressors stemming from climate change. Major challenges to understanding and mitigating these threats include unknown distributions of pathogenic fauna, a lack of baseline estimates of abundance, and the vast spatial extent and ruggedness of the Arctic. To address these important knowledge gaps and detect changes in disease ecology, long-term monitoring programs could be established to collect, analyze, and interpret biological samples collected from wildlife from across the Arctic. A successful program would consider the evolutionary and life histories of the definitive hosts, intermediate hosts (where applicable), and developmental stages of the pathogen. It would further aim to understand the habitat requirements of pathogens through predictive maps and models, thereby expanding the spatial extent of findings to areas not studied or temporally to see where diseases may emerge as the climate warms in the future (Table 21.2).

Effective One Health research programs should be collaborative to expand monitoring across a larger geographic scale (both spatial and temporal) and incorporate diverse skillsets. These collaborative partnerships would also benefit from the inclusion of local community members including hunters, recreationists, and native communities. Likewise, the traditional practice of falconry offers knowledge of raptor health that has been acquired over many centuries of caring for gyrfalcons in falconry (Figure 21.6). A number of research programs currently exist throughout the Arctic, and programs could leverage their research infrastructure and expertise in their fields. For projects that already perform live captures, collecting biological samples for disease identification and quantification could be an easy addition to their protocols and a substantial contribution to disease research. Interpreting agent–host–environmental interactions is difficult even in relatively stable systems, but the additional stochasticity of warmer temperatures and altered ecosystem dynamics (e.g., host species distribution and phenology changes) makes identifying and predicting threats exceedingly complicated.

Fig. 21.6. Master falconer and wildlife veterinarian, Robin W. Radcliffe, with a passage gyrfalcon (*Falco rusticolus*) trapped using centuries old luring techniques made possible by the traditional practice of falconry. Photograph by Travis Booms (Alaska Department of Fish and Game).

Summary

Wildlife in the far north has remained relatively insulated from human interference due to the vastness, remoteness, and ruggedness of the Arctic environment, but anthropogenic climate change may negate some of those protective factors and result in exposure to novel diseases. Research that delineates the threats of changing disease ecology is integral to Arctic conservation, and lessons learned can provide important context for other systems experiencing a changing climate. A proactive approach to conservation is important, because early intervention is less expensive, time intensive, and more successful than attempting to recover critically endangered species. Informed conservation requires a comprehensive understanding of system dynamics and, since the changes occurring rapidly in the Arctic are unlikely to cease, time is of the essence to protect the charismatic species of the north.

LITERATURE CITED

Adde, A., D. Stralberg, T. Logan, C. Lepage, S. Cumming, and M. Darveau. 2020. Projected effects of climate change on the distribution and abundance of breeding waterfowl in eastern Canada. Climatic Change 162:2339–2358.

Aleuy, O. A., and S. Kutz. 2020. Adaptations, life-history traits and ecological mechanisms of parasites to survive extremes and environmental unpredictability in the face of climate change. International Journal for Parasitology: Parasites and Wildlife 12:308–317.

Atkinson, C. T., R. J. Dusek, and J. K. Lease. 2001. Serological responses and immunity to superinfection with avian malaria in experimentally-infected Hawaii Amakihi. Journal of Wildlife Diseases 37:20–27.

Atkinson, C, T., and D. A. LaPointe. 2009. Introduced avian diseases, climate change, and the future of Hawaiian honeycreepers. Journal of Avian Medicine and Surgery 23:53–63.

Atkinson, C. T., K. L. Woods, R. J. Dusek, L.S. Sileo, and W. M. Iko. 1995. Wildlife disease and conservation in Hawaii: pathogenicity of avian malaria (*Plasmodium relictum*) in experimentally infected Iiwi (*Vestiaria coccinea*). Parasitology 111(Supplement S1):S59–S69.

Bagi, Z., H. R. Kraus, and S. Kusza. 2017. A review of the invasive Eurasian collared dove and possible research methods in the future. Balkan Journal of Wildlife Research 4:1–10.

Bal, P., A. I. T. Tulloch, P. F. E. Addison, E. McDonald-Madden, and J. R. Rhodes. 2018. Selecting indicator species for biodiversity management. Frontiers in Environmental Science 16:589–599.

Bech, N., J. Boissier, S. Drovetski, and C. Novoa. 2009. Population genetic structure of rock ptarmigan in the "sky islands" of French Pyrenees: implications for conservation. Animal Conservation 12:138–146.

Bekryaev, R. V., I. V. Polyakov, and V. A. Alexeev. 2010. Role of polar amplification in long-term surface air temperature variations and modern arctic warming. Journal of Climate 23:3888–3906.

Bertran K., N. Busquets, F. X. Abad, J. García de La Fuente, D. Solanes, I. Cordón, T. Costa, R. Dolz, and N. Majó. 2012. Highly (H5N1) and low (H7N2) pathogenic avian influenza virus infection in falcons via nasochoanal route and ingestion of experimentally infected prey. PLoS One. 7(3):e32107.

Booms, T. L., M. Lindgren, and F. Huettmann. 2011. Linking Alaska's predicted climate, gyrfalcon, and ptarmigan distributions in space and time: a unique 200-year perspective. Pages 1–14 *in* R. T. Watson, T. J. Cade, M. Fuller, G. Hunt, and E. Potapov, editors. Proceedings of gyrfalcons and ptarmigan in a changing world conference. The Peregrine Fund, Boise, Idaho, USA.

Bradley, M. J., S. J. Kutz, E. Jenkins, and T. M. O'Hara. 2005. The potential impact of climate change on infectious diseases of Arctic fauna. International Journal of Circumpolar Health 64:468–477.

Buhler, K. J., R. G. Maggi, J. Gailius, T. D. Galloway, N. B. Chilton, R. T. Alisauskas, G. Samelius, É. Bouchard, and E. J. Jenkins. 2020. Hopping species and borders: detection of *Bartonella* spp. in avian nest fleas and Arctic foxes from Nunavut, Canada. Parasites and Vectors 13:469.

Callaghan, T. V., L. O. Björn, Y. Chernov, T. Chapin, T. R. Christensen, B. Huntley, R. A. Ims, M. Johansson, D. Jolly, S. Jonasson et al. 2004. Biodiversity, distributions and adaptations of Arctic species in the context of environmental change. Ambio 33:404–417.

Causey, D., and S. V. Edwards. 2008. Ecology of avian influenza virus in birds. The Journal of Infectious Diseases 197(Supplement 1):S29–S33.

Cooper, J. E., P. T. Redig, and W. Burnham. 1980. Bacterial isolates from the pharynx and cloaca of the peregrine falcon (*Falco peregrinus*) and gyrfalcon (*F. rusticolus*) (bacteria from falcons). Journal of Raptor Research 14:6–9.

Corrigan, L. J., I. J. Winfield, A. R. Hoelzel, and M. C. Lucas. 2011. Dietary plasticity in Arctic charr (*Salvelinus alpinus*) in response to long-term environmental change. Ecology of Freshwater Fish 20:5–13.

Cotter, R. C., D. A. Boag, and C. C. Shank. 1992. Raptor predation on rock ptarmigan (*Lagopus mutus*) in the central Canadian Arctic. Journal of Raptor Research 26:146–151.

Christensen, N. D., Skirnisson K., and O. K. Nielsen. 2015. The parasite fauna of the gyrfalcon (*Falco rusticolus*) in Iceland. Journal of Wildlife Diseases 51:929–933.

Christensen, T., J. Payne, M. Doyle, G. Ibarguchi, J. Taylor, N. M. Schmidt, M. Gill, M. Svoboda, M. Aronsson, C. Behe, et al. 2013. The Arctic Terrestrial Biodiversity Monitoring Plan. Conservation of Arctic Flora and Fauna International Secretariat, CAFF Monitoring Series Report 7, Akureyri, Iceland.

Clausen, B., and F. Gudmundsson. 1981. Causes of mortality among free-ranging gyrfalcons in Iceland. Journal of Wildlife Diseases 17:105–109.

Dobson, A., P. K. Molnár, and S. Kutz. 2015. Climate change and Arctic parasites. Trends in Parasitology 31:181–188.

Dudek, B. M., M. N. Kochert, J. G. Barnes, P. H. Bloom, J. M. Papp, R. W. Gerhold, K. E. Purple, K. V. Jacobson, C. R. Preston, C. R. Vennum, et al. 2018. Prevalence and risk factors of *Trichomonas gallinae* and trichomonosis in golden eagle (*Aquila chrysaetos*) nestlings in western North America. Journal of Wildlife Diseases 54:755–764.

Eberly, L. 1969. Glossary of falconry terms. Raptor Research News 3(3):58–67.

Ehrich, D., J. Henden, R. A. Ims, L. O. Doronina, S. T. Killengren, N. Lecomte, I. G. Pokrovsky, G. Skogstad, A. A. Sokolov, V. A. Sokolov, et al. 2012. The importance of willow thickets for ptarmigan and hares in shrub tundra: the more the better? Oecologia 168:141–151.

Epstein, P. R. 2002. Climate change and infectious disease: stormy weather ahead? Epidemiology 13:373–375.

Franke, A., K. Falk, K. Hawkshaw, S. Ambrose, D. L. Anderson, P. J. Bente, T. L. Booms, K. K. Burnham, J. Ekenstedt, I. Fufachev, et al. 2020. Status and trends of circumpolar peregrine falcon and gyrfalcon populations. Ambio 49:762–783.

Gilg, O., K. M. Kovacs, J. Aars, J. Fort, G. Gauthier, D. Grémillet, R. A. Ims, H. Meltofte, J. Moreau, E. Post, et al. 2012. Climate change and the ecology and evolution of Arctic vertebrates. Annals of the New York Academy of Sciences 1249:166–190.

Gilg, O., B. Sittler, and I. Hanski. 2009. Climate change and cyclic predator–prey population dynamics in the high Arctic. Global Change Biology 15:2634–2652.

Hamilton, M. A., and R. M. Stabler. 1953. Combined trichomoniasis and aspergillosis in a gyrfalcon. Journal of the Colorado-Wyoming Academy of Science 4:58–59.

Hatcher, M. J., J. T. A. Dick, and A. M. Dunn. 2012. Diverse effects of parasites in ecosystems: linking interdependent processes. Frontiers in Environmental Science 10:186–194.

Henderson, M. T., T. L. Booms, D. L. Anderson, K. V. Dhondt, and R. W. Radcliffe. 2021. Gyrfalcon health in Alaska: a temporal and spatial assessment of emergent diseases for an Arctic specialist during rapid climate change. Poster presented at the Raptor Research Foundation annual conference, 9–11 October 2021, Boise, Idaho, USA.

Hoberg, E. P., and D. R. Brooks 2015. Evolution in action: climate change, biodiversity dynamics and emerging infectious disease. Philosophical Transactions of the Royal Society B Biological Sciences 370:20130553.

Hudson, P. J., A. P. Dobson, and D. Newborn. 1992. Do parasites make prey vulnerable to predation? Red grouse and parasites. Journal of Animal Ecology 61:681–692.

Ims, R. A., and E. Fuglei. 2005. Trophic interaction cycles in tundra ecosystems and the impact of climate change. BioScience 55:311–322.

Intergovernmental Panel on Climate Change (IPCC). 2021. Summary for policymakers. *in* V. Masson-Delmotte, P. Zhai, A. Pirani, S. L. Connors, C. Péan, S. Berger, N. Caud, Y. Chen, L. Goldfarb, M. I. Gomis, et al., editors. Climate change 2021: the physical science basis. Contribution of Working Group I to the Sixth Assessment Report of the Intergovernmental Panel on Climate Change. Cambridge University Press, Cambridge, England.

Ip, H. S., M. K. Torchetti, R. Crespo, P. Kohrs, P. DeBruyn, K. G. Mansfield, T. Baszler, L. Badcoe, B. Bodenstein, V. Shearn-Bochsler, and M. L. Killian. 2015. Novel Eurasian highly pathogenic avian influenza A H5 viruses in wild birds, Washington, USA, 2014. Emerging Infectious Diseases 21:886.

Jenkins, E. J., A. M. Veitch, S. J. Kutz, E. P. Hoberg, and L. Polley. 2006. Climate change and the epidemiology of protostrongylid nematodes in northern ecosystems: *Parelaphostrongylus odocoilei* and *Protostrongylus stilesi* in Dall's sheep (*Ovis d. dalli*). Parasitology 132:387–401.

Johnson, D. L., M. T. Henderson, D. L. Anderson, T. L. Booms, and C. T. Williams. 2020. Bayesian stable isotope mixing models effectively characterize the diet of an Arctic raptor. Journal of Animal Ecology 89:2972–2985.

Johnson, J. A., K. K. Burnham, W. A. Burnham, and D. P. Mindell. 2007. Genetic structure among continental and island populations of gyrfalcons. Molecular Ecology 16:3145–3160.

Kafle, P., P. Peller, A. Massolo, E. Hoberg, L. M. Leclerc, M. Tomaselli, and S. Kutz. 2020. Range expansion of muskox lungworms track rapid arctic warming: implications for geographic colonization under climate forcing. Scientific Reports 10:17323.

Kingston, N., J. D., Remple, W. Burnham, R. M. Stabler, and R. B. McGhee. 1976. Malaria in a captively-produced F1 gyrfalcon and in two F1 peregrine falcons. Journal of Wildlife Diseases 12:562–565.

Kutz, S. J., S. Checkley, G. G. Verocai, M. Dumond, E. P. Hoberg, R. Peacock, J. P. Wu, K. Orsel, K. Seegers, A. L. Warren, and A. Abrams. 2013. Invasion, establishment, and range expansion of two parasitic nematodes in the Canadian Arctic. Global Change Biology 19:3254–3262.

Kutz, S. J., E. P. Hoberg, J. Nishi, and L. Polley. 2002. Development of the muskox lungworm, *Umingmak-*

strongylus pallikukensis, in gastropods in the Arctic. Canandian Journal of Zoology 80:1977–1985.

Kutz, S. J., E. P. Hoberg, L. Polley, and E. J. Jenkins. 2005. Global warming is changing the dynamics of Arctic host–parasite systems. Proceedings of the Royal Society B Biological Sciences 272:2571–2576.

Kutz, S. J., E. J. Jenkins, A. M. Veitch, J. Ducrocq, L. Polley, B. Elkin, and S. Lair. 2009. The Arctic as a model for anticipating, preventing, and mitigating climate change impacts on host-parasite interactions. Veterinary Parasitology 163:217–228.

LaPointe, D. A., C. T. Atkinson, and M. D. Samuel. 2012. Ecology and conservation biology of avian malaria. Annals of the New York Academy of Sciences 1249:211–226.

Latham, S. 1615. Lathams Falconry: or, the Faulcons Lure, and Cure: in two books. The first, concerning the ordering and training up of all Hawkes in general; especially the Haggard Faulcon Gentle. The second, teaching approved medicines for the cure of all Diseases in them. Gathered by long practice and experience, and published for the delight of noble mindes, and instruction of young Faulconers in things pertaining to this Princely Art. Thomas Harper, London, England.

Lehikoinen, A., and K. Jaatinen. 2012. Delayed autumn migration in northern European waterfowl. Journal of Ornithology 153:563–570.

Leighton, P. A., J. K. Koffi, Y. Pelcat, L. R. Lindsay, and N. H. Ogden. 2012. Predicting the speed of tick invasion: an empirical model of range expansion for the Lyme disease vector *Ixodes scapularis* in Canada. Journal of Animal Ecology 49:457–464.

Liebezeit, J. E., R. M Cross, and S. Zack. 2012. Assessing climate change vulnerability of breeding birds in Arctic Alaska. A report prepared for the Arctic Landscape Conservation Cooperative. Wildlife Conservation Society, North America Program, Bozeman, Montana, USA.

Lierz, M., H. M. Hafez, R. Klopfleisch, D. Lüschow, C. Prusas, J. P. Teifke, M. Rudolf, C. Grund, D. Kalthoff, T. Mettenleiter, et al. 2007. Protection and virus shedding of falcons vaccinated against highly pathogenic avian influenza A virus (H5N1). Emerging Infectious Diseases 13:1667–1674.

Lierz, M., H. M. Hafez, and O. Krone. 2008. Prevalence of hematozoa in falcons in the United Arab Emirates with respect to the origin of falcon hosts. Journal of Avian Medicine and Surgery 22:208–212.

Lindström, J. 1999. Early development and fitness in birds and mammals. Trends in Ecology and Evolution 14:343–348.

Loiseau, C., R. J. Harrigan, A. J. Cornel, S. L. Guers, M. Dodge, T. Marzec, J. S. Carlson, B. Seppi, and R. N. M. Sehgal. 2012. First evidence and predictions of *Plasmodium* transmission in Alaskan bird populations. PLoS One 7:e44729.

Lowther, P., and R. Johnson. 2020. Rock pigeon (*Columba livia*), version 1.0. *in* S. M. Billerman, editor. Birds of the world. Cornell University, Ithaca, New York, USA. https://birdsoftheworld.org/bow/species/rocpig/cur/introduction. Accessed 1 Mar 2021.

Lynds, M., J. Card, H. Hedstrom, D. Delaney, G. Court, and J. Acorn. 2020. Large winter falcons and their rock pigeon (*Columba livia*) prey at an urban grain terminal in Edmonton, Alberta: an update. Canadian Field-Naturalist 134:205–209.

Malmberg, J. L., L. A. White, and S. Vandewoude. 2021. Bioaccumulation of pathogen exposure in top predators. Trends in Ecology and Evolution 36:411–420.

Meixell, B. W., T. W. Arnold, M. S. Lindberg, M. M. Smith, J. A. Rundstadler, and A. M. Ramey. 2016. Detection, prevalence, and transmission of avian hematozoa in waterfowl at the Arctic/sub-Arctic interface: co-infections, viral infections, and sources of variation. Parasites and Vectors 9:390.

Melin, M., L. Mehtätalo, P. Helle, K. Ikonen, and T. Packalen. 2020. Decline of the boreal willow grouse (*Lagopus lagopus*) has been accelerated by more frequent snow-free springs. Scientific Reports 10:6897.

Min, S., X. Zhang, F. W. Zwiers, and G. C. Hegerl. 2011. Human contribution to more-intense precipitation extremes. Nature 470:378–381.

Moe, B., L. Stempniewicz, D. Jakubas, F. Angelier, O. Chastel, F. Dinessen, G. W. Gabrielsen, F. Hanssen, N. J. Karnovsky, B. Rønning, J. Welcker, et al. 2009. Climate change and phenological responses of two seabird species breeding in the high-Arctic. Marine Ecology Progress Series 393:235–246.

Morin, C. W., B. Stoner-Duncan, K. Winker, M. Scotch, J. J. Hess, J. S. Meschke, K. L. Ebi, and P. M. Rabinowitz. 2018. Avian influenza virus ecology and evolution through a climatic lens. Environment International 119:241–249.

Mørk, T., R. A. Ims, and S. T. Killengreen. 2019. Rodent population cycle as a determinant of gastrointestinal nematode abundance in a low-arctic population of the red fox. International Journal for Parasitology: Parasites and Wildlife 9:36–41.

Mudryk, L. R., J. Dawson, S. E. L. Howell, C. Derksen, T. A. Zagon, and M. Brady. 2021. Impact of 1, 2 and 4 °C of global warming on ship navigation in the Canadian Arctic. Nature Climate Change 11:673–679.

Nemeth, N., D. Gould, R. Bowen, and N. Komar. 2006. Natural and experimental West Nile virus infection in five raptor species. Journal of Wildlife Diseases 42:1–13.

Nielsen, Ó. K. 2003. The impact of food availability on gyrfalcon (*Falco rusticolus*) diet and timing of breeding. Pages 283–302 *in* D. B. A. Thomson, S. M. Redpath and A. H. Fielding, editors. Birds of prey in a changing environment. Scottish Natural Heritage, Joint Nature Conservation Committee, British Ornithologists' Union, Edinburgh, Scotland.

Nielsen, Ó. K., and T. J. Cade. 1990. Seasonal changes in food habits of gyrfalcons in NE-Iceland. Ornis Scandinavica 21:202–211.

Nielsen, Ó. K., and T. J. Cade. 2017. Gyrfalcon and ptarmigan predator–prey relationship. Pages 43–74 *in* D. L. Anderson, C. J. W. McClure and A. Franke, editors. Applied raptor ecology: essentials from gyrfalcon research. The Peregrine Fund, Boise, Idaho, USA.

O'Brien, S. J., and J. F. Evermann. 1988. Interactive influence of infectious disease and genetic diverstiy in natural populations. Trends in Ecology and Evolution 3:254–259.

Ogden, N. H., I. K. Barker, C. M. Francis, A. Heagy, L. R. Lindsay, and K. A. Hobson. 2015. How far north are migrant birds transporting the tick *Ixodes scapularis* in Canada? Insights from stable hydrogen isotope analyses of feathers. Ticks and Tick-borne Diseases 6:715–720.

Ogden, N. H., L. St-Onge, I. K. Barker, S. Brazeau, M. Bigras-Poulin, D. F. Charron, C. M. Francis, A. Heagy, L. R. Lindsay, A. Maarouf, P. Michel, et al. 2008. Risk maps for range expansion of the Lyme disease vector, *Ixodes scapularis*, in Canada now and with climate change. International Journal of Health Geographics 7:24.

Okulewicz, A., and G. Zaleśny. 2005. Biodiversity of Capillariinae. Wiadomości Parazytologiczne 51:9–14.

Pickles, R. A., D. Thornton, R. Feldman, A. Marques, and D. Murray. 2013. Predicting shifts in parasite distribution with climate change: a multitrophic level approach. Global Change Biology 19:2645–2654.

Pilfold, N. W., E. S. Richardson, J. Ellis, E. Jenkins, W. B. Scandrett, A. Hernández-Ortiz, K. Buhler, D. McGeachy, B. Al-Adhami, K. Konecsni, V. A. Lobanov, et al. 2021. Long-term increases in pathogen seroprevalence in polar bears (*Ursus maritimus*) influenced by climate change. Global Change Biology 27:4481–4497.

Raghav, R., and J. Samour. 2019. Inclusion body herpesvirus hepatitis in captive falcons in the Middle East: a review of clinical and pathologic findings. Journal of Avian Medicine and Surgery 33:1–6.

Rahim, M. A., A. O. Bakhiet, and M. F. Hussein. 2013. Aspergillosis in a gyrfalcon (*Falco rusticolus*) in Saudi Arabia. Comparative Clinincal Pathology 22:131–135.

Ramey, A. M., N. J. Hill, T. J. DeLiberto, S. E. J. Gibbs, M. C. Hopkins, A. S. Lang, R. L. Poulson, D. J. Prosser, J. M. Sleeman, D. E. Stallknecht, et al. 2022. Highly pathogenic avian influenza is an emerging disease threat to wild birds in North America. Journal of Wildlife Management 86:e22171.

Ramey, A. M., A. B. Reeves, S. A. Sonsthagen, J. L. TeSlaa, S. Nashold, T. Donnelly, B. Casler, and J. S. Hall. 2015. Dispersal of H9N2 influenza A viruses between East Asia and North America by wild birds. Virology 482:79–83.

Redig, P. T. 1992. Health management of raptors trained for falconry. Pages 258–264 *in* Proceedings of the Association of Avian Veterinarians, 1–5 September 1992, New Orleans, Louisiana, USA.

Redig, P. T., J. E. Cooper, D. Remple, and D. B. Hunter. 1993. Raptor biomedicine. University of Minnesota Press, Saint Paul, USA.

Remple, J. D. 2004. Intracellular hematozoa of raptors: a review and update. Journal of Avian Medicine and Surgery 18:75–88.

Revermann, R., H. Schmid, N. Zbinden, R. Spaar, and B. Schröder. 2012. Habitat at the mountain tops: how long can rock ptarmigan (*Lagopus muta helvetica*) survive rapid climate change in the Swiss Alps? A multi-scale approach. Journal of Ornithology 153:891–905.

Robinson, B. W., T. L. Booms, M. J. Bechard, and D. L. Anderson. 2019. Dietary plasticity in a specialist predator, the gyrfalcon (*Falco rusticolus*): new insights into diet during brood rearing. Journal of Raptor Research 53:115–126.

Romagosa, C. M. 2020. Eurasian collared-dove (*Streptopelia decaocto*). *In* S. M. Billerman, editor. Birds of the world. Cornell University, Ithaca, New York, USA. https://birdsoftheworld.org/bow/species/eucdov/1.0/introduction. Accessed 1 Mar 2021.

Salakij, J., P. Lertwatcharasarakul, C. Kasorndorkbua, and C. Salakij. 2012. *Plasmodium circumflexum* in a shikra (*Accipiter badius*): phylogeny and ultra-structure of the haematozoa. Japanese Journal of Veterinary Research 60:105–109.

Samour, J. H., and J. N. Naldo. 2001. Serratospiculiasis in captive falcons in the Middle East: a review. Journal of Avian Medicine and Surgery 15:2–9.

Schuler, K., M. Claymore, H. Schnitzler, E. Dubovi, T. Rocke, M. J. Perry, D. Bowman, and R. C. Abbott. 2021. Sentinel coyote pathogen survey to assess declining black-footed ferret (*Mustela nigripes*) population in South Dakota, USA. Journal of Wildlife Diseases 57:264–272.

Scridel, D., M. Brambilla, D. R. de Zwaan, N. Froese, S. Wilson, P. Pedrini, and K. Martin. 2021. A genus at risk: predicted current and future distribution of all three *Lagopus* species reveal sensitivity to climate change and efficacy of protected areas. Diversity and Distributions 27:1759–1774.

Seekings, A. H., C. J. Warren, S. S. Thomas, S. Mahmood, J. James, A. M. P. Byrne, S. Watson, C. Bianco, A. Nunez, I. H. Brown, et al. 2021. Highly pathogenic avian influenza virus H5N6 (clade 2.3.4.4b) has a preferable host tropism for waterfowl reflected in its inefficient transmission to terrestrial poultry. Virology 559:74–85.

Sergio, F., T. Caro, D. Brown, B. Claucas, J. Hunter, J. Ketchum, K. McHugh, and F. Hiraldo. 2008. Top predators as conservation tools: ecological rationale, assumptions, and efficacy. Annual Review of Ecology, Evolution, and Systematics 39:1–19.

Shearn-Bochsler, V. I., S. Knowles, and H. S. Ip. 2019. Lethal infection of wild raptors with highly pathogenic avian influenza H5N8 and H5N2 viruses in the USA, 2014–15. Journal of Wildlife Diseases 55:164–168.

Siddig, A. A. H., A. M. Ellison, A. Ochs, C. Villar-Leeman, and M. K. Lau. 2016. How do ecologists select and use indicator species to monitor ecological change? Insights from 14 years of publication in Ecological Indicators. Ecological Indicators 60:223–230.

Skirnisson, K., S. T. Thorarinsdottir, and Ó. K. Nielsen. 2012. The parasite fauna of rock ptarmigan (*Lagopus muta*) in Iceland: prevalence, intensity, and distribution within the host population. Náttúrufræðingurinn 80:33–40.

Smith, M. M., C. Van Hemert, and C. M. Handel. 2019. Evidence of *Culiseta* mosquitoes as vectors for *Plasmodium* parasites in Alaska. Journal of Vector Ecology 44:68–75.

Smith, M. M., C. Van Hemert, and R. Merizon. 2016. Haemosporidian parasite infections in grouse and ptarmigan: prevalence and genetic diversity of blood parasites in resident Alaskan birds. International Journal for Parasitology: Parasites and Wildlife 5:229–239.

Stabler, R. M. 1941. Further studies on trichomoniasis in birds. Auk 58:558–562.

Stabler, R. M. 1956. *Trichomonas gallinae*: a review. Experimental Parasitology 3:368–402.

Stabler, R. M. 1968. *Trichomonas galline* as a factor in the decline of the peregrine falcon. Pages 435–438 *in* J. J. Hickey, editor. Biology and decline of peregrine falcon populations. University of Wisconsin Press, Madison, USA.

Stabler, R. M., and M. A. Hamilton. 1954. Aspergillosis, trichomoniasis, and drug therapy in a gyrfalcon. Auk 71:205–208.

Stenkewitz, U., Ó. K. Nielsen, K. Skírnisson, and G. Stefánsson. 2016. Host–parasite interactions and population dynamics of rock ptarmigan. PLoS One 11:e0165293.

Stout, W. E., A. G. Cassini, J. K. Meece, J. M. Papp, R. N. Rosenfield, and K. D. Reed. 2005. Serological evidence of West Nile virus infection in three wild raptor populations. Avian Diseases 49:371–375.

Stroeve, J., M. M. Holland, W. Meier, T. Scambos, and M. Serreze. 2007. Arctic sea ice decline: faster than forecast. Geophysical Research Letters 34: L09501.

Sturm, M., C. Racine, and K. Tape. 2001. Increasing shrub abundance in the Arctic. Nature 411:546–547.

Tanikawa, T., S. Sakuma, E. Yoshida, R. Tsunekuni, and M. Nakayama. 2021. Comparative susceptibility of the common teal (*Anas crecca*) to infection with high pathogenic avian influenza virus strains isolated in Japan in 2004–2017. Veterinary Microbiology 263:109266.

Tape, K. D., K. Christie, G. Carroll, and J. A. O'Donnell. 2016. Novel wildlife in the Arctic: the influence of changing riparian ecosystems and shrub habitat expansion on snowshoe hares. Global Change Biology 22:208–219.

Tape, K., M. Sturm, and C. Racine. 2006. The evidence for shrub expansion in northern Alaska and the pan-Arctic. Global Change Biology 12:686–702.

Taylor, M. J., R. W. Mannan, J. M. U'Ren, N. P. Garber, R. E. Gallery, and A. E. Arnold. 2019. Age-related variation in the oral microbiome of urban Cooper's hawks (*Accipiter cooperii*). BMC Microbiology 19:47.

Thomas, C. D., A. Cameron, R. E. Green, M. Bakkenes, L. J. Beaumont, Y. C. Collingham, B. F. N. Erasmus, M. F. De Siqueira, A. Grainger, L. Hannah, et al. 2004. Extinction risk from climate change. Nature 427:145–148.

Trainer, D. O., S. D. Folz, and W. M. Samuel. 1968. Capillariasis in the gyrfalcon. Condor 70:276–277.

Uher-Koch, B. D., T. J. Spivey, C. Van Hemert, J. A. Schmutz, J. Kaijun, X.-F. Wan, and A. M. Ramey. 2019. Serologic evidence for influenza a virus exposure in three loon species breeding in Alaska, USA. Journal of Wildlife Diseases 55:862–867.

US Geological Survey (USGS). 2020. Pathology case of the month—gyrfalcon. https://www.usgs.gov/center-news/pathology-case-month-gyrfalcon?qt-news_science_products=1#qt-news_science_products. Accessed 28 October 2021.

Van Hemert, C., J. M. Pearce, and C. M. Handel. 2014. Wildlife health in a rapidly changing north: focus on avian disease. Frontiers in Ecology and the Environment 12:548–555.

Van Hemert, C., T. J. Spivey, B. D. Uher-Koch, T. C. Atwood, R. David, B. W. Meixell, J. W. Hupp, K. Jiang, L. G. Adams, and D. David. 2019. Survey of Arctic Alaskan wildlife for influenza a antibodies: limited evidence for exposure of mammals. Journal of Wildlife Diseases 55:387–398.

Ward, F. P., and D. G. Fairchild. 1972. Air sac parasites of the genus *Serratospiculum* in falcons. Journal of Wildlife Diseases 8:165–168.

Weber, D., P. J. Van Coeverden De Groot, E. Peacock, M. D. Schrenzel, D. A. Perez, S. Thomas, J. M. Shelton, C. K. Else, L. L. Darby, L. Acosta, et al. 2013. Low MHC variation in the polar bear: implications in the face of Arctic warming? Animal Conservation 16:671–683.

Wernery, U., R. Wernery, R. Zachariah, and J. Kinne. 1998. Salmonellosis in relation to chlamydiosis and pox and *Salmonella* infections in captive falcons in the United Arab Emirates. Journal of Veterinary Medicine B 45:577–583.

Wheeler, H. C., T. T. Høye, and J. C. Svenning. 2018. Wildlife species benefitting from a greener Arctic are most sensitive to shrub cover at leading range edges. Global Change Biology 24:212–223.

White, C. M., and H. K. Springer. 1965. Notes on the gyrfalcon in western coastal Alaska. Auk 82:104–105.

Wilkinson, L. C., C. M. Handel, C. Van Hemert, C. Loiseau, and R. N. M. Sehgal. 2016. Avian malaria in a boreal resident species: long-term temporal variability, and increased prevalence in birds with avian keratin disorder. International Journal for Parasitology 46:281–290.

Williams, E. S., D. E. Runde, K. Mills, and L. D. Holler. 1987. Avian cholera in a gyrfalcon (*Falco rusticolus*). Avian Diseases 31:380–382.

Wildlife Diseases Crossing Species and Geographic Barriers

22 Tuberculosis in Free-ranging Wild Animals

Michele A. Miller

Introduction

Tuberculosis (TB) is a term used for a group of diseases caused by a genetically related bacteria in the *Mycobacteria tuberculosis* complex (MTBC) (Gagneux 2018). Infection with these bacteria, which includes *M. tuberculosis* (causative agent of human TB) and *M. bovis* (cause of bovine TB) can occur in a wide range of species, with most mammalian species having some level of susceptibility (Vikas Saket et al. 2017; Fellag et al. 2021; Reis et al. 2021*a*). These species may be further categorized as "reservoir" or "maintenance" hosts, which are capable of maintaining infection without introduction from another species or source (Palmer 2013). Other species are considered "spillover" or "accidental" hosts, which don't maintain spread within the population; infection usually results from direct or indirect contact with a reservoir host. Additional terms that are used to describe TB transmission include "zoonosis," which is TB in humans that has been spread from an animal (i.e., *M. bovis* infection in a human); and the converse, "zooanthroponosis," which is TB in an animal resulting from spread from a human (e.g., *M. tuberculosis* infection in an elephant).

Illustration by Laura Donohue.

The MTBC are a subgroup of more than 120 different species of mycobacteria that can infect birds, reptiles, fish, and mammals (Tortoli 2006; Falkinham 2015; Miller and Lyashchenko 2015). Many mycobacterial species are ubiquitous in the environment and can cause opportunistic infections that may appear like infections with MTBC. There are also environmental sources of MTBC (Fellag et al. 2021). Therefore, it is important to understand how to identify the cause of disease since reporting, management, and control may vary significantly between some of the mycobacterial species. Since many mycobacterial species can infect humans as well as domestic and wild animals, it is essential that understanding, detecting, and controlling tuberculosis uses a One Health approach (Kaneene et al. 2014). For the purposes of this chapter, the focus will be on MTBC infections in free-ranging wildlife and the impact on conservation and management.

History

Tuberculosis is an ancient disease; evidence of infection has been discovered in human mummies and mastodon skeletons (Rothschild and Laub 2006; Zink et al. 2007). Tuberculosis has been found in a bison skeleton from 17,000 years ago as well as in Asian elephants 2,000 years ago. It is believed that the ancestor of the MTBC organisms arose in humans in Africa and spread with migration of people and domestication of animals (Gagneux 2018). There is an evidence-supported theory, based on genetic analyses of mycobacteria in ancient New World human skeletons and seals, that marine mammals may have played a role in transferring ancestral MTBC from Africa to humans in South America (Bos et al. 2014). Evolution of *M. bovis* appears to have occurred in East Africa after the introduction of domesticated cattle and subsequently moved with livestock along trade routes and during colonization (Inlamea et al. 2020; Loiseau et al. 2020; Zimpel et al. 2020). With agricultural intensification, bovine TB became epidemic in Europe during the 18th and 19th centuries and in North America in the late 19th century (Palmer and Waters 2011; Good et al. 2018). Cases of *M. bovis* infection in animals and *M. tuberculosis* in humans have occurred on all continents except Antarctica (MacNeil et al. 2019; Reis et al. 2021*a*).

Etiology

Mycobacteria are gram-positive bacteria with a thick lipid cell wall that allows them to survive both in the environment and resist the host's immune responses. The numbers of MTBC members continue to increase with 13 species or ecotypes currently recognized (Fellag et al. 2021). Some of the bacteria are specifically adapted to their host (i.e., *M. suricattae* in meerkats, *Suricata suricatta*, and *M. mungi* in banded mongooses, *Mungos mungi*), while others, such as *M. bovis*, have a wide species range, including humans (Clarke et al. 2016). Importantly, *M. tuberculosis* can infect animals, including wildlife, especially under human management, such as captive elephants (Mikota et al. 2015); therefore, the infected host species does not provide identification of the specific MTBC organism. Although *M. bovis* and *M. tuberculosis* are found worldwide, the distribution of other MTBC organisms may vary geographically; for example, *M. caprae* and *M. microti* are found mainly in Europe (Rodríguez et al. 2011, Peterhans et al. 2020), and *M. mungi* and *M. suricatti* in southern Africa (Clarke et al. 2016). Members of the MTBC also vary in their ability to cause zoonotic infections: *M. bovis, M. caprae, M. microti, M. pinnipedii*, and *M. orygis* can infect humans, whereas *M. suricattae, M. mungi*, and Dassie bacillus are host specific (Kock et al. 2021)—despite the MTBC bacilli being 99.9% genetically similar (Fellag et al. 2021). Therefore, identification of the specific mycobacterial strain is important in investigating sources of infection and epidemiological contacts, as well as assessing risk for transmission. This requires specialized molecular techniques to distinguish the minor genetic differences between members of the MTBC, which will be discussed in the section on diagnosis.

Demographics and Field Biology

Most mammal species are variably susceptible to infection with MTBC, with *M. bovis* being the most common cause of TB in wildlife and livestock. Although carnivores and perissodactyls have traditionally been viewed as being resistant, more recent studies have discovered that wild canids, felids, mustelids, rodents, bears, rhinoceros, tapirs, and marsupials can become infected with MTBC and even develop TB disease (Bruning-Fann et al. 2001; Cavanagh et al. 2002; Fitzgerald and Kaneene 2012; Miller and Lyashchenko 2015; Bernitz et al. 2021). Since MTBC infections generally take months to years to progress before animals show clinical signs, it is considered a chronic disease, which complicates detection and control. Often it is the older animals in a population that will develop nonspecific signs such as slow progressive weight loss and emaciation, lethargy, inappetence, coughing, dull haircoat, nonhealing wounds, lameness, or even paresis in cases in which bones have become infected (Renwick et al. 2007). In some animals, especially prey species, cases in free-ranging individuals may go undetected due to predation of weaker herd members. Coinfections and malnutrition have also been associated with increased susceptibility, and cases of TB may become more apparent during drought or other suboptimal conditions (Abrantes et al. 2021).

Wildlife that share grazing and water sources with domestic livestock in areas with endemic *M. bovis*–infected populations are also at greater risk of experiencing a spillover and spill back with the wild population (Meunier et al. 2017; VerCauteren et al. 2018; Crispell et al. 2020; Woodroffe et al. 2021). Spillover can also occur in accidental hosts such as coyotes and foxes (Bruning-Fann et al. 2001; Richomme et al. 2020). In countries where there is a high human burden of TB, infected people tending animals can also be a source of infection through aerosols, saliva, or urine contamination of the environment (Ameni et al. 2011; Zachariah et al. 2017). Captive wildlife in zoos, game farms, and ranches are also intermittently infected with *M. bovis* as well as *M. tuberculosis* (Miller and Lyashchenko 2015; Hlokwe et al. 2016). These infections can be caused by intra- or inter-species transmission either associated with wildlife reservoirs or the movement and contact with infected captive animals or human caretakers. Identification of the mycobacterial species causing the infection can be important in helping track the source. For example, investigating *M. tuberculosis* infection in a captive jaguar would lead to closer examination of human contacts resulting in zooanthroponootic disease than if it was caused by *M. bovis* (which could be spread through feeding on an infected carcass). Therefore, it is essential to gather information on land use practices by humans, domestic animals, and wildlife in an area where TB in wildlife is suspected and to employ a One Health approach to investigate outbreaks.

Reports of MTBC-infected wildlife species are abundant in the published literature (Bruning-Fann et al. 2001; Cavanagh et al. 2002; Fitzgerald and Kaneene 2012; Vikas Saket et al. 2017; Domingos et al. 2019; Bernitz et al. 2021; Fellag et al. 2021; Reis et al. 2021*a*). Some wildlife species are important reservoirs of *M. bovis*, which maintain the infection within the population but also facilitate spread to other species as well contaminating the environment. European wild boars (*Sus scofa*) and badgers (*Meles meles*) are both significant wildlife reservoir species for *M. bovis* in Europe, with epidemiological links to infections in cattle and deer (Bhuachalla et al. 2015; Triguero-Ocana et al. 2020). In North America, free-ranging populations of *M. bovis*–infected white-tailed deer (*Odocoileus virginaianus*), elk (*Cervus canadensis*), and wood bison (*Bison bison athabascae*), which are maintenance hosts, have prevented effective regional eradication of the disease (Shury et al. 2015; VerCauteren et al. 2018). Another key reservoir host is the brushtail possum (*Trichosurus vulpecula*) in New Zealand, which is a concern for the farmed deer and cattle industries in that country (Buddle et al. 2015). African buffaloes (*Syncerus caffer*) in South Africa and lechwe antelope (*Kobus leche*) in Zambia have also been implicated in ongoing interspecies transmission of bovine TB (de Garine-Wichatitsky et al. 2010; Hlokwe et al. 2016; Malama et al. 2019).

There are also reservoir hosts of other MTBC organisms. *Mycobacterium pinnipedii* is sporadically found in free-ranging marine mammals (including fur seals, sea lions, and a Hector's dolphin, *Cephalorhynchus hectori*) off the coasts of Australia, New Zealand, and Uruguay (Arbiza et al. 2012; Boardman et al. 2014; Roe et al. 2019). *Mycobacteria mungi* and *M. suricattae* are maintained in very specific populations of banded mongooses in Botswana and meerkats in South Africa, respectively (Clarke et al. 2016). More recently, *M. orygis* infection has been suggested as a threat to endangered greater one-horned rhinoceroses (*Rhinoceros unicornis*) and Asian elephants (*Elephas maximus*) in Southeast Asia (Thapa et al. 2017).

The transmission of MTBC from reservoir hosts is complex and not completely understood. An example of the transmission cycle involving reservoir hosts in North America and Europe is shown in Figure 22.1. A number of factors influence whether a population can maintain and spread MTBC infection. These include a sufficient number of susceptible individuals to continue spread, which usually requires close prolonged contact (more common in social species or herds); whether the host develops disease in a location which allows release of bacteria into aerosols or other secretions; a longer duration of infection without killing the host (often dependent on immune status), which increases the chance of spread; and the capability of the bacteria to remain viable long enough to have contact with the new host.

Suspicion of the presence of TB in a free-ranging wildlife population is usually related to a history of

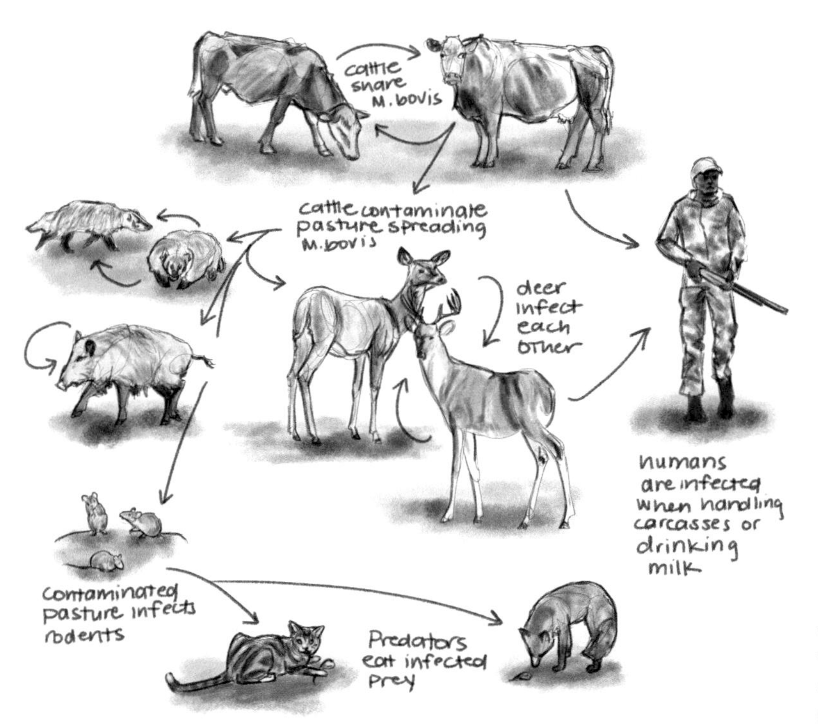

Fig. 22.1. Tuberculosis transmission in North America and Europe. Illustration by Laura Donohue.

TB in livestock or humans in the local area; detection of compatible lesions during postmortem surveillance of hunter-harvested or culled animals; or unexpected changes in appearance, body condition, or behaviour of individuals (de Lisle et al. 2002). Generally, TB is more likely to be found in adult animals with a preponderance of males in some species (such as badgers and elephants), which may be due to differences in behaviour or testosterone effect on immune cells (Graham et al. 2013, Kerr et al. 2019, Dwyer et al. 2020). Risk factors such as drought, increased population density, and aggregation at artificial water and feeding sites have been reported to increase likelihood of finding bovine TB in infected populations (Vicente et al. 2007). Since TB does not present with pathognomic signs, animals showing nonspecific changes, such as loss of body condition, or any individual from an endemically infected population should undergo a complete postmortem examination with collection of tissue samples (frozen and in formalin) for analyses, even in the absence of overt gross pathological changes associated with TB.

Fig. 22.2. Typical large granulomatous abscess of tuberculosis with caseous or liquified centers (Cape buffalo, *Syncerus caffer caffer*, lymph node with advanced *Mycobacterium bovis* infection). Other bacteria can cause similar lesions however, notably Corynebacteria. Photograph by the author.

Signs and Gross Lesions

Often suspicion of TB only arises during postmortem examination. The characteristic gross lesions found in wild ungulates mimic those observed in cattle (Palmer et al. 2019) (Figure 22.2). The hallmark lesion is the granuloma, or tubercle, which is usually a circumscribed yellowish mass, encapsulated by a fibrous capsule and often containing mineralization that feels "gritty" when slicing through the tissue and may have a layered, onion like cross sectional appearance (Domingo et al. 2014). The central area can be necrotic (amorphous, sometimes liquified), caseous ("cheese-like"), or mineralized (Pagán and Ramakrishnan 2018). However, granulomas can be caused by other pathogens (like Corynebacteria) and foreign material and, therefore, are not specific for TB. Granulomas of varying sizes and stages can occur in the same animal; in some cases, granulomas may not be visible but can be detected by palpation of tissue, such as lung. However, granulomas can also coalesce into large masses that distort normal tissue structure. Unlike abscesses caused by some other bacterial infections, TB lesions do not have an odour.

Generally, lesions consistent with bovine TB present as granulomas in the lung and various lymph nodes, especially in the head and chest (Laisse et al. 2011). This is consistent with aerosol transmission, typically in social species. However, more generalized disease with granulomas in multiple organs and abdominal lymph nodes is not uncommon. In cervids with bovine TB, lymph node and lung granulomas often resemble an abscess with a soft caseous core (Zanella et al. 2008; Fitzgerald and Kaneene 2012). In contrast, caseous mineralized lesions occurred primarily in the lymph nodes of the head in wild boars; however, TB granulomas have also been found in the lungs, thoracic and abdominal lymph nodes, and other organs in this species (Naranjo et al. 2008). These lesions resemble those found in other ungulates (Domingo et al. 2014).

Gross lesions in nonungulate species show greater variation in appearance and location. Badgers are typically infected by inhalation of infectious aerosols, with granulomas in the lung and thoracic lymph nodes. However, granulomas are less commonly necrotic or mineralized. Bite wounds contaminated with *M. bovis* appear as draining abscesses or ulcerated lesions (Gormley and Corner 2018*a*). Carnivores, especially lions (*Panthera leo*), tend to develop cavitary lesions in the lung, typically without mineralization, which can be mistaken for pneumonia caused by other organisms (Viljoen et al. 2015). Lymph node granulomas often resemble neoplasia due to the lack of a caseous core. Critically endangered African wild dogs (*Lycaon pictus*) in Kruger National Park (KNP), South Africa, where TB is endemic, were found with generalized TB that appeared as extensively distributed firm masses in multiple lymph nodes, especially in the abdomen (Meiring et al. 2021*a*). However, in other carnivores or omnivores, *M. bovis* has been isolated in the absence of gross lesions in the majority of cases; these included samples from a red fox (*Vulpes vulpes*), coyotes (*Canis latrans*), raccoons (*Procyon lotor*), and a black bear (*Ursus americanus*) in Michigan (Bruning-Fann et al. 2001). Gross lesions in some coyotes were found in the abdominal lymph nodes and were partially mineralized. A rare case of *M. bovis* in a brown bear (*Ursus arctos marsicanus*) in Italy revealed enlarged abdominal lymph nodes that were necrotic (Fico et al. 2019). In carnivores and scavengers, the route of infection is usually oral, through ingestion of infected carcasses (Renwick et al. 2007). These examples show the wide variation in gross pathology between species and support the submission of samples for mycobacterial testing even in the absence of lesions.

In addition to *M. bovis*, *M. microti* can infect humans, as well as domestic and wild mammals (Michelet et al. 2015). Field voles (*Microtus agrestis*), bank voles (*Myodes glareolus*), wood mice (*Apodemus sylvaticus*), and shrews (*Sorex araneus*) have been identified as natural reservoirs of *M. microti* (Cavanagh et al. 2002). Naturally infected field voles, captured in the U.K., had nodular skin lesions with ulceration and granulomas in lymph nodes in the axilla, neck, and abdomen, as well as in the lung (Kipar et al. 2014). The distribution of *M. microti* bacilli in these animals suggest that infected voles may be a transmission risk through direct contact, being ingested, or contaminating the environment. A report describing *M. microti* infection in wild boars in Italy found that it was indistinguishable from *M. bovis* infection, with yellowish-white nodules with a capsule found in the lymph nodes of the head (Boniotti et al. 2014). Like *M. bovis*–infected red foxes in the United States, *M. microti* infection in red foxes in France did not present any TB-like gross lesions (Michelet et al. 2021). Therefore, samples submitted for mycobacterial culture should also undergo genetic speciation and possibly direct polymerase chain reaction (PCR) of samples, since the results can influence the epidemiological investigation (Gavier-Widen et al. 2009, Reis et al. 2021*b*).

Pathogenesis and Pathology

Pathogenic mycobacteria can be transmitted through infectious aerosols produced when the individual coughs or vocalizes, through ingestion of milk or meat from an infected animal, or indirectly, through inhalation or ingestion of mycobacteria present in the environment (Courtenay et al. 2006; Fellag et al. 2021). It is also possible that MTBC organisms present in oral secretions can be spread through biting or contamination of wounds (Gormley and Corner 2018*a*). Once inhaled or ingested, mycobacteria infect macrophages and other phagocytic cells locally, which then travel to the regional lymph nodes. Although phagocytic cells can kill mycobacteria, some bacteria may evade destruction by the host's immune system; these organisms can continue to proliferate, resulting in an attempt by the body to "wall off" the bacilli in a granuloma (Cooper 2009).

Granulomas are considered the hallmark lesion of tuberculosis. As the granuloma matures, fibrosis and necrosis occur (Pagan and Ramakrishnan 2018). Local proliferation as well as spread to other lymph nodes and organs can result in generalized disease, especially in immunocompromised individuals. However, it is important to note that host responses and the resulting pathological changes in MTBC-associated disease are dynamic processes with the immune cell populations and pathogen distribution changing over time, which can result in differences in lesion morphology (Domingo et al. 2014, Palmer 2018).

Histologically, granulomas are comprised of varied cell types including but not limited to epithelial macrophages, multinucleated giant cells, and lymphocytes (Palmer et al. 2019). Characteristics of granulomas in *M. bovis*–infected cattle have been described in different stages of infection; these lesions are similar in infected cervids, wild bovids, and suids (Wangoo et al. 2005). In early infection, irregular clusters of immune cells with little necrosis and a thin fibrous capsule are seen. However, as the granuloma matures, epithelioid macrophages and multinucleated giant cells (a key microscopic finding) surround areas of central necrosis. Peripherally, clusters of lymphocytes and scattered neutrophils extend into the fibrous capsule. Extensive mineralization can also be observed. Low numbers of acid-fast bacilli are present extracellularly within the necrotic center and rarely within the macrophages or multinucleated giant cells (Domingo et al. 2014). In carnivores and other species such as European badgers, the granuloma may appear more proliferative or cavitary with variable necrosis, mineralization, or fibrosis (Viljoen et al. 2015, Gormley and Corner 2018*a*).

Fig. 22.3. (A) Widely disseminated granulomas in the lung of an African elephant (*Loxodonta africana*) caused by *Mycobacterium tuberculosis*. (B) A lung pattern of tuberculosis typically seen in advanced disease when bacteria are inhaled. Photographs by the author.

Diagnosis

Suspicion of tuberculosis in living animals is often based on a known history of pathogen presence in the geographical location or population, along with chronic clinical signs of weight loss or other indications of a "poor doer;" however, many infected individuals may not show evidence of disease (de Lisle et al. 2002). Often, presumptive diagnosis of tuberculosis is only made during postmortem examination when granulomas are discovered in lungs, lymph nodes, or other organs (Figure 22.3). To confirm

MTBC infection, it is crucial that tissues are collected and placed in 10% buffered formalin, and duplicate samples are frozen. For antemortem diagnosis, respiratory or other secretions (such as wound exudate) and tissue biopsies should be used to make impression smears in the field along with additional samples stored in sterile containers and frozen for culture (Lekko et al. 2020).

Definitive diagnosis requires detection and identification of pathogenic mycobacteria in either antemortem or postmortem samples. Rapid screening can be accomplished by direct staining with Ziehl-Neelsen stain of impression smears from tissue or secretions and detection of acid-fast bacilli, although nontuberculous mycobacteria and other organisms can cause a false-positive result (Gcebe and Hlokwe 2017; Thomas et al. 2021). Mycobacterial culture is considered the gold standard for MTBC diagnosis, along with speciation by PCR. However, this can result in false-negative results due to low levels of bacteria, intermittent shedding (in the case of secretions), slow growth of the organisms, and harsh decontamination methods used prior to culture inoculation (Lekko et al. 2020; Goosen et al. 2021). Therefore, various direct PCR methods have been developed to identify MTBC bacilli more rapidly than culture (Bernitz et al. 2021). The disadvantage of using this technique directly on samples is that it cannot distinguish between viable and dead bacteria.

Histopathological examination of formalin-fixed tissues that find characteristic granulomas leads to suspicion of TB, especially if acid-fast organisms are detected, and should undergo further investigation (Thomas et al. 2021). Since not all laboratories are able to perform mycobacterial culture, it is important to check with the laboratory before submitting samples. Specific techniques continue to be improved and have been described in recent reports on wildlife TB diagnosis (Lekko et al. 2020; Bernitz et al. 2021; Goosen et al. 2021).

Since direct detection is not always feasible, other diagnostic tests are being developed for wildlife (Chambers 2013; Lekko et al. 2020; Bernitz et al. 2021; Smith et al. 2021; Thomas et al. 2021). These are predominately immunological or indirect tests. The tuberculin skin test (TST) is an example of using a detectable immune response to mycobacterial components as evidence that infection has occurred (de Lisle et al. 2002; Good and Duignan 2011) (Figure 22.4). Depending on the country or state, the TST is administered and interpretated according to specific guidelines, which may also determine the location of injection as well as whether bovine tuberculin (purified protein derivative) is used solely or in combination with avian tuberculin to assist in detecting cross-reactive responses to environmental mycobacteria. Therefore, prior to performing any TB test in wildlife, consultation with animal health regulatory officials is required to ensure compliance with local regulations.

Although advances have been made, in vitro immunological techniques usually require species-specific reagents and validation and, therefore, are not commonly available. However, serum and whole blood samples should be collected, frozen, and banked from suspect TB cases. In some species, such as cervids, wild suids, elephants, and badgers, sero-

Fig. 22.4. Tuberculin skin test being assessed in immobilized lioness (*Panthera leo*) with advanced bovine tuberculosis. Photograph by the author.

logical tests are available that can detect antibodies to MTBC (Thomas et al. 2021). These assays are particularly useful for population surveillance as well as screening individuals. For wild bovids, MTBC-specific bovine interferon gamma release assays have been used to provide ancillary diagnostic information (Bernitz et al. 2021). However, this requires that a fresh heparinized whole blood sample be processed the same day as collection and, therefore, may not be field friendly for use in remote locations. Novel immunological assays continue to be developed for wildlife and may be available in the future, although this is beyond the scope of this chapter (Smith et al. 2021).

Outcome and Treatment

In general, it is believed that most infections in wildlife species follow a chronic insidious progression to disease and death (de Lisle et al. 2002). However, this process may take months to years, and infected individuals may succumb to other causes, such as predation or loss of fitness (Michel et al. 2006). It is unknown if some individuals may be able to clear or contain infection (similar to humans), if their immune system is not compromised by drought, parasites, or other infections (Ezenwa et al. 2010; Risco et al. 2014). Regardless, in many countries, any suspected case of TB is required to be reported to animal health regulatory agencies (Good and Duignan 2011, Palmer and Waters 2011, Meiring et al. 2018). Officials often determine the outcome, which is usually culling of any test-positive animals. Exceptions for treatment have been made, with specific plans approved by animal health officials, for individuals such as captive elephants, zoo rhinoceros, and nonhuman primates (Mikota et al. 2015; Miller and Lyashchenko 2015). However, effective treatment regimens have not been validated for wildlife species, and treatment requires months-long administration of human drugs that are often cost prohibitive. Any consideration for treatment should be done by consulting with experts in animal TB and public and animal health agencies to tailor a plan. However, in general, treatment is not recommended due to the risk of exposure to other animals and humans.

Management, Financial, Legal, Political, and Social Factors

Recognition of the zoonotic nature of bovine TB has resulted in implementation of control programs in many countries. A national eradication program for bovine TB was implemented in 1917 in the United States, with other countries following their example through the mid-1900s (Olmstead and Rhode 2004; Bezos et al. 2014; Good et al. 2018). Even though bovine TB management plans are regulated in many countries, implementation is poor due to limited resources in low- and middle-income countries (Cosivi et al. 1998; Ayele et al. 2004; Zinsstag et al. 2006; Meiring et al. 2018; Dubey et al. 2020). Despite the increasing evidence of the impact and role of bovine TB in wildlife hosts during the 20th century, few countries have formalized plans for dealing with TB in wildlife. Nepal is one of the few with a national plan for detection and management of TB in elephants (Paudel and Tsubota 2016). The Animal and Plant Health Inspection Service (APHIS) in the United States has developed the Voluntary National Cervid TB Herd Accreditation Program for farmed and captive cervids (USDA APHIS 2020) as well as *Guidelines for Surveillance of Bovine Tuberculosis in Wildlife* (USDA APHIS 2011). However, the relative paucity of knowledge on the epidemiology, pathogenesis, and validated diagnostic tests for TB in wildlife species results in a significant gap in ability to effectively control the disease.

Currently, bovine tuberculosis is a disease listed in the World Organisation for Animal Health (OIE) Terrestrial Animal Health Code, which requires that any cases, including those in wildlife, must be reported by the country's animal health agency (OIE 2021). In addition, each country has differing requirements for notification of national and local animal health authorities. Since many MTBC can cause zoonotic infections, public health agencies may also be involved in management plans in some cases. Generally, management of disease outbreaks associated with *M. bovis*, and occasionally other MTBC, infections in wildlife includes culling of infected individuals, depopulation or quarantine of the susceptible population and restricting intentional movement from the current location (Gormley and Corner 2018*b*; Meiring et al. 2018; Arnot and Michel 2020). In the case of free-ranging wildlife populations, numerous published reports of management and control activities are available (Gortazar et al. 2011; O'Brien et al. 2011*a*, 2011*b*; Fitzgerald and Kaneene 2012; Nugent et al. 2018). Additional actions including habitat management, such as fencing to exclude wildlife access to contaminated water and feeders, and vaccination have been investigated (Gortázar et al. 2015; Smith and Delahay 2018). Since each situation involves different stakeholders, epidemiological factors, and susceptible host populations, management plans should be specifically tailored to each outbreak (More 2019).

In countries with effective bovine TB programs in livestock, the presence of wildlife reservoir host(s) results in costly spillovers and presents a barrier to successful eradication (Zinsstag et al. 2006; McCulloch and Reiss 2017). For example, bovine TB in free-ranging white-tailed deer in Michigan, USA, has been managed by lowering deer densities through hunting, restricted feeding, and using hunter surveillance to assess prevalence and prevent spillover to livestock (O'Brien et al. 2011*a*). The estimated cost of TB-related activities by the state's Department of Natural Resources was over $23 million (USD) between 1997 and 2011. Despite intensive efforts, bovine TB control continues to be an issue in Michigan due to conflicting financial, political, social, and ethical considerations (O'Brien et al. 2011*b*). In contrast, New Zealand has undertaken intensive bovine TB eradication from cattle and deer through lethal control of the wildlife reservoir, brushtail possum *Trichosurus vulpecula*), which was considered acceptable since they were not a native species (Livingstone et al. 2015; Nugent et al. 2018). However, culling of European badgers in the United Kingdom has produced conflicting results and ethical concerns as a method for controlling bovine TB in cattle (McCulloch and Reiss 2017; Downs et al. 2019). Therefore, Bacillus Calmette–Guérin (BCG) vaccination of badgers and improved tracking in cattle have been investigated as additional components of the eradication program (Martin et al. 2020). Vaccination using oral BCG in free-ranging wild boar populations has also been reported to provide a protective effect in piglets (Diez-Delgado et al. 2018). Studies in several species has shown that although vaccination does not provide complete protection against infection, it reduces severity—and, therefore, the likelihood of transmission—and should be considered as a management tool where legally approved (Balseiro et al. 2020).

In some circumstances where TB is endemic in wildlife, culling, vaccination, and depopulation are not feasible nor ethical. For example, bovine TB is endemic in a number of sub-Saharan African wildlife ecosystems, including KNP (Box 22.1) and Hluhluwe-iMfolozi Park (South Africa), Queen Elizabeth National Park (Uganda), Kafue Basin (Zambia) with infection also detected in wildlife in Tarangire National Park and Serengeti (Tanzania), Amboseli (Kenya), Limpopo National Park (Mozambique), and Gonarezhou National Park (Zimbabwe) (Cleaveland et al. 2005; Kalema-Zikusoka et al. 2005; Michel et al. 2006; Munyeme et al. 2010; Tanner et al. 2015). Spillback from wildlife to livestock surrounding these parks has been documented despite attempts to minimize spread, which leads to potential stakeholder conflicts.

Summary

The importance of bovine TB in wildlife is not only a threat to livestock and human health but may have

Box 22.1 Tuberculosis in Kruger National Park, South Africa

Kruger National Park (KNP) is a multihost complex ecosystem in which *M. bovis* has been introduced into free-ranging susceptible species without management interventions. Bovine tuberculosis is considered an "alien" disease, since it was believed to have been introduced to South Africa through importation of infected cattle in the 19th century. Although sporadic cases were reported in wildlife, it wasn't until the 1990s that the disease was considered a threat to wildlife after finding bovine tuberculosis (bTB) in buffalo herds near the southwestern park border, where infected cattle had been present. Since that time, surveys have shown bTB spread from buffalo herds, initially in the south to the northern border by 2006, and across the border to buffalo in Zimbabwe in 2008. During the three decades since bTB was first discovered in KNP, the list of infected species has increased to 19, including black and white rhinoceroses, African elephants, African wild dogs, leopards, lions, banded mongooses, Chacma baboons, warthogs, buffaloes, antelope (greater kudu, common duiker, bushbuck, impala), honey badger, spotted genet, vervet monkey, and more recently, a giraffe and Nile hippopotamus. Although not all species have obvious signs of disease, the presence of multiple reservoir hosts (including buffalos, greater kudu, and lions) ensures the persistence of infection in the ecosystem, although routes of transmission are not always known (Figures 22.5 and 22.6). In addition, environmental changes (such as drought) appear to increase the likelihood of disease development in some species, such as rhinoceros. In contrast, young otherwise healthy *M. bovis*–infected white rhinoceros in bomas (receiving sufficient food and water) appeared to be able to contain and potentially even eliminate the bacteria.

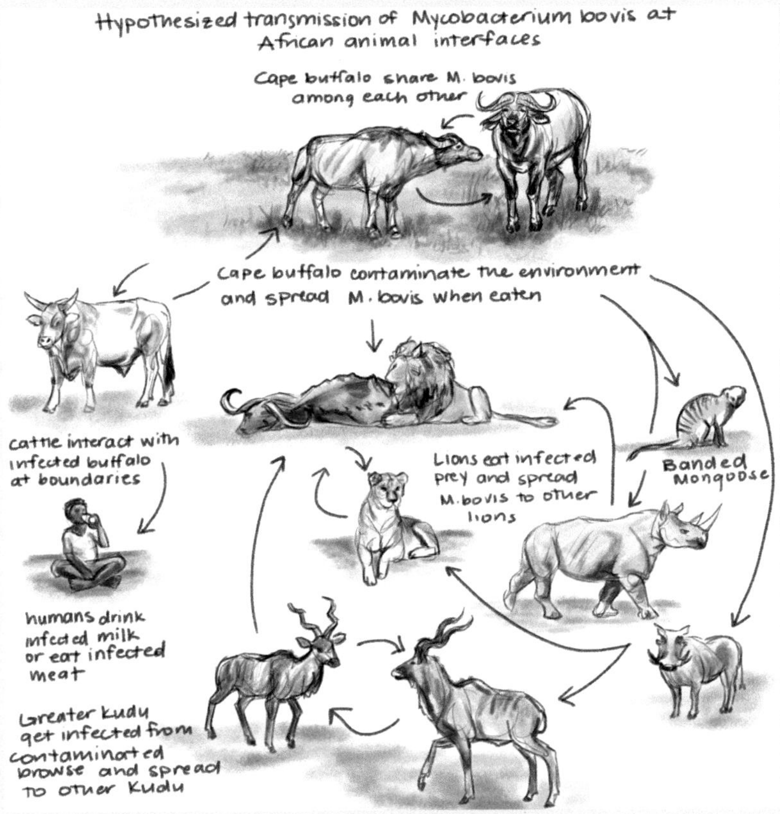

Fig. 22.5. Tuberculosis transmission in Africa. Illustration by Laura Donohue.

(continued)

Fig. 22.6. (A) An adult female greater kudu (*Tragelaphus strepsiceros*) infected with bovine tuberculosis characterized by swollen lymph nodes and loss of body condition. (B) Draining lymph node in a greater kudu infected with *Mycobacterium bovis*. (A) Photograph by the author. (B) Photograph courtesy of Roy Bengis, Skukuza State Veterinarian.

Recent studies suggest that herbivores including rhinoceros, hippopotamus, and antelope may be infected indirectly from environmental sources, whereas carnivores and scavengers can be infected through eating prey. Although lions were initially considered spillover hosts, the discovery that lions have viable mycobacteria in respiratory secretions and appear to transmit *M. bovis* through biting (causing infection and swelling; "elbow hygromas") in other lions has led to including them as a reservoir species in KNP. It is speculated that the high level of bacteria in the system results in spillover to less commonly susceptible hosts, although it is unknown if some of these species can become reservoirs (for example, shedding of bacteria in respiratory samples has been found in African wild dogs). Interestingly, *M. tuberculosis* (human form of tuberculosis) has led to the death of an African elephant bull; the disease was presumably introduced by staff, visitors, or as a result of elephants moving into surrounding communities.

Unlike other game reserves in South Africa, South African National Parks management have elected not to implement disease control interventions. The presence of bTB has resulted in regulatory restrictions in movement of known infected species, such as rhinoceros, from the park without quarantine testing. Therefore, Kruger National Park is a real-world example of how introduced MTBC can impact free-ranging populations of wildlife that have interfaces with domestic animals and humans.

Stellenbosch University and University of Pretoria 2015, Hlokwe et al. 2019, Bernitz et al. 2021, Kerr et al. 2022, Meiring et al. 2021*b*, Miller et al. 2021

a significant impact on conservation and economic effects on wildlife ranching. Premises with documented *M. bovis* infection are often placed under quarantine, which prevents movement of animals completely or, in some cases, only with approved testing. However, there are few validated tests for wildlife. Therefore, reintroduction, translocation for genetic and population management, or animal sales are severely hindered. In addition, the direct effects of disease on fitness of the population are unknown, especially in fragmented habitats. For example, in KNP, the critically endangered Africa wild dog population has a high apparent prevalence of *M. bovis* infection (82%) with associated mortalities (Higgitt et al. 2019; Meiring et al. 2021*a*). In South Asia, TB has been recognized as an emerging threat to endangered species including Asian elephants, greater one-horned rhinoceros, and Hanuman langur (*Presbytis entellus*) (Parmar et al. 2013; Thapa et al. 2017). In areas where wildlife are competing with humans and domestic animals, complex interactions can lead to MTBC spread at the interfaces, which are unlikely to be controlled without significant and costly buy-in and commitment by all stakeholders.

LITERATURE CITED

Abrantes, A. C., J. Serejo, and M. Vieira-Pinto. 2021. The association between Palmer drought severity index data and tuberculosis-like lesions occurrence in Mediterranean hunted wild boars. Animals 11:2060.

Ameni, G., M. Vordermeier, R. Firdessa, A. Aseffa, G. Hewinson, S. V. Gordon, and S. Berg. 2011. *Mycobacterium tuberculosis* infection in grazing cattle in central Ethiopia. The Veterinary Journal 188:359–361.

Arbiza, J., A. Blanc, M. Castro-Ramos, H. Katz, A. P. de León, and M. Clara. 2012. Uruguayan pinnipeds (*Arctocephalus australis* and *Otaria flavescens*): Evidence of influenza virus and *Mycobacterium pinnipedii* infections. New Approaches to Study Marine Mammals 2:151–182.

Arnot, L. F., and A. Michel. 2020. Challenges for controlling bovine tuberculosis in South Africa. Onderstepoort Journal of Veterinary Research 87:a1690.

Ayele, W. Y., S. D. Neill, J. Zinsstag, W. G. Weiss, and I. Pavlik. 2004. Bovine tuberculosis: an old disease but a new threat to Africa. International Journal of Tuberculosis and Lung Disease 8:924–937.

Balseiro, A., J. Thomas, C. Gortázar, and M. A. Risalde. 2020. Development and challenges in animal tuberculosis vaccination. Pathogens 9:472.

Bernitz, N., T. J. Kerr, W. J. Goosen, J. Chileshe, R. L. Higgitt, E. O. Roos, C. Meiring, R. Gumbo, C. De Waal, C. Clarke, and K. Smith. 2021. Review of diagnostic tests for detection of *Mycobacterium bovis* infection in South African wildlife. Frontiers in Veterinary Science 8:588697.

Bezos, J., J. Álvarez, B. Romero, L. de Juan, and L. Domínguez. 2014. Bovine tuberculosis: historical perspective. Research in Veterinary Science 97(Supplement):S3–S4.

Bhuachalla, D. N., L. A. Corner, S. J. More, and E. Gormley E. 2015. The role of badgers in the epidemiology of *Mycobacterium bovis* infection (tuberculosis) in cattle in the United Kingdom and the Republic of Ireland: current perspectives on control strategies. Veterinary Medicine: Research and Reports 6:27.

Boardman, W. S., L. Shephard, I. Bastian, M. Globan, J. A. Fyfe, D. V. Cousins, A. Machado, and L. Woolford. 2014. *Mycobacterium pinnipedii* tuberculosis in a free-ranging Australian fur seal (*Arctocephalus pusillus doriferus*) in South Australia. Journal of Zoo and Wildlife Medicine 45:970–972.

Boniotti, M. B., A. Gaffuri, D. Gelmetti, S. Tagliabue, M. Chiari, A. Mangeli, M. Spisani, C. Nassuato, L. Gibelli, C. Sacchi, and M. Zanoni. 2014. Detection and molecular characterization of *Mycobacterium microti* isolates in wild boar from northern Italy. Journal of Clinical Microbiology 52:2834–2843.

Bos, K. I., K. M. Harkins, A. Herbig, M. Coscolla, N. Weber, I. Comas, S. A. Forrest, J. M. Bryant, S. R. Harris, V. J. Schuenemann, and T. J. Campbell. 2014. Pre-Columbian mycobacterial genomes reveal seals as a source of New World human tuberculosis. Nature 514:494–497.

Bruning-Fann, C. S., S. M. Schmitt, S. D. Fitzgerald, J. S. Fierke, P. D. Friedrich, J. B. Kaneene, K. A. Clarke, K. L. Butler, J. B. Payeur, D. L. Whipple, and T. M. Cooley. 2001. Bovine tuberculosis in free-ranging carnivores from Michigan. Journal of Wildlife Diseases 37:58–64.

Buddle, B. M., G. W. de Lisle, J. F. T. Griffin, and S. A. Hutchings. 2015. Epidemiology, diagnostics, and management of tuberculosis in domestic cattle and deer in New Zealand in the face of a wildlife reservoir. New Zealand Veterinary Journal 63(Supplement 1):19–27.

Cavanagh, R., M. Begon, M. Bennett, T. Ergon, I. M. Graham, P. E. De Haas, C. A. Hart, M. Koedam, K. Kremer, X. Lambin, et al. 2002. *Mycobacterium microti* infection (vole tuberculosis) in wild rodent populations. Journal of Clinical Microbiology 40:3281–3285.

Chambers, M. A. 2013. Review of the diagnosis of tuberculosis in non-bovid wildlife species using immunological methods–an update of published work since 2009. Transboundary and Emerging Diseases 60:14–27.

Clarke, C., P. Van Helden, M. Miller, and S. Parsons. 2016. Animal-adapted members of the *Mycobacterium tuberculosis* complex endemic to the southern African subregion. Journal of the South African Veterinary Association 87:a1322.

Cleaveland, S., T. Mlengeya, R. R. Kazwala, A. Michel, T. Kaare, S. L. Jones, E. Eblate, G. M. Shirima, and C. Packer. 2005. Tuberculosis in Tanzanian wildlife. Journal of Wildlife Diseases 41:446–453.

Cooper, A. M. 2009. Cell-mediated immune responses in tuberculosis. Annual Review of Immunology 27:393–422.

Cosivi, O., J. M. Grange, C. J. Daborn, M. C. Raviglione, T. Fujikura, D. Cousins, R. A. Robinson, H. F. Huchzermeyer, I. de Kantor, and F. X. Meslin. 1998. Zoonotic tuberculosis due to *Mycobacterium bovis* in developing countries. Emerging Infectious Diseases 4:59.

Courtenay, O., L. A. Reilly, F. P. Sweeney, V. Hibberd, S. Bryan, A. Ul-Hassan, C. Newman, D. W. Macdonald, R. J. Delahay, G. J. Wilson, and E. M. H. Wellington. 2006. Is *Mycobacterium bovis* in the environment important for the persistence of bovine tuberculosis? Biology Letters 2:460–462.

Crispell, J., S. Cassidy, K. Kenny, G. McGrath, S. Warde, H. Cameron, G. Rossi, T. MacWhite, P. C. L. White, S. Lycett, R. R. Kao, J. Moriarty, and S. V. Gordon. 2020. *Mycobacterium bovis* genomics reveals transmission of infection between cattle and deer in Ireland. Microbial Genomics 6:mgen000388.

de Garine-Wichatitsky, M., A. Caron, C. Gomo, C. Foggin, K. Dutlow, D. Pfukenyi, E. Lane, S. Le Bel, M. Hofmeyr, T. Hlokwe, and A. Michel. 2010. Bovine tuberculosis in buffaloes, Southern Africa. Emerging Infectious Diseases 16:884–885.

de Lisle, G. W., R. G. Bengis, S. M. Schmitt, and D. J. O Brien. 2002. Tuberculosis in free-ranging wildlife: detection, diagnosis and management. Revue Scientifique et Technique-Office International des Epizooties 21:317–334.

Díez-Delgado, I., I. A. Sevilla, B. Romero, E. Tanner, J. A. Barasona, A. R. White, P. W. W. Lurz, M. Boots, J. de la Fuente, L. Domínguez, et al. 2018. Impact of piglet oral vaccination against tuberculosis in endemic free-ranging wild boar populations. Preventive Veterinary Medicine 155:11–20.

Domingo, M., E. Vidal, and A. Marco. 2014. Pathology of bovine tuberculosis. Research in Veterinary Science 97:20–29.

Domingos, S. C., H. R. C. Júnior, W. Lilenbaum, M. T. Santa Rosa, C. D. Pereira, and L. S. Medeiros. 2019. A systematic review on the distribution of *Mycobacterium bovis* infection among wildlife in the Americas. Tropical Animal Health and Production 51:1801–1805.

Downs, S. H., A. Prosser, A. Ashton, S. Ashfield, L. A. Brunton, A. Brouwer, P. Upton, A. Robertson, C. A. Donnelly, and J. E. Parry. 2019. Assessing effects from four years of industry-led badger culling in England on the incidence of bovine tuberculosis in cattle, 2013–2017. Scientific Reports 9:14666.

Dubey, S., R. V. Singh, B. Gupta, A. Nayak, A. Rai, D. Gupta, R. Patel, D. Soni, and B. M. S. Dhakad. 2020. Bovine tuberculosis and its public health significance/ bovine tuberculosis and its public health significance: a review. Journal of Entomology and Zoology Studies 8:2281–2287.

Dwyer, R. A., C. Witte, P. Buss, W.J. Goosen, and M. Miller. 2020. Epidemiology of tuberculosis in multi-host wildlife systems: implications for black (*Diceros bicornis*) and white (*Ceratotherium simum*) rhinoceros. Frontiers in Veterinary Science 7:580476.

Ezenwa, V. O., R. S. Etienne, G. Luikart, A. Beja-Pereira, and A. E. Jolles. 2010. Hidden consequences of living in a wormy world: nematode-induced immune suppression facilitates tuberculosis invasion in African buffalo. American Naturalist 176:613–624.

Falkinham, J. O, III. 2015. Nontuberculous mycobacterial infections. Pages 538–548 *in* Mukundan, H., M. A. Chambers, W. R. Waters, and M. H. Larsen, editors. Tuberculosis, leprosy and mycobacterial diseases of man and animals: the many hosts of mycobacteria. Commonwealth Agricultural Bureaux International, Boston, Massachusetts, USA.

Fellag, M., A. Loukil, and M. Drancourt. 2021. The puzzle of the evolutionary natural history of tuberculosis. New Microbes and New Infections 41:100712.

Fico, R., A. Mariacher, A. Franco, C. Eleni, E. Ciarrocca, M. L. Pacciarini, and A. Battisti. 2019. Systemic tuberculosis by *Mycobacterium bovis* in a free-ranging Marsican brown bear (*Ursus arctos marsicanus*): a case report. BMC Veterinary Research 15:152.

Fitzgerald, S. D., and J. B. Kaneene. 2012. Wildlife reservoirs of bovine tuberculosis worldwide: hosts, pathology, surveillance, and control. Veterinary Pathology 50:488–499.

Gagneux, S. 2018. Ecology and evolution of *Mycobacterium tuberculosis*. Nature Reviews Microbiology 16:202–213.

Gavier-Widen, D., M. M. Cooke, J. Gallagher, M. A. Chambers, and C. Gortázar. 2009. A review of infection of wildlife hosts with *Mycobacterium bovis* and the diagnostic difficulties of the 'no visible lesion' presentation. New Zealand Veterinary Journal 57:122–131.

Gcebe, N., and T. M. Hlokwe. 2017. Non-tuberculous mycobacteria in South African wildlife: neglected pathogens and potential impediments for bovine tuberculosis diagnosis. Frontiers in Cellular and Infection Microbiology 7:15.

Good, M., D. Bakker, A. Duignan, and D. M. Collins. 2018. The history of in vivo tuberculin testing in bovines: tuberculosis, a "One Health" issue. Frontiers in Veterinary Science 5:59.

Good, M., and A. Duignan. 2011. Perspectives on the history of bovine TB and the role of tuberculin in bovine TB eradication. Veterinary Medicine International 2011:410470.

Goosen, W. J., L. Kleynhans, T. J. Kerr, P. D. van Helden, P. Buss, R. M. Warren, and M. A. Miller. 2021. Improved detection of *Mycobacterium tuberculosis* and *M. bovis* in African wildlife samples using cationic peptide decontamination and mycobacterial culture supplementation. Journal of Veterinary Diagnostic Investigation 2021:10406387211044192.

Gormley, E., and L. Corner. 2018*a*. Pathogenesis of *Mycobacterium bovis* infection: the badger model as a paradigm for understanding tuberculosis in animals. Frontiers in Veterinary Science 4:247.

Gormley, E., and L. A. L. Corner. 2018*b*. Wild animal tuberculosis: stakeholder value systems and management of disease. Frontiers in Veterinary Science 5:327.

Gortázar, C., A. Che Amat, and D. J. O'Brien. 2015. Open questions and recent advances in the control of a multi-host infectious disease: animal tuberculosis. Mammal Review 45:160–175.

Gortazar, C., J. Vicente, M. Boadella, C. Ballesteros, R. C. Galindo, J. Garrido, A. Aranaz, and J. De la Fuente. 2011. Progress in the control of bovine tuberculosis in Spanish wildlife. Veterinary Microbiology 151:170–178.

Graham, J., G. C. Smith, R. J. Delahay, T. Bailey, R. A. McDonald, and D. Hodgson. 2013. Multi-state modelling reveals sex-dependent transmission, progression and severity of tuberculosis in wild badgers. Epidemiology and Infection 141:1429–1436.

Higgitt, R. L., O. L. Van Schalkwyk, L. M. de Klerk-Lorist, P. E. Buss, P. Caldwell, L. Rossouw, T. Manamela, G. A. Hausler, J. Hewlett, E. P. Mitchell, et al. 2019. *Mycobacterium bovis* infection in African wild dogs, Kruger National Park, South Africa. Emerging Infectious Diseases 25:1425–1427.

Hlokwe, T. M., L. M. De Klerk-Lorist, and A. L. Michel. 2016. Wildlife on the move: a hidden tuberculosis threat to conservation areas and game farms through introduction of untested animals. Journal of Wildlife Diseases 52:837–843.

Hlokwe, T. M., A. L. Michel, E. Mitchel, N. Gcebe, and B. Reininghaus. 2019. First detection of *Mycobacterium bovis* infection in giraffe (*Giraffa camelopardalis*) in the Greater Kruger National Park Complex: role and implications. Transboundary and Emerging Diseases 66:2264–2270.

Inlamea, O. F., P. Soares, C. Y. Ikuta, M. B. Heinemann, S. J. Achá, A. Machado, J. S. Ferreira Neto, M. Correia-Neves, and T. Rito. 2020. Evolutionary analysis of *Mycobacterium bovis* genotypes across Africa suggests co-evolution with livestock and humans. PLoS Neglected Tropical Diseases 14:e0008081.

Kalema-Zikusoka, G., R. G. Bengis, A. L. Michel, and M. H. Woodford. 2005. A preliminary investigation of tuberculosis and other diseases in African buffalo (*Syncerus caffer*) in Queen Elizabeth National Park, Uganda. Onderstepoort Journal of Veterinary Research 72:145–151.

Kaneene J. B., R. Miller, J. H. Steele, and C. O. Thoen. 2014. Preventing and controlling zoonotic tuberculosis: a One Health approach. Veterinaria Italiana 50:7–22.

Kerr, T. J., C. R. de Waal, P. E. Buss, J. Hofmeyr, K. P. Lyashchenko, and M. A. Miller. 2019. Seroprevalence of *Mycobacterium tuberculosis* complex in free-ranging African elephants (*Loxodonta africana*) in Kruger National Park, South Africa. Journal of Wildlife Diseases 55:923–927.

Kerr, T. J., W. J. Goosen, R. Gumbo, L.-M. de Klerk-Lorist, O. Pretorius, P. E. Buss, L. Kleynhans, K. P. Lyashchenko, R. M. Warren, P. D. van Helden, and M. A. Miller. 2022. Diagnosis of *Mycobacterium bovis* infection in free-ranging common hippopotamus (*Hippopotamus amphibius*). Transboundary and Emerging Diseases 69:378–384.

Kipar, A., S. J. Burthe, U. Hetzel, M. A. Rokia, S. Telfer, X. Lambin, R. J. Birtles, M. Begon, and M. Bennett. 2014. *Mycobacterium microti* tuberculosis in its maintenance host, the field vole (*Microtus agrestis*) characterization of the disease and possible routes of transmission. Veterinary Pathology 51:903–914.

Kock, R., A. L. Michel, D. Yeboah-Manu, E. I. Azhar, J. B. Torrelles, S. I. Cadmus, L. Brunton, J. M. Chakaya, B. Marais, L. Mboera, et al. 2021. Zoonotic tuberculosis—the changing landscape. International Journal of Infectious Diseases 113(Supplement 1):S68–S72.

Laisse, C. J., D. Gavier-Widén, G. Ramis, C. G. Bila, A. Machado, J. J. Quereda, E. O. Ågren, and P. D. van Helden. 2011. Characterization of tuberculous lesions in naturally infected African buffalo (*Syncerus caffer*). Journal of Veterinary Diagnostic Investigation 23:1022–1027.

Lekko, Y. M., P. T. Ooi, S. Omar, M. Mazlan, S. Z. Ramanoon, S. Jasni, F. F. A. Jesse, and A. Che-Amat. 2020. *Mycobacterium tuberculosis* complex in wildlife: review of current applications of antemortem and postmortem diagnosis. Veterinary World 13:1822.

Livingstone, P. G., N. Hancox, G. Nugent, G. Mackereth, and S.A. Hutchings. 2015. Development of the New Zealand strategy for local eradication of tuberculosis from wildlife and livestock. New Zealand Veterinary Journal 63(Supplement 1):98–107.

Loiseau, C., F. Menardo, A. Aseffa, E. Hailu, B. Gumi, G. Ameni, S. Berg, L. Rigouts, S. Robbe-Austerman, J. Zinsstag, and S. Gagneux. 2020. An African origin for *Mycobacterium bovis*. Evolution, Medicine, and Public Health 2020:49–59.

MacNeil, A., P. Glaziou, C. Sismanidis, S. Maloney, and K. Floyd. 2019. Global epidemiology of tuberculosis and progress toward achieving global targets—2017. Morbidity and Mortality Weekly Reports 68:263–266.

Malama, S., M. Munyeme, and J. B. Muma. 2019. Bovine tuberculosis in Zambia. Pages 445–453 *in* A. B. Dibaba, N. P. J. Kriek, and C. O. Thoen, editors. Tuberculosis in animals: an African perspective. Springer, Cham, Switzerland.

Martin, S. W., J. O'Keeffe, A. W. Byrne, L. E. Rosen, P. W. White, and G. McGrath. 2020. Is moving from targeted culling to BCG-vaccination of badgers (*Meles meles*) associated with an unacceptable increased incidence of cattle herd tuberculosis in the Republic of Ireland? A practical non-inferiority wildlife intervention study in the Republic of Ireland (2011–2017). Preventive Veterinary Medicine 179:105004.

McCulloch, S. P., and M. J. Reiss. 2017. Bovine tuberculosis and badger control in Britain: science, policy and politics. Journal of Agricultural and Environmental Ethics 30:469–484.

Meiring, C., P. D. Helden, and W. J. Goosen. 2018. TB control in humans and animals in South Africa: a perspective on problems and successes. Frontiers in Veterinary Science 5:298.

Meiring, C., R. Higgitt, A. Dippenaar, E. Roos, P. Buss, J. Hewlett, D. Cooper, P. Rogers, L. M. de Klerk-Lorist, L. van Schalkwyk, et al. 2021*a*. Characterizing epidemiological and genotypic features of *Mycobacterium bovis* infection in wild dogs (*Lycaon pictus*). Transboundary and Emerging Diseases 68:3433–3442.

Meiring, C., R. Higgitt, W. J. Goosen, L. van Schalkwyk, L.-M. de Klerk-Lorist, P. Buss, P. D. van Helden, S. D. C. Parsons, M. Moller, and M. Miller. 2021*b*. Shedding of *Mycobacterium bovis* in respiration secretions of free-ranging wild dogs (*Lycaon pictus*): implications for intraspecies transmission. Transboundary and Emerging Diseases 68:2581–2588.

Meunier, N. V., P. Sebulime, R. G. White, and R. Kock. 2017. Wildlife–livestock interactions and risk areas for cross-species spread of bovine tuberculosis. Onderstepoort Journal of Veterinary Research 84:a1221.

Michel, A. L., R. G. Bengis, D. F. Keet, M. Hofmeyr, L. M. de Klerk, P. C. Cross, A. E. Jolles, D. Cooper, I. J. Whyte, P. Buss, and J. Godfroid. 2006. Wildlife tuberculosis in South Africa conservation areas: implications and challenges. Veterinary Microbiology 112:91–100.

Michelet, L., K. de Cruz, G. Zanella, R. Aaziz, T. Bulach, C. Karoui, S. Hénault, G. Joncour, and M. L. Boschiroli., 2015. Infection with *Mycobacterium microti* in animals in France. Journal of Clinical Microbiology 53:981–985.

Michelet, L., C. Richomme, E. Réveillaud, K. De Cruz, J.-L. Moyen, and M. L. Boschiroli. 2020. *Mycobacterium microti* infection in red foxes in France. Microorganisms 9:1257.

Mikota, S. K., K. P. Lyashchenko, L. Lowenstine, D. Agnew, and J. N. Maslow. 2015. Mycobacterial infections in elephants. Pages 259–276 *in* H. Mukundan, M. A. Chambers, W. R. Waters, and M. H. Larsen, editors. Tuberculosis, leprosy and mycobacterial diseases of man and animals: the many hosts of mycobacteria. Commonwealth Agricultural Bureaux International, Boston, Massachusetts, USA.

Miller, M. A., T. J. Kerr, C. R. de Waal, W. J. Goosen, E. M. Streicher, G. Hausler, L. Rossouw, T. Manamela, L. van Schalkwyk, L. Kleynhans, R. Warren, P. van Helden, and P. E. Buss. 2021. *Mycobacterium bovis* infection in free-ranging African elephants. Emerging and Infectious Diseases 27:990–992.

Miller, M., and K. Lyashchenko. 2015. Mycobacterial infections in other zoo animals. Pages 277–297 *in* H. Mukundan, M. A. Chambers, W. R. Waters, and M. H. Larsen, editors. Tuberculosis, leprosy and mycobacterial diseases of man and animals: the many hosts of mycobacteria. Commonwealth Agricultural Bureaux International, Boston, Massachusetts, USA.

More, S. J. 2019. Can bovine TB be eradicated from the Republic of Ireland? Could this be achieved by 2030? Irish Veterinary Journal 72:3.

Munyeme, M., J. B. Muma, V. M. Siamudaala, E. Skjerve, H. M. Munang'andu, and M. Tryland. 2010. Tuberculosis in Kafue lechwe antelopes (*Kobus leche kafuensis*) of the Kafue Basin in Zambia. Preventive Veterinary Medicine 95:305–308.

Naranjo, V., C. Gortazar, J. Vicente, and J. de la Fuente. 2008. Evidence of the role of European wild boar as a reservoir of *Mycobacterium tuberculosis* complex. Veterinary Microbiology 127:1–9.

Nugent, G., A. M. Gormley, D. P. Anderson, and K. Crews. 2018. Roll-back eradication of bovine tuberculosis (TB) from wildlife in New Zealand: concepts, evolving approaches, and progress. Frontiers in Veterinary Science 5:277.

O'Brien, D. J., S. M. Schmitt, S. D. Fitzgerald, and D. E. Berry. 2011*a*. Management of bovine tuberculosis in Michigan wildlife: current status and near term prospects. Veterinary Microbiology 151:179–187.

O'Brien, D. J., S. M. Schmitt, B. A. Rudolph, and G. Nugent. 2011*b*. Recent advances in the management of bovine tuberculosis in free-ranging wildlife. Veterinary Microbiology 151:23–33.

Olmstead, A. L., and P. W. Rhode. 2004. An impossible undertaking: the eradication of bovine tuberculosis in the United States. Journal of Economic History 64:734–772.

Office International des Epizooties (OIE). 2021. World Organisation for Animal Health general disease information sheets. Bovine tuberculosis. https://www.oie.int/fileadmin/Home/eng/Media_Center/docs/pdf/Disease_cards/BOVINE-TB-EN.pdf. Accessed 3 Dec 2021.

Pagan, A., and L. Ramakrishnan. 2018. The formation and function of granulomas. Annual Review of Immunology 36:639–665.

Palmer, M. V. 2013. *Mycobacterium bovis*: characteristics of wildlife reservoir hosts. Transboundary and Emerging Diseases 60(Supplement 1):1–13.

Palmer, M. 2018. Emerging understanding of tuberculosis and the granuloma by comparative analysis in humans, cattle, zebrafish, and nonhuman primates. Veterinary Pathology 55:8–10.

Palmer, M.V., and W. R. Waters. 2011. Bovine tuberculosis and the establishment of an eradication program in the United States: role of veterinarians. Veterinary Medicine International 2011:816345.

Palmer, M. V., J. Wiarda, C. Kanipe, and T. C. Thacker. 2019. Early pulmonary lesions in cattle infected via aerosolized *Mycobacterium bovis*. Veterinary Pathology 56:544–554.

Parmar, S. M., R. G. Jani, F. M. Kapadiya, and D. R. Sutariya. 2013. Status of tuberculosis in the free living hanuman langur (*Presbytis entellus*) of Gujarat state. Indian Veterinary Journal 90:74–75.

Paudel, S., and T. Tsubota. 2016. Tuberculosis in elephants: a zoonotic disease at the human–elephant interface. Japanese Journal of Zoo and Wildlife Medicine 21:65–69.

Peterhans, S., P. Landolt, U. Friedel, F. Oberhänsli, M. Dennler, B. Willi, M. Senn, S. Hinden, K. Kull, A. Kipar, A., et al. 2020. *Mycobacterium microti*: not just a coincidental pathogen for cats. Frontiers in Veterinary Science, 7:590037.

Reis, A. C., B. Ramos, A. C. Pereira, and M. V. Cunha. 2021*a*. Global trends of epidemiological research in livestock tuberculosis for the last four decades. Transboundary and Emerging Diseases 68:333–346.

Reis, A. C., L. Salvador, S. Robbe-Austerman, R. Tenreiro, A. Botelho, T. Albuquerque, and M. V. Cunha. 2021*b*. Whole genome sequencing refines knowledge on the population structure of *Mycobacterium bovis* from a multi-host tuberculosis system. Microorganisms 9:1585.

Renwick, A. R., P. C. L. White, and R. G. Bengis. 2007. Bovine tuberculosis in southern African wildlife: a multi-species host–pathogen system. Epidemiology and Infection 135:529–540.

Richomme, C., E. Réveillaud, J. L. Moyen, P. Sabatier, K. De Cruz, L. Michelet, and M. L. Boschiroli. 2020. *Mycobacterium bovis* infection in red foxes in four animal tuberculosis endemic areas in France. Microorganisms 8:1070.

Risco, D., E. Serrano, P. Fernández-Llario, J. M. Cuesta, P. Gonçalves, W. L. García-Jiménez, R. Martínez, R. Cerrato, R. Velarde, L. Gómez, et al. 2014. Severity of bovine tuberculosis is associated with co-infection with common pathogens in wild boar. PLoS One 9:e110123.

Rodríguez, S., J. Bezos, B. Romero, L. de Juan, J. Álvarez, E. Castellanos, N. Moya, F. Lozano, M. T. Javed, J. L. Sáez-Llorente, and E. Liébana. 2011. *Mycobacterium caprae* infection in livestock and wildlife, Spain. Emerging Infectious Diseases 17:532–535.

Roe, W. D., B. Lenting, A. Kokosinska, S. Hunter, P. J. Duignan, B. Gartrell, L. Rogers, D. M. Collins, G. W. de Lisle, K. Gedye, et al. 2019. Pathology and molecular epidemiology of *Mycobacterium pinnipedii* tuberculosis in native New Zealand marine mammals. PloS One 14:e0212363.

Rothschild, B. M., and R. Laub. 2006. Hyperdisease in the late Pleistocene: validation of an early 20th century hypothesis. Naturwissenschaften 93:557–564.

Shury, T. K., J. S. Nishi, B. T. Elkin, and G. A. Wobeser. 2015. Tuberculosis and brucellosis in wood bison (*Bison bison athabascae*) in northern Canada: a renewed need to develop options for future management. Journal of Wildlife Diseases 51:543–554.

Smith, G. C., and R. J. Delahay. 2018. Modeling as a decision support tool for bovine TB control programs in wildlife. Frontiers in Veterinary Science 5:276.

Smith, K., L. Kleynhans, R. M. Warren, W. J. Goosen, and M. A. Miller. 2021. Cell-mediated immunological biomarkers and their diagnostic application in livestock and wildlife infected with *Mycobacterium bovis*. Frontiers in Immunology 12:483.

Stellenbosch University and University of Pretoria. 2015. Bovine tuberculosis outreach day. https://www.up.ac.za/media/shared/678/btb-outreach-day-booklet-print_03042017.zp114198.pdf. Accessed 4 Dec 2021.

Tanner, M., O. Inlameia, A. Michel, G. Maxlhuza, A. Pondja, J. Fafetine, B. Macucule, M. Zacarias, J. Manguele, I. C. Moiane, et al. 2015. Bovine tuberculosis and brucellosis in cattle and African buffalo in the Limpopo National Park, Mozambique. Transboundary and Emerging Diseases 62:632–638.

Thapa, J., C. Nakajima, K. P. Gairhe, B. Maharjan, S. Paudel, Y. Shah, S. K. Mikota, G. E. Kaufman, D. McCauley, T. Tsubota, et al. 2017. Wildlife tuberculosis: an emerging threat for conservation in South Asia. Pages 73–90 *in* G. A. Lameed, editor. Global exposition of wildlife management. IntechOpen, London, England.

Thomas, J., A Balseiro, C. Gortázar, and M. A. Risalde. 2021. Diagnosis of tuberculosis in wildlife: a systematic review. Veterinary Research 52:1–23.

Tortoli, E. 2006. The new mycobacteria: an update. FEMS Immunology and Medical Microbiology 48:159–178.

Triguero-Ocaña, R., B. Martínez-López, J. Vicente, J. A. Barasona, J. Martínez-Guijosa, and P. Acevedo. 2020. Dynamic network of interactions in the wildlife–livestock interface in Mediterranean Spain: an epidemiological point of view. Pathogens 9:120.

US Department of Agriculture, Animal and Plant Health Inspection Service (USDA APHIS). 2011. Guidelines for surveillance of bovine tuberculosis in wildlife. https://www.aphis.usda.gov/animal_health/animal_diseases/tuberculosis/downloads/wildlife_tb_surv_manual.pdf. Accessed 3 Dec 2021.

US Department of Agriculture, Animal and Plant Health Inspection Service (USDA APHIS). 2020. Cervids: bovine tuberculosis (bTB) in cervids. https://www.aphis.usda.gov/aphis/ourfocus/animalhealth/animal-disease-information/cervid/cervids-bovine-tb. Accessed 3 Dec 2021.

VerCauteren, K. C., M. J. Lavelle, and H. Campa III. 2018. Persistent spillback of bovine tuberculosis from white-tailed deer to cattle in Michigan, USA: status, strategies, and needs. Frontiers in Veterinary Science 5:301.

Vicente, J., U. Höfle, J. M. Garrido, P. Acevedo, R. Juste, M. Barral, and C. Gortazar. 2007. Risk factors associated with the prevalence of tuberculosis-like lesions in fenced wild boar and red deer in south central Spain. Veterinary Research 38:451–464.

Vikas Saket, K., R. Kachhi, and P. Singh. 2017. Tuberculosis in animals and humans: evolution of diagnostics and therapy. Asian Journal of Animal and Veterinary Advances 12:177–188.

Viljoen, I. M., P. van Helden, and R. P. Millar. 2015. *Mycobacterium bovis* infection in the lion (*Panthera leo*): current knowledge, conundrums and research challenges. Veterinary Microbiology 177:252–260.

Wangoo, A., L. Johnson, J. Gough, R. Ackbar, S. Inglut, D. Hicks, Y. Spencer, G. Hewinson, and M. Vordermeier. 2005. Advanced granulomatous lesions in *Mycobacterium bovis*-infected cattle are associated with increased expression of type I procollagen, γδ (WC1+) T cells and CD 68+ Cells. Journal of Comparative Pathology 133:223–234.

Woodroffe, R., C. A. Donnelly, K. Chapman, C. Ham, K. Moyes, N. G. Stratton, and S. J. Cartwright. 2021. Successive use of shared space by badgers and cattle: implications for *Mycobacterium bovis* transmission. Journal of Zoology 314:132–142.

Zachariah, A., J. Pandiyan, G. K. Madhavilatha, S. Mundayoor, B. Chandramohan, P. K. Sajesh, S. Santhosh, and S. K. Mikota. 2017. *Mycobacterium tuberculosis* in wild Asian elephants, southern India. Emerging Infectious Diseases 23:504–506.

Zanella, G., A. Duvauchelle, J. Hars, F. Moutou, and M. L. Boschiroli. 2008. Patterns of lesions of bovine tuberculosis in wild red deer and wild boar. Veterinary Record 163:43–47.

Zimpel, C. K., J. S. L. Patané, A. C. P. Guedes, R. F. de Souza, T. T. Silva-Pereira, N. C. S. Camargo, A. F. de Souza Filho, C. Y. Ikuta, J. S F. Neto, J. C. Setubal, et al. 2020. Global distribution and evolution of *Mycobacterium bovis* lineages. Frontiers in Microbiology 11:843.

Zink, A. R., E. Molnár, N. Motamedi, G. Pálfy, A. Marcsik, and A. G. Nerlich. 2007. Molecular history of tuberculosis from ancient mummies and skeletons. International Journal of Osteoarchaeology 17:380–391.

Zinsstag, J., E. Schelling, F. Roth, and R. Kazwala. 2006. Economics of bovine tuberculosis. Pages 68–83 *in* C.O. Thoen, J. H. Steele, and M. J. Gilsdorf, editors. *Mycobacterium bovis* infection in animals and humans. Second Edition. Blackwell Publishing, Ames, Iowa, USA.

23 Toxoplasmosis and One Health

ELIZABETH VANWORMER, KAREN SHAPIRO

Introduction: *Toxoplasma gondii* Life Cycle and Transmission to Wildlife in Diverse Ecosystems

Toxoplasmosis affects wildlife populations from the tropics to the arctic, impacting the health of terrestrial mammals, aquatic mammals, and avian species. The causative agent of this parasitic disease is *Toxoplasma gondii*, a single-celled protozoan pathogen that is globally distributed. The widespread nature of this pathogen and its ability to infect virtually all warm-blooded animals is likely rooted in its life cycle and different modes of transmission (Figure 23.1). *Toxoplasma* is a particularly remarkable pathogen given its ability to cross ecosystem boundaries and its unique life cycle that includes life stages that can be transmitted to both herbivorous and carnivorous animals.

In the decades that followed the discovery of *Toxoplasma* in 1908 (Ferguson 2009), this zoonotic parasite was predominantly thought to be a pathogen that affected terrestrial hosts, including wildlife, domestic animals, and humans. Only species within the family Felidae (domestic cats and wild felids) can serve as definitive hosts for *Toxoplasma*. In these hosts, the parasite can sexually multiply in the gut, resulting in shedding of oocysts in feces. Oocysts, the

Illustration by Laura Donohue.

Fig. 23.1. *Toxoplasma* infects a broad range of warm-blooded wildlife hosts in diverse environments throughout the world. Illustration by Laura Donohue.

environmental stage of the parasite, are incredibly robust in different matrices, surviving for months or even years under optimal conditions in soil or water (Dumètre and Dardé 2003). This feature of environmental resistance allows oocysts to persist and accumulate in many terrestrial and aquatic habitats, facilitating exposure to a wide range of mammals and birds. While *Toxoplasma* can only form oocysts by sexual reproduction in felids, it is able to multiply asexually in warm-blooded animals (intermediate hosts), and this often results in life-long infection due to its ability to form bradyzoite cysts in tissues such as muscle and brain. The life cycle is then completed when a felid consumes tissues from an intermediate host, such as rodent, bird or ungulate, ingesting parasite cysts in the tissues of its prey (Figure 23.2).

In addition to horizontal transmission from animal to animal through consumption of tissue cysts (carnivorism) or ingestion of oocysts in contaminated food, soil, or water, *Toxoplasma* can also be transmitted vertically—from a pregnant animal to its

Fig. 23.2. *Toxoplasma* is transmitted to wildlife through three main routes: predation or scavenging of intermediate hosts harboring bradyzoite cysts in tissues; ingestion of oocysts in contaminated food, water, or soil; and congenital transmission of tachyzoites that can cross the placenta. Illustration by Laura Donohue.

fetus (Figure 23.2). Tachyzoites are the fast-replicating stage of the parasite that can cross the placenta during pregnancy. While the risk of parasite transmission during pregnancy is thought to be predominantly a concern in naïve female animals that are exposed to *Toxoplasma* for the first time during gestation, evidence suggests that some infected animals can have repeated congenital transmission episodes in consecutive pregnancies over their lifetime (Hide 2016). Data on the importance of vertical transmission of *Toxoplasma* in wildlife are limited given the difficulties of monitoring these events in the field, and we are likely underestimating the effect of congenital toxoplasmosis on wildlife populations.

Vertical transmission of *Toxoplasma* can also help to explain how the parasite persists in environments with no or very few felid definitive hosts (Elmore et al. 2012). High levels of vertical transmission have been observed in diverse mouse species in wild and experimental settings, and *Toxoplasma* infection was maintained in a captive population of mice removed from contact with cats for over 10 years (Hide 2016).

In island or polar environments with no or limited history of free-ranging domestic cat or wild felid populations, migratory birds or mammals can introduce *Toxoplasma* into the food webs where vertical and horizontal transmission both likely contribute to persistence of the parasite (Prestrud et al. 2007, Kutz et al. 2012). On the Arctic island of Svalbard, where Arctic fox (*Vulpes lagopus*) infection and mortality occurs in the absence of free-ranging felid populations, *Toxoplasma* is likely introduced by migratory prey species, such as geese (Prestrud et al. 2007, Sandström et al. 2013). While migrating animals with bradyzoite cysts in their tissues and vertical transmission of tachyzoites can allow *Toxoplasma* to expand to and persist in areas with few or no felids, the ability of *Toxoplasma* to cross terrestrial–aquatic ecosystem boundaries lies in the robust nature of the oocyst stage of the parasite.

The importance of *Toxoplasma* as a waterborne pathogen that can infect a wide range of aquatic animals emerged in the 1990s, nearly a century after the parasite was first discovered. Oocysts in cat feces can become mobilized either directly from feces or from contaminated soil through overland runoff (driven by rainfall events), subsequently washing the parasite into water bodies such as streams, lakes, or coastal waters (Figure 23.3). Once in the aquatic system, oocysts may become concentrated in invertebrates such as shellfish or snails, where they can bioaccumulate until a susceptible host consumes them. Marine turban snails, for example, play a key role in accumulating and transmitting *Toxoplasma* oocysts to southern sea otters (*Enhydra lutris nereis*) in coastal California (Shapiro et al. 2014). Cold-blooded animals, such as fish and shellfish, have also been proposed as mechanical hosts for *Toxoplasma* and serve as a potential source of infection to higher trophic level animals (Arkush et al. 2003, Massie et al. 2010). Cold-blooded animals are not competent intermediate hosts for the parasite, and thus, the oocysts can accumulate in these animals (i.e., as mechanical or paratenic hosts) but will not be able to develop or multiply.

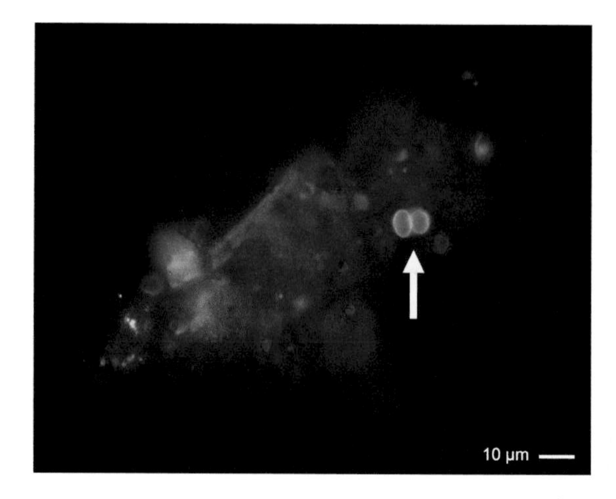

Fig. 23.3. In seawater, *Toxoplasma* oocysts can associate with marine aggregates that impact oocyst movement in coastal waters. In this image, an autofluorescent oocyst (arrow) is embedded in a marine aggregate that also contains phytoplankton (red structures), organic material, and debris. Photograph by Karen Shapiro.

Toxoplasmosis, Wildlife Health, and Implications for Conservation

Following infection with *Toxoplasma*, the manifestation of disease in wild animal hosts ranges from subclinical to severe and fatal toxoplasmosis. In many immunocompetent wildlife hosts, infection is largely asymptomatic, with the parasite encysting in tissues (primarily brain and muscle) for the life of the host. Asymptomatic infection is particularly common in birds and terrestrial mammals that have coevolved with felids, including ungulates and rodents. The ability of *Toxoplasma* to cause life-long infections without overt morbidity and mortality is likely an important reason why exposure (based on serologic testing) is commonly observed in many wildlife populations across the world. Even when *Toxoplasma* infection does not cause clinical illness, the parasite may cause behavioral changes and subclinical impacts on wildlife health. *Toxoplasma*-infected rodents and chimpanzees show attraction to the urine of certain felid predators (Berdoy et al. 2000, Poirotte et al. 2016), and infection in hyenas is linked to increased boldness toward and mortality due to lions (Gering et al. 2021). *Toxoplasma* infection in some wildlife species may reduce fertility (Formenti et al. 2015) or increase the likelihood of being killed by cars (Hollings et al. 2013), but additional research is needed to understand the population-level occurrence and impacts of subclinical toxoplasmosis.

While congenital transmission without evidence of reproductive failure or disease in the offspring has been reported in moose, cetaceans, lemurs, and Australian marsupials (Jardine and Dubey 2002, Resendes et al. 2002, Parameswaran et al. 2009), congenital toxoplasmosis can also cause fetal morbidity and mortality in wildlife. Abortion or miscarriage due to vertical transmission has been reported in free-ranging southern sea otters, captive red ruffed lemurs (*Varecia rubra*), and farmed reindeer (*Rangifer tarandus*) (Dubey et al. 2002, Juan-Sallés et al. 2011, Shapiro et al. 2016, Browning et al. 2021). Fetuses infected in utero may also survive to term and present with disseminated toxoplasmosis that can prove fatal shortly after birth or later in life (Miller et al. 2008). Intriguingly, case reports on wild animals challenge the general dogma that congenital infections only occur in animals who are infected for the first time during pregnancy. The observation that transplacental *Toxoplasma* infection can occur in chronically infected animals with subsequent negative reproductive sequelae warrants further attention due to cascading impacts on conservation of endangered species.

Toxoplasma affects wildlife health at the individual as well as population levels. For some species, our understanding of the severity of toxoplasmosis and potential conservation impact is shaped by observations from very small numbers of captive or wild animals. For example, severe or fatal toxoplasmosis has been reported in endangered hosts including a giant panda (*Ailuropoda melanoleuca*; Ma et al. 2015), a vinaceous Amazon parrot (*Amazona vinacea*; Ferreira et al. 2012), and a group of black-footed ferrets (*Mustela nigripes*; Burns et al. 2003) housed in captivity. On an island near Japan, fatal toxoplasmosis in a free-ranging Amami spiny rat (*Tokudaia osimensis*) raised concern about broader impacts of *Toxoplasma* on the conservation of this endangered rodent species as well as other endemic small mammal hosts (Tokiwa et al. 2019).

For other wildlife species, observations of toxoplasmosis in captive populations as well as free-living ones provide broader evidence for *Toxoplasma* population health and conservation impacts. Severe disease in individual animals or outbreaks of toxoplasmosis leading to mortality events have been reported in captive endangered lemur species (Browning et al. 2021) as well as New World primates including squirrel monkeys, howler monkeys, woolly monkeys, marmosets, and tamarins (Epiphanio et al. 2003; Carme et al. 2009*a*; Salant et al. 2009; Cedillo-Peláez et al. 2011, 2012; Pardini et al. 2015; Santana et al. 2021). Pallas cats (*Otocolobus manul*) in captivity are considered extremely susceptible to *Toxoplasma* infection, but a study of animals in a free-ranging population in Mongolia suggests that they have minimal exposure to the parasite in their natural habitat (Brown et al. 2005). Similarly, fatal toxoplasmosis occurs in sand cats (*Felis margarita*) in captivity, which likely have limited exposure to other cats and oocysts in the wild (Dubey et al. 2010). The severity of disease and high level of mortality in captive Pallas cats and sand cats poses a threat to the success of breeding programs designed to support conservation of wild populations. Deaths and outbreaks due to toxoplasmosis have also been documented in many captive and free-living Australian marsupials including red kangaroos (*Macropus rufus*), black-faced kangaroos (*Macropus fuliginosus melanops*), wallabies, and common wombats (*Vombatus ursinus*) (Dubey et al. 1988, Hartley 2006, Dubey and Crutchley 2008, Carossino et al. 2021). More research is needed to understand whether *Toxoplasma* is a threat to the conservation of marsupial populations (Hillman et al. 2016, Borkens 2021).

The greatest impacts of *Toxoplasma* on free-ranging wildlife populations have been observed in island ecosystems with no native wild felids. In these environments, introduced domestic cats serve as the source of *Toxoplasma* oocysts that can persist in the terrestrial environment and be carried in freshwater runoff to the surrounding sea. *Toxoplasma* infection has caused mortality in threatened and endangered birds and marine mammals in Hawaii and New Zealand. *Toxoplasma*-induced mortality is considered to have caused extinction in the wild of the Hawaiian 'Alala (*Corvus hawaiiensis*), which is the most endangered corvid in the world (Work et al. 2000) and now exists only in a captive breeding population. Toxoplasmosis also occurs in free-ranging endangered Hawaiian geese (Nene, *Branta sandvicensis*) but is not a major cause of mortality at the population level (Work et al. 2016). Hawaiian monk seal (*Neomonachus schauinslandi*) mortality due to toxoplasmosis is an ongoing conservation concern (Barbieri et al. 2016), however, and demonstrates the serious problem created by domestic cat (free-living and feral) fecal contamination of endangered marine mammal ecosystems (Figure 23.4). Similarly, toxoplasmosis poses a significant threat to the health and conservation of endangered Hector's (*Cephalorhynchus hectori*) and Maui dolphins (*Cephalorhynchus hectori* ssp. *maui*) in coastal New Zealand, where domestic cats are the only felid sources of *Toxoplasma* in nearby terrestrial environments (Roe et al. 2013, Roberts et al. 2020).

The question of why some animals develop clinical or fatal toxoplasmosis, while others do not, is not well understood. Host factors such as age, immunocompetence, or coinfection with other pathogens are likely important, as well as pathogen factors such as genotype virulence. Fatal toxoplasmosis in mountain hares (*Lepus timidis*) has been attributed, in part, to lack of an effective cellular immune response (Gustafsson et al. 1997). In an outbreak of toxoplasmosis among marine mammals in the Pacific Northwest, coinfection with the related protozoan parasite *Sarcocystis neurona* was linked to more severe encephalitis and higher mortality (Gibson et al. 2011). In southern sea otters along the central coast of California, death due to toxoplasmosis was linked to specific *Toxoplasma* genotypes (Shapiro et al. 2019*b*). In addition to host immune factors, coinfection, and parasite genetics, an animal's environment and evolutionary history may also shape response to *Toxoplasma* infection. Several groups of animals considered susceptible to the parasite including lemurs, Australian marsupials, and some marine mammal populations evolved in environments without felid hosts prior to the arrival of domestic cats, which introduced *Toxoplasma* oocysts into the environments of these naive hosts (Innes 1997). New World monkeys are also susceptible to severe and fatal toxoplasmosis; although wild felids historically overlapped with their range, the arboreal lifestyle of these primates may have limited contact with oocysts.

Fig. 23.4. Toxoplasmosis is a serious threat to the conservation of Hawaiian monk seals (*Neomonachus schauinslandi*). Oocysts from feral and owned domestic cats can be carried in freshwater runoff from land to sea, where they pose a risk of infection to susceptible marine wildlife. Photograph courtesy of the Kaho'olawe Island Reserve Commission.

Toxoplasmosis and Human Health

Toxoplasma infection is widespread in human populations around the world (Tenter et al. 2000). People can become infected by ingesting water, soil, or food contaminated with oocysts, eating undercooked domestic or wild animal meat containing bradyzoite tissue cysts, or through congenital transmission (Shapiro et al. 2019*a*). Although many individuals with healthy immune systems experience no or mild flu-like symptoms, severe disease and death can occur in individuals with a weakened immune system or babies infected in utero (Tenter et al. 2000). Women who are exposed to *Toxoplasma* for the first time during pregnancy can pass the parasite to their developing fetus, leading to fetal disseminated toxoplasmosis, birth defects, stillbirths, and abortions. Children born with congenital infection may present with neurologic symptoms at birth or develop ocular disease later in life (McLeod et al. 2014). In people with weakened immune systems due to viral infections (AIDS patients) or treatment with immunosuppressive medicines, toxoplasmosis can lead to encephalitis, myocarditis, and death (Luft and Remington 1992). Even in people with healthy immune systems, the genetic type or "genotype" of *Toxoplasma* and the source of exposure may lead to more severe or fatal disease. For example, oocyst-based infections acquired in waterborne outbreaks of toxoplasmosis have caused chorioretinitis and neurologic deficits in immune-competent individuals (Bowie et al. 1997, Vaudaux et al. 2010). Additionally, rapidly expanding research on the more subtle impacts of toxoplasmosis in human populations has identified potential links to changes in behavior, mental health, and neurodegenerative diseases (Johnson and Koshy 2020, Torrey 2021).

A One Health Approach to Understanding and Managing Wildlife Toxoplasmosis

The close connections and interactions among humans, domestic animals, wildlife, and our shared environment require a holistic, One Health approach to understand and reduce the impacts of toxoplasmosis (VanWormer et al. 2013*b*, Aguirre et al. 2019). Although *Toxoplasma* infections in people and domestic animals can be acquired through wildlife food sources including deer, kangaroos, and marine mammals (Tryland et al. 2014, Gaulin et al. 2020, Borkens 2021), a more profound connection that impacts *Toxoplasma* ecology between people and wildlife relates to human population growth and influence on global landscapes. The burden of *Toxoplasma* infections in terrestrial wildlife is correlated with anthropogenic landscape change, with higher levels of exposure in wildlife near areas of higher human density (Wilson et al. 2021; also see chapter 6, figure 6.2). Similarly, *Toxoplasma* infections in aquatic wildlife, such as sea otters, are associated with higher human density in bordering coastal watersheds (Burgess et al. 2018), while muskrats (*Ondatra zibethicus*) living in watersheds with little human development were not exposed to *Toxoplasma* (Ahlers et al. 2020). Anthropogenic (human-mediated) landscape change can result in both a higher total load of *Toxoplasma* oocysts that pollute terrestrial ecosystems, as well as higher likelihood of the parasite flowing from land to aquatic environments via runoff.

As the human population expands, so does that of associated domestic pet and stray cats, increasing the numbers of definitive hosts that can shed oocysts in a given habitat (VanWormer et al. 2013*a*). The contribution of feral cats to *Toxoplasma* oocyst loading in coastal communities was explored by Dabritz et al. (2007). Expanding human settlements also result in removal or degradation of natural landscape features such as vegetated wetlands and replacement with concrete buildings, storm-water canals, and paved roads. The loss of natural cover and replacement with impervious surfaces means that less soil is available to absorb rainfall, and less vegetation is present to act as a filter or sponge for capturing runoff-based pollutants including pathogens like *Toxoplasma* (Shapiro et al. 2010; also see chapter 6, figure 6.4). The end result is a greater force of surface runoff driven by either rainfall or snowmelt that can carry *Toxoplasma* oocysts to aquatic systems (Simon et al. 2013, VanWormer et al. 2016). Climate change can further exacerbate pollution of waterbodies with *Toxoplasma* oocysts, with higher levels of oocyst runoff estimated with increased precipitation (VanWormer et al. 2016). Higher intensity storm events punctuated by longer dry seasons when environmentally resistant oocysts accumulate can lead to higher overall load of the parasite polluting streams, lakes, and nearshore marine waters.

In addition to increasing parasite load and transport across the landscape, human development alters interactions between wild animals and domestic animals that either serve as intermediate hosts for *Toxoplasma* or are the definitive host themselves (domestic cats). Although both wild felids and domestic cats shed oocysts into the environment, the genotypes they shed often differ. Separate domestic and sylvatic cycles of *Toxoplasma* transmission driven by domestic cats and wild felids, respectively, have been identified in North and South America (Carme et al. 2009*b*, Mercier et al. 2011, VanWormer et al. 2014, Shwab et al. 2018). The areas where wild felid and domestic cat home ranges and food webs overlap can allow different genotypes of *Toxoplasma* to be transmitted to new hosts. These interface zones between urban, agricultural, and undeveloped environments may also facilitate opportunities for new parasite strains to emerge through sexual recombination that occurs when felid hosts are coinfected with different *Toxoplasma* genotypes. As some genotypes of *Toxoplasma* are more virulent for humans or wildlife (Carme et al. 2009*b*, Shapiro et al. 2019*b*), overlap between domestic and wild felid definitive hosts, intermediate host prey species, and *Toxoplasma* strains has the potential to shape downstream risks to public and wildlife population health.

Currently, there are limited treatment options for clinical or chronic toxoplasmosis in wildlife. The parasite's ability to encyst and hide from the host immune response contributes to this therapeutic challenge (Miller et al. 2018). Management efforts to reduce the impacts of *Toxoplasma* on free-living wildlife populations focus on reducing the number of oocysts shed into the environment (domestic cat-based approaches) or reducing the movement of oocysts between hosts and habitats (ecosystem-based approaches). Limiting the number of outdoor pet and feral cats benefits wildlife through decreased risk of predation as well as decreased risk of exposure to *Toxoplasma* and other pathogens. One such well-recognized program is Cats Indoors (American Bird Conservancy). Management options for feral cats are contentious, however, and often fiercely debated among cat and wildlife health and welfare stakeholders (Jessup et al. 2023). Experimental vaccines have been developed to reduce oocyst shedding by domestic cats (Mateus-Pinilla et al. 1999, Ramakrishnan et al. 2019), but they are not commercially available and modeling suggests that they may not offer a feasible solution in real-world settings (Bonačić Marinović et al. 2019). In California, a legal approach was implemented to reduce the burden of oocysts in the environment. Rather than managing pet cat populations directly, the law requires labels on cat litter packages asking owners to dispose of litter and cat feces in landfills rather than flushing it in toilets or dumping in yards or at curbside street drains. Strategies that target the ecosystem or landscape level have the potential to be more widely adopted and to reduce broader threats to human and wildlife health from biological and chemical contaminants (see Chapter 6).

Summary

Toxoplasma gondii is one of the most successful and widespread pathogens worldwide. While this zoonotic parasite is often associated with mild or asymptomatic infections, it can also cause morbidity and mortality in a wide range of wildlife species. Preventing toxoplasmosis remains challenging due to the ubiquitous nature of this organism, difficulty in managing definitive hosts (cats) that spread robust oocysts in the environment, and lack of effective vaccines or treatments. Wetland conservation and restoration, introduction of vegetation buffers for storm drain runoff or along streams and rivers, and reductions in impervious surfaces can reduce the flow of *Toxoplasma* oocysts in freshwater runoff from terrestrial to aquatic environments (Shapiro et al. 2010, Daniels et al. 2014; also see chapter 6, figure 6.4). Notably, landscape-based strategies for reducing runoff of *Toxoplasma* from feces to waterways can also decrease the load of broader land-based contaminants such as fertilizers, oil, pesticides, and other fecal pathogens that can pollute aquatic environments, demonstrating a truly One Health approach that can benefit wildlife populations and human health.

LITERATURE CITED

Aguirre, A. A., T. Longcore, M. Barbieri, H. Dabritz, D. Hill, P. N. Klein, C. Lepczyk, E. L. Lilly, R. McLeod, J. Milcarsky, et al. 2019. The One Health approach to toxoplasmosis: epidemiology, control, and prevention strategies. EcoHealth 16:378–390.

Ahlers, A. A., T. M. Wolf, O. Aarrestad, S. K. Windels, B. T. Olson, B. R. Matykiewicz, and J. P. Dubey. 2020. Survey of *Toxoplasma gondii* exposure in muskrats in a relatively pristine ecosystem. Journal of Parasitology 106:346–349.

Arkush, K.D., M. A. Miller, C. M. Leutenegger, I. A. Gardner, A. E. Packham, A. R. Heckeroth, A. M. Tenter, B. C. Barr, and P.A. Conrad. 2003. Molecular and bioassay-based detection of *Toxoplasma gondii* oocyst uptake by mussels (*Mytilus galloprovincialis*). International Journal for Parasitology 33:1087–1097.

Barbieri, M. M., L. Kashinsky, D. S. Rotstein, K. M. Colegrove, K. H. Haman, S. L. Magargal, A. R. Sweeny, A. C. Kaufman, M. E. Grigg, and C. L. Littnan. 2016. Protozoal-related mortalities in endangered Hawaiian monk seals *Neomonachus schauinslandi*. Diseases of Aquatic Organisms 121:85–95.

Berdoy, M., J. P. Webster, and D. W. Macdonald. 2000. Fatal attraction in rats infected with *Toxoplasma gondii*. Proceedings of the Royal Society of London Series B Biological Sciences 267:1591–1594.

Bonačić Marinović, A. A., M. Opsteegh, H. Deng, A. W. M. Suijkerbuijk, P. F. van Gils, and J. van der Giessen. 2019. Prospects of toxoplasmosis control by cat vaccination. Epidemics 30:100380.

Borkens, Y. 2021. *Toxoplasma gondii* in Australian macropods (*Macropodidae*) and its implication to meat consumption. International Journal for Parasitology: Parasites and Wildlife 16:153–162.

Bowie, W. R., A. S. King, D. H. Werker, J. L. Isaac-Renton, A. Bell, S. B. Eng, and S. A. Marion. 1997. Outbreak of toxoplasmosis associated with municipal drinking water. The BC Toxoplasma Investigation Team. Lancet 350:173–177.

Brown, M., M. R. Lappin, J. L. Brown, B. Munkhtsog, and W. F. Swanson. 2005. Exploring the ecologic basis for extreme susceptibility of Pallas' cats (*Otocolobus manul*) to fatal toxoplasmosis. Journal of Wildlife Diseases 41:691–700.

Browning, G. R., C. Singleton, D. Gibson, and I. H. Stalis. 2021. Outcomes of transplacental transmission of *Toxoplasma gondii* from chronically infected female red ruffed lemurs (*Varecia rubra*). Journal of Zoo and Wildlife Medicine 52:1036–1041.

Burgess, T. L., M. T. Tinker, M. A. Miller, J. L. Bodkin, M. J. Murray, J. A. Saarinen, L. M. Nichol, S. Larson, P. A. Conrad, and C. K. Johnson. 2018. Defining the risk landscape in the context of pathogen pollution: *Toxoplasma gondii* in sea otters along the Pacific Rim. Royal Society Open Science 5:171178.

Burns, R., E. S. Williams, D. O'Toole, and J. P. Dubey. 2003. *Toxoplasma gondii* infections in captive black-footed ferrets (*Mustela nigripes*), 1992–1998: clinical signs, serology, pathology, and prevention. Journal of Wildlife Diseases 39:787–797.

Carme, B., D. Ajzenberg, M. Demar, S. Simon, M. L. Dardé, B. Maubert, and B. de Thoisy. 2009*a*. Outbreaks of toxoplasmosis in a captive breeding colony of squirrel monkeys. Veterinary Parasitology 163:132–135.

Carme, B., M. Demar, D. Ajzenberg, and M. L. Dardé. 2009*b*. Severe acquired toxoplasmosis caused by wild cycle of *Toxoplasma gondii*, French Guiana. Emerging Infectious Diseases 15:656–658.

Carossino, M., R. Bauer, M. A. Mitchell, C. O. Cummings, A. C. Stöhr, N. Wakamatsu, K. Harper, I. M. Langohr, K. Schultz, M. S. Mitchell, et al. 2021. Pathologic and immunohistochemical findings in an outbreak of systemic toxoplasmosis in a mob of red kangaroos. Journal of Veterinary Diagnostic Investigation 33:554–565.

Cedillo-Peláez, C., A. Besné-Mérida, D. Espinosa-Aviles, C. P. Rico-Torres, G. Salas-Garrido, and D. Correa. 2012. Distribution of lesions and identification of parasites by immunohistochemistry in cases of acute toxoplasmosis in New World primates and prosimians in captivity in Mexico. International Journal of Infectious Diseases 16:e448.

Cedillo-Peláez, C., C. P. Rico-Torres, C. G. Salas-Garrido, and D. Correa. 2011. Acute toxoplasmosis in squirrel monkeys (*Saimiri sciureus*) in Mexico. Veterinary Parasitology 180:368–371.

Dabritz, H.A., M. A. Miller, E. R. Atwill, I. A. Gardner, C. M. Leutenegger, A. C. Melli, and P. A. Conrad. 2007. Detection of *Toxoplasma gondii*-like oocysts in cat feces and estimates of the environmental oocyst burden. Journal of the American Veterinary Medical Association 231:1676–1684.

Daniels, M. E., J. Hogan, W. A. Smith, S. C. Oates, M. A. Miller, D. Hardin, K. Shapiro, M. Los Huertos, P. A. Conrad, C. Dominik, et al. 2014. Estimating environmental conditions affecting protozoal pathogen removal in surface water wetland systems using a multi-scale, model-based approach. Science of the Total Environment 493:1036–1046.

Dubey, J. P., and C. Crutchley. 2008. Toxoplasmosis in wallabies (*Macropus rufogriseus* and *Macropus eugenii*): blindness, treatment with atovaquone, and isolation of *Toxoplasma gondii*. Journal of Parasitology 94:929–933.

Dubey, J. P., B. Lewis, K. Beam, and B. Abbitt. 2002. Transplacental toxoplasmosis in a reindeer (*Rangifer tarandus*) fetus. Veterinary Parasitology 110:131–135.

Dubey, J. P., J. Ott-Joslin, R. W. Torgerson, M. J. Topper, and J. P. Sundberg. 1988. Toxoplasmosis in black-faced kangaroos (*Macropus fuliginosus melanops*). Veterinary Parasitology 30:97–105.

Dubey, J. P., A. Pas, C. Rajendran, O. C. H. Kwok, L. R. Ferreira, J. Martins, C. Hebel, S. Hammer, and C. Su. 2010. Toxoplasmosis in sand cats (*Felis margarita*) and other animals in the Breeding Centre for Endangered Arabian Wildlife in the United Arab Emirates and Al Wabra Wildlife Preservation, the State of Qatar. Veterinary Parasitology 172:195–203.

Dumètre, A., and M. L. Dardé. 2003. How to detect *Toxoplasma gondii* oocysts in environmental samples? FEMS Microbiology Reviews 27:651–661.

Elmore, S. A., E. J. Jenkins, K. P. Huyvaert, L. Polley, J. J. Root, and C. G. Moore. 2012. *Toxoplasma gondii* in circumpolar people and wildlife. Vector-Borne and Zoonotic Diseases 12:1–9.

Epiphanio, S., I. L. Sinhorini, and J. L. Catão-Dias. 2003. Pathology of toxoplasmosis in captive new world primates. Journal of Comparative Pathology 129:196–204.

Ferguson, D. J. P. 2009. *Toxoplasma gondii*: 1908–2008, homage to Nicolle, Manceaux and Splendore. Memorias do Instituto Oswaldo Cruz 104:133–148.

Ferreira, F. C., Jr., R. V. Donatti, M. V. Romero Marques, R. Ecco, I. S. Preis, H. L. Shivaprasad, D. A. da Rocha Vilela, and N. R. da Silva Martins. 2012. Fatal toxoplasmosis in a vinaceous Amazon parrot (*Amazona vinacea*). Avian Diseases 56:774–777.

Formenti, N., T. Trogu, L. Pedrotti, A. Gaffuri, P. Lanfranchi, and N. Ferrari. 2015. *Toxoplasma gondii* infection in alpine red deer (*Cervus elaphus*): its spread and effects on fertility. PloS One 10:e0142357.

Gaulin, C., D. Ramsay, K. Thivierge, J. Tataryn, A. Courville, C. Martin, P. Cunningham, J. Désilets, D. Morin, and R. Dion. 2020. Acute toxoplasmosis among Canadian deer hunters associated with consumption of undercooked deer meat hunted in the United States. Emerging Infectious Diseases 26:199–205.

Gering, E., Z. M. Laubach, P. S. D. Weber, G. Soboll Hussey, K. D. S. Lehmann, T. M. Montgomery, J. W. Turner, W. Perng, M. O. Pioon, K. E. Holekamp, et al. 2021. *Toxoplasma gondii* infections are associated with costly boldness toward felids in a wild host. Nature Communications 12:3842.

Gibson, A. K., S. Raverty, D. M. Lambourn, J. Huggins, S. L. Magargal, and M. E. Grigg. 2011. Polyparasitism is associated with increased disease severity in *Toxoplasma gondii*-infected marine sentinel species. PLoS Neglected Tropical Diseases 5:e1142.

Gustafsson, K., E. Wattrang, C. Fossum, P. M. Heegaard, P. Lind, and A. Uggla. 1997. *Toxoplasma gondii* infection in the mountain hare (*Lepus timidus*) and domestic rabbit (*Oryctolagus cuniculus*). Journal of Comparative Pathology 117:361–369.

Hartley, M. P. 2006. *Toxoplasma gondii* infection in two common wombats (*Vombatus ursinus*). Australian Veterinary Journal 84:107–109.

Hide, G. 2016. Role of vertical transmission of *Toxoplasma gondii* in prevalence of infection. Expert Review of Anti-infective Therapy 14:335–344.

Hillman, A. E., A. J. Lymbery, and R. C. A. Thompson. 2016. Is *Toxoplasma gondii* a threat to the conservation of free-ranging Australian marsupial populations? International Journal for Parasitology: Parasites and Wildlife 5:17–27.

Hollings, T., M. Jones, N. Mooney, and H. McCallum. 2013. Wildlife disease ecology in changing landscapes: mesopredator release and toxoplasmosis. International Journal for Parasitology: Parasites and Wildlife 2:110–118.

Innes, E. A. 1997. Toxoplasmosis: comparative species susceptibility and host immune response. Comparative Immunology, Microbiology and Infectious Diseases 20:131–138.

Jardine, J. E., and J. P. Dubey. 2002. Congenital toxoplasmosis in a Indo-Pacific bottlenose dolphin (*Tursiops aduncus*). Journal of Parasitology 88:197–199.

Jessup, D. A., C. Lepczyk, S. Hernandez, L. Cherkassky, and R. Gerhold. 2023. Wild birds, cats, and One Health. *in*

B. Speer and Y. van Zeeland, editors. Current therapy in avian medicine and surgery. Second edition. Elsevier, Amsterdam, the Netherlands. In press.

Johnson, H. J., and A. A. Koshy. 2020. Latent toxoplasmosis effects on rodents and humans: how much is real and how much is media hype? mBio 11:e02164-19.

Juan-Sallés, C., M. Mainez, A. Marco, and A. M. M. Sanchís. 2011. Localized toxoplasmosis in a ring-tailed lemur (*Lemur catta*) causing placentitis, stillbirths, and disseminated fetal infection. Journal of Veterinary Diagnostic Investigation 23:1041–1045.

Kutz, S. J., J. Ducrocq, G. G. Verocai, B. M. Hoar, D. D. Colwell, K. B. Beckmen, L. Polley, B. T. Elkin, and E. P. Hoberg. 2012. Parasites in ungulates of Arctic North America and Greenland: a view of contemporary diversity, ecology, and impact in a world under change. Pages 99–252 *in* D. Rollinson and S. I. Hay, editors. Advances in parasitology. Volume 79. Academic Press, Cambridge, Massachusetts, USA.

Luft, B. J., and J. S. Remington. 1992. Toxoplasmic encephalitis in AIDS. Clinical Infectious Diseases 15:211–222.

Ma, H., Z. Wang, C. Wang, C. Li, F. Wei, and Q. Liu. 2015. Fatal *Toxoplasma gondii* infection in the giant panda. Parasite 22:30.

Massie, G.N., M. W. Ware, E. N. Villegas, and M.W. Black. 2010. Uptake and transmission of *Toxoplasma gondii* oocysts by migratory, filter-feeding fish. Veterinary Parasitology 169: 296–303.

Mateus-Pinilla, N. E., J. P. Dubey, L. Choromanski, and R. M. Weigel. 1999. A field trial of the effectiveness of a feline *Toxoplasma gondii* vaccine in reducing *T. gondii* exposure for swine. The Journal of Parasitology 85:855–860.

McLeod, R., J. Lykins, A. G. Noble, P. Rabiah, C. N. Swisher, P. T. Heydemann, D. McLone, D. Frim, S. Withers, F. Clouser, et al. 2014. Management of congenital toxoplasmosis. Current Pediatric Reports 2:166–194.

Mercier, A., D. Ajzenberg, S. Devillard, M. P. Demar, B. de Thoisy, H. Bonnabau, F. Collinet, R. Boukhari, D. Blanchet, S. Simon, et al. 2011. Human impact on genetic diversity of *Toxoplasma gondii*: example of the anthropized environment from French Guiana. Infection, Genetics and Evolution 11:1378–1387.

Miller, M., P. Conrad, E. R. James, A. Packham, S. Toy-Choutka, M. J. Murray, D. Jessup, and M. Grigg. 2008. Transplacental toxoplasmosis in a wild southern sea otter (*Enhydra lutris nereis*). Veterinary Parasitology 153:12–18.

Miller, M. A., K. Shapiro, M. Murray, M. J. Haulena, and S. Raverty. 2018. Protozoan parasites of marine mammals. Pages 425–470 *in* F. M. D. Gulland, L. A. Dierauf, and K. L. Whitman, editors. CRC Handbook of Marine Mammal Medicine. CRC Press, Boca Raton, Florida, USA.

Parameswaran, N., R. M. O'Handley, M. E. Grigg, A. Wayne, and R. C. A. Thompson. 2009. Vertical transmission of *Toxoplasma gondii* in Australian marsupials. Parasitology 136:939–944.

Pardini, L., A. Dellarupe, D. Bacigalupe, M. A. Quiroga, G. Moré, M. Rambeaud, W. Basso, J. M. Unzaga, G. Schares, and M. C. Venturini. 2015. Isolation and molecular characterization of *Toxoplasma gondii* in a colony of captive black-capped squirrel monkeys (*Saimiri boliviensis*). Parasitology International 64:587–590.

Poirotte, C., P. M. Kappeler, B. Ngoubangoye, S. Bourgeois, M. Moussodji, and M. J. E. Charpentier. 2016. Morbid attraction to leopard urine in *Toxoplasma*-infected chimpanzees. Current Biology 26:R98–R99.

Prestrud, K. W., K. Åsbakk, E. Fuglei, T. Mørk, A. Stien, E. Ropstad, M. Tryland, G.W. Gabrielsen, C. Lydersen, K. M. Kovacs, et al. 2007. Serosurvey for *Toxoplasma gondii* in Arctic foxes and possible sources of infection in the high Arctic of Svalbard. Veterinary Parasitology 150:6–12.

Ramakrishnan, C., S. Maier, R. A. Walker, H. Rehrauer, D. E. Joekel, R. R. Winiger, W. U. Basso, M. E. Grigg, A. B. Hehl, P. Deplazes, et al. 2019. An experimental genetically attenuated live vaccine to prevent transmission of *Toxoplasma gondii* by cats. Scientific Reports 9:1474.

Resendes, A. R., S. Almería, J. P. Dubey, E. Obón, C. Juan-Sallés, E. Degollada, F. Alegre, O. Cabezón, S. Pont, and M. Domingo. 2002. Disseminated toxoplasmosis in a Mediterranean pregnant Risso's dolphin (*Grampus griseus*) with transplacental fetal infection. Journal of Parasitology 88:1029–1032.

Roberts, J. O., H. F. E. Jones, and W. D. Roe. 2020. The effects of *Toxoplasma gondii* on New Zealand wildlife: implications for conservation and management. Pacific Conservation Biology 27:208–220.

Roe, W. D., L. Howe, E. J. Baker, L. Burrows, and S. A. Hunter. 2013. An atypical genotype of *Toxoplasma gondii* as a cause of mortality in Hector's dolphins (*Cephalorhynchus hectori*). Veterinary Parasitology 192:67–74.

Salant, H., T. Weingram, D. T. Spira, and T. Eizenberg. 2009. An outbreak of toxoplasmosis amongst squirrel monkeys in an Israeli monkey colony. Veterinary Parasitology 159:24–29.

Sandström, C. A. M., A. G. J. Buma, B. J. Hoye, J. Prop, H. van der Jeugd, B. Voslamber, J. Madsen, and M. J. J. E. Loonen. 2013. Latitudinal variability in the seroprevalence of antibodies against *Toxoplasma gondii* in non-migrant and Arctic migratory geese. Veterinary Parasitology 194:9–15.

Santana, C. H., A. R. de Oliveira, D. O. Dos Santos, S. P. Pimentel, L. D. R. de Souza, L. G. A. Moreira,

H. M. B. Braz, T. P. de Carvalho, C. E. B. Lopes, J. B. S. Oliveira, et al. 2021. Genotyping of *Toxoplasma gondii* in a lethal toxoplasmosis outbreak affecting captive howler monkeys (*Alouatta sp.*). Journal of Medical Primatology 50:99–107.

Shapiro, K., L. Bahia-Oliveira, B. Dixon, A. Dumètre, L. A. deWit, E. VanWormer, and I. Villena. 2019*a*. Environmental transmission of *Toxoplasma gondii*: oocysts in water, soil, and food. Food and Waterborne Parasitology 15:e00049.

Shapiro, K., P. A. Conrad, J. A. K. Mazet, W. W. Wallender, W. A. Miller, and J. L. Largier. 2010. Effect of estuarine wetland degradation on transport of *Toxoplasma gondii* surrogates from land to sea. Applied and Environmental Microbiology 76:6821–6828.

Shapiro, K., C. Krusor, F. F. M. Mazzillo, P. A. Conrad, J. L. Largier, J. A. K. Mazet, and M. W. Silver. 2014. Aquatic polymers can drive pathogen transmission in coastal ecosystems. Proceedings of the Royal Society of London Series B Biological Sciences 281:20141287.

Shapiro, K., M. A. Miller, A. E. Packham, B. Aguilar, P. A. Conrad, E. VanWormer, and M. J. Murray. 2016. Dual congenital transmission of *Toxoplasma gondii* and *Sarcocystis neurona* in a late-term aborted pup from a chronically infected southern sea otter (*Enhydra lutris nereis*). Parasitology 143:276–288.

Shapiro, K., E. VanWormer, A. Packham, E. Dodd, P. A. Conrad, and M. Miller. 2019*b*. Type X strains of *Toxoplasma gondii* are virulent for southern sea otters (*Enhydra lutris nereis*) and present in felids from nearby watersheds. Proceedings of the Royal Society of London Series B Biological Sciences 286:20191334.

Shwab, E. K., P. Saraf, X.-Q. Zhu, D.-H. Zhou, B. M. McFerrin, D. Ajzenberg, G. Schares, K. Hammond-Aryee, P. van Helden, S. A. Higgins, R. W. Gerhold, B. M. Rosenthal, X. Zhao, J. P. Dubey, and C. Su. 2018. Human impact on the diversity and virulence of the ubiquitous zoonotic parasite *Toxoplasma gondii*. Proceedings of the National Academy of Sciences USA 115:E6956–E6963.

Simon, A., A. N. Rousseau, S. Savary, M. Bigras-Poulin, and N. H. Ogden. 2013. Hydrological modelling of *Toxoplasma gondii* oocysts transport to investigate contaminated snowmelt runoff as a potential source of infection for marine mammals in the Canadian Arctic. Journal of Environmental Management 127:150–161.

Tenter, A. M., A. R. Heckeroth, and L. M. Weiss. 2000. *Toxoplasma gondii*: from animals to humans. International Journal for Parasitology 30:1217–1258.

Tokiwa, T., H. Yoshimura, S. Hiruma, Y. Akahori, A. Suzuki, K. Ito, M. Yamamoto, and K. Ike. 2019. *Toxoplasma gondii* infection in Amami spiny rat on Amami-Oshima Island, Japan. International Journal for Parasitology: Parasites and Wildlife 9:244–247.

Torrey, E. F. 2021. Parasites, pussycats, and psychosis: the unknown dangers of human toxoplasmosis. Springer, Cham, Switzerland.

Tryland, M., T. Nesbakken, L. Robertson, D. Grahek-Ogden, and B. T. Lunestad. 2014. Human pathogens in marine mammal meat—a northern perspective. Zoonoses and Public Health 61:377–394.

Vaudaux, J. D., C. Muccioli, E. R. James, C. Silveira, S. L. Magargal, C. Jung, J. P. Dubey, J. L. Jones, M. Z. Doymaz, D. A. Bruckner, et al. 2010. Identification of an atypical strain of *Toxoplasma gondii* as the cause of a waterborne outbreak of toxoplasmosis in Santa Isabel do Ivai, Brazil. Journal of Infectious Diseases 2010:1226–1233.

VanWormer, E., T. E. Carpenter, P. Singh, K. Shapiro, W. W. Wallender, P. A. Conrad, J. L. Largier, M. P. Maneta, and J. A. K. Mazet. 2016. Coastal development and precipitation drive pathogen flow from land to sea: evidence from a *Toxoplasma gondii* and felid host system. Scientific Reports 6:29252.

VanWormer, E., P. A. Conrad, M. A. Miller, A. C. Melli, T. E. Carpenter, and J. A. K. Mazet. 2013*a*. *Toxoplasma gondii*, source to sea: higher contribution of domestic felids to terrestrial parasite loading despite lower infection prevalence. EcoHealth 10:277–289.

VanWormer, E., H. Fritz, K. Shapiro, J. A. K. Mazet, and P. A. Conrad. 2013*b*. Molecules to modeling: *Toxoplasma gondii* oocysts at the human–animal–environment interface. Comparative Immunology, Microbiology and Infectious Diseases 36:217–231.

VanWormer, E., M. A. Miller, P. A. Conrad, M. E. Grigg, D. Rejmanek, T. E. Carpenter, and J. A. K. Mazet. 2014. Using molecular epidemiology to track *Toxoplasma gondii* from terrestrial carnivores to marine hosts: implications for public health and conservation. PLoS Neglected Tropical Diseases 8:e2852.

Wilson, A. G., S. Wilson, N. Alavi, and D. R. Lapen. 2021. Human density is associated with the increased prevalence of a generalist zoonotic parasite in mammalian wildlife. Proceedings of the Royal Society of London Series B Biological Sciences 288:20211724.

Work, T. M., J. G. Massey, B. A. Rideout, C. H. Gardiner, D. B. Ledig, O. C. Kwok, and J. P. Dubey. 2000. Fatal toxoplasmosis in free-ranging endangered 'Alala from Hawaii. Journal of Wildlife Diseases 36:205–212.

Work, T. M., S. K. Verma, C. Su, J. Medeiros, T. Kaiakapu, O. C. Kwok, and J. P. Dubey. 2016. *Toxoplasma gondii* antibody prevalence and two new genotypes of the parasite in endangered Hawaiian geese (Nene: *Branta sandvicensis*). Journal of Wildlife Diseases 52:253–257.

24 Endangered Mountain Gorillas, Ebola, and One Health

KIRSTEN V. GILARDI

Mountain Gorillas: Unique in the World

Gorillas, the world's largest primates, live in equatorial forests spanning the continent of Africa. There are two species, western (*Gorilla gorilla*) and eastern (*Gorilla beringei*) gorillas, each with two subspecies. The International Union for the Conservation of Nature (IUCN) ranks western and eastern gorillas as critically endangered due to habitat loss, poaching, and disease (IUCN 2021).

There are an estimated 1,063 mountain gorillas (*G. b. beringei*), a subspecies of the eastern gorilla (Figure 24.1), remaining in the wild (Hickey et al. 2019). They survive in two separate populations, entirely within protected areas: one in the Virunga Massif (comprising Volcanoes, Virunga, and Mgahinga Gorilla National Parks in Rwanda, Democratic Republic of Congo (DRC), and Uganda, respectively; 01°25′S, 29°32′E) and the other in Bwindi Impenetrable National Park (Uganda; 01°01′S, 29°41′E) (Figure 24.2). The two subpopulations are roughly similar in size, with an estimated minimum count of 604 gorillas (Hickey et al. 2018*a*) in the Virungas and an estimated 459 gorillas (Hickey et al. 2019) in Bwindi.

Mountain gorillas are unique in that a majority of the world population (approximately 60%) is human

Most of the world's mountain gorillas (*Gorilla beringei beringei*) are habituated to the close proximity of humans to facilitate tourism, research, monitoring, and veterinary care. Illustration by Laura Donohue.

Fig. 24.1. Mountain gorillas (*Gorilla beringei beringei*) are a subspecies of eastern gorilla that is classified by the International Union for the Conservation of Nature as endangered. Photograph courtesy of Skyler Bishop for Gorilla Doctors.

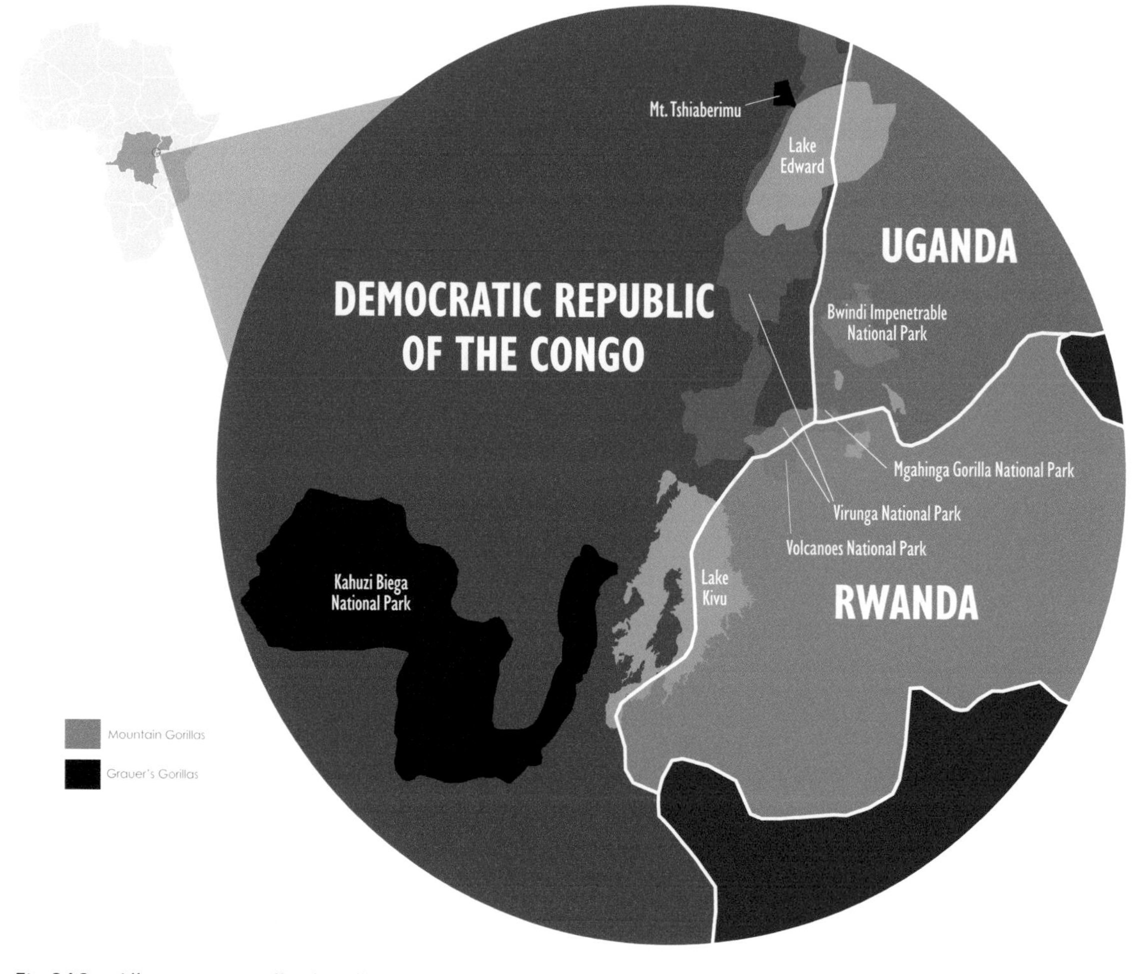

Fig. 24.2. All mountain gorillas (*Gorilla beringei beringei*) live in two protected areas in east-central Africa: in Volcanoes, Virunga, and Mgahinga Gorilla National Parks spanning the borders of Rwanda, Democratic Republic of Congo, and Uganda (respectively) and in Bwindi Impenetrable National Park in Uganda. Grauer's gorillas (*G. b. graueri*), the other subspecies of eastern gorilla, live inside and outside protected areas in eastern Democratic Republic of Congo. Illustration by Gorilla Doctors.

habituated (Williamson and Feistner 2011); in other words, most mountain gorillas in the world have been carefully, gradually, and purposefully made to become accustomed to the close presence of people to facilitate their protection, tourism, and research. For no other species of great ape does such a significant portion of its total world population spend a part of its day, every day, in close proximity to people. It is precisely this unique characteristic of the subspecies, as well as its genetic relatedness to humans, that create the One Health challenges and solutions that are explored further in this chapter.

One could argue that mountain gorillas, due to the high degree of human habituation of the two subpopulations and the fact that they survive in fully protected areas only, which are surrounded by very dense human populations that rely on intensive agriculture, are not truly wild. But it is clear that human habituation has been a key strategy for the conservation of the species. Habituation enables gorilla tourism, which generates essential revenues that fund park protection and management—more than $200 million (USD) annually in Rwanda alone (Maekawa et al. 2013). Tourism dollars also provide livelihood opportunities for communities adjacent to park boundaries (Munanura et al. 2016, Tolbert et al. 2019). Furthermore, habituation allows for ongoing behavioral and ecological research that helps inform management and conservation (Harcourt and Stewart 2007). Best practices around human habituation of great apes for tourism and research have been developed (Macfie and Williamson 2010).

Habituation allows people (trackers and researchers) to get close enough to monitor gorillas daily to detect the presence (or absence) of individual gorillas and observe them for signs of injury or illness. This is important because human habituation is not without consequence. It is critical to recognize that all great apes are susceptible to human pathogens, and the potential for transmission of infectious pathogens from people to great apes is highest where people and great apes are in close proximity. For example, human respiratory pathogens have been confirmed to cause illness in captive (Slater et al. 2014) and wild chimpanzees (*Pan troglodytes*) (e.g., Kaur et al. 2008, Köndgen et al. 2008, Negrey et al. 2019) and in wild western lowland gorillas (Grützmacher et al. 2016).

Mountain gorillas have also proven to be susceptible to human pathogens. In the late 1980s, human measles was the suspected cause of a respiratory illness outbreak in habituated mountain gorillas in Volcanoes National Park, Rwanda (Hastings et al. 1991). Respiratory illness outbreaks are common in mountain gorillas (Spelman et al. 2013). Human metapneumovirus was the primary cause of fatal pneumonia in an adult female mountain gorilla and her infant in 2009 in Volcanoes National Park, Rwanda (Palacios et al. 2011), and respiratory illness in mountain gorillas in Volcanoes National Park in 2012 and 2013 were determined to be caused by human respiratory syncytial virus (Mazet et al. 2020). Previous studies of ecto- and endo-parasites in mountain gorillas have suggested that humans are a source (e.g., Kalema-Zikusoka et al. 2002, Graczyk et al. 2002); more comprehensive and molecular-based work that allows for a closer examination of the extent to which gorillas may (more may not) suffer from parasites acquired from people is currently underway (e.g., Petrželková et al. 2021). While poaching and habitat destruction are significant threats for other gorilla species, close daily proximity with people and the disease risk this poses is a significant conservation threat for the mountain gorilla—the other side of habituation's double-edged sword.

For this reason, One Health—the perspective that animal, human, and environmental health are inextricably linked (Karesh and Cook 2005)—is a highly relevant and critically important approach to mountain gorilla conservation. Ebolaviruses, among the most lethal of all human viruses, provides a compelling example.

Ebola Virus

Ebolavirus is a genus in the Filoviridae family that causes explosive outbreaks of hemorrhagic disease in human communities. First described in 1976, 28 Ebolavirus outbreaks have been documented across Africa (Malvy et al. 2019). Viral genera can comprise one or more species that differ in their genomic, phylogenetic, and phenotypic properties (Kuhn 2021). *Zaire ebolavirus*, also called simply Ebola virus (EBOV), is one of six species within the genus Ebolavirus and one of four species causing human disease in Africa (other species include *Sudan*, *Bundibugyo*, and *Taï Forest ebolavirus*). EBOV can have a case fatality rate up to 90%. The largest Ebolavirus outbreak (caused by EBOV) occurred in West Africa (in Liberia, Sierra Leone, and Guinea) from 2013–2016: more than 28,000 people were infected, and more than 11,000 died (WHO 2021). Ebola virus disease (EVD) is a "hemorrhagic fever" characterized by high fever, abdominal pain, and bloody vomit and diarrhea; the disease progresses rapidly, and death can occur within 72 hours of onset of clinical signs. Disease is the result of viral invasion of endothelial cells lining blood vessels, liver cells, and many immune cells. Viral destruction of endothelial cells leads to depletion of clotting factors, widespread bleeding, and hypovolemic shock.

Ebola viruses are highly contagious, transmitted through direct contact with mucosal surfaces (oral, nasal, and genital) or wounds with bodily fluids and excretions from infected animals or people. Patients suffering from EVD shed large quantities of the virus when they are ill or dying, and live virus can survive in bodily tissues and fluids after patients succumb to their infections. People who are caring for EVD patients or preparing bodies for burial are at great risk for infection. Bats, especially fruit bats, are strongly suspected to be the natural reservoir of Ebola viruses (Leroy et al. 2005, 2009). It is generally assumed that the virus spills over from bats when people or wildlife, including great apes, come into direct contact with an infected bat, or with another infected animal, often as a result of hunting, foraging, or handling bushmeat (Rewar and Mirdha, 2014) (Figure 24.3).

Ebola viruses, in particular *Zaire ebolavirus* and *Taï Forest ebolavirus*, are highly lethal to great apes. *Ebolavirus* is strongly suspected to have caused large die-offs of gorillas and chimpanzees in Gabon and Republic of Congo that coincided with confirmed EVD in other wildlife and humans (Leroy et al. 2004). While clinical signs of EVD in great apes have not been observed, *Ebolavirus* has been confirmed in gorilla and chimpanzee carcasses (Wittman et al. 2007). The actual numbers of great apes that have died due to *Ebolavirus* infections can only be estimated through retrospective data analyses and predictive modelling; all of these analyses have suggested that *Ebolavirus* has caused substantial great ape population declines (Walsh et al. 2003; Bermejo et al. 2006; Caillaud et al. 2006; Devos et al. 2008; Genton et al. 2012, 2015). New tools for detecting the presence of antibodies against EBOV in gorilla and chimpanzee feces may help us to better understand population-level exposure (Reed et al. 2014, Mombo et al. 2020). As with humans, it is assumed that great apes are infected with EBOVs through direct contact with a wildlife reservoir (e.g., bats; Leroy et al. 2005, 2009), with other infected great apes (Caillaud et al. 2006), or with an infected animal carcass.

Ebola Virus and Mountain Gorilla Conservation

The level of human death and suffering caused by EBOV makes it one of the most feared diseases on the planet. The virus poses a very real threat to the conservation of great apes because they are highly susceptible. As more than half the world's mountain gorillas come into close proximity of people daily, they are at higher risk for contracting EVD from infected people as well as from animals in their forest habitat. These highly protected and managed forests comprising the entirety of the mountain gorilla range are surrounded by the highest density of human populations in continental Africa, a situation providing

Fig. 24.3. Ebola virus, an endemic virus of bats, can transmit or spill over to other wildlife and to people who come into direct contact with live or dead bats that are shedding the virus. The virus is highly transmissible, can spread quickly among people and wild animals like gorillas or other great apes, and causes severe disease with high rates of mortality. Illustration by Laura Donohue.

the ideal conditions for EBOV to emerge and impact mountain gorillas.

Ebola virus is a classic example of a One Health disease: its reservoir is wildlife (bats); it is transmitted to humans and other wildlife (in particular, great apes) that come into contact with bats shedding the virus; and the enhanced opportunity for these "spillover" events are the result of our intensifying shared use of forested landscapes. The circumstances surrounding EBOV spillover from wildlife reservoirs is likely the same: people and great apes come into direct contact with Ebola-infected wildlife in the course of utilizing forest habitat for hunting and foraging. Once a person or a great ape is infected, the virus replicates, causing a highly lethal hemorrhagic fever, and EVD is highly transmissible to other people or great apes. The virus is capable of causing high levels or morbidity and mortality to local populations and communities of both humans and great apes.

Ebola virus is of particular concern for mountain gorillas because human outbreaks have occurred literally at the doorsteps of the protected areas in which they live. From 2018 to 2020, a *Zaire ebolavirus* disease outbreak occurred in Ituri and North Kivu Provinces in the eastern DRC. Ultimately, it became the second largest EVD outbreak on record, second only to the 2013–2016 West Africa Outbreak, causing more than 2,000 human deaths. The EVD outbreak zone was directly adjacent to the northern sector of Virunga National Park, including an area called the Mount Tshiaberimu sector that is home to a small population of the other eastern gorilla subspecies, the Grauer's gorillas (*Gorilla beringei graueri*). The outbreak zone was also adjacent to the Uganda border, approximately 60 km from Bwindi Impenetrable National Park in Uganda. While the outbreak zone remained many tens of kilometers from the part of Virunga National Park where mountain gorillas range, it overlapped with various protected areas within the known range of wild Grauer's gorillas, and it literally surrounded the Gorilla Rehabilitation and Conservation Education Center in Kasugho, which houses 14 orphaned Grauer's gorillas.

Thankfully, no mountain gorillas were infected by EBOV during this large human outbreak. A number of One Health elements are in place, or were developed and implemented, to help minimize the threat of EBOV and other infectious agents for gorillas.

Contingency Planning

The 2018–2020 EVD outbreak in the eastern DRC prompted protected area authorities, Gorilla Doctors, and other nongovernmental collaborating organizations to engage in a contingency planning process for minimizing the threat of EVD for mountain gorillas in Virunga National Park. The planning process identified phases of response, key actions for each phase, and responsible parties for implementation, including how Gorilla Doctors would respond to any report of acute death in a gorilla for which an etiology was not immediately obvious (e.g., trauma). Gorilla Doctors also worked with partners to run computer simulation models on what might be expected at the population level should a gorilla become infected with the EBOV, and what options might be available for disease control. Fortunately, at no point during the 2018–2020 EVD outbreak in eastern DRC did a mountain gorilla in Virunga National Park die of EVD. An added benefit of having conducted contingency planning for EVD was that the parks were primed to conduct similar contingency planning for the COVID-19 pandemic and the threat posed by the emergent SARS CoV-2 virus.

Veterinary Monitoring and Clinical Interventions

Since 1986, the Mountain Gorilla Veterinary Project (MGPV, Inc., now partnered with the Karen C. Drayer Wildlife Health Center as "Gorilla Doctors") has provided in situ veterinary care to wild human-habituated mountain gorillas suffering from human-related injury and illness. Human habituation allows for close observation of individual gorillas on a daily basis by trained park personnel and Gorilla Doctors veterinarians. When a gorilla is observed with clinical signs of illness or injury, veterinarians assess the animal's status and prognosis for recovery, and intervene if the injury or illness is deemed life threatening (Figure 24.4). Common reasons for interventions include snare removal (Haggblade et al. 2019), outbreaks of respiratory illness (Spelman et al. 2013), and parasite-related gastrointestinal disease (Muhangi et al. 2021, Petrželková et al. 2021). Gorilla Doctors annually perform between 300 and 400 routine health checks and veterinary monitoring visits to habituated mountain gorillas in all of the parks and conduct between 30 and 40 clinical interventions. This daily gorilla health monitoring enables early detection of clinical signs of potentially devastating outbreaks of disease, like EVD, and the

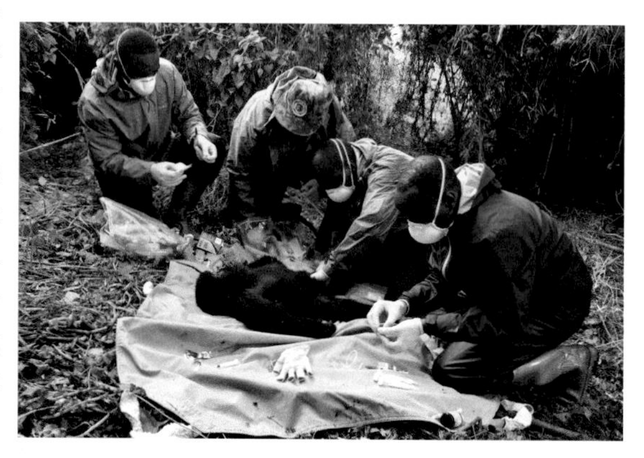

Fig. 24.4. Veterinarians on the staff of Gorilla Doctors (www.gorilladoctors.org) remove a snare from an anesthetized infant mountain gorilla (*Gorilla beringei beringei*). Veterinary care for mountain gorillas has proven to be a significant contributor to their annual population growth rate and, therefore, a critical component of the extreme conservation measures that have brought the mountain gorilla back from the brink of extinction. Photograph by Gorilla Doctors.

established framework and readiness for clinical intervention allows for the potential to intervene to minimize the impacts of disease outbreaks.

Employee Health

Gorilla Doctors has been practicing One Health for many years, long before the term came into common usage, having had the foresight to recognize that the health of the people who come into close proximity of mountain gorillas is critical for preventing the transmission of disease from people to gorillas. Toward that end, Gorilla Doctors worked closely with government partners in Rwanda and the DRC to design and implement an occupational health program for park personnel starting in the early 2000s. In coordination with physicians from the regional hospital, Gorilla Doctors facilitated annual wellness examinations for all park workers, including the performance of routine diagnostics (e.g., bloodwork, fecal exams for parasites, and chest radiographs) and regular treatment for gastrointestinal parasites to improve park workers' nutritional plane and body condition and reduce the incidence of diarrhea (and, therefore, frequency of defecation in the parks). In the early years of this effort, the occupational health program represented park workers' only contact with a physician outside of emergency visits to the regional hospital. Park personnel with underlying health conditions, including human immunodeficiency virus and tuberculosis infections, were identified and referred to the governments' treatment programs. Park personnel diagnosed with poor vision were provided with eyeglasses. Over the years, the program expanded to include quarterly parasite treatment for park workers' immediate family members for their general wellness and to reduce the potential that they would reinfect the park worker(s) living in their households. Ultimately, this occupational health program run by Gorilla Doctors was deemed so critical to park management that the parks hired their own physicians to oversee the program.

Best Practices

The fact that wild great apes having close contact with people may suffer from human-borne infectious disease led the IUCN to publish best-practice recommendations for great ape health monitoring and disease control (Gilardi et al. 2015). Indeed, mountain gorilla conservation and management has been the world's "incubator" for development of many of these best practices, simply by virtue of its long history of human habituation of gorillas for tourism and research purposes.

In addition to international guidelines, Gorilla Doctors works closely with park partners to conduct regular refresher health-surveillance training for park personnel on how to recognize and report clinical signs of illness and injury. Training workshops also provide opportunities to share information about zoonoses, and about the importance of good hygiene for personal and public health protection.

Extreme Conservation

Mountain gorillas are the only great ape whose numbers in the wild are increasing. This can be attributed to "extreme conservation" measures: the current range of the mountain gorilla is entirely protected and managed, and individual gorillas suffering from injury or illness benefit from *in situ* veterinary care (Robbins et al. 2011). Indeed, these conservation measures have resulted in the recent IUCN downlisting of mountain gorillas from critically endangered to endangered (Hickey et al. 2018*b*).

The future for mountain gorillas remains fragile. In addition to the disease threats described in this chapter, full-scale civil war, ongoing rebel activity, and crime surrounding open and black markets for high-value natural resources (e.g., wildlife and bushmeat, precious metals, and natural gas) disrupt civil society and make the east-central part of Africa among the most insecure places in the world (Schouten 2021). Natural events, such as the June 2021 eruption of

Mount Nyiragongo in North Kivu Province, DRC, have the potential to disrupt displace thousands of people and impact protected areas. In the most densely populated part of continental Africa (700 people/km^2), people living adjacent to mountain gorilla parks make only dollars a day at most and rely on subsistence agriculture to feed their families (Maekawa et al. 2013).

Summary

Worldwide, great ape health and conservation is intrinsically linked to the health and well-being of the human communities with which they come into daily contact, and nowhere is this more evident than with the mountain gorillas of east-central Africa. The survival of individual gorillas and their potential to reproduce and grow the population is completely dependent on the sustainability of their forests, the stability of governments, and the health and well-being of the nearby human communities. For these reasons, it is imperative that mountain gorilla conservation efforts continue as a One Health endeavor. This benefits not just the mountain gorilla, but also the human communities with which it shares its ecosystem. Particularly in Rwanda and Uganda where governance is relatively stable, tourism directly and indirectly creates jobs and generates significant revenue—including portions allocated to communities for community-led development projects, like construction of schools of public water sources. Park management helps ensure that the high montane forests remain intact and, therefore, fully functional as sources of freshwater for communities surrounding their boundaries. The increased knowledge of the sharing of pathogens between humans and gorillas facilitates awareness of adults and children around best practices for reducing the risk of communicable disease. And while difficult to measure, there is a tangible pride in all three countries in being the only place in the world where mountain gorillas survive, and a strong sense of purpose in continuing to engage in practices that have led to the recovery of this iconic species from the brink of extinction.

ACKNOWLEDGEMENTS

The author wishes first and foremost to acknowledge all veterinarians, technicians, and administrative staff of the Gorilla Doctors organization, current and past, for their tireless work and dedication since 1986. In particular, Dr. Michael R. Cranfield's leadership of the organization from 1998 to 2018 made Gorilla Doctors a One Health endeavour and success. Governmental partners, donors, and grantors have made all work possible. Lastly, a special thanks to the Karen C. Drayer Wildlife Health Center, University of California Davis School of Veterinary Medicine for giving the author a wholly supportive, enriching, and inspiring professional home.

LITERATURE CITED

Bermejo, M., J. D. Rodríguez-Teijeiro, G. Illera, A. Barroso, C. Vilà, and P. D. Walsh. 2006. Ebola outbreak killed 5000 gorillas. Science 314:1564.

Caillaud, D., F. Levréro, R. Cristescu, S. Gatti, M. Dewas, M. Douadi, A. Gautier-Horn, M. Raymond, and N. Ménard. 2006. Gorilla susceptibility to Ebola virus: the cost of sociality. Current Biology 16:R489–R491.

Devos, C., P. Walsh, E. Arnhem, and M. C. Huynen. 2008. Monitoring population decline: can transect surveys detect the impact of the Ebola virus on apes? Oryx 42:367–374.

Genton, C., R. Cristescu, S. Gatti, F. Levréro, E. Bigot, D. Caillaud, J. S. Pierre, and N. Ménard. 2012. Recovery potential of a western lowland gorilla population following a major Ebola outbreak: results from a ten year study. PLoS One 7:e37106.

Genton, C., A. Pierre, R. Cristescu, F. Levréro, S. Gatti, J.-S. Pierre, N. Ménard, and P. Le Gouar. 2015. How Ebola impacts social dynamics in gorillas: a multistate modelling approach. Journal of Animal Ecology 84:166–176.

Gilardi, K. V., T. R. Gillespie, F. H. Leendertz, E. J. Macfie, D. A. Travis, C. A. Whittier, and E. A. Williamson. 2015. Best practice guidelines for health monitoring and disease control in great ape populations. https://portals.iucn.org/library/node/45793. Accessed 14 Oct 2021.

Graczyk, T. K., J. B. Nizeyi, B. Ssebide, R. C. A. Thompson, C. Read, and M. R. Cranfield. 2002. Anthropozoonotic *Giardia duodenalis* genotype (assemblage) A infections in habitats of free-ranging human-habituated gorillas, Uganda. 2002. Journal of Parasitology 88:905–909.

Grützmacher, K. S., S. Köndgen, V. Keil, A. Todd, A. Feistner, I. Herbinger, K. Petrzelkova, T. Fuh, S. A. Leendertz, S. Calvignac-Spencer, et al. 2016. Co-detection of respiratory syncytial virus in habituated wild western lowland gorillas and humans during a respiratory disease outbreak. EcoHealth 13:499–510.

Haggblade, M., K. Gilardi, J. B. Noheri, M. Cranfield, A. Mudakikwa, and W. A. Smith. 2019. Outcomes of snare-related injuries in endangered mountain gorillas (*Gorilla beringei beringei*) in Rwanda. Journal of Wildlife Diseases 55: 298–303.

Harcourt, A. H., and K. J. Stewart. 2007. Gorilla society: conflict, compromise, and cooperation between the sexes. University of Chicago Press, Chicago, Illinois, USA.

Hastings, B. E., D. Kenny, L. J. Lowenstine, and J. W. Foster.1991. Mountain gorillas and measles: ontogeny of a wildlife vaccination program. Pages 198–205 *in* Proceedings of the American Association of Zoo Veterinarians Annual Conference. American Association of Zoo Veterinarians, Jacksonville, Florida, USA.

Hickey, J. R., A. C. Granjon, L. Vigilant, W. Eckardt, K. V. Gilardi, M. Cranfield, A. Musana, A. B. Masozera, D. Babaasa, F. Ruzigandekwe, et al. 2018*a*. Virunga 2015–2016 surveys: monitoring mountain gorillas, other selected mammals, and illegal activities. Final report. https://igcp.org/content/uploads/2020/09/Virunga-Census-2015-2016-Final-Report-2019-with-French-summary-2019_04_24.pdf. Accessed 14 Oct 2021.

Hickey, J. R., A. Basabose, K. V. Gilardi, D. Greer, S. Nampindo, M. M. Robbins, and T. S. Stoinski. 2018*b*. *Gorilla beringei* ssp. *beringei*. The IUCN Red List of Threatened Species 2018. IUCN Global Species Programme Red List Unit, Cambridge, England.

Hickey, J. R., E. Uzabaho, M. Akantorana, J. Arinaitwe, I. Bakebwa, R. Bitariho, W. Eckardt, K. V. Gilardi, J. Katutu, C. Kayijamahe, et al. 2019. Bwindi–Sarambwe 2018 surveys: monitoring mountain gorillas, other select mammals, and human activities. https://igcp.org/content/uploads/2020/09/Bwindi-Sarambwe-2018-Final-Report-2019_12_16.pdf. Accessed 14 Oct 2021.

International Union for the Conservation of Nature (IUCN). 2021. International Union for the Conservation of Nature Red List. www.iucnredlist.org. Accessed 13 Dec 2021.

Kalema-Zikusoka, G., R. A. Kock, and E. J. Macfie. 2002. Scabies in free-ranging mountain gorilas (*Gorilla beringei beringei*) in Bwindi Impenetrable National Park. Veterinary Record 150:12–15.

Karesh, W. B., and R. A. Cook. 2005. The human–animal link. Foreign Affairs 84:38–50.

Kaur, T., J. Singh, S. Tong, C. Humphrey, D. Clevenger, W. Tan, B. Szekely, Y. Wang, Y. Li, E. A. Muse, et al. 2008. Descriptive epidemiology of fatal respiratory outbreaks and detection of a human-related metapneumovirus in wild chimpanzees (*Pan troglodytes*) at Mahale Mountains National Park, Western Tanzania. American Journal of Primatology 70:755–765.

Köndgen, S., H. Kuhl, P. K. N'Goran, P. D. Walsh, S. Schenk, N. Ernst, R., Biek, P. Formenty, K. Matz-Rensing, B. Schweiger, et al. 2008. Pandemic human viruses cause decline of endangered great apes. Current Biology 18:260–264.

Kuhn, J.H. 2021. Virus taxonomy. Pages 28–37 *in* D. Bamford and M. Zuckerman, editors Encyclopedia of virology. Fourth edition. Elsevier, Amsterdam, The Netherlands.

Leroy, E. M., A. Epelboin, V. Mondonge, X. Pourrut, J.-P. Gonzalez, J.-J. Muyembe-Tamfum, and P. Formenty. 2009. Human Ebola outbreak resulting from direct exposure to fruit bats in Luebo, Democratic Republic of Congo, 2007. Vector-borne and Zoonotic Diseases 9:723–728

Leroy, E. M., B. Kumulungui, X. Pourrut, P. Rouquet, A. Hassanin, P. Yaba, A. Délicat, J. T. Paweska, J.-P. Gonzales, and R. Swanepoel. 2005. Fruit bats as reservoirs of Ebola virus. Nature 438:575–576.

Leroy, E. M., P. Rouquet, P. Formenty, S. Souquiére, A. Kilbourne, J.-M. Froment, M. Bermejo, S. Smit, W. Karesh, R. Swanepoel, et al. 2004. Multiple Ebola virus transmission events and rapid decline of Central African wildlife. Science 303:387–390.

Macfie, E. J., and E. A. Williamson. 2010. Best practice guidelines for great ape tourism. https://www.iucn.org/content/best-practice-guidelines-great-ape-tourism. Accessed 4 Oct 2021.

Maekawa, M., A. Lanjouw, E. Rutagarama, and D. Sharp. 2013. Mountain gorilla tourism generating wealth and peace in post-conflict Rwanda. Natural Resources Forum 37:127–137.

Malvy, D., A. K. McElroy, H. deClerck, S. Günther, and J. van Griensven. 2019. Ebola virus disease. Lancet 393:936–948.

Mazet, J. A. K., B. N. Genovese, L. A. Harris. M. Cranfield, J. B. Noheri, J.-F. Kinani, D. Zimmerman, M. Bahizi, A. Mudakikwa, T. Goldstein, and K. V. K. Gilardi. 2020. Human respiratory syncytial virus detected in mountain gorilla respiratory outbreaks. EcoHealth 17:449–460.

Mombo, I. M., M. Fritz, P. Becquart, F. Leigeois, E. Elguero, L. Boundenga, T. N. Mebaley, F. Pugnolle, G. D. Maganga, and E. M. Leroy. 2020. Detection of Ebola virus antibodies in fecal samples of great apes in Gabon. Viruses 12:1347.

Muhangi, D., C. H. Gardiner, L. Ojok, M. R. Cranfield, K. V. K. Gilardi, A. B. Mudakikwa, and L. J. Lowenstine. 2021. Pathological lesions of the digestive tract in free-ranging mountain gorillas (*Gorilla beringei beringei*). American Journal of Primatology 83:e23290.

Munanura, I. E., K. F. Backman, J. C. Hallo, and R. B. Powell. 2016. Perceptions of tourism revenue sharing impacts on Volcanoes National Park, Rwanda: a sustainable livelihoods framework. Journal of Sustainable Tourism, 24:1709–1726.

Negrey, J., R. Reddy, E. Scully, S. Phillips-Garcia, L. Owens, K. Langergraber, J. Mitani, M. Thompson, R. Wrangham, M. Muller, et al. 2019. Simultaneous outbreaks of respiratory disease in wild chimpanzees caused by distinct viruses of human origin. Emerging Microbes and Infections 8:139–149.

Palacios, G., L. J. Lowenstine, M. R. Cranfield, K. V. K. Gilardi, L. Spelman, M. Lukasik-Braum, J.-F. Kinani, A. Mudakikwa, E. Nyirakaragire, A. V. Bussett, et al. 2011. Human metapneumovirus infection in wild mountain gorillas, Rwanda. Emerging Infectious Diseases 17:711–713.

Petrželková, K. J., C. Uwamahoro, B. Pafčo, B. Červená, P. Samaš, A. Mudakikwa, R. Muvunyi, P. Uwingeli, K. Gilardi, J. Nziza, et al. 2021. Heterogeneity patterns of helminth infections across populations of mountain gorillas (*Gorilla beringei beringei*). Scientific Reports 11:10869.

Reed, P. E., K.N. Cameron, A. U. Ondzie, D. Joly, W. B. Karesh, S. Mulangu, G. Fabozzi, M. Bailey, Z. Shen, N.J. Sullivan, et al. 2014. A new approach for monitoring ebolavirus in wild great apes. PLoS Neglected Tropical Diseases 8:e3143.

Rewar, S. and D. Mirdha. 2014. Transmission of Ebola virus disease: an overview. Annals of Global Health 80:444–451.

Robbins, M. M., M. Gray, K. Fawcett, F. B. Nutter, P. Uwingeli, I. Mburanumwe, E. Kagoda, A. Basabose, T. S. Stoinski, M. Cranfield, et al. 2011. Extreme conservation leads to recovery of the Virunga mountain gorillas. PLoS One 6:e19788.

Schouten, P. 2021 (February 25). Violence is endemic in eastern Congo: what drives it. https://theconversation.com/violence-is-endemic-in-eastern-congo-what-drives-it-156039. Accessed 14 Oct 2021.

Slater O., K. A. Terio, Y. Zhang, D. D. Erdman, E. Schneider, J. M. Kuypers, S. M. Wolinksy, K. J. Kunstman, J. Kunstman, M. J. Kinsel, et al. 2014. Human metapneumovirus infection in chimpanzees, United States. Emerging Infectious Diseases 20:2115–2118.

Spelman, L. H., K. V. K. Gilardi, M. Lukasik-Braum, J.-F. Kinani, E. Nyirakaragire, L. J. Lowenstine, and M. R. Cranfield. 2013. Respiratory disease in mountain gorillas (*Gorilla beringei beringei*) in Rwanda, 1990–2010: outbreaks, clinical course, and medical management. Journal of Zoo and Wildlife Medicine 44:1027–1035.

Tolbert, S., W. Makambo, S. Asuma, A. Musema, and B. Mugabukomeye. 2019. The perceived benefits of protected areas in the Virunga–Bwindi Massif. Environmental Conservation 46(Special Issue 1):76–83.

Walsh, P. D., K. A. Abernethy, M. Bermejo, R. Beyers, P. Wachter, M. E. Akou, B. Huijbregts, D. I. Manbounga, A. K. Toham, A. M. Kilbourn, et al. 2003. Catastrophic ape decline in western equatorial Africa. Nature 422:611–614.

Williamson, E. A., and A. T. C. Feistner. 2011. Habituating primates: processes, techniques, variables and ethics. Pages 33–49 *in* J. M. Setchell and D. J. Curtis, editors. Field and laboratory methods in primatology: a practical guide. Second edition. Cambridge University Press, Cambridge, England.

Wittmann, T. J., R. Biek, A. Hassanin, P. Rouquet, P. Reed, P. Yaba, X. Pourrut, L. A. Real, J. P. Gonzalez, and E. M. Leroy. 2007. Isolates of Zaire ebolavirus from wild apes reveal genetic lineage and recombinants. Proceedings of the National Academy of Sciences USA 104:17123–17127.

World Health Organization (WHO). 2021. Ebola: West Africa, March 2014–2016. https://www.who.int/emergencies/situations/ebola-outbreak-2014-2016-West-Africa. Accessed 13 Dec 2021.

25 Emerging Coronaviruses

A One Health Harbinger

Christine K. Johnson, Janna M. E. Freeman, Tierra Smiley Evans, Diego Montecino-Latorre, Marcela M. Uhart

Introduction

Coronaviruses (CoVs) are enveloped, single-stranded positive-sense RNA viruses that are widespread and cause disease in a range of wild and domesticated animals, as well as humans. In recent years, they have demonstrated their propensity for jumping host species and have caused three notable public health crises. Like many emerging infectious diseases, human coronaviruses (HCoVs) have their origins in wildlife, particularly bats and rodents (Figure 25.1). The most recent zoonotic coronavirus to emerge, severe acute respiratory syndrome coronavirus 2 (SARS-CoV-2), resulted in the coronavirus disease 2019 (COVID-19) global pandemic with an unprecedented toll in human lives lost, societal disruption, and economic cost. In just the first two years, the COVID-19 pandemic has resulted in at least 484.9 million cases and 6.13 million deaths globally (Ritchie et al 2022). This pandemic has revealed substantial weaknesses in our global response to infectious disease outbreaks and highlighted the urgent need for coordinated and proactive multilateral surveillance to facilitate early warning and actionable intelligence to mitigate and ultimately preempt pandemic threats.

A more proactive strategy for pandemic preparedness must involve large-scale and long-term surveillance that includes wildlife, so that cross-species transmission of animal-borne viruses can better inform public health action. Host-jumping capability, or higher virus–host plasticity, has been linked to zoonotic transmission and to human–human transmission of viruses in general (Kreuder Johnson et al. 2015). Evolution and adaptation that include recombination of genetic material between coronaviruses in a single host have been proposed as a significant mechanism underlying host-jumping capabilities. Coronaviruses are also prone to high error rates and lack of error proofing during replication, both of which are common sources of mutations among RNA viruses in general. Cross-species transmission of coronaviruses has included animal–human and human–animal events (Anthony et al. 2017; Graham and Baric 2010; Bashor et al. 2021). These host-jumping traits are not unique to coronaviruses, but the epidemics and pandemics caused by these viruses in the past 20 years are a major concern.

To date, seven coronaviruses with recognized or suspected animal origins that infect humans have been identified (Figure 25.1). Of those seven coronaviruses, four cause mainly mild upper respiratory symptoms in humans; however, HCoV-NL63 and HCoV-HKU1 have been associated with lower respiratory tract disease in children, and HCoV-229E and

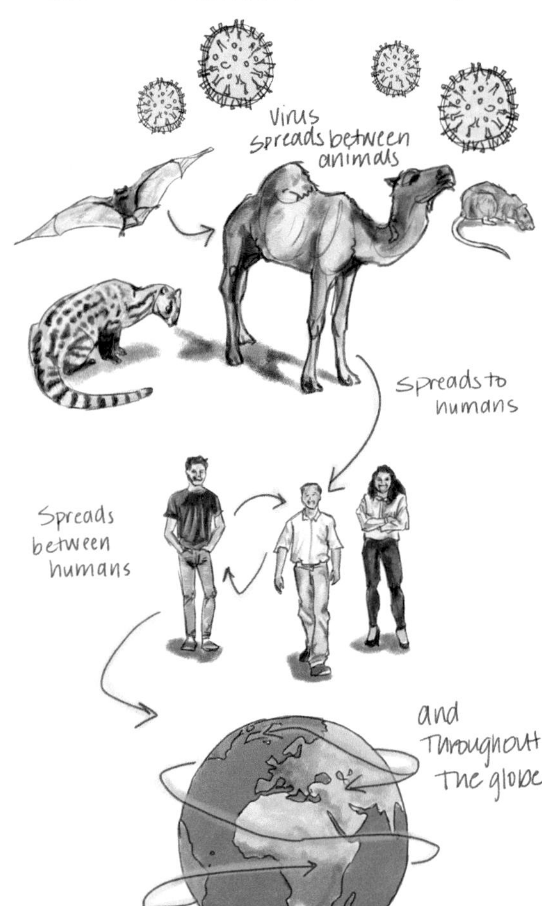

Summary of coronavirus adaptation from animal hosts to humans and spread around the globe. Illustration by Laura Donohue.

HCoV-OC43 have been linked to more severe disease in the elderly (Ye et al. 2020). Three other coronaviruses, SARS-CoV, Middle East respiratory syndrome coronavirus (MERS-CoV), and SARS-CoV-2, have caused severe respiratory disease in humans with high case fatality rates, creating public health emergencies that required regional or global disease control measures. The most recent zoonotic virus to emerge, SARS-CoV-2, has an especially broad range of clinical symptoms with high rates of respiratory transmissibility in advance of symptom onset, facilitating widespread community transmission and rapid progression from initial emergence into a historic global pandemic.

Coronaviruses in Animals

Coronaviruses were initially of veterinary interest as pathogens affecting poultry during the 1930s (Gibbs 1935). Coronavirus infection in wildlife was first reported in ill, captive coyotes (*Canis latrans*; Evermann et al. 1980) and cheetahs (*Acinonyx jubatus*) during the late 1970s and 1980s (Evermann et al. 1988). Subsequent research efforts in nondomestic animal species reported seropositivity in caribou (*Rangifer tarandus*; Elazhary et al. 1981), meadow voles (*Microtus pennsylvanicus*; Descôteaux and Mihok 1986), greylag geese (*Anser anser*), and mallard ducks (*Anas platyrhynchos*; Jonassen et al. 2005). The emergence of SARS-CoV in 2002 and the expansion of molecular techniques to detect genetic material led to rapid growth in surveillance for coronaviruses at the beginning of the 21st century. At present, a filtered review of data available in GenBank (Clark et al. 2016), excluding *Homo sapiens* and experimental exposures, has identified coronavirus nucleotide sequences in 379 unique species, which include 296 mammal species and 83 bird species, representing 22 different taxonomic orders and 67 families (Figure 25.2).* Given interest in the origins of SARS-CoV-2, genomic surveillance for coronaviruses in many species is expected to grow exponentially.

To date, coronaviruses have been detected in 15.1% of Chiroptera species, 1.5% of Rodentia species, 6.6% of Artiodactyla species, and 7.8% of

*The full dataset is available from https://github.com/dmontecino/emerging_coronaviruses and an interactive version of the graph is available from https://drive.google.com/drive/u/0/folders/1s_WGQPaSN_ltnNNalI4RUJwyEieC5K9V by downloading the folder and the HTML file in the same directory and then opening the HTML file.

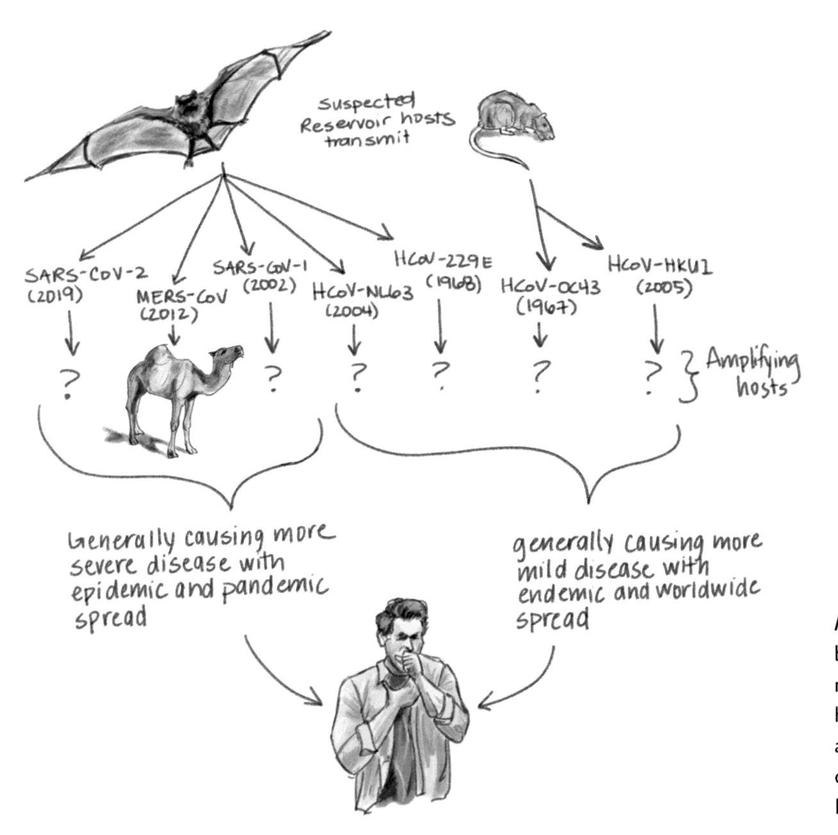

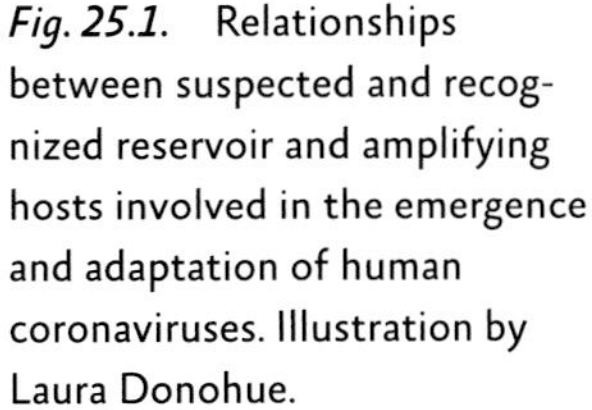

Fig. 25.1. Relationships between suspected and recognized reservoir and amplifying hosts involved in the emergence and adaptation of human coronaviruses. Illustration by Laura Donohue.

Carnivora species (Figure 25.2). Among birds, the highest percentage of species with coronaviruses are Anseriformes (14.7%; even greater than bats), followed by Charadriiformes (4.4%) and Passeriformes (0.2%). Among wild hosts, coronaviruses were most commonly reported in straw-colored fruit bats (*Eidolon helvum*), but this species has likely been the focus of more investigations compared to other wild species. Domesticated species with the most coronavirus detections were chickens (*Gallus domesticus*), pigs (*Sus domesticus*), cats (*Felis catus*), dromedary camels (*Camelus dromedarius*), dogs (*Canis familiaris*), and cattle (*Bos taurus*). Importantly, these animals are intensively surveyed for animal production, food, and public safety purposes.

Of the four coronavirus genera, alpha- and betacoronaviruses (e.g, MERS, SARS-CoV, and SARS-CoV-2) have predominated in mammals, while gamma- and deltacoronaviruses have been most common in birds (Figure 25.2). However, the transmission of coronaviruses between birds and mammals and among different classes of mammals has been described. For example, porcine deltacoronavirus may have originated in birds (Chen et al. 2018) with specific strains of this virus that remain infectious to species of the Aves class (Boley et al. 2020) and have been reported to cause disease in humans (Lednicky et al. 2021). Swine acute diarrhea syndrome coronavirus is an alphacoronavirus that emerged in China in 2016, causing acute and fatal diarrhea in piglets, with likely origins in rhinolophid bats (Zhou et al. 2018). Infection and posterior viral shedding of a specific strain of this virus has been reported in chickens experimentally exposed (Mei et al. 2022). The tendency of coronaviruses to transmit between species in different orders and families underscores the propensity of this group of viruses to adapt to new hosts and habitats.

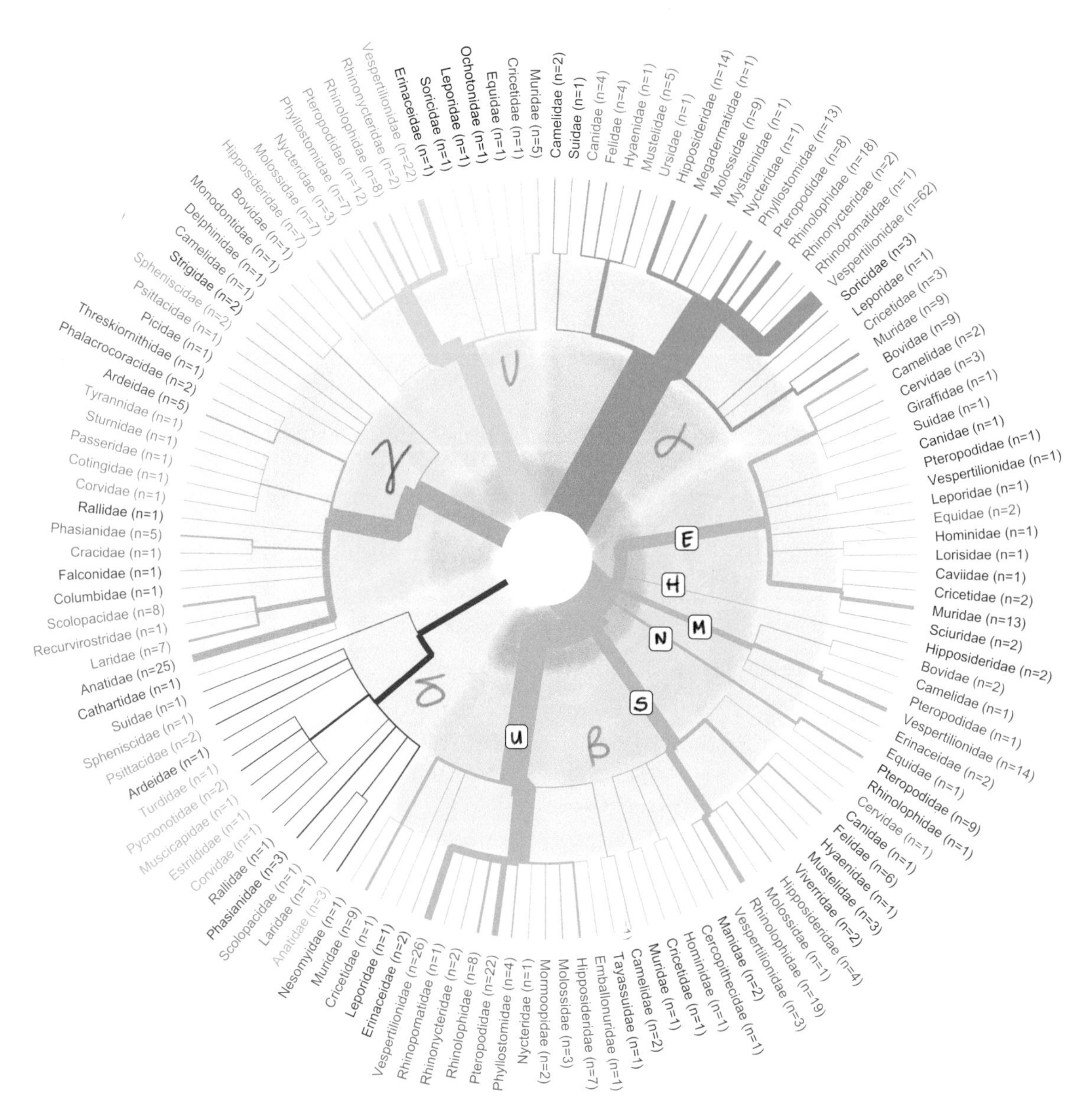

Fig. 25.2. Number of animal species per taxonomic family (*n*) with documented coronavirus nucleotide sequences as reported in GenBank data, grouped by coronavirus genus. Coronaviruses have four genera: alpha- (α, shown in brown), beta- (β, shown in blue), gamma- (γ, shown in purple), and deltacoronavirus (δ, shown in green). The inner two areas divide animal species by coronavirus genus and subgenus (only shown for betacoronavirus). Here E, H, M, N, S, U represents the embeco-, hibeco-, marbeco-, nobeco-, sarbecovirus, and unknown subgenera, respectively. The third, fourth, and fifth levels distribute animal species by their taxonomic class, order, and family, respectively. Branch thickness reflects the number of unique species identified per category. The presence of coronavirus viral nucleotide sequence in animals sampled does not necessarily equate to active infection and does not mean that the host species can serve as a source of infection to other individuals or as maintenance host. Chart by Diego Montecino-Latorre adapted by Laura Donohue.

Emerging Zoonotic Coronaviruses in Humans

Typically, detection of emerging zoonotic viruses occurs when an unusual cluster of human cases is recognized, with the majority of cases going undetected in existing surveillance systems, especially for viruses with a disease spectrum that includes mild or asymptomatic infections. Human–human transmission further complicates efforts to identify index cases, investigate potential animal sources for emerging viruses, and institute effective disease control. Transmission in healthcare settings has facilitated the detection of diseases where healthcare is readily accessible, but diagnostic screening for coronaviruses has been too limited to detect emerging viruses. The remoteness of some settings for animal–human transmission events might have previously limited zoonotic emergence and spread of pathogens, but increased travel and trade through urban centers now enhances the speed and scale of epidemics. Our knowledge of potentially zoonotic coronaviruses in wildlife is largely incomplete, especially because coronaviruses and their hosts are expected to change over time. Wildlife health surveillance capacity is even more limited than public health surveillance, particularly in resource-limited hotspot areas. Nonetheless, this capacity is critical to detecting and characterizing emerging threats from wild animals.

SARS-CoV—A Pandemic Curtailed

The 2002 SARS pandemic caused by a novel coronavirus (SARS-CoV), was the first pandemic of the 21st century. Initial cases were identified in Guangdong Province, China, and spread to 29 countries with 8,422 cases, 916 fatalities, and an 11% case fatality rate in humans (Chan-Yeung and Xu 2003). SARS-CoV is a betacoronavirus in the *Sarbecovirus* subgenus, a group of viruses that generally infect bats. Tracing the origin of SARS-CoV has been especially difficult due to the sporadic nature of early reported cases. Isolation of a SARS-related coronavirus (SARSr-CoV) from masked palm civets (*Paguma larvata*) as well as raccoon dogs (*Nyctereutes procyonoides*) and the detection of SARS-CoV infection in market workers in Guangdong Province provided the first evidence for a zoonotic source of human infection (Guan et al. 2003). After the peak of initial human–human transmission had passed, a cluster of cases was identified in a restaurant serving palm civets. A SARSr-CoV was identified in palm civets at the restaurant where a waitress and a customer eating near the civet cages were infected (Wang et al. 2005). Subsequently, SARSr-CoVs were identified in Chinese horseshoe bats (*Rhinolophus sinicus*), indicating that bats were the most likely source of the progenitor virus for SARS-CoV (Li et al. 2005, Lau et al. 2005). Coronaviruses closely related to SARS-CoV with the ability to use human cellular angiotensin-converting enzyme 2 (ACE2) receptors have since been isolated, providing strong evidence that bats were the original source and raising the possibility that bat coronaviruses can directly infect humans (Wang et al. 2018). Additional Chinese and international wildlife surveillance efforts have led to the discovery of a diverse range of SARSr-CoVs in bats from the southeastern region of the country (Li et al. 2005; Lau et al. 2005; Hu et al. 2017).

Remarkably, the SARS-CoV pandemic was controlled within seven months of its occurrence, enabled by a well-coordinated public health effort aimed at contact tracing and containment (WHO 2004). Nosocomial spreading events were a dominant epidemiologic pattern in this pandemic (Chowell et al. 2015), with infection of healthcare workers by patients more frequent than transmission between family members (Anderson et al. 2004). Peak viral shedding generally occurred 7 to 10 days following the onset of clinical symptoms often corresponding with hospitalization (Anderson et al. 2004), which, along with low rates of asymptomatic infections, facilitated contract tracing and disease control (Wilder-Smith et al. 2005). Controlling the animal sources of SARS-CoV reemergence was also integral. Regulation of wildlife markets, including a ban on the rearing, transport, slaughter, and sale of small wild mammals, particularly civets, was enacted in Guangdong province (Zhong 2004). Following the reemergence of SARS-CoV in late 2003, Chinese authorities ordered the culling of over 10,000 civet cats in animal markets and breeding farms (Watts 2004). Despite these and other measures to control the trade in live wildlife, gaps remain in monitoring and enforcement of regulations (Duffy et al. 2016).

MERS-CoV—A Reemerging Virus of Increasing Concern

Middle East respiratory syndrome coronavirus was first identified in 2012 in Saudi Arabia in a man presenting with symptoms similar to SARS-CoV (Peiris and Perlman 2022). The majority of infections have been reported in Saudi Arabia, other countries in the Arabian Peninsula, and travel-related cases from this region, with limited onward human–human transmission. MERS-CoV causes illness that ranges from asymptomatic to severe, but it has been associated with a 35% case fatality rate in humans, making it more deadly than SARS or SARS-CoV-2 (Dighe et al. 2019, Ye et al. 2020). Self-reported contact with farm animals and livestock led to the suspicion of dromedary camels as the source of human infections, and MERS-CoV was subsequently isolated from camel nasal swabs (Ye et al. 2020, Peiris and Perlman 2022). Transmission to humans often occurs through interactions with camels, which serve as a MERS-CoV amplifying host. Subsequent human–human transmission can initially go undetected if symptoms are mild or general. Outbreaks are most readily identified in healthcare settings or when elderly populations with comorbidities are infected (Peiris and Perlman 2022).

Although first isolated in 2012, it is thought that camels could have hosted MERS-CoVs for decades (Ye et al. 2020, Peiris and Perlman 2022). Evidence suggests that bats are the evolutionary reservoir for MERS-CoV and the source of transmission to camels, possibly repeatedly, over decades (Anthony et al. 2017, Ye et al. 2020). The virus causes only mild illness in dromedary camels (Dighe et al. 2019) with respiratory, fecal, and oral shedding (Ye et al. 2020). South American members of the camelid family, including alpacas (*Vicugna pacos*) and llamas (*Lama glama*), have displayed natural and experimental susceptibility to MERS infection (Dighe et al. 2019).

Strains of MERS-CoV are grouped into distinct clades, with clades A and B found in the Arabian Peninsula and clade C found in western, eastern, and northern Africa (Dighe et al. 2019). Despite seroepidemiologic studies indicating most adult dromedary camels from the Middle East and Africa have past MERS-CoV infections, locally acquired zoonotic infection has only been reported in residents of, or travelers to and from, the Arabian Peninsula. Increasing evidence suggests that the viral strains belonging to clade B, sequestered in the Arabian Peninsula, have greater zoonotic potential than other clades (Peiris and Perlman 2022, Dighe et al. 2019).

Despite containment to localized outbreaks, the high case fatality rate and propensity for human–human transmission, especially in healthcare settings, make this a disease of high concern (Peiris and Perlman 2022, Dighe et al. 2019). Camels are integral to the culture and economy in large parts of the world, so a vaccine for dromedary camels could be important in the prevention of human infections. Gaps in knowledge that must be addressed for an effective vaccination campaign include characterizing MERS-CoV distribution within camel populations (Dighe et al. 2019) and investigating additional potential animal hosts, which could complicate control measures (Ye et al. 2020).

SARS-CoV-2—Emergence of a Pandemic into a Multi-host Pathogen

The world was notified of an outbreak of severe pneumonia in Wuhan, Hubei Province, China, by

Chinese government officials on December 31, 2019 (Worobey et al. 2022). The virus responsible, SARS-CoV-2, has since been identified in every corner of the globe, a result of international spread and subsequent sustained transmission among humans (Li et al. 2020). Factors contributing to the scale of uncontrolled transmission included asymptomatic cases, airborne transmission, and inadequate testing early in the pandemic, which hampered contact tracing efforts. Evaluation of data available in the first year of SARS-CoV-2 transmission suggests that at least one-third of infections were asymptomatic (Oran and Topol 2021). Wide local variations in public health measures have contributed to the epidemiologic trajectories of COVID-19 with waves of new variants evading immunity and infecting susceptible populations. With the widespread infection experienced globally, the likelihood of continued evolution remains high which will continue to challenge existing immunity and preventive approaches, even while the pandemic has prompted unprecedented research advances in therapeutics and vaccine countermeasures.

Alongside research to respond to the COVID-19 pandemic, investigations into the origins of COVID-19 are ongoing. The Huanan Seafood Market has been well characterized as the epicenter for human infections in Wuhan, with epidemiologic evidence pointing to zoonotic transmission involving live wildlife for sale at the market. The earliest documented COVID-19 case was a vendor at the Huanan Seafood Market who became ill on December 10, 2019 (Worobey et al. 2022), and the majority of initial cases had a contact history with this market in central Wuhan (Xiao et al. 2021, Wu et al. 2020). Detection of positive environmental samples clustered in the zone of the Huanan Seafood Market where live wildlife was housed and sold, suggests SARS-CoV-2 was circulating in the market in late 2019. Throughout this period, the market sold animals susceptible to SARS-CoV-2 and capable of transmitting the virus, for example, raccoon dogs, hog badgers (*Arctonyx albogularis*), and red foxes (*Vulpes vulpes*; Worobey et al. 2022, Xiao et al. 2021). Evidence to date geographically links the two SARS-CoV-2 lineages circulating in Wuhan in December 2019 to the seafood market, indicating that the introduction of SARS-CoV-2 into humans could have involved two zoonotic transmission events from a yet-to-be-determined intermediate or amplifying animal hosts at the market (Holmes et al. 2021, Worobey et al. 2022).

Genetically similar SARS-related betacoronaviruses have been found in species of rhinolophid bats across South and East Asia, including *R. pusillus* and *R. malayanus* in China (Zhou et al. 2021), *R. cornutus* in Japan (Murakami et al. 2020), and *R. shameli* in Cambodia (Delaune et al. 2021). Most evolutionary investigations into SARS-CoV-2 emergence have focused on the affinity of the receptor-binding domain to the ACE2 receptor and recent characterization of viruses discovered in *R. malaanus*, *R. pusillus*, and *R. marchalli* identified a receptor-binding domain most genetically similar to SARS-CoV-2 (Temmam et al. 2022).

Human–Animal Transmission of SARS-CoV-2

The spillback of SARS-CoV-2 from humans to animals is a rapidly unfolding story with new human–animal and animal–animal transmission events discovered routinely as new variants emerge and spread worldwide (Table 25.1, Figure 25.3). To date, SARS-CoV-2 spillback from humans to animals has occurred predominantly in settings with close proximity (e.g., households) or in captivity (e.g., zoos, pet shops, and fur animal farms). Spillback of SARS-CoV-2 was initially detected in household dogs and cats linked to infected humans. Signs have varied from asymptomatic disease to severe respiratory distress, and several studies suggest cats are more commonly infected than dogs, mount a longer lasting immune response, and are capable of cat–cat transmission (Dileepan et al. 2021, Gaudreault et al. 2020,

Table 25.1. Natural severe acute respiratory syndrome coronavirus 2 (SARS-CoV-2) infection of animal hosts

Family	Common name (Scientific name)	Interface	Confirmed onward transmission: animal–human leading to (→) human–human	References
Canidae	Domestic dog (*Canis lupus familiaris*)	Pet		USDA APHIS 2022, CSHV 2022
Castoridae	Eurasian beaver (*Castor fiber*)	Beaver breeding center	beaver–beaver[a]	ISID 2021*a*, CSHV 2022
Cervidae	White-tailed deer (*Odocoileus virginianus*)	Wild	deer–deer[b]	Pickering et al., 2022[c]
Cricetidae	Syrian hamster (*Mesocricetus auratus*)	Pet shop, pet shop warehouse	hamster–hamster, hamster–human → human–human	Yen et al. 2022
Felidae	Domestic cat (*Felis catus*)	Pet, feral at mink farm	cat–cat	USDA APHIS 2022, van Aart et al. 2021
	Indian leopard (*Panthera pardus fusca*)	Wild		Mahajan et al., 2022[d]
	Lion (*Panthera leo*)	Zoo, safari park		USDA APHIS 2022
	Tiger (*Panthera tigris*)	Zoo		USDA APHIS 2022
	Puma (*Puma concolor*)	Zoo, wildlife rehabilitation facility		USDA APHIS 2022
	Snow leopard (*Panthera uncia*)	Zoo		USDA APHIS 2022
	Fishing cat (*Prionailurus viverrinus*)	Zoo		USDA APHIS 2022
	Canadian lynx (*Lynx canadensis*)	Zoo		USDA APHIS 2022
Hippopotamidae	Hippopotamus (*Hippopotamus amphibius*)	Zoo		ISID 2021*b*
Hominidae	Western lowland gorilla (*Gorilla gorilla*)	Zoo		USDA APHIS 2022
Hyaenidae	Spotted hyena (*Crocuta crocuta*)	Zoo		USDA APHIS 2022
Mustelidae	American mink (*Neovison vison*)	Commercial farm, escaped from farm, wild	mink–mink, mink–cat, mink–human → human–human	Aguiló-Gisbert et al. 2021, Oude Munnink et al. 2021, van Aart et al. 2022, USDA APHIS 2022
	Ferret (*Mustela furo*)	Pet		USDA APHIS 2022
	Asian small-clawed otter (*Aonyx cinereus*)	Aquarium		USDA APHIS 2022
Procyonidae	South American coati (*Nasua nasua*)	Zoo		USDA APHIS 2022
Viverridae	Binturong (*Arctictis binturong*)	Zoo		USDA APHIS 2022

Confirmed cases of SARS-CoV-2 infection in animal hosts occurring from confirmed or strongly suspected human–animal transmission events over the course of the first 2 years of the coronavirus disease 2019 (COVID-19) pandemic up to February 2022. Cases of SARS-CoV-2 infection were confirmed by PCR viral RNA detection and whole genome sequencing. Data from experimental infections are not shown. Research on SARS-CoV-2 is ongoing, and infections are still being recognized in new hosts.

[a] Beaver–beaver transmission within the breeding facility was suspected.

[b] Evidence strongly suggests deer–deer transmission. Preliminary data on deer–human transmission is pending.

[c] Pickering, B., O. Lung, F. Maguire, P. Kruczkiewicz, J. D. Kotwa, T. Buchanan, M. Gagnier, J. L. Guthrie, C. M. Jardine, A. Marchand-Austin, et al. 2022. Divergent SARS-CoV-2 variant emerges in white-tailed deer with deer-to-human transmission. Nature Microbiology 7:2011–2024.

[d] Mahajan, S., M. Karikalan, V. Chander, A. M. Pawde, G. Saikumar, M. Semmaran, P. S. Lakshmi, M. Sharma, S. Nandi, K. P. Singh, et al. 2022. Detection of SARS-CoV-2 in a free ranging leopard (*Panthera pardus fusca*) in India. European Journal of Wildlife Research 68.

Fig. 25.3. Naturally occurring SARS-CoV-2 transmission from humans to domesticated animal species, captive wild animal species, and free-ranging wild animal species. Cases of human–animal transmission as reported over the course of the first 2 years of the pandemic up to February 2022. Data from experimental infections are not shown. SARS-CoV-2 research is ongoing, and infections are still being recognized in new hosts. Beaver–beaver transmission within the breeding facility is suspected. Onward mink–mink, mink–cat, and mink–human transmission has been shown. Onward hamster–human transmission has been shown. Evidence strongly suggests deer–deer transmission. Preliminary data on deer–human transmission is pending. Illustration by Laura Donohue.

Halfmann et al. 2020). Pet ferrets (*Mustela furo*) and hamsters (*Mesocricetus auratus*) have also been infected by humans (Yen et al. 2022, Račnik et al. 2021). Two events of human–animal transmission and subsequent animal–human transmission involving hamsters were reported in pet shops in Hong Kong in late 2021, with one case of human–human onward transmission (SARS-CoV-2 delta variant) confirmed (Yen et al. 2022).

Spillback of SARS-CoV-2 into wildlife was first noted in zoo animals (Table 25.1); over time, the number of felid species with SARS-CoV-2 increased to include tigers (*Panthera tigris*), lions (*Panthera leo*), puma (*Puma concolor*), and two species of leopards (*Panthera uncia* and *Panthera pardus fusca*) in multiple countries (Table 25.1). Most infections were characterized by mild illness and full recovery, yet a few animals suffered severe disease leading to euthanasia or death. Infections in other carnivores, such as coatis (*Nasua nasua*), otters (*Aonyx cinereus*), hyenas (*Crocuta crocuta*), and binturong (*Arctictis binturong*) were also reported in zoos and aquariums in the United States (Table 25.1). In addition to carnivores, SARS-CoV-2 positive hippopotamus

(*Hippopotamus amphibius*; Belgium), gorillas (*Gorilla gorilla*; USA), and beavers (*Castor fiber*; Mongolia) have been documented in captive settings (Table 25.1).

Transmission from farmworkers to mink (*Neovison vison*) on commercial fur farms across Europe and North America was also documented early in the pandemic (Oude Munnik et al. 2021, Pomorska-Mól et al. 2021). Infections spread quickly. Signs in mink ranged from asymptomatic to severe pneumonia and death (Pomorska-Mól et al. 2021). Mink-acquired mutations have accompanied rapid movement through large, high-density populations (Oude Munnik et al. 2021). Transmission to feral cats (van Aart et al. 2022), escaped mink (Shriner et al. 2021), and suspected transmission from farms to feral mink in Spain (Aguiló-Gisbert et al. 2021) have also been observed. Millions of mink were euthanized early in the pandemic to control the spread of mink-adapted strains to humans and avoid mutations that could render vaccines ineffective.

To date, SARS-CoV-2 infection in free-ranging wildlife has been documented primarily in white-tailed deer (*Odocoileus virginianus*) in North America. In the first two years of the pandemic, infections in deer were confirmed in 14 US states (USDA APHIS 2022) closely aligning with the timing and predominant variant of SARS-CoV-2 circulating in humans (Kuchipudi et al. 2022). Existing data suggest multiple human–deer introductions and in some cases, sustained deer–deer transmission (Kuchipudi et al. 2022). A study conducted on hunted white-tailed deer at the end of 2021 in Ontario, Canada, identified a highly divergent and deer-adapted SARS-CoV-2 virus with evidence of deer–human transmission (Pickering et al., 2022), increasing concerns that white-tailed deer could become a new maintenance host and source host for humans, with implications for viral evolution and potential reemergence in humans. An isolated case of SARS-CoV-2 infection was also reported in a free-ranging Indian leopard cub with systemic disease in a forest reserve in India (Mahajan et al., 2022).

The extent of SARS-CoV-2 transmission in free-ranging wildlife is largely unknown at this time. Activities that enable direct contact with wildlife, such as rehabilitation, research, tourism, and wildlife supplemental feeding, should be monitored along with wild animals adapted to human-habituated settings. Surveillance of SARS-CoV-2 in animals must include investigations into the potential for persistence in animal reservoirs that might inform the future trajectory of SARS-CoV-2 evolution and epidemiology. Banerjee et al. (2021) proposed three criteria to prioritize target animal species for surveillance: receptor homology with human ACE2 sequence, previous reports of infection in a closely related species, and the likelihood of animal–human interactions. These combined factors could inform the risk of new spillover events and better direct future surveillance priorities. Since the onset of the COVID-19 pandemic, only a handful of risk assessments for SARS-CoV-2 spillback to free-ranging terrestrial and marine wildlife have been conducted. In all cases, timely notification of confirmed cases to the relevant national and international authorities with One Health integration of epidemiological veterinary (including both wildlife and domestic) and public health data is highly encouraged.

Implications for Conservation

The impacts of the SARS-CoV-2 pandemic on conservation are still playing out as susceptibility in wildlife and subsequent impacts to populations and ecosystems will not be evident for some time. There are a number of serious implications for species already at conservation risk, such as the great apes, whose endangered status and documented susceptibility to human respiratory pathogens place them at the top of at-risk wild species. Restrictions on travel and tourism bans have already impacted gorillas via revenue loss and poaching (Kalema-Zikusoka et al.

2021). The sustained risk of infection may also impede or pause key interventions, including treatment, research, and rehabilitation that require close human–primate interactions. Fortunately, timely One Health-based planning has thus far succeeded in protecting these highly valued and fragile species (Chapter 24). As tourism reopens and the world moves to the next phase of living with SARS-CoV-2, strict adherence to protocols and COVID-19 screening will remain key preventative barriers.

Further implications including fear and public perceptions of threats carry direct and indirect risks for wildlife conservation and welfare. This was seen with the persecution and retaliatory extermination of bats in response to SARS-CoV-2 emergence. Unfortunately, these recurrent and often misguided reactions stem from public misinformation about the role of wildlife in disease emergence and transmission. The 2017 yellow fever epidemic in Brazil resulted in the killing of countless nonhuman primates, despite the virus being transmitted to humans by mosquitoes, not primates, with the primates themselves already highly susceptible to the disease and important sentinels for the prevention of human cases (Romero 2017). Invariably, the removal of wildlife has little if any positive long-term impact on disease control. Case studies across multiple diseases have shown that the culling of wild animals in the face of disease outbreaks causes a subsequent increase in disease risk, as animals respond with redistribution (Bielby et al. 2016, Amman et al. 2014).

The emergence of zoonotic viruses and pandemic threats have been linked to the exploitation of wildlife and habitat loss, especially as documented for declining and threatened wild animal species (Johnson et al. 2020). Mounting evidence suggests both SARS-CoV and SARS-CoV-2 emerged in the context of wild animal markets where live animals were traded, providing ample opportunity for animal–human disease transmission in crowded, close-contact settings. The wildlife trade has been implicated as a source of various zoonotic infections (IPBES 2020) and is one of the biggest threats to conservation and biodiversity (Smith et al. 2017) particularly in highly biodiverse, low-income countries with poor enforcement capacity and a lack of viable alternatives (Duffy et al. 2016). While contributing to food security and livelihoods in some communities, the wildlife trade is increasingly driven by demand from wealthy societies for luxury items (e.g., fashion, food delicacies, and amulets) or exotic pets (Smith et al. 2017). Unregulated and unsustainable wildlife trade depletes ecosystems of species, disrupts natural ecological balances buffering disease emergence, and entails high-risk practices for disease transmission, such as species mixing, overcrowding, and unsanitary conditions. Regardless of whether the wildlife trade is the source, vehicle, or amplifier for known and unknown zoonotic pathogens, supply chains are poorly characterized and tend to escape sanitary regulations and traceability that apply to the domestic animal trade. Live animal markets where domestic and wild animals mix, are slaughtered, and are dressed in areas of public access with poor sanitation and hygiene have been identified as high-risk sites that should be phased out or transformed (WHO et al. 2021).

Traditional outbreak control methods for zoonotic disease transmission in the wildlife trade have involved mass cullings of animals, such as the elimination of civet cats in markets and farms to control SARS-CoV in China (Watts 2004), mink on Denmark farms (Pomorska-Mól et al. 2021), and pet hamsters in Hong Kong to control SARS-CoV-2. Yet shortly after these interventions, risky human behaviors returned, and the underlying drivers of the wildlife trade remain unaddressed (Duffy et al. 2016). A recent review of the literature failed to identify interventions that predictably reduced the risk of disease emergence in the trade (Stephen et al. 2021). While the scientific community continues to conduct studies that characterize pathogens, few, if any, investigate social, political, financial, and legal tools that could be used to interrupt pathogen spillover or how to effectively prevent and mitigate the emergence of diseases in the trade. The continued emphasis on the wildlife origins of emerging zoonotic pathogens and failure to address the human activities enabling and driving wildlife–human disease transmission presents an ongoing threat to wildlife conservation and public health.

The focus on the negative perception of wildlife, rather than on the essential ecosystem services they provide, unnecessarily places wildlife at odds with human well-being. Complementary efforts between emerging disease surveillance and conservation communities could enable early detection and help address ecological conditions that drive increased risk, while protecting essential ecosystems. Similarly, clear, well-thought-out messaging crafted collaboratively across relevant disciplines should guide public health outreach and education on zoonotic disease.

Summary

Toward a Safer Future

The emergence of multiple coronaviruses in this century is a worrisome trend revealing intrinsic failures in environmental stewardship and disease prevention. The COVID-19 pandemic, in particular, highlighted weaknesses in multilateral response to pandemic threats. Limited scientific understanding of novel coronaviruses in their initial stages of emergence remains a major hurdle for disease mitigation efforts. These events also expose inherent vulnerabilities in existing surveillance systems with inadequate capacity for early detection of zoonotic diseases. Detection of sporadic transmission in people at the point of spillover, or amplification in livestock, remains key to effective control in our increasingly interconnected societies. Strained animal health programs, especially those with limited wildlife veterinary capacity in biodiverse emerging disease hotspots, remain an additional liability for viruses with animal intermediary hosts. Common biosecurity risk-reduction measures have repeatedly failed to halt pathogen transmission at high-risk interfaces for emerging pathogens. International agreements established for conservation purposes can hinder the prompt diagnosis of wildlife with constraints limiting the international movement of wildlife specimens. Furthermore, very few countries have operational, interministerial, well-funded One Health strategies and plans. The lack of attention paid to the novel, particularly wildlife-associated, pathogens in recent national disease prioritization efforts further hinders preparedness and risk reduction activities. For example, only 9.9% (9 of 91) of the assessed countries listed coronaviruses or associated diseases as priorities in their joint external evaluations (2016–2019) despite prior warnings about the significant human and animal health threat they pose (Machalaba et al. 2021).

Wildlife health professionals are uniquely positioned to promote and apply One Health approaches and ecosystem health perspectives to investigating disease risk where wildlife and humans interact. Equally relevant and pressing is the deployment of innovative technology and transdisciplinary efforts to enhance and scale-up surveillance within the context of a cooperative international network to advance disease intelligence. Biodiversity loss, climate change, and emerging infectious diseases with pandemic potential are interlinked, complex, and critical threats that must be addressed cooperatively within a broader recognition of the dependence of human well-being on environmental health.

LITERATURE CITED

Aguiló-Gisbert, J., M. Padilla-Blanco, V. Lizana, E. Maiques, M. Muñoz-Baquero, E. Chillida-Martínez, J. Cardells, and C. Rubio-Guerri. 2021. First description of SARS-CoV-2 infection in two feral American mink (*Neovison vison*) caught in the wild. Animals 11:1422.

Amman, B. R., L. Nyakarahuka, A. K. McElroy, K. A. Dodd, T. K. Sealy, A. J. Schuh, T. R. Shoemaker, S. Balinandi, P. Atimnedi, W. Kaboyo, et al. 2014. Marburgvirus resurgence in Kitaka mine bat population after extermination attempts, Uganda. Emerging Infectious Diseases 20:1761–1764.

Anderson, R. M., C. Fraser, A. C. Ghani, C. A. Donnelly, S. Riley, N. M. Ferguson, G. M. Leung, T. H. Lam, and A. J. Hedley. 2004. Epidemiology, transmission dynamics and control of SARS: the 2002–2003 epidemic. Philosophical Transactions of the Royal Society of London Series B Biological Sciences 359:1091–1105.

Anthony, S. J., C. K. Johnson, D. J. Greig, S. Kramer, X. Che, H. Wells, A. L. Hicks, D. O. Joly, N. D. Wolfe, P. Daszak, et al. 2017. Global patterns in coronavirus diversity. Virus Evolution 3:vex012.

Banerjee, A., K. Mossman, and M. L. Baker. 2021. Zooanthroponotic potential of SARS-CoV-2 and implications of reintroduction into human populations. Cell Host and Microbe 29:160–164.

Bashor, L., R. B. Gagne, A. M. Bosco-Lauth, R. A. Bowen, M. Stenglein, and S. VandeWoude. 2021. SARS-CoV-2 evolution in animals suggests mechanisms for rapid variant selection. Proceedings of the National Academy of Sciences USA 118:e2105253118.

Bielby, J., F. Vial, R. Woodroffe, and C. A. Donnelly. 2016. Localized badger culling increases risk of herd breakdown on nearby, not focal, land. PLOS ONE 11:e0164618.

Boley, P. A., M. A. Alhamo, G. Lossie, K. K. Yadav, M. Vasquez-Lee, L. J. Saif, and S. P. Kenney. 2020. Porcine deltacoronavirus infection and transmission in poultry, United States. Emerging Infectious Diseases 26:255–265.

Chan-Yeung, M., and R.-H. Xu. 2003. SARS: epidemiology. Respirology 8:S9–S14.

Chen, Q., L. Wang, C. Yang, Y. Zheng, P. C. Gauger, T. Anderson, K. M. Harmon, J. Zhang, K.-J. Yoon, R. G. Main, et al. 2018. The emergence of novel sparrow deltacoronaviruses in the United States more closely related to porcine deltacoronaviruses than sparrow deltacoronavirus HKU17. Emerging Microbes and Infections 7:105.

Chowell, G., F. Abdirizak, S. Lee, J. Lee, E. Jung, H. Nishiura, and C. Viboud. 2015. Transmission characteristics of MERS and SARS in the healthcare setting: a comparative study. BMC Medicine 13:210.

Clark, K., I. Karsch-Mizrachi, D. J. Lipman, J. Ostell, and E. W. Sayers. 2016. GenBank. Nucleic Acids Research 44:D67–D72. Accessed 14 Feb 2022.

Complexity Science Hub Vienna (CSHV). 2022. SARS-ANI VIS: a global open access dataset of reported SARS-CoV-2 events in animals. https://vis.csh.ac.at/sars-ani/#infections. Accessed 14 Feb 2022.

Delaune, D., V. Hul, E. A. Karlsson, A. Hassanin, T. P. Ou, A. Baidaliuk, F. Gámbaro, M. Prot, V. T. Tu, S. Chea, et al. 2021. A novel SARS-CoV-2 related coronavirus in bats from Cambodia. Nature Communications 12:6563.

Descôteaux, J.-P., and S. Mihok. 1986. Serologic study on the prevalence of murine viruses in a population of wild meadow voles (*Microtus pennsylvanicus*). Journal of Wildlife Diseases 22:314–319.

Dighe, A., T. Jombart, M. D. Van Kerkhove, and N. Ferguson. 2019. A systematic review of MERS-CoV seroprevalence and RNA prevalence in dromedary camels: implications for animal vaccination. Epidemics 29:100350.

Dileepan, M., D. Di, Q. Huang, S. Ahmed, D. Heinrich, H. Ly, and Y. Liang. 2021. Seroprevalence of SARS-CoV-2 (COVID-19) exposure in pet cats and dogs in Minnesota, USA. Virulence 12:1597–1609.

Duffy, R., F. A. V. St. John, B. Büscher, and D. Brockington. 2016. Toward a new understanding of the links between poverty and illegal wildlife hunting: poverty and illegal wildlife hunting. Conservation Biology 30:14–22.

Elazhary, M. A., J. L. Frechette, A. Silim, and R. S. Roy. 1981. Serological evidence of some bovine viruses in the caribou (*Rangifer tarandus caribou*) in Quebec. J. Wildl. Dis. 17:609–612.

Evermann, J. F., W. Foreyt, L. Maag-Miller, C. W. Leathers, A. J. McKeirnan, and B. LeaMaster. 1980. Acute hemorrhagic enteritis associated with canine coronavirus and parvovirus infections in a captive coyote population. Journal of the American Veterinary Medical Association 177:784–786.

Evermann, J. F., J. L. Heeney, M. E. Roelke, A. J. McKeirnan, and S. J. O'Brien. 1988. Biological and pathological consequences of feline infectious peritonitis virus infection in the cheetah. Archives of Virology 102:155–171.

Gaudreault, N. N., J. D. Trujillo, M. Carossino, D. A. Meekins, I. Morozov, D. W. Madden, S. V. Indran, D. Bold, V. Balaraman, T. Kwon, et al. 2020. SARS-CoV-2 infection, disease and transmission in domestic cats. Emerging Microbes and Infections 9:2322–2332.

Gibbs, C. S. 1935. The etiology of epidemic colds in chickens. Science 81:345–346.

Graham, R. L., and R. S. Baric. 2010. Recombination, reservoirs, and the modular spike: mechanisms of coronavirus cross-species transmission. Journal of Virology 84:3134–3146.

Guan, Y., B. J. Zheng, Y. Q. He, X. L. Liu, Z. X. Zhuang, C. L. Cheung, S. W. Luo, P. H. Li, L. J. Zhang, Y. J. Guan, et al. 2003. Isolation and characterization of viruses related to the SARS coronavirus from animals in southern China. Science 302:276–278.

Halfmann, P. J., M. Hatta, S. Chiba, T. Maemura, S. Fan, M. Takeda, N. Kinoshita, S. Hattori, Y. Sakai-Tagawa, K. Iwatsuki-Horimoto, et al. 2020. Transmission of SARS-CoV-2 in domestic cats. New England Journal of Medicine 383:592–594.

Holmes, E. C., S. A. Goldstein, A. L. Rasmussen, D. L. Robertson, A. Crits -Christoph, J. O. Wertheim, S. J. Anthony, W. S. Barclay, M. F. Boni, P. C. Doherty, et al. 2021. The origins of SARS-CoV-2: a critical review. Cell 184:4848–4856.

Hu, B., L.-P. Zeng, X.-L. Yang, X.-Y. Ge, W. Zhang, B. Li, J.-Z. Xie, X.-R. Shen, Y.-Z. Zhang, N. Wang, et al. 2017. Discovery of a rich gene pool of bat SARS-related coronaviruses provides new insights into the origin of SARS coronavirus. PLOS Pathogens 13:e1006698.

Intergovernmental Science-Policy Platform on Biodiversity and Ecosystem Services (IPBES). 2020. Workshop report on biodiversity and pandemics of the Intergovernmental Platform on Biodiversity and Ecosystem Services (IPBES). https://zenodo.org/record/4147317. Accessed 11 Mar 2022.

International Society for Infectious Diseases (ISID). 2021*a*. Coronavirus disease 2019 update (315): animal, Mongolia, beaver, delta variant, first report. https://promedmail.org/promed-post/?id=8668125. Accessed 14 Feb 2022.

International Society for Infectious Diseases (ISID). 2021*b*. Coronavirus disease 2019 update (418): animal, Belgium, zoo, hippopotamus, first report. https://promedmail.org/promed-post/?id=8700102. Accessed 14 Feb 2022.

Johnson, C. K., P. L. Hitchens, P. S. Pandit, J. Rushmore, T. S. Evans, C. C. W. Young, and M. M. Doyle. 2020. Global shifts in mammalian population trends reveal key predictors of virus spillover risk. Proceedings of the Royal Society B Biological Sciences 287:20192736.

Jonassen, C. M., T. Kofstad, I.-L. Larsen, A. Løvland, K. Handeland, A. Follestad, and A. Lillehaug. 2005. Molecular identification and characterization of novel coronaviruses infecting graylag geese (*Anser anser*), feral pigeons (*Columbia livia*) and mallards (*Anas platyrhynchos*). Journal of General Virology 86:1597–1607.

Kalema-Zikusoka, G., S. Rubanga, A. Ngabirano, and L. Zikusoka. 2021. Mitigating impacts of the COVID-19 Pandemic on gorilla conservation: lessons from Bwindi Impenetrable Forest, Uganda. Frontiers in Public Health 9:655175.

Kreuder Johnson, C., P. L. Hitchens, T. Smiley Evans, T. Goldstein, K. Thomas, A. Clements, D. O. Joly, N. D. Wolfe, P. Daszak, W. B. Karesh, et al. 2015. Spillover and pandemic properties of zoonotic viruses with high host plasticity. Scientific Reports 5:14830.

Kuchipudi, S. V., M. Surendran-Nair, R. M. Ruden, M. Yon, R. H. Nissly, K. J. Vandegrift, R. K. Nelli, L. Li, B. M. Jayarao, C. D. Maranas, et al. 2022. Multiple spillovers from humans and onward transmission of SARS-CoV-2 in white-tailed deer. Proceedings of the National Academy of Sciences USA 119:e2121644119.

Lau, S. K. P., P. C. Y. Woo, K. S. M. Li, Y. Huang, H.-W. Tsoi, B. H. L. Wong, S. S. Y. Wong, S.-Y. Leung, K.-H. Chan, and K.-Y. Yuen. 2005. Severe acute respiratory syndrome coronavirus-like virus in Chinese horseshoe bats. Proceedings of the National Academy of Sciences 102:14040–14045.

Lednicky, J. A., M. S. Tagliamonte, S. K. White, M. A. Elbadry, Md. M. Alam, C. J. Stephenson, T. S. Bonny, J. C. Loeb, T. Telisma, S. Chavannes, et al. 2021. Independent infections of porcine deltacoronavirus among Haitian children. Nature 600:133–137.

Li, Q., X. Guan, P. Wu, X. Wang, L. Zhou, Y. Tong, R. Ren, K. S. M. Leung, E. H. Y. Lau, J. Y. Wong, et al. 2020. Early transmission dynamics in Wuhan, China, of novel coronavirus–infected pneumonia. New England Journal of Medicine 382:1199–1207.

Li, W., Z. Shi, M. Yu, W. Ren, C. Smith, J. H. Epstein, H. Wang, G. Crameri, Z. Hu, H. Zhang, et al. 2005. Bats Are natural reservoirs of SARS-like coronaviruses. Science 310:676–679.

Machalaba, C., M. Uhart, M.-P. Ryser-Degiorgis, and W. B. Karesh. 2021. Gaps in health security related to wildlife and environment affecting pandemic prevention and preparedness, 2007–2020. Bulletin of the World Health Organization 99:342–350B.

Mahajan, S., M. Karikalan, V. Chander, A. M. Pawde, G. Saikumar, M. Semmaran, P. S. Lakshmi, M. Sharma, S. Nandi, K. P. Singh, et al. 2022. Detection of SARS-CoV-2 in a free ranging leopard (*Panthera pardus fusca*) in India. European Journal of Wildlife Research 68.

Mei, X., P. Qin, Y. Yang, M. Liao, Q. Liang, Z. Zhao, F. Shi, B. Wang, and Y. Huang. 2022. First evidence that an emerging mammalian alphacoronavirus is able to infect an avian species. Transboundary and Emerging Diseases 69:e2006–e2019.

Murakami, S., T. Kitamura, J. Suzuki, R. Sato, T. Aoi, M. Fujii, H. Matsugo, H. Kamiki, H. Ishida, A. Takenaka-Uema, et al. 2020. Detection and characterization of bat sarbecovirus phylogenetically related to SARS-CoV-2, Japan. Emerging Infectious Diseases 26:3025–3029.

Oran, D. P., and E. J. Topol. 2021. The proportion of SARS-CoV-2 infections that are asymptomatic: a systematic review. Annals of Internal Medicine 174:655–662.

Oude Munnink, B. B., R. S. Sikkema, D. F. Nieuwenhuijse, R. J. Molenaar, E. Munger, R. Molenkamp, A. van der Spek, P. Tolsma, A. Rietveld, M. Brouwer, et al. 2021.

Transmission of SARS-CoV-2 on mink farms between humans and mink and back to humans. Science 371:172–177.

Peiris, M., and S. Perlman. 2022. Unresolved questions in the zoonotic transmission of MERS. Current Opinion in Virology 52:258–264.

Pickering, B., O. Lung, F. Maguire, P. Kruczkiewicz, J. D. Kotwa, T. Buchanan, M. Gagnier, J. L. Guthrie, C. M. Jardine, A. Marchand-Austin, et al. 2022. Divergent SARS-CoV-2 variant emerges in white-tailed deer with deer-to-human transmission. Nature Microbiology 7:2011–2024.

Pomorska-Mól, M., J. Włodarek, M. Gogulski, and M. Rybska. 2021. Review: SARS-CoV-2 infection in farmed minks—an overview of current knowledge on occurrence, disease and epidemiology. Animal 15:100272.

Račnik, J., A. Kočevar, B. Slavec, M. Korva, K. R. Rus, S. Zakotnik, T. M. Zorec, M. Poljak, M. Matko, O. Z. Rojs, et al. 2021. Transmission of SARS-CoV-2 from human to domestic ferret. Emerging Infectious Diseases 27:2450–2453.

Ritchie, H., E. Mathieu, L. Rodés-Guirao, C. Appel, C. Giattino, E. Ortiz-Ospina, J. Hasell, B. Macdonald, D. Beltekian, and M. Roser. 2022. Coronavirus pandemic (COVID-19). Updated. https://ourworldindata.org/coronavirus. Accessed 14 Feb 2022.

Romero, S. 2017. Brazil yellow fever outbreak spawns alert: stop killing the monkeys. New York Times. 2 May 2017. https://www.nytimes.com/2017/05/02/world/americas/brazil-yellow-fever-monkeys.html. Accessed 14 Feb 2022.

Shriner, S. A., J. W. Ellis, J. J. Root, A. Roug, S. R. Stopak, G. W. Wiscomb, J. R. Zierenberg, H. S. Ip, M. K. Torchetti, and T. J. DeLiberto. 2021. SARS-CoV-2 exposure in escaped mink, Utah, USA. Emerging Infectious Diseases 27:988–990.

Smith, K. M., C. Zambrana-Torrelio, A. White, M. Asmussen, C. Machalaba, S. Kennedy, K. Lopez, T. M. Wolf, P. Daszak, D. A. Travis, et al. 2017. Summarizing US wildlife trade with an eye toward assessing the risk of infectious disease introduction. EcoHealth 14:29–39.

Stephen, C., J. Berezowki, L. P. Carmo, D. de las N. M. Valle, B. Friker, F. M. M. A. de Sousa, and B. Vidondo. 2021. A rapid review of evidence on managing the risk of disease emergence in the wildlife trade. https://web.oie.int/downld/WG/Wildlife/OIE_review_wildlife_trade_March2021.pdf. Accessed 14 Feb 2022.

Temmam, S., K. Vongphayloth, E. Baquero, S. Munier, M. Bonomi, B. Regnault, B. Douangboubpha, Y. Karami, D. Chrétien, D. Sanamxay, et al. 2022. Bat coronaviruses related to SARS-CoV-2 and infectious for human cells. Nature 604:330–336.

US Department of Agriculture Animal and Plant Health Inspection Service (USDA APHIS). 2022. Confirmed Cases of SARS-CoV-2 in Animals in the United States. https://www.aphis.usda.gov/aphis/dashboards/tableau/sars-dashboard. Accessed 14 Feb 2022.

van Aart, A. E., F. C. Velkers, E. A. J. Fischer, E. M. Broens, H. Egberink, S. Zhao, M. Engelsma, R. W. Hakze-van der Honing, F. Harders, M. M. T. Rooij, et al. 2022. SARS-CoV-2 infection in cats and dogs in infected mink farms. Transboundary and Emerging Diseases 69:3001–3007.

Wang, M., M. Yan, H. Xu, W. Liang, B. Kan, B. Zheng, H. Chen, H. Zheng, Y. Xu, E. Zhang, et al. 2005. SARS-CoV infection in a restaurant from palm civet. Emerging Infectious Diseases 11:1860–1865.

Wang, N., S.-Y. Li, X.-L. Yang, H.-M. Huang, Y.-J. Zhang, H. Guo, C.-M. Luo, M. Miller, G. Zhu, A. A. Chmura, E. et al. 2018. Serological evidence of bat SARS-related coronavirus infection in humans, China. Virologica Sinica 33:104–107.

Watts, J. 2004. China culls wild animals to prevent new SARS threat. Lancet 363:134.

Wilder-Smith, A., M. D. Teleman, B. H. Heng, A. Earnest, A. E. Ling, and Y. S. Leo. 2005. asymptomatic SARS coronavirus infection among healthcare workers, Singapore. Emerging Infectious Diseases 11:1142–1145.

World Health Organization (WHO). 2004. Severe acute respiratory syndrome (SARS): report by the Secretariat. https://apps.who.int/iris/handle/10665/20038. Accessed 14 Feb 2022.

World Health Organization (WHO), World Organization for Animal Health (OIE), and United Nations Environment Programme (UNEP). 2021. Reducing public health risks associated with the sale of live wild animals of mammalian species in traditional food markets: interim guidance. https://www.who.int/publications/i/item/WHO-2019-nCoV-Food-safety-traditional-markets-2021.1. Accessed 14 Feb 2022.

Worobey, M., J. I. Levy, L. M. M. Serrano, A. Crits-Christoph, J. E. Pekar, S. A. Goldstein, A. L. Rasmussen, M. U. G. Kraemer, C. Newman, M. P. G. Koopmans, et al. 2022. The Huanan seafood wholesale market in Wuhan was the early epicenter of COVID-19 pandemic. Science 377:951–959.

Wu, F., S. Zhao, B. Yu, Y.-M. Chen, W. Wang, Z.-G. Song, Y. Hu, Z.-W. Tao, J.-H. Tian, Y.-Y. Pei, et al. 2020. A new coronavirus associated with human respiratory disease in China. Nature 579:265–269.

Xiao, X., C. Newman, C. D. Buesching, D. W. Macdonald, and Z.-M. Zhou. 2021. Animal sales from Wuhan wet markets immediately prior to the COVID-19 pandemic. Scientific Reports 11:11898.

Ye, Z.-W., S. Yuan, K.-S. Yuen, S.-Y. Fung, C.-P. Chan, and D.-Y. Jin. 2020. Zoonotic origins of human coronaviruses. International Journal of Biological Sciences 16:1686–1697.

Yen, H.-L., T. H. C. Sit, C. J. Brackman, S. S. Y. Chuk, H. Gu, K. W. S. Tam, P. Y. T. Law, G. M. Leung, M. Peiris, L. L. M. Moon, et al. 2022. Transmission of SARS-CoV-2 delta variant (AY.127) from pet hamsters to humans, leading to onward human-to-human transmission: a case study. Lancet 399:1070–1078.

Zhong, N. 2004. Management and prevention of SARS in China. Philosophical Transactions of the Royal Society of London Series B Biological Sciences 359:1115–1116.

Zhou, H., J. Ji, X. Chen, Y. Bi, J. Li, Q. Wang, T. Hu, H. Song, R. Zhao, Y. Chen, et al. 2021. Identification of novel bat coronaviruses sheds light on the evolutionary origins of SARS-CoV-2 and related viruses. Cell 184:4380–4391.

Zhou, P., H. Fan, T. Lan, X.-L. Yang, W.-F. Shi, W. Zhang, Y. Zhu, Y.-W. Zhang, Q.-M. Xie, S. Mani, et al. 2018. Fatal swine acute diarrhea syndrome caused by an HKU2-related coronavirus of bat origin. Nature 556:255–258.

Conclusion

David A. Jessup, Robin W. Radcliffe, Richard A. Kock

The seeming tsunami of wildlife health challenges and emerging infectious diseases of humans and domestic animals that have appeared in the last 25 years are a bit frightening and perplexing. Frightening because some emergent pathogens seriously threaten human health (zoonotic avian influenza and novel human influenzas, Ebola, and coronaviruses), while other zoonoses persist despite tools to control them being available (rabies, plague, and tuberculosis). From a time when mild levels of disease in wildlife was considered an ecological norm, we now see wildlife disease syndromes that decimate entire populations (avian malaria, sea star wasting, white-nose syndrome, rabbit hemorrhagic disease, distemper and peste des petits ruminants, African swine fever, bighorn pneumonias, chronic wasting disease, the three amphibian diseases, and many of the fish diseases discussed). Other diseases primarily affect agricultural species and wildlife serve as vectors or reservoirs, preventing their elimination from domestic livestock (tuberculosis and brucellosis). Still other wildlife health problems are just chronic, expensive, and inconvenient (bluetongue, lead poisoning, and oil spills). But others, including coral diseases and climate change, appear to be oncoming trains we may not be able to get out of the way of.

These diseases are perplexing because many of them have emerged recently or become much more prevalent for reasons we don't fully understand. It should be clear from reading this book that the health of wildlife and the emergence and spread of diseases is often very clearly linked to climate change, incursion into and degradation of previously pristine wildlife habitats, and biodiversity loss (Figure C.1). All are clearly, or primarily, of anthropogenic origin. This triad is similar to that of One Health in that they are linked and influence one another. For that reason, many of the actions that can be taken to reduce risk of disease emergence, spread, and negative impact are similar to, or the same as, those needed to slow, halt, or reverse biodiversity loss, climate change, and environmental degradation and pollution.

At this point, it is worth revisiting the concept that wildlife diseases occur under unique circumstances of host–pathogen–ecosystem overlap and that they must propagate in animal populations through space and time to pose a serious threat to society or conservation (see Introduction). Pathogens of people and animals (wild and domestic) have existed since time immemorial. But the only way to understand why we seem to be experiencing an unprecedented rate of emergence of diseases in wildlife is to

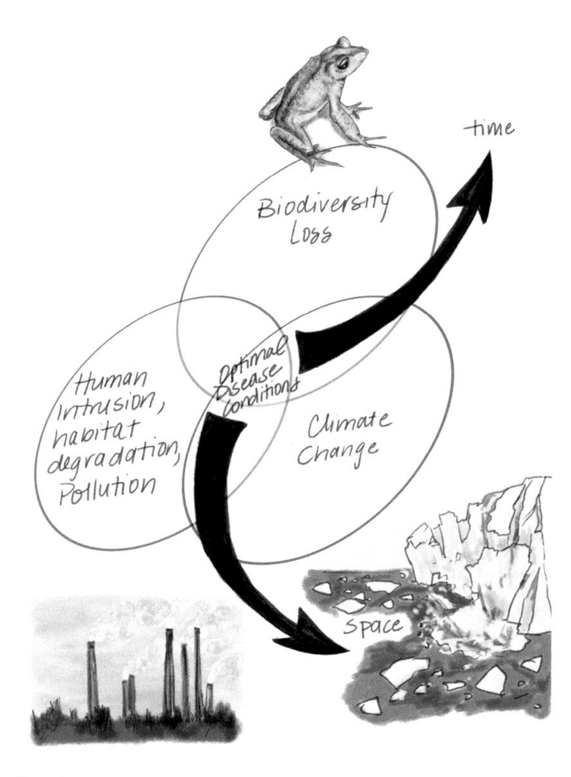

Fig. C.1. Human intrusion into and degradation of wildlife habitats, pollution, and climate change exacerbate health problems that contribute to biodiversity loss, particularly if they persist through space and time. Illustration by Laura Donohue.

recognize that some things have changed. Although some disease-causing organisms are mutating, recombining, infecting new hosts, and jumping species barriers, the emergence of novel agents is not new, but the impacts are now more obvious and much more important to society—with global consequences—given that large human and domestic animal populations now dominate space on earth. These massive "incubators" of disease and their connectivity provide every opportunity for a virus, fungus, bacteria, or other opportunistic parasite to adapt from nature, as they do, and be quickly and widely disseminated. When these emergent or spillback agents, changing environmental conditions, and other stressors tax wildlife population resilience, disease impacts on nature are inevitable. And yes, our ability to diagnose these diseases and identify organisms responsible have increased significantly, but that does not explain much of this "wildlife disease tsunami." It is a bit like blaming global climate change on our increased ability to detect it—that is a cop-out with serious consequences. We should not blame wildlife for these human driven results, rather our lack of knowledge of these processes and impacts, or willingness to apply our knowledge, is responsible. Such thinking trivializes the problem and reveals our ignorance and selfishness, leading to existential consequences.

Should We Intervene, What Are Our Responsibilities?

We introduced this book with Leopold's 1940s observation that wildlife disease, in the context of conservation, has likely been "radically underestimated." After reading these stories of wildlife diseases, and health and ecosystem challenges across a wide array of taxa and environments, one must wonder: what's next, and what should we do? The answer(s) must surely include actions based on sound science, but Leopold also knew that people must reexamine their relationship with both animals and the earth. In the final chapter of *A Sand County Almanac** Leopold outlined his ideas for a land ethic. In short, he defined a new relationship between humans and nature. In its essence, it establishes an ethic that directs people to act on behalf of, and for the mutual benefit of, a community. Importantly, Leopold urged that this "community" must encompass more than people, must include the soils, water, plants, and animals. And he believed that health of the land community was analogous to health of a living organism. He summed up his arguments with a plea: "That land is a community is the basic concept of ecology, but that land is to be loved and respected is an extension of ethics . . . a land ethic, then, reflects the existence of an ecological conscience, and this in turn

* Leopold, A. 1949. A sand county almanac. Oxford University Press, Oxford, England.

reflects a conviction of individual responsibility for the health of the land." Nearly 75 years has passed since Leopold penned those words, and we are still searching for this *ecological conscience*. Could this apparent tsunami of wildlife health challenges be part of the wake-up call needed to move us toward a new social contract with our planet?

Beyond Leopold's holistic views, we each may feel responsibility in this area, and reasons to intervene may stem from professional, personal, social, or moral imperatives. Below we summarize our brief perspectives on how one might view responsibility for, and intervene on behalf of, wildlife health problems.

- Ignorance of wildlife health and disease often serves to bring unwarranted attention to nature as a source of human problems, as history has shown. Fear of the unknown has always stirred human destructive tendencies, and we blame the commons. Wildlife professionals are vital to provide that proportionality of risk for health challenges arising from nature, while promoting a balanced picture of the true drivers of disease emergence, many of which are anthropogenic.
- One size does not fit all. Each wildlife disease and health problem needs careful analysis and a customized response. Some disease processes can, will, and need to run their course. Others may require response, most often because they threaten public health or agriculture, but also at times because of political or social pressures and increasingly because they threaten sensitive wildlife populations. "First do no harm" should be kept in mind as some forms of response can be as detrimental, or more so, than the disease or health problem it is meant to mitigate (cordon fencing of vast portions of the Kalahari Desert to reduce foot and mouth disease transmission risk, for example).
- For those employed by agencies of government, professional responsibility may depend on the policy of your employer, and this may vary markedly. US Department of Interior policy on management of brucellosis is different from that of US Department of Agriculture, as are the laws that established each agency. It is also a matter of degree of policy implementation, for example whether the goal is disease eradication or perhaps to monitor, manage, and research with the goal of developing better and more applicable tools to reduce risk.
- Most academic wildlife health positions carry some expectation of contribution to society beyond teaching and research, but they do not carry with them a legal mandate. There is perhaps more freedom and flexibility in meeting personal and professional goals that serve wildlife health, and many nonprofits are similar.
- A common value with regards to animals, their health, sustainability of their populations, and conservation of their ecosystems is *stewardship*: the responsibility of human beings (individually and collectively), to plan, care for, and manage natural resources. Most natural resource agencies espouse this, and many people see it as a personal as well as a professional responsibility. Societies developing an *ecological conscience*, embracing a *land ethic*, could make this a social, political and even a legal responsibility.
- Even without any formal recognition, most people and almost all cultures recognize we have a generational responsibility to provide for our children and their children and to leave them a healthy natural resource base.
- On a more personal level, most people who have spent time in nature have felt or sensed profound feelings, historically captured well by Henry David Thoreau: "In wilderness is the preservation of the world." More recently the view of planet earth, and a few key photos of it from space, profoundly affected the lives of astronauts and millions of other people. This is formally recognized as "the small blue marble effect" and was one of the major inspirations for "Earth Day."

- Whether one's religious beliefs are theistic, pantheistic, atheistic, or agnostic, it is worth noting that ALL of the worlds organized faiths, and most indigenous cultures, recognize God-like qualities, and the "Spirit in Nature." Many who view Nature as a form of consciousness speculate as to whether emerging wildlife diseases and epidemics, along with other global environmental catastrophes, are a message and warning to us from Nature.
- From a simply utilitarian viewpoint, allowing valuable life forms and natural resources to be destroyed is wasteful and unwise, as there may be vital future needs for them. In many cases, the destruction resulting from wildlife diseases is extremely wasteful, even overwhelming the ability of scavengers to clean up. And to the extent wildlife health problems are caused by human actions, it is often human action that is needed to control or reduce them and avoid them in the future. "You broke it, you fix it."

In summary, it would seem the answer to the questions posed above is a resounding YES, we do have a responsibility to manage, steward, provide for the health of wildlife, the health and sustainability of the ecosystems on which we all depend; to avoid, limit, mitigate or 'treat' wildlife diseases that are traceable to our actions; and to try to optimize the health of animals, people and ecosystems—we return again to the idea of "One Health."

What Can We Do about Wildlife Health and Disease Problems?

To the extent possible, all the authors have attempted to answer this question for the specific disease or health problem in each chapter of this book. We will not go back and review those. We have the power, in some cases, to manipulate or manage the key host, pathogen, and environmental factors (see Introduction), as well as the ability to interrupt how disease processes propagate through space and time (one basis for preventive medicine). The declarations of some that it is "the end of the age of infectious diseases" was premature and conceited. Our technologies are remarkable, but they have also made us complacent and allowed for very unnatural processes to thrive, for example, in our food systems and in changing epidemiological relationships with unintended and unexpected consequences, which we are now confronted with. It's also worth noting that the reason many of the diseases we have discussed are considered important is that they impact people's lives and have some combination of infectivity (R_0), pathogenicity (case fatality rate), or persistence in the host–population–environment that makes them a threat (Table C.1). These factors are not easy to change, but they can show us where we may be more effective in reducing social, biological, ecological, and financial risks and costs. For example, the relatively low pathogenicity and environmental persistence of severe acute respiratory syndrome coronavirus 2 made vaccination and public health precautions a fairly effective response, despite it being highly infectious. Similarly, distemper, peste des petits ruminants, and rinderpest, despite being highly infectious and pathogenic, can be effectively controlled and even eliminated with vaccination and spatial and temporal distancing because these viruses have very poor persistence in the environment. Vaccines have not proven useful for managing brucellosis in wildlife, and the disease—although not highly pathogenic to wildlife—is quite persistent in populations, so management of spatial and temporal separation are currently the best tools to reduce risk of transmission to cattle.

More Generally, What Can Be Done to Reduce Potential Emerging Wildlife Disease Challenges?

- Study wildlife health and disease, research the host, pathogen, and environmental connections, in particular what (if anything) has changed and whether there is a human activity correlate.

Table C.1. Ratings of key disease characteristics for several wildlife health problems

Wildlife disease or health problem	Infectivity (R_0)	Persistence in host, population, or environment	Pathogenicity or virulence (case fatality rate)
Coronavirus disease 2019	High	Low in host and environment, high in populations	Relatively low but variable; however, due to high infectivity, pandemic was responsible for millions of human fatalities worldwide 2019–2022
Chronic wasting disease	Moderate to high	Very high in all three	High for many cervid species, may have population-level effects
African swine fever	Very high	Low in host, high in population and environment	High for all swine
Bovine tuberculosis	High	High in all three	High if untreated in humans, high in most other susceptible species
Rabies	Low	Low in host and environment, high in populations	Very high in people and essentially all susceptible species (mammals)
Highly pathogenic avian influenza	High in some waterfowl and poultry, variable in swine and relatively low in humans with zoonotic forms	Low in environment, variable in host, high in populations	Variable (high to moderate) in poultry, swine and some wild birds, low to moderate with zoonotic forms in humans (main risk in humans is recombination of animal–human viruses)
Brucellosis	High for exposed cattle, bison, elk, and humans	High in all three	Variable, relatively high for disease expression, relatively low for death.
Sea star wasting	Apparently high	Unknown	High
Chytridiomycosis in amphibians	Very high	High	Variable, but high enough to cause multiple amphibian species extinction
Peste des petits ruminants	Very high	Low in host and environment but high in populations	High for many ungulate species, variable in others
Epizootic pneumonia of bighorn sheep	High	High in all three	High in bighorn, relatively low in sheep and goats
Plague in North America	Moderate, requires flea bite, inoculation or inhalation in rare cases	High in all three	High is susceptible species
Rabbit hemorrhagic disease	High	High in all three	High in lagomorphs
White-nose syndrome	High in hibernating and social bat species	High in all three	Moderate to high in most circumstances
Avian malaria	High where competent mosquito vectors live	High in endemic areas, temperature limited	High for most naïve species
Toxoplasmosis	High	High	Moderate to low in many species, high in others
Ebola	High if in close contact, low otherwise	Low in host (most die) and environment, moderate in populations	Very high

The importance of every wildlife disease and health problem differs in various ways and through time, but the most serious are those with some combination of high infectivity; high pathogenicity; and persistence in the host, its populations, and/or the environment. In this table these qualities are rated by the editors for many of the diseases discussed in this book.

Understand proportional risk in terms of pathogen source and drivers of emergence. Preventive medicine is more cost effective, less wasteful and destructive, than crisis management.

- Focus first on the problems we have caused, the most serious ones, and those that can be solved with current knowledge and technology. Develop new and improved technologies.
- Be more thoughtful about land use, remember that planning is a critical part of stewardship, and reduce invasion, degradation, and conversion of wild lands. Where possible, restore wild

lands, reestablishing both biological and ecological communities. Restoration ecology is still a developing science and has significant potential to help heal larger scale health problems.

- Consider the ramifications of, and limit, live animal movements, trade in species and tissues, and unsanitary animal husbandry practices. Some forms of short-term financial gain are high risk for disease emergence (like raising pigs in tropical environments where bats and Nipah virus are endemic or releasing wild pigs on private lands in North America for commercial hunting purposes) and often prove very expensive in the long run.
- Recognize and deal more effectively with the health threats inherent in rapid global transportation and movement of people and their "stuff," especially animals and animal products.
- Recognize "the tragedy of the commons" and the problems inherent in public ownership of wildlife and wild lands where money and politics may overwhelm stewardship. Unfortunately, society tends to care more about wildlife when they are endangered or valuable but not so much otherwise.
- One of the common denominators in many chapters in this book is that global climate change is making, or is expected to make, the problem(s) worse. We need to deal more effectively with climate change NOW (since yesterday isn't possible). This means massive changes in the way energy is generated and consumed. We also need to do much better at managing landscapes, soils, water, and oceans. It means holding politicians and governments responsible to all of society and for assuring our children's and grandchildren's futures. On a personal basis, we can eat a more plant-based diet, fly less often, reduce car use, maximize energy efficiency at home and work, have one less child or adopt or foster, and plant trees—lots and lots of trees.
- Adopt the "precautionary principle." We must be able and willing to act before absolute scientific certainty; if done as management experiments, the learning process may advance further and faster. For example, we are just beginning to understand the profound role of the ocean in our lives—as a sink for carbon and regulator of climate—its tiny diatoms make the oxygen for every other breath we take. We must stop overfishing and filling our waters with plastics and other pollutants, and experiments on how best to do this can serve as both guide and a promissory beginning.

Can We Succeed?

We must succeed! Almost everyone who works in conservation or wildlife management deals with degradation of nature and with disturbing social and political trends; wildlife health specialists, in particular deal with death and dying, often on a very large scale. It can get depressing, at times painful and even overwhelming. It can lead one to give up and believe that it's simply too late. One term for this is "ecogrief." It may even manifest in ways similar to the natural stages of grief that Elisabeth Kübler-Ross identified with death of a loved one: denial, anger, bargaining, depression, and acceptance. Most of us know people stuck in the denial and anger phases, the latter of which occasionally results in violence against perceived perpetrators like large corporations or governments. The bargaining phase may be played out in courts of law. It should be noted that "acceptance" does not mean accepting human actions that cause wildlife disease outbreaks or chronic wildlife health problems, it means accepting the reality of their existence and preventing them in the future. Similarly, as Jane Goodall observes in *The Book of Hope,** hope is not passive, it is not wishful thinking, it is

*Goodall, J., and D. Abrams. 2021. The book of hope: a survival guide for trying times. Caledon Books, New York, New York, USA.

maintaining a positive attitude while working hard for resolution, recovery, and healing in a larger sense. Goodall points to four reasons to maintain hope for stopping biodiversity loss and the climate crisis, to which we would add the most pernicious aspects of the wildlife disease tsunami: the amazing human intellect, the resilience of nature, the power of young people, and the indomitable human spirit. As simplistic as these reasons may sound at first, there is solid thinking and powerful examples that support them. Full exploration of this topic exceeds the scope of this book but suffice it to say there is also no benefit for the individual, society, wildlife, or Nature in giving up. Nobody wins, everybody loses, and there is no real hope without action.

Just Do It!

Talk is cheap. Although reaching a diagnosis (or a consensus) is good, and planning is good too, we need to act to mitigate (treat, manage, prevent) the most serious wildlife health challenges and, perhaps most important of all, understand where the true risk lies for disease emergence and where best to intervene. The time for discussing or arguing what we call it (wildlife disease, wildlife health, or wildlife veterinary medicine), or what framework it fits into (wildlife management, eco-health, conservation medicine, one health, or planetary health), should be behind us. The three Cs should be paramount: cooperation, collaboration, and conservation. We are talking about saving the Earth and its biodiversity, and avoiding biomedical disasters of our own making, here folks. All hands on deck!

Given the pervasive influence of human activities on the emergence of wildlife diseases and the creation of wildlife health challenges, we can't really reduce most risks without social recognition and cooperation. And we can't get that unless people understand that the issues are not only of biological and ecological importance, but also personally relevant to them and their progeny. On a local level, very little of lasting value can be realized unless local people understand and support your efforts. This will very likely require making the science behind wildlife health relevant to the community. Wildlife professionals need to go beyond the scientific and regulatory aspects of the problem to make it real and valuable to people who will be there when we have gone. As several chapters in this book demonstrate, considering the needs of local people and meeting those needs can lead to positive changes in their relationship to animals, ecosystems, and Nature. We must all be teachers, and teaching by example is most effective if we ever hope to adopt an *ecological conscience*.

Our task ahead is huge, and it will not be easy, but it is our *only* small blue marble.

> There is hope if only we can bridge the gap between the clever human brain and the compassionate human heart and act now. Only if we understand, can we care. Only if we care, will we help. Only if we help, shall all be saved.
>
> *Jane Goodall**

* Goodall, J., and D. Abrams. 2021. The book of hope: a survival guide for trying times. Caledon Books, New York, New York, USA.

Index

Page numbers followed by "t" refer to tables.